COLD SPRING HARBOR SYMPOSIA ON QUANTITATIVE BIOLOGY

VOLUME LXXXIV

symposium.cshlp.org

Online access to these articles is free of charge. Please visit our companion website at symposium.cshlp.org.

COLD SPRING HARBOR SYMPOSIA ON QUANTITATIVE BIOLOGY

VOLUME LXXXIV

RNA Control and Regulation

symposium.cshlp.org

Symposium Organizers: Terri Grodzicker, David Stewart, and Bruce Stillman (*Cold Spring Harbor Laboratory*)

Proceedings Editors: David Stewart and Bruce Stillman (*Cold Spring Harbor Laboratory*)

COLD SPRING HARBOR LABORATORY PRESS
2019

COLD SPRING HARBOR SYMPOSIA ON QUANTITATIVE BIOLOGY VOLUME LXXXIV

Published 2019 by Cold Spring Harbor Laboratory Press
International Standard Book Number 978-1-621823-95-7 (cloth)
International Standard Book Number 978-1-621823-92-6 (paper)
International Standard Serial Number 0091-7451
Library of Congress Catalog Card Number 34-8174

Printed in the United States of America
All rights reserved
COLD SPRING HARBOR SYMPOSIA ON QUANTITATIVE BIOLOGY
Founded in 1933 by
REGINALD G. HARRIS
Director of the Biological Laboratory 1924 to 1936

Previous Symposia Volumes

I (1933) Surface Phenomena
II (1934) Aspects of Growth
III (1935) Photochemical Reactions
IV (1936) Excitation Phenomena
V (1937) Internal Secretions
VI (1938) Protein Chemistry
VII (1939) Biological Oxidations
VIII (1940) Permeability and the Nature of Cell Membranes
IX (1941) Genes and Chromosomes: Structure and Organization
X (1942) The Relation of Hormones to Development
XI (1946) Heredity and Variation in Microorganisms
XII (1947) Nucleic Acids and Nucleoproteins
XIII (1948) Biological Applications of Tracer Elements
XIV (1949) Amino Acids and Proteins
XV (1950) Origin and Evolution of Man
XVI (1951) Genes and Mutations
XVII (1952) The Neuron
XVIII (1953) Viruses
XIX (1954) The Mammalian Fetus: Physiological Aspects of Development
XX (1955) Population Genetics: The Nature and Causes of Genetic Variability in Population
XXI (1956) Genetic Mechanisms: Structure and Function
XXII (1957) Population Studies: Animal Ecology and Demography
XXIII (1958) Exchange of Genetic Material: Mechanism and Consequences
XXIV (1959) Genetics and Twentieth Century Darwinism
XXV (1960) Biological Clocks
XXVI (1961) Cellular Regulatory Mechanisms
XXVII (1962) Basic Mechanisms in Animal Virus Biology
XXVIII (1963) Synthesis and Structure of Macromolecules
XXIX (1964) Human Genetics
XXX (1965) Sensory Receptors
XXXI (1966) The Genetic Code
XXXII (1967) Antibodies
XXXIII (1968) Replication of DNA in Microorganisms
XXXIV (1969) The Mechanism of Protein Synthesis
XXXV (1970) Transcription of Genetic Material
XXXVI (1971) Structure and Function of Proteins at the Three-dimensional Level
XXXVII (1972) The Mechanism of Muscle Contraction
XXXVIII (1973) Chromosome Structure and Function
XXXIX (1974) Tumor Viruses
XL (1975) The Synapse
XLI (1976) Origins of Lymphocyte Diversity
XLII (1977) Chromatin
XLIII (1978) DNA: Replication and Recombination
XLIV (1979) Viral Oncogenes
XLV (1980) Movable Genetic Elements
XLVI (1981) Organization of the Cytoplasm
XLVII (1982) Structures of DNA
XLVIII (1983) Molecular Neurobiology
XLIX (1984) Recombination at the DNA Level
L (1985) Molecular Biology of Development
LI (1986) Molecular Biology of Homo sapiens
LII (1987) Evolution of Catalytic Function
LIII (1988) Molecular Biology of Signal Transduction
LIV (1989) Immunological Recognition
LV (1990) The Brain
LVI (1991) The Cell Cycle
LVII (1992) The Cell Surface
LVIII (1993) DNA and Chromosomes
LVIX (1994) The Molecular Genetics of Cancer
LX (1995) Protein Kinesis: The Dynamics of Protein Trafficking and Stability
LXI (1996) Function & Dysfunction in the Nervous System
LXII (1997) Pattern Formation during Development
LXIII (1998) Mechanisms of Transcription
LXIV (1999) Signaling and Gene Expression in the Immune System
LXV (2000) Biological Responses to DNA Damage
LXVI (2001) The Ribosome
LXVII (2002) The Cardiovascular System
LXVIII (2003) The Genome of Homo sapiens
LXIX (2004) Epigenetics
LXX (2005) Molecular Approaches to Controlling Cancer
LXXI (2006) Regulatory RNAs
LXXII (2007) Clocks and Rhythms
LXXIII (2008) Control and Regulation of Stem Cells
LXXIV (2009) Evolution: The Molecular Landscape
LXXV (2010) Nuclear Organization and Function
LXXVI (2011) Metabolism and Disease
LXXVII (2012) The Biology of Plants
LXXVIII (2013) Immunity and Tolerance
LXXIX (2014) Cognition
LXXX (2015) 21st Century Genetics: Genes at Work
LXXXI (2016) Targeting Cancer
LXXXII (2017) Chromosome Segregation and Structure
LXXXIII (2018) Brains and Behavior: Order and Disorder in the Nervous System

***Front cover** (paperback)*: Structure of a postcatalytic spliceosome immediately after exon ligation. RNA components are shown in gold. Image prepared by Max Wilkinson, MRC Laboratory of Molecular Biology, using data from Wilkinson et al. *Science* **358**: 1283–1288 [2017].

Authorization to photocopy items for internal or personal use, or the internal or personal use of specific clients, is granted by Cold Spring Harbor Laboratory Press, provided that the appropriate fee is paid directly to the Copyright Clearance Center (CCC). Write or call CCC at 222 Rosewood Drive, Danvers, MA 01923 (978-750-8400) for information about fees and regulations. Prior to photocopying items for educational classroom use, contact CCC at the above address. Additional information on CCC can be obtained at CCC Online at www.copyright.com.

For a complete catalog of all Cold Spring Harbor Laboratory Press publications, visit our website www.cshlpress.org.

Online access: Please visit our companion website at symposium.cshlp.org. For access issues, please contact Cold Spring Harbor Laboratory Press at subscriptions@cshl.edu.

In warm memory of Kiyoshi Nagai (1949–2019)

Symposium Participants

ABDELMOHSEN, KOTB, National Institutes of Health, Baltimore, Maryland
ABDEL-WAHAB, OMAR, Memorial Sloan Kettering Cancer Center, New York, New York
ADAMS, DEXTER, Stony Brook University, Stony Brook, New York
ADELMAN, KAREN, Harvard Medical School, Boston, Massachusetts
AGGARWAL, DISHA, Cold Spring Harbor Laboratory, Cold Spring Harbor, New York
AGRAWAL, SHIPRA, Nationwide Children's Hospital, Columbus, Ohio
AGUILERA, ANDRÉS, University of Seville, Sevilla, Spain
AKHTAR, ASIFA, MPI of Immunobiology and Epigenetics, Freiburg, Germany
AKIMITSU, NOBUYOSHI, The University of Tokyo, Tokyo, Japan
AKKIPEDDI, SAJAL, Swarthmore College, Swarthmore, Pennsylvania
ALVES, CRISTIANE, Cold Spring Harbor Laboratory, Cold Spring Harbor, New York
ANDERGASSEN, DANIEL, Harvard University, Cambridge, Massachusetts
ANDREASSI, CATIA, University College London, London, United Kingdom
ARAKE DE TACCA, LUISA, University of California, Berkeley, Berkeley, California
ARAVIN, ALEXEI, California Institute of Technology, Pasadena, California
ARISTIZABAL, DAVID, New York University, New York, New York
ARRAIANO, CECILIA, ITQB/Universidade NOVA de Lisboa, Oeiras, Portugal
ATIANAND, MANINJAY, University of Pittsburgh, Pittsburgh, Pennsylvania
BAKER, ALISON, Harvard Medical School, Boston, Massachusetts
BARRON, LINDSEY, UT Health San Antonio, San Antonio, Texas
BARSOUM, JAMES, Arrakis Therapeutics, Waltham, Massachusetts
BARTEL, DAVID, Whitehead Institute/Massachusetts Institute of Technology/Howard Hughes Medical Institute, Cambridge, Massachusetts
BASS, BRENDA, University of Utah, Salt Lake City, Utah
BATISTA, PEDRO, National Cancer Institute, Bethesda, Maryland
BAUMANN, BETHANY, University of Illinois at Chicago, Chicago, Illinois
BAYMURADOV, ULUGBEK, Stanford University, Stanford, California
BECHARA, RAMI, University of Pittsburgh, Pittsburgh, Pennsylvania
BELTRAN, TONI, MRC London Institute of Medical Sciences, London, United Kingdom
BHAT, PRASHANT, California Institute of Technology, Pasadena, California
BHATT, BHARAT, Indian Institute of Science, Bangalore, India
BI, XIANJU, Tsinghua University, Beijing, China
BIBEL, BRIANNA, Cold Spring Harbor Laboratory, Cold Spring Harbor, New York
BICKNELL, ALICIA, Moderna Therapeutics, Cambridge, Massachusetts
BINDEREIF, ALBRECHT, University of Giessen, Giessen, Germany
BISCARINI, SILVIA, Istituto Italiano di Tecnologia, Rome, Italy
BLANCHARD, DANIEL, ThermoFisher Scientific, South San Francisco, California
BLUMBERG, AMIT, Cold Spring Harbor Laboratory, Cold Spring Harbor, New York
BOEHM, STEFANIE, *EMBO Journal*, Heidelberg, Germany
BOFILL-DE ROS, XAVIER, National Cancer Institute, Frederick, Maryland
BONASIO, ROBERTO, University of Pennsylvania, Philadelphia, Pennsylvania
BOREIKAITE, VYTAUTE, MRC Laboratory of Molecular Biology, Cambridge, United Kingdom
BRADLEY, ROBERT, Fred Hutchinson Cancer Research Center, Seattle, Washington
BRADLEY, THOMAS, University of East Anglia, Norwich, United Kingdom
BRENNECKE, JULIUS, IMBA Institute of Molecular Biotechnology GmbH, Vienna, Austria
BRIDGEWATER, HANNAH, University of Warwick, Coventry, United Kingdom
BROCKDORFF, NEIL, University of Oxford, Oxford, United Kingdom
BRODERSEN, PETER, University of Copenhagen, Copenhagen, Denmark
BROOKS, ANGELA, University of California, Santa Cruz, Santa Cruz, California
BROWN, ANISSA, National Institutes of Health, Bethesda, Maryland
CAO, JUN, University of Texas Medical Branch at Galveston, Galveston, Texas
CARDETTI, CAITLYN, Stony Brook University, Stony Brook, New York
CARONE, DAWN, Swarthmore College, Swarthmore, Pennsylvania
CARTEGNI, LUCA, Rutgers University, New York, New York
CASTEELS, TAMARA, CeMM Research Center for Molecular Medicine of the Austrian Academy of Sciences, Vienna, Austria
CERVANTES, MARLENE, University of California, Irvine, Irvine, California
CHAN, JIA JIA, Cancer Science Institute of Singapore, Singapore, Singapore
CHANG, HOWARD, Stanford University School of Medicine, Stanford, California
CHANG, KUNG-CHI, Cold Spring Harbor Laboratory, Cold Spring Harbor, New York
CHEIKHI, AMIN, University of Pittsburgh, Pittsburgh, Pennsylvania
CHEN, LEILEI POLLY, National University of Singapore, Singapore, Singapore
CHEN, LINGLING, CAS, Shanghai Institute of Biochemistry and Cell Biology, Shanghai, China
CHEN, MENGNUO, The University of Hong Kong, Hong Kong SAR
CHEN, PEIWEI, California Institute of Technology, Pasadena, California
CHOI, HEEJIN, National Institutes of Health, Bethesda, Maryland
CHOI, JEONGYOON, University of Oxford, Oxford, United Kingdom
CHOUDHARI, RAMESH, Texas Tech University, El Paso, El Paso, Texas
CISSE, IBRAHIM, Massachusetts Institute of Technology, Cambridge, Massachusetts
CLAYTON, SONIA, The Kinkaid School, Houston, Texas
CONNELL, LAUREEN, *Genes & Development*, Cold Spring Harbor, New York
CONNOLLY, MICHAEL, Novartis Institutes for Biomedical Research, Cambridge, Massachusetts
CONTI, ELENA, Max-Planck-Institut für Biochemie, Martinsfried, Germany
CORRIONERO SAIZ, ANNA, Stoke Therapeutics, Inc., Bedford, Massachusetts
COURCHAINE, EDWARD, Yale University, New Haven, Connecticut
CZECH, BEN, University of Cambridge, Cambridge, United Kingdom
DAI, LISHENG, National Cancer Institute, Frederick, Maryland
DARNELL, JAMES, The Rockefeller University, New York, New York
DARNELL, ROBERT, The Rockefeller University/Howard Hughes Medical Institute/NY Genome Center, New York, New York
DAS, SANKHA SUBHRA, Indian Institute of Technology Kharagpur, Kharagpur, India
DE ZOYSA, MEEMANAGE, University of Rochester Medical Center, Rochester, New York
DEAN, CAROLINE, John Innes Centre, Norwich, United Kingdom
DENLI, AHMET, *Genome Research*, Cold Spring Harbor, New York
DERYUSHEVA, SVETLANA, Carnegie Institution for Science, Baltimore, Maryland
DING, FANGYUAN, California Institute of Technology, Pasadena, California
DING, YILIANG, John Innes Centre, Norwich, United Kingdom
DOBIN, ALEXANDER, Cold Spring Harbor Laboratory, Cold Spring Harbor, New York
DONG, XIANJUN, Harvard Medical School, Boston, Massachusetts

SYMPOSIUM PARTICIPANTS

D'Orazio, Karole, The Johns Hopkins University School of Medicine, Baltimore, Maryland
Dou, Xiaoyang, University of Chicago, Chicago, Illinois
Doudna, Jennifer, University of California, Berkeley/Howard Hughes Medical Institute, Berkeley, California
Dreggors, Liz, Emory University School of Medicine, Atlanta, Georgia
Dreyfuss, Gideon, Howard Hughes Medical Institute/University of Pennsylvania School of Medicine, Philadelphia, Pennsylvania
Duda, Katarzyna, Max Planck Institute, Freiburg, Germany
Duffy, Erin, Harvard Medical School, Boston, Massachusetts
Elchert, Alexandra, National Institutes of Health, Bethesda, Maryland
Elewa, Ahmed, Cold Spring Harbor Laboratory, Cold Spring Harbor, New York
Elguindy, Mahmoud, University of Texas Southwestern, Dallas, Texas
Elkayam, Elad, Cold Spring Harbor Laboratory, Cold Spring Harbor, New York
Enganti, Ramya, Indiana University, Bloomington, Indiana
Erhardt, Sylvia, Heidelberg University, Heidelberg, Germany
Fabry, Martin, University of Cambridge, Cambridge, United Kingdom
Fan, Xiaojuan, Chinese Academy of Sciences, Shanghai, China
Fei, Qili, University of Chicago, Chicago, Illinois
Fernandez, Eric Aria, University of Lausanne, Prilly, Switzerland
Fitzsimmons, Christina, National Cancer Institute, Bethesda, Maryland
Flynn, Ryan, Stanford University, Stanford, California
Flynt, Alex, University of Southern Mississippi, Hattiesburg, Mississippi
Fonseca Cardenas, Alejandro, Pontificia Universidad Catolica de Chile, Santiago, Chile
Frendewey, David, Regeneron Pharmaceuticals, Inc., New York, New York
Fu, Xiang-Dong, University of California, San Diego, La Jolla, California
Gaffen, Sarah, University of Pittsburgh, Pittsburgh, Pennsylvania
Gainetdinov, Ildar, University of Massachusetts Medical School, Worcester, Massachusetts
Gandhi, Minakshi, German Cancer Research Center, Heidelberg, Germany
Garcia, David, University of Oregon, Eugene, Oregon
Garg, Salil, Massachusetts Institute of Technology, Cambridge, Massachusetts
Gasser, Susan, Friedrich Miescher Institute for Biomedical Research, Basel, Switzerland
Gennarino, Vincenzo, Columbia University Medical Center, New York, New York
Genzor, Pavol, National Institute of Diabetes and Digestive and Kidney Diseases/National Institutes of Health, Bethesda, Maryland
George, Rani, Harvard Medical School, Dana-Farber Cancer Institute, Boston, Massachusetts
Ghalei, Homa, Emory University School of Medicine, Atlanta, Georgia
Gingeras, Thomas, Cold Spring Harbor Laboratory, Cold Spring Harbor, New York
Gladfelter, Amy, University of North Carolina, Chapel Hill, North Carolina
Godneeva, Baira, Institute of Molecular Genetics RAS, Moscow, Russia
Goh, Sho, Genome Institute of Singapore, Singapore, Singapore
Goh, Yeek Teck, Genome Institute of Singapore, Singapore, Singapore
Gokhale, Nandan, Duke University, Durham, North Carolina
Gopinath, Srikar, Institute for Stem Cell Science and Regenerative Medicine, Bangalore, India
Gordon, Katrina, University of Edinburgh, Edinburgh, United Kingdom
Greer, Eric, Harvard Medical School/Boston Children's Hospital, Boston, Massachusetts
Grewal, Shiv, National Institutes of Health, Bethesda, Maryland
Grodzicker, Terri, Cold Spring Harbor Laboratory, Cold Spring Harbor, New York
Gu, Shuo, National Cancer Institute, Frederick, Maryland
Guo, Xuecui, MeiraGTx, New York, New York
Gutbrod, Michael, Cold Spring Harbor Laboratory, Cold Spring Harbor, New York
Guzman Gundermann, Daniel, New York University, New York, New York

Haase, Astrid, National Institute of Diabetes and Digestive and Kidney Diseases/National Institutes of Health, Bethesda, Maryland
Haberman, Nejc, Imperial College London, London, United Kingdom
Hacisuleyman, Ezgi, The Rockefeller University, New York, New York
Hafner, Markus, National Institutes of Health, Bethesda, Maryland
Hale, Caryn, Rockefeller University, New York, New York
Hallacli, Erinc, Brigham and Women's Hospital, Boston, Massachusetts
Hammarskjold, Marie-Louise, University of Virginia, Charlottesville, Virginia
Hammell, Christopher, Cold Spring Harbor Laboratory, Cold Spring Harbor, New York
Han, Cecil, Georgetown University, Washington, D.C.
Hang, Jing, Peking University Third Hospital, Beijing, China
Hao, Qinyu, University of Illinois at Urbana-Champaign, Urbana, Illinois
Harandi, Omid, Rubius Therapeutics, Cambridge, Massachusetts
Harlow, Matt, Dana-Farber Cancer Institute, Boston, Massachusetts
Hartenian, Ella, UC Berkeley, Berkeley, California
He, Chuan, The University of Chicago, Chicago, Illinois
Hertz, Laura, National Institutes of Health, Bethesda, Maryland
Hirose, Tetsuro, Hokkaido University, Sapporo, Japan
Hoek, Tim, Hubrecht Institute, Utrecht, Netherlands
Holling, Aaron, Roswell Park Comprehensive Cancer Center, Buffalo, New York
Hong, Dawon, Dankook University, Yongin-si, South Korea
Hong, Lingzi, Case Western Reserve University, Cleveland, Ohio
Hung, Lee-Hsueh, University of Giessen, Giessen, Germany
Hwang, Hun-Way, University of Pittsburgh, Pittsburgh, Pennsylvania
Ipsaro, Jonathan, Cold Spring Harbor Laboratory, Cold Spring Harbor, New York
Irmady, Krithi, The Rockefeller University, New York, New York
Jaffrey, Samie, Weill Cornell Medical College, New York, New York
Jaganathan, Kishore, Illumina Inc., Foster City, California
Jakimo, Alan, Hofstra University, Hempstead, New York
Jalloh, Binta, Emory University, Atlanta, Georgia
Jensen, Christina, Stanford University, Stanford, California
Jensen, Mads Aaboe, Roche, Hørsholm, Denmark
Jeong, Jiwon, Dankook University, Yongin-si, South Korea
Joshua-Tor, Leemor, Howard Hughes Medical Institute/Cold Spring Harbor Laboratory, Cold Spring Harbor, New York
Josipovic, Natasa, University of Göttingen, Göttingen, Germany
Jung, Soo-Jin, Institute for Basic Science, Seoul, South Korea
Kadlec, Jan, CNRS/Institut de Biologie Structurale, Grenoble, France
Kamenova, Ivanka, *Nature Protocols*, London, United Kingdom
Kang, Jian, National University of Singapore, Singapore, Singapore
Kang, Minjeong, Kaist, Daejeon, South Korea
Kawamoto, Sachiyo, National Institutes of Health, Bethesda, Maryland
Kazlauskiene, Migle, University of Zürich, Zürich, Switzerland
Keinath, Melissa, Carnegie Institution for Science, Baltimore, Maryland
Khanwalkar, Urjeet, Rubius Therapeutics, Cambridge, Massachusetts
Khateb, Mamdoh, National Institutes of Health, Bethesda, Maryland
Khuperkar, Deepak, Hubrecht Institute, Utrecht, Netherlands
Kim, Dongwan, Seoul National University, Seoul Gwanak-gu, South Korea
Kim, Haedong, Seoul National University, Seoul, South Korea
Kim, Sujin, KAIST, Daejeon, South Korea
Kim, Yoosik, KAIST, Daejeon, South Korea
Kim, Young, Stony Brook University, Stony Brook, New York
Kim, Young Jin, Cold Spring Harbor Laboratory, Cold Spring Harbor, New York
King, Benjamin, University of Maine, Orono, Maine
Kneuss, Emma, University of Cambridge, Cambridge, United Kingdom
Knupp, David, University of Nevada, Reno, Reno, Nevada
Koh, Wei Qian Casslynn, A*STAR, Singapore, Singapore
Konstantinidou, Parthena, National Institutes of Health, Bethesda, Maryland
Kornblihtt, Alberto, University of Buenos Aires, Buenos Aires, Argentina
Kotaja, Noora, University of Turku, Turku, Finland
Krainer, Adrian, Cold Spring Harbor Laboratory, Cold Spring Harbor, New York

SYMPOSIUM PARTICIPANTS

KRAJEWSKA, MALGORZATA, Dana-Farber Cancer Institute, Boston, Massachusetts
KRETOV, DMITRY, Boston University School of Medicine, Boston, Massachusetts
KU, JAYOUNG, Korea Advanced Institute of Science and Technology, Daejeon, South Korea
KU, YONGSUK, Korea Advanced Institute of Science and Technology, Daejeon, South Korea
KUBICA, NEIL, Arrakis Therapeutics, Waltham, Massachusetts
KWAK, YEONUI, Cornell University, Ithaca, New York
KWOK, ZHI HAO, Cancer Science Institute of Singapore, Singapore, Singapore
KWON, JUNSU, National University Singapore, Singapore, Singapore
LAGRANGE, THIERRY, CNRS, Perpignan, France
LAI, ERIC, Memorial Sloan Kettering Cancer Center, New York, New York
LAKSHMANAN, VAIRAVAN, Institute for Stem Cell Biology and Regenerative Medicine, Bangalore, India
LAN, PENGFEI, Shanghai Jiao Tong University School of Medicine, Shanghai, China
LATIFKAR, ARASH, Cornell University, Ithaca, New York
LAU, NELSON, Boston University School of Medicine, Boston, Massachusetts
LEE, HEUIRAN, University of Ulsan College of Medicine, Seoul, South Korea
LEE, JE, Cold Spring Harbor Laboratory, Cold Spring Harbor, New York
LEE, JEANNIE, Massachusetts General Hospital, Boston, Massachusetts
LEE, JONG-SUN, UT Southwestern Medical Center, Dallas, Texas
LEE, JOO-HYUNG, University of Texas Health Science at Houston, Houston, Texas
LEE, MIHYE, Soonchunhyang University, Cheonan, South Korea
LEE, SEUNG KYU, National Institute on Aging/National Institutes of Health, Baltimore, Maryland
LEE, SHIH-HAN, Memorial Sloan Kettering Cancer Center, New York, New York
LEE, SUNGYUL, Seoul National University, Seoul, South Korea
LEE, YOUNG-SUK, Seoul National University, Seoul, South Korea
LEHMANN, RUTH, NYU School of Medicine, New York, New York
LEHTINIEMI, TIINA, University of Turku, Turku, Finland
LEI, MING, Shanghai Jiao Tong University School of Medicine, Shanghai, China
LI, HUABING, Shanghai JiaoTong University School of Medicine, Shanghai, China
LI, QIN, Stanford University, Stanford, California
LI, WENBO, University of Texas Health Science Center Houston, Houston, Texas
LI, XIAOXIA, Cleveland Clinic/Lerner Research Institute, Cleveland, Ohio
LI, ZHIZHONG, Fudan University, Shanghai, China
LICHINCHI, GIANLUIGI, Roche Innovation Center Copenhagen, Hørsholm, Denmark
LIM, PYUNG OK, DGIST, Daegu, South Korea
LIN, XIANZHI, Cedars-Sinai Medical Center, Los Angeles, California
LIN, XUN, Stony Brook University, Stony Brook, New York
LIN, YI, UT Southwestern Medical Center, Dallas, Texas
LING, JONATHAN, Johns Hopkins University School of Medicine, Baltimore, Maryland
LIU, ALISON, Wiley, Hoboken, New Jersey
LIU, HAIBIN, National Cancer Institute, Frederick, Maryland
LIU, JUN, University of Chicago, Chicago, Illinois
LIU, JUN-JIE, University of California Berkeley, Berkeley, California
LJUNGMAN, MATS, University of Michigan, Ann Arbor, Michigan
LOPEZ-MEJIA, ISABEL, University of Lausanne, Lausanne, Switzerland
LU, SYDNEY, Memorial Sloan Kettering Cancer Center, New York, New York
LUO, YICHENG, California Institute of Technology, Pasadena, California
LURAIN, KAYCEE, Ribometrix, Durham, North Carolina
LY, MICHAEL, University of California, Berkeley, Berkeley, California
LYNCH, KRISTEN, University of Pennsylvania, Philadelphia, Pennsylvania
MA, JINBIAO, Fudan University, Shanghai, China
MA, WEIRUI, Memorial Sloan Kettering Cancer Center, New York, New York
MACRAE, RHIANNON, The Broad Institute, Belmont, Massachusetts
MAHAT, JAY, Massachusetts Institute of Technology, Cambridge, Massachusetts
MANGILET, ANCHILIE, University of Oldenburg, Oldenburg, Germany
MANGONE, MARCO, Arizona State University, Tempe, Arizona
MANLEY, JAMES, Columbia University, New York, New York
MANZOUROLAJDAD, AMIRHOSSEIN, National Institutes of Health, Bethesda, Maryland
MAO, STEVE, Science/American Association for the Advancement of Science, Washington, D.C.
MAQUAT, LYNNE, University of Rochester Medical Center, Rochester, New York
MARLIN ANDREWS, CELINE, National Institutes of Health, Bethesda, Maryland
MARRAN, KRISTA, Ribometrix, Durham, North Carolina
MARTIENSSEN, ROBERT, Cold Spring Harbor Laboratory/Howard Hughes Medical Institute, Cold Spring Harbor, New York
MARX, STEPHEN, National Institutes of Health/National Institute of Child Health and Human Development, Bethesda, Maryland
MAUGER, DAVID, Moderna Therapeutics, Cambridge, Massachusetts
MAYR, CHRISTINE, Memorial Sloan Kettering Cancer Center, New York, New York
MCKNIGHT, STEVEN, UT Southwestern Medical Center, Dallas, Texas
MENDELL, JOSHUA, UT Southwestern Medical Center/Howard Hughes Medical Institute, Dallas, Texas
MEYER, CINDY, The Rockefeller University, New York City, New York
MIKEDIS, MARIA, Whitehead Institute, Cambridge, Massachusetts
MISHRA, VIBHOR, Howard Hughes Medical Institute/Indiana University, Bloomington, Indiana
MOHAMMED, JAAVED, Stanford University, Palo Alto, California
MONAHAN, JACK, EMBL-EBI, Cambridge, United Kingdom
MONIAN, PRASHANT, Wave Life Sciences, Cambridge, Massachusetts
MONTEZ, MIGUEL, Institute of Biochemistry and Biophysics, Warsaw, Poland
MUCKENFUSS, LENA, University of Zürich, Zürich, Switzerland
MUKHERJEE, POOJA, CERBM GIE IGBMC, Illkirch, France
MUNOZ, ADRIANA, Cold Spring Harbor Laboratory, Cold Spring Harbor, New York
NARLIKAR, GEETA, University of California, San Francisco, San Francisco, California
NECHOOSHTAN, GAL, Cold Spring Harbor Laboratory, Woodbury, New York
NEUGEBAUER, KARLA, Yale University, New Haven, Connecticut
NILSEN, TIMOTHY, Case Western Reserve University, Cleveland, Ohio
NNADI, CHIDINMA, Johns Hopkins University, Baltimore, Maryland
OLEJNICZAK, SCOTT, Roswell Park Comprehensive Cancer Center, Buffalo, New York
OLTHOF, ANOUK, University of Connecticut, Storrs, Connecticut
ON, KIN, Cold Spring Harbor Laboratory, Cold Spring Harbor, New York
ONOGUCHI-MIZUTANI, RENA, The University of Tokyo, Tokyo, Japan
OUYANG, ZHENGQING, The Jackson Laboratory for Genomic Medicine, Farmington, Connecticut
PALENCIA, ANDRÉS, INSERM, Grenoble, France
PARK, ANNSEA, Yale School of Medicine, New Haven, Connecticut
PARKER, ROY, University of Colorado, Boulder/Howard Hughes Medical Institute, Boulder, Colorado
PASSMORE, LORI, MRC Laboratory of Molecular Biology, Cambridge, United Kingdom
PATRICK, KRISTIN, Texas A&M Health Science Center, Bryan, Texas
PIKAARD, CRAIG, Howard Hughes Medical Institute/Indiana University, Bloomington, Indiana
PIWECKA, MONIKA, Max Delbrück Center for Molecular Medicine, Berlin, Germany
POLASKI, JACOB, Fred Hutchinson Cancer Research Center, Seattle, Washington
POMAVILLE, MONICA, Dana-Farber Cancer Institute, Boston, Massachusetts
POTTER, KRISTINE, State University of New York, Canton, Canton, New York
PRAJAPAT, MAHENDRA, National Institutes of Health/National Cancer Institute, Bethesda, Maryland
PREMSRIRUT, PREM, Mirimus, Inc., Brooklyn, New York
PROUDFOOT, NICHOLAS, University of Oxford, Oxford, United Kingdom
QIU, CHEN, National Institute of Environmental Health Sciences/National Institutes of Health, Research Triangle Park, North Carolina

RAESCH, FELIX, Max Planck Institute for Developmental Biology, Tübingen, Germany
RAMBOUT, XAVIER, University of Rochester Medical Center, Rochester, New York
RANDO, OLIVER, UMass Medical School, Worcester, Massachusetts
RAY, ATREYEI, University of Wisconsin—Milwaukee, Milwaukee, Wisconsin
REKOSH, DAVID, University of Virginia, Charlottesville, Virginia
REN, JIE, Cold Spring Harbor Laboratory, Cold Spring Harbor, New York
REN, WENQING, Wistar Institute, Philadelphia, Pennsylvania
ROCHE, BENJAMIN, Cold Spring Harbor Laboratory, Cold Spring Harbor, New York
RODRIGUEZ, ANTONY, Baylor College of Medicine, Houston, Texas
ROZHKOV, NIKOLAY, Cold Spring Harbor Laboratory, Cold Spring Harbor, New York
RUBIEN, JACK, Swarthmore College, Swarthmore, Pennsylvania
SACCHI, NICOLETTA, Roswell Park Cancer Institute, Buffalo, New York
SALA, LAURA, National Institutes of Health/National Cancer Institute, Bethesda, Maryland
SARKAR, DEBINA, University of Otago, Dunedin, New Zealand
SARMA, KAVITHA, The Wistar Institute, Philadelphia, Pennsylvania
SARUULDALAI, ENKHJIN, National Cancer Center, Goyang-si, Gyeonggi-do, South Korea
SAVAN, RAM, University of Washington, Seattle, Washington
SCHAENING BURGOS, CASSANDRA, Massachusetts Institute of Technology, Cambridge, Massachusetts
SCHAFFER, AMOS, Bar Ilan University, Ramat Gan, Israel
SCHORN, ANDREA, Cold Spring Harbor Laboratory, Cold Spring Habor, New York
SEVER, RICHARD, Cold Spring Harbor Laboratory Press, Cold Spring Harbor, New York
SEYDOUX, GERALDINE, Howard Hughes Medical Institute/Johns Hopkins University School of Medicine, Baltimore, Maryland
SHANG, RENFU, Memorial Sloan Kettering Cancer Center, New York, New York
SHARP, PHILLIP, Massachusetts Institute of Technology, Cambridge, Massachusetts
SHEN, WEIPING, National Institutes of Health, Baltimore, Maryland
SHEN, XIULONG, UT Southwestern Medical Center, Dallas, Texas
SHETTY, SUNIL, University of Basel, Basel, Switzerland
SHI, HAILING, The University of Chicago, Chicago, Illinois
SHI, XIANLE, Icahn School of Medicine at Mount Sinai, New York, New York
SHORE, DAVID, University of Geneva, Geneva, Switzerland
SINGER, ROBERT, Albert Einstein College of Medicine, Bronx, New York
SINGH, JASLEEN, Indiana University/Howard Hughes Medical Institute, Bloomington, Indiana
SKOURTI-STATHAKI, KONSTANTINA, University of Edinburgh, Edinburgh, United Kingdom
SO, BYUNG RAN, University of Pennsylvania, Philadelphia, Pennsylvania
SONG, YANGYANG, National University of Singapore, Singapore, Singapore
SPARMANN, ANKE, Springer Nature, Berlin, Germany
SPECTOR, DAVID, Cold Spring Harbor Laboratory, Cold Spring Harbor, New York
STEIN, CHAD, Harvard Medical School, Boston, Massachusetts
STEINBERGER, JUTTA, McGill University, Montreal, Quebec, Canada
STEITZ, JOAN, Yale University/Howard Hughes Medical Institute, New Haven, Connecticut
STEWART, DAVID, Cold Spring Harbor Laboratory, Cold Spring Harbor, New York
STILLMAN, BRUCE, Cold Spring Harbor Laboratory, Cold Spring Harbor, New York
SU, SHUAIKUN, National Institute on Aging, Baltimore, Maryland
SUN, QINYU, University of Illinois at Urbana-Champaign, Urbana, Illinois
SUN, SHUYING, Johns Hopkins University School of Medicine, Baltimore, Maryland
SUN, TAO, Stanford University, Stanford, California
SUN, YU, Yale University, New Haven, Connecticut
SUSSMAN, HILLARY, *Genome Research*, Cold Spring Harbor, New York
SUZAWA, MASATAKA, The University of Tokyo, Tokyo, Japan
SVEJSTRUP, JESPER, Francis Crick Institute, London, United Kingdom
SWIFT, MICHAEL, Stanford University, Stanford, California
SWIGUT, TOMASZ, Stanford University, Stanford, California
SYED, FAROOQ, Indiana University School of Medicine, Indianapolis, Indiana
SZEWCZAK, LARA, *Cell*, Cambridge, Massachusetts
TAN, MENG HOW, Agency for Science Technology and Research, Singapore, Singapore
TANG, SZE JING, National University of Singapore, Singapore, Singapore
TAY, YVONNE, National University of Singapore, Singapore, Singapore
TELESNITSKY, ALICE, University of Michigan, Ann Arbor, Michigan
TENG, CHONG, Donald Danforth Plant Science Center, Saint Louis, Missouri
THOMAS, JAMES, Fred Hutchinson Cancer Research Center, Seattle, Washington
TILGNER, HAGEN, Weill Cornell Medicine, New York, New York
TJIAN, ROBERT, University of California, Berkeley, Berkeley, California
TO, JENN, Bayer Crop Science, Chesterfield, Missouri
TORRES-PADILLA, MARIA-ELENA, Institute of Epigenetics & Stem Cells, Münich, Germany
TSAI, MIAO-CHIH, Cell Press, Cambridge, Massachusetts
ULITSKY, IGOR, Weizmann Institute, Rehovot, Israel
UNGERLEIDER, NATHAN, Tulane University School of Medicine, New Orleans, Louisiana
VALLE-GARCIA, DAVID, Harvard Medical School, Boston, Massachusetts
VAN KEUREN-JENSEN, KENDALL, TGen, Tempe, Arizona
VIDIGAL, JOANA, National Institutes of Health/National Cancer Institute, Bethesda, Maryland
VIRTANEN, ANDERS, Uppsala University, Uppsala, Sweden
WALTON, ELEANOR, Abcam, Cambridge, United Kingdom
WANG, JIE, Springer Nature, Shanghai, China
WANG, JINGJING, University of Colorado Boulder, Boulder, Colorado
WANG, YANG, Chinese Academy of Sciences, Shanghai, China
WANG, YONGBO, School of Basic Medical Sciences, Fudan University, Shanghai, China
WARE, DOREEN, Cold Spring Harbor Laboratory/U.S. Department of Agriculture/Agricultural Research Service, Cold Spring Harbor, New York
WATERTON, SEAN, Massachusetts Institute of Technology, Cambridge, Massachusetts
WEEKS, KEVIN, University of North Carolina, Chapel Hill, North Carolina
WEI, GUIFENG, University of Oxford, Oxford, United Kingdom
WEI, JIANGBO, The University of Chicago, Chicago, Illinois
WEI, LIAN-HUAN, Yale University, New Haven, Connecticut
WEICK, EVA-MARIA, Memorial Sloan Kettering Institute, New York, New York
WELDON, CARIKA, The University of Oxford, Oxford, United Kingdom
WENG, ZHIPING, University of Massachusetts Medical School, Worcester, Massachusetts
WHIPPLE, AMANDA, Massachusetts Institute of Technology, Cambridge, Massachusetts
WILKINSON, MAX, MRC Laboratory of Molecular Biology, Cambridge, United Kingdom
WILUSZ, JEREMY, University of Pennsylvania, Philadelphia, Pennsylvania
WITHERS, JOHANNA, Yale University, New Haven, Connecticut
WITKOWSKI, JAN, Cold Spring Harbor Laboratory, Cold Spring Harbor, New York
WU, JIAN, Shanghai Jiao Tong University School of Medicine, Shanghai, China
WU, PEI-HSUAN (XUAN), University of Massachusetts Medical School, Worcester, Massachusetts
WU, QIUSHUANG, Stowers Institute for Medical Research, Kansas City, Missouri
WU, XIAOLI, Cold Spring Harbor Laboratory, Cold Spring Harbor, New York
WU, XIAOYUN, Broad Institute, Cambridge, Massachusetts
WULF, MADALEE, New England Biolabs, Ipswich, Massachusetts
WYSOCKA, JOANNA, Stanford University, Stanford, California
XIANG, KEHUI, Whitehead Institute, Cambridge, Massachusetts
XIAO, MEISHENG, University of Pennsylvania Perelman School of Medicine, Philadelphia, Pennsylvania
XIE, MINGYI, University of Florida, Gainesville, Florida
XIONG, FENG, UT Health/Texas Medical Center, Houston, Texas
XU, GUANG, Chinese Academy of Sciences, Shanghai, China

SYMPOSIUM PARTICIPANTS

Yan, Qingqing, Wistar Institute, Philadelphia, Pennsylvania
Yang, Acong, National Cancer Institute, Frederick, Maryland
Yang, Li, CAS-MPG Partner Institute/Computational Biology, Shanghai, China
Yang, Ze, Stanford University, Stanford, California
Yeh, Johannes, Cold Spring Harbor Laboratory, Cold Spring Harbor, New York
Yu, Allen, Cold Spring Harbor Laboratory, Cold Spring Harbor, New York
Yu, Lulu, National Cancer Institute, Frederick, Maryland
Zali, Narges, Cold Spring Harbor Laboratory, Cold Spring Harbor, New York
Zhan, Huichun, Stony Brook University, Stony Brook, New York
Zhan, Xiechao, Tsinghua University, Beijing, China
Zhang, Bin, National University of Singapore, Singapore, Singapore
Zhang, Feng, The Broad Institute of MIT and Harvard, Cambridge, Massachusetts
Zhang, Xiaoou, UMass Medical School, Worcester, Massachusetts
Zhao, Yixin, Cold Spring Harbor Laboratory, New York, New York
Zheng, Zhi-ming (Thomas), National Cancer Institute, Frederick, Maryland
Zhou, Chan, Mass General Hospital, Harvard Medical School, Boston, Massachusetts
Zhou, Xiaoming, UT Southwestern Medical Center, Dallas, Texas
Zhu, Pan, John Innes Centre, Norwich, United Kingdom
Zinad, Hany, Newcastle University, Newcastle Upon Tyne, United Kingdom
Zlotorynski, Eytan, *Nature Reviews Molecular Cell Biology*, London, United Kingdom

Row 1: J. Mendell, C. Mayr, J. Manley, L. Maquat, J. Steitz, R. Parker; X. Dong, E. Hallacli
Row 2: S. Jaffrey; A. Aguilera, M. Montez; B. Jalloh, R. Lehmann
Row 3: R. Macrae, Z. Lippman; M. Prajapat, L. Sala
Row 4: S. McKnight, J. Steitz; M. Ljungman, K. Abdelmohsen

Row 1: M. De Zoysa, C. Schaening Burgos; A. Mangilet, P. Sharp
Row 2: R. Darnell, J. Darnell, R. Singer; A. Krainer
Row 3: C. Dean, C. Pikaard; A. Kornblihtt, J. Svejstrop; E-M. Weick
Row 4: G. Narlikar, J. Wysocka; D. Stewart, M.A. Jensen

Row 1: J. Manley, N. Proudfoot; K. Fejes Toth, J. Brennecke
Row 2: T. Bradley, S.S. Das; S. Mao, M-E. Torres-Padilla; J. Liu, X. Dou
Row 3: E. Walton, A. Brown; S. Grewal, N. Proudfoot
Row 4: C. Arraiano, L. Maquat, N. Brockdorff; X-D. Fu, K. Lynch

Row 1: A. Krainer, R. Bradley; P. Brodersen, E. Lai; X. Wu, Q. Hao
Row 2: J. Lee, K. Sarma; M. Montez, J. Thomas, V. Mishra
Row 3: G. Dreyfuss, J. Wilusz; Symposium Posters
Row 4: L. Chen, X-D. Fu; D. Bartel, Z-M. Zheng

Row 1: A. Akhtar, C. Mayr; V. Lakshmanan, S. Gopinath
Row 2: T. Nilsen, D. Rekosh; R. Singer, J. Doudna; Z-M. Zheng, D. Rekosh
Row 3: D. Shore, S. Gasser, B. Stillman, R. Parker; C. Weldon, L. Szewczak
Row 4: L. Szewczak, J. Doudna, K. Adelman; M. Piwecka, M. Xiao

Row 1: M. Mangone, A. Denli; J. Witkowski, R. Darnell
Row 2: C. Meyer, S. Boehm; F. Zhang; M. Hafner, C. Weldon
Row 3: D. Spector, M. Gandhi; A. Virtanen, A. Stenlund
Row 4: A. Gladfelter, K. Weeks; C. Arraiano, B. Jalloh

Row 1: A. Bindereif, D. Frendewey; R. Lehmann, G. Seydoux
Row 2: G. Narlikar, K. Neugebauer; E. Greer, O. Rando
Row 3: Beach Picnic
Row 4: S. McKnight, S. Gasser; B. Bass, E. Conti

Foreword

The Cold Spring Harbor Symposia on Quantitative Biology series is now in its 84th year, having been initiated by then-Director Reginald Harris back in 1933 when the Symposium lasted a full month! The Cold Spring Harbor Symposia bring together scientists to present and evaluate new data and ideas in rapidly moving areas of biological research. Each year, a topic is chosen at a stage at which general and intensive scrutiny and review are warranted. Many previous Cold Spring Harbor Symposia have addressed different aspects of RNA biology, including Nucleic Acids and Nucleoproteins in 1947, Transcription of Genetic Material in 1970, Mechanisms of Transcription in 1998, The Ribosome in 2001, and Regulatory RNAs in 2006. The enormous progress in the field in the past 15 years led us to conclude that the time was past due for another Symposium squarely focused on RNA. In fact, this is only the second Symposium in its entire history to include RNA in the title of the meeting, which we think conveys how central this molecule is in so many areas of the biology of the cell and increasingly to how we think about treating disease. Topics addressed at the 2019 Symposium included RNA-based structures; RNA modifications; nuclear localization of RNA; quality control and editing; RNA and gene regulation; cotranscriptional splicing; intron–exon boundaries; alternative polyadenylation; transposon control; small noncoding RNAs; long noncoding RNAs; RNA and development; membraneless organelles; phase separation; RNA-based diseases; and novel RNA functions.

The Symposium attracted more than 400 participants and provided an extraordinary five-day synthesis of current understanding in the field. Opening night talks setting the scene for later sessions included Roy Parker (HHMI/University of Colorado Boulder) on RNP granules in health and disease, Christine Mayr (Memorial Sloan Kettering Cancer Center) on the regulation of 3′ UTR–mediated protein–protein interactions, Joshua Mendell (UT Southwestern Medical Center) on the regulation and function of noncoding RNAs in mammalian physiology and disease, and Adrian Krainer (Cold Spring Harbor Laboratory), who addressed targeted modulation of splicing or nonsense-mediated mRNA decay (NMD) for disease therapy. Jennifer Doudna (HHMI/University of California Berkeley) delivered a fascinating Dorcas Cummings lecture on "Editing the Code of Life" for the Laboratory's friends and neighbors. Rising to the challenging task of condensing more than 50 talks over the prior five days, Gideon Dreyfuss (HHMI/University of Pennsylvania School of Medicine) provided a masterly summary of the state of the field at the conclusion of the Symposium. Interviews by participating editors, including Steve Mao, Richard Sever, Anke Sparmann, Lara Szewczak, Carika Weldon, and Jan Witkowski, were conducted throughout the Symposium to provide a snapshot of the state of current research and are available on the CSHL Leading Strand channel (https://www.youtube.com/user/LeadingStrand). Transcripts of these Symposium conversations are provided here.

We thank Val Pakaluk, Mary Smith, Ed Campodonico, and his staff in the Meetings & Courses Program for their assistance in organizing and running the Symposium and John Inglis and his staff at Cold Spring Harbor Laboratory Press, particularly Inez Sialiano, Maria Smit, Kathleen Bubbeo, and Denise Weiss, for publishing the printed and online versions of the Symposium proceedings. Photographer Connie Brukin captured candid snapshots throughout the meeting.

Symposium Organizers
Terri Grodzicker, David Stewart, and Bruce Stillman
Cold Spring Harbor Laboratory

Symposium Editors
David Stewart and Bruce Stillman
Cold Spring Harbor Laboratory

Sponsors

Financial support from the corporate sponsors of our meetings program is essential for these Symposia to remain a success and we are most grateful for their continued support:

Corporate Benefactors
Estée Lauder Companies
Regeneron
Thermo Fisher Scientific

Corporate Sponsors
Agilent Technologies
Bayer
Bristol-Myers Squibb Company
Calico Labs
Celgene
Genentech, Inc.
Merck
New England BioLabs
Pfizer

Corporate Partners
Alexandria Real Estate
Enzo Life Sciences
Gilead Sciences
Lundbeck
Novartis Institutes for Biomedical Research
Sanofi

Contents

Symposium Participants — vii
Foreword — xix

Global Analyses and Structures

Zipcode Binding Protein 1 (ZBP1; IGF2BP1): A Model for Sequence-Specific RNA Regulation   *Jeetayu Biswas, Leti Nunez, Sulagna Das, Young J. Yoon, Carolina Eliscovich, and Robert H. Singer* — 1

Pre-mRNA Splicing in the Nuclear Landscape   *Tucker J. Carrocci and Karla M. Neugebauer* — 11

Recognition of Poly(A) RNA through Its Intrinsic Helical Structure   *Terence T.L. Tang and Lori A. Passmore* — 21

Subcellular Spatial Transcriptomes: Emerging Frontier for Understanding Gene Regulation   *Furqan M. Fazal and Howard Y. Chang* — 31

Gene Regulation

Transcriptional Coactivator PGC-1α Binding to Newly Synthesized RNA via CBP80: A Nexus for Co- and Posttranscriptional Gene Regulation   *Xavier Rambout, Hana Cho, and Lynne E. Maquat* — 47

Mechanistic Dissection of RNA-Binding Proteins in Regulated Gene Expression at Chromatin Levels   *Jia-Yu Chen, Do-Hwan Lim, and Xiang-Dong Fu* — 55

Linking RNA Processing and Function   *Run-Wen Yao, Chu-Xiao Liu, and Ling-Ling Chen* — 67

Attenuation of Eukaryotic Protein-Coding Gene Expression via Premature Transcription Termination   *Deirdre C. Tatomer and Jeremy E. Wilusz* — 83

Regulation of RNA Functions

3′ UTRs Regulate Protein Functions by Providing a Nurturing Niche during Protein Synthesis   *Christine Mayr* — 95

The THO Complex as a Paradigm for the Prevention of Cotranscriptional R-Loops   *Rosa Luna, Ana G. Rondón, Carmen Pérez-Calero, Irene Salas-Armenteros, and Andrés Aguilera* — 105

U1 snRNP Telescripting Roles in Transcription and Its Mechanism   *Chao Di, Byung Ran So, Zhiqiang Cai, Chie Arai, Jingqi Duan, and Gideon Dreyfuss* — 115

Functional and Mechanistic Interplay of Host and Viral Alternative Splicing Regulation during Influenza Infection   *Matthew G. Thompson and Kristen W. Lynch* — 123

Chromatin and RNA

Small RNA Function in Plants: From Chromatin to the Next Generation   *Jean-Sébastien Parent, Filipe Borges, Atsushi Shimada, and Robert A. Martienssen* — 133

A Nuclear RNA Degradation Pathway Helps Silence Polycomb/H3K27me3-Marked Loci in *Caenorhabditis elegans*   *Anna Mattout, Dimos Gaidatzis, Véronique Kalck, and Susan M. Gasser* — 141

To Process or to Decay: A Mechanistic View of the Nuclear RNA Exosome   *Mahesh Lingaraju, Jan M. Schuller, Sebastian Falk, Piotr Gerlach, Fabien Bonneau, Jérôme Basquin, Christian Benda, and Elena Conti* — 155

Long Noncoding RNAs in Development and Regeneration of the Neural Lineage   *Hadas Hezroni, Rotem Ben Tov Perry, and Igor Ulitsky* — 165

Small RNAs and Defense Systems

How Complementary Targets Expose the microRNA 3′ End for Tailing and Trimming during Target-Directed microRNA Degradation   *Paulina Pawlica, Jessica Sheu-Gruttadauria, Ian J. MacRae, and Joan A. Steitz* ... 179

Dicer's Helicase Domain: A Meeting Place for Regulatory Proteins   *Sarah R. Hansen, Adedeji M. Aderounmu, Helen M. Donelick, and Brenda L. Bass* ... 185

Reconstitution of siRNA Biogenesis In Vitro: Novel Reaction Mechanisms and RNA Channeling in the RNA-Directed DNA Methylation Pathway   *Jasleen Singh and Craig S. Pikaard* ... 195

Membraneless Bodies and Phase Separation

RNP Granule Formation: Lessons from P-Bodies and Stress Granules   *Giulia Ada Corbet and Roy Parker* ... 203

Biophysical Properties of HP1-Mediated Heterochromatin   *Serena Sanulli, John D. Gross, and Geeta J. Narlikar* ... 217

Architectural RNAs for Membraneless Nuclear Body Formation   *Tomohiro Yamazaki, Shinichi Nakagawa, and Tetsuro Hirose* ... 227

Summary

Myriad RNAs and RNA-Binding Proteins Control Cell Functions, Explain Diseases, and Guide New Therapies   *Byung Ran So and Gideon Dreyfuss* ... 239

Dorcas Cummings Lecture ... 243

Dorcas Cummings Lecture   *Jennifer Doudna* ... 245

Conversations at the Symposium ... 251

Karen Adelman ... 253
Andrés Aguilera ... 256
David Bartel ... 259
Ling-Ling Chen ... 262
Caroline Dean ... 264
Susan Gasser ... 266
Samie Jaffrey ... 268
Leemor Joshua-Tor ... 271
Alberto Kornblihtt ... 274
Adrian Krainer ... 276
Lynne Maquat ... 279
Karla Neugebauer ... 282
Nicholas Proudfoot ... 285
Oliver Rando ... 288
Phillip Sharp ... 291
Maria-Elena Torres-Padilla ... 294
Igor Ulitsky ... 296
Jeremy Wilusz ... 299
Feng Zhang ... 302

Author Index ... 305

Subject Index ... 307

Zipcode Binding Protein 1 (ZBP1; IGF2BP1): A Model for Sequence-Specific RNA Regulation

Jeetayu Biswas,[1] Leti Nunez,[1,5] Sulagna Das,[1,5] Young J. Yoon,[1,2] Carolina Eliscovich,[1,3] and Robert H. Singer[1,2,4]

[1]Department of Anatomy and Structural Biology, [2]Department of Neuroscience, [3]Department of Medicine, Albert Einstein College of Medicine, Bronx, New York 10461, USA
[4]Howard Hughes Medical Institute, Janelia Research Campus, Ashburn, Virginia 20147, USA
Correspondence: robert.singer@einsteinmed.org

The fate of an RNA, from its localization, translation, and ultimate decay, is dictated by interactions with RNA binding proteins (RBPs). β-actin mRNA has functioned as the classic example of RNA localization in eukaryotic cells. Studies of β-actin mRNA over the past three decades have allowed understanding of how RBPs, such as ZBP1 (IGF2BP1), can control both RNA localization and translational status. Here, we summarize studies of β-actin mRNA and focus on how ZBP1 serves as a model for understanding interactions between RNA and their binding protein(s). Central to the study of RNA and RBPs were technological developments that occurred along the way. We conclude with a future outlook highlighting new technologies that may be used to address still unanswered questions about RBP-mediated regulation of mRNA during its life cycle, within the cell.

Changes in protein synthesis define all aspects of a cell's life. Although most cells carry identical copy numbers of their DNA, differences in transcription and translation define both cellular fate and moment to moment cellular decisions. Therefore, the regulation of RNA is central to the control of protein synthesis (for review, see Vera et al. 2016). The timing and localization of protein synthesis is highly regulated at the level of RNA by RNA binding proteins (RBPs).

In eukaryotes, RBPs shuttle between the nucleus and cytoplasm participating in all aspects of gene expression. RBPs can direct the processes of splicing, polyadenylation, export, localization, translation, and decay (for review, see Glisovic et al. 2008). RBPs allow for rapid spatiotemporal regulation of transcripts. Although long-term changes in steady state RNA abundance can occur during differentiation, there are several instances when more rapid modulation of transcripts is necessary. By storing transcripts in a translationally inactive state until stimulated, cellular RBPs can rapidly initiate localized translation.

RNA localization patterns appear across organisms and cell types (for reviews, see Martin and Ephrussi 2009; Buxbaum et al. 2015). Asymmetric distribution of poly(A) RNAs was first described in ascidian eggs (Jeffery et al. 1983). Shortly thereafter, specific RNAs were observed to be localized to the animal and vegetal pole in maternal *Xenopus* eggs (Rebagliati et al. 1985). In chicken embryonic myoblasts both tritiated and biotinylated probes were used to observe cytoskeletal mRNA localization (Singer and Ward 1982; Lawrence and Singer 1986). Of the three RNAs tested (actin, vimentin, and tubulin), actin had the most distinct localization pattern. Approximately 95% of myoblasts had a nonrandom distribution of β-actin mRNA, in which RNAs were concentrated at the cell extremities.

Key to the observation of RNA localization was the development and optimization of RNA fluorescence in situ hybridization (RNA FISH) (Singer and Ward 1982; Lawrence and Singer 1985). Improvements to both the detectors and fluorophores allowed for visualization and tracking of single RNA molecules within cells (Bertrand et al. 1998; Femino et al. 2003). Building on the advent of single-molecule RNA FISH, advances in microfluidics have allowed for multiplexed reprobing of samples and transcriptome-wide studies of RNA localization in both cells and tissues (Levsky et al. 2002; Lécuyer et al. 2007; Eng et al. 2019). By increasing the throughput of approaches to study RNA localization, it has now been possible to appreciate the breadth of RNAs that undergo this process. For example, RNA localization was found to be prevalent in the developing *Drosophila* embryo (Lécuyer et al. 2007). RNA FISH against 3370 genes determined that >70% of the RNAs tested showed different patterns of localization, not only emphasizing the prevalence of RNA localization patterns but also the different mechanisms by which localization patterns are encoded. In neurons, more than one-half of the transcripts queried by RNA sequencing were localized in the neurite and nearly one-half of the localized mRNAs led to neurite localized proteins (Zappulo et al. 2017).

---

[5]These authors contributed equally to this work.

© 2019 Biswas et al. This article is distributed under the terms of the Creative Commons Attribution-NonCommercial License, which permits reuse and redistribution, except for commercial purposes, provided that the original author and source are credited.

Profiling of RNAs from different subcellular compartments shows that RNA localization confers function on the subcellular scale. Within the cytoplasm, localization and local translation of mRNAs to the mitochondria and endoplasmic reticulum (ER) ensures correct expression of mitochondrial and secretory proteins (Garcia et al. 2010; Gadir et al. 2011; Jan et al. 2014; Williams et al. 2014). In the migrating cell, β-actin mRNA localization to the leading edge allows for directional movement in response to guidance cues (Lawrence and Singer 1986; Shestakova et al. 2001; Katz et al. 2012).

RNA transport and localization provide the means for gene expression regulation, in which local sites can determine what proteins are made in response to spatiotemporal cues. In highly polarized neurons, there are clear advantages to local control of gene expression. As dendrites and axons can extend hundreds of microns away from the cell body, the process of RNA localization and local protein synthesis is particularly relevant to neuronal function (Zappulo et al. 2017). Synapses that mediate transmission of information from one neuron to another have the capacity to undergo long-term cytoskeletal remodeling, growing or shrinking the size of the dendritic spine in response to firing (Hotulainen and Hoogenraad 2010). This synaptic modulation requires newly translated proteins and underlies higher cognitive functions such as learning and memory (for review, see Costa-Mattioli et al. 2009).

In the developing axon growth cone, RNA localization has been shown to be involved in growth cone pathfinding, allowing the axon to twist and turn as it navigates chemical gradients and forms immature synapses (Eom et al. 2003). In the mature neuron, dynamic RNA localization occurs in response to neural stimulation of specific dendrites, with RNAs localizing to their base and new protein synthesis occurring in the specifically stimulated spines (Yoon et al. 2016). Therefore, not only does the timing of translation play a key role when a dendrite is stimulated but also the proper RNA location at a specific synapse is critical.

MECHANISMS OF RNA LOCALIZATION

It is now appreciated that RNA localization, facilitated by RBPs, is a highly conserved mechanism to spatially confine protein synthesis, amplify local protein concentration, or even direct integration into macromolecular complexes (for reviews, see Glisovic et al. 2008; Buxbaum et al. 2015). Compartment-specific targeting of mRNA involves recognition of a short nucleotide sequence known as a *cis*-acting localization element by an RBP called the *trans*-acting factor. Many well-characterized RBPs bind sequence specifically, whereas others have been shown to recognize specific stem–loop structure or nonspecifically bind to single- or double-stranded RNA. The direct association between RBPs and *cis*-acting localization elements leads to the formation of a ribonucleoprotein (RNP) complex that travels along the cytoskeleton with the help of motor proteins (for review, see Eliscovich et al. 2013). Identification of canonical *trans*-acting factors have been based on (i) their ability to interact specifically with their target mRNAs, and (ii) where loss of function or loss of expression results in mislocalization of the mRNA. Disruptions in the formation of transport RNPs can have significant consequences, especially in neurons in which localization is necessary for the synthesis of proteins and consequently synaptic plasticity.

Several models have been developed to understand how RNAs can become localized within a cell (for review, see Buxbaum et al. 2015). RNA molecules can passively diffuse until they reach an anchoring point (diffusion and entrapment). Different examples of anchors include cytoskeletal proteins or RBPs (Beach et al. 1999; Farina et al. 2003). RNAs can also be locally protected from degradation or conversely selectively degraded, leading to the accumulation of RNA in areas with lower decay rates (Zaessinger et al. 2006; Tadros et al. 2007). Additionally, RNAs can be localized by motor proteins, often through their interaction with an RBP that serves as an adapter (Long et al. 2000; Song et al. 2015). Although each mechanism of localization can act alone, they can also be combined to perform biological functions. By combining directed transport with local entrapment, it has been shown that β-actin mRNA can rapidly localize to a stimulated dendritic spine where new proteins are then synthesized (Yoon et al. 2016). Key to both simple and complex modes of RNA localization are RBPs.

ZBP1: THE RBP THAT CONTROLS β-ACTIN mRNA LOCALIZATION

Since the initial observation of polarized RNA within cells, β-actin mRNA localization has provided a model system for understanding the mechanisms of RNA localization within eukaryotic cells (Fig. 1; for reviews, see Eliscovich et al. 2013; Buxbaum et al. 2015; Eliscovich and Singer 2017; Das et al. 2019). Biochemical methodologies, structural analysis, and imaging-based approaches have analyzed how a *cis*-acting element and *trans*-acting factors act together to ensure the cytoplasmic fate of the mRNA once it is transcribed in the nucleus.

Cytoplasmic β-actin mRNA localization was initially observed using in situ hybridization in chicken embryonic skeletal myoblasts and fibroblasts (Lawrence and Singer 1986). Later reporter plasmids expressing different regions of the 3′ UTR of β-actin mRNA demonstrated that a 54-nt *cis*-acting "zipcode" element was responsible for the localization of β-actin mRNA to the cellular periphery (Kislauskis et al. 1994). This 54-nt zipcode sequence is conserved from chicken to mouse and human β-actin 3′ UTRs. Zipcode Binding Protein 1 (ZBP; also called IGF2BP1), the key RBP that binds the 54-nt zipcode in the nucleus and regulates localization and translational repression of β-actin mRNA in the cytoplasm (Farina et al. 2003; Oleynikov and Singer 2003; Hüttelmaier et al. 2005; Pan et al. 2007; Wu et al. 2015), was identified by UV cross-linking and affinity purification (Ross et al. 1997). Subsequently, a number of RBPs homologous to ZBP1 were also discovered in human, mouse, fly, and frog (Yisraeli 2005). A timeline highlighting ZBP's role as a critical

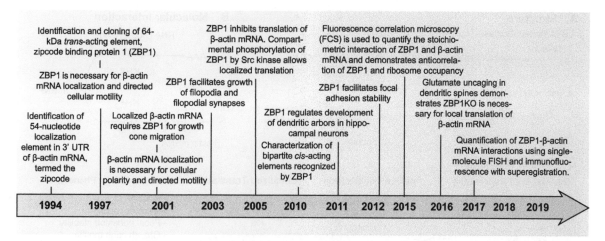

Figure 1. Timeline of ZBP1 and β-actin mRNA discoveries. Over the past two decades, significant work has revealed that the zipcode binding protein-1 (ZBP1) is involved in localization of mRNA. Presented here is a summary of important findings that have led to the current understanding of how ZBP1 acts to regulate β-actin mRNA.

RBP for β-actin mRNA localization in different cell types has been illustrated in Figure 1.

RNA reporter assays (Kislauskis et al. 1994) and biochemical (Farina et al. 2003) and structural characterization of the 54-nt zipcode led to the identification of a minimal 28-nt consensus bipartite element that is specifically recognized by ZBP1 (Chao et al. 2010; Patel et al. 2012). The structural studies also showed that ZBP1KH34 (third and fourth hnRNP K-homology domains) specifically binds the bipartite β-actin 3′-UTR element, with ZBP1KH4 and ZBP1KH3 recognizing β-actin 5′-CGGAC-3′ and 5′-(C/A)CA(C/U)-3′ sequences, respectively (Chao et al. 2010; Patel et al. 2012). For each RNA consensus sequence to bind to the two protein binding sites on opposite ends of ZBP1KH34, the RNA must loop around the protein. The distortion sequesters the stop codon of β-actin mRNA and likely contributes to translational repression (Figs. 2 and 3A–D; Chao et al. 2010; Wu et al. 2015).

ZBP1 CONTROLS LOCAL β-ACTIN TRANSLATION

Both the zipcode sequence within the 3′ UTR of β-actin and the KH34 domains of ZBP1 are necessary for the formation of the ZBP1-β-actin mRNA complex (Fig. 2A,B). Regulation of β-actin mRNA fate is dependent on these associations.

In the cytoplasm, formation of the ZBP1-β-actin mRNA complex sterically inhibits the large ribosomal subunit from binding the β-actin mRNA, thereby preventing translation (Hüttelmaier et al. 2005; Wu et al. 2015). Motorized movement of this complex along cytoskeletal filaments leads to peripheral localization (e.g., leading edge in fibroblasts, dendrites or axonal cone in neurons) and accounts for the nonrandom distribution of β-actin mRNA observed in migrating or stimulated cells (Bassell et al. 1998; Mukherjee et al. 2019). Src kinase–dependent phosphorylation of Tyr-396 in ZBP1 leads to disassembly of the ZBP1-β-actin mRNA complex at the periphery. Subsequent release of β-actin mRNA permits binding of ribosomal subunits and leads to localized translation (Fig. 2C; Hüttelmaier et al. 2005). Thus, ZBP1 can both localize and control the translation of β-actin mRNA, suggesting that RBPs can perform complex roles within the cells.

A similar pattern of events occurs in yeast when *ASH1* mRNA localizes to the bud tip in a She2-dependent manner (Long et al. 2000). Motor-dependent localization along cytoskeletal filaments and phosphorylation-dependent translation initiation in both yeast and mammalian cells suggests there exist common mechanisms by which RBPs can localize RNAs and regulate their translation.

Although ZBP1's regulation of β-actin mRNA is well characterized, it is likely that multiple RBPs bind to and regulate a given RNA during its entire life cycle. ZBP1 binding to the β-actin 3′ UTR is in coordination with other proteins, such as ZBP2 (human hnRNP protein K-homology splicing regulator protein, KHSRP). Binding of ZBP2/KHSRP in the nucleus facilitates nuclear ZBP1 association and further cytoplasmic localization in fibroblasts (Gu et al. 2002; Farina et al. 2003). Therefore, the binding of an RBP to its target can be facilitated or hindered by cofactors. Perturbations of mRNA–RBP interactions leads to functional consequences in multiple cell types. For example, in fibroblasts, the absence of the mRNA zipcode or ZBP1 results in loss of polarity, focal adhesions, and random mRNA movement (Shestakova et al. 2001; Katz et al. 2012). Similarly in neurons, loss of either the zipcode or ZBP1 affects synaptic formation and dendritic branching (Perycz et al. 2011). The neuronal observations are consistent with β-actin protein being an important cytoskeletal component involved in growth cone migration and facilitating synapse formation. Loss of the zipcode or ZBP1 in forebrain neurons is associated with decreased growth cone migration (Zhang et al. 2001; Welshhans and Bassell 2011). Hippocampal neurons, in

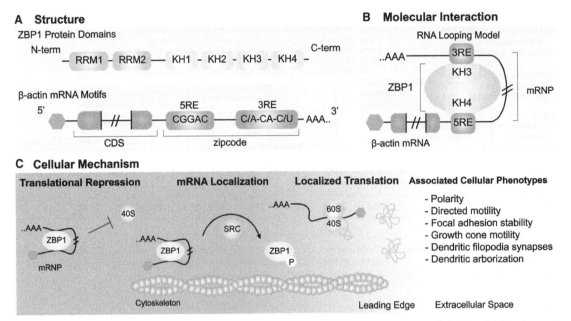

Figure 2. Summary of the molecular interaction of ZBP1 with β-actin mRNA, which determines cellular fate. (*A*) The interaction of ZBP1 with cognate mRNA; β-actin is dependent on structural motifs. ZBP1 contains two RNA recognition motifs (RRM) (purple) and four hnRNP-K homology (KH) (green) domains. β-actin mRNA contains a bipartite localization element in the 3′ UTR, named the zipcode (red). (*B*) Interaction between the zipcode and ZBP1 has been proposed to occur by RNA looping, thereby forming the ribonucleoprotein complex (mRNP). (*C*) Formation of the mRNP complex has molecular and cellular consequences. The ZBP1-β-actin mRNP prevents ribosomal subunits (e.g., 40S) from binding β-actin mRNA, thus resulting in translational repression. Active transport of the ZBP1-β-actin mRNP along cytoskeletal filaments (motors not depicted) allows localization of β-actin mRNA. Disassembly of the ZBP1-β-actin mRNP is facilitated by Src-dependent phosphorylation of ZBP1 at the leading edge. Release of β-actin mRNA ultimately allows for ribosomal subunit assembly and local translation to occur. Localized translation of β-actin mRNA at the leading edge is associated with cellular phenotypes including polarity, directed motility, and focal adhesion stability in fibroblasts and growth cone motility, dendritic filopodia synapses, and dendritic arborization in neurons.

contrast, have notable changes to dendritic arborization, with a severe reduction in branching and filopodia formation (Fig. 3I; Eom et al. 2003; Perycz et al. 2011). Taken together, these studies demonstrate the importance of these conserved and coordinated molecular events underlying localized translation. However, the role of ZBP1 extends beyond translation by functioning as an adapter between the RNA and cytoskeletal motors and, thus, trafficking RNAs to their final destinations (Song et al. 2015).

RNA TRANSPORT IN DENDRITES

The dendritic arbor represents a complex maze for mRNA transport and trafficking. Real-time imaging and tracking of mRNAs allows understanding of the basic rules for transport behavior. Endogenous and reporter mRNAs containing zipcodes in their 3′ UTRs exhibit similar velocities during directed motion, consistent with the *cis*-acting element(s) and the *trans*-acting factor(s) being the critical determinants of RNP transport. With neuronal stimulation, the β-actin mRNP can become unmasked and release a translatable pool of β-actin mRNA in distal dendrites (Buxbaum et al. 2014). To determine whether mRNP localization and unmasking could be controlled, glutamate uncaging delivered neurotransmitter to a subset of dendritic spines. Upon localized stimulation, endogenous β-actin mRNAs localized to the stimulated synapses, resulting in an accumulation of mRNAs over 2 hours. This local capture of β-actin mRNPs was blocked in neurons from the ZBP1 knockout mice, suggesting the role of ZBP1 as the targeting protein (Fig. 3H; Yoon et al. 2016).

When the mRNP particles are in motion, they move processively (0.5–2.0 μm/sec), move in a series of short distances (few microns) intervened by short pauses (<10 sec), or remain corralled (diffusion within a small volume of space). The directed movement is indicative of motor-driven transport along microtubule tracks, with instantaneous velocities ranging from 0.5 to 5 μm/sec (Yoon et al. 2016; Das et al. 2018). Dendritic mRNPs can move in either direction or switch directions—depending on the combined force of the bound motors and the orientation of the microtubules. Collectively, all dendritic mRNAs exhibit bidirectional motion, but with a slight bias toward the anterograde, which allows them to be delivered to the distal dendrite to participate in local translation when needed.

mRNA localization in dendrites supports a "local entrapment" model as a general mechanism of how local activity can capture mRNAs that are cruising along the dendrites. This represents a "sushi belt" where β-actin mRNA(s) patrol through multiple synapses like a circling conveyor belt until they are captured by the recently activated synapses and anchored to the base of the spine (Doyle and Kiebler 2011). To determine the fate of the

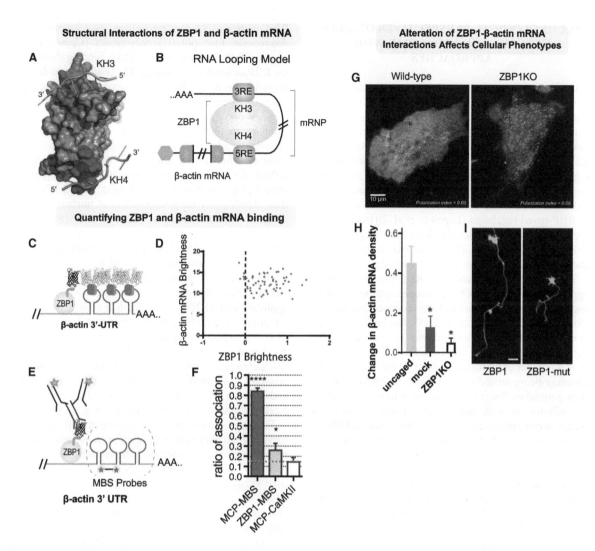

Figure 3. Exploring RBP–RNA interactions by studying ZBP1-β-actin mRNA. Here, we highlight studies that have contributed to our understanding of ZBP1-β-actin mRNA structural interactions (*A,B*), stoichiometry of binding (*C–F*), and how alterations of their interactions lead to respective changes in cellular phenotype (*G–I*). ZBP1 has two main domains (KH34) that are reported to interact with the zipcode element in the 3′ UTR of β-actin (*A*), which is proposed to form an RNA loop when bound by ZBP1 (*B*). The stoichiometric interaction of ZBP1 with β-actin mRNA has revealed a 1:1 binding. Two parallel imaging approaches have been utilized to confirm the stoichiometric associations of the ZBP1-β-actin complex: fluctuation correlation microscopy (FCS) (*C,D*) and fluorescence in situ hybridization and immunofluorescence (FISH-IF) (*E,F*). Using the bacteriophage MS2 system to label β-actin mRNA with stem-loops (MBS), capsid proteins are labeled with fluorescent proteins. Simultaneous expression of ZBP1 with a fluorescent tag (*C*). Using live tracking of single particles in FCS videos, the fluorescent intensity of both particles is used to evaluate stoichiometry (*D*). Alternatively, fixed images from FISH-IF can be analyzed to determine stoichiometric association. Using a similar approach, ZBP1 is expressed with a fluorescent tag. To increase the signal from IF, staining is performed against thee fluorescent tag. FISH probes are used against the stem-loop sequence (MBS) (*E*) and spatial association of signals are used to determine stoichiometry (*F*). Alteration of the ZBP1-β-actin mRNA complex leads to a variety of cellular defects. In fibroblasts, deletion of ZBP1 leads to loss of polarization. In neurons, ZBP1 is important for localization of β-actin mRNA in dendrites. The localization of β-actin mRNA occurs in response to glutamate stimulation (photoactivable uncaging of glutamate). Loss of ZBP1 results in a reduction of β-actin mRNA in response to glutamate release (*H*). Neurons also show alterations in outgrowth in response to mutations of the ZBP1 phosphorylation site (Hüttelmaier et al. 2005). (*A*, Modified, with permission, from Patel et al. 2012, © Cold Spring Harbor Laboratory Press; *D*, modified from Wu et al. 2015, © Elsevier; *F*, modified from Eliscovich et al. 2017; *H*, adapted from Yoon et al. 2016.)

localized mRNP, β-actin translation reporters showed that after an mRNA localizes to an activated spine, it persists there for hours undergoing multiple rounds of stimulation and translation to generate a pool of new β-actin proteins that can be incorporated into the expanding dendritic spine structure. Future studies are required to validate whether this "sushi belt" model of mRNA transport is applicable to other dendritically localized mRNAs along the entire length of the dendrite.

To determine the conservation of ZBP1-dependent local translation for transcripts other than β-actin, transcriptome-wide profiling of ZBP1 interacting transcripts can subsequently be evaluated for physiologically relevant interactions.

TECHNOLOGICAL PERSPECTIVES ON RBPs: GENOME- AND PROTEOME-WIDE APPROACHES

Given the complex nature of RNA–protein interactions, it is now appreciated that few transcripts act with a single RBP, and few RBPs act upon a single transcript. To determine the breadth of interactions development of genome-wide approaches has been essential. Although each technique focuses on one aspect of RNA regulation (binding to a target, downstream effects on stability, translation regulation), the integration of multiple approaches provides the opportunity to uncover the breadth and depth of RBP-based regulation (Lapointe et al. 2018). Here we highlight a few techniques that have been used to study ZBP1 and related RBPs and hypothesize a few possible themes that may result from their integration.

To determine the targets of an RBP, the most notable approaches are RIP (RNA immunoprecipitation) and its successor CLIP (UV cross-linking immunoprecipitation) (for review, see Wheeler et al. 2018). Antibodies are used to isolate the RBP of interest, and next-generation sequencing is used to identify the fragments that are isolated. However, a number of limitations exist with this approach, the foremost being related to the indirect isolation of RBPs with antibodies. Varying degrees of antibody specificity and affinity are used, making the comparison of CLIP studies across groups, cell lines, and experiments challenging. Recently an effort lead by the ENCODE project has characterized commercial antibodies and optimized the CLIP protocol (Van Nostrand et al. 2016). With the optimized protocol, hundreds of RBPs have been studied in HepG2 and K562 cells (Van Nostrand et al. 2016). However, the process of UV cross-linking, antibody-based immunoprecipitation as well as most of the downstream processing steps have known associated biases. Complementary approaches to CLIP have been necessary to validate the putative targets that come from this approach.

One such promising in vitro approach performs highly multiplexed measurements of protein ON and OFF rates with thousands of randomly synthesized RNA targets. Named RNA arrays on high-throughput sequences (RNA-HiTS), this approach utilizes the microscope, microfluidics, and flow cells of DNA sequencers. Thousands of RNA sequences are synthesized during sequencing, and individual cluster locations are annotated and then interrogated by flowing in fluorescent RBPs (for review, see Denny and Greenleaf 2019). By determining the landscape of binding affinities in vitro, rules by which RBPs recognize the sequence and structures of their targets can be interrogated. This allows for refinement of CLIP and other in vivo data, motifs found by RNA-HiTS can be used to identify both false positives and false negative such as putative targets that may not be highly expressed in the cell type used.

Two antibody independent approaches to study RBPs were recently developed, RNA tagging in yeast and TRIBE (targets or RNA binding proteins identified by editing) in *Drosophila*. Both approaches utilize enzymes fused to a RBP of interest to deposit covalent marks on RNA targets. RNA tagging utilizes yeast genetics to endogenously fuse poly(U) polymerase (*Caenorhabditis elegans* PUP-2 which lacks an RNA binding domain) to the RBP of interest. "U"-tailed RNAs are then isolated by poly(A) selection, after which paired-end sequencing libraries are generated. Poly(U) tail length for RNA targets in yeast correlated with both affinity for the RBP and cellular function (Lapointe et al. 2015). TRIBE fuses *Drosophila* adenosine deaminase (ADAR) to an RBP of interest. After FACS sorting of cells expressing the ADAR–RBP fusion protein, standard RNA sequencing library preparation is then performed. Edited RNAs are then identified by the heterogeneity of nucleotides at individual positions along a transcript (McMahon et al. 2016; Rahman et al. 2018; Xu et al. 2018).

Whereas neither RNA tagging nor TRIBE provide the nucleotide level resolution of CLIP, they may select for longer-lived RNA–protein interactions and provide methods complementary to CLIP. The recent adaptation and application of TRIBE in mammalian cells (Biswas et al. 2019b) allowed for the profiling of mammalian RBPs such as ZBP1 and its family members (J Biswas, R Rahman, V Gupta, et al., in prep.). Data comparing mammalian TRIBE to CLIP suggests that TRIBE can avoid the biases and false positives associated with antibody immunoprecipitation and allow for all RBP binding sites to be discovered within a cell (Biswas et al. 2019b; J Biswas, et al., in prep.).

To determine the proteins associated with a specific transcript, biochemical techniques such as complementary biotinylated oligonucleotide-based isolation of RNA followed by mass spectrometry have classically been used (Ross et al. 1997). Recent advances in proximity-based labeling techniques using the biotin ligases have emerged as a powerful complementary approach to map RNA–protein interactions. One of the most studied biotin ligases is BirA, a bifunctional protein expressed in *Escherichia coli*, which mediates biotinylation of a specific lysine residue of a subunit of acetyl-CoA carboxylase. Using a mutated BirA, which causes promiscuous biotinylation, proximity-dependent biotin identification (BioID) was developed, which labels any proximal proteins within a radius of ~10 nm (Kim and Roux 2016). The β-actin mRNA associated proteome was identified by performing BioID upon β-actin mRNA in MEFs (Mukherjee et al. 2019). Besides ZBP1, this approach identified novel regulators of β-actin mRNA localization including FUBP3. Additionally, this approach determined which factors were constitutively loaded upon the RNA and which exchanged during the process of serum starvation or stimulation in which localization to the leading edge is decreased or increased respectively. BioID has been extended to map large-scale RNP interactome by analyzing the biotinylation profile of 119 proteins associated at different stages of the mRNA life cycle (Youn et al. 2018). However, one caveat to BioID is that it requires biotinylation over 16–24 h and therefore captures multiple interactions occurring over a long period of time. For more rapid labeling, TurboID (Branon et al. 2018), ascorbate peroxidase (APEX) (Lam et al. 2015), and RNA–protein interaction detection (RaPID) (Ramanathan et al. 2018) are the techniques of choice, which can capture transient interactions in situ, in different subcellular compartments. An engineered APEX, which oxidizes biotin-phenol, generates short-lived radicals that can covalently react with tyrosine and other electron-rich amino acids as well as the amino group on guanosine (Kim and Roux 2016). The optimization of the enzyme tags and the ability to biotinylate both RNA and proteins have extended the capability of APEX to map RNA–protein interactions in cells with subcellular precision (Kaewsapsak et al. 2017; Fazal et al. 2019). More recently, proximity biotinylation by APEX has been used to profile the RNAs associated with different organelles and stress granules (Fazal et al. 2019; Padrón et al. 2019). Future studies may allow for the profiling of RNA granule components using APEX-based strategies.

IMAGING-BASED APPROACHES TO STUDY RNA–PROTEIN INTERACTIONS IN SITU

The study of an mRNA and its binding protein(s) by ensemble biochemical approaches (Mili and Steitz 2004) lacks spatial information. Given the wealth of RNA–protein interactions now defined by the aforementioned techniques, follow-up experiments are required to determine in situ RNA–protein interactions as well as subcellular RNA localization. Imaging-based approaches therefore can provide spatial information currently inaccessible by other methods, and this is particularly critical for the understanding of RNA localization.

High-precision imaging—namely, "super registration"—is capable of determining whether two molecules are physically interacting in situ or simply in proximity by random chance using standard wide-field microscopy (Fig. 3E,F; Eliscovich et al. 2017). This imaging technique corrects the chromatic aberration uniquely intrinsic to individual commercial microscope objectives so that two molecules can be superimposed within 10 nm precision. By precisely localizing the mRNA and protein using single-molecule FISH to detect mRNA and immunofluorescence (IF) against a RBP, a significant fraction of proteins biochemically defined to bind β-actin mRNA were then shown to not interact with the RNA in situ. Therefore, super registration can complement information from biochemical interactions and visualize physical contacts in situ, with high precision in both neurons and cell lines (Eliscovich et al. 2017; Mukherjee et al. 2019).

Direct RNA–protein interactions within living cells can be interrogated with multicolor fluorescence correlation spectroscopy. By exciting a femtomolar volume within the cell, the association of different molecules can be correlated over time at a specific point in space. This approach can be used alongside super registration microscopy to determine live cell associations of RNA (labeled with an MBS array) with fluorescently tagged RBPs (Fig. 3C,D; Wu et al. 2015). Future developments may combine this approach with super-resolution STED imaging (Lanzanò et al. 2017) to allow for interactions to be measured within a dense environment—for instance, in an individual dendritic spine.

Although single-molecule imaging often interrogates individual genes or proteins, recent advances in highly multiplexed RNA FISH can now localize thousands of RNA species inside the cell. By using iterative RNA FISH and multiplex barcoding of the FISH probes, techniques such as SeqFISH (Eng et al. 2019) and MERFISH (Xia et al. 2019) allow transcriptome-wide localization of RNAs within both cells and tissues. These approaches have found novel transcripts localized to cellular protrusions, expanding our understanding of RNA localization (Eng et al. 2019). Future applications of these techniques may follow the entire breadth of ZBP1 targets (as defined by CLIP or TRIBE) in rapidly changing environments such as the migrating fibroblast, developing growth cone, or stimulated synapse. Although both approaches have been combined with single color IF, recent multicolored fluorescence imaging of barcoded antibodies has allowed for spatial profiling of more than a dozen different proteins (Guo et al. 2019). Combined barcoding of RNA and protein may allow for future colocalization of the two molecules at a massively parallel scale within cells.

Instead of using oligonucleotide barcoding, other approaches have performed direct sequencing of RNAs in situ. Approaches such as in situ transcriptome accessibility sequencing have begun to correlate in situ sequencing truncation events with the location of RBP binding sites. Using this approach, it was discovered that *Drosophila* ZBP1 (dIMP1) binding sites were found within the Act5C mRNA more localized to the periphery of retinal cells (Fürth et al. 2019). As evidenced by the above technologies, further development of high-throughput protein and RNA imaging will allow for RNA–protein interactions to be directly imaged in situ at a massively parallel scale.

CONCLUSIONS—BIOLOGICAL PERSPECTIVES ON RBPs

As evidenced above, work on β-actin-ZBP1 interactions has developed models for how, when, and where RNAs become localized inside a cell. Future technical developments are required to allow the dimensions of RNA tracking to be expanded to include the entire life of an RNA as well as the entirety of its protein interactions. Tracking of RNA fate from transcription to decay is possible with the MS2 system (Tutucci et al. 2018), and high-throughput methods allow a plethora of RBP interactions to be profiled. Missing from the current studies are in vivo dynamics of RNA–protein interactions—how are regulatory factors exchanged during a RNA's lifetime? Genome-wide studies like CLIP show that many RBPs interact with the same transcript, often at the same site. How does this multitude of interactions direct the fate of the transcript? Is there functional redundancy among RBP family members or RBPs from different families (Conway et al. 2016; Biswas et al. 2019a)? How do the targets of an RBP change as cellular identity changes during differentiation or reprogramming? Studies have yet to clearly define how changes in RBP expression or stoichiometry affect the process of finding RNA targets. These open questions will require high-throughput, time-resolved, nondestructive measurements of RNA–protein interactions, ideally within single cells.

ACKNOWLEDGMENTS

We thank members of the Singer lab past and present for their helpful discussions and contributions to the work presented in this perspective. J.B. and L.N. were supported with funding from a Medical Scientist Training Program (MSTP) Training Grant T32GM007288. J.B. was additionally supported by predoctoral fellowship F30CA214009. C.E., Y.J.Y., and R.H.S. were supported by the National Institutes of Health (NIH) grant NS083085. Y.J.Y. was also supported by NIH grant MH120496 and R.H.S by NIH grant U01DA047729.

REFERENCES

Bassell GJ, Zhang H, Byrd AL, Femino AM, Singer RH, Taneja KL, Lifshitz LM, Herman IM, Kosik KS. 1998. Sorting of β-actin mRNA and protein to neurites and growth cones in culture. *J Neurosci* **18:** 251–265. doi:10.1523/JNEUROSCI.18-01-00251.1998

Beach DL, Salmon ED, Bloom K. 1999. Localization and anchoring of mRNA in budding yeast. *Curr Biol* **9:** 569–578. doi:10.1016/S0960-9822(99)80260-7

Bertrand E, Chartrand P, Schaefer M, Shenoy SM, Singer RH, Long RM. 1998. Localization of *ASH1* mRNA particles in living yeast. *Mol Cell* **2:** 437–445. doi:10.1016/S1097-2765(00)80143-4

Biswas J, Patel VL, Bhaskar V, Chao JA, Singer RH, Eliscovich C. 2019a. The structural basis for RNA selectivity by the IMP family of RNA-binding proteins. *Nat Commun* **10:** 4440. doi:10.1038/s41467-019-12193-7

Biswas J, Rahman R, Gupta V, Rosbash M, Singer RH. 2019b. MS2-TRIBE evaluates protein-RNA interactions and nuclear organization of transcription by RNA editing. bioRxiv doi:10.1101/829606

Branon TC, Bosch JA, Sanchez AD, Udeshi ND, Svinkina T, Carr SA, Feldman JL, Perrimon N, Ting AY. 2018. Efficient proximity labeling in living cells and organisms with TurboID. *Nat Biotechnol* **36:** 880–887. doi:10.1038/nbt.4201

Buxbaum AR, Wu B, Singer RH. 2014. Single β-actin mRNA detection in neurons reveals a mechanism for regulating its translatability. *Science* **343:** 419–422. doi:10.1126/science.1242939

Buxbaum AR, Haimovich G, Singer RH. 2015. In the right place at the right time: visualizing and understanding mRNA localization. *Nat Rev Mol Cell Biol* **16:** 95–109. doi:10.1038/nrm3918

Chao JA, Patskovsky Y, Patel V, Levy M, Almo SC, Singer RH. 2010. ZBP1 recognition of β-actin zipcode induces RNA looping. *Genes Dev* **24:** 148–158. doi:10.1101/gad.1862910

Conway AE, Van Nostrand EL, Pratt GA, Aigner S, Wilbert ML, Sundararaman B, Freese P, Lambert NJ, Sathe S, Liang TY, et al. 2016. Enhanced CLIP uncovers IMP protein-RNA targets in human pluripotent stem cells important for cell adhesion and survival. *Cell Rep* **15:** 666–679. doi:10.1016/j.celrep.2016.03.052

Costa-Mattioli M, Sossin WS, Klann E, Sonenberg N. 2009. Translational control of long-lasting synaptic plasticity and memory. *Neuron* **61:** 10–26. doi:10.1016/j.neuron.2008.10.055

Das S, Moon HC, Singer RH, Park HY. 2018. A transgenic mouse for imaging activity-dependent dynamics of endogenous Arc mRNA in live neurons. *Sci Adv* **4:** eaar3448. doi:10.1126/sciadv.aar3448

Das S, Singer RH, Yoon YJ. 2019. The travels of mRNAs in neurons: do they know where they are going? *Curr Opin Neurobiol* **57:** 110–116. doi:10.1016/j.conb.2019.01.016

Denny SK, Greenleaf WJ. 2019. Linking RNA sequence, structure, and function on massively parallel high-throughput sequencers. *Cold Spring Harb Perspect Biol* **11:** a032300. doi:10.1101/cshperspect.a032300

Doyle M, Kiebler MA. 2011. Mechanisms of dendritic mRNA transport and its role in synaptic tagging: mechanisms of dendritic mRNA transport. *EMBO J* **30:** 3540–3552. doi:10.1038/emboj.2011.278

Eliscovich C, Singer RH. 2017. RNP transport in cell biology: the long and winding road. *Curr Opin Cell Biol* **45:** 38–46. doi:10.1016/j.ceb.2017.02.008

Eliscovich C, Buxbaum AR, Katz ZB, Singer RH. 2013. mRNA on the move: the road to its biological destiny. *J Biol Chem* **288:** 20361–20368. doi:10.1074/jbc.R113.452094

Eliscovich C, Shenoy SM, Singer RH. 2017. Imaging mRNA and protein interactions within neurons. *Proc Natl Acad Sci* **114:** E1875–E1884. doi:10.1073/pnas.1621440114

Eng C-HL, Lawson M, Zhu Q, Dries R, Koulena N, Takei Y, Yun J, Cronin C, Karp C, Yuan G-C, et al. 2019. Transcriptome-scale super-resolved imaging in tissues by RNA seqFISH. *Nature* **568:** 235–239. doi:10.1038/s41586-019-1049-y

Eom T, Antar LN, Singer RH, Bassell GJ. 2003. Localization of a β-actin messenger ribonucleoprotein complex with zipcode-binding protein modulates the density of dendritic filopodia and filopodial synapses. *J Neurosci* **23:** 10433–10444. doi:10.1523/JNEUROSCI.23-32-10433.2003

Farina KL, Hüttelmaier S, Musunuru K, Darnell R, Singer RH. 2003. Two ZBP1 KH domains facilitate β-actin mRNA localization, granule formation, and cytoskeletal attachment. *J Cell Biol* **160:** 77–87. doi:10.1083/jcb.200206003

Fazal FM, Han S, Parker KR, Kaewsapsak P, Xu J, Boettiger AN, Chang HY, Ting AY. 2019. Atlas of subcellular RNA localization revealed by APEX-Seq. *Cell* **178:** 473–490.e26. doi:10.1016/j.cell.2019.05.027

Femino AM, Fogarty K, Lifshitz LM, Carrington W, Singer RH. 2003. Visualization of single molecules of mRNA in situ. *Methods Enzymol* **361:** 245–304. doi:10.1016/S0076-6879(03)61015-3

Fürth D, Hatini V, Lee JH. 2019. In situ transcriptome accessibility sequencing (INSTA-seq). bioRxiv doi:10.1101/722819

Gadir N, Haim-Vilmovsky L, Kraut-Cohen J, Gerst JE. 2011. Localization of mRNAs coding for mitochondrial proteins in the yeast *Saccharomyces cerevisiae*. *RNA* **17:** 1551–1565. doi:10.1261/rna.2621111

Garcia M, Delaveau T, Goussard S, Jacq C. 2010. Mitochondrial presequence and open reading frame mediate asymmetric localization of messenger RNA. *EMBO Rep* **11:** 285–291. doi:10.1038/embor.2010.17

Glisovic T, Bachorik JL, Yong J, Dreyfuss G. 2008. RNA-binding proteins and post-transcriptional gene regulation. *FEBS Lett* **582:** 1977–1986. doi:10.1016/j.febslet.2008.03.004

Gu W, Pan F, Zhang H, Bassell GJ, Singer RH. 2002. A predominantly nuclear protein affecting cytoplasmic localization of β-actin mRNA in fibroblasts and neurons. *J Cell Biol* **156:** 41–52. doi:10.1083/jcb.200105133

Guo S-M, Veneziano R, Gordonov S, Li L, Danielson E, de Arce KP, Park D, Kulesa AB, Wamhoff E-C, Blainey PC, et al. 2019. Multiplexed and high-throughput neuronal fluorescence imaging with diffusible probes. *Nat Commun* **10:** 4377. doi:10.1038/s41467-019-12372-6

Hotulainen P, Hoogenraad CC. 2010. Actin in dendritic spines: connecting dynamics to function. *J Cell Biol* **189:** 619–629. doi:10.1083/jcb.201003008

Hüttelmaier S, Zenklusen D, Lederer M, Dictenberg J, Lorenz M, Meng X, Bassell GJ, Condeelis J, Singer RH. 2005. Spatial regulation of β-actin translation by Src-dependent phosphorylation of ZBP1. *Nature* **438:** 512–515. doi:10.1038/nature04115

Jan CH, Williams CC, Weissman JS. 2014. Principles of ER cotranslational translocation revealed by proximity-specific ribosome profiling. *Science* **346:** 1257521. doi:10.1126/science.1257521

Jeffery WR, Tomlinson CR, Brodeur RD. 1983. Localization of actin messenger RNA during early ascidian development. *Dev Biol* **99:** 408–417. doi:10.1016/0012-1606(83)90290-7

Kaewsapsak P, Shechner DM, Mallard W, Rinn JL, Ting AY. 2017. Live-cell mapping of organelle-associated RNAs via proximity biotinylation combined with protein-RNA cross-linking. *Elife* **6:** e29224. doi:10.7554/eLife.29224

Katz ZB, Wells AL, Park HY, Wu B, Shenoy SM, Singer RH. 2012. β-Actin mRNA compartmentalization enhances focal adhesion stability and directs cell migration. *Genes Dev* **26:** 1885–1890. doi:10.1101/gad.190413.112

Kim DI, Roux KJ. 2016. Filling the void: proximity-based labeling of proteins in living cells. *Trends Cell Biol* **26:** 804–817. doi:10.1016/j.tcb.2016.09.004

Kislauskis EH, Zhu X, Singer RH. 1994. Sequences responsible for intracellular localization of β-actin messenger RNA also affect cell phenotype. *J Cell Biol* **127:** 441–451. doi:10.1083/jcb.127.2.441

Lam SS, Martell JD, Kamer KJ, Deerinck TJ, Ellisman MH, Mootha VK, Ting AY. 2015. Directed evolution of APEX2 for electron microscopy and proximity labeling. *Nat Methods* **12:** 51–54. doi:10.1038/nmeth.3179

Lanzanò L, Scipioni L, Di Bona M, Bianchini P, Bizzarri R, Cardarelli F, Diaspro A, Vicidomini G. 2017. Measurement of nanoscale three-dimensional diffusion in the interior of living cells by STED-FCS. *Nat Commun* **8:** 1–10. doi:10.1038/s41467-017-00117-2

Lapointe CP, Wilinski D, Saunders HAJ, Wickens M. 2015. Protein-RNA networks revealed through covalent RNA marks. *Nat Methods* **12:** 1163–1170. doi:10.1038/nmeth.3651

Lapointe CP, Stefely JA, Jochem A, Hutchins PD, Wilson GM, Kwiecien NW, Coon JJ, Wickens M, Pagliarini DJ. 2018. Multi-omics reveal specific targets of the RNA-binding protein Puf3p and its orchestration of mitochondrial biogenesis. *Cell Syst* **6:** 125–135.e6. doi:10.1016/j.cels.2017.11.012

Lawrence JB, Singer RH. 1985. Quantitative analysis of in situ hybridization methods for the detection of actin gene expression. *Nucleic Acids Res* **13:** 1777–1799. doi:10.1093/nar/13.5.1777

Lawrence JB, Singer RH. 1986. Intracellular localization of messenger RNAs for cytoskeletal proteins. *Cell* **45:** 407–415. doi:10.1016/0092-8674(86)90326-0

Lécuyer E, Yoshida H, Parthasarathy N, Alm C, Babak T, Cerovina T, Hughes TR, Tomancak P, Krause HM. 2007. Global analysis of mRNA localization reveals a prominent role in organizing cellular architecture and function. *Cell* **131:** 174–187. doi:10.1016/j.cell.2007.08.003

Levsky JM, Shenoy SM, Pezo RC, Singer RH. 2002. Single-cell gene expression profiling. *Science* **297:** 836–840. doi:10.1126/science.1072241

Long RM, Gu W, Lorimer E, Singer RH, Chartrand P. 2000. She2p is a novel RNA-binding protein that recruits the Myo4p–She3p complex to *ASH1* mRNA. *EMBO J* **19:** 6592–6601. doi:10.1093/emboj/19.23.6592

Martin KC, Ephrussi A. 2009. mRNA localization: gene expression in the spatial dimension. *Cell* **136:** 719–730. doi:10.1016/j.cell.2009.01.044

McMahon AC, Rahman R, Jin H, Shen JL, Fieldsend A, Luo W, Rosbash M. 2016. TRIBE: hijacking an RNA-editing enzyme to identify cell-specific targets of RNA-binding proteins. *Cell* **165:** 742–753. doi:10.1016/j.cell.2016.03.007

Mili S, Steitz JA. 2004. Evidence for reassociation of RNA-binding proteins after cell lysis: implications for the interpretation of immunoprecipitation analyses. *RNA* **10:** 1692–1694. doi:10.1261/rna.7151404

Mukherjee J, Hermesh O, Eliscovich C, Nalpas N, Franz-Wachtel M, Maček B, Jansen R-P. 2019. β-actin mRNA interactome mapping by proximity biotinylation. *Proc Natl Acad Sci* **116:** 12863–12872. doi:10.1073/pnas.1820737116

Oleynikov Y, Singer RH. 2003. Real-time visualization of ZBP1 association with β-actin mRNA during transcription and localization. *Curr Biol* **13:** 199–207. doi:10.1016/S0960-9822(03)00044-7

Padrón A, Iwasaki S, Ingolia NT. 2019. Proximity RNA labeling by APEX-seq reveals the organization of translation initiation complexes and repressive RNA granules. *Mol Cell* **75:** 875–887.e5. doi:10.1016/j.molcel.2019.07.030

Pan F, Hüttelmaier S, Singer RH, Gu W. 2007. ZBP2 facilitates binding of ZBP1 to β-actin mRNA during transcription. *Mol Cell Biol* **27:** 8340–8351. doi:10.1128/MCB.00972-07

Patel VL, Mitra S, Harris R, Buxbaum AR, Lionnet T, Brenowitz M, Girvin M, Levy M, Almo SC, Singer RH, et al. 2012. Spatial arrangement of an RNA zipcode identifies mRNAs under post-transcriptional control. *Genes Dev* **26:** 43–53. doi:10.1101/gad.177428.111

Perycz M, Urbanska AS, Krawczyk PS, Parobczak K, Jaworski J. 2011. Zipcode binding protein 1 regulates the development of dendritic arbors in hippocampal neurons. *J Neurosci* **31:** 5271–5285. doi:10.1523/JNEUROSCI.2387-10.2011

Rahman R, Xu W, Jin H, Rosbash M. 2018. Identification of RNA-binding protein targets with HyperTRIBE. *Nat Protoc* **13:** 1829. doi:10.1038/s41596-018-0020-y

Ramanathan M, Majzoub K, Rao DS, Neela PH, Zarnegar BJ, Mondal S, Roth JG, Gai H, Kovalski JR, Siprashvili Z, et al. 2018. RNA–protein interaction detection in living cells. *Nat Methods* **15:** 207–212. doi:10.1038/nmeth.4601

Rebagliati MR, Weeks DL, Harvey RP, Melton DA. 1985. Identification and cloning of localized maternal RNAs from *Xenopus* eggs. *Cell* **42:** 769–777. doi:10.1016/0092-8674(85)90273-9

Ross AF, Oleynikov Y, Kislauskis EH, Taneja KL, Singer RH. 1997. Characterization of a β-actin mRNA zipcode-binding protein. *Mol Cell Biol* **17:** 2158–2165. doi:10.1128/MCB.17.4.2158

Shestakova EA, Singer RH, Condeelis J. 2001. The physiological significance of β-actin mRNA localization in determining cell polarity and directional motility. *Proc Natl Acad Sci* **98:** 7045–7050. doi:10.1073/pnas.121146098

Singer RH, Ward DC. 1982. Actin gene expression visualized in chicken muscle tissue culture by using in situ hybridization with a biotinated nucleotide analog. *Proc Natl Acad Sci* **79:** 7331–7335. doi:10.1073/pnas.79.23.7331

Song T, Zheng Y, Wang Y, Katz Z, Liu X, Chen S, Singer RH, Gu W. 2015. Specific interaction of KIF11 with ZBP1 regulates the transport of β-actin mRNA and cell motility. *J Cell Sci* **128:** 1001–1010. doi:10.1242/jcs.161679

Tadros W, Goldman AL, Babak T, Menzies F, Vardy L, Orr-Weaver T, Hughes TR, Westwood JT, Smibert CA, Lipshitz HD. 2007. SMAUG is a major regulator of maternal mRNA destabilization in *Drosophila* and its translation is activated by the PAN GU kinase. *Dev Cell* **12:** 143–155. doi:10.1016/j.devcel.2006.10.005

Tutucci E, Vera M, Biswas J, Garcia J, Parker R, Singer RH. 2018. An improved MS2 system for accurate reporting of the mRNA life cycle. *Nat Methods* **15:** 81–89. doi:10.1038/nmeth.4502

Van Nostrand EL, Pratt GA, Shishkin AA, Gelboin-Burkhart C, Fang MY, Sundararaman B, Blue SM, Nguyen TB, Surka C, Elkins K, et al. 2016. Robust transcriptome-wide discovery of RNA-binding protein binding sites with enhanced CLIP (eCLIP). *Nat Methods* **13:** 508–514. doi:10.1038/nmeth.3810

Vera M, Biswas J, Senecal A, Singer RH, Park HY. 2016. Single-cell and single-molecule analysis of gene expression regulation. *Annu Rev Genet* **50:** 267–291. doi:10.1146/annurev-genet-120215-034854

Welshhans K, Bassell GJ. 2011. Netrin-1-induced local β-actin synthesis and growth cone guidance requires zipcode binding protein 1. *J Neurosci* **31:** 9800–9813. doi:10.1523/JNEUROSCI.0166-11.2011

Wheeler EC, Nostrand ELV, Yeo GW. 2018. Advances and challenges in the detection of transcriptome-wide protein–RNA interactions. *Wiley Interdiscip Rev RNA* **9:** e1436. doi:10.1002/wrna.1436

Williams CC, Jan CH, Weissman JS. 2014. Targeting and plasticity of mitochondrial proteins revealed by proximity-specific ribosome profiling. *Science* **346:** 748–751. doi:10.1126/science.1257522

Wu B, Buxbaum AR, Katz ZB, Yoon YJ, Singer RH. 2015. Quantifying protein-mRNA interactions in single live cells. *Cell* **162:** 211–220. doi:10.1016/j.cell.2015.05.054

Xia C, Fan J, Emanuel G, Hao J, Zhuang X. 2019. Spatial transcriptome profiling by MERFISH reveals subcellular RNA compartmentalization and cell cycle-dependent gene expression. *Proc Natl Acad Sci* **116:** 19490–19499. doi:10.1073/pnas.1912459116

Xu W, Rahman R, Rosbash M. 2018. Mechanistic implications of enhanced editing by a HyperTRIBE RNA-binding protein. *RNA* **24:** 173–182. doi:10.1261/rna.064691.117

Yisraeli JK. 2005. VICKZ proteins: a multi-talented family of regulatory RNA-binding proteins. *Biol Cell* **97:** 87–96. doi:10.1042/BC20040151

Yoon YJ, Wu B, Buxbaum AR, Das S, Tsai A, English BP, Grimm JB, Lavis LD, Singer RH. 2016. Glutamate-induced RNA localization and translation in neurons. *Proc Natl Acad Sci* **113:** E6877–E6886. doi:10.1073/pnas.1614267113

Youn J-Y, Dunham WH, Hong SJ, Knight JDR, Bashkurov M, Chen GI, Bagci H, Rathod B, MacLeod G, Eng SWM, et al. 2018. High-density proximity mapping reveals the subcellular organization of mRNA-associated granules and bodies. *Mol Cell* **69:** 517–532.e11. doi:10.1016/j.molcel.2017.12.020

Zaessinger S, Busseau I, Simonelig M. 2006. Oskar allows nanos mRNA translation in *Drosophila* embryos by preventing its deadenylation by Smaug/CCR4. *Development* **133:** 4573–4583. doi:10.1242/dev.02649

Zappulo A, van den Bruck D, Mattioli CC, Franke V, Imami K, McShane E, Moreno-Estelles M, Calviello L, Filipchyk A, Peguero-Sanchez E, et al. 2017. RNA localization is a key determinant of neurite-enriched proteome. *Nat Commun* **8:** 583. doi:10.1038/s41467-017-00690-6

Zhang HL, Eom T, Oleynikov Y, Shenoy SM, Liebelt DA, Dictenberg JB, Singer RH, Bassell GJ. 2001. Neurotrophin-induced transport of a β-actin mRNP complex increases β-actin levels and stimulates growth cone motility. *Neuron* **31:** 261–275. doi:10.1016/S0896-6273(01)00357-9

Pre-mRNA Splicing in the Nuclear Landscape

TUCKER J. CARROCCI AND KARLA M. NEUGEBAUER

Department of Molecular Biophysics and Biochemistry, Yale University, New Haven, Connecticut 06520, USA

Correspondence: karla.neugebauer@yale.edu

Eukaryotic gene expression requires the cumulative activity of multiple molecular machines to synthesize and process newly transcribed pre-messenger RNA. Introns, the noncoding regions in pre-mRNA, must be removed by the spliceosome, which assembles on the pre-mRNA as it is transcribed by RNA polymerase II (Pol II). The assembly and activity of the spliceosome can be modulated by features including the speed of transcription elongation, chromatin, post-translational modifications of Pol II and histone tails, and other RNA processing events like 5′-end capping. Here, we review recent work that has revealed cooperation and coordination among co-transcriptional processing events and speculate on new avenues of research. We anticipate new mechanistic insights capable of unraveling the relative contribution of coupled processing to gene expression.

Eukaryotic pre-messenger RNA (pre-mRNA) is processed in the nuclear milieu by multiple molecular machines working in concert to affect gene expression. Pre-mRNA processing starts concurrently with transcription elongation (i.e., co-transcriptionally) and likely proceeds until the mRNA is packaged for export to the cytoplasm (Fig. 1; Beyer and Osheim 1988; Baurén and Wieslander 1994). A major component of pre-mRNA processing is RNA splicing, which excises noncoding, intervening regions (introns) from a transcript to generate mRNA. Introns can be plentiful in eukaryotic genes, and the selective removal of introns can significantly impact gene expression by altering transcript stability, coding potential, or localization. Some of the first evidence for co-transcriptional processing was the transcription-dependent recruitment of splicing factors to chromatin (Sass and Pederson 1984). In the past 10 years, global analyses have revealed that co-transcriptional removal of introns is conserved from yeast to humans (Carrillo Oesterreich et al. 2010; Ameur et al. 2011; Khodor et al. 2011, 2012; Schmidt et al. 2011; Girard et al. 2012; Tilgner et al. 2012; Windhager et al. 2012; Nojima et al. 2015; Pai et al. 2017).

An increasingly prominent theme in molecular biology is that the complexes responsible for synthesizing and modifying mRNA cross-regulate to fine-tune gene outputs. For instance, the presence of an intron in transgenes is positively correlated with transcriptional activity (Brinster et al. 1988), intron–exon boundaries are associated with active chromatin marks (Bieberstein et al. 2012), and splicing factors stimulate in vitro transcription reactions (Fong and Zhou 2001). This evidence suggests coordination between transcription and splicing. It is also well known that RNA polymerase II (Pol II) speed influences alternative splicing and intron retention in all examined species (Schor et al. 2009, 2013; Aslanzadeh et al. 2018; Saldi et al. 2018; Godoy Herz et al. 2019). Moreover, transcripts with multiple introns tend to have either all the introns removed or all the introns retained in *Schizosaccharomyces pombe*, suggesting coordination or cross-regulation between individual splicing events (Herzel et al. 2018). This may also be true in insect and mammalian cells, because coordination among adjacent introns has been well documented (Tilgner et al. 2015, 2018; Kim et al. 2017; Pai et al. 2017; Drexler et al. 2020). Furthermore, pre-mRNAs that are spliced are also more efficiently cleaved for polyadenylation, suggesting a relationship between the two machineries (Davidson and West 2013; Herzel et al. 2018). At the other end of the gene, we know that capping of the pre-mRNA 5′ end occurs in a rapid and coordinated fashion with transcription initiation (Hagler and Shuman 1992; Rasmussen and Lis 1993) and that promoter identity plays a role in alternative splicing and even RNA half-life (Cramer et al. 1997; Trcek et al. 2011; Fiszbein et al. 2019).

Over the years, major advances in understanding coupled processing events have emerged from new technologies. Here, we review recent progress on understanding coupled and coordinated pre-mRNA splicing focusing on the yeast *Saccharomyces cerevisiae*. Budding yeast is a powerful system for uncovering the mechanisms of splicing and transcription coordination, because although only 300 genes have introns, these encoded intron-containing mRNAs account for 25%–30% of the transcripts in yeast cells (Ares et al. 1999; Lopez and Seraphin 1999). Indeed, the fitness of yeast depends on having introns as the physical presence of introns in genes can promote survival in starvation conditions (Parenteau et al. 2019). The yeast spliceosome has also served as a useful model for understanding the effect of splicing factor mutations identified in human diseases including various hematopoietic malignancies and retinitis pigmentosa (Tang et al. 2016; Carrocci et al. 2017; Ruzickova and Stanek 2017).

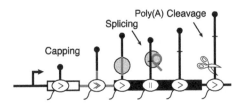

Figure 1. Eukaryotic mRNA is processed concurrently with transcription to modify the transcript output. Processing can include 5′-end capping with 7-methylguanosine, splicing, and polyadenylation cleavage. The rate of elongation is also non-uniform along the gene body.

THE MECHANISM OF PRE-mRNA SPLICING

Pre-mRNA splicing is catalyzed by the spliceosome, a single-turnover enzyme composed of the U1, U2, U4, U5, and U6 small nuclear ribonucleoproteins (snRNP) and a number of protein-only factors (Lerner et al. 1980). Early work identified several splicing factors as temperature-sensitive mutants in yeast (Hartwell 1967; Hartwell et al. 1970). The spliceosome defines the intron through recognition of conserved sequences located in the intron: the 5′ splice site (5′ss), the branchpoint sequence (bps), and the 3′ splice site (3′ss). Splice site recognition is aided by the activity of non-snRNP factors, such as Mud2 and branchpoint binding protein (BBP), which promote pairing between the 5′ss and 3′ss (Abovich and Rosbash 1997; Lacadie et al. 2006). The intronic sequences are obligate components of the splicing reaction and form part of the spliceosome active site during catalysis. Splicing occurs in two sequential trans-esterification reactions (Fig. 2; Moore and Sharp 1993). The catalytic steps of splicing are ATP-independent, but ATP is required for the activity of a number of ATPases that structurally and compositionally remodel the spliceosome during assembly and catalysis (Liu and Cheng 2015). In the first reaction, the branchpoint adenosine acts as the nucleophile to cleave the phosphodiester backbone at the 5 exon–intron boundary and generate a free 5′ exon and lariat intron–3′ exon intermediate. During the second step, the 3′OH of the 5′ exon acts as the nucleophile to excise the lariat intron and ligate the exons together.

The spliceosome can use a variety of splice sites to varying degrees of efficiency and the conservation of splice sites differs greatly among organisms (Will and Luhrmann 2011; Qin et al. 2016). Splice sites in the yeast *S. cerevisiae* are very well conserved and rarely diverge from the consensus sequence, whereas metazoan splice sites are more degenerate. This is in part due to the presence of splicing regulatory proteins in metazoans that recognize splicing regulatory elements and influence splice site usage (Zhong et al. 2009). In yeast, splicing regulation has been attributed to Npl3 and Nam8. Npl3 is an SR-like protein, which is required for proper spliceosome assembly in the co-transcriptional context, and *NPL3* mutations lead to intron retention (Kress et al. 2008). Interestingly, splice site usage can be further modified by the surrounding sequence context. In the context of a minigene reporter system, Wong et al. (2018) showed that the spliceosome can accommodate different 5′ss sequences depending on the origin of the intron. In yeast, Nam8 is a poly(U) binding protein that binds near the 5′ss and enhances 5′ss recognition by the U1 snRNP in a manner analogous to TIA-1 in humans (Puig et al. 1999; Förch et al. 2000; Spingola and Ares 2000; Qiu et al. 2011).

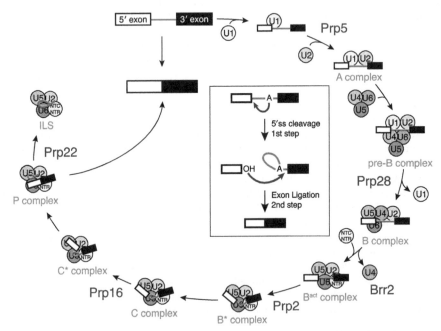

Figure 2. Splicing occurs in two sequential chemical steps (*inset*) to excise the intron (red) and ligate the coding exons. The spliceosome assembles in an ordered manner from preassembled components responsible for identifying the splice sites, forming the active site, and catalysis. The spliceosome is then released and the components recycled for the next round of splicing. Splicing complex names are shown in blue, and helicases that help mediate major transitions in the splicing cycle are shown in gray.

The steps of spliceosome assembly have been extensively characterized using in vitro biochemistry and genetics (Fig. 2) (Will and Luhrmann 2011). More recently, the molecular architecture of many intermediate complexes along the splicing reaction have been revealed using cryo-EM (reviewed extensively in Fica and Nagai 2017; Shi 2017; Wilkinson et al. 2019). Initially, the 5′ end of the U1 snRNA base pairs with the 5′ss found at the 5′ boundary of the intron to form the spliceosome E complex. Association of the U2 snRNP with the bps/3′ss region converts the E complex to the A complex. A complex is converted to the pre-B complex upon addition of the U4/U6.U5 tri-snRNP and subsequently to the B complex upon exchange of base pairing at the 5′ss from the U1 snRNA to the U6 snRNA and release of the U1 snRNP. Unwinding of U4/U6 base pairing and release of the U4 snRNA permits the formation of the U2/U6 catalytic active site and marks formation of the B^{act} complex. The spliceosome is then heavily remodeled to bring together the U2/U6 duplex and the U2/bps duplex (B* complex). Splicing factors then activate the spliceosome to promote the catalytic steps in the C and C* complexes to ligate the exons and form the postcatalytic P complex. The P complex spliceosome is then released from the mRNA and disassembled. Splicing is energetically costly and the spliceosome must assemble anew from its constituent components for every intron. This means that in mammalian cells, where genes contain on average eight introns, eight spliceosomes must assemble and disassemble for every pre-mRNA synthesized. In yeast, the majority of genes lack introns (Spingola et al. 1999); yet, the fact that 35% of transcripts must be spliced generates a strong demand for spliceosomal components. Reduced demand for splicing leads to a "hungry spliceosome," which can act on cryptic splice sites in RNA that are not normally recognized by the spliceosome (Munding et al. 2013; Talkish et al. 2019).

CO-TRANSCRIPTIONAL ASSEMBLY OF THE SPLICEOSOME

Studies from a number of laboratories and in many species have shown that pre-mRNA splicing occurs concurrently with transcription elongation by Pol II (Carrillo Oesterreich et al. 2010; Ameur et al. 2011; Khodor et al. 2011, 2012; Schmidt et al. 2011; Girard et al. 2012; Tilgner et al. 2012; Windhager et al. 2012; Nojima et al. 2015; Pai et al. 2017). If splicing occurs co-transcriptionally, then so must spliceosome assembly. This scenario raises the possibility that components of the spliceosome would associate with the nascent RNA potentially as soon as the corresponding RNA element is synthesized (e.g., the U1 snRNP would bind the RNA as soon as Pol II has transcribed the 5′ss). Additional possibilities include the potential for Pol II itself and/or components of chromatin to help recruit spliceosomal components. The work of our laboratory and others established that splicing factors associate with gene bodies in vivo in a manner that mirrors the stepwise assembly pathway defined in vitro (Kotovic et al. 2003; Gornemann et al. 2005; Tardiff et al. 2006; Hoskins et al. 2011). Because splicing occurs co-transcriptionally, splicing factors are adjacent to the DNA axis and can cross-link to the underlying chromatin (Fig. 3A). These "splicing factor ChIP" experiments showed that U1 snRNP signal peaks in the region downstream from the 5′ exon–intron boundary and decays downstream

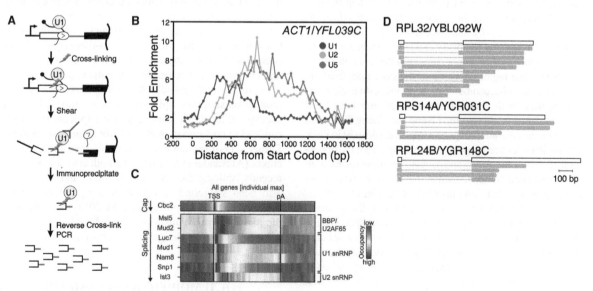

Figure 3. Monitoring spliceosome assembly co-transcriptionally. (*A*) Schematic describing a splicing factor ChIP assay. Splicing factors (i.e., U1) are formaldehyde-cross-linked to the chromatin, and chromatin is subsequently fragmented. Amplification of fragments that coimmunoprecipitated infer the association of the targeted splicing factor along the gene body. (*B*) Representative spliceosome assembly profile for *ACT1/YFL039C*. U1 associates first and departs first. U2 addition is followed by U5, consistent with the ordered assembly model. (*C*) Splicing factor cross-linking immunoprecipitation identifying splicing factor association with the nascent RNA. (*D*) Representative long reads from yeast nascent RNA sequenced on the Pacific Biosciences RSII platform showing rapid splicing of the intron while Pol II is within ~200 nt of the 3′ss. Data reproduced from Oesterreich et al. (2016). (*B*, Reproduced, with permission from Tardiff et al. 2006; *C*, reproduced, with permission, from Baejen et al. 2014.)

from the intron–3′exon boundary (Fig. 3B) (Kotovic et al. 2003; Tardiff et al. 2006). Furthermore, U2 and U5 snRNPs peaked in the downstream exon, consistent with the addition of U2 and U5 subsequent to both U1 binding and synthesis of the 3′ss. ChIP data were later extended by high resolution cross-linking and immunoprecipitation (CLIP) experiments that identified the binding sites of many splicing factors on the mRNA substrate (Fig. 3C). For example, it was unclear how Mud2—similar to metazoan U2AF that binds the polypyrimidine tract at the 3′ss—might participate in gene expression in yeast, which generally lack a polypyrimidine tract. Mud2 CLIP shows that this protein binds along the entire length of the intron, displaying a different role in mRNA processing than suspected from protein homology (Baejen et al. 2014).

Broadly speaking, the ChIP and CLIP profiles in budding yeast were consistent with spliceosome assembly during transcription, but the degree to which splicing completed co-transcriptionally was contentious, with some studies concluding that exon ligation could not complete before cleavage at the poly(A) site (Tardiff et al. 2006). However, our recent work has suggested that splicing can occur rapidly relative to transcription of the 3′ss and is often upstream of previously identified ChIP peaks (Tardiff et al. 2006; Oesterreich et al. 2016; Wallace and Beggs 2017). This discrepancy is likely in part due to the low resolution of ChIP data compared to more recent custom sequencing strategies like single-molecule intron tracking and long-read sequencing, which identify the position of Pol II with single-nucleotide resolution while simultaneously determining the transcript splicing status. Spliced products can be identified while Pol II is still in close proximity to the 3′ss (Fig. 3D), suggesting Pol II and the spliceosome are physically close and corroborating earlier measurements of efficient co-transcriptional splicing with high-density tiling arrays (Carrillo Oesterreich et al. 2010). Taken together, these findings suggest that although the chemistry of RNA splicing may occur quickly, spliceosomal components may be retained in downstream gene regions because of the slow release of the post-catalytic spliceosome. This is an interesting observation because both chemical steps of splicing have been shown to be reversible in vitro (Tseng and Cheng 2008) and might suggest that an analogous complex could exist in vivo.

Simultaneous transcription and splicing afford the opportunity for cross-regulation and demand coupling for accurate gene expression. Splice site sequences are short and consensus sequences alone are insufficient to identify introns because of the presence of near-cognates in the genome. How can co-transcriptional processing aid the spliceosome in selecting the correct sites? Co-transcriptional splicing reduces the number of potential splice sites to only the subset of those in the transcript that have been transcribed at the time of spliceosome assembly. This idea led to the "first come, first served" model of splice usage which states that the splice sites that are transcribed first are used by the spliceosome (Kuhne et al. 1983). The Kornblihtt laboratory subsequently showed that "first come, first served" could refer to splicing commitment and not necessarily the completion of intron removal in the order of synthesis (de la Mata et al. 2010). Concordantly, changes in the rate of Pol II elongation can lead to changes in the outcomes of splicing even in budding yeast (Howe et al. 2003; Braberg et al. 2013; Oesterreich et al. 2016; Aslanzadeh et al. 2018). Interestingly, both faster and slower elongation leads to changes, suggesting that the transcription and splicing machinery are finely tuned to work concurrently.

Pol II elongation along gene bodies in human cells is nonuniform, with elongation rates slower over exons than introns (Veloso et al. 2014). In budding and fission yeasts, high-resolution elongation rate measurements are difficult to make and have not been accomplished; instead, a single study established an average elongation rate in budding yeast of 1.5 kb/min (Mason and Struhl 2005). Reduction in Pol II elongation (i.e., transcriptional pausing) has been proposed as an important determinant of splicing, particularly when excising introns with nonconsensus splice sites (Aslanzadeh et al. 2018). Changes in the overall rate of transcription has broad effects on splicing efficiency, indicating tuning between the rates of two processes (Oesterreich et al. 2016). Indeed, reduction in elongation rate either through mutation of Pol II or through drug treatment can rescue middle exon inclusion in multi-intron yeast genes (Howe et al. 2003). Terminal exon pausing has been identified in short endogenous genes that are spliced efficiently co-transcriptionally but the molecular features that contribute to pausing remain unclear (Carrillo Oesterreich et al. 2010). Understanding the functional consequences of pausing in terminal exons is difficult, because we currently lack the tools to manipulate pausing per se; changing the overall rate of transcription is not the same as modulating elongation at particular gene locations. Theoretically, pausing should play a role in the extent of transcripts that are spliced co-transcriptionally, but the rapid appearance of spliced products relative to transcription progress suggests that pausing may not be absolutely required for splicing. Dissecting the contribution of transcriptional pausing to efficient co-transcriptional processing will require the development of new tools to manipulate pausing on genes in their native context.

Perturbation of pre-mRNA splicing also feeds back to modify transcription. Splicing has been proposed to be a checkpoint for Pol II pausing because splicing inhibition can lead to an increase in Pol II ChIP signal over introns, further supporting communication between the two machines (Alexander et al. 2010; Chathoth et al. 2014). More recently, splicing factors, such as Mud2 and Npl3, have also been suggested to act in transcription elongation (Dermody et al. 2008; Minocha et al. 2018). Nevertheless, the molecular mechanism by which Mud2 or Npl3 contributes to efficient elongation remains unknown.

POL II MODIFICATIONS AND THE SPLICEOSOME

The major proposed mechanism of communication between elongating Pol II and complexes that modify the nascent RNA is the Pol II carboxy-terminal domain (CTD) (Custodio and Carmo-Fonseca 2016; Harlen and Churchman 2017; Wallace and Beggs 2017). The CTD is highly conserved among eukaryotes and contains a species-specific number of heptad repeats of the sequence Tyr1–Ser2–Pro3–Thr4–Ser5–Pro6–Ser7 (26 repeats in *S. cerevisiae* and 52 in humans). Although these repeats may act as a platform to assist in the recruitment of splicing factors to the nascent RNA, the CTD alone does not specify recruitment; for example, snRNPs are not recruited to intronless genes in yeast (Kotovic et al. 2003; Gornemann et al. 2005). Further, a CTD domain is not strictly required for splicing. Splicing occurs in the absence of transcription in vitro, and even in the context of an elongating polymerase there is no significant change in the efficiency of splicing between complexes that do and do not have a CTD (Lin et al. 1985; Natalizio et al. 2009). Eukaryotic genes including the *S. pombe* U6 gene that are transcribed by RNA polymerase III, which lacks a CTD, can be efficiently spliced by the spliceosome (Tani and Ohshima 1989). Nevertheless, phosphorylation of Pol II stimulates pre-mRNA splicing (Hirose et al. 1999), suggesting that splicing factor recruitment is modulated by kinases and phosphatases that dynamically post-translationally modify the CTD as Pol II transcribes along the body of the gene. Importantly, mNET-seq data from human cells has shown that Ser5P antibodies precipitate intermediates of the splicing reaction including free 5′ exons and lariat intron–3′ exon intermediates (Mayer et al. 2015; Nojima et al. 2015, 2018). This indicates that nascent RNAs that are undergoing splicing are attached to Pol II molecules with this modification. Indeed, antibodies targeting the Ser5P phospho-epitope immunoprecipitate splicing factors alongside the polymerase (Harlen et al. 2016). Additionally, pre-mRNAs transcribed by Pol II are more efficiently processed in vitro than pre-mRNAs transcribed by another polymerase lacking a CTD (Das et al. 2006). Together, these data suggest that the modified CTD may interact with the spliceosome but that other features may also significantly contribute to splicing outcomes.

CONNECTIONS TO THE mRNA CAP

The 5′ end of mRNA is capped with an inverted, methylated guanosine nucleotide that plays a major role in the life of the transcript, including mRNA splicing, 3′-end formation, export, and overall stability (Ramanathan et al. 2016). The cap is one of the first modifications installed on eukaryotic mRNA and the effects of the cap are mediated through interaction with the cap-binding complex (CBC). The CBC binds to the nascent RNA early during transcription and may help promote escape from the promoter (Lidschreiber et al. 2013). The CBC has been shown to be important, but not essential, for efficient pre-mRNA splicing in both mammalian and yeast systems (Fresco and Buratowski 1996; Schwer and Shuman 1996). Defects in mRNA capping lead to the accumulation of unspliced precursors. The CBC helps to promote E complex formation by interacting directly with U1 to recruit the snRNP and stabilize it at the 5′ss to promote spliceosome assembly (Fig. 4A,B; Lewis et al. 1996; Gornemann et al. 2005; Larson and Hoskins 2017; Li et al. 2019). Furthermore, the CBC has been proposed to play a role in the stable recruitment of the tri-snRNP (Pabis et al. 2013). Therefore, interactions with the 5′ cap of the mRNA have been proposed to be a major determinant of how the transcript will be processed. Interestingly, Ser5 phosphorylation of the Pol II CTD has also been shown to be important for RNA capping (Fabrega et al. 2003). The lethality associated with Ser5A Pol II CTD mutant in *S. pombe* can be bypassed by fusion of the mammalian capping enzyme to Pol II (Schwer et al. 2012). Therefore, whether the Ser5 phosphomark is directly required for splicing or whether it contributes indirectly by promoting efficient capping remains unclear.

CHROMATIN AND SPLICING

The co-transcriptional nature of splicing places the spliceosome in proximity to chromatin—the array of nucleosomes made up of DNA wrapped around eight histones. Some early indication of the impact of chromatin on the regulation of splicing came from work in which the integration of the adenovirus genome at different genomic locations resulted in altered splicing outcomes (Adami and Babiss 1991). In metazoans, nucleosomes are found more frequently at exons than introns, and it is proposed

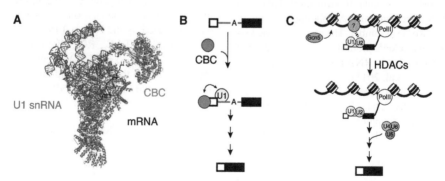

Figure 4. Cap-binding complex aids in U1 recruitment. (*A*) The structure of the spliceosome E complex showed physical interactions with CBC and the U1 snRNP component Snp1 (U1-70K in humans; pdb 6N7P) (Li et al. 2019). (*B*) Capping of the nascent RNA recruits CBC, which in turn can nucleate spliceosome assembly and splicing. (*C*) Gcn5-dependent histone acetylation recruits an unknown factor to chromatin to modulate U2 recruitment. Deacetylation facilitates subsequent steps in splicing (Gunderson et al. 2011).

that they act as landmarks of exons (Kogan and Trifonov 2005; Spies et al. 2009; Tilgner et al. 2009). Nucleosome positioning can act as a transcriptional barrier that might aid in the definition of exons and subsequent assembly of the spliceosome through transient polymerase pausing (Hodges et al. 2009). Independent of nucleosome position, gene architecture—namely, the arrangement of introns and exons in any given gene—has a profound effect on the distribution of histone post-translational modifications (PTMs), which likely impacts the splicing-dependent amplification of gene output. Specifically, H3K4me3 and other active marks are best aligned to the first 5′ss in genes, even better than alignment to transcription start sites; importantly, these peaks can be moved in the gene by changing its architecture (Bieberstein et al. 2012). This suggests that components of the splicing machinery enhance gene output in part through an uncharacterized interaction with histone modifying enzymes. Consistent with this idea, perturbations of histone deacetylases (HDACs) shift alternative splicing patterns globally (Hnilicova et al. 2011). Specific and physiologically relevant examples of histone PTM alterations that determine alternative splicing patterns include the changed chromatin landscape of the *NCAM* gene in stimulated or differentiating neurons (Schor et al. 2009, 2013). It is very difficult to disentangle the roles of chromatin, transcription rates, and splicing, which seem to have arrows pointing in all directions. This experimental obstacle may explain why relatively few specific examples of regulation, such as the targeting of an alternative splicing regulator to chromatin sites, have emerged (Sims et al. 2007; Luco and Misteli 2011; Kfir et al. 2015). Nevertheless, these seminal studies in metazoans show the complex roles the chromatin landscape can play in splicing regulation.

Although budding and fission yeasts essentially lack alternative splicing, chromatin plays a role in splicing efficiency. The recent high-throughput method epistatic mini array profile (E-MAP) has revealed links between splicing and chromatin (Braberg et al. 2013). For example, Npl3 was found to promote H2B monoubiquitination through interactions with Bre1 in budding yeast (Moehle et al. 2012), and the chromatin remodeler SWI/SNF was linked to spliceosome activation in fission yeast (Patrick et al. 2015). Histone post-translational modifications have also been proposed to affect pre-mRNA splicing by directly modulating spliceosome assembly. The activity of the histone acetyltransferase Gcn5 promotes co-transcriptional U2 recruitment to the nascent RNA (Gunderson and Johnson 2009; Gunderson et al. 2011). More recently, genetic interactions between components of the U2 snRNP and the rare histone variant, H2A.Z, have been identified, pointing to a role in promoting splicing of introns with nonconsensus splice sites (Neves et al. 2017; Nissen et al. 2017). Furthermore, the histone methyltransferase Set2 and its corresponding H3K36 methylation mark has also been shown to play a role in the recruitment of splicing factors to the nascent transcript (Sorenson et al. 2016; Leung et al. 2019). Thus, although the molecular underpinnings of these linkages are unknown, the case for cross-regulation between chromatin and splicing is well-substantiated in the budding yeast system. Future work will be necessary to understand how this network of interactions translates into gene-specific splicing efficiencies.

OUTLOOK: PRE-mRNA SPLICING IN THE NUCLEAR LANDSCAPE

The nuclear landscape in metazoan cells is dominated by chromosomes and membraneless organelles, such as nucleoli, Cajal bodies, speckles, and others (Mao et al. 2011). Currently the field is engaged in exploring how these biomolecular condensates contribute to gene expression. It is becoming hard to imagine what nucleoplasm is, when even individual genes can be thought of as their own organelles with Pol II, RNA, and DNA- and RNA-binding proteins concentrated there (Herzel et al. 2017; Hnisz et al. 2017). Components of the transcription and splicing machinery are enriched in intrinsically disordered regions (IDRs) (Courchaine et al. 2016; Herzel et al. 2017), providing a logic for how small local condensates could form when nascent RNA is present. Indeed, SR proteins and the CTD of Pol II have some of the most extensive intrinsically disordered regions of all proteins. The concentration of SR proteins is high at transcription sites, and phosphorylation changes on the Pol II CTD during elongation could play a role in promoting splicing by recruiting snRNPs and other splicing factors, creating a high local concentration for the removal of multiple introns (Neugebauer and Roth 1997; Galganski et al. 2017; Herzel et al. 2017; Guo et al. 2019). At present, it is unclear how this understanding can be extended to budding and fission yeasts which often lack these long IDRs. Nevertheless, P-bodies were discovered in yeast (Sheth and Parker 2003), and yeast RNA has been shown to mediate condensation (Van Treeck et al. 2018). Yeast nucleoli are enormous, but speckles and Cajal bodies have not been observed. Therefore, importance of biomolecular condensation in yeast nuclei and its role in protein-coding gene expression is currently a frontier awaiting exploration.

ACKNOWLEDGMENTS

We thank Dr. Margaret Rodgers and Kirsten Reimer for their insightful comments on the manuscript. This work was supported by the National Institutes of Health (NIH) (GM134572 to T.J.C., R01 HL133406 to K.M.N., and R01 GM112766 to K.M.N.). Its contents are solely the responsibility of the authors and do not necessarily represent the official views of the NIH.

REFERENCES

Abovich N, Rosbash M. 1997. Cross-intron bridging interactions in the yeast commitment complex are conserved in mammals. *Cell* **89:** 403–412. doi:10.1016/S0092-8674(00)80221-4

Adami G, Babiss LE. 1991. DNA template effect on RNA splicing: two copies of the same gene in the same nucleus are processed differently. *EMBO J* **10:** 3457–3465. doi:10.1002/j.1460-2075.1991.tb04910.x

Alexander RD, Innocente SA, Barrass JD, Beggs JD. 2010. Splicing-dependent RNA polymerase pausing in yeast. *Mol Cell* **40:** 582–593. doi:10.1016/j.molcel.2010.11.005

Ameur A, Zaghlool A, Halvardson J, Wetterbom A, Gyllensten U, Cavelier L, Feuk L. 2011. Total RNA sequencing reveals nascent transcription and widespread co-transcriptional splicing in the human brain. *Nat Struct Mol Biol* **18:** 1435–1440. doi:10.1038/nsmb.2143

Ares M Jr, Grate L, Pauling MH. 1999. A handful of intron-containing genes produces the lion's share of yeast mRNA. *RNA* **5:** 1138–1139. doi:10.1017/S1355838299991379

Aslanzadeh V, Huang Y, Sanguinetti G, Beggs JD. 2018. Transcription rate strongly affects splicing fidelity and cotranscriptionality in budding yeast. *Genome Res* **28:** 203–213. doi:10.1101/gr.225615.117

Baejen C, Torkler P, Gressel S, Essig K, Soding J, Cramer P. 2014. Transcriptome maps of mRNP biogenesis factors define pre-mRNA recognition. *Mol Cell* **55:** 745–757. doi:10.1016/j.molcel.2014.08.005

Baurén G, Wieslander L. 1994. Splicing of Balbiani ring 1 gene pre-mRNA occurs simultaneously with transcription. *Cell* **76:** 183–192. doi:10.1016/0092-8674(94)90182-1

Beyer AL, Osheim YN. 1988. Splice site selection, rate of splicing, and alternative splicing on nascent transcripts. *Genes Dev* **2:** 754–765. doi:10.1101/gad.2.6.754

Bieberstein NI, Carrillo Oesterreich F, Straube K, Neugebauer KM. 2012. First exon length controls active chromatin signatures and transcription. *Cell Rep* **2:** 62–68. doi:10.1016/j.celrep.2012.05.019

Braberg H, Jin H, Moehle EA, Chan YA, Wang S, Shales M, Benschop JJ, Morris JH, Qiu C, Hu F, et al. 2013. From structure to systems: high-resolution, quantitative genetic analysis of RNA polymerase II. *Cell* **154:** 775–788. doi:10.1016/j.cell.2013.07.033

Brinster RL, Allen JM, Behringer RR, Gelinas RE, Palmiter RD. 1988. Introns increase transcriptional efficiency in transgenic mice. *Proc Natl Acad Sci* **85:** 836–840. doi:10.1073/pnas.85.3.836

Carrillo Oesterreich F, Preibisch S, Neugebauer KM. 2010. Global analysis of nascent RNA reveals transcriptional pausing in terminal exons. *Mol Cell* **40:** 571–581. doi:10.1016/j.molcel.2010.11.004

Carrocci TJ, Zoerner DM, Paulson JC, Hoskins AA. 2017. SF3b1 mutations associated with myelodysplastic syndromes alter the fidelity of branchsite selection in yeast. *Nucleic Acids Res* **45:** 4837–4852. doi:10.1093/nar/gkw1349

Chathoth KT, Barrass JD, Webb S, Beggs JD. 2014. A splicing-dependent transcriptional checkpoint associated with prespliceosome formation. *Mol Cell* **53:** 779–790. doi:10.1016/j.molcel.2014.01.017

Courchaine EM, Lu A, Neugebauer KM. 2016. Droplet organelles? *EMBO J* **35:** 1603–1612. doi:10.15252/embj.201593517

Cramer P, Pesce CG, Baralle FE, Kornblihtt AR. 1997. Functional association between promoter structure and transcript alternative splicing. *Proc Natl Acad Sci* **94:** 11456–11460. doi:10.1073/pnas.94.21.11456

Custodio N, Carmo-Fonseca M. 2016. Co-transcriptional splicing and the CTD code. *Crit Rev Biochem Mol Biol* **51:** 395–411. doi:10.1080/10409238.2016.1230086

Das R, Dufu K, Romney B, Feldt M, Elenko M, Reed R. 2006. Functional coupling of RNAP II transcription to spliceosome assembly. *Genes Dev* **20:** 1100–1109. doi:10.1101/gad.1397406

Davidson L, West S. 2013. Splicing-coupled 3′ end formation requires a terminal splice acceptor site, but not intron excision. *Nucleic Acids Res* **41:** 7101–7114. doi:10.1093/nar/gkt446

de la Mata M, Lafaille C, Kornblihtt AR. 2010. First come, first served revisited: factors affecting the same alternative splicing event have different effects on the relative rates of intron removal. *RNA* **16:** 904–912. doi:10.1261/rna.1993510

Dermody JL, Dreyfuss JM, Villen J, Ogundipe B, Gygi SP, Park PJ, Ponticelli AS, Moore CL, Buratowski S, Bucheli ME. 2008. Unphosphorylated SR-like protein Npl3 stimulates RNA polymerase II elongation. *PLoS One* **3:** e3273. doi:10.1371/journal.pone.0003273

Drexler HL, Choquet K, Churchman LS. 2020. Splicing kinetics and coordination revealed by direct nascent RNA sequencing through nanopores. *Mol Cell* **77:** 985–998 e988. doi:10.1016/j.molcel.2019.11.017

Fabrega C, Shen V, Shuman S, Lima CD. 2003. Structure of an mRNA capping enzyme bound to the phosphorylated carboxy-terminal domain of RNA polymerase II. *Mol Cell* **11:** 1549–1561. doi:10.1016/S1097-2765(03)00187-4

Fica SM, Nagai K. 2017. Cryo-electron microscopy snapshots of the spliceosome: structural insights into a dynamic ribonucleoprotein machine. *Nat Struct Mol Biol* **24:** 791–799. doi:10.1038/nsmb.3463

Fiszbein A, Krick KS, Begg BE, Burge CB. 2019. Exon-mediated activation of transcription starts. *Cell* **179:** 1551–1565 e1517. doi:10.1016/j.cell.2019.11.002

Fong YW, Zhou Q. 2001. Stimulatory effect of splicing factors on transcriptional elongation. *Nature* **414:** 929–933. doi:10.1038/414929a

Förch P, Puig O, Kedersha N, Martínez C, Granneman S, Séraphin B, Anderson P, Valcárcel J. 2000. The apoptosis-promoting factor TIA-1 is a regulator of alternative pre-mRNA splicing. *Mol Cell* **6:** 1089–1098. doi:10.1016/S1097-2765(00)00107-6

Fresco LD, Buratowski S. 1996. Conditional mutants of the yeast mRNA capping enzyme show that the cap enhances, but is not required for, mRNA splicing. *RNA* **2:** 584–596.

Galganski L, Urbanek MO, Krzyzosiak WJ. 2017. Nuclear speckles: molecular organization, biological function and role in disease. *Nucleic Acids Res* **45:** 10350–10368. doi:10.1093/nar/gkx759

Girard C, Will CL, Peng J, Makarov EM, Kastner B, Lemm I, Urlaub H, Hartmuth K, Luhrmann R. 2012. Post-transcriptional spliceosomes are retained in nuclear speckles until splicing completion. *Nat Commun* **3:** 994. doi:10.1038/ncomms1998

Godoy Herz MA, Kubaczka MG, Brzyzek G, Servi L, Krzyszton M, Simpson C, Brown J, Swiezewski S, Petrilli E, Kornblihtt AR. 2019. Light regulates plant alternative splicing through the control of transcriptional elongation. *Mol Cell* **73:** 1066–1074 e1063. doi:10.1016/j.molcel.2018.12.005

Gornemann J, Kotovic KM, Hujer K, Neugebauer KM. 2005. Cotranscriptional spliceosome assembly occurs in a stepwise fashion and requires the cap binding complex. *Mol Cell* **19:** 53–63. doi:10.1016/j.molcel.2005.05.007

Gunderson FQ, Johnson TL. 2009. Acetylation by the transcriptional coactivator Gcn5 plays a novel role in co-transcriptional spliceosome assembly. *PLoS Genet* **5:** e1000682. doi:10.1371/journal.pgen.1000682

Gunderson FQ, Merkhofer EC, Johnson TL. 2011. Dynamic histone acetylation is critical for cotranscriptional spliceosome assembly and spliceosomal rearrangements. *Proc Natl Acad Sci* **108:** 2004–2009. doi:10.1073/pnas.1011982108

Guo YE, Manteiga JC, Henninger JE, Sabari BR, Dall'Agnese A, Hannett NM, Spille JH, Afeyan LK, Zamudio AV, Shrinivas K, et al. 2019. Pol II phosphorylation regulates a switch between transcriptional and splicing condensates. *Nature* **572:** 543–548. doi:10.1038/s41586-019-1464-0

Hagler J, Shuman S. 1992. A freeze-frame view of eukaryotic transcription during elongation and capping of nascent mRNA. *Science* **255:** 983–986. doi:10.1126/science.1546295

Harlen KM, Churchman LS. 2017. The code and beyond: transcription regulation by the RNA polymerase II carboxy-terminal domain. *Nat Rev Mol Cell Biol* **18:** 263–273. doi:10.1038/nrm.2017.10

Harlen KM, Trotta KL, Smith EE, Mosaheb MM, Fuchs SM, Churchman LS. 2016. Comprehensive RNA polymerase II interactomes reveal distinct and varied roles for each phospho-CTD residue. *Cell Rep* **15:** 2147–2158. doi:10.1016/j.celrep.2016.05.010

Hartwell LH. 1967. Macromolecule synthesis in temperature-sensitive mutants of yeast. *J Bacteriol* **93:** 1662–1670. doi:10.1128/JB.93.5.1662-1670.1967

Hartwell LH, McLaughlin CS, Warner JR. 1970. Identification of ten genes that control ribosome formation in yeast. *Mol Gen Genet* **109:** 42–56. doi:10.1007/BF00334045

Herzel L, Ottoz DSM, Alpert T, Neugebauer KM. 2017. Splicing and transcription touch base: co-transcriptional spliceosome assembly and function. *Nat Rev Mol Cell Biol* **18:** 637–650. doi:10.1038/nrm.2017.63

Herzel L, Straube K, Neugebauer KM. 2018. Long-read sequencing of nascent RNA reveals coupling among RNA processing events. *Genome Res* **28:** 1008–1019. doi:10.1101/gr.232025.117

Hirose Y, Tacke R, Manley JL. 1999. Phosphorylated RNA polymerase II stimulates pre-mRNA splicing. *Genes Dev* **13:** 1234–1239. doi:10.1101/gad.13.10.1234

Hnilicova J, Hozeifi S, Duskova E, Icha J, Tomankova T, Stanek D. 2011. Histone deacetylase activity modulates alternative splicing. *PLoS One* **6:** e16727. doi:10.1371/journal.pone.0016727

Hnisz D, Shrinivas K, Young RA, Chakraborty AK, Sharp PA. 2017. A phase separation model for transcriptional control. *Cell* **169:** 13–23. doi:10.1016/j.cell.2017.02.007

Hodges C, Bintu L, Lubkowska L, Kashlev M, Bustamante C. 2009. Nucleosomal fluctuations govern the transcription dynamics of RNA polymerase II. *Science* **325:** 626–628. doi:10.1126/science.1172926

Hoskins AA, Friedman LJ, Gallagher SS, Crawford DJ, Anderson EG, Wombacher R, Ramirez N, Cornish VW, Gelles J, Moore MJ. 2011. Ordered and dynamic assembly of single spliceosomes. *Science* **331:** 1289–1295. doi:10.1126/science.1198830

Howe KJ, Kane CM, Ares M Jr. 2003. Perturbation of transcription elongation influences the fidelity of internal exon inclusion in *Saccharomyces cerevisiae*. *RNA* **9:** 993–1006. doi:10.1261/rna.5390803

Kfir N, Lev-Maor G, Glaich O, Alajem A, Datta A, Sze SK, Meshorer E, Ast G. 2015. SF3B1 association with chromatin determines splicing outcomes. *Cell Rep* **11:** 618–629. doi:10.1016/j.celrep.2015.03.048

Khodor YL, Rodriguez J, Abruzzi KC, Tang CH, Marr MT II, Rosbash M. 2011. Nascent-seq indicates widespread cotranscriptional pre-mRNA splicing in *Drosophila*. *Genes Dev* **25:** 2502–2512. doi:10.1101/gad.178962.111

Khodor YL, Menet JS, Tolan M, Rosbash M. 2012. Cotranscriptional splicing efficiency differs dramatically between *Drosophila* and mouse. *RNA* **18:** 2174–2186. doi:10.1261/rna.034090.112

Kim SW, Taggart AJ, Heintzelman C, Cygan KJ, Hull CG, Wang J, Shrestha B, Fairbrother WG. 2017. Widespread intra-dependencies in the removal of introns from human transcripts. *Nucleic Acids Res* **45:** 9503–9513. doi:10.1093/nar/gkx661

Kogan S, Trifonov EN. 2005. Gene splice sites correlate with nucleosome positions. *Gene* **352:** 57–62. doi:10.1016/j.gene.2005.03.004

Kotovic KM, Lockshon D, Boric L, Neugebauer KM. 2003. Cotranscriptional recruitment of the U1 snRNP to intron-containing genes in yeast. *Mol Cell Biol* **23:** 5768–5779. doi:10.1128/MCB.23.16.5768-5779.2003

Kress TL, Krogan NJ, Guthrie C. 2008. A single SR-like protein, Npl3, promotes pre-mRNA splicing in budding yeast. *Mol Cell* **32:** 727–734. doi:10.1016/j.molcel.2008.11.013

Kuhne T, Wieringa B, Reiser J, Weissmann C. 1983. Evidence against a scanning model of RNA splicing. *EMBO J* **2:** 727–733. doi:10.1002/j.1460-2075.1983.tb01492.x

Lacadie SA, Tardiff DF, Kadener S, Rosbash M. 2006. In vivo commitment to yeast cotranscriptional splicing is sensitive to transcription elongation mutants. *Genes Dev* **20:** 2055–2066. doi:10.1101/gad.1434706

Larson JD, Hoskins AA. 2017. Dynamics and consequences of spliceosome E complex formation. *Elife* **6:** e27592. doi:10.7554/eLife.27592

Lerner MR, Boyle JA, Mount SM, Wolin SL, Steitz JA. 1980. Are snRNPs involved in splicing? *Nature* **283:** 220–224. doi:10.1038/283220a0

Leung CS, Douglass SM, Morselli M, Obusan MB, Pavlyukov MS, Pellegrini M, Johnson TL. 2019. H3K36 methylation and the chromodomain protein Eaf3 are required for proper cotranscriptional spliceosome assembly. *Cell Rep* **27:** 3760–3769 e3764. doi:10.1016/j.celrep.2019.05.100

Lewis JD, Izaurralde E, Jarmolowski A, McGuigan C, Mattaj IW. 1996. A nuclear cap-binding complex facilitates association of U1 snRNP with the cap-proximal 5′ splice site. *Genes Dev* **10:** 1683–1698. doi:10.1101/gad.10.13.1683

Li X, Liu S, Zhang L, Issaian A, Hill RC, Espinosa S, Shi S, Cui Y, Kappel K, Das R, et al. 2019. A unified mechanism for intron and exon definition and back-splicing. *Nature* **573:** 375–380. doi:10.1038/s41586-019-1523-6

Lidschreiber M, Leike K, Cramer P. 2013. Cap completion and C-terminal repeat domain kinase recruitment underlie the initiation-elongation transition of RNA polymerase II. *Mol Cell Biol* **33:** 3805–3816. doi:10.1128/MCB.00361-13

Lin RJ, Newman AJ, Cheng SC, Abelson J. 1985. Yeast mRNA splicing in vitro. *J Biol Chem* **260:** 14780–14792.

Liu YC, Cheng SC. 2015. Functional roles of DExD/H-box RNA helicases in Pre-mRNA splicing. *J Biomed Sci* **22:** 54. doi:10.1186/s12929-015-0161-z

Lopez PJ, Seraphin B. 1999. Genomic-scale quantitative analysis of yeast pre-mRNA splicing: implications for splice-site recognition. *RNA* **5:** 1135–1137. doi:10.1017/S135583829999091X

Luco RF, Misteli T. 2011. More than a splicing code: integrating the role of RNA, chromatin and non-coding RNA in alternative splicing regulation. *Curr Opin Genet Dev* **21:** 366–372. doi:10.1016/j.gde.2011.03.004

Mao YS, Zhang B, Spector DL. 2011. Biogenesis and function of nuclear bodies. *Trends Genet* **27:** 295–306. doi:10.1016/j.tig.2011.05.006

Mason PB, Struhl K. 2005. Distinction and relationship between elongation rate and processivity of RNA polymerase II in vivo. *Mol Cell* **17:** 831–840. doi:10.1016/j.molcel.2005.02.017

Mayer A, di Iulio J, Maleri S, Eser U, Vierstra J, Reynolds A, Sandstrom R, Stamatoyannopoulos JA, Churchman LS. 2015. Native elongating transcript sequencing reveals human transcriptional activity at nucleotide resolution. *Cell* **161:** 541–554. doi:10.1016/j.cell.2015.03.010

Minocha R, Popova V, Kopytova D, Misiak D, Huttelmaier S, Georgieva S, Strasser K. 2018. Mud2 functions in transcription by recruiting the Prp19 and TREX complexes to transcribed genes. *Nucleic Acids Res* **46:** 9749–9763. doi:10.1093/nar/gky640

Moehle EA, Ryan CJ, Krogan NJ, Kress TL, Guthrie C. 2012. The yeast SR-like protein Npl3 links chromatin modification to mRNA processing. *PLoS Genet* **8:** e1003101. doi:10.1371/journal.pgen.1003101

Moore MJ, Sharp PA. 1993. Evidence for two active sites in the spliceosome provided by stereochemistry of pre-mRNA splicing. *Nature* **365:** 364–368. doi:10.1038/365364a0

Munding EM, Shiue L, Katzman S, Donohue JP, Ares M Jr. 2013. Competition between pre-mRNAs for the splicing machinery drives global regulation of splicing. *Mol Cell* **51:** 338–348. doi:10.1016/j.molcel.2013.06.012

Natalizio BJ, Robson-Dixon ND, Garcia-Blanco MA. 2009. The carboxyl-terminal domain of RNA polymerase II is not sufficient to enhance the efficiency of pre-mRNA capping or splicing in the context of a different polymerase. *J Biol Chem* **284:** 8692–8702. doi:10.1074/jbc.M806919200

Neugebauer KM, Roth MB. 1997. Distribution of pre-mRNA splicing factors at sites of RNA polymerase II transcription. *Genes Dev* **11:** 1148–1159. doi:10.1101/gad.11.9.1148

Neves LT, Douglass S, Spreafico R, Venkataramanan S, Kress TL, Johnson TL. 2017. The histone variant H2A.Z promotes efficient cotranscriptional splicing in *S. cerevisiae*. *Genes Dev* **31:** 702–717. doi:10.1101/gad.295188.116

Nissen KE, Homer CM, Ryan CJ, Shales M, Krogan NJ, Patrick KL, Guthrie C. 2017. The histone variant H2A.Z promotes splicing of weak introns. *Genes Dev* **31:** 688–701. doi:10.1101/gad.295287.116

Nojima T, Gomes T, Grosso ARF, Kimura H, Dye MJ, Dhir S, Carmo-Fonseca M, Proudfoot NJ. 2015. Mammalian NET-seq reveals genome-wide nascent transcription coupled to RNA processing. *Cell* **161:** 526–540. doi:10.1016/j.cell.2015.03.027

Nojima T, Rebelo K, Gomes T, Grosso AR, Proudfoot NJ, Carmo-Fonseca M. 2018. RNA polymerase II phosphorylated on CTD serine 5 interacts with the spliceosome during co-transcriptional splicing. *Mol Cell* **72:** 369–379 e364. doi:10.1016/j.molcel.2018.09.004

Oesterreich FC, Herzel L, Straube K, Hujer K, Howard J, Neugebauer KM. 2016. Splicing of nascent RNA coincides with intron exit from RNA polymerase II. *Cell* **165:** 372–381. doi:10.1016/j.cell.2016.02.045

Pabis M, Neufeld N, Steiner MC, Bojic T, Shav-Tal Y, Neugebauer KM. 2013. The nuclear cap-binding complex interacts with the U4/U6.U5 tri-snRNP and promotes spliceosome assembly in mammalian cells. *RNA* **19:** 1054–1063. doi:10.1261/rna.037069.112

Pai AA, Henriques T, McCue K, Burkholder A, Adelman K, Burge CB. 2017. The kinetics of pre-mRNA splicing in the Drosophila genome and the influence of gene architecture. *Elife* **6:** e32537. doi:10.7554/eLife.32537

Parenteau J, Maignon L, Berthoumieux M, Catala M, Gagnon V, Abou Elela S. 2019. Introns are mediators of cell response to starvation. *Nature* **565:** 612–617. doi:10.1038/s41586-018-0859-7

Patrick KL, Ryan CJ, Xu J, Lipp JJ, Nissen KE, Roguev A, Shales M, Krogan NJ, Guthrie C. 2015. Genetic interaction mapping reveals a role for the SWI/SNF nucleosome remodeler in spliceosome activation in fission yeast. *PLoS Genet* **11:** e1005074. doi:10.1371/journal.pgen.1005074

Puig O, Gottschalk A, Fabrizio P, Seraphin B. 1999. Interaction of the U1 snRNP with nonconserved intronic sequences affects 5′ splice site selection. *Genes Dev* **13:** 569–580. doi:10.1101/gad.13.5.569

Qin D, Huang L, Wlodaver A, Andrade J, Staley JP. 2016. Sequencing of lariat termini in *S. cerevisiae* reveals 5′ splice sites, branch points, and novel splicing events. *RNA* **22:** 237–253. doi:10.1261/rna.052829.115

Qiu ZR, Schwer B, Shuman S. 2011. Determinants of Nam8-dependent splicing of meiotic pre-mRNAs. *Nucleic Acids Res* **39:** 3427–3445. doi:10.1093/nar/gkq1328

Ramanathan A, Robb GB, Chan SH. 2016. mRNA capping: biological functions and applications. *Nucleic Acids Res* **44:** 7511–7526. doi:10.1093/nar/gkw551

Rasmussen EB, Lis JT. 1993. In vivo transcriptional pausing and cap formation on three *Drosophila* heat shock genes. *Proc Natl Acad Sci* **90:** 7923–7927. doi:10.1073/pnas.90.17.7923

Ruzickova S, Stanek D. 2017. Mutations in spliceosomal proteins and retina degeneration. *RNA Biol* **14:** 544–552. doi:10.1080/15476286.2016.1191735

Saldi T, Fong N, Bentley DL. 2018. Transcription elongation rate affects nascent histone pre-mRNA folding and 3′ end processing. *Genes Dev* **32:** 297–308. doi:10.1101/gad.310896.117

Sass H, Pederson T. 1984. Transcription-dependent localization of U1 and U2 small nuclear ribonucleoproteins at major sites of gene activity in polytene chromosomes. *J Mol Biol* **180:** 911–926. doi:10.1016/0022-2836(84)90263-8

Schmidt U, Basyuk E, Robert MC, Yoshida M, Villemin JP, Auboeuf D, Aitken S, Bertrand E. 2011. Real-time imaging of cotranscriptional splicing reveals a kinetic model that reduces noise: implications for alternative splicing regulation. *J Cell Biol* **193:** 819–829. doi:10.1083/jcb.201009012

Schor IE, Rascovan N, Pelisch F, Allo M, Kornblihtt AR. 2009. Neuronal cell depolarization induces intragenic chromatin modifications affecting NCAM alternative splicing. *Proc Natl Acad Sci* **106:** 4325–4330. doi:10.1073/pnas.0810666106

Schor IE, Fiszbein A, Petrillo E, Kornblihtt AR. 2013. Intragenic epigenetic changes modulate NCAM alternative splicing in neuronal differentiation. *EMBO J* **32:** 2264–2274. doi:10.1038/emboj.2013.167

Schwer B, Shuman S. 1996. Conditional inactivation of mRNA capping enzyme affects yeast pre-mRNA splicing in vivo. *RNA* **2:** 574–583.

Schwer B, Sanchez AM, Shuman S. 2012. Punctuation and syntax of the RNA polymerase II CTD code in fission yeast. *Proc Natl Acad Sci* **109:** 18024–18029. doi:10.1073/pnas.1208995109

Sheth U, Parker R. 2003. Decapping and decay of messenger RNA occur in cytoplasmic processing bodies. *Science* **300:** 805–808. doi:10.1126/science.1082320

Shi Y. 2017. Mechanistic insights into precursor messenger RNA splicing by the spliceosome. *Nat Rev Mol Cell Biol* **18:** 655–670. doi:10.1038/nrm.2017.86

Sims RJ III, Millhouse S, Chen CF, Lewis BA, Erdjument-Bromage H, Tempst P, Manley JL, Reinberg D. 2007. Recognition of trimethylated histone H3 lysine 4 facilitates the recruitment of transcription postinitiation factors and pre-mRNA splicing. *Mol Cell* **28:** 665–676. doi:10.1016/j.molcel.2007.11.010

Sorenson MR, Jha DK, Ucles SA, Flood DM, Strahl BD, Stevens SW, Kress TL. 2016. Histone H3K36 methylation regulates pre-mRNA splicing in *Saccharomyces cerevisiae*. *RNA Biol* **13:** 412–426. doi:10.1080/15476286.2016.1144009

Spies N, Nielsen CB, Padgett RA, Burge CB. 2009. Biased chromatin signatures around polyadenylation sites and exons. *Mol Cell* **36:** 245–254. doi:10.1016/j.molcel.2009.10.008

Spingola M, Ares M. 2000. A yeast intronic splicing enhancer and Nam8p are required for Mer1p-activated splicing. *Mol Cell* **6:** 329–338. doi:10.1016/S1097-2765(00)00033-2

Spingola M, Grate L, Haussler D, Ares M. 1999. Genome-wide bioinformatic and molecular analysis of introns in *Saccharomyces cerevisiae*. *RNA* **5:** 221–234. doi:10.1017/S135583829981682

Talkish J, Igel H, Perriman RJ, Shiue L, Katzman S, Munding EM, Shelansky R, Donohue JP, Ares M Jr. 2019. Rapidly evolving protointrons in *Saccharomyces* genomes revealed by a hungry spliceosome. *PLoS Genet* **15:** e1008249. doi:10.1371/journal.pgen.1008249

Tang Q, Rodriguez-Santiago S, Wang J, Pu J, Yuste A, Gupta V, Moldon A, Xu YZ, Query CC. 2016. SF3B1/Hsh155 HEAT motif mutations affect interaction with the spliceosomal ATPase Prp5, resulting in altered branch site selectivity in pre-mRNA splicing. *Genes Dev* **30:** 2710–2723. doi:10.1101/gad.291872.116

Tani T, Ohshima Y. 1989. The gene for the U6 small nuclear RNA in fission yeast has an intron. *Nature* **337:** 87–90. doi:10.1038/337087a0

Tardiff DF, Lacadie SA, Rosbash M. 2006. A genome-wide analysis indicates that yeast pre-mRNA splicing is predominantly posttranscriptional. *Mol Cell* **24:** 917–929. doi:10.1016/j.molcel.2006.12.002

Tilgner H, Nikolaou C, Althammer S, Sammeth M, Beato M, Valcarcel J, Guigo R. 2009. Nucleosome positioning as a determinant of exon recognition. *Nat Struct Mol Biol* **16:** 996–1001. doi:10.1038/nsmb.1658

Tilgner H, Knowles DG, Johnson R, Davis CA, Chakrabortty S, Djebali S, Curado J, Snyder M, Gingeras TR, Guigo R. 2012. Deep sequencing of subcellular RNA fractions shows splicing to be predominantly co-transcriptional in the human genome but inefficient for lncRNAs. *Genome Res* **22:** 1616–1625. doi:10.1101/gr.134445.111

Tilgner H, Jahanbani F, Blauwkamp T, Moshrefi A, Jaeger E, Chen F, Harel I, Bustamante CD, Rasmussen M, Snyder MP. 2015. Comprehensive transcriptome analysis using synthetic long-read sequencing reveals molecular co-association of distant splicing events. *Nat Biotechnol* **33:** 736–742. doi:10.1038/nbt.3242

Tilgner H, Jahanbani F, Gupta I, Collier P, Wei E, Rasmussen M, Snyder M. 2018. Microfluidic isoform sequencing shows widespread splicing coordination in the human transcriptome. *Genome Res* **28:** 231–242. doi:10.1101/gr.230516.117

Trcek T, Larson DR, Moldon A, Query CC, Singer RH. 2011. Single-molecule mRNA decay measurements reveal promoter-regulated mRNA stability in yeast. *Cell* **147:** 1484–1497. doi:10.1016/j.cell.2011.11.051

Tseng CK, Cheng SC. 2008. Both catalytic steps of nuclear pre-mRNA splicing are reversible. *Science* **320:** 1782–1784. doi:10.1126/science.1158993

Van Treeck B, Protter DSW, Matheny T, Khong A, Link CD, Parker R. 2018. RNA self-assembly contributes to stress granule formation and defining the stress granule transcriptome. *Proc Natl Acad Sci* **115:** 2734–2739. doi:10.1073/pnas.1800038115

Veloso A, Kirkconnell KS, Magnuson B, Biewen B, Paulsen MT, Wilson TE, Ljungman M. 2014. Rate of elongation by RNA polymerase II is associated with specific gene features and epigenetic modifications. *Genome Res* **24:** 896–905. doi:10.1101/gr.171405.113

Wallace EWJ, Beggs JD. 2017. Extremely fast and incredibly close: cotranscriptional splicing in budding yeast. *RNA* **23:** 601–610. doi:10.1261/rna.060830.117

Wilkinson ME, Charenton C, Nagai K. 2019. RNA splicing by the spliceosome. *Annu Rev Biochem* doi:10.1146/annurev-biochem-091719-064225

Will CL, Luhrmann R. 2011. Spliceosome structure and function. *Cold Spring Harb Perspect Biol* **3:** a003707. doi:10.1101/cshperspect.a003707

Windhager L, Bonfert T, Burger K, Ruzsics Z, Krebs S, Kaufmann S, Malterer G, L'Hernault A, Schilhabel M, Schreiber S, et al. 2012. Ultrashort and progressive 4sU-tagging reveals key characteristics of RNA processing at nucleotide resolution. *Genome Res* **22:** 2031–2042. doi:10.1101/gr.131847.111

Wong MS, Kinney JB, Krainer AR. 2018. Quantitative activity profile and context dependence of all human 5′ splice sites. *Mol Cell* **71:** 1012–1026 e1013. doi:10.1016/j.molcel.2018.07.033

Zhong XY, Wang P, Han J, Rosenfeld MG, Fu XD. 2009. SR proteins in vertical integration of gene expression from transcription to RNA processing to translation. *Mol Cell* **35:** 1–10. doi:10.1016/j.molcel.2009.06.016

Recognition of Poly(A) RNA through Its Intrinsic Helical Structure

Terence T.L. Tang and Lori A. Passmore

MRC Laboratory of Molecular Biology, Cambridge CB2 0QH, United Kingdom
Correspondence: passmore@mrc-lmb.cam.ac.uk

The polyadenosine (poly(A)) tail, which is found on the 3′ end of almost all eukaryotic messenger RNAs (mRNAs), plays an important role in the posttranscriptional regulation of gene expression. Shortening of the poly(A) tail, a process known as deadenylation, is thought to be the first and rate-limiting step of mRNA turnover. Deadenylation is performed by the Pan2–Pan3 and Ccr4–Not complexes that contain highly conserved exonuclease enzymes Pan2, and Ccr4 and Caf1, respectively. These complexes have been extensively studied, but the mechanisms of how the deadenylase enzymes recognize the poly(A) tail were poorly understood until recently. Here, we summarize recent work from our laboratory demonstrating that the highly conserved Pan2 exonuclease recognizes the poly(A) tail, not through adenine-specific functional groups, but through the conformation of poly(A) RNA. Our biochemical, biophysical, and structural investigations suggest that poly(A) forms an intrinsic base-stacked, single-stranded helical conformation that is recognized by Pan2, and that disruption of this structure inhibits both Pan2 and Caf1. This intrinsic structure has been shown to be important in poly(A) recognition in other biological processes, further underlining the importance of the unique conformation of poly(A).

Almost all mature eukaryotic messenger RNAs (mRNAs) carry a 3′ polyadenosine (poly(A)) tail. As a nascent mRNA emerges from RNA polymerase II (Pol II), it is specifically recognized, cleaved, and polyadenylated by the cleavage and polyadenylation factor (CPF in yeast, CPSF in mammals) (Proudfoot 2011). The length of the mature poly(A) tail ranges from ~70 nt in yeast (Brown and Sachs 1998) to ~250 nt in mammalian cells (Brawerman 1981). The poly(A) tail is required for the export of mature mRNA into the cytoplasm (Fuke and Ohno 2008), where it is bound by the cytoplasmic poly(A)-binding protein (Pab1 in yeast, PABPC1 in mammals) (Kuhn and Wahle 2004). In the cytoplasm, poly(A) tails are shortened: In general, the poly(A) tail length of cytoplasmic mRNAs is much shorter than that synthesized in the nucleus (Chang et al. 2014; Lima et al. 2017; Eisen et al. 2020).

The poly(A) tail regulates posttranscriptional gene expression through multiple mechanisms (Fig. 1). First, the poly(A) tail increases the efficiency of translation initiation. This is thought to occur through protein–protein interactions between the cytoplasmic poly(A)-binding protein on the poly(A) tail and the eIF4G subunit of the cap-binding complex at the 5′ cap (Sachs 1990; Tarun and Sachs 1996; Hentze 1997), effectively circularizing the transcript. A circularized form of mRNA has been directly observed by atomic force microscopy in vitro (Wells et al. 1998), but controversy remains regarding whether or not the circularized form of mRNA is prevalent in cells (Pierron and Weil 2018; Vicens et al. 2018).

Second, the poly(A) tail is important for mRNA stability; shortening of the poly(A) tail, or deadenylation, is the first and rate-limiting step of mRNA decay for most eukaryotic transcripts (Chen and Shyu 2011). Once the poly(A) tail is removed, decapping and further 3′–5′ or 5′–3′ degradation can occur. Thus, deadenylation is an important step in the posttranscriptional regulation of gene expression because it determines transcript half-life (Wilson and Treisman 1988; Meyer et al. 2004; Parker and Song 2004). Indeed, deadenylation has been implicated in physiological processes such as development (Nakamura et al. 2004; Morita et al. 2007; Neely et al. 2010) and tumorigenesis (Faraji et al. 2016). In eukaryotes, deadenylation is primarily performed by two highly conserved multiprotein complexes, Pan2–Pan3 and Ccr4–Not, that shorten the poly(A) tail in a 3′–5′ direction. Within these complexes, the exonucleases that carry out deadenylation are Pan2/PAN2, and Ccr4/CNOT6/CNOT6L and Caf1/CNOT7, respectively.

Deadenylase complexes are thought to degrade only the poly(A) tail and not the transcript body. This specificity is thought to arise through the intrinsic specificity of the exonuclease enzymes for adenosines, as well as through specific interactions of the poly(A) tail with subunits of the conserved deadenylase complexes. Here, we review the molecular basis of poly(A) specificity in deadenylation. These data reveal the importance of the unique intrinsic conformation of the poly(A) tail that is also exploited for its recognition in several other biological processes.

POLY(A) RECOGNITION BY DEADENYLASE COMPLEXES

The activities of the *Schizosaccharomyces pombe* and *Homo sapiens* Ccr4–Not complexes have been shown to

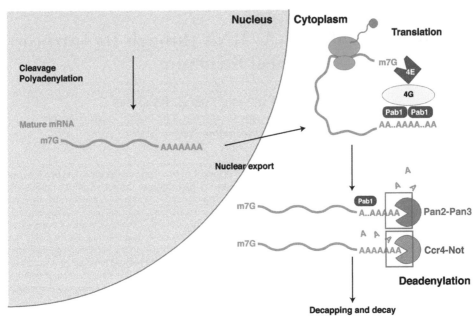

Figure 1. Overview of the role of poly(A) in posttranscriptional regulation of gene expression. The poly(A) tail is added to the 3′ end of nascent transcripts by the cleavage and polyadenylation machinery in the eukaryotic nucleus. The polyadenylated mature mRNA can be exported into the cytoplasm, where it is bound by poly(A)-binding protein to increase the efficiency of translation. Cytoplasmic mRNAs also undergo turnover and decay; the first step of mRNA turnover is the shortening of the poly(A) tail, a process known as deadenylation. Once the poly(A) tail is shortened, translation is inhibited and the transcript undergoes decapping and degradation.

be specific for adenosines in in vitro deadenylation assays as the complexes stall upon encountering non-A stretches upstream of the poly(A) tail (Stowell et al. 2016; Raisch et al. 2019). The molecular basis of poly(A) recognition by Ccr4 has been elucidated through a crystal structure of human CNOT6L bound to single-stranded poly(A) DNA (Fig. 2A; Wang et al. 2010). In this structure, poly(A) DNA is bound in the active site cleft of the heart-shaped CNOT6L enzyme and the scissile phosphate group points into the base of the cleft toward the active site residues. The interactions between Ccr4 and poly(A) suggest that the specificity for adenine is determined by a hydrogen bond between the carboxyl oxygen of Asn412 and the N6 amine group of the penultimate adenine (A_{-1}), as well as a stacking interaction between the aromatic adenine base and Phe484 (Fig. 2B). Nonetheless, this study did not address whether or not other subunits of the Ccr4–Not complex contribute to the recognition of the poly(A) tail, and how Caf1, the other exonuclease of the Ccr4–Not complex, specifies for poly(A).

Within the other major deadenylase complex, Pan2–Pan3, several domains and motifs contribute to poly(A) specificity. Pan2–Pan3 is composed of one Pan2 and two Pan3 molecules (Jonas et al. 2014; Schäfer et al. 2014, 2019; Wolf et al. 2014). The Pan3 subunit binds to the carboxy-terminal domain of cytoplasmic poly(A)-binding protein via a polypeptide stretch known as the PABP-interacting motif 2 (PAM2 motif), thereby recruiting the Pan2–Pan3 complex to the poly(A) tail (Siddiqui et al. 2007). Moreover, Pan3 contains an amino-terminal zinc finger domain that specifically binds poly(A) (Wolf et al. 2014). Thus, Pan3 contributes to the recognition of

poly(A) RNA, but it was unclear whether the exonuclease domain of Pan2 also contained intrinsic specificity for adenosine.

SPECIFICITY OF Pan2 AND Caf1 EXONUCLEASES

To determine the nucleotide specificities of the DEDD exonucleases Caf1 and Pan2, we used in vitro biochemical assays with recombinant Pan2–Pan3 or with Ccr4–Not containing a catalytic mutant of Ccr4 (such that Caf1 was the only active nuclease) (Tang et al. 2019). Both enzymes showed a preference for poly(A) when incubated with fluorescently labeled RNA substrates containing poly(A) tails with varied 3′-terminal nucleotides (A_{30}-U_3, -C_3, -G_3) (Fig. 3A,B). Caf1 shows strict specificity for poly(A) and is inhibited by all non-A nucleotides, whereas Pan2 is substantially inhibited only by guanosines at the end of a poly(A) tail (Fig. 3A,B). These specificities likely prevent the 3′ untranslated region (UTR) of the transcript from being degraded by deadenylase complexes. Notably, both deadenylase complexes are generally inhibited when they reach the end of the poly(A) tail.

NON-A SEQUENCES IN POLY(A) TAILS

In cells, the poly(A) tail can be modified by the addition of other nucleotides. After deadenylation, shortened oligo(A) tails can be marked by an oligo(U) tail deposited by terminal uridyl transferases (TUTases) to label specific transcripts for degradation (Rissland and Norbury 2009;

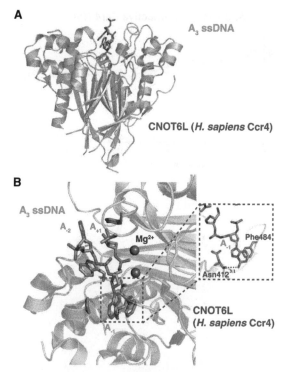

Figure 2. Recognition of poly(A) by *Homo sapiens* CNOT6L. A crystal structure of the Ccr4 deadenylase bound to poly(A) DNA (PDB: 3NGO) revealed the molecular determinants of poly(A) specificity of CNOT6L (Ccr4 in other organisms). (*A*) An overview of the CNOT6L structure (purple, shown as cartoon) bound to poly(A) DNA (green, shown as sticks). (*B*) A close-up view of poly(A) DNA in the CNOT6L active site. The scissile phosphate group is coordinated by two metal ions (shown as spheres) in the active site, which in turn is lined by hydrophobic residues (shown as sticks). (*Inset*) Interactions of the adenosine 5′ to the scissile phosphate (A_{-1}) with the carbonyl oxygen of Asn412 and the phenyl ring of Phe484.

Lim et al. 2014). Recent studies have also identified that non-A nucleotides can be incorporated throughout mammalian poly(A) tails at a low frequency (Chang et al. 2014; Legnini et al. 2019) by the noncanonical poly(A) polymerases TENT4A (PAPD7) and TENT4B (PAPD5) (Lim et al. 2018). The presence of guanosines within the poly(A) tail correlates with increased transcript half-life, suggesting that these modifications may affect transcript stability (Chang et al. 2014).

The lack of inhibitory effect by uracils on Pan2–Pan3 and Ccr4–Not suggests that the oligo(U) tail alone does not impede deadenylation (Tang et al. 2019). Nonetheless, an oligo(U) tail is thought to recruit RNA-binding proteins, such as the Lsm complex, that could block deadenylation (Song and Kiledjian 2007). In contrast, the presence of guanosines within the poly(A) tail inhibits both Pan2–Pan3 and Ccr4–Not (Fig. 3A,B). This in vitro result agrees with the observation that guanylated poly(A) tails correlate positively with transcript half-life, suggesting that guanylation could be a mechanism by which transcripts are selectively stabilized by inhibition of deadenylation. It remains unclear whether the incorporation of non-A nucleotides into the poly(A) tail is a regulated or stochastic process, and how significant this process is in regulating gene expression in a global or transcript-specific manner.

POLY(A) RECOGNITION BY Pan2

We used X-ray crystallography to further investigate DEDD deadenylase specificity. Previous crystal structures of the carboxy-terminal half of Pan2 in the absence of RNA had revealed that it consists of a pseudo–ubiquitin hydrolase (UCH) domain and an exonuclease (Exo) domain with the two domains forming a contiguous structural unit (UCH-Exo) (Jonas et al. 2014; Schäfer et al. 2014). We determined the molecular basis of poly(A) recognition by the Pan2 exonuclease from a crystal structure of a UCH-Exo catalytic mutant from *Saccharomyces cerevisiae* bound to oligo(A) RNA (Fig. 3C; Tang et al. 2019). Pan2 does not undergo any major conformational changes upon substrate binding, suggesting that the UCH-Exo domains are rigid. Although one of the metal-coordinating residues was mutated to prevent RNA degradation, the RNA scissile phosphate bond faced the key catalytic residues within the active site, consistent with productive RNA binding. Furthermore, this structure revealed the contacts between Pan2 and oligo(A) RNA in the active site, including a π-stacking interaction between the terminal adenine and the phenyl group of Y975, as well as putative hydrogen bonds between amino acid residues (F913, N1019, Y1046, S1048, and L1049) and the ribophosphate backbone (Fig. 3C, inset). Surprisingly, apart from the stacking interaction of the terminal adenine, there were no interactions between Pan2 and the adenine bases, raising the question of how Pan2 specifically recognizes poly(A).

THE INTRINSIC STRUCTURE OF POLY(A)

Within the crystal structure, oligo(A) formed a single-stranded A-form-like helix in the Pan2 active site, where each adenine base was π-stacked in an offset parallel manner onto adjacent bases (Fig. 3C, inset). This suggested that Pan2 may recognize the shape of the RNA, instead of directly binding functional groups specific to adenine.

The in vitro conformation of poly(A) RNA has been extensively investigated by circular dichroism (CD) (Brahms et al. 1966; Hashizume and Imahori 1967), temperature jump studies (Dewey and Turner 1979), and crystallography (Suck et al. 1976). More recently, the conformation has been further studied by atomic force microscopy (Smith et al. 1997), optical tweezers (Seol et al. 2007), nuclear magnetic resonance (NMR) (Isaksson et al. 2004), and protein nanopores (Lin et al. 2010). From these studies, it had been proposed that poly(A) can form a single-stranded A-form helix at physiological pH in solution, with the adenine bases stacked in a roughly parallel orientation (Saenger 1984; Bloomfield et al. 1999). The helical conformation adopted by oligo(A) in the Pan2 active site (Fig. 3D) is similar to that hypothesized by Saenger et al. (1975) derived from the stacked configuration of two adenosines in a crystal structure of

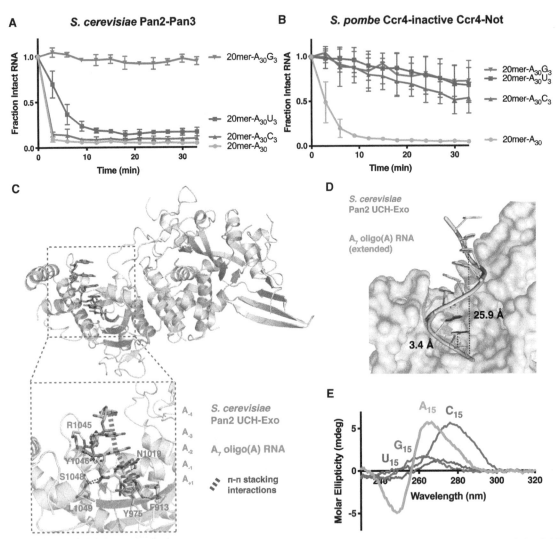

Figure 3. Nucleotide specificity of DEDD deadenylases. (*A,B*) In vitro deadenylation by *Saccharomyces cerevisiae* Pan2–Pan3 (*A*) or *Schizosaccharomyces pombe* Ccr4-inactive Ccr4–Not (*B*) on RNA substrates with a 30-nt poly(A) tail harboring different 3′ ribonucleotides. The disappearance of the substrate was quantified by intensity measurements of the band corresponding to the intact, fluorescently labeled RNA. Data were normalized to that at time = 0 and individual points are connected by straight lines for clarity. Assays were performed in triplicate; the points represent the mean, and the error bars represent standard deviation. (Reproduced from Tang et al. 2019, Figs. 1B and 2C.) (*C*) Crystal structure of the Pan2 UCH and exonuclease (Exo) domains (blue, shown as cartoon) bound to a poly(A) RNA (green, shown as sticks) (PDB: 6R9J). (*Inset*) Protein–RNA interactions in the Pan2 active site. Amino acids involved in the interaction are shown as sticks and labeled in blue. Putative hydrogen bonds are indicated with black dashed lines, and π–π stacking interactions are indicated with thick gray dashes. (*D*) Extended oligo(A) helix bound to Pan2, modeled by duplication and superposition of the observed A_5. Distances are shown in Ångstroms. (*E*) Circular dichroism spectra of A_{15} (green), U_{15} (purple), C_{15} (blue), and G_{15} (red) RNAs. (Reproduced from Tang et al. 2019, Fig. 7B.)

A_3 RNA. Given the lack of base-specific contacts between adenine functional groups and the protein, our data suggested that Pan2 recognizes the intrinsic structure formed by oligo(A).

To assess if poly(A) RNA forms an intrinsic structure in solution in the absence of protein, we used CD to study 15-mer polyribonucleotides. CD spectra are sensitive to higher-order chiral structures formed by a macromolecule. Poly(A) is unique in forming a signature peak (265-nm) and trough (250-nm) structure, which cannot be found with other polyribonucleotides (Fig. 3E). Interestingly, poly(C) adopts a different structure with a peak at 278 nm, presumably corresponding to a previously solved crystal structure of poly(C) RNA (Akinrimisi et al. 1963; Arnott et al. 1976) with different characteristic helical parameters relative to poly(A). Thus, in vitro, poly(A) RNA forms a unique structure in solution compared to other polyribonucleotides.

THE STACKED, HELICAL STRUCTURE OF POLY(A) IS IMPORTANT IN DEADENYLATION

In vitro deadenylation assays showed that Pan2 was not strongly inhibited by uracils (Us) or cytosines (Cs) at the end of a poly(A) tail (Fig. 3A,B), but these nucleotides did not show the characteristic CD signature of helical poly(A), either alone (Fig. 3E) or in the context of oligo (A) (Fig. 4A). If Pan2 recognizes the unique helical structure of poly(A) in its active site, how does it remove these non-A nucleotides? In crystal structures of Pan2 bound to different oligonucleotides, we observed that oligo(A) RNA containing two Us or Cs forms a similar stacked, helical structure to poly(A) in the active site (Fig. 4B; Tang et al. 2019). Thus, C- and U-containing RNAs can form the π-stacking interactions necessary for the helical conformation while bound to the Pan2 active site, further suggesting that Pan2 specifically recognizes the formation of a poly(A) helix-like structure. As these C- and U-containing RNAs do not adopt an intrinsic poly(A)-like structure, the formation of this structure in the Pan2 active site likely comes at an entropic cost, leading to a small reduction in Pan2 activity on these substrates.

In contrast, the presence of guanosines (Gs) disrupts the stacked, helical structure of poly(A) as the crystal structure revealed that the G-containing RNA is unstacked in the Pan2 active site (Fig. 4B; Tang et al. 2019). This unstacking likely disrupts the correct recognition of the ribophosphate backbone and in particular the scissile phosphate bond, leading to the inhibition of deadenylation. Although previous studies of dinucleotides have predicted that guanosines have energetically favorable stacking interactions with each other and with adenosines (Friedman and Honig 1995; Brown et al. 2015), the configurations of these stacked guanosine dinucleotides cannot be accommodated within an A-form-like single-stranded RNA helix. This leads to unstacking within the context of poly(A) and disruption of ideal helical geometry.

Pan2 recognition of the stacked, helical form of poly(A) was further tested using modified nucleotides that inhibit stacking. Dihydrouracil (DHU) is a uracil analog that contains the same functional groups as uracil, except for a C–C single bond between C5 and C6 instead of a C=C double bond. As such, DHU is nonplanar and disrupts stacking interactions between adjacent bases. The introduction of two DHUs into a poly(A) tail strongly inhibits Pan2 activity relative to two uracils, which only cause a slight stall in deadenylation (Fig. 4C; Tang et al. 2019). This supports the finding that disruption of base-stacking inhibits Pan2, and that Pan2 requires its substrate to adopt a poly(A)-like stacked, helical conformation.

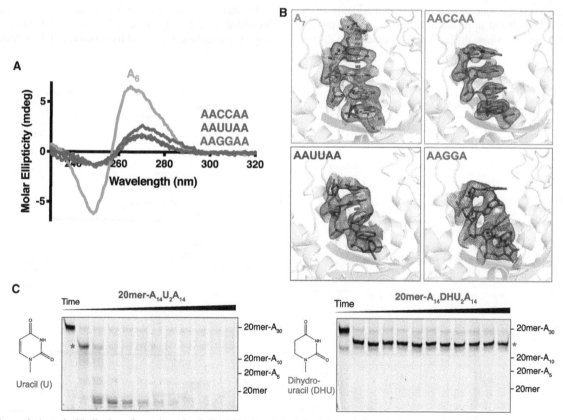

Figure 4. A stacked helical conformation of poly(A) RNA is required for Pan2 activity. (*A*) Circular dichroism spectra of A_6 (green), AAUUAA (purple), AACCAA (blue), and AAGGAA (red) RNAs. (*B*) Crystal structures of Pan2 UCH-Exo (shown as light blue cartoon) bound to oligo(A) RNAs (shown as green sticks) interrupted by two uracils (purple, PDB: 6R9P), cytosines (blue, PDB: 6R9Q), or guanosines (red, PDB: 6R9O). The electron density shown represents feature-enhanced maps (FEM) contoured to 1.8 σ. π-stacking interactions between adjacent bases, if any, are denoted in thick dashed lines in gray. (*C*) In vitro deadenylation assays of Pan2 UCH-Exo on fluorescently labeled RNAs with 30-nt poly(A) tails interrupted either by two uracils (*left*, purple) or two dihydrouracils (*right*, pink). The chemical structures of uracil and dihydrouracil are shown alongside each assay for reference. The red asterisk indicates the poly(A) tail length at which stalling occurs (-UUAAA or -DDAAAA). (Reproduced from Tang et al. 2019, Fig, 6C,D.)

EFFECT OF ADENINE MODIFICATIONS ON DEADENYLATION

Why does the planar guanine disrupt stacking of a poly(A) helix? Guanines and adenines differ in the position of an amino group and the addition of a carbonyl group around the purine moiety (Fig. 5A). This affects the electronic distribution within the aromatic ring system through inductive and resonance effects. The distribution of functional groups in adenine likely enables electrostatic complementarity when adenines are stacked on top of each other in a single-stranded helix. To test this hypothesis, we designed RNA substrates with poly(A) tails interrupted by two purines (Ps), which lack the amine group on C6 (Fig. 5B), as well as two 2-aminopurines (2APs), which lack the amine group on C6 and contain an additional amine group on C2, compared to adenine (Fig. 5C). These were then tested in deadenylation assays with Pan2 and compared to an RNA containing a pure poly(A) tail (Fig. 5A; TTL Lang, LA Passmore, unpubl. data). The introduction of purines or 2APs results in a strong inhibition of Pan2 exonuclease activity (Fig. 5B,C; TTL Lang, LA Passmore, unpubl.). This is in agreement with an important role for the electrostatic distribution of functional groups in forming the stacked, helical conformation of poly(A). If the functional groups within adenine are removed or altered in position, as in the case of purines, 2APs, and guanines, the poly(A)-like structure is disrupted, which in turn inhibits Pan2. Thus, the electronic distribution around the central purine of adenine is unique and likely enables its intrinsic stacked, helical structure.

Adenine can be chemically modified in vivo by the addition of a methyl group onto the N6 amine to form N^6-methyladenosine (m^6A) (Shi et al. 2019). This modification can be specifically recognized by diverse RNA-binding proteins, regulating processes such as RNA degradation (Du et al. 2016) and splicing (Xiao et al. 2016). Adenine can also be deaminated at the N6 position to inosine (Alseth et al. 2014), which has been reported in mRNAs (Paul and Bass 1998). Deamination of adenine has been implicated in numerous human diseases such as psychiatric disorders and cancers (Slotkin and Nishikura 2013). Importantly, these modifications can cause subtle changes in the electronic distribution of adenine.

To test whether these modifications affect the stacked, helical conformation of poly(A) and thus deadenylation by Pan2, we introduced two m^6As or two inosines into the poly(A) tail and tested the substrates in deadenylation assays with Pan2 (Fig. 5D,E: TTL Lang, LA Passmore, unpubl. data). We observed that m^6A had almost no effect on deadenylation by Pan2 relative to the unmodified poly(A) tail (Fig. 5D). In contrast, when Pan2 encounters inosines, there is a stall in deadenylation (Fig. 5E).

To date, m^6A and inosine nucleotides have not been identified in the poly(A) tail; however, limitations in sequencing techniques would have precluded their detec-

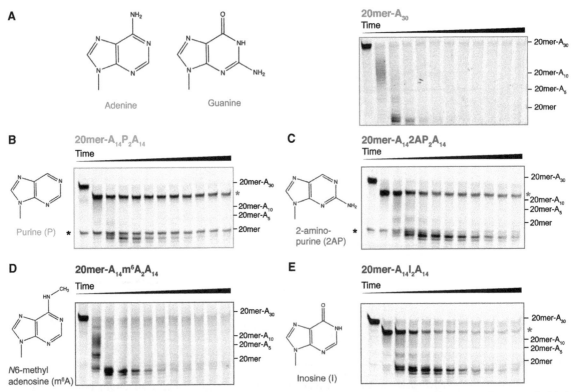

Figure 5. Effect of modified nucleotides on Pan2 activity. In vitro deadenylation assays of Pan2 UCH-Exo on fluorescently labeled RNAs with 30-nt poly(A) tails (*A*) or 30-nt poly(A) tails interrupted by two purine (*B*, orange), two 2-aminopurine (*C*, magenta), two N6-methyladenosine (*D*, dark purple), or two inosine (*E*, gray) nucleotides. The chemical structures of the corresponding bases are shown beside each assay for reference. The red asterisk indicates the poly(A) tail length at which stalling occurs. The black asterisk indicates a contaminating nucleic acid band.

tion. We expect that the increased sensitivity and improvements in sequencing techniques, such as long-read nanopore sequencing, will be able to detect such modifications if they occur within the poly(A) tail (Liu et al. 2019). Overall, our results show that the formation of the helical conformation is dependent on the electronic distribution in adenine, enabling electrostatic complementarity upon adenine base-stacking.

THE POLY(A) STRUCTURE IN OTHER BIOLOGICAL PROCESSES

Although a structure of poly(A) had been previously proposed, it had never been directly observed, and the biological significance of the helical conformation was unclear. Together, our data show that the intrinsic single-stranded helix of poly(A) is exploited by Pan2 for specificity and recognition, and that disruption of this structure is sufficient to inhibit Pan2 (Fig. 6A).

Caf1 is a DEDD-family exonuclease in the Ccr4–Not complex with structural homology with the Pan2 exonuclease domain. Thus, we hypothesized that Caf1 would also recognize the stacked, helical structure of poly(A). We were unable to obtain crystals of the DEDD deadenylase Caf1 in complex with oligo(A) RNA, but we could model the poly(A) helix into the active site of a previously determined structure of *S. pombe* Caf1 (Andersen et al. 2009). This showed that the poly(A) helix can be accommodated in the Caf1 active site, forming plausible contacts between the ribophosphate backbone and side chains of

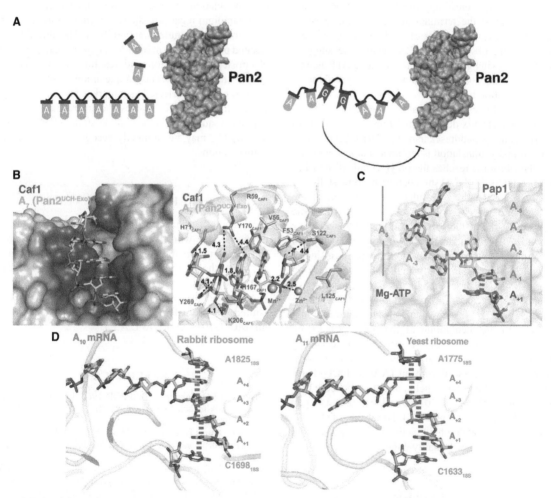

Figure 6. Role of the stacked, single-stranded poly(A) helix in biological processes. (*A*) Proposed model for the recognition of poly(A) RNA by Pan2 through its stacked, single-stranded, helical conformation. Disruption of this conformation inhibits Pan2. (*B*) Proposed model of recognition of poly(A) (shown as green sticks) by Caf1 (shown as pink surface and as cartoon; PDB: 3G0Z). The Caf1 active site (*left*, surface) is colored by proximity to the poly(A) substrate, showing that poly(A) is more buried compared to in the Pan2 active site. Putative protein–RNA contacts between amino acid side chains (labeled, shown as sticks) and the poly(A) substrate are shown in the *right* panel. Possible hydrogen bonds are shown as black dashed lines and are labeled with distance, shown in Ångstroms. (Reproduced from Tang et al. 2019, Supplemental Figs. 4A,B.) (*C*) Structure of yeast poly(A) polymerase Pap1 (shown as khaki surface) bound to poly(A) RNA and an incoming ATP (shown as green sticks) (PDB: 2Q66). The 3′-terminal adenine appears to stack against the adenine of the incoming ATP; the π-stacking interaction is depicted with a thick gray dash. (*D*) Structures of oligo(A) RNA (shown in sticks, green) bound to rabbit ribosome (*left*, shown as orange cartoon; PDB: 6SGC) or yeast ribosome (*right*, shown as orange cartoon; PDB: 6T7T). The oligo(A) RNA is stacked between an adenine and cytosine (shown as orange sticks) of the 18S rRNA. π-stacking interactions are shown as thick gray dashes.

Caf1 (Fig. 6B). Similar to Pan2, Caf1 is strongly inhibited by DHU relative to uracils, suggesting that it may also recognize the stacked, helical conformation of poly(A) (Tang et al. 2019). However, our modeling shows that there are also putative base-specific contacts, consistent with our observation that Caf1 shows greater nucleotide specificity compared to Pan2 (Tang et al. 2019).

Adenine stacking has also been observed in the context of polyadenylation, the process whereby adenosines are processively added by a poly(A) polymerase to the 3' end of a nascent transcript (Kumar et al. 2019). A structure of yeast poly(A) polymerase in complex with ATP and oligo (A) has been determined, providing a model for how adenosines are added to the 3' end of an elongating poly(A) tail (Balbo and Bohm 2007). Most adenines are flipped out to form base-specific contacts with the Pap1 active site (A_{-2}, A_{-3}, A_{-4}, A_{-5}) but, interestingly, the incoming ATP appears to π-stack against the terminal adenine of the existing poly(A) tail (A_{-1}). This mode of stacking is reminiscent of the π-stacking observed in the Pan2 active site, suggesting that the stacking geometry of an incoming ATP against the existing 3' adenosine contributes to the specificity of adenosine addition by Pap1.

Two recent studies have revealed a role for the single-stranded poly(A) helix during translation of a poly(A) tail by the ribosome (Chandrasekaran et al. 2019; Tesina et al. 2020). Normally, translation is terminated at a stop codon before the ribosome reaches the poly(A) tail. If the ribosome encounters a poly(A) stretch, the ribosome stalls and a quality control pathway results in degradation of the transcript and disassembly of the stalled ribosome (Doma and Parker 2006; Shao et al. 2013; Juszkiewicz and Hegde 2017; Sundaramoorthy et al. 2017). In both rabbit and yeast ribosomes, poly(A) RNA forms a single-stranded RNA helix in the A-site of the ribosome, stacking between 18S rRNA bases A1825 and C1698, leading to a structural rearrangement in the decoding center of the ribosome (Fig. 6D). This contributes to ribosome stalling. Analogous to the recognition of the poly(A) helix by Pan2, disruption of the poly(A) helix in the decoding center by the introduction of guanosines—for instance, when poly(A) is replaced with an $(AAG)_n$ tract—leads to disruption of the poly(A) helix, and the ribosome thus does not stall (Arthur et al. 2015; Juszkiewicz and Hegde 2017; Sundaramoorthy et al. 2017). The ability of poly(A) to form a single-stranded helix is thereby recognized and exploited in translational quality control.

CONCLUSION

The recent work described here shows that proteins can recognize the sequence of RNA, particularly poly(A), through the intrinsic conformation of single-stranded RNA. This is reminiscent of the mechanism whereby DNA binding proteins (such as the Trp repressor) use indirect readout to recognize specific DNA sequences (Otwinowski et al. 1988). Poly(A) RNA likely interconverts between helical and unstructured conformations in solution. In the Pan2 active site, this helical conformation is stabilized by contacts with the ribophosphate backbone. Interestingly, the characteristic CD signature of adenosine stacking can be observed with oligo(A) polyribonucleotides as short as A_3 (TTL Lang, LA Passmore, unpubl. data). Proteins can also recognize the poly(A) tail via base-specific interactions. For instance, in the poly(A) polymerase active site, the terminal adenosines within the existing poly(A) tail contact the protein via specific functional groups (Balbo and Bohm 2007). Similarly, the RNA recognition motif domains (RRMs) of the cytoplasmic poly(A) binding protein recognize poly(A) RNA through interactions with base-specific groups with the RNA in an extended, nonhelical conformation (Deo et al. 1999; Safaee et al. 2012).

The single-stranded helical structure of poly(A) is important in deadenylation, translation quality control, and polyadenylation. The studies described here uncover a new paradigm regarding the recognition of the characteristic structure of single-stranded RNA. The ubiquity of the stacked poly(A) helix in biology suggests that, because the electrostatic distribution and resulting conformation of single-stranded poly(A) is unique among polyribonucleotides, it has been selected as a marker for a correctly processed transcript. This raises the interesting possibility that the sequence of other single-stranded RNAs, such as poly(C), may be indirectly recognized through their conformations.

ACKNOWLEDGMENTS

We thank members of the Passmore laboratory for discussions. This work was supported by a Herchel Smith PhD Studentship from the University of Cambridge (to T.T.L.T.); the European Union's Horizon 2020 research and innovation programme (ERC grant no. 725685, to L.A.P.); and the Medical Research Council, as part of United Kingdom Research and Innovation (MRC grant no. MC_U105192715, to L.A.P.).

REFERENCES

Akinrimisi EO, Sander C, Ts'o POP. 1963. Properties of helical polycytidylic acid. *Biochemistry* **2:** 340–344. doi:10.1021/bi00902a028

Alseth I, Dalhus B, Bjørås M. 2014. Inosine in DNA and RNA. *Curr Opin Genet Dev* **26:** 116–123. doi:10.1016/j.gde.2014.07.008

Andersen KR, Jonstrup AT, Van LB, Brodersen DE. 2009. The activity and selectivity of fission yeast Pop2p are affected by a high affinity for Zn^{2+} and Mn^{2+} in the active site. *RNA* **15:** 850–861. doi:10.1261/rna.1489409

Arnott S, Chandrasekaran R, Leslie AGW. 1976. Structure of the single-stranded polyribonucleotide polycytidylic acid. *J Mol Biol* **106:** 735–748. doi:10.1016/0022-2836(76)90262-X

Arthur L, Pavlovic-Djuranovic S, Smith-Koutmou K, Green R, Szczesny P, Djuranovic S. 2015. Translational control by lysine-encoding A-rich sequences. *Sci Adv* **1:** e1500154. doi:10.1126/sciadv.1500154

Balbo PB, Bohm A. 2007. Mechanism of poly(A) polymerase: structure of the enzyme–MgATP–RNA ternary complex and kinetic analysis. *Structure* **15:** 1117–1131. doi:10.1016/j.str.2007.07.010

Bloomfield VA, Crothers DM, Tinoco I. 1999. *Nucleic acids: structures, properties, and functions*. University Science Books, Sausalito, CA.

Brahms J, Michelson AM, Van Holde KE. 1966. Adenylate oligomers in single- and double-strand conformation. *J Mol Biol* 15: 467–488. doi:10.1016/S0022-2836(66)80122-5

Brawerman G. 1981. The role of the poly(A) sequence in mammalian messenger RNA. *CRC Crit Rev Biochem* 10: 1–38. doi:10.3109/10409238109114634

Brown CE, Sachs AB. 1998. Poly(A) tail length control in *Saccharomyces cerevisiae* occurs by message-specific deadenylation. *Mol Cell Biol* 18: 6548–6559. doi:10.1128/MCB.18.11.6548

Brown RF, Andrews CT, Elcock AH. 2015. Stacking free energies of all DNA and RNA nucleoside pairs and dinucleoside-monophosphates computed using recently revised AMBER parameters and compared with experiment. *J Chem Theory Comput* 11: 2315–2328. doi:10.1021/ct501170h

Chandrasekaran V, Juszkiewicz S, Choi J, Puglisi JD, Brown A, Shao S, Ramakrishnan V, Hegde RS. 2019. Mechanism of ribosome stalling during translation of a poly(A) tail. *Nat Struct Mol Biol* 26: 1132–1140. doi:10.1038/s41594-019-0331-x

Chang H, Lim J, Ha M, Kim VN. 2014. TAIL-seq: genome-wide determination of poly(A) tail length and 3′ end modifications. *Mol Cell* 53: 1044–1052. doi:10.1016/j.molcel.2014.02.007

Chen C-Y, Shyu A-B. 2011. Mechanisms of deadenylation-dependent decay. *Wiley Interdiscip Rev RNA* 2: 167–183. doi:10.1002/wrna.40

Deo RC, Bonanno JB, Sonenberg N, Burley SK. 1999. Recognition of polyadenylate RNA by the poly(A)-binding protein. *Cell* 98: 835–845. doi:10.1016/S0092-8674(00)81517-2

Dewey TG, Turner DH. 1979. Laser temperature-jump study of stacking in adenylic acid polymers. *Biochemistry* 18: 5757–5762. doi:10.1021/bi00593a002

Doma MK, Parker R. 2006. Endonucleolytic cleavage of eukaryotic mRNAs with stalls in translation elongation. *Nature* 440: 561–564. doi:10.1038/nature04530

Du H, Zhao Y, He J, Zhang Y, Xi H, Liu M, Ma J, Wu L. 2016. YTHDF2 destabilizes m^6A-containing RNA through direct recruitment of the CCR4–NOT deadenylase complex. *Nat Commun* 7: 12626. doi:10.1038/ncomms12626

Eisen TJ, Eichhorn SW, Subtelny AO, Lin KS, McGeary SE, Gupta S, Bartel DP. 2020. The dynamics of cytoplasmic mRNA metabolism. *Mol Cell* 77: 786–799.e10. doi:10.1016/j.molcel.2019.12.005

Faraji F, Hu Y, Yang HH, Lee MP, Winkler GS, Hafner M, Hunter KW. 2016. Post-transcriptional control of tumor cell autonomous metastatic potential by CCR4–NOT deadenylase CNOT7. *PLoS Genet* 12: e1005820. doi:10.1371/journal.pgen.1005820

Friedman RA, Honig B. 1995. A free energy analysis of nucleic acid base stacking in aqueous solution. *Biophys J* 69: 1528–1535. doi:10.1016/S0006-3495(95)80023-8

Fuke H, Ohno M. 2008. Role of poly(A) tail as an identity element for mRNA nuclear export. *Nucleic Acids Res* 36: 1037–1049. doi:10.1093/nar/gkm1120

Hashizume H, Imahori K. 1967. Circular dichroism and conformation of natural and synthetic polynucleotides. *J Biochem* 61: 738–749. doi:10.1093/oxfordjournals.jbchem.a128608

Hentze MW. 1997. eIF4G: a multipurpose ribosome adapter? *Science* 275: 500–501. doi:10.1126/science.275.5299.500

Isaksson J, Acharya S, Barman J, Cheruku P, Chattopadhyaya J. 2004. Single-stranded adenine-rich DNA and RNA retain structural characteristics of their respective double-stranded conformations and show directional differences in stacking pattern. *Biochemistry* 43: 15996–16010. doi:10.1021/bi048221v

Jonas S, Christie M, Peter D, Bhandari D, Loh B, Huntzinger E, Weichenrieder O, Izaurralde E. 2014. An asymmetric PAN3 dimer recruits a single PAN2 exonuclease to mediate mRNA deadenylation and decay. *Nat Struct Mol Biol* 21: 599–608. doi:10.1038/nsmb.2837

Juszkiewicz S, Hegde RS. 2017. Initiation of quality control during poly(A) translation requires site-specific ribosome ubiquitination. *Mol Cell* 65: 743–750.e4. doi:10.1016/j.molcel.2016.11.039

Kuhn U, Wahle E. 2004. Structure and function of poly(A) binding proteins. *Biochim Biophys Acta* 1678: 67–84. doi:10.1016/j.bbaexp.2004.03.008

Kumar A, Clerici M, Muckenfuss LM, Passmore LA, Jinek M. 2019. Mechanistic insights into mRNA 3′-end processing. *Curr Opin Struct Biol* 59: 143–150. doi:10.1016/j.sbi.2019.08.001

Legnini I, Alles J, Karaiskos N, Ayoub S, Rajewsky N. 2019. FLAM-seq: full-length mRNA sequencing reveals principles of poly(A) tail length control. *Nat Methods* 16: 879–886. doi:10.1038/s41592-019-0503-y

Lim J, Ha M, Chang H, Kwon SC, Simanshu DK, Patel DJ, Kim VN. 2014. Uridylation by TUT4 and TUT7 marks mRNA for degradation. *Cell* 159: 1365–1376. doi:10.1016/j.cell.2014.10.055

Lim J, Kim D, Lee Y-S, Ha M, Lee M, Yeo J, Chang H, Song J, Ahn K, Kim VN. 2018. Mixed tailing by TENT4A and TENT4B shields mRNA from rapid deadenylation. *Science* 361: 701–704. doi:10.1126/science.aam5794

Lima SA, Chipman LB, Nicholson AL, Chen Y-H, Yee BA, Yeo GW, Coller J, Pasquinelli AE. 2017. Short poly(A) tails are a conserved feature of highly expressed genes. *Nat Struct Mol Biol* 24: 1057–1063. doi:10.1038/nsmb.3499

Lin J, Kolomeisky A, Meller A. 2010. Helix-coil kinetics of individual polyadenylic acid molecules in a protein channel. *Phys Rev Lett* 104: 158101. doi:10.1103/PhysRevLett.104.158101

Liu H, Begik O, Lucas MC, Ramirez JM, Mason CE, Wiener D, Schwartz S, Mattick JS, Smith MA, Novoa EM. 2019. Accurate detection of m^6A RNA modifications in native RNA sequences. *Nat Commun* 10: 4079. doi:10.1038/s41467-019-11713-9

Meyer S, Temme C, Wahle E. 2004. Messenger RNA turnover in eukaryotes: pathways and enzymes. *Crit Rev Biochem Mol Biol* 39: 197–216. doi:10.1080/10409230490513991

Morita M, Suzuki T, Nakamura T, Yokoyama K, Miyasaka T, Yamamoto T. 2007. Depletion of mammalian CCR4b deadenylase triggers elevation of the $p27^{Kip1}$ mRNA level and impairs cell growth. *Mol Cell Biol* 27: 4980–4990. doi:10.1128/MCB.02304-06

Nakamura T, Yao R, Ogawa T, Suzuki T, Ito C, Tsunekawa N, Inoue K, Ajima R, Miyasaka T, Yoshida Y, et al. 2004. Oligo-astheno-teratozoospermia in mice lacking Cnot7, a regulator of retinoid X receptor β. *Nat Genet* 36: 528–533. doi:10.1038/ng1344

Neely GG, Kuba K, Cammarato A, Isobe K, Amann S, Zhang L, Murata M, Elmén L, Gupta V, Arora S, et al. 2010. A global *in vivo Drosophila* RNAi screen identifies NOT3 as a conserved regulator of heart function. *Cell* 141: 142–153. doi:10.1016/j.cell.2010.02.023

Otwinowski Z, Schevitz RW, Zhang R-G, Lawson CL, Joachimiak A, Marmorstein RQ, Luisi BF, Sigler PB. 1988. Crystal structure of *trp* repressor/operator complex at atomic resolution. *Nature* 335: 321–329. doi:10.1038/335321a0

Parker R, Song H. 2004. The enzymes and control of eukaryotic mRNA turnover. *Nat Struct Mol Biol* 11: 121–127. doi:10.1038/nsmb724

Paul MS, Bass BL. 1998. Inosine exists in mRNA at tissue-specific levels and is most abundant in brain mRNA. *EMBO J* 17: 1120–1127. doi:10.1093/emboj/17.4.1120

Pierron G, Weil D. 2018. Re-viewing the 3D organization of mRNPs. *Mol Cell* 72: 603–605. doi:10.1016/j.molcel.2018.10.044

Proudfoot NJ. 2011. Ending the message: poly(A) signals then and now. *Genes Dev* 25: 1770–1782. doi:10.1101/gad.17268411

Raisch T, Chang CT, Levdansky Y, Muthukumar S, Raunser S, Valkov E. 2019. Reconstitution of recombinant human CCR4–NOT reveals molecular insights into regulated deadenylation. *Nat Commun* 10: 3173. doi:10.1038/s41467-019-11094-z

Rissland OS, Norbury CJ. 2009. Decapping is preceded by 3′ uridylation in a novel pathway of bulk mRNA turnover. *Nat Struct Mol Biol* **16:** 616–623. doi:10.1038/nsmb.1601

Sachs A. 1990. The role of poly(A) in the translation and stability of mRNA. *Curr Opin Cell Biol* **2:** 1092–1098. doi:10.1016/0955-0674(90)90161-7

Saenger W. 1984. *Principles of nucleic acid structure*. Springer-Verlag, New York.

Saenger W, Riecke J, Suck D. 1975. A structural model for the polyadenylic acid single helix. *J Mol Biol* **93:** 529–534. doi:10.1016/0022-2836(75)90244-2

Safaee N, Kozlov G, Noronha AM, Xie JW, Wilds CJ, Gehring K. 2012. Interdomain allostery promotes assembly of the poly(A) mRNA complex with PABP and eIF4G. *Mol Cell* **48:** 375–386. doi:10.1016/j.molcel.2012.09.001

Schäfer IB, Rode M, Bonneau F, Schüssler S, Conti E. 2014. The structure of the Pan2–Pan3 core complex reveals cross-talk between deadenylase and pseudokinase. *Nat Struct Mol Biol* **21:** 591–598. doi:10.1038/nsmb.2834

Schäfer IB, Yamashita M, Schuller JM, Schüssler S, Reichelt P, Strauss M, Conti E. 2019. Molecular basis for poly(A) RNP architecture and recognition by the Pan2–Pan3 deadenylase. *Cell* **177:** 1619–1631.e21. doi:10.1016/j.cell.2019.04.013

Seol Y, Skinner GM, Visscher K, Buhot A, Halperin A. 2007. Stretching of homopolymeric RNA reveals single-stranded helices and base-stacking. *Phys Rev Lett* **98:** 158103. doi:10.1103/PhysRevLett.98.158103

Shao S, von der Malsburg K, Hegde RS. 2013. Listerin-dependent nascent protein ubiquitination relies on ribosome subunit dissociation. *Mol Cell* **50:** 637–648. doi:10.1016/j.molcel.2013.04.015

Shi H, Wei J, He C. 2019. Where, when, and how: context-dependent functions of RNA methylation writers, readers, and erasers. *Mol Cell* **74:** 640–650. doi:10.1016/j.molcel.2019.04.025

Siddiqui N, Mangus DA, Chang T-C, Palermino J-M, Shyu A-B, Gehring K. 2007. Poly(A) nuclease interacts with the C-terminal domain of polyadenylate-binding protein domain from poly(A)-binding protein. *J Biol Chem* **282:** 25067–25075. doi:10.1074/jbc.M701256200

Slotkin W, Nishikura K. 2013. Adenosine-to-inosine RNA editing and human disease. *Genome Med* **5:** 105. doi:10.1186/gm508

Smith BL, Gallie DR, Le H, Hansma PK. 1997. Visualization of poly(A)-binding protein complex formation with poly(A) RNA using atomic force microscopy. *J Struct Biol* **119:** 109–117. doi:10.1006/jsbi.1997.3864

Song M-G, Kiledjian M. 2007. 3′ Terminal oligo U-tract-mediated stimulation of decapping. *RNA* **13:** 2356–2365. doi:10.1261/rna.765807

Stowell JAW, Webster MW, Kögel A, Wolf J, Shelley KL, Passmore LA. 2016. Reconstitution of targeted deadenylation by the Ccr4–not complex and the YTH domain protein Mmi1. *Cell Rep* **17:** 1978–1989. doi:10.1016/j.celrep.2016.10.066

Suck D, Manor PC, Saenger W. 1976. The structure of a trinucleoside diphosphate: adenylyl-(3′,5′)-adenylyl-(3′,5′)-adenosine hexahydrate. *Acta Cryst* **32:** 1727–1737. doi:10.1107/S0567740876006316

Sundaramoorthy E, Leonard M, Mak R, Liao J, Fulzele A, Bennett EJ. 2017. ZNF598 and RACK1 regulate mammalian ribosome-associated quality control function by mediating regulatory 40S ribosomal ubiquitylation. *Mol Cell* **65:** 751–760.e4. doi:10.1016/j.molcel.2016.12.026

Tang TTL, Stowell JAW, Hill CH, Passmore LA. 2019. The intrinsic structure of poly(A) RNA determines the specificity of Pan2 and Caf1 deadenylases. *Nat Struct Mol Biol* **26:** 433–442. doi:10.1038/s41594-019-0227-9

Tarun SZ J, Sachs AB. 1996. Association of the yeast poly(A) tail binding protein with translation initiation factor eIF-4G. *EMBO J* **15:** 7168–7177. doi:10.1002/j.1460-2075.1996.tb01108.x

Tesina P, Lessen LN, Buschauer R, Cheng J, Wu C-C, Berninghausen O, Buskirk AR, Becker T, Beckmann R, Green R. 2020. Molecular mechanism of translational stalling by inhibitory codon combinations and poly(A) tracts. *EMBO J* **39:** e103365. doi:10.15252/embj.2019103365

Vicens Q, Kieft JS, Rissland OS. 2018. Revisiting the closed-loop model and the nature of mRNA 5′–3′ communication. *Mol Cell* **72:** 805–812. doi:10.1016/j.molcel.2018.10.047

Wang H, Morita M, Yang X, Suzuki T, Yang W, Wang J, Ito K, Wang Q, Zhao C, Bartlam M, et al. 2010. Crystal structure of the human CNOT6L nuclease domain reveals strict poly(A) substrate specificity. *EMBO J* **29:** 2566–2576. doi:10.1038/emboj.2010.152

Wells SE, Hillner PE, Vale RD, Sachs AB. 1998. Circularization of mRNA by eukaryotic translation initiation factors. *Mol Cell* **2:** 135–140. doi:10.1016/S1097-2765(00)80122-7

Wilson T, Treisman R. 1988. Removal of poly(A) and consequent degradation of *c-fos* mRNA facilitated by 3′ AU-rich sequences. *Nature* **336:** 396–399. doi:10.1038/336396a0

Wolf J, Valkov E, Allen MD, Meineke B, Gordiyenko Y, McLaughlin SH, Olsen TM, Robinson CV, Bycroft M, Stewart M, et al. 2014. Structural basis for Pan3 binding to Pan2 and its function in mRNA recruitment and deadenylation. *EMBO J* **33:** 1514–1526. doi:10.15252/embj.201488373

Xiao W, Adhikari S, Dahal U, Chen Y-S, Hao Y-J, Sun B-F, Sun H-Y, Li A, Ping X-L, Lai W-Y, et al. 2016. Nuclear m^{6}A reader YTHDC1 regulates mRNA splicing. *Mol Cell* **61:** 507–519. doi:10.1016/j.molcel.2016.01.012

Subcellular Spatial Transcriptomes: Emerging Frontier for Understanding Gene Regulation

FURQAN M. FAZAL[1] AND HOWARD Y. CHANG[1,2]

[1]*Center for Personal Dynamic Regulomes, Stanford University, Stanford, California 94305, USA*
[2]*Howard Hughes Medical Institute, Stanford University School of Medicine, Stanford, California 94305, USA*

Correspondence: howchang@stanford.edu

RNAs are trafficked and localized with exquisite precision inside the cell. Studies of candidate messenger RNAs have shown the vital importance of RNA subcellular location in development and cellular function. New sequencing- and imaging-based methods are providing complementary insights into subcellular localization of RNAs transcriptome-wide. APEX-seq and ribosome profiling as well as proximity-labeling approaches have revealed thousands of transcript isoforms are localized to distinct cytotopic locations, including locations that defy biochemical fractionation and hence were missed by prior studies. Sequences in the 3′ and 5′ untranslated regions (UTRs) serve as "zip codes" to direct transcripts to particular locales, and it is clear that intronic and retrotransposable sequences within transcripts have been co-opted by cells to control localization. Molecular motors, nuclear-to-cytosol RNA export, liquid–liquid phase separation, RNA modifications, and RNA structure dynamically shape the subcellular transcriptome. Location-based RNA regulation continues to pose new mysteries for the field, yet promises to reveal insights into fundamental cell biology and disease mechanisms.

A eukaryotic cell is highly organized, with biomolecules localizing to specific regions of the cell that are integral to their function. For more than three decades, evidence has been accumulating to suggest that the RNAs for thousands of genes show pronounced subcellular localization, and that this localization is an essential mechanism for post-transcriptional regulation. RNA localization influences RNA folding, editing, splicing, degradation, translation, binding partner, catalytic activity, and even the fate of the protein that is encoded. Some of the earliest experiments examining the localization of messenger RNAs (mRNAs) were performed in *Xenopus* and *Drosophila* eggs, and were followed by similar demonstrations in yeast, mammalian neurons, and in developing *Drosophila* embryos. Such studies have revealed that sequences within (i.e., *cis* elements) RNAs, also termed "zip codes," direct the localization of mRNAs, typically by recruiting proteins (i.e., *trans* factors).

In this review, we summarize the history of mRNA localization studies and focus on exciting new developments in the last decade to track the localization of thousands of transcripts within cells using either sequencing- or imaging-based approaches. We identify how new techniques are starting to systematically dissect the *cis* and *trans* regulators of RNA localization. Although it now appears that RNA subcellular localization is the norm rather than the exception for both coding and noncoding RNAs (Wilk et al. 2016) and is broadly conserved evolutionarily (Benoit Bouvrette et al. 2018), our understanding of the extent, importance, and regulation of subcellular spatial transcriptomics continues to be limited. Furthermore, the relevant techniques toolkit for such RNA studies lags behind those developed for subcellular spatial proteomics, for which we have detailed information for more than 10,000 human protein-coding genes with subcellular resolution (Uhlen et al. 2010; Thul et al. 2017) across many tissues (Uhlen et al. 2015). In contrast, even today we do not have a good map or atlas of RNA subcellular localization, although promising new technological developments (Chen et al. 2015; Shah et al. 2016; Fazal et al. 2019) are making such milestones within reach.

BRIEF HISTORY OF RNA LOCALIZATION STUDIES

Early studies in RNA localization focused on easy-to-image cells such as the relatively large *Xenopus* (Rebagliati et al. 1985; Weeks and Melton 1987) and *Drosophila* eggs. Such initial studies established mRNA localization as a way to regulate protein expression and have shown that sequences within the transcript, particularly in the 3′ untranslated region (UTR), can direct localization of transcripts; these findings have since been extended to mammalian systems. For example, the *Vg1* mRNA in *Xenopus* was found to localize to one (the vegetal) pole, whereas the *bicoid* mRNA in *Drosophila* egg cell was shown to localize to the anterior pole and to require the protein Staufen for localization (Fig. 1; St Johnston et al. 1991).

In mammals, one of the most well-studied transcripts is *β-actin* mRNA, which localizes to the leading edge of chicken embryo fibroblasts and to the growth cones of

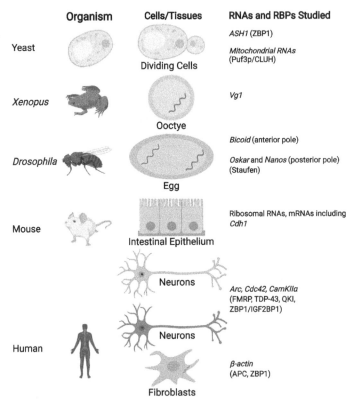

Figure 1. Model systems to study RNA localization. RNA-binding proteins involved are shown in parentheses.

developing neurons. β-actin has been shown to contain a 54-nucleotide (nt) zip code region in the 3′ UTR that is essential for localization (Kislauskis et al. 1994) in a translation-independent manner. This localization is, in turn, regulated by the protein factors IGF2BP1 (ZBP1) and ZBP2. ZBP1 controls local translation of β-actin by sequestering the transcript until it reaches the periphery of the cell, where the phosphorylation of ZBP1 releases the mRNA and permits its translation (Huttelmaier et al. 2005). Other RNAs have similarly been shown to localize to cellular protrusions and to require adenomatous polyposis coil (APC) protein (Mili et al. 2008; Baumann et al. 2020). Likewise, the protein fragile X mental retardation protein (FMRP) functions as a translational regulator of localized RNAs in many systems, including neurons and fibroblasts (Mili et al. 2008). FMRP functions by binding to and repressing the translation of mRNAs and is mediated by recognition of RNA secondary structure. Upon reaching their destination, FMRP release of the RNAs triggers local translation, as in the case of axons. Findings from many studies have converged on the hypothesis that the mRNAs are transported along with retinol-binding proteins (RBPs) in the cytosol as translationally repressed RNA granules (Anderson and Kedersha 2006). Supporting studies have shown that the cytoskeleton and its associated molecular motors play an integral role in this mRNA transport (Wang et al. 2016).

The lessons learned to dissect β-actin mRNA transport have since been extended to other mammalian systems, particularly neurons that are ideal systems to study localization defects because of the vast distances metabolites need to be transported. Neurons need to coordinate functions between the cell nucleus and the axons and dendrites, which can be >1 m apart. Neurons also need to dynamically regulate their proteomes in response to changing environments, and it is now clear that local translation of mRNAs in dendrites is widespread and essential. It is now thought that RNA localization is the primary determinant of the proteome of neurites, rather than transport of corresponding proteins (Zappulo et al. 2017). Furthermore, many essential RBPs whose processing is dysregulated in neuronal disorders have been shown to bind hundreds of RNAs and to be involved in their localization. Two such RBPs are FMRP, whose loss of function results in fragile X syndrome and autism, and TDP-43, whose dysregulation is associated with amyotrophic lateral sclerosis (ALS) (Neumann et al. 2006; Sreedharan et al. 2008). TDP-43, which regulates RNA metabolism through many mechanisms, will form cytoplasmic messenger ribonucleoprotein (mRNP) granules that undergo microtubule-dependent transport in neurons (Fig. 2; Alami et al. 2014).

MECHANISMS OF RNA LOCALIZATION

Role of *cis* Elements, including Retrotransposable Elements and Features in the 3′ UTR

Studies focusing on specific RNAs such as β-actin have revealed that mRNAs can localize to subcellular locales independent of translation and are guided by internal zip

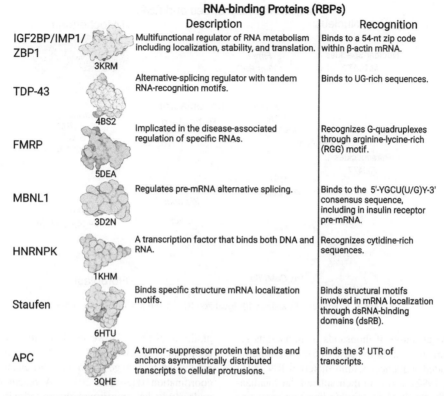

Figure 2. Some representative RNA binding proteins implicated in mRNA localization. Structures are generated from Protein Data Bank (PDB) entries.

code sequences, particularly in 3′ UTRs. Surprisingly, even smaller RNAs such as microRNAs (miRNAs) can contain sequence elements within that direct them to subcellular locales (Hwang et al. 2007), as do many long noncoding RNAs (lncRNAs) (Batista and Chang 2013). However, for the vast majority of transcripts, the zip codes responsible for localization to specific organelles and biological condensates remain unknown, although newly developed transcriptome-wide approaches are laying the foundation for identifying more *cis*- localization elements. For example, in neurons, hundreds of genes, including *Cdc42*, have transcript isoforms that localize differently between neurites and the soma based on sequence differences in 3′ UTRs (Ciolli Mattioli et al. 2019). Similarly, the endoplasmic reticulum (ER) is known to recruit transcripts directly in a translation-independent matter (Pyhtila et al. 2008), and a recent transcriptome-wide study has identified a sequence termed SECRETE that can recruit mRNAs encoding secretory/membrane proteins to the ER. The SECRETE sequence, comprising a ≥10-nt triplet repeat, occurs in both prokaryotes and eukaryotes (Cohen-Zontag et al. 2019).

A robust approach to identify zip codes within transcripts, which has been particularly fruitful for lncRNAs, is to identify a transcript(s) that localizes to a specific locale and then to systematically test whether sequences within are necessary and sufficient to direct localization (Fig. 3). Such an approach has identified an ∼600-nt region in human cells that is required for localization of the lncRNA *MALAT1* to nuclear speckles (Miyagawa et al. 2012). Similarly, sequences within the lncRNA *Xist* called A-repeats, located near the 5′ end, are responsible for localization to the nuclear periphery (Wutz et al. 2002), likely secondary to the ability of this RNA element to induce facultative heterochromatinization (Chaumeil et al. 2006). Another lncRNA *Firre* has a 156-nt repeating RNA domain (RRD), recognized by the protein hnRNPU, that aids in localizing it to chromatin (Hacisuleyman et al. 2014); hnRNPU binding is also required for the proper localization of *Xist* (Hasegawa et al. 2010). Recently several studies, including computation and experimental approaches, have revealed that sequences derived from transposable elements, which are present in many mRNAs and lncRNAs, contribute to the nuclear retention of many lncRNAs (Carlevaro-Fita et al. 2019). Although such studies show the widespread occurrence of zip code sequences, systematic high-throughput experiments are needed to identify the *cis* elements necessary for the observed extensive RNA subcellular localization transcripts.

Protein Factors That Interact with mRNAs and lncRNAs

RNA localization is thought to be orchestrated by RNA-binding proteins that can recognize sequence motifs or RNA structural features, including single-stranded regions or stem loops. Although we know of a few RBPs mediating localization, including Staufen and Puf3, how the cell co-

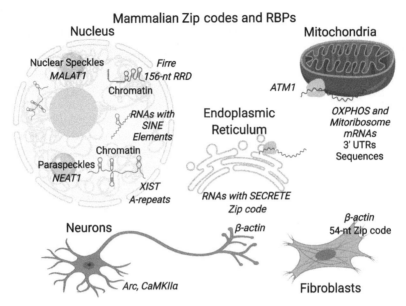

Figure 3. Examples of locales with localized RNAs and associated zip codes.

ordinates the localization of thousands of transcripts remains poorly understood. In the case of Staufen, binding to double-stranded sequences within maternal RNAs (St Johnston et al. 1992) results in their subcellular localization, whereas for the Puf3 protein binding to a sequence motif (Zhu et al. 2009) within some nuclear-encoded RNAs results in their recruitment to the mitochondria in yeast (Saint-Georges et al. 2008; Gadir et al. 2011). Future work on mapping RNA–protein interactions (Ramanathan et al. 2019) will likely be crucial in discovering RBPs essential for localization.

Active Transport and the Role of Molecular Motors

Many studies suggest mRNA localization in the cytosol is facilitated by the underlying cytoskeleton network, although the relative contributions of individual players remain unclear. However, we do know that molecular motors operating on microtubules as well as actin filaments participate in RNA transport (Fig. 4; Maday et al. 2014), including myosin motors that walk on actin filaments and kinesins and cytoplasmic dynein that move on microtubules. For example, early studies established that the localization of *oskar* mRNA in *Drosophila* oocytes to the posterior pole requires the cytoskeleton (Erdelyi et al. 1995), with subsequent studies implicating kinesin-1 (Zimyanin et al. 2008). Similar studies of other RNAs in yeast have implicated myosin V (Bertrand et al. 1998).

Furthermore, this RNA localization process can be dynamically regulated through active transport, as shown in intestinal epithelial cells where mRNAs strongly localize (Moor et al. 2017). The RNA, packaged in RNPs, can be transported bidirectionally along microtubules by plus-end-directed kinesins (Kanai et al. 2004) and minus-end-directed dynein motors (Hirokawa et al. 2010). Kinesins typically transport RNAs toward the cell periphery, whereas dynein transports RNAs toward the cell center (retrograde transport). However, how the different motors cooperate is unclear, and in fact, different motor types are known to engage cargo and participate in tug-of-war coordination (Hancock 2014). A recent transcriptome-wide study has confirmed that hundreds of transcripts rely on microtubule-based transport to get their cytosolic destinations (Fazal et al. 2019), and continued progress is being made in understanding the transport of RNAs, as shown in reconstituted in vitro systems (Baumann et al. 2020) and inside living cells (Krauss et al. 2009).

Splicing, Intron Retention, and Nuclear Export

The nucleus of a eukaryotic cell is enveloped by a double lipid bilayer that serves as the gateway for mRNAs exiting to the cytosol. The export of RNAs through the nuclear pore complexes (NPCs) spanning the envelope has been extensively studied (Muller-McNicoll and Neugebauer 2013; Katahira 2015), with translocation through the pore thought to be diffusive, and with only a fraction (one-third or less) of mammalian mRNAs that interact with the NPC eventually exiting (Ma et al. 2013). Importantly, this export process can vary depending on the type of RNA species (mRNAs, ribosomal RNAs [rRNAs], micrmiRNAs, transfer RNAs [tRNAs], etc.) in question (Muller-McNicoll and Neugebauer 2013; Katahira 2015). Furthermore, mRNAs can move bidirectionally through the pore, and not all pores are created equally (Grunwald and Singer 2010; Siebrasse et al. 2012). The NPCs are hypothesized to show considerable heterogeneity (Colon-Ramos et al. 2003), with specialized NPCs mediating the transport of mRNAs from distinct genomic loci with nuclei to specific regions of the cytosol and, therefore, optimizing nuclear export and facilitating subsequent translation (Brown and Silver 2007).

Within the nucleus, splicing has a profound influence on nuclear export (Kim-Ha et al. 1993; Hachet and Eph-

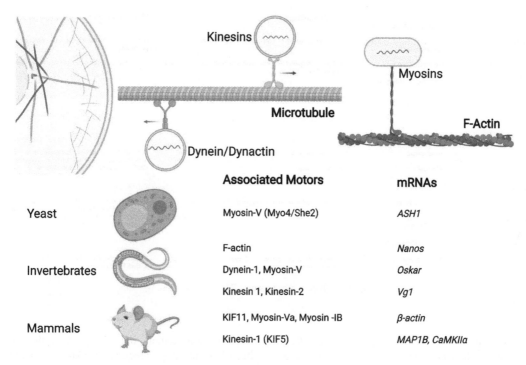

Figure 4. Molecular motors implicated in RNA transport.

russi 2004), with pre-mRNAs recruiting splicing factors along with the conserved mRNA export machinery (TREX, transcription/export complex). TREX is recruited to the 5′ end of transcripts and accounts for the export of mRNAs through the pore (Cheng et al. 2006). Likewise, the deposition of exon-junction complex (EJC) during splicing is essential for the localization of developmentally important transcripts (Braunschweig et al. 2013), including *oskar* mRNA in *Drosophila* (Ghosh et al. 2012). Alternative splicing provides yet another opportunity for the cell to influence RNA localization, as has been shown recently where isoform-specific localization to neurites is guided by alternative last exons (ALEs) (Taliaferro et al. 2016). Furthermore, partial splicing of transcripts results in their nuclear retention, which partially explains why many lncRNAs that are substantially less-efficiently spliced relative to mRNAs are nuclear (Zuckerman and Ulitsky 2019). By retaining some introns ("detained introns") in polyadenylated transcripts that are only excised before export, cells use nuclear retention in mRNAs and a constant nuclear-export rate to reduce cytoplasmic gene expression noise due to bursty transcription-related noise (Bahar Halpern et al. 2015). Such detained introns are widespread, enriched in UTRs and noncoding RNAs, and thought to functionally tune transcriptomes (Braunschweig et al. 2014).

MODERN APPROACHES TO STUDY RNA LOCALIZATION

Tracking Single RNAs

Currently, there are two general approaches to map RNA subcellular localization: imaging- and sequencing-based. Imaging-based approaches over the last two decades have yielded insights into the dynamics of single RNAs in cells, revealing their complicated history. One of the early studies focused on the dynamics of *ASH1* mRNA in yeast, and its 3′ UTR localization using the MS2 RNA-hairpin system (Fig. 5; Bertrand et al. 1998). Since then, the MS2 system has been optimized and applied extensively to study mRNA localization and transcription in living cells (Darzacq et al. 2007), and complementary approaches have been developed (Wu et al. 2016, 2019; Braselmann et al. 2018; Chen et al. 2019a; Wan et al. 2019) to study localization and local translation. Similar studies have revealed the intricate dynamics of nuclear pore mRNA export (Grunwald and Singer 2010; Siebrasse et al. 2012; Chen et al. 2017), and the trafficking of mRNAs to membrane-less organelles (MLOs) such as stress granules (Nelles et al. 2016). In addition to live-cell imaging, in situ hybridization approaches (Lawrence et al. 1989), which have evolved to use fluorescent in situ hybridization (FISH) labeling (Femino et al. 1998), provide complementary information with routine single-molecule sensitivity (Raj et al. 2008). The FISH-based approach has recently been extended to study RNA localization for hundreds of thousands of RNAs simultaneously, as discussed below.

Transcriptome-Wide Imaging Technologies

An exciting development in the field is the new approaches that finally enable imaging of hundreds and even thousands of RNAs within fixed cells. An early study was based on in situ RNA sequencing using complementary DNA amplicons, which permitted thousands of

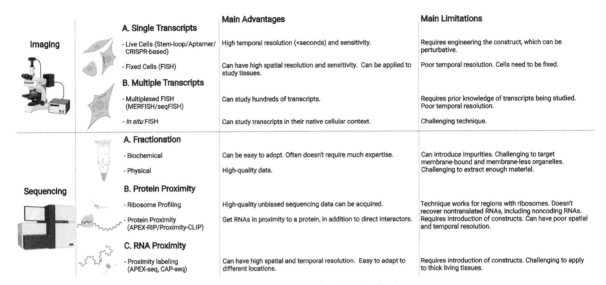

Figure 5. Techniques to study RNA localization.

RNAs to be simultaneously interrogated (Lee et al. 2014). However, although this technology is promising (Ke et al. 2013; Lee et al. 2014) and improvements continue to be made (Fürth et al. 2019), so far this challenging approach has not been widely adopted. Instead, visualizing many RNAs using sequential FISH is at the forefront of high-throughput localization studies, and two groups have mainly advanced this approach. In one iteration, called MERFISH developed by the Zhuang laboratory, the locations of RNAs in fixed cells are interrogated by performing sequential FISH through multiple rounds of hybridization of DNA oligonucleotides ("oligos") to the complementary RNAs of interest. This sequential approach uses an error-correcting scheme to design and select for hybridization oligos, such that some errors in the binding of DNA oligos to the complementary RNA molecules can be tolerated and correctly decoded. Although the high density of RNAs in a cell puts a limit on how many transcripts can be resolved and their relative abundances (Chen et al. 2015), in practice, hundreds to thousands of transcripts in individual cells can be interrogated. Further advances, including the integration of other techniques such as expansion microscopy, have further aided throughput (Xia et al. 2019). Recently the MERFISH approach has also been extended to carry out phenotypic screening in cells (Emanuel et al. 2017), as shown by a study identifying positive and negative regulators of the nuclear-speckle localization of the lncRNA *MALAT1* (Wang et al. 2019a).

The second sequential FISH approach, advanced by Cai and coworkers, called SeqFISH (Lubeck et al. 2014; Shah et al. 2016), allows multiplexed imaging of hundreds of genes through signal amplification and error-correction schemes, similar to MERFISH. Excitingly, SeqFISH facilitates mapping the subcellular localization of thousands of RNAs, including nascent transcripts (Shah et al. 2018) and splice isoforms. However, the limitations of both MERFISH and SeqFISH include working with fixed and not live cells. Furthermore, unlike sequencing that can be unbiased, the techniques require prior knowledge of the transcripts being targeted, as oligos can be designed to image those transcripts. In the near term, the unique advantage of imaging-based approaches in simultaneously interrogating many cells makes them especially well suited in exploring RNA heterogeneity across cells and in tissues (Moffitt et al. 2018), thereby distinguishing cell-types based on the RNAs they express (Eng et al. 2019).

Transcriptome-Wide Sequencing Technologies

The advent of next-generation sequencing technologies has ushered in a new revolution in biology, including in the investigation of RNA localization. Biochemical fractionation protocols coupled with RNA sequencing have, for example, been applied to study nuclear-versus-cytosol RNA dynamics (Djebali et al. 2012; Benoit Bouvrette et al. 2018) and to determine RNAs being actively translated through polyribosome profiling. In recent years, fractionation protocols have also been developed to determine the RNAs in challenging locations such as the membrane-less nucleolus and stress granules (Khong et al. 2017). Likewise, physical/mechanical separation of long neuronal cells, typically through microdissection (Cajigas et al. 2012), has been productive in determining their transcriptomes. Other techniques such as laser capture microscopy (LCM) have also enabled careful dissection of both single cells and subcellular locations (Nichterwitz et al. 2016) within and been applied to study rapid changes in RNA localization in mouse intestinal epithelial cells in response to food gradients (Moor et al. 2017).

An innovative approach to study the RNAs in cytosolic locales such as the endoplasmic reticulum membrane (ERM) and outer mitochondrial membrane (OMM) has been through proximity-specific ribosome profiling, in which ribosomes in specific locations undergo proximity biotinylation (Roux et al. 2012). These ribosomes with a biotin tag can subsequently be isolated through streptavidin-biotin pulldown, and the RNAs they are bound to and translating profiled by sequencing them. Such ribosome profiling experiments have revealed the RNAs bound to ribosomes in the ER in yeast and humans (Jan et al. 2014) and at the outer surface of the mitochondria in yeast (Williams et al. 2014; Costa et al. 2018).

Despite these existing sequencing technologies, many critical locations within the cell, including membrane-bound and membrane-less organelles, continue to be difficult, if not impossible, to interrogate. Furthermore, unlike live-cell-imaging approaches, these sequencing-based approaches are generally not well suited to study the dynamics of transcript localization. However, a new approach discussed below using proximity labeling of RNAs in living cells provides an opportunity to investigate RNA subcellular spatial dynamics (Fazal et al. 2019; Padron et al. 2019), albeit currently at a bulk rather than single-cell scale.

Subcellular Transcriptomics through Proximity Labeling

A recent approach to determine the RNAs at subcellular locales, called APEX-seq, yields an unbiased transcriptome that can be applied to study membrane-less and membrane-bound organelles. APEX-seq leverages an engineered enzyme called APEX2 (ascorbate peroxidase, version 2) that can be targeted to specific cellular locales by fusing it to a protein or peptide that is known to localize to the desired location. Upon providing the reagents biotin-phenol and hydrogen peroxide, APEX2 generates biotin-phenoxy radicals that result in the spatial tagging of nearby metabolites within cells with a biotin tag (Rhee et al. 2013). For example, when plasmids containing APEX2, which itself is around the size of green fluorescent protein (GFP), are fused to the nuclear localization sequence (NLS) and introduced into cells, APEX2 localizes to the nucleus and permits tagging of metabolites there. These metabolites include proteins (Rhee et al. 2013), RNAs, DNA, and small molecules; in APEX-seq, the labeled RNAs are isolated and enriched for using streptavidin-biotin pulldown, followed by RNA sequencing. Excitingly, APEX RNA labeling can achieve high spatial (∼10-nm) and temporal (∼1-min) resolution in almost any location of interest, including in MLOs such as the nucleolus (Fazal et al. 2019) and stress granules (Markmiller et al. 2018), as well as the membrane-bound ER (Kaewsapsak et al. 2017) and outer mitochondrial membrane (Fazal et al. 2019). Furthermore, APEX-based approaches have been applied to different model systems, including in mice, worms, and flies, as well as in cultured neurons (Hung et al. 2016). Likewise, fusing APEX to dCas9 (Gao et al. 2018; Myers et al. 2018; Qiu et al. 2019) allows targeting APEX to any genomic locus and obtaining the interacting proteins and RNAs.

In the initial demonstrations of obtaining subcellular RNAs using proximity labeling, APEX was used to label proteins, which were then cross-linked with RNAs nearby. These biotin-labeled proteins were then enriched by streptavidin-biotin pulldown, and the cross-linked RNAs were released and sequenced. Using this cross-linking approach, called APEX-RIP (Kaewsapsak et al. 2017) and proximity-CLIP (Benhalevy et al. 2018), APEX labeling has been used to determine the RNAs in the cytosol, nucleus, and mitochondria. In APEX-RIP, formaldehyde cross-linking is performed, whereas in proximity-CLIP UV cross-linking and metabolic labeling is used to improve specificity.

In contrast, the more straightforward APEX-seq approach entailing direct RNA labeling (Zhou et al. 2019) has generated subcellular transcriptomes of many organelles in human cells (Fazal et al. 2019; Padron et al. 2019). By targeting APEX to multiple subcellular locales in the nucleus and cytosol, APEX-seq has revealed that thousands of RNAs show robust subcellular localization (Fazal et al. 2019). Independently, Ingolia and coworkers have used APEX-seq to examine RNAs in proximity to the 7-methylguanosine (m^7G) cap-binding protein eIF4E1, while also obtaining subcellular proteomic information. In addition, changes in RNA localization upon heat shock and stress granule assembly on the timescale of minutes were tracked (Padron et al. 2019).

APEX-seq (Fazal et al. 2019) has revealed that the RNA transcripts for thousands of genes localize to specific locales within cells, including in the nucleolus, nuclear lamina, nuclear pore, OMM, and ERM. Moreover, APEX-seq detected many transcripts with distinct isoforms showing differential subcellular localization. In addition to providing a map of subcellular RNA localization, APEX-seq has confirmed the role of the nuclear pore in mRNA surveillance and shown that the location of mature RNA transcripts within the nucleus is connected with the underlying genome architecture. For example, transcripts found at the nuclear lamina are enriched for genes found in DNA lamina-associated domains (LADs), as well as transcripts containing retrotransposable elements such as the short interspersed nuclear elements (SINEs) and long interspersed nuclear elements (LINEs). APEX-seq also revealed two modes of mRNA localization to the OMM: ribosome-dependent (i.e., requiring translation) and RNA-dependent. Transcripts coding for mitochondrial proteins that localize to the OMM independent of translation were found to have shorter 3′ UTRs and shorter poly(A) tails. RNA localization to the OMM depends on active transport, as shown by time course experiments showing mislocalization of transcripts within minutes of adding nocodazole, a microtubule depolymerizer.

APEX-seq, in conjunction with approaches to identify proteins interacting with specific RNAs (Chu et al. 2015; Ramanathan et al. 2018, 2019; Mukherjee et al. 2019; Han et al. 2020), is likely to emerge as a powerful approach to identify both localized RNAs and their corresponding RBP partners. Likewise, new approaches to spatial tagging of RNAs are continually being invented. One such method is based on spatially restricted nucleobase oxidation, which uses localized fluorophores (Li et al. 2017). Another approach uses an enzyme to add uridine residues to RNAs in specific locations in *Caenorhabditis elegans*, including in the mitochondria and ER (Medina-Muñoz et al. 2019). A third approach, called CAP-seq, uses light-activated, proximity-dependent photo-oxidation of RNA (Wang et al. 2019b).

Machine-Learning Approaches

Subcellular RNA-seq data, including from the nucleus, cytosol, and the ER, provide rich data sets to identify the sequencing elements involved in RNA localization. Applying machine-learning approaches, including deep-learning algorithms (LeCun et al. 2015), to these data sets is likely to provide new insights into the sequence-determinants of localization (Fig. 6). For example, bioinformatics approaches have identified transposable elements as being important for the nuclear retention of many lncRNAs (Carlevaro-Fita et al. 2019). Furthermore, a statistical analysis has shown that the transposable element Alu has a strong preference for being in the 3′ UTR of transcripts that are overrepresented in the nucleus, Golgi, and mitochondria (Chen et al. 2018). Similarly, a deep-neural-network approach to predict lncRNA localization as nuclear or cytosolic directly from transcript sequences had modest success, approaching an accuracy of 72% (Gudenas and Wang 2018). In that study, the feature set for learning included sequences as k-mers, known RNA-binding protein motif sites, as well as genomic characteristics of the RNAs such as whether they were intergenic, antisense, or sense lncRNAs. Likewise, another group used k-mers along with other features to obtain an accuracy of 59% for localization of lncRNAs (Cao et al. 2018), although another group in a different context claimed 87% accuracy using 8-mer nucleotide segments along with other features (Su et al. 2018). Another model called RNATracker has used a convolutional neural network to classify RNA localization; so far, success has been modest (Yan et al. 2019). A recent computational approach called RNA-GPS (Wu et al. 2020), which uses an APEX-seq data set (Fazal et al. 2019) comprising the subcellular transcriptome of eight locations, uses k-mers as features to obtain an overall accuracy of 70%. RNA-GPS implicates transcript splicing as an important process influencing localization

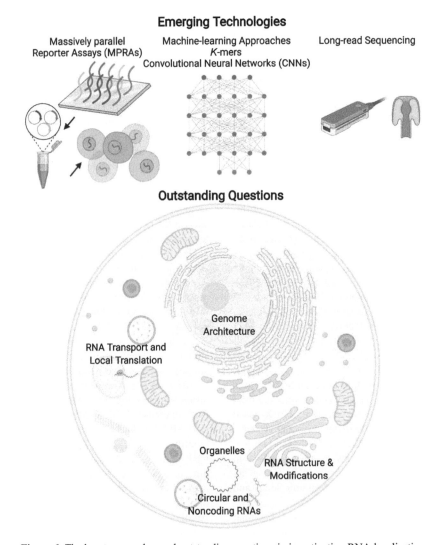

Figure 6. The latest approaches and outstanding questions in investigating RNA localization.

for organelles within both the nucleus and cytosol. In summary, although such approaches are in their infancy, they should provide candidates sequences that can be directly tested for their localization potential, and the corresponding interacting RBP identified (Wu et al. 2020).

Massively Parallel Reporter Assays

An experimental strategy to identify and test for zip code sequences within cells is the use of massively parallel reporter assay (MPRAs), in which tens of thousands of sequences, typically 75–200 nt in length, can be interrogated. Using MPRAs, along with machine-learning models, particularly convolutional neural networks (CNNs) (Movva et al. 2019), is likely to facilitate the rapid discovery of zip code sequences. Two groups recently used high-throughput screens to identify *cis*-acting RNA localization elements that promote nuclear retention. Rinn and coworkers tested and designed more than 10,000 oligos derived from 38 human lncRNAs with known both nuclear and cytosolic localization. Similarly, the Ullitsky group used approximately 5500 oligos gathered from 37 lncRNAs as well as some mRNAs. Both these studies were performed by introducing these oligos into an RNA and assessing its change in nuclear retention by nuclear-cytosolic fractionation following by sequencing. Through the MPRA experiments, the Ulltisky group found a cytosine-rich element, RCCTCCC (R = A/G), derived from an antisense Alu element, which they named SIRLOIN (SINE-derived nuclear localization element), that promotes nuclear retention (Lubelsky and Ulitsky 2018). By screening the binding sites of more than 100 RBPs using publicly available RBP-binding data sets (Van Nostrand et al. 2016), they identified the heterogeneous nuclear ribonucleoprotein K (HNRNPK) as binding to and nuclear-retaining SIRLOIN-containing RNAs. The Rinn group found a similar motif contributing to nuclear retention. The role of SINE elements in nuclear retention of *MALAT1* lncRNA through HNRNPK recruitment was recently confirmed by another study (Nguyen et al. 2020).

MPRA-based screens are likely going to be a powerful way to screen for zip code components, including sequence motifs and structural elements. Concomitantly, MPRA experimental and computational strategies continue to improve, and it is now possible to test more than 100 million sequences (de Boer et al. 2019).

Long-Read Sequencing

Next-generation-sequencing (NGS) approaches, including using the Illumina platform, continue to transform in biology. However, a significant limitation continues to be the relatively short sequencing reads (typically <300 basepair [bp]) generated. Fortunately, the latest third-generation approaches such as Oxford Nanopore Technologies (ONT) and Pacific Biosciences (PacBio) provide much longer-read sequencing reads (>1000 bp) and can sequence RNA directly without having to reverse transcribe it to make complementary DNA (cDNA) (Garalde et al. 2018). By being able to generate full-length transcript sequences, in addition to yielding RNA modification (Soneson et al. 2019; Workman et al. 2019), these techniques can reveal the landscape of variation in splicing isoforms, poly(A)-tail-length (Legnini et al. 2019), and RNA modifications (Workman et al. 2019). Exciting future studies will undoubtedly implement these approaches to explore transcript-isoform localization differences and dissect the role of RNA modifications in localization. Previous studies indeed identify the abundant N6-methyladenosine (m^{6}A) modification to be important for facilitating the nuclear export of mRNAs, with modified transcripts "fast-tracked" to the cytosol for translation (Lesbirel and Wilson 2019).

SOME OUTSTANDING QUESTIONS IN THE FIELD

Although RNA localization studies have a rich history spanning more than three decades, many critical issues in the field remain unanswered. The central question continues to persist: Why do cells localize their RNA contents? In some cell types, such as neuronal cells in which the distances involved for transporting biomolecules are vast, it is easy to rationalize that actively transporting mRNAs to their destination to be locally translated to make proteins would be convenient and efficient. However, it remains unclear why RNA subcellular localization is ubiquitously observed in almost all cell types, including ones in which the process of diffusion should be fast (seconds or less). To address the question of why cells localize their RNA contents, we must first explain the following questions.

Relative Contribution of Translation- versus RNA-Dependent mRNA Localization

A vital issue in the field is ascertaining to what extent the observed subcellular RNA localization is translation-dependent, and whether RNAs can be transported, particularly actively by molecular motors, with the ribosome engaged in the translation of the mRNA. It was generally accepted that the transport of mRNAs occurs through mRNPs that are translationally repressed until they get to their destination (Fig. 6). Furthermore, translating mRNAs interact with RNP granules dynamically, whereas nontranslating mRNAs can form stable associations (Moon et al. 2019). However, recent studies have begun to question this assumption, including imaging experiments that have revealed that active transport of mRNAs can occur after the mRNA has started translation and entered the polysome state (Wang et al. 2016; Moon et al. 2019). Similarly, APEX-seq has revealed that many nuclear RNAs destined for the mitochondria begin the process of translation elsewhere, such as in the cytosol, and then the translating-ribosome complex comprising of the nascent peptide being synthesized, RNA, and ribosome is directed to the mitochondria. APEX-seq experiments (Fazal et al. 2019) also implicated the cytoskeleton and its associated motors as being necessary for this transport, suggesting that engagement of mRNA with transport motors and translating ribosomes can co-occur. These studies also indicate that some observed mRNA localization is a consequence of translation, as has been suggested for mitochondria in yeast (Eliyahu et al. 2010). In contrast, other RNAs were found to localize to the mitochondria independent of translation and to be preferentially coding for mitoribosome and oxidative phosphorylation proteins.

Understanding how RNAs find their destination continues to be a fascinating problem that will require imaging, sequencing, and biophysical insights. Cells rely on different approaches to transport mRNAs, and future studies will likely also focus on understanding how organelles are optimized and regulated to control the localization of transcripts to them (Tsuboi et al. 2019).

Role of RNA Modifications and Structure

RNAs are extensively modified within cells, and there exist more than a hundred types of chemical modifications (Roundtree et al. 2017a), some of which are likely to be important in specifying RNA localization. For example, the abundant epitranscriptomic modification m^6A has been shown to influence the nuclear export of RNAs, with the m^6A-binding protein YTHDC1 mediating this process (Roundtree et al. 2017b). RBPs such as FMRP have been identified as m^6A readers that promote export (Edens et al. 2019), and RNA modifications are known to be involved in forming and localizing to phase-separated, membrane-less granules under stress conditions. Furthermore, m^6A-modified mRNAs are enriched in stress granules (SGs), and the m^6A-binding YTHDF protein is critical for SG formation (Fu and Zhuang 2019; Ries et al. 2019). Likewise, changes in the poly(A)-tail length at the end of 3′ UTRs have been implicated with RNA-localization changes (Fazal et al. 2019). Thus, although evidence for widespread involvement of modifications in RNA localization remains limited, these multiple observations in different systems warrant future investigation.

In addition to RNA modification, the secondary and tertiary structures of RNAs undoubtedly guide RNA localization patterns. RNA structure within cells varies across different cellular locations (Sun et al. 2019), and many RBPs such as Staufen interact with structural elements in RNAs (Bevilacqua et al. 2016). Furthermore, structured RNAs (Langdon et al. 2018; Maharana et al. 2018) in different subcellular locations show different propensities for forming liquid–liquid phase-separated condensates and organelles, including nuclear speckles, paraspeckles, Cajal bodies, nuclear stress bodies, and even heterochromatin (Sanulli et al. 2019). Thus structure-mapping studies should complement localization studies in identifying *cis* elements directing RNA localization.

How RNAs Influence the Genome Architecture

Genomic DNA is highly organized in three-dimensional space, and RNA has long been known to be an essential regulator of chromatin (Nickerson et al. 1989). RNA binding seems to promote CTCF-dependent chromatin looping and thus is vital for the organization of the genome into megabase structures called topologically associated domains (TADs) (Saldaña-Meyer et al. 2019). Furthermore, RNAse treatment to degrade RNAs, as well as transcriptional inhibition, affects both the structure and formation of DNA TADS (Barutcu et al. 2019). Likewise, disruption of the RNA-binding domain of CTCF, including through mutations (Hansen et al. 2019), has a global effect on chromatin binding, gene expression, and the formation of chromatin loops. However, the exact identity of RNAs in each genomic neighborhood within the nucleus that modulates the underlying processes of transcription, splicing, and genome organization remains unclear. A recently developed technology to perform RNA-directed chromosome conformation may aid in solving this mystery (Mumbach et al. 2019).

Recent studies suggest RNAs can act as structural scaffolds for organizing chromatin domains, including the lncRNA *Firre* that maintains the H3K27m3 chromatin state of the inactive X chromosome in female cells and makes contact with several autosomes (Thakur et al. 2019). Other RNAs such as *MALAT1* and *NEAT1* have also been shown to have scaffolding roles within the nucleus, particularly within the nuclear speckles and paraspeckles respectively. Another RNA *Xist*, required for transcriptional silencing of the X chromosome, is brought to the nuclear lamina as part of its function (Chen et al. 2016).

APEX-seq in the nucleus revealed a correlation between the location of mature, polyadenylated transcripts, and the underlying genome architecture (Fazal et al. 2019). For example, the lamina transcriptome was found to be enriched for genes found in lamina-associated domains (LADs), and the nucleolus transcriptome is enriched for genes found in nucleolus-associated domains (NADs). LADs, DNA regions near the lamina, comprise 30%–40% of the genome and contain thousands of genes that are generally lowly expressed. In summary, there seems to be an intimate connection between subnuclear RNA localization and the underlying genome organization and regulation that warrants further investigation.

How RNAs Localize to Organelles

How cells orchestrate the localization of hundreds of RNAs to a subcellular location continues to remain a mystery. Locales such as the ERM and OMM are known to have more than a thousand transcripts localizing there. Recently, APEX-seq (Fazal et al. 2019) revealed the landscape of RNA localization and local translation to the outside of the mitochondria, identifying both translation-dependent and translation-independent mechanisms of RNA localization. For reasons not clear, the translation-independent transcripts had shorter 3′ UTRs and shorter poly(A)-tail lengths. Similarly, the RBP CLUH is known to bind a subset of mRNAs for nuclear-encoded mitochondrial proteins in mammals (Gao et al. 2014). Nonetheless, the localization mechanism of transcripts to the mammalian mitochondria remains opaque and will undoubtedly be an active area of future investigation.

In yeast, where RNA localization to the mitochondria is better understood, it has been speculated that the mitochondrial proteins translated near the mitochondria are of prokaryotic origin, whereas accessory proteins are often translated in free cytoplasmic polysomes (Garcia et al. 2007; Marc et al. 2002). Furthermore, although the localization of proteins to the mitochondria is aided by specific amino acids in the translated nascent peptide, called mitochondria-targeting sequences (MTSs), sequences in the 3′ UTR of the corresponding RNA have also been shown to be essential for local translation. For example, in yeast, either the MTS or the 3′ UTR was sufficient to independently target ATM1 mRNA to the vicinity of the mitochondria (Corral-Debrinski et al. 2000). Also, some RBPs such as the Puf family of proteins in yeast control the localization of hundreds of transcripts, particularly Puf3 that associates with transcripts encoding proteins localizing to the mitochondria (Hogan et al. 2008).

In addition to the ER and mitochondria, many locations in cells concentrate RNAs, including MLOs present in both the nucleus and cytosol. Interestingly, the MLOs' nucleolus and stress bodies are known to phase separate and are tuned and regulated by the concentration of proteins and RNAs within them. Furthermore, long RNAs with stable secondary structures that bind RNA binding proteins are particularly good at promoting phase separation, including in nuclear locations such as paraspeckles. Understanding how RNAs are specifically targeted to MLOs and membrane-bound organelles continues to be a fascinating, unanswered question.

How Nonpolyadenylated RNAs, including Circular RNAs, Localize within Cells

Cells contain many different RNA species, and extending our current understanding of mRNA localization to other RNA species, including tRNAs and circular RNAs (circRNAs), will be important in understanding the regulation of these molecules. circRNAs have received a lot of interest in recent years, and it is known that they can code for proteins (Jeck and Sharpless 2014) and show asymmetric subcellular localization (Saini et al. 2019). Initial studies, for example, suggest circRNAs localize differently relative to other RNAs in neuronal projections (Saini et al. 2019). In addition, immunogenic circRNAs that are sensed as foreign are localized to distinct locations in the cytoplasm compared to endogenous circRNAs (Chen et al. 2019b). Thus, the mechanisms of circRNA localization, often without the benefit of 5′ or 3′ UTRs present on linear mRNAs, are likely to shed new light on RNA localization and circRNA functions.

CONCLUSION

Subcellular RNA localization is an essential but underappreciated aspect of gene regulation. This review focuses on the eukaryotic cell, but even prokaryotic cells are known to have highly localized RNAs (Nevo-Dinur et al. 2011). In prokaryotes, RNAs are directed to specific locations such as the inner membrane, although whether this localization is exclusively translation-dependent or not remains an open question (Moffitt et al. 2016). With the advent of high-throughput imaging and sequencing approaches, it is now possible to comprehensive interrogate the transcriptomes of subcellular locations in different cell types and model systems. Exciting future studies will undoubtedly map out the regulatory code guiding localization, and explain why organisms ubiquitously use such mechanisms.

ACKNOWLEDGMENTS

F.M.F. acknowledges funding from the Arnold O. Beckman postdoctoral fellowship, and by a National Institutes of Health (NIH) K99/R00 award from the National Human Genome Research Institute (NHGRI) (HG010910). H.Y.C. is supported by RM1-HG007735, R35-CA209919, and R01-HG004361. H.Y.C. is an Investigator of the Howard Hughes Medical Institute. We apologize to colleagues for the exclusion of references because of space constraints. Some figures were created with BioRender as part of an academic license.

REFERENCES

Alami NH, Smith RB, Carrasco MA, Williams LA, Winborn CS, Han SSW, Kiskinis E, Winborn B, Freibaum BD, Kanagaraj A, et al. 2014. Axonal transport of TDP-43 mRNA granules is impaired by ALS-causing mutations. *Neuron* **81:** 536–543. doi:10.1016/j.neuron.2013.12.018

Anderson P, Kedersha N. 2006. RNA granules. *J Cell Biol* **172:** 803–808. doi:10.1083/jcb.200512082

Bahar Halpern K, Caspi I, Lemze D, Levy M, Landen S, Elinav E, Ulitsky I, Itzkovitz S. 2015. Nuclear retention of mRNA in mammalian tissues. *Cell Rep* **13:** 2653–2662. doi:10.1016/j.celrep.2015.11.036

Barutcu AR, Blencowe BJ, Rinn JL. 2019. Differential contribution of steady-state RNA and active transcription in chromatin organization. *EMBO Rep* **20:** e48068. doi:10.15252/embr.201948068

Batista PJ, Chang HY. 2013. Long noncoding RNAs: cellular address codes in development and disease. *Cell* **152:** 1298–1307. doi:10.1016/j.cell.2013.02.012

Baumann S, Komissarov A, Gili M, Ruprecht V, Wieser S, Maurer SP. 2020. A reconstituted mammalian APC-kinesin complex selectively transports defined packages of axonal mRNAs. *Sci Adv* **6:** eaaz1588. doi:10.1126/sciadv.aaz1588

Benhalevy D, Anastasakis DG, Hafner M. 2018. Proximity-CLIP provides a snapshot of protein-occupied RNA elements in subcellular compartments. *Nat Methods* **15:** 1074–1082. doi:10.1038/s41592-018-0220-y

Benoit Bouvrette LP, Cody NAL, Bergalet J, Lefebvre FA, Diot C, Wang X, Blanchette M, Lecuyer E. 2018. CeFra-seq reveals broad asymmetric mRNA and noncoding RNA distribution profiles in *Drosophila* and human cells. *RNA* **24:** 98–113. doi:10.1261/rna.063172.117

Bertrand E, Chartrand P, Schaefer M, Shenoy SM, Singer RH, Long RM. 1998. Localization of ASH1 mRNA particles in living yeast. *Mol Cell* **2:** 437–445. doi:10.1016/S1097-2765(00)80143-4

Bevilacqua PC, Ritchey LE, Su Z, Assmann SM. 2016. Genome-wide analysis of RNA secondary structure. *Annu Rev Genet* **50:** 235–266. doi:10.1146/annurev-genet-120215-035034

Braselmann E, Wierzba AJ, Polaski JT, Chrominski M, Holmes ZE, Hung ST, Batan D, Wheeler JR, Parker R, Jimenez R, et al. 2018. A multicolor riboswitch-based platform for imaging of RNA in live mammalian cells. *Nat Chem Biol* **14:** 964–971. doi:10.1038/s41589-018-0103-7

Braunschweig U, Gueroussov S, Plocik AM, Graveley BR, Blencowe BJ. 2013. Dynamic integration of splicing within gene regulatory pathways. *Cell* **152:** 1252–1269. doi:10.1016/j.cell.2013.02.034

Braunschweig U, Barbosa-Morais NL, Pan Q, Nachman EN, Alipanahi B, Gonatopoulos-Pournatzis T, Frey B, Irimia M, Blencowe BJ. 2014. Widespread intron retention in mammals functionally tunes transcriptomes. *Genome Res* **24:** 1774–1786. doi:10.1101/gr.177790.114

Brown CR, Silver PA. 2007. Transcriptional regulation at the nuclear pore complex. *Curr Opin Genet Dev* **17:** 100–106. doi:10.1016/j.gde.2007.02.005

Cajigas IJ, Tushev G, Will TJ, tom Dieck S, Fuerst N, Schuman EM. 2012. The local transcriptome in the synaptic neuropil revealed by deep sequencing and high-resolution imaging. *Neuron* **74:** 453–466. doi:10.1016/j.neuron.2012.02.036

Cao Z, Pan X, Yang Y, Huang Y, Shen HB. 2018. The lncLocator: a subcellular localization predictor for long non-coding RNAs based on a stacked ensemble classifier. *Bioinformatics* **34:** 2185–2194. doi:10.1093/bioinformatics/bty085

Carlevaro-Fita J, Polidori T, Das M, Navarro C, Zoller TI, Johnson R. 2019. Ancient exapted transposable elements promote nuclear enrichment of human long noncoding RNAs. *Genome Res* **29:** 208–222. doi:10.1101/gr.229922.117

Chaumeil J, Le Baccon P, Wutz A, Heard E. 2006. A novel role for Xist RNA in the formation of a repressive nuclear compartment into which genes are recruited when silenced. *Genes Dev* **20:** 2223–2237. doi:10.1101/gad.380906

Chen KH, Boettiger AN, Moffitt JR, Wang S, Zhuang X. 2015. RNA imaging. Spatially resolved, highly multiplexed RNA profiling in single cells. *Science* **348:** aaa6090. doi:10.1126/science.aaa6090

Chen CK, Blanco M, Jackson C, Aznauryan E, Ollikainen N, Surka C, Chow A, Cerase A, McDonel P, Guttman M. 2016. Xist recruits the X chromosome to the nuclear lamina to enable chromosome-wide silencing. *Science* **354:** 468–472. doi:10.1126/science.aae0047

Chen M, Ma Z, Wu X, Mao S, Yang Y, Tan J, Krueger CJ, Chen AK. 2017. A molecular beacon-based approach for live-cell imaging of RNA transcripts with minimal target engineering at the single-molecule level. *Sci Rep* **7:** 1550. doi:10.1038/s41598-017-01740-1

Chen K, Wang Y, Sun J. 2018. A statistical analysis on transcriptome sequences: the enrichment of Alu-element is associated with subcellular location. *Biochem Biophys Res Commun* **499:** 397–402. doi:10.1016/j.bbrc.2018.03.024

Chen X, Zhang D, Su N, Bao B, Xie X, Zuo F, Yang L, Wang H, Jiang L, Lin Q, et al. 2019a. Visualizing RNA dynamics in live cells with bright and stable fluorescent RNAs. *Nat Biotechnol* **37:** 1287–1293. doi:10.1038/s41587-019-0249-1

Chen YG, Chen R, Ahmad S, Verma R, Kasturi SP, Amaya L, Broughton JP, Kim J, Cadena C, Pulendran B, et al. 2019b. N^6-methyladenosine modification controls circular RNA immunity. *Mol Cell* **76:** 96–109 e109. doi:10.1016/j.molcel.2019.07.016

Cheng H, Dufu K, Lee CS, Hsu JL, Dias A, Reed R. 2006. Human mRNA export machinery recruited to the 5′ end of mRNA. *Cell* **127:** 1389–1400. doi:10.1016/j.cell.2006.10.044

Chu C, Zhang QC, da Rocha ST, Flynn RA, Bharadwaj M, Calabrese JM, Magnuson T, Heard E, Chang HY. 2015. Systematic discovery of Xist RNA binding proteins. *Cell* **161:** 404–416. doi:10.1016/j.cell.2015.03.025

Ciolli Mattioli C, Rom A, Franke V, Imami K, Arrey G, Terne M, Woehler A, Akalin A, Ulitsky I, Chekulaeva M. 2019. Alternative 3′ UTRs direct localization of functionally diverse protein isoforms in neuronal compartments. *Nucleic Acids Res* **47:** 2560–2573. doi:10.1093/nar/gky1270

Cohen-Zontag O, Baez C, Lim LQJ, Olender T, Schirman D, Dahary D, Pilpel Y, Gerst JE. 2019. A secretion-enhancing *cis* regulatory targeting element (SECReTE) involved in mRNA localization and protein synthesis. *PLoS Genet* **15:** e1008248. doi:10.1371/journal.pgen.1008248

Colon-Ramos DA, Salisbury JL, Sanders MA, Shenoy SM, Singer RH, Garcia-Blanco MA. 2003. Asymmetric distribution of nuclear pore complexes and the cytoplasmic localization of β2-tubulin mRNA in *Chlamydomonas reinhardtii*. *Dev Cell* **4:** 941–952. doi:10.1016/S1534-5807(03)00163-1

Corral-Debrinski M, Blugeon C, Jacq C. 2000. In yeast, the 3′ untranslated region or the presequence of ATM1 is required for the exclusive localization of its mRNA to the vicinity of mitochondria. *Mol Cell Biol* **20:** 7881–7892. doi:10.1128/MCB.20.21.7881-7892.2000

Costa EA, Subramanian K, Nunnari J, Weissman JS. 2018. Defining the physiological role of SRP in protein-targeting efficiency and specificity. *Science* **359:** 689–692. doi:10.1126/science.aar3607

Darzacq X, Shav-Tal Y, de Turris V, Brody Y, Shenoy SM, Phair RD, Singer RH. 2007. In vivo dynamics of RNA polymerase II transcription. *Nat Struct Mol Biol* **14:** 796–806. doi:10.1038/nsmb1280

de Boer CG, Vaishnav ED, Sadeh R, Abeyta EL, Friedman N, Regev A. 2019. Deciphering eukaryotic gene-regulatory logic with 100 million random promoters. *Nat Biotechnol* **38:** 56–65. doi:10.1038/s41587-019-0315-8

Djebali S, Davis CA, Merkel A, Dobin A, Lassmann T, Mortazavi A, Tanzer A, Lagarde J, Lin W, Schlesinger F, et al. 2012. Landscape of transcription in human cells. *Nature* **489:** 101–108. doi:10.1038/nature11233

Edens BM, Vissers C, Su J, Arumugam S, Xu Z, Shi H, Miller N, Rojas Ringeling F, Ming GL, He C, et al. 2019. FMRP modulates neural differentiation through m^{6}A-dependent mRNA nuclear export. *Cell Rep* **28:** 845–854 e845. doi:10.1016/j.celrep.2019.06.072

Eliyahu E, Pnueli L, Melamed D, Scherrer T, Gerber AP, Pines O, Rapaport D, Arava Y. 2010. Tom20 mediates localization of mRNAs to mitochondria in a translation-dependent manner. *Mol Cell Biol* **30:** 284–294. doi:10.1128/MCB.00651-09

Emanuel G, Moffitt JR, Zhuang X. 2017. High-throughput, image-based screening of pooled genetic-variant libraries. *Nat Methods* **14:** 1159–1162. doi:10.1038/nmeth.4495

Eng CL, Lawson M, Zhu Q, Dries R, Koulena N, Takei Y, Yun J, Cronin C, Karp C, Yuan GC, et al. 2019. Transcriptome-scale super-resolved imaging in tissues by RNA seqFISH. *Nature* **568:** 235–239. doi:10.1038/s41586-019-1049-y

Erdelyi M, Michon AM, Guichet A, Glotzer JB, Ephrussi A. 1995. Requirement for *Drosophila* cytoplasmic tropomyosin in *oskar* mRNA localization. *Nature* **377:** 524–527. doi:10.1038/377524a0

Fazal FM, Han S, Parker KR, Kaewsapsak P, Xu J, Boettiger AN, Chang HY, Ting AY. 2019. Atlas of subcellular RNA localization revealed by APEX-seq. *Cell* **178:** 473–490 e426. doi:10.1016/j.cell.2019.05.027

Femino AM, Fay FS, Fogarty K, Singer RH. 1998. Visualization of single RNA transcripts in situ. *Science* **280:** 585–590. doi:10.1126/science.280.5363.585

Fu Y, Zhuang X. 2019. m6A-binding YTHDF proteins promote stress granule formation by modulating phase separation of stress granule proteins. bioRxiv doi:10.1101/694455

Fürth D, Hatini V, Lee JH. 2019. In situ transcriptome accessibility sequencing (INSTA-seq). bioRxiv doi:10.1101/722819

Gadir N, Haim-Vilmovsky L, Kraut-Cohen J, Gerst JE. 2011. Localization of mRNAs coding for mitochondrial proteins in the yeast *Saccharomyces cerevisiae*. *RNA* **17:** 1551–1565. doi:10.1261/rna.2621111

Gao J, Schatton D, Martinelli P, Hansen H, Pla-Martin D, Barth E, Becker C, Altmueller J, Frommolt P, Sardiello M, et al. 2014. CLUH regulates mitochondrial biogenesis by binding mRNAs of nuclear-encoded mitochondrial proteins. *J Cell Biol* **207:** 213–223. doi:10.1083/jcb.201403129

Gao XD, Tu LC, Mir A, Rodriguez T, Ding Y, Leszyk J, Dekker J, Shaffer SA, Zhu LJ, Wolfe SA, et al. 2018. C-BERST: defining subnuclear proteomic landscapes at genomic elements with dCas9-APEX2. *Nat Methods* **15:** 433–436. doi:10.1038/s41592-018-0006-2

Garalde DR, Snell EA, Jachimowicz D, Sipos B, Lloyd JH, Bruce M, Pantic N, Admassu T, James P, Warland A, et al. 2018. Highly parallel direct RNA sequencing on an array of nanopores. *Nat Methods* **15:** 201–206. doi:10.1038/nmeth.4577

Garcia M, Darzacq X, Delaveau T, Jourdren L, Singer RH, Jacq C. 2007. Mitochondria-associated yeast mRNAs and the biogenesis of molecular complexes. *Mol Biol Cell* **18:** 362–368. doi:10.1091/mbc.e06-09-0827

Ghosh S, Marchand V, Gaspar I, Ephrussi A. 2012. Control of RNP motility and localization by a splicing-dependent structure in *oskar* mRNA. *Nat Struct Mol Biol* **19:** 441–449. doi:10.1038/nsmb.2257

Grunwald D, Singer RH. 2010. In vivo imaging of labelled endogenous β-actin mRNA during nucleocytoplasmic transport. *Nature* **467:** 604–607. doi:10.1038/nature09438

Gudenas BL, Wang L. 2018. Prediction of LncRNA subcellular localization with deep learning from sequence features. *Sci Rep* **8:** 16385. doi:10.1038/s41598-018-34708-w

Hachet O, Ephrussi A. 2004. Splicing of *oskar* RNA in the nucleus is coupled to its cytoplasmic localization. *Nature* **428:** 959–963. doi:10.1038/nature02521

Hacisuleyman E, Goff LA, Trapnell C, Williams A, Henao-Mejia J, Sun L, McClanahan P, Hendrickson DG, Sauvageau M, Kelley DR, et al. 2014. Topological organization of multichromosomal regions by the long intergenic noncoding RNA Firre. *Nat Struct Mol Biol* **21:** 198–206. doi:10.1038/nsmb.2764

Han S, Zhao BS, Myers SA, Carr SA, He C, Ting AY. 2020. RNA-protein interaction mapping via MS2 or Cas13-based APEX targeting. bioRxiv doi:10.1101/968297

Hancock WO. 2014. Bidirectional cargo transport: moving beyond tug of war. *Nat Rev Mol Cell Biol* **15:** 615–628. doi:10.1038/nrm3853

Hansen AS, Hsieh TS, Cattoglio C, Pustova I, Saldaña-Meyer R, Reinberg D, Darzacq X, Tjian R. 2019. Distinct classes of chromatin loops revealed by deletion of an RNA-binding region in CTCF. *Mol Cell* **76:** 395–411 e313. doi:10.1016/j.molcel.2019.07.039

Hasegawa Y, Brockdorff N, Kawano S, Tsutui K, Tsutui K, Nakagawa S. 2010. The matrix protein hnRNP U is required for chromosomal localization of Xist RNA. *Dev Cell* **19:** 469–476. doi:10.1016/j.devcel.2010.08.006

Hirokawa N, Niwa S, Tanaka Y. 2010. Molecular motors in neurons: transport mechanisms and roles in brain function, development, and disease. *Neuron* **68:** 610–638. doi:10.1016/j.neuron.2010.09.039

Hogan DJ, Riordan DP, Gerber AP, Herschlag D, Brown PO. 2008. Diverse RNA-binding proteins interact with functionally related sets of RNAs, suggesting an extensive regulatory system. *PLoS Biol* **6:** e255. doi:10.1371/journal.pbio.0060255

Hung V, Udeshi ND, Lam SS, Loh KH, Cox KJ, Pedram K, Carr SA, Ting AY. 2016. Spatially resolved proteomic mapping in living cells with the engineered peroxidase APEX2. *Nat Protoc* **11:** 456–475. doi:10.1038/nprot.2016.018

Huttelmaier S, Zenklusen D, Lederer M, Dictenberg J, Lorenz M, Meng X, Bassell GJ, Condeelis J, Singer RH. 2005. Spatial regulation of β-actin translation by Src-dependent phosphorylation of ZBP1. *Nature* **438:** 512–515. doi:10.1038/nature04115

Hwang HW, Wentzel EA, Mendell JT. 2007. A hexanucleotide element directs microRNA nuclear import. *Science* **315:** 97–100. doi:10.1126/science.1136235

Jan CH, Williams CC, Weissman JS. 2014. Principles of ER cotranslational translocation revealed by proximity-specific ribosome profiling. *Science* **346:** 1257521. doi:10.1126/science.1257521

Jeck WR, Sharpless NE. 2014. Detecting and characterizing circular RNAs. *Nat Biotechnol* **32:** 453–461. doi:10.1038/nbt.2890

Kaewsapsak P, Shechner DM, Mallard W, Rinn JL, Ting AY. 2017. Live-cell mapping of organelle-associated RNAs via proximity biotinylation combined with protein-RNA cross-linking. *Elife* **6:** e29224. doi:10.7554/eLife.29224

Kanai Y, Dohmae N, Hirokawa N. 2004. Kinesin transports RNA: isolation and characterization of an RNA-transporting granule. *Neuron* **43:** 513–525. doi:10.1016/j.neuron.2004.07.022

Katahira J. 2015. Nuclear export of messenger RNA. *Genes (Basel)* **6:** 163–184. doi:10.3390/genes6020163

Ke R, Mignardi M, Pacureanu A, Svedlund J, Botling J, Wahlby C, Nilsson M. 2013. In situ sequencing for RNA analysis in preserved tissue and cells. *Nat Methods* **10:** 857–860. doi:10.1038/nmeth.2563

Khong A, Matheny T, Jain S, Mitchell SF, Wheeler JR, Parker R. 2017. The stress granule transcriptome reveals principles of mRNA accumulation in stress granules. *Mol Cell* **68:** 808–820 e805. doi:10.1016/j.molcel.2017.10.015

Kim-Ha J, Webster PJ, Smith JL, Macdonald PM. 1993. Multiple RNA regulatory elements mediate distinct steps in localization of *oskar* mRNA. *Development* **119:** 169–178.

Kislauskis EH, Zhu X, Singer RH. 1994. Sequences responsible for intracellular localization of β-actin messenger RNA also affect cell phenotype. *J Cell Biol* **127:** 441–451. doi:10.1083/jcb.127.2.441

Krauss J, Lopez de Quinto S, Nusslein-Volhard C, Ephrussi A. 2009. Myosin-V regulates *oskar* mRNA localization in the *Drosophila* oocyte. *Curr Biol* **19:** 1058–1063. doi:10.1016/j.cub.2009.04.062

Langdon EM, Qiu Y, Ghanbari Niaki A, McLaughlin GA, Weidmann CA, Gerbich TM, Smith JA, Crutchley JM, Termini CM, Weeks KM, et al. 2018. mRNA structure determines specificity of a polyQ-driven phase separation. *Science* **360:** 922–927. doi:10.1126/science.aar7432

Lawrence JB, Singer RH, Marselle LM. 1989. Highly localized tracks of specific transcripts within interphase nuclei visualized by in situ hybridization. *Cell* **57:** 493–502. doi:10.1016/0092-8674(89)90924-0

LeCun Y, Bengio Y, Hinton G. 2015. Deep learning. *Nature* **521:** 436–444. doi:10.1038/nature14539

Lee JH, Daugharthy ER, Scheiman J, Kalhor R, Yang JL, Ferrante TC, Terry R, Jeanty SS, Li C, Amamoto R, et al. 2014. Highly multiplexed subcellular RNA sequencing in situ. *Science* **343:** 1360–1363. doi:10.1126/science.1250212

Legnini I, Alles J, Karaiskos N, Ayoub S, Rajewsky N. 2019. FLAM-seq: full-length mRNA sequencing reveals principles of poly(A) tail length control. *Nat Methods* **16:** 879–886. doi:10.1038/s41592-019-0503-y

Lesbirel S, Wilson SA. 2019. The m^6A methylase complex and mRNA export. *Biochim Biophys Acta* **1862:** 319–328. doi:10.1016/j.bbagrm.2018.09.008

Li Y, Aggarwal MB, Nguyen K, Ke K, Spitale RC. 2017. Assaying RNA localization in situ with spatially restricted nucleobase oxidation. *ACS Chem Biol* **12:** 2709–2714. doi:10.1021/acschembio.7b00519

Lubeck E, Coskun AF, Zhiyentayev T, Ahmad M, Cai L. 2014. Single-cell in situ RNA profiling by sequential hybridization. *Nat Methods* **11:** 360–361. doi:10.1038/nmeth.2892

Lubelsky Y, Ulitsky I. 2018. Sequences enriched in Alu repeats drive nuclear localization of long RNAs in human cells. *Nature* **555:** 107–111. doi:10.1038/nature25757

Ma J, Liu Z, Michelotti N, Pitchiaya S, Veerapaneni R, Androsavich JR, Walter NG, Yang W. 2013. High-resolution three-dimensional mapping of mRNA export through the nuclear pore. *Nat Commun* **4:** 2414. doi:10.1038/ncomms3414

Maday S, Twelvetrees AE, Moughamian AJ, Holzbaur EL. 2014. Axonal transport: cargo-specific mechanisms of motility and regulation. *Neuron* **84:** 292–309. doi:10.1016/j.neuron.2014.10.019

Maharana S, Wang J, Papadopoulos DK, Richter D, Pozniakovsky A, Poser I, Bickle M, Rizk S, Guillen-Boixet J, Franzmann TM, et al. 2018. RNA buffers the phase separation behavior of prion-like RNA binding proteins. *Science* **360:** 918–921. doi:10.1126/science.aar7366

Marc P, Margeot A, Devaux F, Blugeon C, Corral-Debrinski M, Jacq C. 2002. Genome-wide analysis of mRNAs targeted to yeast mitochondria. *EMBO Rep* **3:** 159–164. doi:10.1093/embo-reports/kvf025

Markmiller S, Soltanieh S, Server KL, Mak R, Jin W, Fang MY, Luo EC, Krach F, Yang D, Sen A, et al. 2018. Context-dependent and disease-specific diversity in protein interactions within stress granules. *Cell* **172:** 590–604 e513. doi:10.1016/j.cell.2017.12.032

Medina-Muñoz HC, Lapointe CP, Porter DF, Wickens M. 2019. Records of RNA localization through covalent tagging. bioRxiv doi:10.1101/785816

Mili S, Moissoglu K, Macara IG. 2008. Genome-wide screen reveals APC-associated RNAs enriched in cell protrusions. *Nature* **453:** 115–119. doi:10.1038/nature06888

Miyagawa R, Tano K, Mizuno R, Nakamura Y, Ijiri K, Rakwal R, Shibato J, Masuo Y, Mayeda A, Hirose T, et al. 2012. Identification of *cis*- and *trans*-acting factors involved in the localization of MALAT-1 noncoding RNA to nuclear speckles. *RNA* **18:** 738–751. doi:10.1261/rna.028639.111

Moffitt JR, Pandey S, Boettiger AN, Wang S, Zhuang X. 2016. Spatial organization shapes the turnover of a bacterial transcriptome. *Elife* **5:** e13065. doi:10.7554/eLife.13065

Moffitt JR, Bambah-Mukku D, Eichhorn SW, Vaughn E, Shekhar K, Perez JD, Rubinstein ND, Hao J, Regev A, Dulac C, et al. 2018. Molecular, spatial, and functional single-cell profiling of the hypothalamic preoptic region. *Science* **362:** eaau5324. doi:10.1126/science.aau5324

Moon SL, Morisaki T, Khong A, Lyon K, Parker R, Stasevich TJ. 2019. Multicolour single-molecule tracking of mRNA interactions with RNP granules. *Nat Cell Biol* **21:** 162–168. doi:10.1038/s41556-018-0263-4

Moor AE, Golan M, Massaa EE, Lemze D, Weizman T, Shenhav R, Baydatch S, Mizrahi O, Winkler R, Golani O, et al. 2017. Global mRNA polarization regulates translation efficiency in the intestinal epithelium. *Science* **357:** 1299–1303. doi:10.1126/science.aan2399

Movva R, Greenside P, Marinov GK, Nair S, Shrikumar A, Kundaje A. 2019. Deciphering regulatory DNA sequences and noncoding genetic variants using neural network models of massively parallel reporter assays. *PLoS One* **14:** e0218073. doi:10.1371/journal.pone.0218073

Mukherjee J, Hermesh O, Eliscovich C, Nalpas N, Franz-Wachtel M, Macek B, Jansen RP. 2019. β-Actin mRNA interactome mapping by proximity biotinylation. *Proc Natl Acad Sci* **116:** 12863–12872. doi:10.1073/pnas.1820737116

Muller-McNicoll M, Neugebauer KM. 2013. How cells get the message: dynamic assembly and function of mRNA-protein complexes. *Nat Rev Genet* **14:** 275–287. doi:10.1038/nrg3434

Mumbach MR, Granja JM, Flynn RA, Roake CM, Satpathy AT, Rubin AJ, Qi Y, Jiang Z, Shams S, Louie BH, et al. 2019. HiChIRP reveals RNA-associated chromosome conformation. *Nat Methods* **16:** 489–492. doi:10.1038/s41592-019-0407-x

Myers SA, Wright J, Peckner R, Kalish BT, Zhang F, Carr SA. 2018. Discovery of proteins associated with a predefined genomic locus via dCas9-APEX-mediated proximity labeling. *Nat Methods* **15:** 437–439. doi:10.1038/s41592-018-0007-1

Nelles DA, Fang MY, O'Connell MR, Xu JL, Markmiller SJ, Doudna JA, Yeo GW. 2016. Programmable RNA tracking in live cells with CRISPR/Cas9. *Cell* **165:** 488–496. doi:10.1016/j.cell.2016.02.054

Neumann M, Sampathu DM, Kwong LK, Truax AC, Micsenyi MC, Chou TT, Bruce J, Schuck T, Grossman M, Clark CM, et al. 2006. Ubiquitinated TDP-43 in frontotemporal lobar degeneration and amyotrophic lateral sclerosis. *Science* **314:** 130–133. doi:10.1126/science.1134108

Nevo-Dinur K, Nussbaum-Shochat A, Ben-Yehuda S, Amster-Choder O. 2011. Translation-independent localization of mRNA in *E. coli*. *Science* **331:** 1081–1084. doi:10.1126/science.1195691

Nguyen TM, Kabotyanski EB, Reineke LC, Shao J, Xiong F, Lee JH, Dubrulle J, Johnson H, Stossi F, Tsoi PS, et al. 2020. The SINEB1 element in the long non-coding RNA Malat1 is necessary for TDP-43 proteostasis. *Nucleic Acids Res* **48:** 2621–2642. doi:10.1093/nar/gkz1176

Nichterwitz S, Chen G, Aguila Benitez J, Yilmaz M, Storvall H, Cao M, Sandberg R, Deng Q, Hedlund E. 2016. Laser capture microscopy coupled with Smart-seq2 for precise spatial transcriptomic profiling. *Nat Commun* **7:** 12139. doi:10.1038/ncomms12139

Nickerson JA, Krochmalnic G, Wan KM, Penman S. 1989. Chromatin architecture and nuclear RNA. *Proc Natl Acad Sci* **86:** 177–181. doi:10.1073/pnas.86.1.177

Padron A, Iwasaki S, Ingolia NT. 2019. Proximity RNA labeling by APEX-seq reveals the organization of translation initiation complexes and repressive RNA granules. *Mol Cell* **75:** 875–887 e875. doi:10.1016/j.molcel.2019.07.030

Pyhtila B, Zheng T, Lager PJ, Keene JD, Reedy MC, Nicchitta CV. 2008. Signal sequence- and translation-independent mRNA localization to the endoplasmic reticulum. *RNA* **14:** 445–453. doi:10.1261/rna.721108

Qiu W, Xu Z, Zhang M, Zhang D, Fan H, Li T, Wang Q, Liu P, Zhu Z, Du D, et al. 2019. Determination of local chromatin interactions using a combined CRISPR and peroxidase APEX2 system. *Nucleic Acids Res* **47:** e52. doi:10.1093/nar/gkz134

Raj A, van den Bogaard P, Rifkin SA, van Oudenaarden A, Tyagi S. 2008. Imaging individual mRNA molecules using multiple singly labeled probes. *Nat Methods* **5:** 877–879. doi:10.1038/nmeth.1253

Ramanathan M, Majzoub K, Rao DS, Neela PH, Zarnegar BJ, Mondal S, Roth JG, Gai H, Kovalski JR, Siprashvili Z, et al. 2018. RNA-protein interaction detection in living cells. *Nat Methods* **15:** 207–212. doi:10.1038/nmeth.4601

Ramanathan M, Porter DF, Khavari PA. 2019. Methods to study RNA-protein interactions. *Nat Methods* **16:** 225–234. doi:10.1038/s41592-019-0330-1

Rebagliati MR, Weeks DL, Harvey RP, Melton DA. 1985. Identification and cloning of localized maternal RNAs from *Xenopus* eggs. *Cell* **42:** 769–777. doi:10.1016/0092-8674(85)90273-9

Rhee HW, Zou P, Udeshi ND, Martell JD, Mootha VK, Carr SA, Ting AY. 2013. Proteomic mapping of mitochondria in living cells via spatially restricted enzymatic tagging. *Science* **339:** 1328–1331. doi:10.1126/science.1230593

Ries RJ, Zaccara S, Klein P, Olarerin-George A, Namkoong S, Pickering BF, Patil DP, Kwak H, Lee JH, Jaffrey SR. 2019. m(6)A enhances the phase separation potential of mRNA. *Nature* **571:** 424–428. doi:10.1038/s41586-019-1374-1

Roundtree IA, Evans ME, Pan T, He C. 2017a. Dynamic RNA modifications in gene expression regulation. *Cell* **169:** 1187–1200. doi:10.1016/j.cell.2017.05.045

Roundtree IA, Luo GZ, Zhang Z, Wang X, Zhou T, Cui Y, Sha J, Huang X, Guerrero L, Xie P, et al. 2017b. YTHDC1 mediates nuclear export of N^6-methyladenosine methylated mRNAs. *Elife* **6:** e31311. doi:10.7554/eLife.31311

Roux KJ, Kim DI, Raida M, Burke B. 2012. A promiscuous biotin ligase fusion protein identifies proximal and interacting proteins in mammalian cells. *J Cell Biol* **196:** 801–810. doi:10.1083/jcb.201112098

Saini H, Bicknell AA, Eddy SR, Moore MJ. 2019. Free circular introns with an unusual branchpoint in neuronal projections. *Elife* **8:** e47809. doi:10.7554/eLife.47809

Saint-Georges Y, Garcia M, Delaveau T, Jourdren L, Le Crom S, Lemoine S, Tanty V, Devaux F, Jacq C. 2008. Yeast mitochondrial biogenesis: a role for the PUF RNA-binding protein Puf3p in mRNA localization. *PLoS One* **3:** e2293. doi:10.1371/journal.pone.0002293

Saldaña-Meyer R, Rodriguez-Hernaez J, Escobar T, Nishana M, Jacome-Lopez K, Nora EP, Bruneau BG, Tsirigos A, Furlan-Magaril M, Skok J, et al. 2019. RNA interactions are essential for CTCF-mediated genome organization. *Mol Cell* **76:** 412–422 e415. doi:10.1016/j.molcel.2019.08.015

Sanulli S, Trnka MJ, Dharmarajan V, Tibble RW, Pascal BD, Burlingame AL, Griffin PR, Gross JD, Narlikar GJ. 2019. HP1 reshapes nucleosome core to promote phase separation of heterochromatin. *Nature* **575:** 390–394. doi:10.1038/s41586-019-1669-2

Shah S, Lubeck E, Zhou W, Cai L. 2016. In situ transcription profiling of single cells reveals spatial organization of cells in the mouse hippocampus. *Neuron* **92:** 342–357. doi:10.1016/j.neuron.2016.10.001

Shah S, Takei Y, Zhou W, Lubeck E, Yun J, Eng CL, Koulena N, Cronin C, Karp C, Liaw EJ, et al. 2018. Dynamics and spatial genomics of the nascent transcriptome by intron seqFISH. *Cell* **174:** 363–376 e316. doi:10.1016/j.cell.2018.05.035

Siebrasse JP, Kaminski T, Kubitscheck U. 2012. Nuclear export of single native mRNA molecules observed by light sheet fluorescence microscopy. *Proc Natl Acad Sci* **109:** 9426–9431. doi:10.1073/pnas.1201781109

Soneson C, Yao Y, Bratus-Neuenschwander A, Patrignani A, Robinson MD, Hussain S. 2019. A comprehensive examination of Nanopore native RNA sequencing for characterization of complex transcriptomes. *Nat Commun* **10:** 3359. doi:10.1038/s41467-019-11272-z

Sreedharan J, Blair IP, Tripathi VB, Hu X, Vance C, Rogelj B, Ackerley S, Durnall JC, Williams KL, Buratti E, et al. 2008. TDP-43 mutations in familial and sporadic amyotrophic lateral sclerosis. *Science* **319:** 1668–1672. doi:10.1126/science.1154584

St Johnston D, Beuchle D, Nusslein-Volhard C. 1991. *Staufen*, a gene required to localize maternal RNAs in the *Drosophila* egg. *Cell* **66:** 51–63. doi:10.1016/0092-8674(91)90138-O

St Johnston D, Brown NH, Gall JG, Jantsch M. 1992. A conserved double-stranded RNA-binding domain. *Proc Natl Acad Sci* **89:** 10979–10983. doi:10.1073/pnas.89.22.10979

Su ZD, Huang Y, Zhang ZY, Zhao YW, Wang D, Chen W, Chou KC, Lin H. 2018. iLoc-lncRNA: predict the subcellular location of lncRNAs by incorporating octamer composition into general PseKNC. *Bioinformatics* **34:** 4196–4204.

Sun L, Fazal FM, Li P, Broughton JP, Lee B, Tang L, Huang W, Kool ET, Chang HY, Zhang QC. 2019. RNA structure maps across mammalian cellular compartments. *Nat Struct Mol Biol* **26:** 322–330. doi:10.1038/s41594-019-0200-7

Taliaferro JM, Vidaki M, Oliveira R, Olson S, Zhan L, Saxena T, Wang ET, Graveley BR, Gertler FB, Swanson MS, et al. 2016. Distal alternative last exons localize mRNAs to neural projections. *Mol Cell* **61:** 821–833. doi:10.1016/j.molcel.2016.01.020

Thakur J, Fang H, Llagas T, Disteche CM, Henikoff S. 2019. Architectural RNA is required for heterochromatin organization. bioRxiv doi:10.1101/78435v1

Thul PJ, Akesson L, Wiking M, Mahdessian D, Geladaki A, Ait Blal H, Alm T, Asplund A, Björk L, Breckels LM, et al. 2017. A subcellular map of the human proteome. *Science* **356:** eaal3321. doi:10.1126/science.aal3321

Tsuboi T, Viana MP, Xu F, Yu J, Chanchani R, Arceo XG, Tutucci E, Choi J, Chen YS, Singer RH, et al. 2019. Mitochondrial volume fraction and translation speed impact mRNA localization and production of nuclear-encoded mitochondrial proteins. bioRxiv doi:10.1101/529289

Uhlen M, Oksvold P, Fagerberg L, Lundberg E, Jonasson K, Forsberg M, Zwahlen M, Kampf C, Wester K, Hober S, et al. 2010. Towards a knowledge-based Human Protein Atlas. *Nat Biotechnol* **28:** 1248–1250. doi:10.1038/nbt1210-1248

Uhlen M, Fagerberg L, Hallstrom BM, Lindskog C, Oksvold P, Mardinoglu A, Sivertsson A, Kampf C, Sjostedt E, Asplund A, et al. 2015. Proteomics. Tissue-based map of the human proteome. *Science* **347:** 1260419. doi:10.1126/science.1260419

Van Nostrand EL, Pratt GA, Shishkin AA, Gelboin-Burkhart C, Fang MY, Sundararaman B, Blue SM, Nguyen TB, Surka C, Elkins K, et al. 2016. Robust transcriptome-wide discovery of RNA-binding protein binding sites with enhanced CLIP (eCLIP). *Nat Methods* **13:** 508–514. doi:10.1038/nmeth.3810

Wan Y, Zhu N, Lu Y, Wong PK. 2019. DNA transformer for visualizing endogenous RNA dynamics in live cells. *Anal Chem* **91:** 2626–2633. doi:10.1021/acs.analchem.8b02826

Wang C, Han B, Zhou R, Zhuang X. 2016. Real-time imaging of translation on single mRNA transcripts in live cells. *Cell* **165:** 990–1001. doi:10.1016/j.cell.2016.04.040

Wang C, Lu T, Emanuel G, Babcock HP, Zhuang X. 2019a. Imaging-based pooled CRISPR screening reveals regulators of lncRNA localization. *Proc Natl Acad Sci* **116:** 10842–10851. doi:10.1073/pnas.1903808116

Wang P, Tang W, Li Z, Zou Z, Zhou Y, Li R, Xiong T, Wang J, Zou P. 2019b. Mapping spatial transcriptome with light-activated proximity-dependent RNA labeling. *Nat Chem Biol* **15:** 1110–1119. doi:10.1038/s41589-019-0368-5

Weeks DL, Melton DA. 1987. A maternal mRNA localized to the vegetal hemisphere in *Xenopus* eggs codes for a growth factor related to TGF-β. *Cell* **51:** 861–867. doi:10.1016/0092-8674(87)90109-7

Wilk R, Hu J, Blotsky D, Krause HM. 2016. Diverse and pervasive subcellular distributions for both coding and long noncoding RNAs. *Genes Dev* **30:** 594–609. doi:10.1101/gad.276931.115

Williams CC, Jan CH, Weissman JS. 2014. Targeting and plasticity of mitochondrial proteins revealed by proximity-specific ribosome profiling. *Science* **346:** 748–751. doi:10.1126/science.1257522

Workman RE, Tang AD, Tang PS, Jain M, Tyson JR, Razaghi R, Zuarte PC, Gilpatrick T, Payne A, Quick J, et al. 2019. Nanopore native RNA sequencing of a human poly(A) transcriptome. *Nat Methods* **16:** 1297–1305. doi:10.1038/s41592-019-0617-2

Wu B, Eliscovich C, Yoon YJ, Singer RH. 2016. Translation dynamics of single mRNAs in live cells and neurons. *Science* **352:** 1430–1435. doi:10.1126/science.aaf1084

Wu J, Zaccara S, Khuperkar D, Kim H, Tanenbaum ME, Jaffrey SR. 2019. Live imaging of mRNA using RNA-stabilized fluorogenic proteins. *Nat Methods* **16:** 862–865. doi:10.1038/s41592-019-0531-7

Wu KE, Parker KR, Fazal FM, Chang H, Zou J. 2020. RNA-GPS predicts high-resolution RNA subcellular localization and highlights the role of splicing. *RNA* doi:10.1261/rna.074161.119

Wutz A, Rasmussen TP, Jaenisch R. 2002. Chromosomal silencing and localization are mediated by different domains of Xist RNA. *Nat Genet* **30:** 167–174. doi:10.1038/ng820

Xia C, Fan J, Emanuel G, Hao J, Zhuang X. 2019. Spatial transcriptome profiling by MERFISH reveals subcellular RNA compartmentalization and cell cycle-dependent gene expression. *Proc Natl Acad Sci* **116:** 19490–19499. doi:10.1073/pnas.1912459116

Yan Z, Lecuyer E, Blanchette M. 2019. Prediction of mRNA subcellular localization using deep recurrent neural networks. *Bioinformatics* **35:** i333–i342. doi:10.1093/bioinformatics/btz337

Zappulo A, van den Bruck D, Ciolli Mattioli C, Franke V, Imami K, McShane E, Moreno-Estelles M, Calviello L, Filipchyk A, Peguero-Sanchez E, et al. 2017. RNA localization is a key determinant of neurite-enriched proteome. *Nat Commun* **8:** 583. doi:10.1038/s41467-017-00690-6

Zhou Y, Wang G, Wang P, Li Z, Yue T, Wang J, Zou P. 2019. Expanding APEX2 substrates for proximity-dependent labeling of nucleic acids and proteins in living cells. *Angew Chem Int Ed Engl* **58:** 11763–11767. doi:10.1002/anie.201905949

Zhu D, Stumpf CR, Krahn JM, Wickens M, Hall TM. 2009. A 5′ cytosine binding pocket in Puf3p specifies regulation of mitochondrial mRNAs. *Proc Natl Acad Sci* **106:** 20192–20197. doi:10.1073/pnas.0812079106

Zimyanin VL, Belaya K, Pecreaux J, Gilchrist MJ, Clark A, Davis I, St Johnston D. 2008. In vivo imaging of *oskar* mRNA transport reveals the mechanism of posterior localization. *Cell* **134:** 843–853. doi:10.1016/j.cell.2008.06.053

Zuckerman B, Ulitsky I. 2019. Predictive models of subcellular localization of long RNAs. *RNA* **25:** 557–572. doi:10.1261/rna.068288.118

Transcriptional Coactivator PGC-1α Binding to Newly Synthesized RNA via CBP80: A Nexus for Co- and Posttranscriptional Gene Regulation

Xavier Rambout,[1,2] Hana Cho,[1,2] and Lynne E. Maquat[1,2]

[1]Department of Biochemistry and Biophysics, School of Medicine and Dentistry,
[2]Center for RNA Biology, University of Rochester, Rochester, New York 14642, USA
Correspondence: lynne_maquat@urmc.rochester.edu

Mammalian cells have many quality-control mechanisms that regulate protein-coding gene expression to ensure proper transcript synthesis, processing, and translation. Should a step in transcript metabolism fail to fulfill requisite spatial, temporal, or structural criteria, including the proper acquisition of RNA-binding proteins, then that step will halt, fail to proceed to the next step, and ultimately result in transcript degradation. Quality-control mechanisms constitute a continuum of processes that initiate in the nucleus and extend to the cytoplasm. Here, we present published and unpublished data for protein-coding genes whose expression is activated by the transcriptional coactivator PGC-1α. We show that PGC-1α movement from chromatin, to which it is recruited by DNA-binding proteins, to CBP80 at the 5′ cap of nascent transcripts begins a series of co- and posttranscriptional quality- and quantity-control steps that, in total, ensure proper gene expression.

CBP80

Cap-binding protein 80 (CBP80) and CBP20 constitute the heterodimer called the cap-binding complex (CBC). CBP80, via CBP20, binds the m^7G 5′ cap of nascent RNAs, such as precursors to mRNAs (pre-mRNAs), as they are synthesized by RNA polymerase II (RNAPII) (Lejeune et al. 2002; Listerman et al. 2006; Glover-Cutter et al. 2008). Whereas CBP20 directly binds the cap, CBP80 increases the affinity of CBP20 for the cap and also serves as a docking platform for numerous RNA processing factors (for review, see Gonatopoulos-Pournatzis and Cowling 2014). Over the years, the CBC has been found to bind, remodel, and regulate the fate of protein-coding transcripts during their synthesis and processing in the nucleus, export from the nucleus to the cytoplasm and, as our laboratory has found in studies of mammalian cells, cytoplasmic "pioneer" rounds of translation that, depending on the mRNA, may be coupled to nonsense-mediated mRNA decay (NMD) (for reviews, see Gonatopoulos-Pournatzis and Cowling 2014; Kurosaki et al. 2019). For mRNAs that are not degraded by NMD, the CBC is subsequently replaced by another cap-binding protein, eukaryotic translation initiation factor 4E (eIF4E) (Lejeune et al. 2002), which typically supports the bulk of cellular translation (Pelletier and Sonenberg 2019).

Our laboratory's continuing interest in the CBC and, in particular, CBP80 function began with our finding that CBP80 promotes NMD. The best-characterized NMD targets harbor at least one exon-junction complex (EJC), deposited as a consequence of pre-mRNA splicing (Le Hir et al. 2000), on their 3′ untranslated region (3′ UTR).

3′-UTR EJC-dependent NMD initiates while the 5′ cap is still bound by the CBC during what we have called pioneer rounds of translation (Ishigaki et al. 2001; Hwang et al. 2010). These rounds occur on the cytoplasmic side of the nuclear envelope, as soon as newly synthesized mRNAs can be accessed by cytoplasmic ribosomes (Trcek et al. 2013). Depending on the length of the open reading frame (ORF) and efficiency of translation initiation, pioneer rounds can involve more than one associated ribosome translating the ORF, which is consistent with the decay steps of NMD requiring inhibition of further translation initiation events (Isken et al. 2008; Hoek et al. 2019).

Our studies have shown that CBP80 promotes NMD not only as a translation initiation factor but also as a choreographer of mRNA–protein rearrangements. Should a 3′-UTR EJC remain downstream from a ribosome after translation termination, as is often the scenario when translation terminates at a premature termination codon (PTC), then CBP80 escorts the essential NMD factor UPF1, which is an ATP-dependent RNA helicase, together with the UPF1 kinase, SMG1, first to the translation termination factors bound at the PTC and subsequently to the 3′-UTR EJC (Hosoda et al. 2005; Hwang et al. 2010). Details of these and other steps involving cytoplasmic functions of CBP80 and the CBC have been recently reviewed (Kurosaki et al. 2019).

The CBC also functions in the nucleus, where it promotes gene transcription, pre-mRNA splicing, and nuclear mRNA export. Of these functions, how the CBC promotes transcription is not well-characterized, especially in higher eukaryotes. To date, CBC-dependent transcription re-initiation has been established for *Saccharomyces cerevisiae*

© 2019 Rambout et al. This article is distributed under the terms of the Creative Commons Attribution-NonCommercial License, which permits reuse and redistribution, except for commercial purposes, provided that the original author and source are credited.

Published by Cold Spring Harbor Laboratory Press; doi:10.1101/sqb.2019.84.040212

(Lahudkar et al. 2011; Li et al. 2016a) and *Drosophila melanogaster* (Kachaev et al. 2019). Additionally, CBC-dependent transcription elongation has been shown for *S. cerevisiae* (Hossain et al. 2013; Lidschreiber et al. 2013) and mammalian cells (Lenasi et al. 2011). Notably, CBC-mediated control of transcription appears to be gene-specific, and gene-specific molecular determinants of CBC-mediated re-initiation of transcription have been identified (Lahudkar et al. 2011; Li et al. 2016a; Kachaev et al. 2019). However, how the CBC recruits elongation factors, such as the positive transcription elongation factor P-TEFb and the cyclin-dependent kinase CDK12, on specific genes is unknown (Lenasi et al. 2011; Hossain et al. 2013; Lidschreiber et al. 2013).

THE CBP80-BINDING MOTIF OF PGC-1 PROTEINS

Given our long-standing interest in the mechanistic continuum of nuclear and cytoplasmic RNA processes, we aimed to extend our largely cytoplasmic studies of mammalian-cell CBP80 to the nucleus. To begin, we interrogated a previously published yeast two-hybrid screen for human cDNAs encoding proteins that directly interact with human CBP80 (Kataoka et al. 1995). Among CBP80-binding partners identified were nuclear cap-binding protein (NCBP)-interacting protein 1 (NIP1) and NIP2, which were later renamed, respectively, CBP20 and PPARγ coactivator 1beta (PGC-1β, in which PPAR is an abbreviation for peroxisome proliferator-activated receptor). PGC-1 proteins PGC-1β and PGC-1α are established transcriptional coactivators recruited to particular DNA promoters and enhancers via numerous transcription factors, including nuclear receptors, so as to regulate many aspects of metabolism and the anti-inflammatory response (see, e.g., Correia et al. 2015; Martínez-Redondo et al. 2015). However, the molecular mechanisms governing their transcriptional activities have been understudied in view of their importance to cellular and organismal physiology.

There are obvious gaps in our current understanding of how DNA-associated and/or RNA-associating proteins in mammalian cells coordinate the various steps involved in protein-coding gene expression (Rambout et al. 2016, 2018). To address gaps that involve CBP80, we began by characterizing the physical interaction between CBP80 and PGC-1β (Cho et al. 2018). Using GST pull-downs, we mapped the CBP80-binding domain of PGC-1β to its extreme carboxyl terminus. We then solved the X-ray crystal structure of this terminus in complex with cap-bound CBC to reveal that a 9-amino acid α-helix of PGC-1β was positioned on a lipped shelf consisting of the two amino-terminal-most α-helices of CBP80 (Fig. 1A; modified from Cho et al. 2018). We named the α-helix of PGC-1β the CBP80-binding motif (CBM). In vitro fluorescence polarization competition assays showed that all five residues of the PGC-1β CBM directly contacting CBP80 in the X-ray crystal structure are required for CBP80 binding (Cho et al. 2018). Given that these residues are conserved in PGC-1β and PGC-1α of every mammalian species we examined, and considering that a number of PGC-1α target genes—both protein-coding genes and their associated enhancers—had been identified in cells of interest to us (i.e., mouse C2C12 myoblasts [MBs]) (Aguilo et al. 2016), we set out to determine if PGC-1α serves as an important transcriptional cofactor for, initially, the CBC-dependent control of PGC-1α-responsive gene expression.

Prior to our work, PGC-1α was known to promote gene transcription through at least two distinct mechanisms. In one mechanism, PGC-1α tethers via its amino-terminal activation domain histone acetyl transferases, such as the CREB-binding protein/E1A-binding protein CBP/p300 and the proto-oncogene tyrosine-protein kinase SRC-1, presumably to loosen local chromatin compaction and facilitate the formation of the transcription pre-initiation complex at target-gene transcription start sites (Puigserver et al. 1999; Wallberg et al. 2003). In a second mechanism, the carboxy-terminal arginine-serine rich (RS) domain of PGC-1α directly binds the MED1 subunit of the master transcriptional regulator Mediator (Wallberg et al. 2003; Chen et al. 2009). Although Mediator has been shown to play many roles in transcription (Allen and Taatjes 2015), the mechanism by which PGC-1α binding to MED1 promotes gene transcription is unclear. In addition to its transcriptional coactivator function, PGC-1α acts as a transcriptional repressor, as illustrated by its binding to and inhibition of the transcription factor XBP1s (Lee et al. 2018), and also as an alternative splicing factor (Monsalve et al. 2000; Martínez-Redondo et al. 2016).

THE PGC-1α CBM PROMOTES TARGET-GENE TRANSCRIPTION

Consistent with the emerging view that chromatin-associated RNA-binding proteins can regulate gene transcription (Xiao et al. 2019), we showed that PGC-1α recruitment via its CBM to the CBC at the 5′ cap of nascent pre-mRNAs and their associated enhancer RNAs (eRNAs), is critical for the activation of the majority of its target genes. These conclusions are supported by three lines of evidence. First, the co-immunoprecipitation (co-IP) of C2C12-MB pre-mRNAs produced by PGC-1α-activated genes with FLAG-tagged wild-type (WT) PGC-1α (FLAG-PGC-1α(WT)) is lost when RNase H is directed to cleave 6 nt downstream from their 5′ caps (Fig. 1B–D; modified from Cho et al. 2018). Second, the co-IP of CBP80 as well as PGC-1α up-regulated mRNAs and eRNAs with FLAG-PGC-1α(WT) is lost when the CBM of FLAG-PGC-1α(WT) is deleted (PGC-1α(ΔCBM)) (Fig. 1B–E; modified from Cho et al. 2018; see Cho et al. 2018 for data on eRNAs). Third, ∼80% of genes responsive to PGC-1α-mediated transcriptional activation require the PGC-1α CBM, as evidenced from transcriptome-wide RNA-seq data comparing WT C2C12 MBs, C2C12 MBs in which cellular PGC-1α is knocked down (PGC-1α-KD C2C12 MBs), and C2C12 MBs in which cellular PGC-1α was replaced by FLAG-PGC-1α(WT)

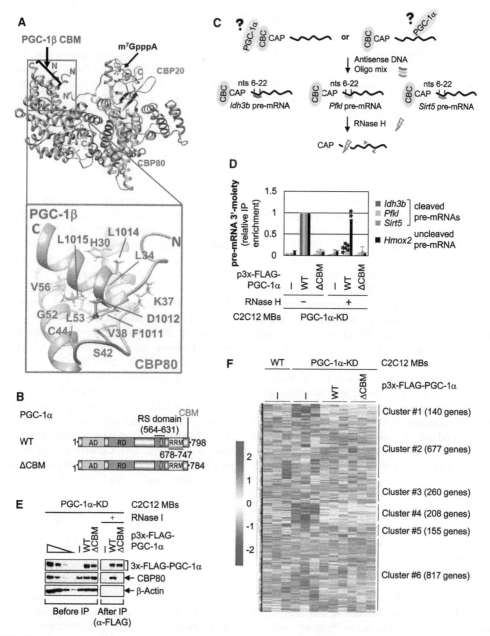

Figure 1. CBP80 binding to the carboxy-terminal end of the PGC-1 family of transcriptional coactivators can promote target-gene transcription (modified from Cho et al. 2018). (*A*) Crystal structure at 2.68 Å resolution of the carboxy-terminal end of human PGC-1β (golden brown) in association with the CBP80 (green)–CBP20 (blue) CBC heterodimer bound to m^{7}GpppA, which mimics the m^{7}G cap linked 5′-5′ to the first transcribed adenine nucleotide of an RNAPII-synthesized transcript. The CBP80–PGC-1β interface, including the 9-amino acid α-helix that we defined as the CBP80-binding motif (CBM), is enlarged at the bottom. (*B*) Diagram of human PGC-1α (WT) and PGC-1α(ΔCBM), denoting some of the known or putative functional regions. (RD) Repression domain, (AD) activation domain, (RS) serine-arginine-rich domain, (RRM) RNA-recognition motif, (CBM) CBP80-binding motif. Numbers pertain to amino acids. (*C*) Schematic illustrating the strategy used to show that PGC-1α binds the 5′ cap of three PGC-1α-activated transcripts. Annealing of a mix of three DNA oligonucleotides, each specific to the 5′ end of one of the transcripts, directed RNase H-mediated transcript cleavage so as to separate the 5′ cap (CAP) and associated CBC from each transcript body. (*D*) Histogram representation of RT-qPCR quantitations of PGC-1α-responsive pre-mRNA levels after, relative to before, anti-FLAG (α-FLAG) immunoprecipitation (IP) using lysates of PGC-1α-KD C2C12 MBs transiently transfected with plasmid (p) encoding FLAG alone (−) or the specified FLAG-tagged PGC-1α (FLAG-PGC-1α) variant. IPs were performed in the presence of the oligonucleotide mix as described in *C*, either in the presence (+) or absence (−) of RNase H. Values after anti-FLAG-PGC-1α(WT) IP in the absence of RNase H relative to before IP are set to 1. Results are means ± S.D. $n = 2$. (∗∗) $P < 0.01$, compared to no RNase H treatment by a two-tailed unpaired Student's *t*-test. (*E*) Western blots of lysates of PGC-1α-KD C2C12 MBs transiently transfected with plasmid encoding FLAG alone (−) or the specified FLAG-tagged PGC-1α variant before or after anti-FLAG IP, the latter in the presence (+) of RNase I. β-actin serves to control for variations in loading and IP specificity. Here, as in subsequent western blots, the *top left* wedge denotes threefold serial dilutions of samples to provide semiquantitative analyses. (*F*) Heatmap and hierarchical clustering of the expression of protein-coding genes using RNA sequencing (RNA-seq) analyses of the specified cells: wild-type (WT) C2C12 MBs expressing FLAG alone (−), or PGC-1α-knockdown (KD) C2C12 MBs expressing FLAG alone (−), FLAG-PGC-1α(WT), or FLAG-PGC-1α(ΔCBM). Only genes whose expression was significantly regulated following PGC-1α KD and significantly rescued by re-expression of FLAG-PGC-1α(WT) and/or FLAG-PGC-1α(ΔCBM) are shown. The color key represents row-scaled average expression values ($n = 3$). (*A–F*, Modified from Cho et al. 2018.)

or FLAG-PGC-1α(ΔCBM) (Fig. 1F; modified from Cho et al. 2018).

All of these data indicate that once PGC-1α is recruited to the promoter, or enhancer, of a PGC-1α-activated gene, its interaction with CBP80 at the 5′ cap of the associated nascent transcript is key to gene expression (Cho et al. 2018). Should the nascent transcript fail to undergo 5′ capping or otherwise fail to acquire the CBC, then the synthesis of full-length pre-mRNA and, therefore, the production of mRNA fail to occur (Cho et al. 2018). Thus, the PGC-1α CBM can be viewed as key to one of the many steps involved in the quality control of PGC-1α-activated gene expression, functioning as a type of pre-mRNA surveillance. Whether PGC-1α CBM-mediated transcriptional activation is a consequence of up-regulated transcription re-initiation or up-regulated transcription elongation has yet to be resolved. Given increasing evidence that overcoming promoter-proximal pausing is central to productive transcription initiation in higher eukaryotes (Core and Adelman 2019), we are currently testing the hypothesis that the PGC-1α CBM, via its binding to CBP80 at the 5′ cap of nascent transcripts, promotes the release of RNAPII from promoter-proximal pausing on PGC-1α-activated genes. Additional open questions pertain to the functions of the carboxy-terminal RS domain and the RNA-recognition motif (RRM) of PGC-1α, both of which we found are required for the stable RNase-insensitive co-IP of CBP80 with PGC-1α and, consequently, the association of PGC-1α with its target transcripts (Cho et al. 2018).

THE PGC-1α CBM PREVENTS CYTOPLASMIC ACCUMULATION OF INTRON 1–CONTAINING mRNAs DERIVING FROM PGC-1α CBM-ACTIVATED GENES

Largely concomitant with pre-mRNA synthesis are the steps of pre-mRNA splicing. Another of our interests is how transcription factors and cofactors "imprint" nascent protein-coding transcripts to regulate their subsequent fate during splicing (Rambout et al. 2018). Beyond pre-mRNA splicing, previous work has focused on how transcript imprinting by the erythroblast transformation-specific transcription factor ERG conditionally regulates RNA metabolic steps that extend out into the cytoplasm so as to affect cytoplasmic mRNA stability (Rambout et al. 2016).

Returning to our long-standing interest in how NMD degrades newly spliced mRNAs on which the CBC has yet to be replaced by eIF4E (Lejeune et al. 2002) led us to test if the PGC-1α CBM participates in NMD. This seemed to be reasonable considering that (i) PGC-1α remains associated with CBC-bound mRNAs produced by PGC-1α-responsive genes (Cho et al. 2018) and, as discussed above, (ii) 3′-UTR EJC-dependent NMD degrades cytoplasmic CBC-bound mRNAs.

In possible support of PGC-1α function during NMD, one purpose of NMD is to function as a quality-control pathway that eliminates incompletely or improperly spliced transcripts (Jacob and Smith 2017). As an example, should an mRNA be incompletely spliced and exported to the cytoplasm, and should the retained intronic sequences introduce a frameshift mutation or nonsense codon that produces a PTC, the resulting intron-containing mRNA is likely to be degraded by NMD. Of course, the farther the retained intronic sequences reside relative to the 3′ UTR, the greater the likelihood that these retained sequences result in a PTC upstream of 3′-UTR EJC. It follows that retained intron 1 sequences are most likely to trigger NMD.

The CBC promotes recognition of the cap-proximal 5′-splice site (Lewis et al. 1996) so as to expedite the splicing of intron 1 (Bittencourt et al. 2008; Laubinger et al. 2008; Li et al. 2016b). And PGC-1α has been shown to function in pre-mRNA splicing, albeit alternative pre-mRNA splicing (Monsalve et al. 2000; Martínez-Redondo et al. 2016). Therefore, we began to investigate if the PGC-1α CBM functions in splicing and, possibly, NMD by analyzing the cytoplasmic-to-nuclear (C/N) ratios of intron 1–retained *Idh3b*, *Pfkl*, and *Sirt5* transcripts (Fig. 1), which we had shown are bound by PGC-1α at their 5′ cap (Cho et al. 2018). We used WT C2C12 MBs expressing FLAG alone, or PGC-1α-KD C2C12 MBs expressing either FLAG alone, FLAG-PGC-1α(WT), or FLAG-PGC-1α(ΔCBM). As MB-fractionation controls, nucleoporin NUP160 and the glycolytic protein GAPDH were detectable only in nuclear and cytoplasmic fractions, respectively (Fig. 2A). Consistent with these results, *U6* small nuclear (sn)RNA and *Gapdh* mRNA were enriched in nuclear and cytoplasmic fractions, respectively (Fig. 2B).

We previously reported that, relative to either WT C2C12 MBs or PGC1α-KD C2C12 MBs expressing FLAG-PGC-1α(WT), PGC1α-KD MBs expressing either FLAG alone or FLAG-PGC-1α(ΔCBM) produce abnormally low levels of normally spliced *Idh3b*, *Pfkl*, and *Sirt5* mRNAs (Fig. 2C; adapted from Cho et al. 2018). However, we report here that accompanying these low levels are increased C/N ratios of the corresponding intron 1–containing mRNAs (Fig. 2D), without detectable changes in the C/N ratios of either intron 2– or intron 3–containing mRNAs (data not shown).

A PGC-1α CBM-dependent decrease in the C/N ratio of intron 1–containing mRNAs could be the result of increased mRNA stability in the nucleus, impaired nuclear mRNA export, and/or decreased mRNA stability in the cytoplasm (e.g., increased NMD). Supporting the hypothesis that the PGC-1α CBM promotes NMD, inhibiting NMD by siRNA-mediated KD of the key NMD factor UPF1 (Fig. 2E) increased the cytoplasmic levels of each of the three intron 1–containing mRNAs, normalized to the non-NMD target *Gapdh* mRNA, in PGC1α-KD MBs expressing FLAG-PGC-1α(WT) relative to PGC1α-KD MBs expressing FLAG-PGC-1α(ΔCBM) (Fig. 2F). In other words, intron 1–containing mRNAs deriving from PGC-1α-activated genes and whose cap is not bound by PGC-1α appear to evade the NMD surveillance mechanism and, as a result, accumulate in the cytoplasm. Altogether, our results indicate that the PGC1α CBM (i) is required for efficient intron 1 splicing and (ii)

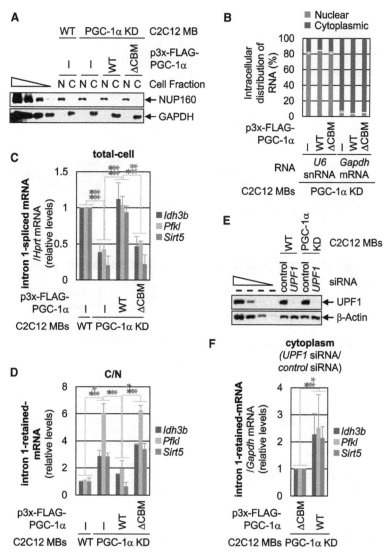

Figure 2. CBP80 binding to PGC-1α reduces the cytoplasmic accumulation of intron 1–containing transcripts deriving from PGC-1α-activated genes. (*A*) Western blots of nuclear and cytoplasmic fractions from the same number of WT C2C12 MBs expressing FLAG alone (−) or PGC-1α-KD C2C12 MBs expressing FLAG alone (−) or the denoted FLAG-PGC-1α variant. NUP160 and GAPDH are nuclear pore and cytoplasmic constituents, respectively. (N) Nucleus, (C) cytoplasm. (*B*) Stacked histogram representation of RT-qPCR quantitations of the relative distribution of *U6* snRNA (largely nuclear) and *Gapdh* mRNA (largely cytoplasmic) in the nuclear (blue) and cytoplasmic (red) fraction of cells treated as in *A*. Results are means ± S.D. $n = 2$. (*C*) Histogram representation of RT-qPCR quantitations of total-cell levels of three intron 1–spliced mRNAs deriving from PGC-1α CBM-activated genes using cells treated as in *A*. Results are means ± S.D. $n = 3$. (∗) $P < 0.05$, (∗∗) $P < 0.01$ by a two-tailed unpaired Student's *t*-test. (*D*) Histogram representation of RT-qPCR quantitations of cytoplasmic-to-nuclear (C/N) ratios of three intron 1–retained mRNAs deriving from PGC-1α CBM-activated genes using cells treated as in *A*. Results are means ± S.D. $n = 2$. (∗) $P < 0.05$, (∗∗) $P < 0.01$ by a two-tailed unpaired Student's *t*-test. (*E*) Western blot of lysates of WT or PGC-1α-KD C2C12 MBs transiently transfected with either control or *UPF1* siRNA, demonstrating effective UPF1 KD to ≤10% of normal. (*F*) Histogram representation of RT-qPCR quantitations of the cytoplasmic levels of three intron 1–retained mRNAs deriving from PGC-1α CBM-activated genes, normalized to the cytoplasmic level of *Gapdh* mRNA, using PGC-1α-KD C2C12 MBs transiently transfected with *UPF1* or control siRNA and, subsequently, FLAG-PGC-1α(WT) or FLAG-PGC-1α(ΔCBM). Values for PGC-1α-KD C2C12 MBs expressing FLAG-PGC-1α(ΔCBM) are set to 1. Results are means ± S.D. $n = 2$. (∗) $P < 0.05$, (∗∗) $P < 0.01$ by a two-tailed unpaired Student's *t*-test. Methods from which these data derive have been published elsewhere (Gong et al. 2009; Cho et al. 2018). (*C*, Modified from Cho et al. 2018.)

in cases where cytoplasmic intron 1–containing mRNAs are nonetheless produced, permits if not promotes their degradation by NMD. We do not exclude the possibility that PGC1α via its CBM functions in other steps of gene expression, such as nuclear mRNA stability and export, that may also contribute to prevent the accumulation of intron 1–containing mRNAs.

CONCLUSION

Although NMD is arguably the best-characterized mechanism by which nuclear and cytoplasmic RNA processing machineries are coordinated so as to control the quality of protein-coding gene expression, it is clear there are additional connections to be discovered and

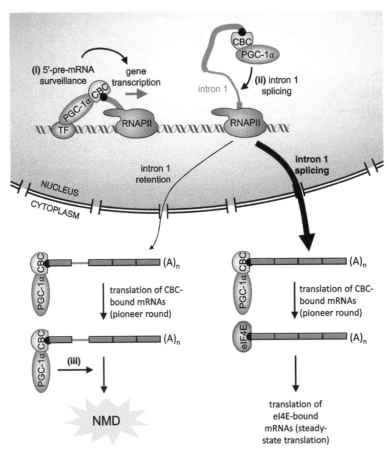

Figure 3. Model for PGC-1α–CBP80-dependent coupling of co- and posttranscriptional steps of gene expression. PGC-1α binding to CBP80 of the CBC promotes: (step i) target-gene transcription provided the nascent transcript has undergone proper 5′-end capping and CBC binding; (step ii) intron 1 splicing; and (step iii) NMD of cytoplasmic mRNAs in which intron 1 was mistakenly retained.

characterized. Work done by ourselves and others has shaped the current understanding of how CBP80 functions as a nexus for regulating co- and posttranscriptional RNA processes, forming a continuum that begins with transcription initiation in the nucleus and ends only when the CBC is replaced by eIF4E in the cytoplasm.

We recently showed that binding of the transcriptional coactivator PGC-1α to CBP80 promotes the transcription of the majority of PGC-1α-activated genes (Cho et al. 2018). We see this interaction not only as a cue for increased gene transcription but also as a checkpoint that surveys the integrity of pre-mRNA 5′-end capping and CBC acquisition (Fig. 3, step i). Beyond this function, we provide evidence that PGC-1α binding to CBP80 promotes intron 1 splicing (Fig. 3, step ii) and the NMD of intron 1–containing mRNAs (Fig. 3, step iii) deriving from PGC-1α CBM-activated genes. Thus, our data indicate that PGC-1α and CBP80 not only promote the transcription of target genes, but they also ensure that only 5′-end-capped and properly spliced mRNAs accumulate in the cytoplasm where they can be translated into functional proteins. The failure of eIF4E to co-immunoprecipitate with PGC-1α (Cho et al. 2018) indicates that PGC-1α dissociates from the mRNA cap together with the CBC shortly after nuclear export and therefore does not take part in steady-state mRNA translation.

The mechanism by which PGC-1α binding to CBP80 promotes gene transcription is currently unknown (see above), as is the mechanism by which PGC-1α binding to CBP80 facilitates NMD. Possibilities for the latter include (i) enhanced translation initiation during the pioneer round of translation, (ii) increased recruitment of UPF1 to an NMD target, and/or (iii) differential recruitment of other positive or negative regulators of NMD. It should be noted that our RNA-seq data indicate that some PGC-1α-responsive genes are repressed rather than activated by the CBM (Fig. 1F, gene cluster #6; modified from Cho et al. 2018). Analyses of previously published data deriving from chromatin immunoprecipitation coupled to sequencing (ChIP-seq) show that FLAG-tagged PGC-1α is recruited to the promoter of at least some of these genes in C2C12 MBs (Baresic et al. 2014), raising the possibility that CBM-mediated gene repression may be direct. However, whether this repression is through inhibition of gene transcription or through coupling to pre-mRNA splicing and/or mRNA decay (e.g., NMD) remains to be tested.

Future studies will undoubtedly determine if other transcriptional coactivators function in pre-mRNA surveillance mechanisms. Whereas the CBM is conserved in all three members of the PGC-1 family of proteins, we failed to identify CBM-like motifs in other transcriptional coactivators or DNA-binding transcription factors (Cho et al. 2018). However, we recently reviewed the numerous ways by which transcription factors directly control alternative pre-mRNA splicing (Rambout et al. 2018). Thus, we anticipate that additional gene-specific chromatin-acting mediators of pre-mRNA surveillance pathways will be identified in the future.

ACKNOWLEDGMENTS

This work was supported by the American Heart Association (AHA) Postdoctoral Fellowships 18POST3 3960339 to X.R. and 16POST27260273 to H.C. and by the National Institutes of Health (NIH) grant R01 GM059514 to L.E.M.

REFERENCES

Aguilo F, Li S, Balasubramaniyan N, Sancho A, Benko S, Zhang F, Vashisht A, Rengasamy M, Andino B, Chen C-H, et al. 2016. Deposition of 5-methylcytosine on enhancer RNAs enables the coactivator function of PGC-1α. *Cell Rep* 14: 479–492. doi:10.1016/j.celrep.2015.12.043

Allen BL, Taatjes DJ. 2015. The mediator complex: a central integrator of transcription. *Nat Rev Mol Cell Biol* 16: 155–166. doi:10.1038/nrm3951

Baresic M, Salatino S, Kupr B, van Nimwegen E, Handschin C. 2014. Transcriptional network analysis in muscle reveals AP-1 as a partner of PGC-1α in the regulation of the hypoxic gene program. *Mol Cell Biol* 34: 2996–3012. doi:10.1128/MCB .01710-13

Bittencourt D, Dutertre M, Sanchez G, Barbier J, Gratadou L, Auboeuf D. 2008. Cotranscriptional splicing potentiates the mRNA production from a subset of estradiol-stimulated genes. *Mol Cell Biol* 28: 5811–5824. doi:10.1128/MCB.02231-07

Chen W, Yang Q, Roeder RG. 2009. Dynamic interactions and cooperative functions of PGC-1α and MED1 in TRα-mediated activation of the brown-fat-specific *UCP-1* gene. *Mol Cell* 35: 755–768. doi:10.1016/j.molcel.2009.09.015

Cho H, Rambout X, Gleghorn ML, Nguyen PQT, Phipps CR, Miyoshi K, Myers JR, Kataoka N, Fasan R, Maquat LE. 2018. Transcriptional coactivator PGC-1α contains a novel CBP80-binding motif that orchestrates efficient target gene expression. *Genes Dev* 32: 555–567. doi:10.1101/gad.309773.117

Core L, Adelman K. 2019. Promoter-proximal pausing of RNA polymerase II: a nexus of gene regulation. *Genes Dev* 33: 960–982. doi:10.1101/gad.325142.119

Correia JC, Ferreira DM, Ruas JL. 2015. Intercellular: local and systemic actions of skeletal muscle PGC-1s. *Trends Endocrinol Metab* 26: 305–314. doi:10.1016/j.tem.2015.03.010

Glover-Cutter K, Kim S, Espinosa J, Bentley DL. 2008. RNA polymerase II pauses and associates with pre-mRNA processing factors at both ends of genes. *Nat Struct Mol Biol* 15: 71–78. doi:10.1038/nsmb1352

Gonatopoulos-Pournatzis T, Cowling VH. 2014. Cap-binding complex (CBC). *Biochem J* 457: 231–242. doi:10.1042/B J20131214

Gong C, Kim YK, Woeller CF, Tang Y, Maquat LE. 2009. SMD and NMD are competitive pathways that contribute to myogenesis: effects on PAX3 and myogenin mRNAs. *Genes Dev* 23: 54–66. doi:10.1101/gad.1717309

Hoek TA, Khuperkar D, Lindeboom RGH, Sonneveld S, Verhagen BMP, Boersma S, Vermeulen M, Tanenbaum ME. 2019. Single-molecule imaging uncovers rules governing nonsense-mediated mRNA decay. *Mol Cell* 75: 324–339.e11. doi:10 .1016/j.molcel.2019.05.008

Hosoda N, Kim YK, Lejeune F, Maquat LE. 2005. CBP80 promotes interaction of Upf1 with Upf2 during nonsense-mediated mRNA decay in mammalian cells. *Nat Struct Mol Biol* 12: 893–901. doi:10.1038/nsmb995

Hossain MA, Chung C, Pradhan SK, Johnson TL. 2013. The yeast cap binding complex modulates transcription factor recruitment and establishes proper histone H3K36 trimethylation during active transcription. *Mol Cell Biol* 33: 785–799. doi:10.1128/MCB.00947-12

Hwang J, Sato H, Tang Y, Matsuda D, Maquat LE. 2010. UPF1 association with the cap-binding protein, CBP80, promotes nonsense-mediated mRNA decay at two distinct steps. *Mol Cell* 39: 396–409. doi:10.1016/j.molcel.2010.07.004

Ishigaki Y, Li X, Serin G, Maquat LE. 2001. Evidence for a pioneer round of mRNA translation: mRNAs subject to nonsense-mediated decay in mammalian cells are bound by CBP80 and CBP20. *Cell* 106: 607–617. doi:10.1016/S0092-8674(01)00475-5

Isken O, Kim YK, Hosoda N, Mayeur GL, Hershey JW, Maquat LE. 2008. Upf1 phosphorylation triggers translational repression during nonsense-mediated mRNA decay. *Cell* 133: 314–327. doi:10.1016/j.cell.2008.02.030

Jacob AG, Smith CWJ. 2017. Intron retention as a component of regulated gene expression programs. *Hum Genet* 136: 1043–1057. doi:10.1007/s00439-017-1791-x

Kachaev ZM, Lebedeva LA, Shaposhnikov AV, Moresco JJ, Yates JR III, Schedl P, Shidlovskii YV. 2019. Paip2 cooperates with Cbp80 at an active promoter and participates in RNA polymerase II phosphorylation in *Drosophila*. *FEBS Lett* 593: 1102–1112. doi:10.1002/1873-3468.13391

Kataoka N, Ohno M, Moda I, Shimura Y. 1995. Identification of the factors that interact with NCBP, an 80 kDa nuclear cap binding protein. *Nucleic Acids Res* 23: 3638–3641. doi:10 .1093/nar/23.18.3638

Kurosaki T, Popp MW, Maquat LE. 2019. Quality and quantity control of gene expression by nonsense-mediated mRNA decay. *Nat Rev Mol Cell Biol* 20: 406–420. doi:10.1038/s41580-019-0126-2

Lahudkar S, Shukla A, Bajwa P, Durairaj G, Stanojevic N, Bhaumik SR. 2011. The mRNA cap-binding complex stimulates the formation of pre-initiation complex at the promoter via its interaction with Mot1p in vivo. *Nucleic Acids Res* 39: 2188–2209. doi:10.1093/nar/gkq1029

Laubinger S, Sachsenberg T, Zeller G, Busch W, Lohmann JU, Ratsch G, Weigel D. 2008. Dual roles of the nuclear cap-binding complex and SERRATE in pre-mRNA splicing and microRNA processing in *Arabidopsis thaliana*. *Proc Natl Acad Sci* 105: 8795–8800. doi:10.1073/pnas.0802493105

Lee J, Salazar Hernández MA, Auen T, Mucka P, Lee J, Ozcan U. 2018. PGC-1α functions as a co-suppressor of XBP1s to regulate glucose metabolism. *Mol Metab* 7: 119–131. doi:10 .1016/j.molmet.2017.10.010

Le Hir H, Moore MJ, Maquat LE. 2000. Pre-mRNA splicing alters mRNP composition: evidence for stable association of proteins at exon–exon junctions. *Genes Dev* 14: 1098–1108.

Lejeune F, Ishigaki Y, Li X, Maquat LE. 2002. The exon junction complex is detected on CBP80-bound but not eIF4E-bound mRNA in mammalian cells: dynamics of mRNP remodeling. *EMBO J* 21: 3536–3545. doi:10.1093/emboj/cdf345

Lenasi T, Peterlin BM, Barboric M. 2011. Cap-binding protein complex links pre-mRNA capping to transcription elongation and alternative splicing through positive transcription elongation factor b (P-TEFb). *J Biol Chem* 286: 22758–22768. doi:10.1074/jbc.M111.235077

Lewis JD, Izaurralde E, Jarmolowski A, McGuigan C, Mattaj IW. 1996. A nuclear cap-binding complex facilitates association of U1 snRNP with the cap-proximal 5′ splice site. *Genes Dev* 10: 1683–1698. doi:10.1101/gad.10.13.1683

Li T, De Clercq N, Medina DA, Garre E, Sunnerhagen P, Perez-Ortin JE, Alepuz P. 2016a. The mRNA cap-binding protein Cbc1 is required for high and timely expression of genes by promoting the accumulation of gene-specific activators at promoters. *Biochim Biophys Acta* **1859:** 405–419. doi:10.1016/j.bbagrm.2016.01.002

Li Z, Jiang D, Fu X, Luo X, Liu R, He Y. 2016b. Coupling of histone methylation and RNA processing by the nuclear mRNA cap-binding complex. *Nat Plants* **2:** 16015. doi:10.1038/nplants.2016.15

Lidschreiber M, Leike K, Cramer P. 2013. Cap completion and C-terminal repeat domain kinase recruitment underlie the initiation-elongation transition of RNA polymerase II. *Mol Cell Biol* **33:** 3805–3816. doi:10.1128/MCB.00361-13

Listerman I, Sapra AK, Neugebauer KM. 2006. Cotranscriptional coupling of splicing factor recruitment and precursor messenger RNA splicing in mammalian cells. *Nat Struct Mol Biol* **13:** 815–822. doi:10.1038/nsmb1135

Martínez-Redondo V, Pettersson AT, Ruas JL. 2015. The hitchhiker's guide to PGC-1α isoform structure and biological functions. *Diabetologia* **58:** 1969–1977. doi:10.1007/s00125-015-3671-z

Martínez-Redondo V, Jannig PR, Correia JC, Ferreira DM, Cervenka I, Lindvall JM, Sinha I, Izadi M, Pettersson-Klein AT, Agudelo LZ, et al. 2016. Peroxisome proliferator-activated receptor γ coactivator-1α isoforms selectively regulate multiple splicing events on target genes. *J Biol Chem* **291:** 15169–15184. doi:10.1074/jbc.M115.705822

Monsalve M, Wu Z, Adelmant G, Puigserver P, Fan M, Spiegelman BM. 2000. Direct coupling of transcription and mRNA processing through the thermogenic coactivator PGC-1. *Mol Cell* **6:** 307–316. doi:10.1016/S1097-2765(00)00031-9

Pelletier J, Sonenberg N. 2019. The organizing principles of eukaryotic ribosome recruitment. *Annu Rev Biochem* **88:** 307–335. doi:10.1146/annurev-biochem-013118-111042

Puigserver P, Adelmant G, Wu Z, Fan M, Xu J, O'Malley B, Spiegelman BM. 1999. Activation of PPARγ coactivator-1 through transcription factor docking. *Science* **286:** 1368–1371. doi:10.1126/science.286.5443.1368

Rambout X, Detiffe C, Bruyr J, Mariavelle E, Cherkaoui M, Brohée S, Demoitié P, Lebrun M, Soin R, Lesage B, et al. 2016. The transcription factor ERG recruits CCR4-NOT to control mRNA decay and mitotic progression. *Nat Struct Mol Biol* **23:** 663–672. doi:10.1038/nsmb.3243

Rambout X, Dequiedt F, Maquat LE. 2018. Beyond transcription: roles of transcription factors in pre-mRNA splicing. *Chem Rev* **118:** 4339–4364. doi:10.1021/acs.chemrev.7b00470

Trcek T, Sato H, Singer RH, Maquat LE. 2013. Temporal and spatial characterization of nonsense-mediated mRNA decay. *Genes Dev* **27:** 541–551. doi:10.1101/gad.209635.112

Wallberg AE, Yamamura S, Malik S, Spiegelman BM, Roeder RG. 2003. Coordination of p300-mediated chromatin remodeling and TRAP/mediator function through coactivator PGC-1α. *Mol Cell* **12:** 1137–1149. doi:10.1016/S1097-2765(03)00391-5

Xiao R, Chen J-Y, Liang Z, Luo D, Chen G, Lu ZJ, Chen Y, Zhou B, Li H, Du X, et al. 2019. Pervasive chromatin-RNA binding protein interactions enable RNA-based regulation of transcription. *Cell* **178:** 107–121.e18. doi:10.1016/j.cell.2019.06.001

Mechanistic Dissection of RNA-Binding Proteins in Regulated Gene Expression at Chromatin Levels

JIA-YU CHEN, DO-HWAN LIM, AND XIANG-DONG FU

Department of Cellular and Molecular Medicine, Institute of Genomic Medicine, University of California, San Diego, La Jolla, California 92093, USA

Correspondence: xdfu@ucsd.edu

Eukaryotic genomes are known to prevalently transcribe diverse classes of RNAs, virtually all of which, including nascent RNAs from protein-coding genes, are now recognized to have regulatory functions in gene expression, suggesting that RNAs are both the products and the regulators of gene expression. Their functions must enlist specific RNA-binding proteins (RBPs) to execute their regulatory activities, and recent evidence suggests that nearly all biochemically defined chromatin regions in the human genome, whether defined for gene activation or silencing, have the involvement of specific RBPs. Interestingly, the boundary between RNA- and DNA-binding proteins is also melting, as many DNA-binding proteins traditionally studied in the context of transcription are able to bind RNAs, some of which may simultaneously bind both DNA and RNA to facilitate network interactions in three-dimensional (3D) genome. In this review, we focus on RBPs that function at chromatin levels, with particular emphasis on their mechanisms of action in regulated gene expression, which is intended to facilitate future functional and mechanistic dissection of chromatin-associated RBPs.

The traditional view of transcription is to produce either structural RNAs or protein-coding mRNAs. Specific structural RNAs are assembled into various RNA machines to catalyze specific biochemical reactions, and protein-coding RNAs are processed in the nucleus (such as capping, splicing, and polyadenylation) and then exported to the cytoplasm to translate into proteins. The advent of deep-sequencing technologies has now revealed that mammalian genomes are far more active in transcription (Djebali et al. 2012), generating a large repertoire of regulatory RNAs, including long noncoding RNAs (lncRNAs) (Long et al. 2017), repeat-derived RNAs (Johnson and Straight 2017), and enhancer RNAs (eRNAs) (Li et al. 2016). Even protein-coding genes are producing various smaller RNA species that are either cause or consequence of regulated gene expression as a result of divergent or convergent transcription and transcription pausing and pause release (Wissink et al. 2019). Most regulatory RNAs are predominantly retained in the nucleus (Li and Fu 2019), where they may modulate gene expression at different steps of transcription on specific transcription units or genomic loci (Skalska et al. 2017), remodel chromatin structures and dynamics (Bohmdorfer and Wierzbicki 2015), and mediate long-distance genomic interactions (Schoenfelder and Fraser 2019), together contributing to the organization of the three-dimensional (3D) genome.

RNA molecules contain a series of single- and double-stranded regions that enable them to interact with DNA, RNA, and protein, thus providing versatile structural modules that are distinct from those in proteins to mediate network interactions. Through functional dissection of specific RNA metabolism pathways, a large number of RNA-binding proteins (RBPs) have been characterized, which often process unique structural motifs for direct contact with RNA sequences, base compositions and modifications, polynucleotide backbone, double-stranded regions, or RNA tertiary structures. However, recent global surveys of RBPs reveal ~1500 RBPs encoded by mammalian genomes, many of which do not carry canonical RNA-binding domains (Hentze et al. 2018). It is particularly interesting to note that many DNA-binding proteins are also able to directly bind RNA through either the same or distinct nucleic acid recognition motif(s), which are collectively termed DNA/RNA-binding proteins (DRBPs) (Hudson and Ortlund 2014). Consequently, many traditional DNA-binding transcription factors (TFs) may also function as RBPs in mammalian cells. These DRBPs are exemplified by many zinc-finger proteins, which often contain multiple fingers in the same polypeptides with divided tasks in interacting with DNA, RNA, and/or protein.

Given prevalent transcription activities in mammalian genomes, our recent large-scale chromatin immunoprecipitation sequencing (ChIP-seq) analysis of RBPs reveals that nearly all biochemically defined chromatin regions (based on RNA production, chromatin marks, and accessible chromatin regions) in the human genome involve specific RBPs, and a significant fraction of these nuclear RBPs appear to directly participate in transcriptional control (Xiao et al. 2019). In this review, we focus on RBPs that function at chromatin levels. We highlight recent advances in detecting RBP–chromatin interactions and in dissecting their mechanisms in transcriptional control and co-transcription RNA processing through acting on selective "hotspots" on chromatin to aid in future research

---

© 2019 Chen et al. This article is distributed under the terms of the Creative Commons Attribution License, which permits unrestricted reuse and redistribution provided that the original author and source are credited.

Published by Cold Spring Harbor Laboratory Press; doi:10.1101/sqb.2019.84.039222

to (i) understand a suspected function of an RBP on chromatin, (ii) probe the regulatory activity of a chromatin-associated RNA through identifying and characterizing its associated RBPs, (iii) dissect a specific chromatin activity that may involve both regulatory RNAs and RBPs, or (iv) deduce global DNA–RNA–protein networks in 3D genome critical for specific biological processes. Because of limited space, we select specific examples to illustrate how to experimentally approach the function and mechanism of chromatin-associated RBPs, rather than trying to be comprehensive in covering all related literature on regulatory RNAs and RBPs. Readers are directed to the outstanding reviews on such topics cited above.

STRATEGIES TO DETECT CHROMATIN-ASSOCIATED RBPs

Chromatin-associated RBPs can be detected either on an individual basis or at the genome-wide scale. If the experimental goal is to explore a suspected function of a specific RBP on chromatin (Fig. 1A), the first step is to perform ChIP-seq if specific antibody is available or through genomic tagging using the CRISPR technology to determine the binding pattern of the RBP of interest on chromatin, essentially treating the RBP under investigation as a candidate TF. Options for genomic tagging include in-frame insertion of a GFP, FLAG, or SPY tag to the endogenous gene (Zakeri et al. 2012; Kimple et al. 2013). Multiple variations of ChIP-seq may be chosen, including ChIP-exonuclease (ChIP-exo) to increase the resolution (Rhee and Pugh 2011) or ChIPmentation to improve the robustness in library construction (Schmidl et al. 2015). More sophisticated variations include CUT&RUN and CUT&Tag, which use a Protein A-Micrococcal Nuclease or a Protein-A-Tn5 fusion protein to recognize chromatin-bound antibody (Skene and Henikoff 2017; Kaya-Okur et al. 2019). These techniques would avoid cross-linking and the harsh sonication step as in standard ChIP-seq. Deduced RBP-binding peaks can lead to functional and mechanistic studies through motif identification (Landt et al. 2012), Gene Ontology (GO)-term and Kyoto Encyclopedia of Genes and Genomes (KEGG) enrichment analyses (Kanehisa and Goto 2000; The Gene Ontology Consortium 2019), and cobinding analyses by using existing ChIP-seq data for known TFs and other RBPs (Xiao et al. 2019).

Before mechanistic dissection, two important questions may be addressed. The first is to determine whether the RBP of interest binds chromatin in an RNA-dependent manner. This can be addressed by using RNase A to treat permeabilized cells before ChIP-seq or using a drug, such as α-amanitin, to block transcription to determine the dependence on nascent RNA production. The second ques-

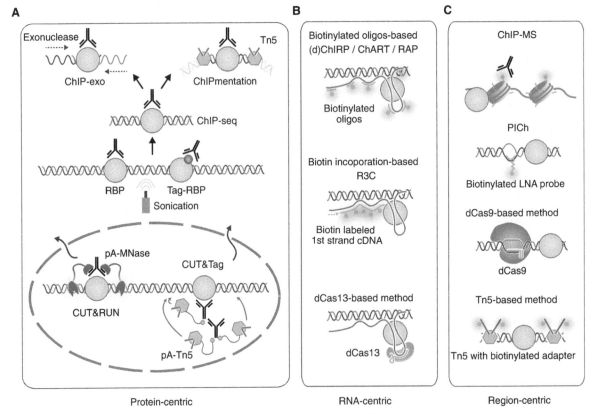

Figure 1. Strategies to detect chromatin-associated RNA-binding proteins (RBPs). (*A*) Protein-centric approaches to explore whether a specific RBP directly acts on chromatin. (*B*) RNA-centric approaches to identify both RBPs and associated DNA regions. (*C*) Region-centric approaches to profile associated RBPs.

tion is to identify specific RNAs that might mediate the interaction of the RBP with chromatin, which may not be as straightforward as it might sound. One approach is to sequence RNA, rather than DNA, in the IPed sample by RNA immunoprecipitation (RIP) (Gilbert and Svejstrup 2006) or formaldehyde RIP (fRIP) (Hendrickson et al. 2016). As these techniques do not differentiate direct from indirect binding, a better choice would be cross-linking immunoprecipitation (CLIP) (Lee and Ule 2018). Still, the problem is twofold: RBPs may bind RNAs both on and beyond chromatin and the RNA-binding profiles for most RBPs rarely match with those detected by ChIP-seq (Ji et al. 2013; Van Nostrand et al. 2018). One potential solution to this problem is to perform CLIP on biochemically enriched chromatin fractions, which may provide critical insights into RNA-guided interactions with DNA. This may be amenable with *cis*-acting RNAs, but it is quite challenging to link RBP–chromatin interactions mediated by *trans*-acting RNAs. However, this is readily approachable if a study begins with a specific chromatin-associated noncoding RNA and the goal is to understand the function and mechanism of such potential regulatory RNA on chromatin by identifying its interaction with DNA and then searching for potential RBPs involved (Fig. 1B). For this experimental goal, established RNA capture strategies could be used to identify the associated DNA and proteins, as exemplified by a set of related methods, such as ChIRP (Chu et al. 2011; Quinn et al. 2014), CHART (Simon et al. 2011), RAP (Engreitz et al. 2013), R3C (Zhang et al. 2014), and a potential dCas13-based approach similar to that using a dCas9-based strategy to target a specific genomic loci (Tsui et al. 2018). Additionally, RNA-linked chromatin architecture could be approached with HiChIRP (Mumbach et al. 2019), and RNA–RNA interactions with hiCLIP (Sugimoto et al. 2015). These technologies enable the elucidation of potential DNA-, RNA-, and protein-mediated network interactions in 3D genome.

Given that all active chromatin regions are suspected to involve regulatory RNAs and RBPs, it would require strategies to identify specific RNAs and RBPs at a specific genomic locus or genome-wide that are linked to a specific epigenetic event or a regulated gene expression program (Fig. 1C). A specific nucleic acid probe (Dejardin and Kingston 2009) or a dCas9-based strategy (Tsui et al. 2018) have been developed to detect locus-specific interactions by capture followed by genomic or proteomic profiling. To identify potential regulatory RNAs and RBPs associated with an epigenetic event, IP-coupled chromatin proteomic profiling would provide a general approach. This strategy has been applied to specific histone-marked genomic regions, uncovering multiple RBPs in complex with specific histone modification events (Ji et al. 2015). Conversely, regulated gene expression may enlist proteins that interact with nascent RNAs from various genomic regions, which can be approached by ethynyl uridine (EU) labeling to enable nascent RNAs to react with azide-biotin for streptavidin enrichment, which has uncovered many noncanonical RBPs that are well-known TFs and chromatin remodelers (Bao et al. 2018). In fact, EU may be combined with 4-thiouridine (4sU) labeling to enhance protein–RNA cross-linking (Huang et al. 2018). We may also envision a general strategy to systematically identify chromatin-associated RBPs by loading a biotin-labeled adaptor to Tn5 (Lai et al. 2018) to access all open chromatin regions followed by streptavidin enrichment and proteomic profiling. This approach would enable unbiased survey of annotated RBPs on chromatin in different cell types.

DEFINING THE FUNCTIONAL IMPACT OF CHROMATIN-ASSOCIATED RBPs ON GENE EXPRESSION

Chromatin-associated RBPs may have direct roles in transcription or mediate cotranscriptional processing or both. It is thus critical to measure their functional impacts before investigating the mechanism of their actions. RNA-seq following knockdown has been typically used to quickly assess the functional consequence; however, the data do not necessarily reflect regulated gene expression at the level of transcription because steady state RNA is the collective consequence of transcription, RNA processing, and stability as well as indirect effects induced by knockdown of a specific RBP. To determine potential impact on transcription, the most straightforward assay is global nuclear run-on (GRO-seq), which measures nascent RNA production (Core et al. 2008).

To help differentiate between direct versus indirect effects, one may determine whether RBP–chromatin interactions are linked to target genes the RBP binds, assuming that bound target genes are more affected than unbound genes. However, if the RBP under investigation preferentially binds intergenic regions, such as enhancers, it would be important to link individual binding events to likely target genes (Yao et al. 2015). Most studies infer the closest genes as targets for enhancers, which is reasonable for metagene analysis, but there are numerous exceptions to this assumption, as many enhancers may engage in long-distance interactions with target gene promoters through DNA looping, thus skipping the nearest neighboring genes (Schoenfelder and Fraser 2019). As transcription is a multistep process from the assembly of preinitiation complex (PIC) to transcription pausing and pause release to productive elongation to termination; various high-throughput technologies for analyzing different portions of nascent RNAs have been developed for mechanistic dissection. Readers are referred to a recent thorough review on these technologies and their applications to addressing specific mechanistic questions (Wissink et al. 2019).

If an RBP is suspected to play a direct role in transcription, it is often informative to survey its impact on the behavior of specific RNA polymerases. Using Pol II as an example, the carboxy-terminal domain (CTD) of the largest subunit is posttranslationally modified, and specific modification events have been linked to different steps in transcription according to the so-called CTD code (Hsin and Manley 2012). Therefore, quantifying those modification events and mapping them to chromatin by ChIP-seq in response to RBP knockdown are often informative to pinpoint a specific transcription step(s) being regulated (Takeuchi et al. 2018). Moreover, altered transcription is frequently linked to modified histones according to the histone code hypothesis (Jenuwein and Allis 2001), which can also be used to characterize the regulation of the epigenetic landscape by a specific chromatin-associated RBP.

Chromatin-bound RBPs are not necessarily involved in transcription, but rather in coupling transcription with downstream RNA processing events, which can affect RNA fates by multiple mechanisms. A battery of high-throughput technologies may be used to pinpoint changes in RNA fate. RNA-seq is again a powerful strategy to obtain the first approximation on differential gene expression. Resultant high-density reads can be aligned to the reference genome to deduce alternative splicing events by using rMATS (Shen et al. 2014). Various strategies to sequence the 3' end of mRNAs can be used to quantify changes in steady state mRNAs as well as evaluate potential alternative polyadenylation (Zhou et al. 2014). Cotranscriptional RNA modifications can be determined by mapping specific modification events, such as m6A and ΨU, in mRNAs (Limbach and Paulines 2017). Functional impacts can be evaluated with 4sU-based methods in pulse-chase experiments for RNA stability (Tani and Akimitsu 2012), with subcellular fractionation and RNA sequencing (Frac-seq) for RNA export by performing separate RNA-seq analyses on cytoplasmic versus nuclear RNAs (Sterne-Weiler et al. 2013), and with ribosome profiling (Ribo-seq) for translational control (Ingolia et al. 2009). As an RBP may perform any of those functions independent of their association with chromatin, it has been a great challenge to determine whether their chromatin-binding activities contribute to specific functional impacts.

ACTIONS AND MECHANISMS OF RBPs ON CHROMATIN

Below, we discuss specific RBPs that have been characterized to some mechanistic details to illustrate how RBPs may be coupled with regulatory RNAs to control gene expression via their actions on chromatin.

RBPs as Part of RNA Polymerase Complexes

Multiple RBPs are known to be either part of the Pol II holoenzyme or PIC that contains Pol II. In fact, the active center cleft of Pol II has been found to be able to bind B2 noncoding RNA transcribed from a repeat (SINE) element (Kettenberger et al. 2006), which can potently inhibit transcription initiation (Espinoza et al. 2004). One of the Pol II subunit POLR2G (aka RPB7) contains a putative RNA-binding domain, which can bind DNA and RNA with similar affinity (Meka et al. 2005). As part of Pol II, this subunit sits close to the RNA exit channel and may play a critical role in transcription elongation. Interestingly, this Pol II subunit forms a heterodimer with POLR2D (aka RPB4), which has been recognized to modulate 3'-end formation of a subset of mRNAs in yeast, suggesting that it may play a critical role in polyadenylation-coupled transcription termination (Runner et al. 2008). By coIP, the RBP SFPQ has been reported to tightly associate with Pol II and modulate its phosphorylation in the Ser2 positions to influence transcription elongation (Takeuchi et al. 2018). Collectively, these findings illustrate that the Pol II complex has the capacity to bind RNA, either through its own active site or an RNA-binding subunit or a tightly associated RBP, thus rendering the Pol II machinery a direct target for modulation by regulatory RNAs.

Promoters as Hotspots for RBP Actions

Initiation of transcription requires the accessibility of gene promoters to DNA-binding TFs, and a recent large-scale survey of RBPs reveals that gene promoters are also the most predominant hotspots for RBPs (Xiao et al. 2019). Although it is entirely conceivable that various downstream RNA processing events may begin via promoter-associated RBPs (see below), increasing evidence suggests that RBPs may have direct roles in facilitating chromatin accessibility to aid in transcription initiation. The formation of non-B DNA structure, such as Z-DNA, is known to contribute to the formation of nucleosome-free chromatin regions (Liu et al. 2006; Mulholland et al. 2012) to license transcription activation (Fig. 2A; Shin et al. 2016). Interestingly, ADAR1, a double-stranded RBP functioning in RNA editing (Nishikura 2010), appears to have the capacity to bind Z-DNA to enhance gene expression (Oh et al. 2002). Many gene promoters contain CpG islands. G-rich sequences have been suggested to form G-quadruplex (G4), and the RBP HNRNPA1 appears to help unfold such G4 DNA, thus altering the chromatin accessibility (Fig. 2B; Paramasivam et al. 2009). Interestingly, the opposite C-rich strand has also been postulated to form a four-stranded DNA referred to as "i-motif," a structure that a recent study suggests does form in the cell (Zeraati et al. 2018), and HNRNPLL and several other RBPs appear to bind and unfold this motif to activate transcription (Fig. 2B; Kang et al. 2014; Abou Assi et al. 2018). The exposed single-stranded DNA (ssDNA) region at promoters is a potential platform for RBPs to act on, especially by those RBPs containing a KH domain or RRM motif that has been long recognized to also bind ssDNA (Fig. 2B; Maris et al. 2005; Valverde et al. 2008). These findings together suggest that RBPs may participate in the formation and resolution of various non-B DNA structures to modulate chromatin accessibility, thus modulating transcription.

Most gene promoters in mammalian genomes are now known to be regulated by transcription pausing and pause release in promoter-proximal regions (Core and Adelman 2019). A large number of gene promoters contain CpG islands, which promote the formation of R-loop, a three-stranded RNA/DNA structure in which nascent RNA anneals back to template DNA (Ginno et al. 2012). The ability of nascent RNA to invade into duplex DNA is greatly enhanced by the high propensity of nontemplate DNA to form G4-like structure, and, thus, R-loops are tightly associated with GC-skewed (G-rich sequence in nontemplate and C-rich sequence in template DNA) pro-

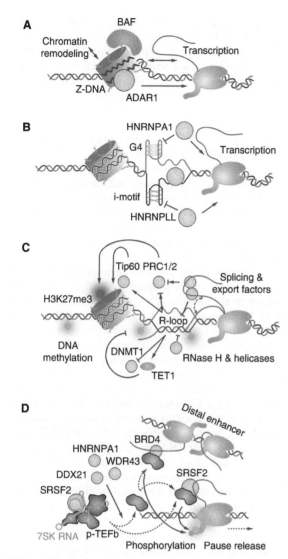

Figure 2. Mechanisms for RNA-binding proteins (RBPs) acting on promoters. (*A*) Mutual influence of Z-DNA formation and transcription, and regulation by the RNA editing enzyme ADAR1. (*B*) Recognition of G4, i-motif, or single-stranded DNA (ssDNA) by RBPs to regulate transcription. (*C*) The formation and resolution of R-loops, which can play positive roles in the recruitment of chromatin remodelers and DNA modification enzymes. (*D*) Multiple mechanisms for transcriptional pause release by relocating p-TEFb from 7SK complex to paused Pol II.

moter regions (Chen et al. 2017b). R-loop formation is likely part of the mechanism for Pol II pausing (Chen et al. 2017b), which has been thought to repress gene expression and induce genome instability (Skourti-Stathaki and Proudfoot 2014). However, recent studies show that R-loop formation is also linked to gene activation, perhaps by facilitating the recruitment of chromatin remodeler, such as Tip60 (Fig. 2C; Chen et al. 2015), thereby enhancing TF binding (Boque-Sastre et al. 2015). Because RNA is a key participant in R-loop formation, various RBPs have been shown to modulate R-loop formation and/or resolution, which creates opportunities for RBPs to positively or negatively modulate transcription (Cristini et al. 2018; Wang et al. 2018). For example, the RNA helicases DDX21 and DDX5 have been suggested to help resolve R-loop to promote transcription (Fig. 2C; Argaud et al. 2019; Mersaoui et al. 2019). Pre-mRNA splicing and RNA export factors appear to help pull nascent RNAs out from R-loops to facilitate RNA processing and transport, thus preventing R-loop accumulation (Fig. 2C; Santos-Pereira and Aguilera 2015). Transient R-loop formation may also lead to sustainable changes in gene promoters by repulsing DNA methyltransferases DNMT1 (Grunseich et al. 2018) and recruiting DNA demethylase TET1 (Arab et al. 2019), together converting them to the hypomethylated state for gene activation (Fig. 2C). On the other hand, a recent study (Alecki et al. 2019) suggests that Polycomb complex 2 (PRC2) can help RNA invade into DNA to enhance R-loop formation and that Polycomb complex 1 (PRC1) can bind R-loop, together facilitating H3K27me3 deposition to silence gene expression (Fig. 2C).

The release of paused Pol II is a major step in transcriptional control, which is regulated by the P-TEFb complex, consisting of cyclin T and CDK9 kinase, to phosphorylate NELF, DSIF, and Pol II Ser2 (Saunders et al. 2006). Interestingly, P-TEFb is part of an inhibitory complex containing the 7SK noncoding RNA, which is associated with gene promoters, and releasing and relocating P-TEFb from the 7SK complex to the Pol II complex has been recognized to play a key role in Pol II pause release (Fig. 2D; Core and Adelman 2019). An increasing number of RBPs have been shown to be involved in this process. The RBP HNRNPA1 appears to promote the disassociation of P-TEFb from 7SK complex, thus tripping the kinase in promoter-proximal regions (Barrandon et al. 2007; Van Herreweghe et al. 2007). SR proteins, which have been extensively characterized as splicing commitment factors, are also part of the 7SK complex, which help extract P-TEFb from the 7SK complex and relocate this critical Pol II CTD kinase to nascent RNA to activate transcription (Ji et al. 2013). DDX21 has also been shown to release p-TEFb from the 7SK complex via its helicase activity (Calo et al. 2015), and more recently, another RBP WDR43 has been found to activate transcription by releasing P-TEFb from the 7SK complex (Bi et al. 2019). These findings suggest multiple mechanisms for releasing P-TEFb from the 7SK complex through coordinated actions of regulatory RNAs and RBPs. In fact, nascent RNA-induced P-TEFb release may also be part of the mechanism for eRNAs to activate transcription (Schröder et al. 2012; Chen et al. 2017a; Rahnamoun et al. 2018).

RBPs to Facilitate Heterochromatin Formation, Spreading, and Maintenance

Gene silencing results from the formation of heterochromatin, but, counterintuitively, both the formation and maintenance of heterochromatin appear to depend on ongoing transcription. Heterochromatin can be further classified into facultative or constitutive heterochromatin, which are respectively decorated with H3K27me3 and H3K9me2/3. Interestingly, both lncRNAs and small RNAs have been shown to play critical roles in establishing heterochromatin (Li and Fu 2019). Exemplary analysis of X inactivation has provided critical insights into facultative heterochromatin formation and spreading, which is mediated by the lncRNA Xist (Wutz 2011). Through RNA pull-down coupled with mass spectrometric analysis, Xist has been shown to interact with multiple RBPs, particularly SPEN/SHARP, which appears to synergize with PRC2 in depositing H3K27me3 on targeted DNA regions (McHugh et al. 2015). Interestingly, all subunits of PRC2 have the capacity to directly bind RNA, thus contributing to both PRC2 recruitment and PRC2 spreading (Yan et al. 2019). PRC2 also interacts with nascent mRNAs; however, the functional consequence is still under active debate. The RBP RBFox2 has been shown to couple nascent RNA production with the recruitment of PRC2 as part of the feedback mechanism to maintain the bivalency for a subset of gene promoters (Wei et al. 2016), which are particularly prevalent in stem cells (Bernstein et al. 2006). The recently elucidated roles of Polycomb complexes in R-loop formation and recognition is consistent with RNA-guided H3K27me3 deposition in gene promoter regions (Alecki et al. 2019). On the other hand, when nascent RNAs are of sufficient abundance, they are able to evict PRC2 to prevent H3K27me3 deposition and thus help maintain genes in the highly active state (Kaneko et al. 2013; Beltran et al. 2016, 2019; Wang et al. 2017).

Repeat-derived small RNAs are well known to mediate the formation of constitutive heterochromatin (Matzke and Mosher 2014; Holoch and Moazed 2015; Johnson and Straight 2017). Briefly, in fission yeast, repeat-derived transcripts from active retrotransposons are amplified by RNA-dependent RNA polymerase and processed by Dicer to generate endo-siRNAs. In *Drosophila* germline cells, piRNAs are generated and amplified by the "ping-pong" mechanism. These endo-siRNAs or piRNAs are loaded on RNA-induced transcriptional silencing complex to target nascent homologous transcripts on chromatin, which help recruit H3K9me2/3 methyltransferases (SUV39) to deposit the histone marks to establish constitutive heterochromatin. Importantly, this process involves numerous RNA-mediated interactions, including direct interactions of the constitutive heterochromatin factors themselves with RNA to facilitate both the formation and spreading of heterochromatin (Muchardt et al. 2002; Johnson et al. 2017). It is thus conceivable that additional RBPs may be involved to fine-tune various steps, as illustrated by the role of the RBP Vigilin/HDLBP in binding hyperedited RNAs or other unstructured RNAs to enhance SUV39H1 recruitment (Zhou et al. 2008). Therefore, both RNAs and RBPs are instrumental to heterochromatin formation and maintenance, particularly in centromeric and pericentromeric regions, which is known to be critical for chromosome alignment during mitosis (Simon et al. 2015).

CONNECTING TRANSCRIPTION TO DOWNSTREAM RNA METABOLISM EVENTS

The primary purpose for RBPs to associate with chromatin has been thought to facilitate cotranscriptional processing events from RNA modification to intron removal to polyadenylation to RNA export (Proudfoot et al. 2002; Bentley 2014). Although cotranscriptional RNA processing has been well-documented, the mechanisms are still poorly understood. A popular idea is that specific RBPs or RNA processing machineries may ride with elongating Pol II to facilitate cotranscriptional RNA processing. The CTD of the largest Pol II subunit is thought to play a key role in this process by providing a docking platform for various RNA processing machineries. However, whereas depletion of CTD did show profound impacts on capping, alternative splicing, and alternative polyadenylation (Fong and Bentley 2001), it remains to be determined whether the CTD is required for Pol II to interact with various RNA processing machineries. To our knowledge, this has only been documented with capping enzymes (McCracken et al. 1997).

Interestingly, promoters have been reported to dictate downstream events from splicing (Cramer et al. 1997; Moldón et al. 2008) to RNA stability (Bregman et al. 2011; Trcek et al. 2011) to RNA export (Xiao et al. 2019), and even translation in the cytoplasm (Zid and O'Shea 2014). These promoter-dependent RNA metabolic steps have been convincingly shown by promoter swap in budding yeast genome, but such strategy has not been applied to mammals. It is conceivable that specific promoter-associated RBPs may be switched to nascent RNAs according to the so-called recruitment model (Naftelberg et al. 2015), but to date, specific RBPs critical for such promoter-dependent RNA processing events have not yet been identified. Alternatively, different promoters may equip the Pol II machinery with different factors to influence elongation speed, thereby creating different windows of opportunities for positive and negative regulators to recognize emerging RNA signals to facilitate specific RNA processing events (Bentley 2014; Fong et al. 2014). This has been referred to as the kinetic model (Naftelberg et al. 2015), but the mechanism for this attractive model has remained poorly understood, which requires the identification of specific RNA processing regulators that recognize specific nascent RNA elements in a Pol II elongation speed-dependent manner.

Another popular idea is for RBPs to interact with various modified histones, thereby coupling specific epigenetic features to the regulation of RNA processing. Indeed, many RBPs have been identified to associate with histone modification events (Ji et al. 2015), and some potential H3K36me3 readers have been reported to mediate alternative splicing (Luco et al. 2010; Guo et al. 2014). However, evidence has remained relatively thin to support modified histones as a widespread coupling mechanism between transcription and cotranscriptional RNA processing, and, therefore, the epigenetic control of cotranscriptional RNA processing still largely remains as an attractive hypothesis. Interestingly, the converse scenario has also been documented by which cotranscriptional splicing appears also to influence specific histone modification events, such as H3K4me3 in gene promoters (Bieberstein et al. 2012) and H3K36me3 in gene bodies (de Almeida et al. 2011; Kim et al. 2011).

ORGANIZING 3D GENOME FOR REGULATED GENE EXPRESSION PROGRAMS

Chromatin-associated RBPs may play larger roles in 3D genome beyond their functions in modulating local transcriptional and cotranscriptional activities (Fig. 3), as regulatory RNAs have been increasingly recognized to provide multivalent interactions to coordinate chromosomal interactions (Li and Fu 2019). Chromosomes can be segregated into active A compartments and inactive B compartments (Lieberman-Aiden et al. 2009). Recent studies show that RNA-dependent oligomerization of the nuclear matrix–associated RBP HNRNPU plays key regulatory roles at the chromosome level (Nozawa et al. 2017; Fan et al. 2018). Deletion of HNRNPU weakens the boundary between A and B compartments, leading to A-to-B switches and overall chromosome condensation (Nozawa et al. 2017; Fan et al. 2018). HNRNPU appears to work with various chromatin-associated RNAs to help organize 3D genome, including some lncRNAs, such as FIRRE (Hacisuleyman et al. 2014), Xist (Hasegawa et al. 2010), and double-stranded viral RNAs (Cao et al. 2019). As 3D genome involves numerous long-range interactions, chromatin-associated RNAs and RBPs may bridge such interactions (Kim and Shendure 2019). CTCF, one of the best-known high-order chromatin organizers, is able to bind DNA and RNA (Sun et al. 2013; Saldaña-Meyer et al. 2014), which has recently been shown to be mediated by distinct zinc fingers (Hansen et al. 2019; Saldaña-Meyer et al. 2019). As zinc fingers are also involved in protein–protein interactions, CTCF appears to form oligomers through the same zinc finger for RNA binding, and disruption of this domain greatly impairs specific long-range genomic interactions (Hansen et al. 2019; Saldaña-Meyer et al. 2019), suggesting that CTCF may mediate critical RNA–protein–DNA interaction networks in 3D genome.

More recently, another zinc-finger-containing TF YY1 is recognized to also play a broad role in mediating long-distance genomic interactions between promoters and enhancers (Weintraub et al. 2017). Interestingly, YY1 is also able to bind RNA, which appears to be required for its efficient targeting to specific promoters and enhancers (Sigova et al. 2015). A recent large-scale cobinding analysis between TFs and RBPs on chromatin reveals that YY1 colocalizes with the RBP RBM25 on chromatin in which RBM25 appears to direct YY1 to target genomic loci (Xiao et al. 2019). Thus, unlike CTCF, which can bind both DNA and RNA, the genome organization function of YY1 is mediated through its partnership with a specific RBP. It is attempting to speculate that various TFs may functionally interact with specific RBPs to increase the

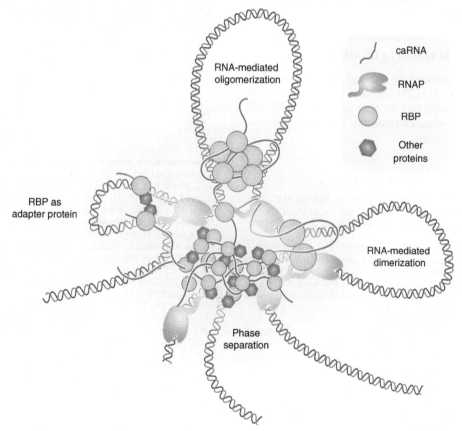

Figure 3. RNAs and RNA-binding proteins (RBPs) in 3D genome organization. Depicted are RNA-mediated dimerization (i.e., YY1), oligomerization (i.e., CTCF and HNRNPU), and genomic targeting (i.e., RBM25) for different TFs, together contributing to multivalent interactions to drive phase separation and formation of transcription hubs.

specificity and/or efficiency in genomic targeting, which is in line with the emerging concept for the formation of transcription hubs that result from multivalent interactions to induce liquid–liquid phase transition for gene activation in the nucleus (Hnisz et al. 2017; Boija et al. 2018; Sabari et al. 2018; Cramer 2019). RBPs may play critical roles in this process (Saha et al. 2016; Maharana et al. 2018), as RBPs are highly enriched with intrinsically disordered regions (Castello et al. 2012), which have been shown to be key driving forces for phase separation (Molliex et al. 2015; Lin et al. 2017). Therefore, 3D genome is likely orchestrated by network interactions among DNA, RNA, TFs, and RBPs to drive cell type–specific gene expression programs (Guo et al. 2019).

CONCLUSION

We have particularly focused in this review on diverse functions and mechanisms of chromatin-associated RBPs in regulated gene expression. An important message is that RNAs are no longer just products, but are also regulators of gene expression. Their regulatory functions are executed by specific RBPs, together contributing to network interactions in 3D genome. Such RNA and RBP-mediated interactions are critical for both gene activation and silencing, which calls for future research in this direction to understand how functional genome is organized and dynamically regulated by RNAs and RBPs in development and disease.

ACKNOWLEDGMENTS

X.-D.F. is supported by National Institutes of Health (NIH) grants GM131796, HG004659, and DK098808, and J.-Y.C. is supported by K99 grant DK120952. We thank members of the Fu laboratory for their critical reading of this manuscript.

REFERENCES

Abou Assi H, Garavis M, Gonzalez C, Damha MJ. 2018. i-Motif DNA: structural features and significance to cell biology. *Nucleic Acids Res* **46**: 8038–8056. doi:10.1093/nar/gky735

Alecki C, Chiwara V, Sanz LA, Grau D, Pérez OA, Armache K-J, Chédin F, Francis NJ. 2019. RNA strand invasion activity of the Polycomb complex PRC2. bioRxiv doi:10.1101/635722

Arab K, Karaulanov E, Musheev M, Trnka P, Schafer A, Grummt I, Niehrs C. 2019. GADD45A binds R-loops and recruits TET1 to CpG island promoters. *Nat Genet* **51**: 217–223. doi:10.1038/s41588-018-0306-6

Argaud D, Boulanger MC, Chignon A, Mkannez G, Mathieu P. 2019. Enhancer-mediated enrichment of interacting JMJD3-DDX21 to ENPP2 locus prevents R-loop formation and promotes transcription. *Nucleic Acids Res* **47**: 8424–8438. doi:10.1093/nar/gkz560

Bao X, Guo X, Yin M, Tariq M, Lai Y, Kanwal S, Zhou J, Li N, Lv Y, Pulido-Quetglas C, et al. 2018. Capturing the interactome of newly transcribed RNA. *Nat Methods* **15**: 213–220. doi:10.1038/nmeth.4595

Barrandon C, Bonnet F, Nguyen VT, Labas V, Bensaude O. 2007. The transcription-dependent dissociation of P-TEFb-HEXIM1-7SK RNA relies upon formation of hnRNP-7SK RNA complexes. *Mol Cell Biol* **27**: 6996–7006. doi:10.1128/MCB.00975-07

Beltran M, Yates CM, Skalska L, Dawson M, Reis FP, Viiri K, Fisher CL, Sibley CR, Foster BM, Bartke T, et al. 2016. The interaction of PRC2 with RNA or chromatin is mutually antagonistic. *Genome Res* **26**: 896–907. doi:10.1101/gr.197632.115

Beltran M, Tavares M, Justin N, Khandelwal G, Ambrose J, Foster BM, Worlock KB, Tvardovskiy A, Kunzelmann S, Herrero J, et al. 2019. G-tract RNA removes Polycomb repressive complex 2 from genes. *Nat Struct Mol Biol* **26**: 899–909. doi:10.1038/s41594-019-0293-z

Bentley DL. 2014. Coupling mRNA processing with transcription in time and space. *Nat Rev Genet* **15**: 163–175. doi:10.1038/nrg3662

Bernstein BE, Mikkelsen TS, Xie X, Kamal M, Huebert DJ, Cuff J, Fry B, Meissner A, Wernig M, Plath K, et al. 2006. A bivalent chromatin structure marks key developmental genes in embryonic stem cells. *Cell* **125**: 315–326. doi:10.1016/j.cell.2006.02.041

Bi X, Xu Y, Li T, Li X, Li W, Shao W, Wang K, Zhan G, Wu Z, Liu W, et al. 2019. RNA targets ribogenesis factor WDR43 to chromatin for transcription and pluripotency control. *Mol Cell* **75**: 102–116 e109. doi:10.1016/j.molcel.2019.05.007

Bieberstein NI, Carrillo Oesterreich F, Straube K, Neugebauer KM. 2012. First exon length controls active chromatin signatures and transcription. *Cell Rep* **2**: 62–68. doi:10.1016/j.celrep.2012.05.019

Bohmdorfer G, Wierzbicki AT. 2015. Control of chromatin structure by long noncoding RNA. *Trends Cell Biol* **25**: 623–632. doi:10.1016/j.tcb.2015.07.002

Boija A, Klein IA, Sabari BR, Dall'Agnese A, Coffey EL, Zamudio AV, Li CH, Shrinivas K, Manteiga JC, Hannett NM, et al. 2018. Transcription factors activate genes through the phase-separation capacity of their activation domains. *Cell* **175**: 1842–1855 e1816. doi:10.1016/j.cell.2018.10.042

Boque-Sastre R, Soler M, Oliveira-Mateos C, Portela A, Moutinho C, Sayols S, Villanueva A, Esteller M, Guil S. 2015. Head-to-head antisense transcription and R-loop formation promotes transcriptional activation. *Proc Natl Acad Sci* **112**: 5785–5790. doi:10.1073/pnas.1421197112

Bregman A, Avraham-Kelbert M, Barkai O, Duek L, Guterman A, Choder M. 2011. Promoter elements regulate cytoplasmic mRNA decay. *Cell* **147**: 1473–1483. doi:10.1016/j.cell.2011.12.005

Calo E, Flynn RA, Martin L, Spitale RC, Chang HY, Wysocka J. 2015. RNA helicase DDX21 coordinates transcription and ribosomal RNA processing. *Nature* **518**: 249–253. doi:10.1038/nature13923

Cao L, Liu S, Li Y, Yang G, Luo Y, Li S, Du H, Zhao Y, Wang D, Chen J, et al. 2019. The nuclear matrix protein SAFA surveils viral RNA and facilitates immunity by activating antiviral enhancers and super-enhancers. *Cell Host Microbe* **26**: 369–384 e368. doi:10.1016/j.chom.2019.08.010

Castello A, Fischer B, Eichelbaum K, Horos R, Beckmann BM, Strein C, Davey NE, Humphreys DT, Preiss T, Steinmetz LM, et al. 2012. Insights into RNA biology from an atlas of mammalian mRNA-binding proteins. *Cell* **149**: 1393–1406. doi:10.1016/j.cell.2012.04.031

Chen PB, Chen HV, Acharya D, Rando OJ, Fazzio TG. 2015. R loops regulate promoter-proximal chromatin architecture and cellular differentiation. *Nat Struct Mol Biol* **22**: 999–1007. doi:10.1038/nsmb.3122

Chen FX, Xie P, Collings CK, Cao K, Aoi Y, Marshall SA, Rendleman EJ, Ugarenko M, Ozark PA, Zhang A, et al. 2017a. PAF1 regulation of promoter-proximal pause release via enhancer activation. *Science* **357**: 1294–1298. doi:10.1126/science.aan3269

Chen L, Chen JY, Zhang X, Gu Y, Xiao R, Shao C, Tang P, Qian H, Luo D, Li H, et al. 2017b. R-ChIP using inactive RNase H reveals dynamic coupling of R-loops with transcriptional pausing at gene promoters. *Mol Cell* **68**: 745–757 e745. doi:10.1016/j.molcel.2017.10.008

Chu C, Qu K, Zhong FL, Artandi SE, Chang HY. 2011. Genomic maps of long noncoding RNA occupancy reveal principles of RNA-chromatin interactions. *Mol Cell* **44:** 667–678. doi:10.1016/j.molcel.2011.08.027

Core L, Adelman K. 2019. Promoter-proximal pausing of RNA polymerase II: a nexus of gene regulation. *Genes Dev* **33:** 960–982. doi:10.1101/gad.325142.119

Core LJ, Waterfall JJ, Lis JT. 2008. Nascent RNA sequencing reveals widespread pausing and divergent initiation at human promoters. *Science* **322:** 1845–1848. doi:10.1126/science.1162228

Cramer P. 2019. Organization and regulation of gene transcription. *Nature* **573:** 45–54. doi:10.1038/s41586-019-1517-4

Cramer P, Pesce CG, Baralle FE, Kornblihtt AR. 1997. Functional association between promoter structure and transcript alternative splicing. *Proc Natl Acad Sci* **94:** 11456–11460. doi:10.1073/pnas.94.21.11456

Cristini A, Groh M, Kristiansen MS, Gromak N. 2018. RNA/DNA hybrid interactome identifies DXH9 as a molecular player in transcriptional termination and R-loop-associated DNA damage. *Cell Rep* **23:** 1891–1905. doi:10.1016/j.celrep.2018.04.025

de Almeida SF, Grosso AR, Koch F, Fenouil R, Carvalho S, Andrade J, Levezinho H, Gut M, Eick D, Gut I, et al. 2011. Splicing enhances recruitment of methyltransferase HYPB/Setd2 and methylation of histone H3 Lys36. *Nat Struct Mol Biol* **18:** 977–983. doi:10.1038/nsmb.2123

Dejardin J, Kingston RE. 2009. Purification of proteins associated with specific genomic loci. *Cell* **136:** 175–186. doi:10.1016/j.cell.2008.11.045

Djebali S, Davis CA, Merkel A, Dobin A, Lassmann T, Mortazavi A, Tanzer A, Lagarde J, Lin W, Schlesinger F, et al. 2012. Landscape of transcription in human cells. *Nature* **489:** 101–108. doi:10.1038/nature11233

Engreitz JM, Pandya-Jones A, McDonel P, Shishkin A, Sirokman K, Surka C, Kadri S, Xing J, Goren A, Lander ES, et al. 2013. The Xist lncRNA exploits three-dimensional genome architecture to spread across the X chromosome. *Science* **341:** 1237973. doi:10.1126/science.1237973

Espinoza CA, Allen TA, Hieb AR, Kugel JF, Goodrich JA. 2004. B2 RNA binds directly to RNA polymerase II to repress transcript synthesis. *Nat Struct Mol Biol* **11:** 822–829. doi:10.1038/nsmb812

Fan H, Lv P, Huo X, Wu J, Wang Q, Cheng L, Liu Y, Tang QQ, Zhang L, Zhang F, et al. 2018. The nuclear matrix protein HNRNPU maintains 3D genome architecture globally in mouse hepatocytes. *Genome Res* **28:** 192–202. doi:10.1101/gr.224576.117

Fong N, Bentley DL. 2001. Capping, splicing, and 3′ processing are independently stimulated by RNA polymerase II: different functions for different segments of the CTD. *Genes Dev* **15:** 1783–1795. doi:10.1101/gad.889101

Fong N, Kim H, Zhou Y, Ji X, Qiu J, Saldi T, Diener K, Jones K, Fu XD, Bentley DL. 2014. Pre-mRNA splicing is facilitated by an optimal RNA polymerase II elongation rate. *Genes Dev* **28:** 2663–2676. doi:10.1101/gad.252106.114

Gilbert C, Svejstrup JQ. 2006. RNA immunoprecipitation for determining RNA-protein associations in vivo. *Curr Protoc Mol Biol* **Chapter 27:** Unit 27 24. doi:10.1002/0471142727.mb2704s75

Ginno PA, Lott PL, Christensen HC, Korf I, Chedin F. 2012. R-loop formation is a distinctive characteristic of unmethylated human CpG island promoters. *Mol Cell* **45:** 814–825. doi:10.1016/j.molcel.2012.01.017

Grunseich C, Wang IX, Watts JA, Burdick JT, Guber RD, Zhu Z, Bruzel A, Lanman T, Chen K, Schindler AB, et al. 2018. Senataxin mutation reveals how R-loops promote transcription by blocking DNA methylation at gene promoters. *Mol Cell* **69:** 426–437 e427. doi:10.1016/j.molcel.2017.12.030

Guo R, Zheng L, Park JW, Lv R, Chen H, Jiao F, Xu W, Mu S, Wen H, Qiu J, et al. 2014. BS69/ZMYND11 reads and connects histone H3.3 lysine 36 trimethylation-decorated chromatin to regulated pre-mRNA processing. *Mol Cell* **56:** 298–310. doi:10.1016/j.molcel.2014.08.022

Guo YE, Manteiga JC, Henninger JE, Sabari BR, Dall'Agnese A, Hannett NM, Spille JH, Afeyan LK, Zamudio AV, Shrinivas K, et al. 2019. Pol II phosphorylation regulates a switch between transcriptional and splicing condensates. *Nature* **572:** 543–548. doi:10.1038/s41586-019-1464-0

Hacisuleyman E, Goff LA, Trapnell C, Williams A, Henao-Mejia J, Sun L, McClanahan P, Hendrickson DG, Sauvageau M, Kelley DR, et al. 2014. Topological organization of multichromosomal regions by the long intergenic noncoding RNA Firre. *Nat Struct Mol Biol* **21:** 198–206. doi:10.1038/nsmb.2764

Hansen AS, Hsieh T-HS, Cattoglio C, Pustova I, Saldaña-Meyer R, Reinberg D, Darzacq X, Tjian R. 2019. Distinct classes of chromatin loops revealed by deletion of an RNA-binding region in CTCF. *Mol Cell* **76:** 395–411.e13. doi:10.1016/j.molcel.2019.07.039.

Hasegawa Y, Brockdorff N, Kawano S, Tsutui K, Tsutui K, Nakagawa S. 2010. The matrix protein hnRNP U is required for chromosomal localization of Xist RNA. *Dev Cell* **19:** 469–476. doi:10.1016/j.devcel.2010.08.006

Hendrickson DG, Kelley DR, Tenen D, Bernstein B, Rinn JL. 2016. Widespread RNA binding by chromatin-associated proteins. *Genome Biol* **17:** 28. doi:10.1186/s13059-016-0878-3

Hentze MW, Castello A, Schwarzl T, Preiss T. 2018. A brave new world of RNA-binding proteins. *Nat Rev Mol Cell Biol* **19:** 327–341. doi:10.1038/nrm.2017.130

Hnisz D, Shrinivas K, Young RA, Chakraborty AK, Sharp PA. 2017. A phase separation model for transcriptional control. *Cell* **169:** 13–23. doi:10.1016/j.cell.2017.02.007

Holoch D, Moazed D. 2015. RNA-mediated epigenetic regulation of gene expression. *Nat Rev Genet* **16:** 71–84. doi:10.1038/nrg3863

Hsin JP, Manley JL. 2012. The RNA polymerase II CTD coordinates transcription and RNA processing. *Genes Dev* **26:** 2119–2137. doi:10.1101/gad.200303.112

Huang R, Han M, Meng L, Chen X. 2018. Transcriptome-wide discovery of coding and noncoding RNA-binding proteins. *Proc Natl Acad Sci* **115:** E3879–E3887. doi:10.1073/pnas.1718406115

Hudson WH, Ortlund EA. 2014. The structure, function and evolution of proteins that bind DNA and RNA. *Nat Rev Mol Cell Biol* **15:** 749–760. doi:10.1038/nrm3884

Ingolia NT, Ghaemmaghami S, Newman JR, Weissman JS. 2009. Genome-wide analysis in vivo of translation with nucleotide resolution using ribosome profiling. *Science* **324:** 218–223. doi:10.1126/science.1168978

Jenuwein T, Allis CD. 2001. Translating the histone code. *Science* **293:** 1074–1080. doi:10.1126/science.1063127

Ji X, Zhou Y, Pandit S, Huang J, Li H, Lin CY, Xiao R, Burge CB, Fu XD. 2013. SR proteins collaborate with 7SK and promoter-associated nascent RNA to release paused polymerase. *Cell* **153:** 855–868. doi:10.1016/j.cell.2013.04.028

Ji X, Dadon DB, Abraham BJ, Lee TI, Jaenisch R, Bradner JE, Young RA. 2015. Chromatin proteomic profiling reveals novel proteins associated with histone-marked genomic regions. *Proc Natl Acad Sci* **112:** 3841–3846. doi:10.1073/pnas.1502971112

Johnson WL, Straight AF. 2017. RNA-mediated regulation of heterochromatin. *Curr Opin Cell Biol* **46:** 102–109. doi:10.1016/j.ceb.2017.05.004

Johnson WL, Yewdell WT, Bell JC, McNulty SM, Duda Z, O'Neill RJ, Sullivan BA, Straight AF. 2017. RNA-dependent stabilization of SUV39H1 at constitutive heterochromatin. *Elife* **6:** e25299. doi:10.7554/eLife.25299

Kanehisa M, Goto S. 2000. KEGG: Kyoto Encyclopedia of Genes and Genomes. *Nucleic Acids Res* **28:** 27–30. doi:10.1093/nar/28.1.27

Kaneko S, Son J, Shen SS, Reinberg D, Bonasio R. 2013. PRC2 binds active promoters and contacts nascent RNAs in embryonic stem cells. *Nat Struct Mol Biol* **20:** 1258–1264. doi:10.1038/nsmb.2700

Kang HJ, Kendrick S, Hecht SM, Hurley LH. 2014. The transcriptional complex between the BCL2 i-motif and hnRNP LL is a molecular switch for control of gene expression that can be modulated by small molecules. *J Am Chem Soc* **136:** 4172–4185. doi:10.1021/ja4109352

Kaya-Okur HS, Wu SJ, Codomo CA, Pledger ES, Bryson TD, Henikoff JG, Ahmad K, Henikoff S. 2019. CUT&Tag for efficient epigenomic profiling of small samples and single cells. *Nat Commun* **10:** 1930. doi:10.1038/s41467-019-09982-5

Kettenberger H, Eisenfuhr A, Brueckner F, Theis M, Famulok M, Cramer P. 2006. Structure of an RNA polymerase II-RNA inhibitor complex elucidates transcription regulation by noncoding RNAs. *Nat Struct Mol Biol* **13:** 44–48. doi:10.1038/nsmb1032

Kim S, Shendure J. 2019. Mechanisms of interplay between transcription factors and the 3D genome. *Mol Cell* **76:** 306–319. doi:10.1016/j.molcel.2019.08.010

Kim S, Kim H, Fong N, Erickson B, Bentley DL. 2011. Pre-mRNA splicing is a determinant of histone H3K36 methylation. *Proc Natl Acad Sci* **108:** 13564–13569. doi:10.1073/pnas.1109475108

Kimple ME, Brill AL, Pasker RL. 2013. Overview of affinity tags for protein purification. *Curr Protoc Protein Sci* **73:** 9.9.1–9.9.23. doi:10.1002/0471140864.ps0909s73

Lai B, Tang Q, Jin W, Hu G, Wangsa D, Cui K, Stanton BZ, Ren G, Ding Y, Zhao M, et al. 2018. Trac-looping measures genome structure and chromatin accessibility. *Nat Methods* **15:** 741–747. doi:10.1038/s41592-018-0107-y

Landt SG, Marinov GK, Kundaje A, Kheradpour P, Pauli F, Batzoglou S, Bernstein BE, Bickel P, Brown JB, Cayting P, et al. 2012. ChIP-seq guidelines and practices of the ENCODE and modENCODE consortia. *Genome Res* **22:** 1813–1831. doi:10.1101/gr.136184.111

Lee FCY, Ule J. 2018. Advances in CLIP technologies for studies of protein–RNA interactions. *Mol Cell* **69:** 354–369. doi:10.1016/j.molcel.2018.01.005

Li X, Fu XD. 2019. Chromatin-associated RNAs as facilitators of functional genomic interactions. *Nat Rev Genet* **20:** 503–519. doi:10.1038/s41576-019-0135-1

Li W, Notani D, Rosenfeld MG. 2016. Enhancers as non-coding RNA transcription units: recent insights and future perspectives. *Nat Rev Genet* **17:** 207–223. doi:10.1038/nrg.2016.4

Lieberman-Aiden E, van Berkum NL, Williams L, Imakaev M, Ragoczy T, Telling A, Amit I, Lajoie BR, Sabo PJ, Dorschner MO, et al. 2009. Comprehensive mapping of long-range interactions reveals folding principles of the human genome. *Science* **326:** 289–293. doi:10.1126/science.1181369

Limbach PA, Paulines MJ. 2017. Going global: the new era of mapping modifications in RNA. *Wiley Interdiscip Rev RNA* **8:** e1367. doi:10.1002/wrna.1367

Lin Y, Currie SL, Rosen MK. 2017. Intrinsically disordered sequences enable modulation of protein phase separation through distributed tyrosine motifs. *J Biol Chem* **292:** 19110–19120. doi:10.1074/jbc.M117.800466

Liu H, Mulholland N, Fu H, Zhao K. 2006. Cooperative activity of BRG1 and Z-DNA formation in chromatin remodeling. *Mol Cell Biol* **26:** 2550–2559. doi:10.1128/MCB.26.7.2550-2559.2006

Long Y, Wang X, Youmans DT, Cech TR. 2017. How do lncRNAs regulate transcription? *Sci Adv* **3:** eaao2110. doi:10.1126/sciadv.aao2110

Luco RF, Pan Q, Tominaga K, Blencowe BJ, Pereira-Smith OM, Misteli T. 2010. Regulation of alternative splicing by histone modifications. *Science* **327:** 996–1000. doi:10.1126/science.1184208

Maharana S, Wang J, Papadopoulos DK, Richter D, Pozniakovsky A, Poser I, Bickle M, Rizk S, Guillen-Boixet J, Franzmann TM, et al. 2018. RNA buffers the phase separation behavior of prion-like RNA binding proteins. *Science* **360:** 918–921. doi:10.1126/science.aar7366

Maris C, Dominguez C, Allain FH. 2005. The RNA recognition motif, a plastic RNA-binding platform to regulate post-transcriptional gene expression. *FEBS J* **272:** 2118–2131. doi:10.1111/j.1742-4658.2005.04653.x

Matzke MA, Mosher RA. 2014. RNA-directed DNA methylation: an epigenetic pathway of increasing complexity. *Nat Rev Genet* **15:** 394–408. doi:10.1038/nrg3683

McCracken S, Fong N, Rosonina E, Yankulov K, Brothers G, Siderovski D, Hessel A, Foster S, Shuman S, Bentley DL. 1997. 5′-Capping enzymes are targeted to pre-mRNA by binding to the phosphorylated carboxy-terminal domain of RNA polymerase II. *Genes Dev* **11:** 3306–3318. doi:10.1101/gad.11.24.3306

McHugh CA, Chen CK, Chow A, Surka CF, Tran C, McDonel P, Pandya-Jones A, Blanco M, Burghard C, Moradian A, et al. 2015. The Xist lncRNA interacts directly with SHARP to silence transcription through HDAC3. *Nature* **521:** 232–236. doi:10.1038/nature14443

Meka H, Werner F, Cordell SC, Onesti S, Brick P. 2005. Crystal structure and RNA binding of the Rpb4/Rpb7 subunits of human RNA polymerase II. *Nucleic Acids Res* **33:** 6435–6444. doi:10.1093/nar/gki945

Mersaoui SY, Yu Z, Coulombe Y, Karam M, Busatto FF, Masson JY, Richard S. 2019. Arginine methylation of the DDX5 helicase RGG/RG motif by PRMT5 regulates resolution of RNA:DNA hybrids. *EMBO J* **38:** e100986. doi:10.15252/embj.2018100986

Moldón A, Malapeira J, Gabrielli N, Gogol M, Gómez-Escoda B, Ivanova T, Seidel C, Ayté J. 2008. Promoter-driven splicing regulation in fission yeast. *Nature* **455:** 997–1000. doi:10.1038/nature07325

Molliex A, Temirov J, Lee J, Coughlin M, Kanagaraj AP, Kim HJ, Mittag T, Taylor JP. 2015. Phase separation by low complexity domains promotes stress granule assembly and drives pathological fibrillization. *Cell* **163:** 123–133. doi:10.1016/j.cell.2015.09.015

Muchardt C, Guillemé M, Seeler JS, Trouche D, Dejean A, Yaniv M. 2002. Coordinated methyl and RNA binding is required for heterochromatin localization of mammalian HP1α. *EMBO Rep* **3:** 975–981. doi:10.1093/embo-reports/kvf194

Mulholland N, Xu Y, Sugiyama H, Zhao K. 2012. SWI/SNF-mediated chromatin remodeling induces Z-DNA formation on a nucleosome. *Cell Biosci* **2:** 3. doi:10.1186/2045-3701-2-3

Mumbach MR, Granja JM, Flynn RA, Roake CM, Satpathy AT, Rubin AJ, Qi Y, Jiang Z, Shams S, Louie BH, et al. 2019. HiChIRP reveals RNA-associated chromosome conformation. *Nat Methods* **16:** 489–492. doi:10.1038/s41592-019-0407-x

Naftelberg S, Schor IE, Ast G, Kornblihtt AR. 2015. Regulation of alternative splicing through coupling with transcription and chromatin structure. *Annu Rev Biochem* **84:** 165–198. doi:10.1146/annurev-biochem-060614-034242

Nishikura K. 2010. Functions and regulation of RNA editing by ADAR deaminases. *Annu Rev Biochem* **79:** 321–349. doi:10.1146/annurev-biochem-060208-105251

Nozawa RS, Boteva L, Soares DC, Naughton C, Dun AR, Buckle A, Ramsahoye B, Bruton PC, Saleeb RS, Arnedo M, et al. 2017. SAF-A regulates interphase chromosome structure through oligomerization with chromatin-associated RNAs. *Cell* **169:** 1214–1227 e1218. doi:10.1016/j.cell.2017.05.029

Oh DB, Kim YG, Rich A. 2002. Z-DNA-binding proteins can act as potent effectors of gene expression in vivo. *Proc Natl Acad Sci* **99:** 16666–16671. doi:10.1073/pnas.262672699

Paramasivam M, Membrino A, Cogoi S, Fukuda H, Nakagama H, Xodo LE. 2009. Protein hnRNP A1 and its derivative Up1 unfold quadruplex DNA in the human KRAS promoter: implications for transcription. *Nucleic Acids Res* **37:** 2841–2853. doi:10.1093/nar/gkp138

Proudfoot NJ, Furger A, Dye MJ. 2002. Integrating mRNA processing with transcription. *Cell* **108:** 501–512. doi:10.1016/S0092-8674(02)00617-7

Quinn JJ, Ilik IA, Qu K, Georgiev P, Chu C, Akhtar A, Chang HY. 2014. Revealing long noncoding RNA architecture and functions using domain-specific chromatin isolation by RNA purification. *Nat Biotechnol* **32:** 933–940. doi:10.1038/nbt.2943

Rahnamoun H, Lee J, Sun Z, Lu H, Ramsey KM, Komives EA, Lauberth SM. 2018. RNAs interact with BRD4 to promote enhanced chromatin engagement and transcription activation. *Nat Struct Mol Biol* **25:** 687–697. doi:10.1038/s41594-018-0102-0

Rhee HS, Pugh BF. 2011. Comprehensive genome-wide protein-DNA interactions detected at single-nucleotide resolution. *Cell* **147:** 1408–1419. doi:10.1016/j.cell.2011.11.013

Runner VM, Podolny V, Buratowski S. 2008. The Rpb4 subunit of RNA polymerase II contributes to cotranscriptional recruitment of 3′ processing factors. *Mol Cell Biol* **28:** 1883–1891. doi:10.1128/MCB.01714-07

Sabari BR, Dall'Agnese A, Boija A, Klein IA, Coffey EL, Shrinivas K, Abraham BJ, Hannett NM, Zamudio AV, Manteiga JC, et al. 2018. Coactivator condensation at super-enhancers links phase separation and gene control. *Science* **361:** eaar3958. doi:10.1126/science.aar3958

Saha S, Weber CA, Nousch M, Adame-Arana O, Hoege C, Hein MY, Osborne-Nishimura E, Mahamid J, Jahnel M, Jawerth L, et al. 2016. Polar positioning of phase-separated liquid compartments in cells regulated by an mRNA competition mechanism. *Cell* **166:** 1572–1584 e1516. doi:10.1016/j.cell.2016.08.006

Saldaña-Meyer R, González-Buendía E, Guerrero G, Narendra V, Bonasio R, Recillas-Targa F, Reinberg D. 2014. CTCF regulates the human p53 gene through direct interaction with its natural antisense transcript, Wrap53. *Genes Dev* **28:** 723–734. doi:10.1101/gad.236869.113

Saldaña-Meyer R, Rodriguez-Hernaez J, Escobar T, Nishana M, Jácome-López K, Nora EP, Bruneau BG, Tsirigos A, Furlan-Magaril M, Skok J, et al. 2019. RNA interactions are essential for CTCF-mediated genome organization. *Mol Cell* **76:** 412–422.e5. doi:10.1016/j.molcel.2019.08.015.

Santos-Pereira JM, Aguilera A. 2015. R loops: new modulators of genome dynamics and function. *Nat Rev Genet* **16:** 583–597. doi:10.1038/nrg3961

Saunders A, Core LJ, Lis JT. 2006. Breaking barriers to transcription elongation. *Nat Rev Mol Cell Biol* **7:** 557–567. doi:10.1038/nrm1981

Schmidl C, Rendeiro AF, Sheffield NC, Bock C. 2015. ChIPmentation: fast, robust, low-input ChIP-seq for histones and transcription factors. *Nat Methods* **12:** 963–965. doi:10.1038/nmeth.3542

Schoenfelder S, Fraser P. 2019. Long-range enhancer–promoter contacts in gene expression control. *Nat Rev Genet* **20:** 437–455. doi:10.1038/s41576-019-0128-0

Schröder S, Cho S, Zeng L, Zhang Q, Kaehlcke K, Mak L, Lau J, Bisgrove D, Schnölzer M, Verdin E, et al. 2012. Two-pronged binding with bromodomain-containing protein 4 liberates positive transcription elongation factor b from inactive ribonucleoprotein complexes. *J Biol Chem* **287:** 1090–1099. doi:10.1074/jbc.M111.282855

Shen S, Park JW, Lu ZX, Lin L, Henry MD, Wu YN, Zhou Q, Xing Y. 2014. rMATS: robust and flexible detection of differential alternative splicing from replicate RNA-Seq data. *Proc Natl Acad Sci* **111:** E5593–5601. doi:10.1073/pnas.1419161111

Shin SI, Ham S, Park J, Seo SH, Lim CH, Jeon H, Huh J, Roh TY. 2016. Z-DNA-forming sites identified by ChIP-Seq are associated with actively transcribed regions in the human genome. *DNA Res* **23:** 477–486. doi:10.1093/dnares/dsw031

Sigova AA, Abraham BJ, Ji X, Molinie B, Hannett NM, Guo YE, Jangi M, Giallourakis CC, Sharp PA, Young RA. 2015. Transcription factor trapping by RNA in gene regulatory elements. *Science* **350:** 978–981. doi:10.1126/science.aad3346

Simon MD, Wang CI, Kharchenko PV, West JA, Chapman BA, Alekseyenko AA, Borowsky ML, Kuroda MI, Kingston RE. 2011. The genomic binding sites of a noncoding RNA. *Proc Natl Acad Sci* **108:** 20497–20502. doi:10.1073/pnas.1113536108

Simon L, Voisin M, Tatout C, Probst AV. 2015. Structure and function of centromeric and pericentromeric heterochromatin in *Arabidopsis thaliana*. *Front Plant Sci* **6:** 1049. doi:10.3389/fpls.2015.01049

Skalska L, Beltran-Nebot M, Ule J, Jenner RG. 2017. Regulatory feedback from nascent RNA to chromatin and transcription. *Nat Rev Mol Cell Biol* **18:** 331–337. doi:10.1038/nrm.2017.12

Skene PJ, Henikoff S. 2017. An efficient targeted nuclease strategy for high-resolution mapping of DNA binding sites. *Elife* **6:** e21856. doi:10.7554/eLife.21856

Skourti-Stathaki K, Proudfoot NJ. 2014. A double-edged sword: R loops as threats to genome integrity and powerful regulators of gene expression. *Genes Dev* **28:** 1384–1396. doi:10.1101/gad.242990.114

Sterne-Weiler T, Martinez-Nunez RT, Howard JM, Cvitovik I, Katzman S, Tariq MA, Pourmand N, Sanford JR. 2013. Frac-seq reveals isoform-specific recruitment to polyribosomes. *Genome Res* **23:** 1615–1623. doi:10.1101/gr.148585.112

Sugimoto Y, Vigilante A, Darbo E, Zirra A, Militti C, D'Ambrogio A, Luscombe NM, Ule J. 2015. hiCLIP reveals the in vivo atlas of mRNA secondary structures recognized by Staufen 1. *Nature* **519:** 491–494. doi:10.1038/nature14280

Sun S, Del Rosario BC, Szanto A, Ogawa Y, Jeon Y, Lee JT. 2013. Jpx RNA activates Xist by evicting CTCF. *Cell* **153:** 1537–1551. doi:10.1016/j.cell.2013.05.028

Takeuchi A, Iida K, Tsubota T, Hosokawa M, Denawa M, Brown JB, Ninomiya K, Ito M, Kimura H, Abe T, et al. 2018. Loss of *Sfpq* causes long-gene transcriptopathy in the brain. *Cell Rep* **23:** 1326–1341. doi:10.1016/j.celrep.2018.03.141

Tani H, Akimitsu N. 2012. Genome-wide technology for determining RNA stability in mammalian cells: historical perspective and recent advantages based on modified nucleotide labeling. *RNA Biol* **9:** 1233–1238. doi:10.4161/rna.22036

The Gene Ontology Consortium. 2019. The Gene Ontology Resource: 20 years and still GOing strong. *Nucleic Acids Res* **47:** D330–D338. doi:10.1093/nar/gky1055

Trcek T, Larson DR, Moldón A, Query CC, Singer RH. 2011. Single-molecule mRNA decay measurements reveal promoter-regulated mRNA stability in yeast. *Cell* **147:** 1484–1497. doi:10.1016/j.cell.2011.11.051

Tsui C, Inouye C, Levy M, Lu A, Florens L, Washburn MP, Tjian R. 2018. dCas9-targeted locus-specific protein isolation method identifies histone gene regulators. *Proc Natl Acad Sci* **115:** E2734–E2741. doi:10.1073/pnas.1718844115

Valverde R, Edwards L, Regan L. 2008. Structure and function of KH domains. *FEBS J* **275:** 2712–2726. doi:10.1111/j.1742-4658.2008.06411.x

Van Herreweghe E, Egloff S, Goiffon I, Jady BE, Froment C, Monsarrat B, Kiss T. 2007. Dynamic remodelling of human 7SK snRNP controls the nuclear level of active P-TEFb. *EMBO J* **26:** 3570–3580. doi:10.1038/sj.emboj.7601783

Van Nostrand EL, Freese P, Pratt GA, Wang X, Wei X, Xiao R, Blue SM, Chen J-Y, Cody NAL, Dominguez D, et al. 2018. A large-scale binding and functional map of human RNA binding proteins. bioRxiv doi:10.1101/179648

Wang X, Paucek RD, Gooding AR, Brown ZZ, Ge EJ, Muir TW, Cech TR. 2017. Molecular analysis of PRC2 recruitment to DNA in chromatin and its inhibition by RNA. *Nat Struct Mol Biol* **24:** 1028–1038. doi:10.1038/nsmb.3487

Wang IX, Grunseich C, Fox J, Burdick J, Zhu Z, Ravazian N, Hafner M, Cheung VG. 2018. Human proteins that interact with RNA/DNA hybrids. *Genome Res* **28:** 1405–1414. doi:10.1101/gr.237362.118

Wei C, Xiao R, Chen L, Cui H, Zhou Y, Xue Y, Hu J, Zhou B, Tsutsui T, Qiu J, et al. 2016. RBFox2 binds nascent RNA to globally regulate Polycomb complex 2 targeting in mammalian genomes. *Mol Cell* **62:** 875–889. doi:10.1016/j.molcel.2016.04.013

Weintraub AS, Li CH, Zamudio AV, Sigova AA, Hannett NM, Day DS, Abraham BJ, Cohen MA, Nabet B, Buckley DL, et al. 2017. YY1 is a structural regulator of enhancer-promoter loops. *Cell* **171:** 1573–1588 e1528. doi:10.1016/j.cell.2017.11.008

Wissink EM, Vihervaara A, Tippens ND, Lis JT. 2019. Nascent RNA analyses: tracking transcription and its regulation. *Nat Rev Genet* **20:** 705–723. doi:10.1038/s41576-019-0159-6.

Wutz A. 2011. Gene silencing in X-chromosome inactivation: advances in understanding facultative heterochromatin formation. *Nat Rev Genet* **12**: 542–553. doi:10.1038/nrg3035

Xiao R, Chen JY, Liang Z, Luo D, Chen G, Lu ZJ, Chen Y, Zhou B, Li H, Du X, et al. 2019. Pervasive chromatin-RNA binding protein interactions enable RNA-based regulation of transcription. *Cell* **178**: 107–121 e118. doi:10.1016/j.cell.2019.06.001

Yan J, Dutta B, Hee YT, Chng WJ. 2019. Towards understanding of PRC2 binding to RNA. *RNA Biol* **16**: 176–184. doi:10.1080/15476286.2019.1565283

Yao L, Berman BP, Farnham PJ. 2015. Demystifying the secret mission of enhancers: linking distal regulatory elements to target genes. *Crit Rev Biochem Mol Biol* **50**: 550–573. doi:10.3109/10409238.2015.1087961

Zakeri B, Fierer JO, Celik E, Chittock EC, Schwarz-Linek U, Moy VT, Howarth M. 2012. Peptide tag forming a rapid covalent bond to a protein, through engineering a bacterial adhesin. *Proc Natl Acad Sci* **109**: E690–E697. doi:10.1073/pnas.1115485109

Zeraati M, Langley DB, Schofield P, Moye AL, Rouet R, Hughes WE, Bryan TM, Dinger ME, Christ D. 2018. I-motif DNA structures are formed in the nuclei of human cells. *Nat Chem* **10**: 631–637. doi:10.1038/s41557-018-0046-3

Zhang H, Zeitz MJ, Wang H, Niu B, Ge S, Li W, Cui J, Wang G, Qian G, Higgins MJ, et al. 2014. Long noncoding RNA-mediated intrachromosomal interactions promote imprinting at the Kcnq1 locus. *J Cell Biol* **204**: 61–75. doi:10.1083/jcb.201304152

Zhou J, Wang Q, Chen LL, Carmichael GG. 2008. On the mechanism of induction of heterochromatin by the RNA-binding protein vigilin. *RNA* **14**: 1773–1781. doi:10.1261/rna.1036308

Zhou Y, Li HR, Huang J, Jin G, Fu XD. 2014. Multiplex analysis of polyA-linked sequences (MAPS): an RNA-seq strategy to profile poly(A+) RNA. *Methods Mol Biol* **1125**: 169–178. doi:10.1007/978-1-62703-971-0_15

Zid BM, O'Shea EK. 2014. Promoter sequences direct cytoplasmic localization and translation of mRNAs during starvation in yeast. *Nature* **514**: 117–121. doi:10.1038/nature13578

Linking RNA Processing and Function

RUN-WEN YAO,[1] CHU-XIAO LIU,[1] AND LING-LING CHEN[1,2]

[1]*State Key Laboratory of Molecular Biology, Shanghai Key Laboratory of Molecular Andrology, CAS Center for Excellence in Molecular Cell Science, Shanghai Institute of Biochemistry and Cell Biology, University of Chinese Academy of Sciences, Chinese Academy of Sciences, Shanghai 200031, China*
[2]*School of Life Science and Technology, ShanghaiTech University, Shanghai 201210, China*
Correspondence: linglingchen@sibcb.ac.cn

RNA processing is critical for eukaryotic mRNA maturation and function. It appears there is no exception for other types of RNAs. Long noncoding RNAs (lncRNAs) represent a subclass of noncoding RNAs, have sizes of >200 nucleotides (nt), and participate in various aspects of gene regulation. Although many lncRNAs are capped, polyadenylated, and spliced just like mRNAs, others are derived from primary transcripts of RNA polymerase II and stabilized by forming circular structures or by ending with small nucleolar RNA–protein complexes. Here we summarize the recent progress in linking the processing and function of these unconventionally processed lncRNAs; we also discuss how directional RNA movement is achieved using the radial flux movement of nascent precursor ribosomal RNA (pre-rRNA) in the human nucleolus as an example.

Surprisingly, the human genome is pervasively transcribed (>80%), and >98% of this transcriptional output represents non-protein-coding RNAs (ncRNAs) (Derrien et al. 2012; Mudge et al. 2013). Long ncRNAs (lncRNAs), which are longer than 200 nucleotides (nt) and lack protein-coding potential, have emerged as a major class of eukaryotic regulatory transcripts involved in multiple layers of gene expression. Statistics from Human GENCODE Release version 28 suggest that the human genome contains more than 16,000 lncRNA genes, but other estimates for the number of lncRNA genes exceed 100,000 in humans (Zhao et al. 2016).

Over the past decades, the study of lncRNA biogenesis and regulation has greatly improved our understanding of the overall regulatory RNA diversity and function, ranging from the classical mRNA-like lncRNAs (e.g., *Xist* [Engreitz et al. 2013; Simon et al. 2013; Chu et al. 2015; McHugh et al. 2015; Minajigi et al. 2015; Chen et al. 2016; Creamer and Lawrence 2017], *CCAT1-L* [Xiang et al. 2014], *NORAD* [Lee et al. 2016; Tichon et al. 2016, 2018; Munschauer et al. 2018]) to other types of unconventionally formatted linear (e.g., *NEAT1* [Clemson et al. 2009; Sunwoo et al. 2009; Souquere et al. 2010; Adriaens et al. 2016; Wang et al. 2018], *MALAT1* [Wilusz et al. 2008; Tripathi et al. 2010; Nakagawa et al. 2012; Zhang et al. 2012; Arun et al. 2016; Malakar et al. 2017], *sno-lncRNAs* [Yin et al. 2012; Xing et al. 2017], *SPAs* [Wu et al. 2016; Lykke-Andersen et al. 2018]) and circular RNAs (Salzman et al. 2012; Hansen et al. 2013; Jeck et al. 2013; Memczak et al. 2013; Zhang et al. 2013). Compared to microRNAs, which are ∼22-nt RNAs that mainly direct post-transcriptional repression of mRNA targets in eukaryotes (Bartel 2018), lncRNAs exhibit a surprisingly wide range of sizes, shapes, and functions (Wu et al. 2017; Kopp and Mendell 2018; Uszczynska-Ratajczak et al. 2018; Carlevaro-Fita and Johnson 2019; Yao et al. 2019a). It is now well-known that lncRNAs participate in the regulation of genetic flow of protein expression from chromatin organization to transcription regulation in the nucleus to modulation of mRNA stability, translation, and post-translation in the cytoplasm. These diverse functional potentials depend on the processing, subcellular localization, and formation of structural modules of individual lncRNAs to partner with associated proteins, which may undergo rapid changes depending on local or cellular environments.

Most annotated lncRNAs transcribed from genomic intergenic regions (lincRNAs) by RNA polymerase II (Pol II) are capped, polyadenylated, and spliced just like mRNAs (Cabili et al. 2011; Derrien et al. 2012; Djebali et al. 2012; Quinn and Chang 2016). Although exhibiting tissue- or cell type–specific expression (Cabili et al. 2011; Goff and Rinn 2015; Ulitsky 2016), they differ from mRNAs by being less evolutionarily conserved and less abundant and containing fewer exons. lincRNAs are generally more nuclear localized than their mRNA counterparts (Derrien et al. 2012; Cabili et al. 2015), in part because some lincRNAs are transcribed by deregulated Pol II, weakly spliced (Mele et al. 2017), inefficiently polyadenylated, and degraded by the nuclear exosome on chromatin (Schlackow et al. 2017). Thus, in principle, functional lincRNAs must escape this targeted nuclear surveillance process to accumulate to high levels in specific cell types.

In addition to lincRNAs, recent studies have uncovered other types of lncRNAs that are processed from primary Pol II transcripts and are stabilized by distinct mechanisms. In this review, we briefly summarize our current understanding of the biogenesis of these unconventionally processed lncRNAs and emphasize the importance of

---

© 2019 Yao et al. This article is distributed under the terms of the Creative Commons Attribution-NonCommercial License, which permits reuse and redistribution, except for commercial purposes, provided that the original author and source are credited.

Published by Cold Spring Harbor Laboratory Press; doi:10.1101/sqb.2019.84.039495

linking their processing to functions in innate immunity and in the regulation of structure and function of human nuclear subdomains. To be functional, these different types of lncRNAs need to be translocated to different subcellular compartmentations. Although mechanisms of RNA trafficking remain incompletely understood, we discuss some insights into this question using the radial flux nature of nascent precursor ribosomal RNA (pre-rRNA) trafficking in the nucleolus as a model.

MALAT1, *NEAT1*, AND NONPOLYADENYLATED TRANSCRIPTOMES

Earlier studies from the Spector laboratory reported alternative 3′-end processing of *MALAT1* (Wilusz et al. 2008) and *NEAT1_long* (the long isoform of *NEAT1* [Sunwoo et al. 2009]), which are two abundant nuclear-enriched lncRNAs that are localized to nuclear speckles and paraspeckles, respectively (Hutchinson et al. 2007; Clemson et al. 2009; Sasaki et al. 2009; Sunwoo et al. 2009). These RNAs are processed at their 3′ ends by recognition and cleavage of tRNA-like structures by RNase P (which is best known to process the 5′ ends of tRNAs) (Wilusz et al. 2008; Sunwoo et al. 2009). RNase P cleavage leads to the formation of mature 3′ ends, which are subsequently protected by a conserved stable U·A·U triple-helical RNA structure (· denotes the Hoogsteen face and - denotes the Watson–Crick face) (Wilusz et al. 2012; Brown et al. 2012). A similar triple-helical structure, but not formed by RNase P processing and called a nuclear retention element (NRE), has also been found at the 3′ end of the *PAN* lncRNA, which is expressed from Kaposi's sarcoma–associated herpesvirus (KSHV), and in RNAs from other viruses (Mitton-Fry et al. 2010; Tycowski et al. 2012). In addition to RNase P–mediated 3′ processing of lncRNAs, a group of lncRNA transcripts containing miRNAs (lnc-pri-miRNAs) use Microprocessor cleavage to terminate transcription, resulting in unstable lncRNAs without 3′-end poly(A) tails (Dhir et al. 2015).

Inspired by alternative 3′-end processing of *MALAT1* and *NEAT1*, we began to explore the nonpolyadenylated (poly(A)) transcriptomes in human cells (Yang et al. 2011). In this method, total RNAs collected from human cells were incubated with oligo(dT) beads to select poly(A) RNAs. The unbound, flow-through RNAs from the oligo(dT) beads were collected and subjected to multiple rounds of rRNA depletion. Both poly(A) RNAs and poly(A)-/ribo-RNAs were then subjected to RNA-seq library preparation and sequencing (Yang et al. 2011). In addition to the replication-dependent histone mRNAs, this non-poly(A) RNA-seq unexpectedly identified hundreds of abundant RNA signals that did not align fully to annotated genes but derived from either introns and exons, which were termed as "excised introns" and "excised exons" (Yang et al. 2011). Later, these excised "exons" were identified as circular RNAs produced from backsplicing of exons of pre-mRNAs (Salzman et al. 2012; Zhang et al. 2014a); "excised introns" were characterized as circular intronic RNAs (ciRNAs) (Zhang et al. 2013) and snoRNA-ended lncRNAs (*sno-lncRNAs*) (Yin et al. 2012). Further, *SPA* (5′ snoRNA capped and 3′ polyadenylated) lncRNAs were uncovered using fibrillarin (a key protein component of Box C/D snoRNP complex)—RNA immunoprecipitation (RIP) and RNA-seq (Wu et al. 2016).

IDENTIFICATION OF DIFFERENT TYPES OF CIRCULAR RNAs FROM NON-POLY(A) TRANSCRIPTOMES

Covalently closed circular RNAs were first observed by electron microscopy in plants and eukaryotic cells in the 1970s (Sanger et al. 1976; Hsu and Cocaprados 1979). This was followed by observations of RNA circles with scrambled exons, which were thought to be "by-products" of aberrant splicing with little functional potential (Nigro et al. 1991; Cocquerelle et al. 1992; Capel et al. 1993).

Genome-wide studies of rRNA-depleted and non-poly (A) transcriptomes, as well as rRNA-depleted and RNase R–enriched transcriptomes, together with specific bioinformatics algorithms for circular RNA detection (Kristensen et al. 2019), have revealed widespread expression of circular RNAs in metazoans (Yang et al. 2011; Gardner et al. 2012; Salzman et al. 2012, 2013; Jeck et al. 2013; Memczak et al. 2013; Zhang et al. 2013, 2014a; Guo et al. 2014; Westholm et al. 2014; Ivanov et al. 2015) and in plants (Wang et al. 2014; Lu et al. 2015). These include two types of circular RNAs derived from the primary Pol II transcripts.

The first type of ciRNAs derive from lariat introns in mammalian cells, and their formation depends on a consensus sequence containing a 7-nt GU-rich motif near the 5′ splice site and an 11-nt C-rich motif at the branchpoint site that inhibits debranching. The resulting RNA circles are covalently ligated through a 2′, 5′-phosphodiester bond at the joining site and lack the linear part stretching from the 3′ end of the intron to the branchpoint (Fig. 1A). Hundreds of ciRNAs were identified in human cells including the human embryonic carcinoma cell line PA1 and the human embryonic stem cell (hESC) H9 line. Some abundantly expressed ciRNAs including *ci-ankrd52* and *ci-sirt7* were found to primarily accumulate in the nucleus and interact with the elongating Pol II complex. Depleting these ciRNAs led to decreased transcription of the parental *ANKRD52* or *SIRT7* genes. These results suggest a *cis*-regulatory role of intronic noncoding sequences on their parent coding genes (Fig. 1A; Zhang et al. 2013). Stable intronic RNAs derived from lariats were also found in oocytes of *Xenopus tropicalis* (Gardner et al. 2012; Talhouarne and Gall 2014) and *Drosophila melanogaster* (Jia Ng et al. 2018).

Of note, different from higher eukaryotic cells, it has recently seen that excised introns in yeast can be stabilized as linear forms that can regulate cell growth (Morgan et al. 2019; Parenteau et al. 2019). Thirty-four excised linear introns were found in *Saccharomyces cerevisiae* that remain associated with components of the spliceosome. These differ from classical spliceosomal introns by con-

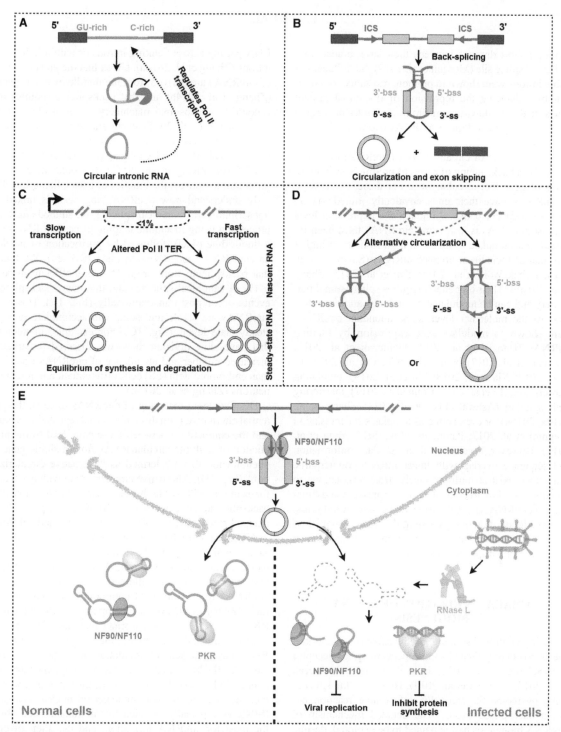

Figure 1. Production, structure, and degradation of circular RNAs. (*A*) A model for circular intronic RNA (ciRNA) processing. ciRNAs are derived from lariat introns and are covalently ligated via a 2′,5′-phosphodiester bond. Their formation depends on a consensus sequence near the 5′ splice site and containing a 7-nt GU-rich motif near the 5′ splice site and an 11-nt C-rich motif at the branchpoint site. (*B*) A model for circular RNA (circRNA) processing. circRNAs are produced by spliceosome-catalyzed backsplicing of exon(s). Generally, flanking intronic complementary sequences (ICSs; most are *Alu* elements) promote exon circularization, whose efficiency can be regulated by competition between ICSs across introns. (*C*) Nascent circRNA production correlates with Pol II elongation rate. In cells, the efficiency of backsplicing from pre-mRNA is relatively low, but because of the stability, circRNAs can accumulate to achieve high abundance at the steady state levels. (*D*) A model of alternative circularization. Alternative formation of inverted repeated ICS pairs and the competition between the pairs can lead to multiple circRNAs produced from a single gene locus. (The red dashed arrows indicate potential competition between different ICS pairs.) (*E*) Linking circRNA processing and structure to innate immunity regulation. Under normal conditions (*left*), NF90/NF110 promotes circRNA production by stabilizing ICSs flanking intronic RNA pairs to juxtapose the circRNA-forming exon(s). Many examined circRNAs form intramolecular short imperfect double-stranded RNAs (dsRNAs) that act as inhibitors of innate immune dsRNA receptor PKR and NF90/NF110. Upon viral infection (*right*), circRNAs are globally reduced at the production level because of export of NF90/NF110, as well as at the steady state level because of the rapid turnover upon RNase L activation. This global reduction of circRNAs may free PKR and NF90/NF110 to be engaged in antiviral immune responses. Misregulation of this process is found in patients with autoimmune disease including systemic lupus erythematosus (SLE).

taining a short distance between their lariat branchpoint and the 3′ splice site (Morgan et al. 2019). Such linear but stable introns were shown to promote resistance to starvation by enhancing the repression of ribosomal protein genes that are downstream from the nutrient-sensing TORC1 and protein kinase A (PKA) pathways (Morgan et al. 2019).

A second type of circular RNAs (circRNAs) are produced from backsplicing of exons of pre-mRNAs in thousands of gene loci in eukaryotes. For these, exons from pre-mRNAs have their ends covalently joined via 3′,5′-phosphodiester bonds. Although the expression level of most circRNAs is low, some circRNAs have been reported to accumulate to levels as high as or even higher than that of their linear cognate mRNAs (Salzman et al. 2013; Rybak-Wolf et al. 2015; You et al. 2015; Zhang et al. 2016b). Their biogenesis requires spliceosomal machinery and can be modulated by both cis- and trans-factors. At the molecular level, some abundant circRNAs were shown to modulate gene expression by titrating miRNAs (Hansen et al. 2013; Memczak et al. 2013; Piwecka et al. 2017) or proteins (Chen et al. 2017; Li et al. 2017; Xia et al. 2018; Liu et al. 2019), regulating transcription (Li et al. 2015; Conn et al. 2017), interfering with splicing (Ashwal-Fluss et al. 2014; Zhang et al. 2014a, 2016a), or even acting as templates for translation (Legnini et al. 2017; Pamudurti et al. 2017; Yang et al. 2017). However, because of their circular conformation and sequence overlap with linear mRNA counterparts, challenges exist at multiple levels, from annotations to functional studies, to understand the expression and functions of circRNAs (Li et al. 2018). In the past several years, we have focused on understanding the regulation of their life cycles, which led us to uncover important functions in innate immune responses (see below).

CHARACTERIZATION OF circRNA BIOGENESIS

During exon backsplicing, a downstream splice-donor site is covalently ligated to an upstream splice-acceptor site (Salzman et al. 2012; Jeck et al. 2013; Memczak et al. 2013; Salzman et al. 2013). How does the spliceosome overcome the sterically unfavorable reaction between backspliced exons? The detailed mechanism of circRNA biogenesis had remained to be explored, despite a noted association with *Alu* elements or other complementary sequences in flanking introns of circle-forming exons (Capel et al. 1993; Jeck et al. 2013). By developing a computational algorithm (CIRCexplorer) (Zhang et al. 2014a), we identified thousands of circRNAs in non-poly(A) data sets derived from H9 and HeLa cells [described as "excised exons" (Yang et al. 2011)]. Using circular RNA recapitulation assays that contain flanking intronic complementary sequences in circle-forming vectors, we have shown that exon circularization is in general promoted by flanking intronic complementary sequences (ICSs), most being *Alu* elements in human (Fig. 1B). The efficiency of exon circularization is regulated by competition between RNA pairing across flanking introns or within individual introns (Zhang et al. 2014a). Given that the great majority of circRNAs are derived from the middle exons of genes (Zhang et al. 2014a) and that backsplicing requires the canonical spliceosomal machinery (Ashwal-Fluss et al. 2014; Starke et al. 2015; Wang and Wang 2015; Liang et al. 2017), it has been proposed that backsplicing events compete with linear RNA splicing (Fig. 1B; Ashwal-Fluss et al. 2014; Zhang et al. 2014a). It would therefore be important to address the kinetics of backsplicing in cells.

To understand how circRNA biogenesis is linked to transcription and canonical splicing, we studied circRNA processing using metabolic tagging of nascent RNAs via 4-thiouridine (4sU), followed by purification of labeled nascent RNAs and RNA-seq (4sUDRB-seq). We found that backsplicing from pre-mRNA is inefficient in cells (<1% of canonical splicing) and that many backsplicing events occur post-transcriptionally (Fig. 1C). However, circRNAs are stable and some can accumulate to high steady state levels (Fig. 1C). For example, in neurons that have slow division rates, we observed remarkably increased number and abundance of circRNAs upon human embryonic stem (ES) cells differentiation to forebrain neurons (Zhang et al. 2016b).

Despite slow processing of circRNAs in cells, the potential alternative formation of inverted repeated ICS pairs and the competition between these pairs lead to the production of multiple circular RNAs from a single gene, a phenomenon that we termed as "alternative circularization" (Fig. 1D). Alternative circularization widely occurs; for example, >50% of highly expressed circRNA (mapped backsplice junction reads ≥ 0.1 RPM [reads per million mapped reads]) gene loci in multiple examined cell lines can produce more than one circRNA (Zhang et al. 2014a; Zhang et al. 2016a). We have annotated patterns of alternative circularization, including different types of alternative backsplicing and alternative splicing events, in circRNAs from a range of commonly used human cell lines (Zhang et al. 2016a). Compared to linear cognate RNAs, circRNAs exhibit distinct patterns of alternative backsplicing and alternative splicing. Quantification of RNA pairing capacity of orientation-opposite ICSs across circRNA-flanking introns using a complementary sequence index revealed that among all types of complementary sequences, short interspersed nuclear elements (SINEs)—especially *Alu* elements in human—contribute the most for circRNA formation, and that their diverse distribution across species results in the increased complexity of circRNA expression during species evolution (Dong et al. 2017). These findings together reveal the complexity of post-transcriptional regulation in mammalian transcriptomes.

PRODUCTION, STRUCTURE, AND DEGRADATION OF circRNAs REGULATE INNATE IMMUNE RESPONSES

Because of overlapping sequences, it has been challenging to study individual circRNA function due to inadequate methods in distinguishing exons in circRNAs from those in linear cognate mRNAs (Li et al. 2018). We speculated that understanding the life cycle difference between circular and linear RNAs might provide some insights into their functions.

Although having the same *cis*-elements, expression levels of circRNAs from the same loci exhibit cell type– and tissue-specific patterns, indicating the participation of protein factors in circRNA biogenesis. To identify such factors, we applied a genome-wide siRNA screening that targets all human unique genes with an in-house-developed efficient dual-color circular/linear RNA expression reporter. In this vector, backsplicing produces a translatable circular *mCherry* and canonical splicing produces a translatable linear *egfp*. This screening identified 103 proteins that have an impact on mCherry expression but not EGFP (Li et al. 2017). In addition to protein candidates that have known roles in splicing regulation, we identified multiple factors related to host immune responses involved in circRNA production (Li et al. 2017). One such factor is the human interleukin enhancer binding factor 3 (ILF3), whose gene produces multiple mRNA isoforms by alternative splicing to express factors including nuclear factor 90 (NF90) and nuclear factor 110 (NF110). iCLIP and mutagenesis experiments confirmed that NF90/NF110 directly bind to inverted-repeated *Alu* elements juxtaposing circRNA-forming exon(s) to promote circRNA production by stabilizing intronic RNA pairs in the nucleus (Fig. 1E). Interestingly, mature circRNAs as a group were found to associate with NF90/NF110 in the cytoplasm (Li et al. 2017). It is known that upon viral infection, NF90/NF110 are rapidly exported to the cytoplasm, where they participate in innate immunity (Harashima et al. 2010). After infection, we observed a global reduction of nascent circRNA production, which was consistent with a deassociation of NF90/NF110 from circRNA–protein complexes for antiviral activity. Consistently, overexpression of endogenous circRNAs facilitated viral infection of human cells (Li et al. 2017). These findings indicated that circRNA biogenesis is unfavorable for innate immune responses, thus linking endogenous circRNA production to innate immunity regulation (Fig. 1E; Li et al. 2017).

Remarkably, not only is nascent circRNA production limited globally, but the steady state level of circRNAs is also globally and rapidly (with a turnover of ∼1 h) reduced upon poly(I:C) treatment to mimic pathogenic dsRNAs or the encephalomyocarditis virus (EMCV) infection (Liu et al. 2019). This global reduction of steady state circRNAs is catalyzed by RNase L, an endonuclease that becomes activated upon the activation of innate immune response to cleave viral and host mRNAs as one way to limit viral spread (Han et al. 2014; Huang et al. 2014). Structural analysis by an optimized in cell SHAPE-MaP assay (Smola et al. 2015) revealed that endogenous circRNAs tend to form 16- to 26-bp imperfect RNA duplexes and act as inhibitors of a group of nucleic acids receptors with antiviral activities including the IFN-inducible isoform of adenosine deaminase acting on RNA 1 (ADAR1), ADAR1p150, NF90/NF110, and double-stranded RNA (dsRNA)-activated protein kinase (PKR) (Liu et al. 2019). Compared to other examined proteins, circRNAs exhibit the highest binding preference for PKR and regulate PKR activation. PKR undergoes autophosphorylation and activation by long dsRNAs (>33-bp), but this activation is blocked by short dsRNAs (16- to 33-bp) (Zheng and Bevilacqua 2004). circRNAs, but not their linear cognate mRNAs, are inhibitors of the dsRNA-induced activation of PKR in a sequence-independent, but dsRNA structure–dependent manner. Depleting RNase L in cells resulted in delayed PKR activation, whereas introducing dsRNA-containing circRNAs into cells rendered them susceptible to EMCV infection (Liu et al. 2019). This regulation is physiologically important, and misregulation has been observed to be related to the autoimmune disease systemic lupus erythematosus (SLE). For example, augmented PKR phosphorylation and circRNA reduction were found in peripheral blood mononuclear cells (PBMCs) derived from patients of SLE. Importantly, introducing the dsRNA-containing circRNA, but not their linear cognate mRNAs, into PBMCs or T cells derived from SLE patients attenuated the aberrant PKR activation cascade.

Collectively, by studying processing, structure and degradation of circRNAs, we have discovered that endogenous circRNAs as a group can dampen innate immune responses (Fig. 1E; Li et al. 2017; Liu et al. 2019). These findings are consistent with a recent report that patients with a genetic defect leading to the accumulation of intron lariat–derived RNA circles are more susceptible to viral infections (Zhang et al. 2018). Different from endogenous circRNAs, there are conflicting reports on whether exogenously produced circular RNAs themselves trigger immune responses (Chen et al. 2017; Chen et al. 2019; Wesselhoeft et al. 2019). Future studies are warranted to clarify the modes of action of in vitro–synthesized versus endogenous circular RNAs in the regulation and application in the innate immunity.

IDENTIFICATION OF snoRNA-ENDED LONG NONCODING RNAs

Although it is generally believed that most introns or intron fragments are unstable (Rodriguez-Trelles et al. 2006), the RNA-seq identified a number of nonannotated non-poly(A) RNA signals that mapped to intronic regions in hES H9 and HeLa cells (Yang et al. 2011). Intriguingly, careful analysis of these nonpolyadenylated reads revealed that some excised introns are stabilized by small nucleolar ribonucleoprotein complexes (snoRNPs) at each end, which were named as *sno-lncRNA*s (snoRNA ended long noncoding RNAs) (Fig. 2A). snoRNAs are a family of conserved nuclear RNAs (70–200 nt) that function in the modification of small nuclear RNAs (snRNAs) or processing of ribosomal RNAs (Kiss 2001). Binding of core proteins cotranscriptionally is essential to protect the termini of mature snoRNAs from exonucleolytic degradation (Samarsky et al. 1998; Kufel et al. 2000). In many higher eukaryotes, the great majority of snoRNAs are

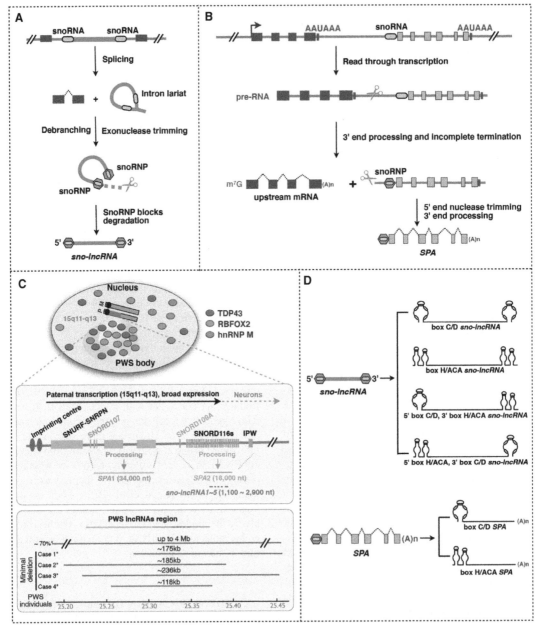

Figure 2. The biogenesis and diversity of snoRNA-ended lncRNAs. (*A*) A model for snoRNA-ended long noncoding RNA (*sno-lncRNA*) processing. Typically, when one intron contains two snoRNA genes, the cotranscriptional formation of snoRNPs at the ends can protect the intronic sequences from exonuclease trimming after debranching during splicing, leading to the formation of sno-lncRNAs. (*B*) A model for 5′-end snoRNP-capped, 3′-polyadenylated (*SPA*) lncRNA processing. Like that of *sno-lncRNA*, *SPA* processing requires an intact snoRNA at the 5′ end. A weak poly(A) signal located downstream from the coding region of mRNA is also necessary for generating readthrough transcripts of pre-*SPAs*. (*C*) snoRNA-ended lncRNAs are derived from the critical region deleted in PWS patients. (*Top*) *sno-lncRNAs* and *SPAs* derived from the PWS deletion region (human 15q11-q13) interact with multiple RBPs including TDP43, RBFOX2, and hnRNP M in the nucleus of human ES cells. (*Bottom*) Minimal chromosome deletions reported in four PWS individuals (Cases 1–4) (Sahoo et al. 2008; de Smith et al. 2009; Duker et al. 2010; Bieth et al. 2015). (*D*) Subtypes of snoRNA-ended lncRNAs. *sno-lncRNAs*, and *SPAs* can be ended with either box C/D or box H/ACA snoRNAs at their ends.

produced from introns, and usually one intron only contains one snoRNA (Weinstein and Steitz 1999; Dieci et al. 2009). When one intron contains two snoRNA genes, the sequences between the snoRNAs are not degraded after splicing, leading to the accumulation of lncRNAs flanked by snoRNA sequences but lacking 5′ caps and 3′ poly(A) tails (Fig. 2A; Yin et al. 2012). Dozens of *sno-lncRNAs* have been found in mammalian genomes, whose expression is species-specific, because of species-specific alternative splicing that results in single snoRNA or two snoRNAs in one intron (Zhang et al. 2014b; Xing and Chen 2018).

Further application of fibrillarin–RIP and RNA-seq identified 5′-end snoRNP-capped, 3′ polyadenylated

(*SPA*) lncRNAs. *SPA* processing is associated with kinetic competition of the 5′ to 3′ exonuclease XRN2 and Pol II elongation speed downstream from polyadenylation signals in the nucleus. Following cleavage/polyadenylation of an upstream gene, the downstream uncapped *pre-SPA* RNA is trimmed by XRN2 until this exonuclease reaches the cotranscriptionally assembled snoRNP complex (Fig. 2B). More recent study has revealed that the 5′ to 3′ exonuclease activity of the cytoplasmic nonsense-mediated decay pathway mediated by SMG6-mediated endonucleolytic cleavage can also trigger the production of a box C/D snoRD86-containing *SPA*s in the cytoplasm (*c-SPA*s) (Lykke-Andersen et al. 2018).

INVOLVEMENT OF snoRNA-ENDED lncRNAs IN PRADER–WILLI SYNDROME

Among the snoRNA-ended lncRNAs, we found that five *sno-lncRNA*s (sizes are 1000—3000 nt, named *sno-lncRNA1* to *sno-lncRNA5*) and two *SPA*s (35,000 nt for *SPA1* and 16,000 nt for *SPA2*) are conspicuously missing in Prader–Willi syndrome, a neurodevelopmental genetic disorder with elusive molecular causes (Cassidy et al. 2012; Yin et al. 2012; Chamberlain 2013; Wu et al. 2016; Aman et al. 2018). In normal hESCs, these abundant PWS region snoRNA-ended lncRNAs accumulate in *cis* to form one 1- to ~2-µm³ RNA–protein puncta per nucleus. Importantly, these lncRNAs interact with different RNA binding proteins (RBPs) including TAR DNA-binding protein 43 (TDP43), heterogeneous nuclear ribonucleoprotein M (hnRNPM), and RBP fox-1 homolog 2 (RBFOX2) in normal hESCs and other examined human cell lines. Super resolution microscopy and iCLIP further showed strong preferences between different sno-ended lncRNAs and their interacting RBPs: *SPA1* preferred to interact with TDP43, *sno-lncRNA*s preferred to bind RBFOX2, whereas *SPA2* did not have a preference to these examined RBPs. Generation of a PWS cellular model in hESCs by depleting these lncRNAs using CRISPR–Cas9 revealed the mislocalization of their associated RBPs and an altered pattern of alternative splicing of more than 300 mRNAs without affecting the expression level of individual corresponding mRNAs (Fig. 2C). Importantly, some genes with altered splicing regulation are associated with synaptosome and neurotrophin signaling pathways (Yin et al. 2012; Wu et al. 2016). These studies together indicate that missing of these snoRNA-related lncRNAs is linked to PWS pathogenesis (Fig. 2C).

Interestingly, the lncRNAs *116HG* from the Prader–Willi locus in mouse also form similar cloud-like nuclear accumulations (Powell et al. 2013; Coulson et al. 2018), but currently existing PWS mouse models do not fully recapitulate the PWS patient phenotypes (Ding et al. 2008), highlighting the importance of developing human PWS cellular models in studying the pathogenesis of PWS in the future.

Future work is warranted to examine whether hESCs lacking all these lncRNAs would have any measurable effect on neuronal functions, especially hypothalamic neurons, including not only the morphological phenotypes but also alterations in transcriptomes and alternative splicing patterns. In addition, the contributions of *sno-lncRNA*s and *SPA*s to the molecular and disease phenotype at this locus should also be pursued by identifying additional proteins, DNAs and RNAs that are associated with individual *sno-lncRNA*s and *SPA*s in hESCs and differentiated neurons. Nevertheless, these findings expand the diversity of lncRNAs and provide previously unappreciated insights into PWS pathogenesis.

REGULATION OF RNA POLYMERASE I TRANSCRIPTION BY *SLERT* IN THE HUMAN NUCLEOLUS

There are two main classes of snoRNAs: box C/D snoRNAs and box H/ACA snoRNAs in eukaryotes according to their conserved sequence motifs (Reichow et al. 2007). Thus, four different subtypes of *sno-lncRNA*s that each contain the same or different box C/D or box H/ACA snoRNPs at the ends might be expected to exist (Fig. 2D; Zhang et al. 2014b). In addition to the PWS region *sno-lncRNA*s that are capped by a box C/D snoRNA at each end (Yin et al. 2012), *SLERT* (snoRNA-ended lncRNA enhances pre-ribosomal RNA transcription) is a *sno-lncRNA* that contains box H/ACA snoRNAs at its ends and is highly expressed in multiple human cell lines (Fig. 3A; Xing et al. 2017). Of note, *sno-lncRNA*s containing a box C/D or a box H/ACA snoRNP at each end were also observed in human cells (Zhang et al. 2014b). Similarly, in addition to the 5′ box C/D snoRNA-ended PWS region *SPA*s and *c-SPA*s (Wu et al. 2016; Lykke-Andersen et al. 2018), box H/ACA snoRNA-ended *SPA*s were also detected in several human cell lines (Luo, We, Chen, et al., unpubl. data) (Fig. 2D).

SLERT is generated from the intron of the transforming growth factor beta regulator 4 (*TBRG4*) gene locus and alters RNA polymerase I (Pol I) transcription of ribosomal RNAs (rRNAs) (Xing et al. 2017). Alternative splicing leading to skipping of exons 4 and 5 of the *TBRG4* locus results in two box H/ACA snoRNAs embedded within one intron, which subsequently forms *SLERT* (Fig. 3A). Unlike PWS region *sno-lncRNA*s, *SLERT* does not localize to its own transcription site but instead accumulates in the nucleolus, which is the largest nuclear subdomain in which rRNA biogenesis takes place. Translocation of *SLERT* from its transcription site to the nucleolus depends on its two box H/ACA snoRNAs (Fig. 3B).

Once localized in the nucleolus, *SLERT* directly interacts with the DEAD-box family protein 21 (DDX21) via a 143-nt-long internal region within the two snoRNAs. DDX21 is a DEAD-box RNA helicase that is involved in multiple steps of ribosome biogenesis by contacting both rRNA and snoRNAs and is thought to modulate rRNA transcription, processing, and modification (Holmstrom et al. 2008; Calo et al. 2015; Sloan et al. 2015). Applying super-resolution structured illumination microscopy (SIM) to examine DDX21 localization in live and fixed human cells in detail, we found that DDX21 is

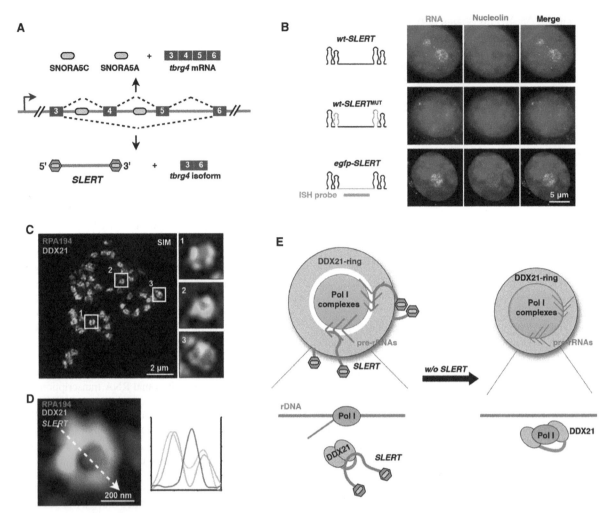

Figure 3. The biogenesis and function of *SLERT*. (*A*) Alternative splicing of the *TBRG4* locus. Alternative splicing of *tbrg4* pre-mRNA generates either two snoRNAs and the *tbrg4* mRNA (*top*) or a snoRNA-ended lncRNA that enhances preribosomal RNA transcription (*SLERT*) and a short isoform of *tbrg4* (rapidly degraded) with skipped exons 4 and 5. (*B*) *SLERT* requires its box H/ACA snoRNA ends to be translocated to the nucleolus. Each indicated *SLERT* or *SLERT* mutants were expressed in HeLa cells, followed by co-staining of WT-*SLERT*, WT-*SLERT*^{MUT}, or *egfp-SLERT* (green) and nucleolar marker nucleolin (red). (*C*) The Pol I complexes are located within DDX21 rings, shown by SIM in PA1 cells. (*D*) *SLERT* interacts with DDX21 rings but not Pol I complexes. A representative image of co-staining DDX21 (green), RPA194 (red), and *SLERT* (blue) by SIM and a plot profile of the image are shown. (*E*) *SLERT* modulates DDX21 in Pol I transcription regulation. In the nucleolus, DDX21 ring-shaped arrangement surrounds multiple Pol I complexes and inhibits Pol I activity. Binding by *SLERT* allosterically alters individual DDX21 molecules, leading to the reduced interaction between DDX21 and Pol I, which subsequently allows the Pol I complexes to occupy the actively transcribed rDNAs. (*A–E*, Adapted and modified from Xing et al. 2017.)

largely enriched in the nucleolus and form dozens of ring-like structures encircling Pol I complexes (Fig. 3C). Further biochemical analyses showed that DDX21 interacts with subunits of Pol I complexes and prevents them loading onto rDNAs in an RNA helicase activity–independent manner and subsequently inhibits pre-rRNA transcription.

Depleting one snoRNA of *SLERT* by CRSIPR–Cas9 abolished *SLERT* expression in examined human cells and led to suppressed Pol I transcription. Under SIM observation, *SLERT* does not directly interact with Pol I, but instead specifically localizes to individual DDX21 rings (Fig. 3D). Mechanistically, *SLERT* binding to DDX21 alters the conformation of individual DDX21 molecules, suppressing the interaction between DDX21 and Pol I, therefore releasing Pol I to be engaged to rDNAs for active pre-RNA transcription (Fig. 3E).

ULTRA-STRUCTURE ORGANIZATION OF THE HUMAN NUCLEOLUS

The primary function of the nucleolus serves as the site of rRNA biogenesis. It has been well-established that the mammalian nucleolus is highly organized and is comprised of three morphologically distinct subregions and named fibrillar centers (FCs), dense fibrillar components (DFCs), and the granular component (GC), shown by the electron density under electron microscopy (EM) (Fig. 4A; Koberna et al. 2002; Boisvert et al. 2007). Recent single-molecule images have further revealed the tripartite

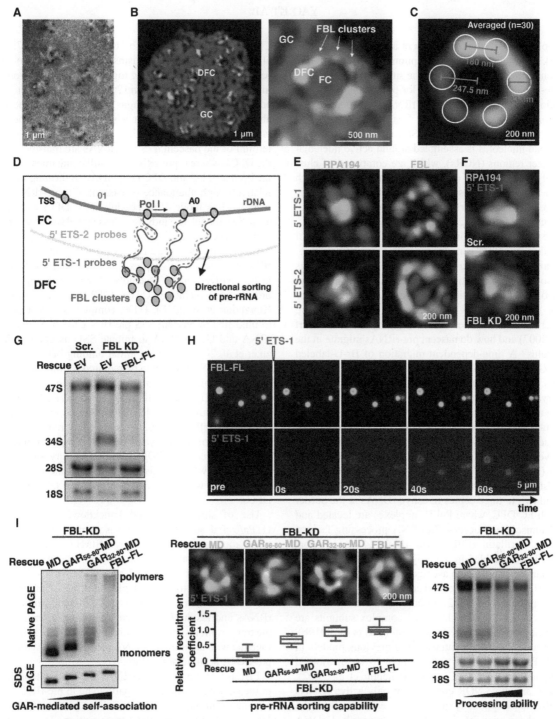

Figure 4. Ultra-structure organization of the human nucleolus and nascent pre-rRNA translocation is required for proper pre-rRNA processing and function. (*A*) Electron microscopy shows the localization of nucleolus in thin-sectioned permeabilized HeLa cells. (*B*) Representative SIM images of nucleoli and three nucleolar subregions in live HeLa cells. (*C*) Cross-correlation of aligned and averaged images shows that the max-cross sections of individual DFC regions contain six FBL clusters. The minimum distance between two clusters is ∼180 nm; the diameter of FBL cluster is ∼133 nm; the distance between the center of individual clusters and the center of DFC is ∼247.5 nm; the detached distance between two adjacent PF clusters is ∼180 nm. (*D*) A schematic of RNA smFISH probes to detect transcribing pre-rRNA. The 5′ ETS-1 probe detects nt 1–414 of pre-rRNAs; the 5′ ETS-2 probe detects nt 498–977 of pre-rRNA. (*E*) The 5′ ETS-1 probe detects 47S pre-rRNAs that are largely distributed outside of the FC, whereas the 5′ ETS-2 probe detects pre-rRNAs that are mainly located at the FC/DFC border, shown by SIM. (*F*) FBL knockdown (KD) results in impaired localization of nascent pre-rRNAs shown by the highest colocalization signal between RPA194 and 5′ ETS-1. (*G*) FBL KD results in aberrant accumulation of 47S and 34S pre-rRNAs, accompanied by reduced 28S and 18S rRNAs. (*H*) Cy3-labeled 5′ ETS-1 (magenta) is sorted to the mNeonGreen-FBL-FL droplets (green) in vitro. (*I*) The IDR length in the GAR domain of FBL promotes FBL self-association, which confers the capability of pre-rRNA sorting and processing in which FBL is involved. (*Left*) Increased GAR domain length in FBL mutants leads to augmented FBL self-association shown by increased multimerization. (*Middle*) Increased IDR length in the GAR domain of FBL mutants promotes translocation of pre-rRNAs into the DFC region. Twenty cells were analyzed under each condition by boxplot. (*Right*) Increased IDR length in the GAR domain of FBL positively correlates with proper 47S pre-rRNA processing, shown by northern blots. (*A*, Adapted from Koberna et al. 2002; *B–I*, adapted and modified from Yao et al. 2019b.)

nucleolar organization (Szczurek et al. 2016; Khan et al. 2018). Continuous Pol I transcription at the FC produces nascent pre-rRNAs and causes the subsequent radial flux of pre-rRNAs through the DFC for processing and modification to produce 28S and 18S rRNAs, and then into the GC for ribosome assembly and finally into the nucleoplasm (Boisvert et al. 2007; McStay and Grummt 2008). A human nucleolus is assembled around active nucleolar organizer regions (NORs), which are composed of clusters of tandem repeats of ribosomal DNA (rDNA), each with a long intergenic spacer (IGS) of ∼30 kb and a preribosomal RNA (pre-rRNA) coding region of ∼14 kb (McStay and Grummt 2008).

The SIM observation that DDX21 forms dozens of ring-like structures surrounding Pol I complexes (Fig. 3C) raised an intriguing question—how are DDX21-rings positioned in the tripartite organization of the mammalian nucleolus? Other questions remaining to be addressed include where does Pol I transcription occur (Mais and Scheer 2001; Cheutin et al. 2002; Huang 2002; Boisvert et al. 2007) and how do nascent pre-rRNAs migrate in the nucleolus? A time-dependent migration of Br-U-labeled pre-rRNAs from FCs to DFCs was observed in HeLa cells (Thiry et al. 2000), but the underlying mechanism of this observation had remained unclear.

We applied CRISPR–Cas9-mediated knock-in of fluorescently tagged proteins to visualize FC, DFC, and GC subnuclear organization under SIM that allowed us to uncover previously uncharacterized nucleolar ultra-structures (Fig. 4B; Yao et al. 2019b). A human nucleolus consists of dozens of FC/DFC units that are assembled around two to three copies of active rDNAs at the border of each FC/DFC, where Pol I complexes are located and Pol I transcription occurs. Pre-rRNA processing factors, such as fibrillarin (FBL) form 18 to 24 clusters that are further assembled into a polyhedron-like shell of the DFC surrounding FC. On average, each spherical PF cluster is ∼133 nm in diameter and each DFC region contains ∼628 nm outer and ∼362 nm inner diameters (Fig. 4C). Consistent with this model, Pol I complex subunits are enriched at the FC border as clusters (Yao et al. 2019b), which are in striking contrast to the previous models in which Pol I complexes were thought to be distributed throughout the entire FC region (Cheutin et al. 2002). Further, an active NOR contains not only active rDNAs but also transcriptionally inert rDNAs, consistent with the earlier observation of discontinuous transcribed rDNA clusters in rDNA spreads (McKnight and Miller 1976; McStay and Grummt 2008). Collectively, these observations represent a substantial advance in our understanding of nucleolar spatial organization.

NASCENT pre-rRNA TRANSLOCATION IS REQUIRED FOR PROPER pre-rRNA PROCESSING AND FUNCTION

An increasing number of RNAs have emerged as important modulators of nuclear structure and function by acting in *cis* or in *trans* (Bergmann and Spector 2014; Chen 2016; Engreitz et al. 2016; Tomita et al. 2017; Yao et al. 2019a). What mechanisms does the cell use to keep nascent RNAs from sticking together while also promoting directional sorting in *trans*? It has been a challenge to address this question in single cells because of the low abundance of most nascent RNAs that usually undergo rapid processing. Remarkably, we observed that the 5′ termini of the nascent 47S pre-rRNAs are translocated to the DFCs, whereas pre-rRNAs are still being transcribed at the border of FC and DFC (Fig. 4D,E), indicating that the relatively high abundance of nascent 47S pre-rRNA and its radial flux mode of processing in FC/DFCs can be an attractive model to study how nascent RNA directional sorting is achieved.

Earlier studies in yeast revealed that the eukaryotic pre-rRNA processing is complex. The early step was thought to be involved in the assembly of small subunit (SSU) processomes including UTPa, UTPb, and U3 snoRNP (Barandun et al. 2018). UTPa complexes were reported to bind first to nascent 35S pre-rRNA to chaperone pre-rRNA and U3 snoRNA to initiate SSU assembly (Hunziker et al. 2016). Such an initial binding between UTPa and the 5′ ETS appeared to be required for the subsequent recruitment of the UTPb complex and the U3 snoRNP components (Henras et al. 2015; Sharma and Lafontaine 2015; Barandun et al. 2018). These studies in yeast might not directly reflect processing of nascent pre-rRNA in the human nucleolus since yeast cells have bipartite nucleoli containing merged FC/DFCs and GC (Thiry and Lafontaine 2005), which is distinct from tripartite nucleoli in human cells, where pre-rRNA processing takes place in DFCs.

Use of shRNA-mediated knockdown of factors in UTPa/b and snoRNP complexes combined with structured illumination microscopy (SIM)-based screening have revealed that FBL plays a key role in promoting the movement of the 5′ end of 47S pre-rRNA from its transcription site to DFC. Depletion of FBL dramatically blocks the translocation of 5′ ends of transcribing pre-rRNAs and results in the accumulation of pre-rRNAs at the transcription sites at the border of FC/DFCs (Fig. 4F), together with aberrant pre-rRNA processing shown as 34S pre-rRNA accumulation (Fig. 4G).

FBL is well-known as an rRNA 2′-O-methyltransferase in the snoRNP particles that participate in pre-rRNA processing (Tollervey et al. 1991; Tollervey et al. 1993; Tafforeau et al. 2013) and Pol I transcription (Tessarz et al. 2014; Loza-Muller et al. 2015). Its amino-terminus contains a glycine- and arginine-rich (GAR) domain highly enriched with intrinsically disordered regions (IDRs) that are required for *Xenopus laevis* FBL phase separation (Feric et al. 2016); its carboxyl terminus contains an RNA binding domain (RBD) and the α domain for methyltransferase activity (MD domain).

How does FBL control the movement of the 5′ end of 47S pre-rRNA to the DFC? First, distinct from other components in snoRNP complexes, FBL specifically and strongly interacts with the 5′ end of 47S pre-rRNA as revealed by PAR-CLIP (Kishore et al. 2013) and by in vitro binding and phase-separation assays (Yao et al.

2019b). These observations indicate an additional function of FBL in nascent pre-rRNA sorting beyond its classical role in U3 snoRNP. Second, the GAR domain is necessary and sufficient for human FBL self-aggregation into droplets at physiological concentrations (Yao et al. 2019b). Importantly, both in vitro and in vivo lines of evidence suggest a model whereby the binding and sorting of nascent 47S pre-rRNA utilizes different FBL domains: The MD domain binds to the 5′ end of 47S pre-rRNA, and the RNA-interacting FBL then moves toward the DFC by GAR domain self-association (Fig. 4H–I). Such pre-rRNA sorting strongly correlates with FBL self-association via the strength of IDRs in FBL and is required for proper pre-rRNA processing in cells (Fig. 4I). For example, rescue of FBL KD in cells with FBL mutants containing different length of GARs but with intact MD domain shows the enhanced formation of FBL polymers, enhanced 5′-end pre-rRNA sorting to DFC, and increased proper production of 47S pre-rRNA (Fig. 4I). Importantly, this sorting and translocation process is required for proper ribosome production, thus representing yet another example illustrating the importance of linking RNA processing to function.

More broadly, as many RBPs associated with nascent pre-mRNA processing events are known to contain structurally disordered regions (Banani et al. 2017; Gueroussov et al. 2017; Shin and Brangwynne 2017; Ying et al. 2017), a similar liquid–liquid phase-separation controlled nascent rRNA sorting mechanism is likely used by the cell to keep other types of nascent RNAs from unnecessary or unwanted self-aggregation.

CONCLUSION

It has been well-established that Pol II transcription, nascent pre-mRNA splicing, capping, polyadenylation, mRNA export, and surveillance are seamlessly integrated (Moore and Proudfoot 2009). Studies of unconventionally processed lncRNAs including circular RNAs (Fig. 1), snoRNA-related RNAs (Figs. 2 and 3), and the "housekeeping" pre-rRNAs (Fig. 4) have revealed no exception for ncRNAs: Throughout the maturation process of each district class of RNA, transcription and processing are crucially important for subcellular localization and function.

One may argue that all examples illustrated so far are unconventionally processed ncRNAs. Indeed, despite these examples, it has been well-studied that Pol II transcription and the 3′-end alternative processing of the NEAT1 lncRNA act together to modulate paraspeckle morphology and function (Mao et al. 2011; Naganuma et al. 2012; Hirose et al. 2014; Wang et al. 2018; Yamazaki et al. 2018). Emerging studies have also revealed both cis- and trans-factors are required for the subcellular localization of Pol II–transcribed mRNA-like lincRNAs and functions (Hacisuleyman et al. 2014; Zhang et al. 2014b; Lubelsky and Ulitsky 2018; Shukla et al. 2018).

Recent studies using the most robustly available methods have greatly advanced our understanding of cellular functions of lncRNAs (Kopp and Mendell 2018; Yao et al. 2019a). Beyond linking RNA processing and function, how lncRNAs are structured and what structural conformations lncRNAs adopt for their interacting partners have remained mysteries. Their large sizes and flexible nature have endowed them with previously underappreciated functional potentials; however, this has also presented unexpected experimental challenges to understanding their regulation at multiple levels. Future studies aimed at understanding in greater detail the regulation of expression, subcellular localization patterns, interaction partners, and conformational information of lncRNAs, as well as understanding how the biogenesis and turnover of RNAs are linked to their subcellular localization, will provide greater insight into their cellular functions.

REFERENCES

Adriaens C, Standaert L, Barra J, Latil M, Verfaillie A, Kalev P, Boeckx B, Wijnhoven PW, Radaelli E, Vermi W, et al. 2016. p53 induces formation of *NEAT1* lncRNA-containing paraspeckles that modulate replication stress response and chemosensitivity. *Nat Med* 22: 861–868. doi:10.1038/nm.4135

Aman LCS, Manning KE, Whittington JE, Holland AJ. 2018. Mechanistic insights into the genetics of affective psychosis from Prader–Willi syndrome. *Lancet Psychiatry* 5: 370–378. doi:10.1016/S2215-0366(18)30009-9

Arun G, Diermeier S, Akerman M, Chang KC, Wilkinson JE, Hearn S, Kim Y, MacLeod AR, Krainer AR, Norton L, et al. 2016. Differentiation of mammary tumors and reduction in metastasis upon *Malat1* lncRNA loss. *Genes Dev* 30: 34–51. doi:10.1101/gad.270959.115

Ashwal-Fluss R, Meyer M, Pamudurti NR, Ivanov A, Bartok O, Hanan M, Evantal N, Memczak S, Rajewsky N, Kadener S. 2014. circRNA biogenesis competes with pre-mRNA splicing. *Mol Cell* 56: 55–66. doi:10.1016/j.molcel.2014.08.019

Banani SF, Lee HO, Hyman AA, Rosen MK. 2017. Biomolecular condensates: organizers of cellular biochemistry. *Nat Rev Mol Cell Biol* 18: 285–298. doi:10.1038/nrm.2017.7

Barandun J, Hunziker M, Klinge S. 2018. Assembly and structure of the SSU processome—a nucleolar precursor of the small ribosomal subunit. *Curr Opin Struct Biol* 49: 85–93. doi:10.1016/j.sbi.2018.01.008

Bartel DP. 2018. Metazoan microRNAs. *Cell* 173: 20–51. doi:10.1016/j.cell.2018.03.006

Bergmann JH, Spector DL. 2014. Long non-coding RNAs: modulators of nuclear structure and function. *Curr Opin Cell Biol* 26: 10–18. doi:10.1016/j.ceb.2013.08.005

Bieth E, Eddiry S, Gaston V, Lorenzini F, Buffet A, Conte Auriol F, Molinas C, Cailley D, Rooryck C, Arveiler B, et al. 2015. Highly restricted deletion of the SNORD116 region is implicated in Prader–Willi syndrome. *Eur J Hum Genet* 23: 252–255. doi:10.1038/ejhg.2014.103

Boisvert FM, van Koningsbruggen S, Navascues J, Lamond AI. 2007. The multifunctional nucleolus. *Nat Rev Mol Cell Biol* 8: 574–585. doi:10.1038/nrm2184

Brown JA, Valenstein ML, Yario TA, Tycowski KT, Steitz JA. 2012. Formation of triple-helical structures by the 3′-end sequences of MALAT1 and MENβ noncoding RNAs. *Proc Natl Acad Sci* 109: 19202–19207. doi:10.1073/pnas.1217338109

Cabili MN, Trapnell C, Goff L, Koziol M, Tazon-Vega B, Regev A, Rinn JL. 2011. Integrative annotation of human large intergenic noncoding RNAs reveals global properties and specific subclasses. *Genes Dev* 25: 1915–1927. doi:10.1101/gad.17446611

Cabili MN, Dunagin MC, McClanahan PD, Biaesch A, Padovan-Merhar O, Regev A, Rinn JL, Raj A. 2015. Localization and abundance analysis of human lncRNAs at single-cell and single-molecule resolution. *Genome Biol* **16:** 20. doi:10.1186/s13059-015-0586-4

Calo E, Flynn RA, Martin L, Spitale RC, Chang HY, Wysocka J. 2015. RNA helicase DDX21 coordinates transcription and ribosomal RNA processing. *Nature* **518:** 249–253. doi:10.1038/nature13923

Capel B, Swain A, Nicolis S, Hacker A, Walter M, Koopman P, Goodfellow P, Lovellbadge R. 1993. Circular transcripts of the testis-determining gene *Sry* in adult-mouse testis. *Cell* **73:** 1019–1030. doi:10.1016/0092-8674(93)90279-Y

Carlevaro-Fita J, Johnson R. 2019. Global positioning system: understanding long noncoding RNAs through subcellular localization. *Mol Cell* **73:** 869–883. doi:10.1016/j.molcel.2019.02.008

Cassidy SB, Schwartz S, Miller JL, Driscoll DJ. 2012. Prader–Willi syndrome. *Genet Med* **14:** 10–26. doi:10.1038/gim.0b013e31822bead0

Chamberlain SJ. 2013. RNAs of the human chromosome 15q11-q13 imprinted region. *Wiley Interdiscip Rev RNA* **4:** 155–166. doi:10.1002/wrna.1150

Chen LL. 2016. Linking long noncoding RNA localization and function. *Trends Biochem Sci* **41:** 761–772. doi:10.1016/j.tibs.2016.07.003

Chen CK, Blanco M, Jackson C, Aznauryan E, Ollikainen N, Surka C, Chow A, Cerase A, McDonel P, Guttman M. 2016. Xist recruits the X chromosome to the nuclear lamina to enable chromosome-wide silencing. *Science* **354:** 468–472. doi:10.1126/science.aae0047

Chen YG, Kim MV, Chen XQ, Batista PJ, Aoyama S, Wilusz JE, Iwasaki A, Chang HY. 2017. Sensing self and foreign circular RNAs by intron identity. *Mol Cell* **67:** 228–238. doi:10.1016/j.molcel.2017.05.022

Chen YG, Chen R, Ahmad S, Verma R, Kasturi SP, Amaya L, Broughton JP, Kim J, Cadena C, Pulendran B, et al. 2019. N^6-methyladenosine modification controls circular RNA immunity. *Mol Cell* **76:** 96–109. doi:10.1016/j.molcel.2019.07.016

Cheutin T, O'Donohue MF, Beorchia A, Vandelaer M, Kaplan H, Defever B, Ploton D, Thiry M. 2002. Three-dimensional organization of active rRNA genes within the nucleolus. *J Cell Sci* **115:** 3297–3307.

Chu C, Zhang QC, da Rocha ST, Flynn RA, Bharadwaj M, Calabrese JM, Magnuson T, Heard E, Chang HY. 2015. Systematic discovery of Xist binding proteins. *Cell* **161:** 404–416. doi:10.1016/j.cell.2015.03.025

Clemson CM, Hutchinson JN, Sara SA, Ensminger AW, Fox AH, Chess A, Lawrence JB. 2009. An architectural role for a nuclear noncoding RNA: NEAT1 RNA is essential for the structure of paraspeckles. *Mol Cell* **33:** 717–726. doi:10.1016/j.molcel.2009.01.026

Cocquerelle C, Daubersies P, Majerus MA, Kerckaert JP, Bailleul B. 1992. Splicing with inverted order of exons occurs proximal to large introns. *EMBO J* **11:** 1095–1098. doi:10.1002/j.1460-2075.1992.tb05148.x

Conn VM, Hugouvieux V, Nayak A, Conos SA, Capovilla G, Cildir G, Jourdain A, Tergaonkar V, Schmid M, Zubieta C, et al. 2017. A circRNA from *SEPALLATA3* regulates splicing of its cognate mRNA through R-loop formation. *Nat Plants* **3:** 17053. doi:10.1038/nplants.2017.53

Coulson RL, Powell WT, Yasui DH, Dileep G, Resnick J, LaSalle JM. 2018. Prader–Willi locus *Snord116* RNA processing requires an active endogenous allele and neuron-specific splicing by *Rbfox3*/NeuN. *Hum Mol Genet* **27:** 4051–4060. doi:10.1093/hmg/ddy296

Creamer KM, Lawrence JB. 2017. XIST RNA: a window into the broader role of RNA in nuclear chromosome architecture. *Philos Trans R Soc Lond B Biol Sci* **372:** 20160360. doi:10.1098/rstb.2016.0360.

Derrien T, Johnson R, Bussotti G, Tanzer A, Djebali S, Tilgner H, Guernec G, Martin D, Merkel A, Knowles DG, et al. 2012. The GENCODE v7 catalog of human long noncoding RNAs: analysis of their gene structure, evolution, and expression. *Genome Res* **22:** 1775–1789. doi:10.1101/gr.132159.111

de Smith AJ, Purmann C, Walters RG, Ellis RJ, Holder SE, Van Haelst MM, Brady AF, Fairbrother UL, Dattani M, Keogh JM, et al. 2009. A deletion of the HBII-85 class of small nucleolar RNAs (snoRNAs) is associated with hyperphagia, obesity and hypogonadism. *Hum Mol Genet* **18:** 3257–3265. doi:10.1093/hmg/ddp263

Dhir A, Dhir S, Proudfoot NJ, Jopling CL. 2015. Microprocessor mediates transcriptional termination of long noncoding RNA transcripts hosting microRNAs. *Nat Struct Mol Biol* **22:** 319–327. doi:10.1038/nsmb.2982

Dieci G, Preti M, Montanini B. 2009. Eukaryotic snoRNAs: a paradigm for gene expression flexibility. *Genomics* **94:** 83–88. doi:10.1016/j.ygeno.2009.05.002

Ding F, Li HH, Zhang S, Solomon NM, Camper SA, Cohen P, Francke U. 2008. SnoRNA *Snord116* (*Pwcr1/MBII-85*) deletion causes growth deficiency and hyperphagia in mice. *PLoS One* **3:** e1709. doi:10.1371/journal.pone.0001709

Djebali S, Davis CA, Merkel A, Dobin A, Lassmann T, Mortazavi A, Tanzer A, Lagarde J, Lin W, Schlesinger F, et al. 2012. Landscape of transcription in human cells. *Nature* **489:** 101–108. doi:10.1038/nature11233

Dong R, Ma XK, Chen LL, Yang L. 2017. Increased complexity of circRNA expression during species evolution. *RNA Biol* **14:** 1064–1074. doi:10.1080/15476286.2016.1269999

Duker AL, Ballif BC, Bawle EV, Person RE, Mahadevan S, Alliman S, Thompson R, Traylor R, Bejjani BA, Shaffer LG, et al. 2010. Paternally inherited microdeletion at 15q11.2 confirms a significant role for the SNORD116 C/D box snoRNA cluster in Prader–Willi syndrome. *Eur J Hum Genet* **18:** 1196–1201. doi:10.1038/ejhg.2010.102

Engreitz JM, Pandya-Jones A, McDonel P, Shishkin A, Sirokman K, Surka C, Kadri S, Xing J, Goren A, Lander ES, et al. 2013. The Xist lncRNA exploits three-dimensional genome architecture to spread across the X chromosome. *Science* **341:** 1237973. doi:10.1126/science.1237973

Engreitz JM, Ollikainen N, Guttman M. 2016. Long non-coding RNAs: spatial amplifiers that control nuclear structure and gene expression. *Nat Rev Mol Cell Biol* **17:** 756–770. doi:10.1038/nrm.2016.126

Feric M, Vaidya N, Harmon TS, Mitrea DM, Zhu L, Richardson TM, Kriwacki RW, Pappu RV, Brangwynne CP. 2016. Coexisting liquid phases underlie nucleolar subcompartments. *Cell* **165:** 1686–1697. doi:10.1016/j.cell.2016.04.047

Gardner EJ, Nizami ZF, Talbot CC, Gall JG. 2012. Stable intronic sequence RNA (sisRNA), a new class of noncoding RNA from the oocyte nucleus of *Xenopus tropicalis*. *Genes Dev* **26:** 2550–2559. doi:10.1101/gad.202184.112

Goff LA, Rinn JL. 2015. Linking RNA biology to lncRNAs. *Genome Res* **25:** 1456–1465. doi:10.1101/gr.191122.115

Gueroussov S, Weatheritt RJ, O'Hanlon D, Lin ZY, Narula A, Gingras AC, Blencowe BJ. 2017. Regulatory expansion in mammals of multivalent hnRNP assemblies that globally control alternative splicing. *Cell* **170:** 324–339.e323. doi:10.1016/j.cell.2017.06.037

Guo JU, Agarwal V, Guo HL, Bartel DP. 2014. Expanded identification and characterization of mammalian circular RNAs. *Genome Biol* **15:** 409. doi:10.1186/s13059-014-0409-z

Hacisuleyman E, Goff LA, Trapnell C, Williams A, Henao-Mejia J, Sun L, McClanahan P, Hendrickson DG, Sauvageau M, Kelley DR, et al. 2014. Topological organization of multichromosomal regions by the long intergenic noncoding RNA Firre. *Nat Struct Mol Biol* **21:** 198–206. doi:10.1038/nsmb.2764

Han YC, Donovan J, Rath S, Whitney G, Chitrakar A, Korennykh A. 2014. Structure of human RNase L reveals the basis for regulated RNA decay in the IFN response. *Science* **343:** 1244–1248. doi:10.1126/science.1249845

Hansen TB, Jensen TI, Clausen BH, Bramsen JB, Finsen B, Damgaard CK, Kjems J. 2013. Natural RNA circles function as efficient microRNA sponges. *Nature* **495:** 384–388. doi:10.1038/nature11993

Harashima A, Guettouche T, Barber GN. 2010. Phosphorylation of the NFAR proteins by the dsRNA- dependent protein kinase

PKR constitutes a novel mechanism of translational regulation and cellular defense. *Genes Dev* **24:** 2640–2653. doi:10.1101/gad.1965010

Henras AK, Plisson-Chastang C, O'Donohue MF, Chakraborty A, Gleizes PE. 2015. An overview of pre-ribosomal RNA processing in eukaryotes. *Wiley Interdiscip Rev RNA* **6:** 225–242. doi:10.1002/wrna.1269

Hirose T, Virnicchi G, Tanigawa A, Naganuma T, Li R, Kimura H, Yokoi T, Nakagawa S, Benard M, Fox AH, et al. 2014. NEAT1 long noncoding RNA regulates transcription via protein sequestration within subnuclear bodies. *Mol Biol Cell* **25:** 169–183. doi:10.1091/mbc.e13-09-0558

Holmstrom TH, Mialon A, Kallio M, Nymalm Y, Mannermaa L, Holm T, Johansson H, Black E, Gillespie D, Salminen TA, et al. 2008. c-Jun supports ribosomal RNA processing and nucleolar localization of RNA helicase DDX21. *J Biol Chem* **283:** 7046–7053. doi:10.1074/jbc.M709613200

Hsu MT, Cocaprados M. 1979. Electron-microscopic evidence for the circular form of RNA in the cytoplasm of eukaryotic cells. *Nature* **280:** 339–340. doi:10.1038/280339a0

Huang S. 2002. Building an efficient factory: where is pre-rRNA synthesized in the nucleolus? *J Cell Biol* **157:** 739–741. doi:10.1083/jcb.200204159

Huang H, Zeqiraj E, Dong BH, Jha BK, Duffy NM, Orlicky S, Thevakumaran N, Talukdar M, Pillon MC, Ceccarelli DF, et al. 2014. Dimeric structure of pseudokinase RNase L bound to 2–5 Å reveals a basis for interferon-induced antiviral activity. *Mol Cell* **53:** 221–234. doi:10.1016/j.molcel.2013.12.025

Hunziker M, Barandun J, Petfalski E, Tan D, Delan-Forino C, Molloy KR, Kim KH, Dunn-Davies H, Shi Y, Chaker-Margot M, et al. 2016. UtpA and UtpB chaperone nascent pre-ribosomal RNA and U3 snoRNA to initiate eukaryotic ribosome assembly. *Nat Commun* **7:** 12090. doi:10.1038/ncomms12090

Hutchinson JN, Ensminger AW, Clemson CM, Lynch CR, Lawrence JB, Chess A. 2007. A screen for nuclear transcripts identifies two linked noncoding RNAs associated with SC35 splicing domains. *BMC Genomics* **8:** 39. doi:10.1186/1471-2164-8-39

Ivanov A, Memczak S, Wyler E, Torti F, Porath HT, Orejuela MR, Piechotta M, Levanon EY, Landthaler M, Dieterich C, et al. 2015. Analysis of intron sequences reveals hallmarks of circular RNA biogenesis in animals. *Cell Rep* **10:** 170–177. doi:10.1016/j.celrep.2014.12.019

Jeck WR, Sorrentino JA, Wang K, Slevin MK, Burd CE, Liu JZ, Marzluff WF, Sharpless NE. 2013. Circular RNAs are abundant, conserved, and associated with ALU repeats. *RNA* **19:** 141–157. doi:10.1261/rna.035667.112

Jia Ng SS, Zheng RT, Osman I, Pek JW. 2018. Generation of *Drosophila* sisRNAs by independent transcription from cognate introns. *iScience* **4:** 68–75. doi:10.1016/j.isci.2018.05.010

Khan S, Verma NC, Chethana, Nandi CK. 2018. Carbon dots for single-molecule imaging of the nucleolus. *ACS Applied Nano Mater* **1:** 483–487. doi:10.1021/acsanm.7b00175

Kishore S, Gruber AR, Jedlinski DJ, Syed AP, Jorjani H, Zavolan M. 2013. Insights into snoRNA biogenesis and processing from PAR-CLIP of snoRNA core proteins and small RNA sequencing. *Genome Biol* **14:** R45. doi:10.1186/gb-2013-14-5-r45

Kiss T. 2001. Small nucleolar RNA-guided post-transcriptional modification of cellular RNAs. *EMBO J* **20:** 3617–3622. doi:10.1093/emboj/20.14.3617

Koberna K, Malinsky J, Pliss A, Masata M, Vecerova J, Fialova M, Bednar J, Raska I. 2002. Ribosomal genes in focus: new transcripts label the dense fibrillar components and form clusters indicative of "Christmas trees" in situ. *J Cell Biol* **157:** 743–748. doi:10.1083/jcb.200202007

Kopp F, Mendell JT. 2018. Functional classification and experimental dissection of long noncoding RNAs. *Cell* **172:** 393–407. doi:10.1016/j.cell.2018.01.011

Kristensen LS, Andersen MS, Stagsted LVW, Ebbesen KK, Hansen TB, Kjems J. 2019. The biogenesis, biology and characterization of circular RNAs. *Nat Rev Genet* **20:** 675–691. doi:10.1038/s41576-019-0158-7

Kufel J, Allmang C, Chanfreau G, Petfalski E, Lafontaine DL, Tollervey D. 2000. Precursors to the U3 small nucleolar RNA lack small nucleolar RNP proteins but are stabilized by La binding. *Mol Cell Biol* **20:** 5415–5424. doi:10.1128/MCB.20.15.5415-5424.2000

Lee S, Kopp F, Chang TC, Sataluri A, Chen B, Sivakumar S, Yu H, Xie Y, Mendell JT. 2016. Noncoding RNA *NORAD* regulates genomic stability by sequestering PUMILIO proteins. *Cell* **164:** 69–80. doi:10.1016/j.cell.2015.12.017

Legnini I, Di Timoteo G, Rossi F, Morlando M, Briganti F, Sthandier O, Fatica A, Santini T, Andronache A, Wade M, et al. 2017. Circ-ZNF609 is a circular RNA that can be translated and functions in myogenesis. *Mol Cell* **66:** 22–37. doi:10.1016/j.molcel.2017.02.017

Li ZY, Huang C, Bao C, Chen L, Lin M, Wang XL, Zhong GL, Yu B, Hu WC, Dai LM, et al. 2015. Exon-intron circular RNAs regulate transcription in the nucleus. *Nat Struct Mol Biol* **22:** 256–264. doi:10.1038/nsmb.2959

Li X, Liu CX, Xue W, Zhang Y, Jiang S, Yin QF, Wei J, Yao RW, Yang L, Chen LL. 2017. Coordinated circRNA biogenesis and function with NF90/NF110 in viral infection. *Mol Cell* **67:** 214–227. doi:10.1016/j.molcel.2017.05.023

Li X, Yang L, Chen LL. 2018. The biogenesis, functions, and challenges of circular RNAs. *Mol Cell* **71:** 428–442. doi:10.1016/j.molcel.2018.06.034

Liang D, Tatomer DC, Luo Z, Wu H, Yang L, Chen LL, Cherry S, Wilusz JE. 2017. The output of protein-coding genes shifts to circular RNAs when the pre-mRNA processing machinery is limiting. *Mol Cell* **68:** 940–954. doi:10.1016/j.molcel.2017.10.034

Liu CX, Li X, Nan F, Jiang S, Gao X, Guo SK, Xue W, Cui YG, Dong KG, Ding HH, et al. 2019. Structure and degradation of circular RNAs regulate PKR activation in innate immunity. *Cell* **177:** 865–880. doi:10.1016/j.cell.2019.03.046

Loza-Muller L, Rodríguez-Corona U, Sobol M, Rodríguez-Zapata LC, Hozak P, Castano E. 2015. Fibrillarin methylates H2A in RNA polymerase I trans-active promoters in *Brassica oleracea*. *Front Plant Sci* **6:** 976. doi:10.3389/fpls.2015.00976

Lu TT, Cui LL, Zhou Y, Zhu CR, Fan DL, Gong H, Zhao Q, Zhou CC, Zhao Y, Lu DF, et al. 2015. Transcriptome-wide investigation of circular RNAs in rice. *RNA* **21:** 2076–2087. doi:10.1261/rna.052282.115

Lubelsky Y, Ulitsky I. 2018. Sequences enriched in Alu repeats drive nuclear localization of long RNAs in human cells. *Nature* **555:** 107–111. doi:10.1038/nature25757

Lykke-Andersen S, Ardal BK, Hollensen AK, Damgaard CK, Jensen TH. 2018. Box C/D snoRNP autoregulation by a *cis*-acting snoRNA in the *NOP56* pre-mRNA. *Mol Cell* **72:** 99–111.e115. doi:10.1016/j.molcel.2018.08.017

Mais C, Scheer U. 2001. Molecular architecture of the amplified nucleoli of *Xenopus* oocytes. *J Cell Sci* **114:** 709–718.

Malakar P, Shilo A, Mogilevsky A, Stein I, Pikarsky E, Nevo Y, Benyamini H, Elgavish S, Zong X, Prasanth KV, et al. 2017. Long noncoding RNA MALAT1 promotes hepatocellular carcinoma development by SRSF1 upregulation and mTOR activation. *Cancer Res* **77:** 1155–1167. doi:10.1158/0008-5472.CAN-16-1508

Mao YS, Sunwoo H, Zhang B, Spector DL. 2011. Direct visualization of the co-transcriptional assembly of a nuclear body by noncoding RNAs. *Nat Cell Biol* **13:** 95–101. doi:10.1038/ncb2140

McHugh CA, Chen CK, Chow A, Surka CF, Tran C, McDonel P, Pandya-Jones A, Blanco M, Burghard C, Moradian A, et al. 2015. The Xist lncRNA interacts directly with SHARP to silence transcription through HDAC3. *Nature* **521:** 232–236. doi:10.1038/nature14443

McKnight SL, Miller OL Jr. 1976. Ultrastructural patterns of RNA synthesis during early embryogenesis of *Drosophila melanogaster*. *Cell* **8:** 305–319. doi:10.1016/0092-8674(76)90014-3

McStay B, Grummt I. 2008. The epigenetics of rRNA genes: from molecular to chromosome biology. *Annu Rev Cell Dev Biol* **24:** 131–157. doi:10.1146/annurev.cellbio.24.110707.175259

Mele M, Mattioli K, Mallard W, Shechner DM, Gerhardinger C, Rinn JL. 2017. Chromatin environment, transcriptional regulation, and splicing distinguish lincRNAs and mRNAs. *Genome Res* **27:** 27–37. doi:10.1101/gr.214205.116

Memczak S, Jens M, Elefsinioti A, Torti F, Krueger J, Rybak A, Maier L, Mackowiak SD, Gregersen LH, Munschauer M, et al. 2013. Circular RNAs are a large class of animal RNAs with regulatory potency. *Nature* **495:** 333–338. doi:10.1038/nature11928

Minajigi A, Froberg J, Wei C, Sunwoo H, Kesner B, Colognori D, Lessing D, Payer B, Boukhali M, Haas W, et al. 2015. Chromosomes. A comprehensive Xist interactome reveals cohesin repulsion and an RNA-directed chromosome conformation. *Science* **349:** aab2276. doi:10.1126/science.aab2276.

Mitton-Fry RM, DeGregorio SJ, Wang J, Steitz TA, Steitz JA. 2010. Poly(A) tail recognition by a viral RNA element through assembly of a triple helix. *Science* **330:** 1244–1247. doi:10.1126/science.1195858

Moore MJ, Proudfoot NJ. 2009. Pre-mRNA processing reaches back to transcription and ahead to translation. *Cell* **136:** 688–700. doi:10.1016/j.cell.2009.02.001

Morgan JT, Fink GR, Bartel DP. 2019. Excised linear introns regulate growth in yeast. *Nature* **565:** 606–611. doi:10.1038/s41586-018-0828-1

Mudge JM, Frankish A, Harrow J. 2013. Functional transcriptomics in the post-ENCODE era. *Genome Res* **23:** 1961–1973. doi:10.1101/gr.161315.113

Munschauer M, Nguyen CT, Sirokman K, Hartigan CR, Hogstrom L, Engreitz JM, Ulirsch JC, Fulco CP, Subramanian V, Chen J, et al. 2018. The *NORAD* lncRNA assembles a topoisomerase complex critical for genome stability. *Nature* **561:** 132–136. doi:10.1038/s41586-018-0453-z

Naganuma T, Nakagawa S, Tanigawa A, Sasaki YF, Goshima N, Hirose T. 2012. Alternative 3′-end processing of long noncoding RNA initiates construction of nuclear paraspeckles. *EMBO J* **31:** 4020–4034. doi:10.1038/emboj.2012.251

Nakagawa S, Ip JY, Shioi G, Tripathi V, Zong X, Hirose T, Prasanth KV. 2012. Malat1 is not an essential component of nuclear speckles in mice. *RNA* **18:** 1487–1499. doi:10.1261/rna.033217.112

Nigro JM, Cho KR, Fearon ER, Kern SE, Ruppert JM, Oliner JD, Kinzler KW, Vogelstein B. 1991. Scrambled exons. *Cell* **64:** 607–613. doi:10.1016/0092-8674(91)90244-S

Pamudurti NR, Bartok O, Jens M, Ashwal-Fluss R, Stottmeister C, Ruhe L, Hanan M, Wyler E, Perez-Hernandez D, Ramberger E, et al. 2017. Translation of circRNAs. *Mol Cell* **66:** 9–21. doi:10.1016/j.molcel.2017.02.021

Parenteau J, Maignon L, Berthoumieux M, Catala M, Gagnon V, Abou Elela S. 2019. Introns are mediators of cell response to starvation. *Nature* **565:** 612–617. doi:10.1038/s41586-018-0859-7

Piwecka M, Glazar P, Hernandez-Miranda LR, Memczak S, Wolf SA, Rybak-Wolf A, Filipchyk A, Klironomos F, Jara CAC, Fenske P, et al. 2017. Loss of a mammalian circular RNA locus causes miRNA deregulation and affects brain function. *Science* **357:** eaam8526. doi:10.1126/science.aam8526

Powell WT, Coulson RL, Crary FK, Wong SS, Ach RA, Tsang P, Alice Yamada N, Yasui DH, Lasalle JM. 2013. A Prader–Willi locus lncRNA cloud modulates diurnal genes and energy expenditure. *Hum Mol Genet* **22:** 4318–4328. doi:10.1093/hmg/ddt281

Quinn JJ, Chang HY. 2016. Unique features of long non-coding RNA biogenesis and function. *Nat Rev Genet* **17:** 47–62. doi:10.1038/nrg.2015.10

Reichow SL, Hamma T, Ferre-D'Amare AR, Varani G. 2007. The structure and function of small nucleolar ribonucleoproteins. *Nucleic Acids Res* **35:** 1452–1464. doi:10.1093/nar/gkl1172

Rodriguez-Trelles F, Tarrio R, Ayala FJ. 2006. Origins and evolution of spliceosomal introns. *Annu Rev Genet* **40:** 47–76. doi:10.1146/annurev.genet.40.110405.090625

Rybak-Wolf A, Stottmeister C, Glazar P, Jens M, Pino N, Giusti S, Hanan M, Behm M, Bartok O, Ashwal-Fluss R, et al. 2015. Circular RNAs in the mammalian brain are highly abundant, conserved, and dynamically expressed. *Mol Cell* **58:** 870–885. doi:10.1016/j.molcel.2015.03.027

Sahoo T, del Gaudio D, German JR, Shinawi M, Peters SU, Person RE, Garnica A, Cheung SW, Beaudet AL. 2008. Prader–Willi phenotype caused by paternal deficiency for the HBII-85 C/D box small nucleolar RNA cluster. *Nat Genet* **40:** 719–721. doi:10.1038/ng.158

Salzman J, Gawad C, Wang PL, Lacayo N, Brown PO. 2012. Circular RNAs are the predominant transcript isoform from hundreds of human genes in diverse cell types. *PLoS One* **7:** e30733. doi:10.1371/journal.pone.0030733

Salzman J, Chen RE, Olsen MN, Wang PL, Brown PO. 2013. Cell-type specific features of circular rna expression. *PLoS Genet* **9:** e1003777. doi:10.1371/journal.pgen.1003777

Samarsky DA, Fournier MJ, Singer RH, Bertrand E. 1998. The snoRNA box C/D motif directs nucleolar targeting and also couples snoRNA synthesis and localization. *EMBO J* **17:** 3747–3757. doi:10.1093/emboj/17.13.3747

Sanger HL, Klotz G, Riesner D, Gross HJ, Kleinschmidt AK. 1976. Viroids are single-stranded covalently closed circular RNA molecules existing as highly base-paired rod-like structures. *Proc Natl Acad Sci* **73:** 3852–3856. doi:10.1073/pnas.73.11.3852

Sasaki YT, Ideue T, Sano M, Mituyama T, Hirose T. 2009. MENε/β noncoding RNAs are essential for structural integrity of nuclear paraspeckles. *Proc Natl Acad Sci* **106:** 2525–2530. doi:10.1073/pnas.0807899106

Schlackow M, Nojima T, Gomes T, Dhir A, Carmo-Fonseca M, Proudfoot NJ. 2017. Distinctive patterns of transcription and RNA processing for human lincRNAs. *Mol Cell* **65:** 25–38. doi:10.1016/j.molcel.2016.11.029

Sharma S, Lafontaine DLJ. 2015. 'View from a bridge': a new perspective on eukaryotic rRNA base modification. *Trends Biochem Sci* **40:** 560–575. doi:10.1016/j.tibs.2015.07.008

Shin Y, Brangwynne CP. 2017. Liquid phase condensation in cell physiology and disease. *Science* **357:** eaaf4382. doi:10.1126/science.aaf4382

Shukla CJ, McCorkindale AL, Gerhardinger C, Korthauer KD, Cabili MN, Shechner DM, Irizarry RA, Maass PG, Rinn JL. 2018. High-throughput identification of RNA nuclear enrichment sequences. *EMBO J* **37:** e98452. doi:10.15252/embj.201798452

Simon MD, Pinter SF, Fang R, Sarma K, Rutenberg-Schoenberg M, Bowman SK, Kesner BA, Maier VK, Kingston RE, Lee JT. 2013. High-resolution Xist binding maps reveal two-step spreading during X-chromosome inactivation. *Nature* **504:** 465–469. doi:10.1038/nature12719

Sloan KE, Leisegang MS, Doebele C, Ramirez AS, Simm S, Safferthal C, Kretschmer J, Schorge T, Markoutsa S, Haag S, et al. 2015. The association of late-acting snoRNPs with human pre-ribosomal complexes requires the RNA helicase DDX21. *Nucleic Acids Res* **43:** 553–564. doi:10.1093/nar/gku1291

Smola MJ, Rice GM, Busan S, Siegfried NA, Weeks KM. 2015. Selective 2′-hydroxyl acylation analyzed by primer extension and mutational profiling (SHAPE-MaP) for direct, versatile and accurate RNA structure analysis. *Nat Protoc* **10:** 1643–1669. doi:10.1038/nprot.2015.103

Souquere S, Beauclair G, Harper F, Fox A, Pierron G. 2010. Highly ordered spatial organization of the structural long noncoding NEAT1 RNAs within paraspeckle nuclear bodies. *Mol Biol Cell* **21:** 4020–4027. doi:10.1091/mbc.e10-08-0690

Starke S, Jost I, Rossbach O, Schneider T, Schreiner S, Hung LH, Bindereif A. 2015. Exon circularization requires canonical splice signals. *Cell Rep* **10:** 103–111. doi:10.1016/j.celrep.2014.12.002

Sunwoo H, Dinger ME, Wilusz JE, Amaral PP, Mattick JS, Spector DL. 2009. MEN ε/β nuclear-retained non-coding RNAs are up-regulated upon muscle differentiation and are essential components of paraspeckles. *Genome Res* **19:** 347–359. doi:10.1101/gr.087775.108

Szczurek A, Xing J, Birk UJ, Cremer C. 2016. Single molecule localization microscopy of mammalian cell nuclei on the nanoscale. *Front Genet* **7:** 114. doi:10.3389/fgene.2016.00114

Tafforeau L, Zorbas C, Langhendries JL, Mullineux ST, Stamatopoulou V, Mullier R, Wacheul L, Lafontaine DL. 2013. The complexity of human ribosome biogenesis revealed by systematic nucleolar screening of pre-rRNA processing factors. *Mol Cell* **51:** 539–551. doi:10.1016/j.molcel.2013.08.011

Talhouarne GJS, Gall JG. 2014. Lariat intronic RNAs in the cytoplasm of *Xenopus tropicalis* oocytes. *RNA* **20:** 1476–1487. doi:10.1261/rna.045781.114

Tessarz P, Santos-Rosa H, Robson SC, Sylvestersen KB, Nelson CJ, Nielsen ML, Kouzarides T. 2014. Glutamine methylation in histone H2A is an RNA-polymerase-I-dedicated modification. *Nature* **505:** 564–568. doi:10.1038/nature12819

Thiry M, Lafontaine DL. 2005. Birth of a nucleolus: the evolution of nucleolar compartments. *Trends Cell Biol* **15:** 194–199. doi:10.1016/j.tcb.2005.02.007

Thiry M, Cheutin T, O'Donohue MF, Kaplan H, Ploton D. 2000. Dynamics and three-dimensional localization of ribosomal RNA within the nucleolus. *RNA* **6:** 1750–1761. doi:10.1017/S1355838200001564

Tichon A, Gil N, Lubelsky Y, Havkin Solomon T, Lemze D, Itzkovitz S, Stern-Ginossar N, Ulitsky I. 2016. A conserved abundant cytoplasmic long noncoding RNA modulates repression by Pumilio proteins in human cells. *Nat Commun* **7:** 12209. doi:10.1038/ncomms12209

Tichon A, Perry RB, Stojic L, Ulitsky I. 2018. SAM68 is required for regulation of Pumilio by the NORAD long noncoding RNA. *Genes Dev* **32:** 70–78. doi:10.1101/gad.309138.117

Tollervey D, Lehtonen H, Carmo-Fonseca M, Hurt EC. 1991. The small nucleolar RNP protein NOP1 (fibrillarin) is required for pre-rRNA processing in yeast. *EMBO J* **10:** 573–583. doi:10.1002/j.1460-2075.1991.tb07984.x

Tollervey D, Lehtonen H, Jansen R, Kern H, Hurt EC. 1993. Temperature-sensitive mutations demonstrate roles for yeast fibrillarin in pre-rRNA processing, pre-rRNA methylation, and ribosome assembly. *Cell* **72:** 443–457. doi:10.1016/0092-8674(93)90120-F

Tomita S, Abdalla MO, Fujiwara S, Yamamoto T, Iwase H, Nakao M, Saitoh N. 2017. Roles of long noncoding RNAs in chromosome domains. *Wiley Interdiscip Rev RNA* **8:** e1384. doi:10.1002/wrna.1384.

Tripathi V, Ellis JD, Shen Z, Song DY, Pan Q, Watt AT, Freier SM, Bennett CF, Sharma A, Bubulya PA, et al. 2010. The nuclear-retained noncoding RNA MALAT1 regulates alternative splicing by modulating SR splicing factor phosphorylation. *Mol Cell* **39:** 925–938. doi:10.1016/j.molcel.2010.08.011

Tycowski KT, Shu MD, Borah S, Shi M, Steitz JA. 2012. Conservation of a triple-helix-forming RNA stability element in noncoding and genomic RNAs of diverse viruses. *Cell Rep* **2:** 26–32. doi:10.1016/j.celrep.2012.05.020

Ulitsky I. 2016. Evolution to the rescue: using comparative genomics to understand long non-coding RNAs. *Nat Rev Genet* **17:** 601–614. doi:10.1038/nrg.2016.85

Uszczynska-Ratajczak B, Lagarde J, Frankish A, Guigo R, Johnson R. 2018. Towards a complete map of the human long non-coding RNA transcriptome. *Nat Rev Genet* **19:** 535–548. doi:10.1038/s41576-018-0017-y

Wang Y, Wang ZF. 2015. Efficient backsplicing produces translatable circular mRNAs. *RNA* **21:** 172–179. doi:10.1261/rna.048272.114

Wang PL, Bao Y, Yee MC, Barrett SP, Hogan GJ, Olsen MN, Dinneny JR, Brown PO, Salzman J. 2014. Circular RNA is expressed across the eukaryotic tree of life. *PLoS One* **9:** e90859. doi:10.1371/journal.pone.0090859

Wang Y, Hu SB, Wang MR, Yao RW, Wu D, Yang L, Chen LL. 2018. Genome-wide screening of *NEAT1* regulators reveals cross-regulation between paraspeckles and mitochondria. *Nat Cell Biol* **20:** 1145–1158. doi:10.1038/s41556-018-0204-2

Weinstein LB, Steitz JA. 1999. Guided tours: from precursor snoRNA to functional snoRNP. *Curr Opin Cell Biol* **11:** 378–384. doi:10.1016/S0955-0674(99)80053-2

Wesselhoeft RA, Kowalski PS, Parker-Hale FC, Huang YX, Bisaria N, Anderson DG. 2019. RNA circularization diminishes immunogenicity and can extend translation duration in vivo. *Mol Cell* **74:** 508–520. doi:10.1016/j.molcel.2019.02.015

Westholm JO, Miura P, Olson S, Shenker S, Joseph B, Sanfilippo P, Celniker SE, Graveley BR, Lai EC. 2014. Genome-wide analysis of *Drosophila* circular RNAs reveals their structural and sequence properties and age-dependent neural accumulation. *Cell Rep* **9:** 1966–1980. doi:10.1016/j.celrep.2014.10.062

Wilusz JE, Freier SM, Spector DL. 2008. 3′ end processing of a long nuclear-retained noncoding RNA yields a tRNA-like cytoplasmic RNA. *Cell* **135:** 919–932. doi:10.1016/j.cell.2008.10.012

Wilusz JE, JnBaptiste CK, Lu LY, Kuhn CD, Joshua-Tor L, Sharp PA. 2012. A triple helix stabilizes the 3′ ends of long noncoding RNAs that lack poly(A) tails. *Genes Dev* **26:** 2392–2407. doi:10.1101/gad.204438.112

Wu H, Yin QF, Luo Z, Yao RW, Zheng CC, Zhang J, Xiang JF, Yang L, Chen LL. 2016. Unusual processing generates SPA LncRNAs that sequester multiple RNA binding proteins. *Mol Cell* **64:** 534–548. doi:10.1016/j.molcel.2016.10.007

Wu H, Yang L, Chen LL. 2017. The diversity of long noncoding RNAs and their generation. *Trends Genet* **33:** 540–552. doi:10.1016/j.tig.2017.05.004

Xia P, Wang S, Ye BQ, Du Y, Li C, Xiong Z, Qu Y, Fan ZS. 2018. A circular RNA protects dormant hematopoietic stem cells from DNA sensor cGAS-mediated exhaustion. *Immunity* **48:** 688–701. doi:10.1016/j.immuni.2018.03.016

Xiang JF, Yin QF, Chen T, Zhang Y, Zhang XO, Wu Z, Zhang S, Wang HB, Ge J, Lu X, et al. 2014. Human colorectal cancer-specific *CCAT1-L* lncRNA regulates long-range chromatin interactions at the *MYC* locus. *Cell Res* **24:** 513–531. doi:10.1038/cr.2014.35

Xing YH, Chen LL. 2018. Processing and roles of snoRNA-ended long noncoding RNAs. *Crit Rev Biochem Mol Biol* **53:** 596–606. doi:10.1080/10409238.2018.1508411

Xing YH, Yao RW, Zhang Y, Guo CJ, Jiang S, Xu G, Dong R, Yang L, Chen LL. 2017. *SLERT* regulates DDX21 rings associated with Pol I transcription. *Cell* **169:** 664–678.e616. doi:10.1016/j.cell.2017.04.011

Yamazaki T, Souquere S, Chujo T, Kobelke S, Chong YS, Fox AH, Bond CS, Nakagawa S, Pierron G, Hirose T. 2018. Functional domains of NEAT1 architectural lncRNA induce paraspeckle assembly through phase separation. *Mol Cell* **70:** 1038–1053.e1037. doi:10.1016/j.molcel.2018.05.019

Yang L, Duff MO, Graveley BR, Carmichael GG, Chen LL. 2011. Genomewide characterization of non-polyadenylated RNAs. *Genome Biol* **12:** R16. doi:10.1186/gb-2011-12-2-r16

Yang Y, Fan XJ, Mao MW, Song XW, Wu P, Zhang Y, Jin YF, Yang Y, Chen LL, Wang Y, et al. 2017. Extensive translation of circular RNAs driven by N^6-methyladenosine. *Cell Res* **27:** 626–641. doi:10.1038/cr.2017.31

Yao RW, Wang Y, Chen LL. 2019a. Cellular functions of long noncoding RNAs. *Nat Cell Biol* **21:** 542–551. doi:10.1038/s41556-019-0311-8

Yao RW, Xu G, Wang Y, Shan L, Luan PF, Wang Y, Wu M, Yang LZ, Xing YH, Yang L, et al. 2019b. Nascent Pre-rRNA sorting via phase separation drives the assembly of dense fibrillar components in the human nucleolus. *Mol Cell* **76:** 767–783. doi:10.1016/j.molcel.2019.08.014

Yin QF, Yang L, Zhang Y, Xiang JF, Wu YW, Carmichael GG, Chen LL. 2012. Long noncoding RNAs with snoRNA ends. *Mol Cell* **48:** 219–230. doi:10.1016/j.molcel.2012.07.033

Ying Y, Wang XJ, Vuong CK, Lin CH, Damianov A, Black DL. 2017. Splicing activation by Rbfox requires self-aggregation through its tyrosine-rich domain. *Cell* **170:** 312–323.e310. doi:10.1016/j.cell.2017.06.022

You XT, Vlatkovic I, Babic A, Will T, Epstein I, Tushev G, Akbalik G, Wang MT, Glock C, Quedenau C, et al. 2015. Neural circular RNAs are derived from synaptic genes and regulated by development and plasticity. *Nat Neurosci* **18:** 603–610. doi:10.1038/nn.3975

Zhang B, Arun G, Mao YS, Lazar Z, Hung G, Bhattacharjee G, Xiao X, Booth CJ, Wu J, Zhang C, et al. 2012. The lncRNA *Malat1* is dispensable for mouse development but its transcription plays a cis-regulatory role in the adult. *Cell Rep* **2:** 111–123. doi:10.1016/j.celrep.2012.06.003

Zhang Y, Zhang XO, Chen T, Xiang JF, Yin QF, Xing YH, Zhu SS, Yang L, Chen LL. 2013. Circular intronic long noncoding RNAs. *Mol Cell* **51:** 792–806. doi:10.1016/j.molcel.2013.08.017

Zhang XO, Wang HB, Zhang Y, Lu XH, Chen LL, Yang L. 2014a. Complementary sequence-mediated exon circularization. *Cell* **159:** 134–147. doi:10.1016/j.cell.2014.09.001

Zhang XO, Yin QF, Wang HB, Zhang Y, Chen T, Zheng P, Lu X, Chen LL, Yang L. 2014b. Species-specific alternative splicing leads to unique expression of *sno-lncRNAs*. *BMC Genomics* **15:** 287. doi:10.1186/1471-2164-15-287

Zhang XO, Dong R, Zhang Y, Zhang JL, Luo Z, Zhang J, Chen LL, Yang L. 2016a. Diverse alternative back-splicing and alternative splicing landscape of circular RNAs. *Genome Res* **26:** 1277–1287. doi:10.1101/gr.202895.115

Zhang Y, Xue W, Li X, Zhang J, Chen SY, Zhang JL, Yang L, Chen LL. 2016b. The biogenesis of nascent circular RNAs. *Cell Rep* **15:** 611–624. doi:10.1016/j.celrep.2016.03.058

Zhang SY, Clark NE, Freije CA, Pauwels E, Taggart AJ, Okada S, Mandel H, Garcia P, Ciancanelli MJ, Biran A, et al. 2018. Inborn errors of RNA lariat metabolism in humans with brainstem viral infection. *Cell* **172:** 952–965. doi:10.1016/j.cell.2018.02.019

Zhao Y, Li H, Fang S, Kang Y, Wu W, Hao Y, Li Z, Bu D, Sun N, Zhang MQ, et al. 2016. NONCODE 2016: an informative and valuable data source of long non-coding RNAs. *Nucleic Acids Res* **44:** D203–D208. doi:10.1093/nar/gkv1252

Zheng XF, Bevilacqua PC. 2004. Activation of the protein kinase PKR by short double-stranded RNAs with single-stranded tails. *RNA* **10:** 1934–1945. doi:10.1261/rna.7150804

Attenuation of Eukaryotic Protein-Coding Gene Expression via Premature Transcription Termination

DEIRDRE C. TATOMER AND JEREMY E. WILUSZ

Department of Biochemistry and Biophysics, University of Pennsylvania Perelman School of Medicine, Philadelphia, Pennsylvania 19104, USA

Correspondence: wilusz@pennmedicine.upenn.edu

A complex network of RNA transcripts is generated from eukaryotic genomes, many of which are processed in unexpected ways. Here, we highlight how premature transcription termination events at protein-coding gene loci can simultaneously lead to the generation of short RNAs and attenuate production of full-length mRNA transcripts. We recently showed that the Integrator (Int) complex can be selectively recruited to protein-coding gene loci, including *Drosophila* metallothionein A (MtnA), where the IntS11 RNA endonuclease cleaves nascent transcripts near their 5′ ends. Such premature termination events catalyzed by Integrator can repress the expression of some full-length mRNAs by more than 100-fold. Transcription at small nuclear RNA (snRNA) loci is likewise terminated by Integrator cleavage, but protein-coding and snRNA gene loci have notably distinct dependencies on Integrator subunits. Additional mechanisms that attenuate eukaryotic gene outputs via premature termination have been discovered, including by the cleavage and polyadenylation machinery in a manner controlled by U1 snRNP. These mechanisms appear to function broadly across the transcriptome. This suggests that synthesis of full-length transcripts is not always the default option and that premature termination events can lead to a variety of transcripts, some of which may have important and unexpected biological functions.

It is well established that nascent transcripts must be extensively processed in order to generate a mature RNA that is stable and functional. For transcripts derived from protein-coding genes, they typically must be capped at their 5′ ends, spliced to remove intronic sequences, and modified at their 3′ ends by the addition of a poly(A) tail. Each of these canonical pre-mRNA processing steps is extensively regulated, ensuring quality control as well as enabling the generation of a diverse set of functional RNAs (for reviews, see Moore and Proudfoot 2009; Braunschweig et al. 2013; Tian and Manley 2017; Hentze et al. 2018). Besides well-studied transcripts that undergo all of the canonical pre-mRNA processing steps (Fig. 1A), a growing number of transcripts that lack a standard 5′ cap, a poly(A) tail, or both of these terminal structures have been identified, with some accumulating to high levels and having key cellular roles (for reviews, see Wilusz and Spector 2010; Zhang et al. 2014; Wilusz 2016; Kiledjian 2018). For example, metazoan replicative histone mRNAs are efficiently translated despite ending in a stem–loop structure (Fig. 1B; for review, see Marzluff et al. 2008); the MALAT1 and MEN β (also known as NEAT1_2) nuclear-retained long noncoding RNAs have 3′ terminal triple helices (Wilusz et al. 2008, 2012; Brown et al. 2012, 2014) and play key roles in cancer metastasis (Ji et al. 2003; Arun et al. 2016) and nuclear paraspeckle formation (Clemson et al. 2009; Sasaki et al. 2009; Sunwoo et al. 2009; Mao et al. 2011), respectively; and circular RNAs have covalently linked ends and, in some cases, regulate microRNA activity, the immune system, or other cellular pathways (for reviews, see Li et al. 2018; Wilusz 2018; Kristensen et al. 2019; Patop et al. 2019).

Because nonpolyadenylated RNAs and circular RNAs are structurally distinct from canonical mRNAs, they are subjected to different biogenesis and post-transcriptional control mechanisms as well as likely bound by unique factors. A major goal of our laboratory has thus been to understand how the fates of noncanonical RNAs are controlled. This has led us in a variety of scientific directions and led to a number of unexpected findings, including mechanisms that control (i) circular RNA levels and localization (Liang and Wilusz 2014; Kramer et al. 2015; Liang et al. 2017; Huang et al. 2018), (ii) non-AUG translation (Kearse et al. 2019), and (iii) transcription elongation (Tatomer et al. 2019), which will be the focus of this manuscript. We will summarize how a high-throughput RNAi screening effort revealed that the Integrator (Int) complex is a potent inhibitor of the transcription of many protein-coding genes. This is because the IntS11 RNA endonuclease cleaves nascent transcripts and catalyzes premature transcription termination. Additional mechanisms that attenuate eukaryotic gene outputs via premature termination have been discovered and may likewise be widespread across the transcriptome (for a review, see Kamieniarz-Gdula and Proudfoot 2019). This suggests that synthesis of full-length transcripts is likely not the default option. Instead, premature cleavage events by the Integrator complex, the cleavage and polyadenylation (CPA) machinery, and likely other endonucleases need to be actively suppressed for full-length transcripts to be generated.

© 2019 Tatomer and Wilusz. This article is distributed under the terms of the Creative Commons Attribution-NonCommercial License, which permits reuse and redistribution, except for commercial purposes, provided that the original author and source are credited.

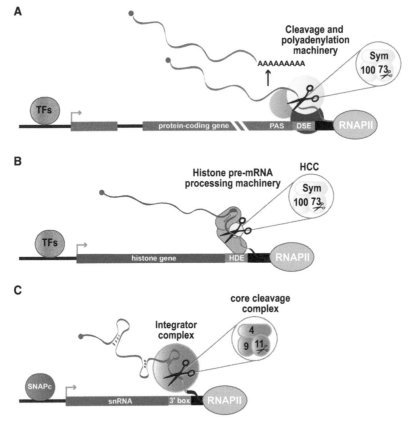

Figure 1. RNA 3′-end processing mechanisms. Proper processing of the RNA 3′ end is critical for the generation of a mature, functional transcript. (*A*) For most protein-coding genes, the cleavage and polyadenylation machinery are recruited to a polyadenylation signal (PAS) with help from a downstream sequence element (DSE). The Cpsf73 endonuclease (denoted 73) cleaves the nascent transcript, which is then modified by the addition of a poly(A) tail. (*B*) At replication-dependent histone genes, nascent transcripts are cleaved by Cpsf73 (which is part of the histone cleavage complex, HCC) between a stem–loop structure and a purine-rich histone downstream element (HDE). During histone mRNA 3′-end processing, the HDE interacts with the U7 snRNP and the stem–loop is recognized by stem–loop binding protein (SLBP). Cleavage is not followed by addition of a poly(A) tail. (*C*) Nascent snRNA transcripts are recognized by the Integrator complex and cleaved upstream of a 3′ box sequence by the IntS11 endonuclease (denoted 11), which is homologous to Cpsf73 and is part of a core cleavage complex within Integrator.

RNAi SCREENING REVEALED THE INTEGRATOR COMPLEX AS A POTENT NEGATIVE REGULATOR OF *DROSOPHILA* MtnA TRANSCRIPTION

To identify novel factors that control RNA fates, we worked with Sara Cherry's group to set up genome-scale RNAi screens in *Drosophila* cells. In particular, we examined the outputs of inducible reporter mRNAs that ended in distinct 3′ terminal sequences—for example, eGFP or mCherry mRNAs ending in a poly(A) tail, the histone stem–loop structure, or the MALAT1 triple helix (Fig. 2A). By quantitating the levels of the encoded fluorescent proteins using automated microscopy, we reasoned that this approach should enable the identification of (i) general factors that regulate all of the reporter RNAs (e.g., general transcription and translation control factors), (ii) factors that regulate all of the nonpolyadenylated RNAs, and (iii) factors that are specific for particular 3′ ends (e.g., stem–loop binding protein for the histone stem–loop [Wang et al. 1996]). Among other observations, these efforts surprisingly revealed that nearly all 14 subunits of the Int complex (for a review, see Baillat and Wagner 2015) were among the most potent negative regulators of the reporters (Fig. 2B,C; Tatomer et al. 2019). When many Integrator subunits were individually depleted, we observed increased levels of the reporter RNAs and encoded reporter proteins, and these effects were observed regardless of the open reading frame or 3′ end processing mechanism that was present downstream from the inducible metallothionein A (MtnA) promoter (Fig. 2B,C). This suggested that the Integrator complex inhibits the transcriptional output of the MtnA promoter.

The >1 MDa Integrator complex is conserved across metazoans, interacts with RNA polymerase II (RNAPII), and consists of 14 subunits, but few of the subunits have identifiable paralogs within eukaryotic genomes (Baillat et al. 2005; Egloff et al. 2010; Baillat and Wagner 2015). Strikingly, many Integrator subunits are devoid of known protein domains, with the most common domains being α-helical repeats (e.g., HEAT, ARM, TPR, or VWA domains) that may function as protein–protein interaction surfaces. Notable exceptions are Integrator subunits 11 (IntS11) and 9 (IntS9) that are homologous to Cpsf73

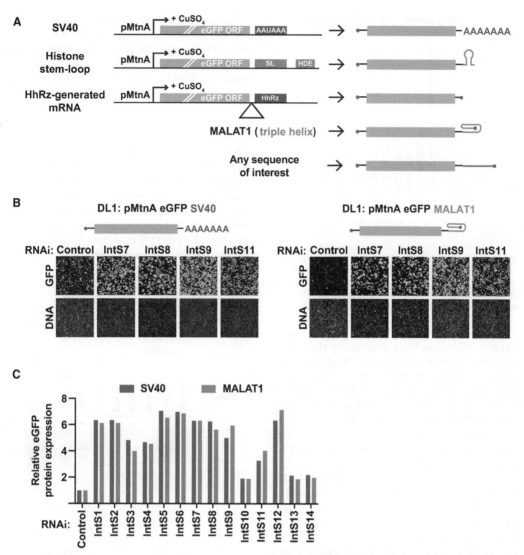

Figure 2. The Integrator complex inhibits expression of reporter mRNAs generated from the MtnA promoter. (*A*) A set of reporter plasmids was generated that encode fluorescent proteins and are processed by distinct RNA 3′-end processing mechanisms. All reporters are driven by the *Drosophila* MtnA promoter, which is rapidly induced when the intracellular concentration of heavy metals (e.g., copper or cadmium) is increased. The SV40 polyadenylation signal enables generation of an mRNA ending in a poly(A) tail, the histone mRNA processing signals enable generation of an mRNA ending in the histone stem–loop, and the self-cleaving hammerhead ribozyme (HhRz) enables generation of nonpolyadenylated reporter RNAs with defined 3′ ends. Insertion of the MALAT1 3′ end upstream of the HhRz sequence enables generation of an mRNA ending in a triple helix. (*B,C*) *Drosophila* DL1 cells stably maintaining an eGFP reporter ending in the SV40 polyadenylation signal or the MALAT1 triple helix were treated with double-stranded RNAs (dsRNAs) for 3 d to induce RNAi and depletion of the indicated Integrator subunits. $CuSO_4$ was added for the last 6 h. (*B*) Representative images of eGFP and DNA (Hoechst 33342). (*C*) The integrated eGFP intensity (amount of eGFP signal in each well divided by the number of cells) was quantified. Integrator depletion resulted in similar effects on eGFP protein levels regardless of the mRNA 3′-end processing mechanism.

and Cpsf100, which bind one another and function as a zinc-dependent endonuclease that is critical for cleaving the 3′ ends of mRNAs prior to the addition of the poly(A) tail (Fig. 1; Baillat et al. 2005; Shi and Manley 2015). The IntS4, IntS9, and IntS11 proteins interact with one another (Fig. 1C; Wu et al. 2017; Albrecht et al. 2018) and, like the Cpsf73, Cpsf100, Symplekin complex (Fig. 1A; Sullivan et al. 2009), can catalyze cleavage of RNA. In particular, IntS11 has been well-established to cleave the 3′ ends of nascent small nuclear RNA (snRNA) transcripts, thereby releasing the processed snRNA from RNAPII so that it can function in pre-mRNA splicing as part of the spliceosome (Fig. 1C; Baillat et al. 2005). Until fairly recently, snRNA cleavage was the only known function for Integrator, but recent work has suggested that this complex may also bind to a subset of protein-coding, noncoding, and enhancer loci to control their expression (Cazalla et al. 2011; Gardini et al. 2014; Stadelmayer et al. 2014; Lai et al. 2015; Skaar et al. 2015; Xie et al. 2015; Barbieri et al. 2018; Rubtsova et al. 2019).

We, therefore, examined binding of Integrator subunits to the endogenous MtnA locus using chromatin immuno-

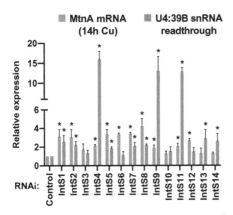

Figure 3. The Integrator complex regulates the outputs of endogenous snRNA and MtnA genes. *Drosophila* DL1 cells were treated with dsRNAs for 3 d to induce RNAi and depletion of the indicated Integrator subunits. RT-qPCR was used to quantify expression of endogenous MtnA mRNA after 14 h of $CuSO_4$ treatment. Northern blotting was used to quantify readthrough transcription downstream from the U4:39B snRNA. Data are shown as mean ± SD, $N \geq 3$. (*) $P < 0.05$.

precipitation (ChIP)-quantitative polymerase chain reaction (qPCR) in collaboration with Eric J. Wagner's group. The MtnA locus encodes a metal chelator, and its promoter is rapidly induced when the intracellular concentration of heavy metals (e.g., copper or cadmium) is increased (Günther et al. 2012). Much more significant binding of Integrator subunits to the 5′ end of the MtnA locus was noted during copper stress (the same conditions used for the genome-scale RNAi screens) as compared to cadmium stress (Tatomer et al. 2019). Consistent with these ChIP-qPCR results, depletion of Integrator subunits resulted in increased levels of MtnA pre-mRNA and mRNA during copper (Fig. 3) but not cadmium stress. Integrator can thus display context-specific binding and regulation, including helping with homeostatic control of intracellular heavy metal levels via direct control of MtnA gene expression.

THE INTEGRATOR COMPLEX CLEAVES NASCENT MtnA TRANSCRIPTS TO TRIGGER PREMATURE TRANSCRIPTION TERMINATION

Individual depletion of many Integrator subunits, including noncatalytic subunits that currently lack a known function, resulted in increased MtnA expression (Fig. 3; Tatomer et al. 2019). Nevertheless, to reveal the underlying molecular mechanism, we first examined whether the well-established RNA endonuclease activity of IntS11 is required for the Integrator complex to limit the output from the MtnA promoter. Endogenous IntS11 was depleted from *Drosophila* cells using RNAi coupled to expression of either a wild-type or catalytically dead (E203Q) RNAi-resistant IntS11 transgene. These knockdown-rescue experiments revealed a clear requirement for the IntS11 RNA endonuclease activity (Tatomer et al. 2019) and sparked a search for where Integrator may be cleaving RNAs derived from the MtnA locus.

We reasoned that the RNA products of Integrator cleavage may be unstable because of the lack of a poly(A) tail and further noted that many RNA exosome core components and cofactors (including Rrp40 and Mtr4) had scored as positive regulators of the MtnA promoter in the genome-scale RNAi screens. We thus hypothesized that Integrator may cleave nascent MtnA transcripts to prematurely terminate transcription and that these cleaved transcripts are then targeted for rapid degradation by the RNA exosome. To test this model, the exosome-associated RNA helicase Mtr4 (Lubas et al. 2011) was depleted from cells during copper stress (Fig. 4, lane 4). This resulted in a reduction in full-length MtnA mRNA expression that was coupled to increased expression of several small RNAs, including prominent transcripts with lengths of ∼85 and ∼110 nt (Fig. 4, lane 4; Tatomer et al. 2019). These small RNAs (i) were capped at their 5′ ends, (ii) had

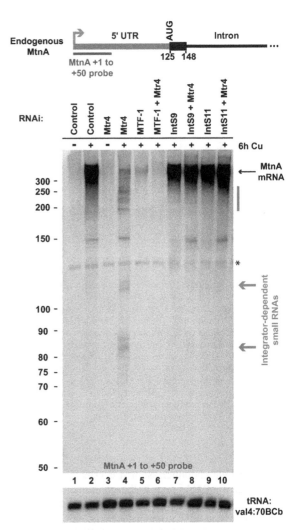

Figure 4. The Integrator complex catalyzes premature transcription termination by cleaving nascent MtnA RNAs. Northern blotting was used to analyze RNAs generated from the endogenous MtnA locus in *Drosophila* DL1 cells treated with the indicated dsRNAs for 3 d and $CuSO_4$ for the last 6 h. Full-length MtnA mRNA (black arrow) and Integrator-dependent small RNAs (orange) are indicated. *, Nonspecific band.

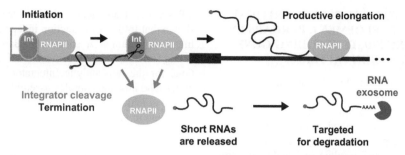

Figure 5. Model for Integrator-dependent premature transcription termination. After transcription initiation, RNA polymerase II can transition into productive elongation to generate the mature mRNA. Alternatively, the IntS11 endonuclease within the Integrator complex can cleave the nascent RNA, enabling transcription termination and degradation of the short RNA by the RNA exosome.

the same transcription start site (TSS) as full-length MtnA mRNA and a similar requirement for the MTF-1 transcription factor (Fig. 4, lane 6), as well as (iii) had detectable oligoadenylation, a mark known to facilitate RNA degradation by the RNA exosome. These features are all consistent with premature transcription termination products. Upon then co-depleting Mtr4 and Integrator subunits, these small RNAs were eliminated and full-length MtnA mRNA expression was restored (Fig. 4, lanes 8 and 10). Generation of the MtnA small RNAs is thus dependent on the presence of Integrator and, in particular, catalytically active IntS11.

In total, these data indicate that the Integrator complex can be recruited to the MtnA locus, where it cleaves nascent RNAs and facilitates premature transcription termination (Fig. 5; Tatomer et al. 2019). Integrator cleavage limits production of full-length MtnA mRNA, likely as a homeostatic mechanism to fine-tune copper levels in cells. This is particularly important as copper is required for the function of a subset of enzymes and must be maintained in a narrow concentration range (for a review, see Festa and Thiele 2011).

THE INTEGRATOR COMPLEX CATALYZES PREMATURE TRANSCRIPTION TERMINATION AT MANY PROTEIN-CODING GENES

Given that Integrator cleavage potently attenuates MtnA transcription, we next addressed whether additional protein-coding genes are similarly regulated. The IntS9 subunit was depleted from *Drosophila* cells (in copper stress conditions) and RNA-seq was used to identify mRNA expression changes across the transcriptome. This revealed 409 and 49 genes that were up- and down-regulated, respectively, upon IntS9 depletion (fold change >1.5 and $P < 0.001$) (Tatomer et al. 2019). Integrator thus predominately inhibits protein-coding gene expression in *Drosophila* cells, which is in contrast to previous reports in other systems that suggested that Integrator predominantly stimulates protein-coding gene expression (Gardini et al. 2014; Stadelmayer et al. 2014). Furthermore, we find that Integrator can function as a very potent inhibitor of gene expression, with some genes being repressed by more than 25-fold by Integrator (e.g., Pepck1 and Hml are repressed by ~100- and 35-fold, respectively), including in unstressed cells.

To validate these results, seven mRNAs that had differing magnitudes of fold change upon IntS9 depletion were selected for further analyses, which included reverse transcription (RT)-qPCR, ChIP-qPCR, and northern blotting to identify small RNAs (Tatomer et al. 2019). The results at these loci largely mirrored what was observed at the MtnA locus: (i) expression of the pre-mRNAs increased upon depletion of many Integrator subunits (including noncatalytic subunits), indicative of a transcriptional effect, (ii) Integrator subunits were bound at the 5′ ends of these loci, but not at loci whose expression was unchanged upon Integrator depletion, (iii) the IntS11 endonuclease was required for regulation of these loci, and (iv) prematurely terminated small RNAs could be detected using northern blotting. Interestingly, the prematurely terminated small RNAs were of defined lengths and often 50–110 nt, roughly mirroring the sizes of cleavage products observed at the MtnA locus.

At snRNA gene loci, a conserved but relatively degenerate 3′ box sequence is required for Integrator cleavage (Fig. 1C; Hernandez 1985), but similar sequences are not immediately obvious at any of the protein-coding transcripts we have examined in detail. Introducing deletions into the MtnA 5′ UTR did not alter the cleavage product sizes (Tatomer et al. 2019), suggesting that Integrator may cleave nascent mRNAs at a set distance from the TSS in a manner independent of local DNA/RNA sequence, perhaps at positions of RNA polymerase II pausing/stalling or nucleosomes. Indeed, collaborative work with the laboratories of Karen Adelman and Eric J. Wagner suggests a role for RNA polymerase II pausing/stalling in dictating Integrator cleavage sites (Elrod et al. 2019). It will be very informative in the future to understand how exactly cleavage sites are selected, especially those downstream from classically defined promoter-proximal pausing sites. Regardless of the underlying details, our work (which built upon that of others [Skaar et al. 2015; Shah et al. 2018; Gómez-Orte et al. 2019]) has now revealed that the Integrator complex can potently attenuate the expression of many protein-coding genes via catalyzing premature transcription termination (Elrod et al. 2019; Tatomer et al. 2019).

THE INTEGRATOR COMPLEX IS CRITICAL FOR PROPER DEVELOPMENT, PERHAPS BY CONTROLLING PROTEIN-CODING GENE EXPRESSION PATTERNS

Considering that Integrator directly controls the expression of snRNAs, enhancer RNAs, and many protein-coding genes, it is perhaps not surprising that developmental phenotypes have been observed when Integrator subunits are knocked down/out or mutated. Early work showed that mutation of IntS4 or IntS7 resulted in lethality in *Drosophila* (Rutkowski and Warren 2009; Ezzeddine et al. 2011), and a mutation in IntS6 was subsequently shown to result in embryonic defects/lethality in zebrafish (Kapp et al. 2013). Focusing specifically on the brain, depletion of Integrator subunits is associated with excess immature neuroblasts in *Drosophila* (Zhang et al. 2019) and cortical neuron migration defects in mice (van den Berg et al. 2017). Recent work has even extended these observations to humans as six individuals with severe neurodevelopmental delay have been shown to carry biallelic mutations in Integrator subunits (Oegema et al. 2017).

However, the underlying molecular mechanisms responsible for these phenotypes remain largely unclear. Slight increases in unprocessed snRNAs have been observed when Integrator subunits are depleted or mutated, but the levels of mature snRNAs have generally not been found to be significantly different (Ezzeddine et al. 2011; Oegema et al. 2017; Tatomer et al. 2019). There is no denying that functional snRNAs are critical for life, but we hypothesize that Integrator's role in fine-tuning protein-coding transcription may also be critical for normal development, and that perturbation of this tuning process may lead to the phenotypes that have been observed. For example, Integrator may catalyze premature transcription termination at genes that control differentiation processes, thereby preventing expression of the encoded proteins until they are needed. Alternatively, Integrator may be needed to repress embryonic genes in fully differentiated cells to enforce cell identity. Future studies that map Integrator target genes in different tissues as well as during differentiation programs will provide unprecedented insights into how this complex functions across the transcriptome in vivo.

SPECIALIZED INTEGRATOR SUBCOMPLEXES

Endonucleolytic cleavage by IntS11 is critical for Integrator regulation at snRNA and protein-coding genes, but our RNAi screening data indicate that these loci have different dependencies on Integrator subunits (Tatomer et al. 2019). IntS4, IntS9, and IntS11, which comprise the Integrator cleavage module (Fig. 1C; Albrecht et al. 2018), are most important for snRNA processing, whereas the noncatalytic subunits play only minor roles (Fig. 3). In contrast, large increases in mRNA expression were observed when many of the noncatalytic subunits were depleted (Fig. 3; Tatomer et al. 2019). This may suggest that distinct Integrator complexes are recruited to snRNA and mRNA loci. Consistent with this idea, recent work has shown that IntS13 can function independently from other Integrator subunits at enhancers (Barbieri et al. 2018). Going forward, it will be critical to define molecular functions for the noncatalytic Integrator subunits, to better understand how the complex assembles, and to characterize the stoichiometry of individual Integrator subunits that are bound to target gene loci.

ATTENUATION OF EUKARYOTIC PROTEIN-CODING GENE EXPRESSION IS LIKELY MORE WIDESPREAD THAN HAS BEEN APPRECIATED

Attenuation of protein-coding gene expression via premature termination is well-established in bacteria (e.g., the tryptophan operon) (reviewed in Merino and Yanofsky 2005; Naville and Gautheret 2010), but has generally received much less attention in eukaryotes (for reviews, see Wright 1993; Kamieniarz-Gdula and Proudfoot 2019). Nevertheless, there are clear examples, some of which date back more than 40 years (Evans et al. 1979), of how regulated premature termination can modulate gene outputs in eukaryotes. In *Saccharomyces cerevisiae*, the Nrd1–Nab3–Sen1 (NNS) complex uses the Sen1 helicase to pull nascent transcripts out of the RNA polymerase II active site, thereby catalyzing premature termination of some protein-coding genes, including Nrd1 itself and genes involved in nucleotide biosynthesis (Steinmetz et al. 2001; Arigo et al. 2006; Jenks et al. 2008; Kuehner and Brow 2008; Thiebaut et al. 2008; Vasiljeva et al. 2008; Bresson et al. 2017). Notably, the NNS complex also catalyzes termination at snRNA loci, suggesting parallels with the Integrator complex (which is absent from yeast species) even though the underlying molecular mechanisms of transcription termination are distinct. In *Schizosaccharomyces pombe*, the MTREC (Mtl1-Red1 core) complex likewise triggers premature termination and promotes heterochromatin formation (Lee et al. 2013; Chalamcharla et al. 2015).

In metazoans, short capped transcripts can be detected from the 5′ ends of many genes (Kapranov et al. 2007; Preker et al. 2008; Nechaev et al. 2010), likely a result of premature transcription termination events. Indeed, single-molecule footprinting (Krebs et al. 2017), ChIP-seq (Erickson et al. 2018), and fluorescence recovery after photobleaching (FRAP) experiments (Steurer et al. 2018) all suggest high levels of RNA polymerase II turnover near the 5′ ends of genes. Besides Integrator, premature termination events can be driven by other complexes, such as the cleavage and polyadenylation machinery (e.g., PCF11 [Kamieniarz-Gdula et al. 2019]), the Microprocessor (Wagschal et al. 2012), and the decapping/Xrn2 machinery (Brannan et al. 2012). Nevertheless, the exact number of genes that are controlled by premature termination events is not yet clear as many prematurely terminated transcripts are rapidly degraded by the RNA exosome (Preker et al. 2008; Almada et al. 2013; Ntini et al. 2013).

WIDESPREAD PREMATURE CLEAVAGE AND POLYADENYLATION EVENTS CAN OCCUR, BUT ARE OFTEN PREVENTED BY U1 snRNP

The widespread potential for premature cleavage and polyadenylation events has been revealed in recent years via a set of analyses that began in the laboratory of Gideon Dreyfuss with a very unrelated line of scientific inquiry (for a review, see Venters et al. 2019). It was long known that stoichiometric amounts of U1, U2, U4, U5, and U6 small nuclear ribonucleoproteins (snRNPs) are required to assemble a catalytically active spliceosome (for a review, see Shi 2017), yet these snRNPs are expressed at varying levels in cells, with U1 snRNP often being most abundant. To explore the functional significance of this observation, the Dreyfuss laboratory used antisense morpholino oligonucleotides (AMOs) to functionally deplete U1 snRNA and then examined effects on the transcriptomes of human HeLa cells, mouse 3T3 cells, and *Drosophila* S2 cells (Kaida et al. 2010; Berg et al. 2012; Oh et al. 2017). Transcripts from some protein-coding genes showed accumulation of one or more introns, consistent with general splicing inhibition. However, transcripts from the majority of genes showed a strikingly distinct and unexpected pattern: the transcripts extended several kilobases from the TSS into the first intron and then abruptly ended (Fig. 6). This could have been due to U1 snRNP functioning as a positive elongation factor (e.g., at roadblocks that cause RNA polymerase II stalling), but 3′ RACE (rapid amplification of cDNA ends) revealed that these transcripts were, in fact, prematurely terminated and polyadenylated because of the use of intronic polyadenylation signals (Kaida et al. 2010; Berg et al. 2012; Oh et al. 2017). Notably, AMOs complementary to U2 snRNA or treatment with spliceostatin A, a small-molecule inhibitor of pre-mRNA splicing that binds the SF3b complex (Kaida et al. 2007), did not result in similar premature cleavage and polyadenylation patterns. This indicated that U1 snRNP has a critical role in protecting transcripts from premature cleavage that is in addition to and independent of its role in splicing (Kaida et al. 2010; Berg et al. 2012).

How does U1 snRNP control polyadenylation site usage, especially those in introns? A simple (but incomplete) model is that U1 binding to canonical 5′ splice sites blocks usage of nearby cryptic polyadenylation signals. Mutation of the 5′ splice site in a minigene construct inhibited splicing and resulted in increased levels of cleavage and polyadenylation in the downstream intron, but treatment with the U1 AMO increased the level of intronic termination events from this mutated construct even further (Kaida et al. 2010). Therefore, base-pairing of U1 snRNP to both canonical and noncanonical (or cryptic) 5′ splice sites (Engreitz et al. 2014; Oh et al. 2017) is required for complete protection from premature cleavage

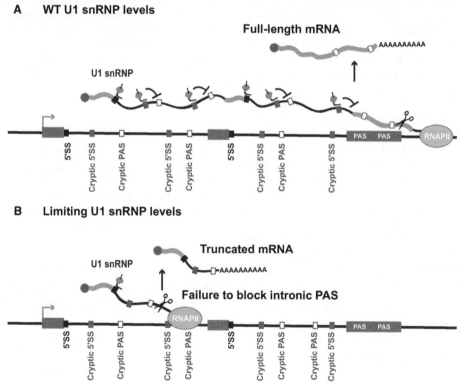

Figure 6. U1 binding to nascent transcripts controls polyadenylation signal usage. Besides canonical 5′ splice sites (5′SS) at exon–intron junctions and a polyadenylation signal (PAS) at the annotated 3′ end of the gene, there are often a number of cryptic 5′SS and PAS throughout nascent protein-coding transcripts. (*A*) Under standard conditions when U1 snRNP levels are high, U1 binds to canonical and cryptic 5′SS and blocks cleavage at nearby PAS. This results in the generation of full-length mRNAs. (*B*) In contrast, when U1 snRNP levels are limited, 5′SS tend to not be bound by U1, resulting in cleavage at a nearby PAS. Transcription is prematurely terminated and the generated short polyadenylated transcripts are often rapidly degraded.

and polyadenylation events (Fig. 6). At these binding sites, U1 appears to interact with a number of proteins within the cleavage and polyadenylation machinery (Boelens et al. 1993; Lutz et al. 1996; Gunderson et al. 1998; Awasthi and Alwine 2003), but the cleavage reaction is blocked, perhaps because of the lack of key factors that activate the cleavage reaction (e.g., CFIm68 and PABN1) (So et al. 2019).

Additional key insights into the underlying mechanism were obtained by modulating the amount of AMO that was added to cells. When U1 activity was approximately fully inhibited, cleavage and polyadenylation often occurred in the first intron with little or no transcription beyond that point. In contrast, when U1 activity was inhibited by 10%–50%, the position of the termination event was shifted further downstream from the TSS (e.g., proximal rather than distal polyadenylation sites were used in the 3′ UTR) (Berg et al. 2012). Suppression of cleavage and polyadenylation by U1 is thus a 5′–3′ directional process and, in general, the smaller the decrease in functional U1 levels, the greater the distance from the TSS to the polyadenylation signal that is selected for 3′-end processing. These data further suggested that 3′-end processing happens at the first actionable polyadenylation signal that is not "protected" by U1 snRNP bound nearby (within ∼3.5 kb). This protective role for U1 has been referred to as "telescripting" as it is necessary for nascent transcripts to be extended over long lengths (Berg et al. 2012).

Uncontrolled premature cleavage and polyadenylation likely would result in rampant transcriptional attrition and the inability to transcribe functional mRNAs. It thus makes sense that the cell would want to only change the outputs of a subset of genes when U1 levels are limiting. In fact, it was recently shown that some genes do not undergo changes in their cleavage and polyadenylation patterns when U1 is limiting (Oh et al. 2017). Remarkably, these unaffected genes are smaller than average (median length of 14.2 kb, compared to a median length of 22.8 kb for all expressed genes) and often encode proteins with functions related to cell-stress responses or basic cellular processes required for survival, including transcription, splicing, translation, and signaling. The overall expression of some of these genes can even be up-regulated upon U1 inhibition. In contrast, genes subjected to premature cleavage and polyadenylation when U1 is limiting are longer than average (median length of 39 kb) and often encode proteins related to cell cycle progression, DNA replication, or developmental processes (Oh et al. 2017). It thus appears that evolution has broadly selected against intron expansion in genes that are crucial for cell survival under adverse conditions. This likely ensures that these genes can be more rapidly induced (as fewer nucleotides need to be transcribed) with minimal potential for premature cleavage and polyadenylation events.

CONCLUSION

Here, we have highlighted two widespread mechanisms by which premature transcription termination can be used to attenuate metazoan gene outputs—one that is catalyzed by the Integrator complex and the other by the cleavage and polyadenylation machinery. These are near certainly not the only two mechanisms by which premature termination is catalyzed (and other mechanisms have already been identified as discussed above), but they nicely highlight how termination can shape the outputs of specific genes and thus the transcriptome. For example, U1 binding sites are depleted in the antisense direction of divergent RNAPII promoters, thus enabling cleavage and polyadenylation of antisense transcripts shortly after initiation and reinforcing promoter directionality (Almada et al. 2013; Ntini et al. 2013). Likewise, Integrator cleavage events can down-regulate the expression of some protein-coding genes by more than 100-fold. Controlling the recruitment/activity of the Integrator complex thus can likely function as a very efficient on/off gene expression switch. It is now critical for the field to reveal the underlying details of how these premature termination complexes are recruited to specific gene loci and how their cleavage activities are controlled. Potential cross talk between the premature termination pathways is unclear at this point, but seems likely to exist. In many cases, premature termination events lead to RNAs that are rapidly degraded, but it will be very interesting to determine if any of these transcripts are stable and what their biological functions are.

In total, it is becoming increasingly clear that synthesis of full-length transcripts is not necessarily the default option, and that premature cleavage events need to be actively suppressed. This raises the exciting possibility that there are many regulatory steps and fates for nascent transcripts that the field has largely not considered in the past. In addition to revealing general trends, each of these noncanonical RNAs may have its own unique biology that awaits discovery.

ACKNOWLEDGMENTS

We thank Sara Cherry, Gideon Dreyfuss, Karen Adelman, Eric Wagner, and all members of the Wilusz laboratory for discussions. This work was supported by National Institutes of Health (NIH) grants R35-GM119735 (to J.E.W.) and K99-GM131028 (to D.C.T.). J.E.W. is a Rita Allen Foundation Scholar.

REFERENCES

Albrecht TR, Shevtsov SP, Wu Y, Mascibroda LG, Peart NJ, Huang KL, Sawyer IA, Tong L, Dundr M, Wagner EJ. 2018. Integrator subunit 4 is a 'Symplekin-like' scaffold that associates with INTS9/11 to form the Integrator cleavage module. *Nucleic Acids Res* **46:** 4241–4255. doi:10.1093/nar/gky100

Almada AE, Wu X, Kriz AJ, Burge CB, Sharp PA. 2013. Promoter directionality is controlled by U1 snRNP and polyadenylation signals. *Nature* **499:** 360–363. doi:10.1038/nature12349

Arigo JT, Carroll KL, Ames JM, Corden JL. 2006. Regulation of yeast NRD1 expression by premature transcription termination. *Mol Cell* **21:** 641–651. doi:10.1016/j.molcel.2006.02.005

Arun G, Diermeier S, Akerman M, Chang KC, Wilkinson JE, Hearn S, Kim Y, MacLeod AR, Krainer AR, Norton L, et al.

2016. Differentiation of mammary tumors and reduction in metastasis upon Malat1 lncRNA loss. *Genes Dev* **30:** 34–51. doi:10.1101/gad.270959.115

Awasthi S, Alwine JC. 2003. Association of polyadenylation cleavage factor I with U1 snRNP. *RNA* **9:** 1400–1409. doi:10.1261/rna.5104603

Baillat D, Wagner EJ. 2015. Integrator: surprisingly diverse functions in gene expression. *Trends Biochem Sci* **40:** 257–264. doi:10.1016/j.tibs.2015.03.005

Baillat D, Hakimi MA, Näär AM, Shilatifard A, Cooch N, Shiekhattar R. 2005. Integrator, a multiprotein mediator of small nuclear RNA processing, associates with the C-terminal repeat of RNA polymerase II. *Cell* **123:** 265–276. doi:10.1016/j.cell.2005.08.019

Barbieri E, Trizzino M, Welsh SA, Owens TA, Calabretta B, Carroll M, Sarma K, Gardini A. 2018. Targeted enhancer activation by a subunit of the integrator complex. *Mol Cell* **71:** 103–116 e107. doi:10.1016/j.molcel.2018.05.031

Berg MG, Singh LN, Younis I, Liu Q, Pinto AM, Kaida D, Zhang Z, Cho S, Sherrill-Mix S, Wan L, et al. 2012. U1 snRNP determines mRNA length and regulates isoform expression. *Cell* **150:** 53–64. doi:10.1016/j.cell.2012.05.029

Boelens WC, Jansen EJ, van Venrooij WJ, Stripecke R, Mattaj IW, Gunderson SI. 1993. The human U1 snRNP-specific U1A protein inhibits polyadenylation of its own pre-mRNA. *Cell* **72:** 881–892. doi:10.1016/0092-8674(93)90577-D

Brannan K, Kim H, Erickson B, Glover-Cutter K, Kim S, Fong N, Kiemele L, Hansen K, Davis R, Lykke-Andersen J, et al. 2012. mRNA decapping factors and the exonuclease Xrn2 function in widespread premature termination of RNA polymerase II transcription. *Mol Cell* **46:** 311–324. doi:10.1016/j.molcel.2012.03.006

Braunschweig U, Gueroussov S, Plocik AM, Graveley BR, Blencowe BJ. 2013. Dynamic integration of splicing within gene regulatory pathways. *Cell* **152:** 1252–1269. doi:10.1016/j.cell.2013.02.034

Bresson S, Tuck A, Staneva D, Tollervey D. 2017. Nuclear RNA decay pathways aid rapid remodeling of gene expression in yeast. *Mol Cell* **65:** 787–800 e785. doi:10.1016/j.molcel.2017.01.005

Brown JA, Valenstein ML, Yario TA, Tycowski KT, Steitz JA. 2012. Formation of triple-helical structures by the 3′-end sequences of MALAT1 and MENβ noncoding RNAs. *Proc Natl Acad Sci* **109:** 19202–19207. doi:10.1073/pnas.1217338109

Brown JA, Bulkley D, Wang J, Valenstein ML, Yario TA, Steitz TA, Steitz JA. 2014. Structural insights into the stabilization of MALAT1 noncoding RNA by a bipartite triple helix. *Nat Struct Mol Biol* **21:** 633–640. doi:10.1038/nsmb.2844

Cazalla D, Xie M, Steitz JA. 2011. A primate herpesvirus uses the integrator complex to generate viral microRNAs. *Mol Cell* **43:** 982–992. doi:10.1016/j.molcel.2011.07.025

Chalamcharla VR, Folco HD, Dhakshnamoorthy J, Grewal SI. 2015. Conserved factor Dhp1/Rat1/Xrn2 triggers premature transcription termination and nucleates heterochromatin to promote gene silencing. *Proc Natl Acad Sci* **112:** 15548–15555. doi:10.1073/pnas.1522127112

Clemson CM, Hutchinson JN, Sara SA, Ensminger AW, Fox AH, Chess A, Lawrence JB. 2009. An architectural role for a nuclear noncoding RNA: NEAT1 RNA is essential for the structure of paraspeckles. *Mol Cell* **33:** 717–726. doi:10.1016/j.molcel.2009.01.026

Egloff S, Szczepaniak SA, Dienstbier M, Taylor A, Knight S, Murphy S. 2010. The integrator complex recognizes a new double mark on the RNA polymerase II carboxyl-terminal domain. *J Biol Chem* **285:** 20564–20569. doi:10.1074/jbc.M110.132530

Elrod ND, Henriques T, Huang KL, Tatomer DC, Wilusz JE, Wagner EJ, Adelman K. 2019. The integrator complex attenuates promoter-proximal transcription at protein-coding genes. *Mol Cell* **76:** 738–752 e737. doi:10.1016/j.molcel.2019.10.034

Engreitz JM, Sirokman K, McDonel P, Shishkin AA, Surka C, Russell P, Grossman SR, Chow AY, Guttman M, Lander ES.
2014. RNA–RNA interactions enable specific targeting of noncoding RNAs to nascent pre-mRNAs and chromatin sites. *Cell* **159:** 188–199. doi:10.1016/j.cell.2014.08.018

Erickson B, Sheridan RM, Cortazar M, Bentley DL. 2018. Dynamic turnover of paused Pol II complexes at human promoters. *Genes Dev* **32:** 1215–1225. doi:10.1101/gad.316810.118

Evans R, Weber J, Ziff E, Darnell JE. 1979. Premature termination during adenovirus transcription. *Nature* **278:** 367–370. doi:10.1038/278367a0

Ezzeddine N, Chen J, Waltenspiel B, Burch B, Albrecht T, Zhuo M, Warren WD, Marzluff WF, Wagner EJ. 2011. A subset of *Drosophila* integrator proteins is essential for efficient U7 snRNA and spliceosomal snRNA 3′-end formation. *Mol Cell Biol* **31:** 328–341. doi:10.1128/MCB.00943-10

Festa RA, Thiele DJ. 2011. Copper: an essential metal in biology. *Curr Biol* **21:** R877–R883. doi:10.1016/j.cub.2011.09.040

Gardini A, Baillat D, Cesaroni M, Hu D, Marinis JM, Wagner EJ, Lazar MA, Shilatifard A, Shiekhattar R. 2014. Integrator regulates transcriptional initiation and pause release following activation. *Mol Cell* **56:** 128–139. doi:10.1016/j.molcel.2014.08.004

Gómez-Orte E, Sáenz-Narciso B, Zheleva A, Ezcurra B, de Toro M, López R, Gastaca I, Nilsen H, Sacristán MP, Schnabel R, et al. 2019. Disruption of the *Caenorhabditis elegans* Integrator complex triggers a non-conventional transcriptional mechanism beyond snRNA genes. *PLoS Genet* **15:** e1007981. doi:10.1371/journal.pgen.1007981

Gunderson SI, Polycarpou-Schwarz M, Mattaj IW. 1998. U1 snRNP inhibits pre-mRNA polyadenylation through a direct interaction between U1 70K and poly(A) polymerase. *Mol Cell* **1:** 255–264. doi:10.1016/S1097-2765(00)80026-X

Günther V, Lindert U, Schaffner W. 2012. The taste of heavy metals: gene regulation by MTF-1. *Biochim Biophys Acta* **1823:** 1416–1425. doi:10.1016/j.bbamcr.2012.01.005

Hentze MW, Castello A, Schwarzl T, Preiss T. 2018. A brave new world of RNA-binding proteins. *Nat Rev Mol Cell Biol* **19:** 327–341. doi:10.1038/nrm.2017.130

Hernandez N. 1985. Formation of the 3′ end of U1 snRNA is directed by a conserved sequence located downstream of the coding region. *EMBO J* **4:** 1827–1837. doi:10.1002/j.1460-2075.1985.tb03857.x

Huang C, Liang D, Tatomer DC, Wilusz JE. 2018. A length-dependent evolutionarily conserved pathway controls nuclear export of circular RNAs. *Genes Dev* **32:** 639–644. doi:10.1101/gad.314856.118

Jenks MH, O'Rourke TW, Reines D. 2008. Properties of an intergenic terminator and start site switch that regulate *IMD2* transcription in yeast. *Mol Cell Biol* **28:** 3883–3893. doi:10.1128/MCB.00380-08

Ji P, Diederichs S, Wang W, Böing S, Metzger R, Schneider PM, Tidow N, Brandt B, Buerger H, Bulk E, et al. 2003. MALAT-1, a novel noncoding RNA, and thymosin β4 predict metastasis and survival in early-stage non–small cell lung cancer. *Oncogene* **22:** 8031–8041. doi:10.1038/sj.onc.1206928

Kaida D, Motoyoshi H, Tashiro E, Nojima T, Hagiwara M, Ishigami K, Watanabe H, Kitahara T, Yoshida T, Nakajima H, et al. 2007. Spliceostatin A targets SF3b and inhibits both splicing and nuclear retention of pre-mRNA. *Nat Chem Biol* **3:** 576–583. doi:10.1038/nchembio.2007.18

Kaida D, Berg MG, Younis I, Kasim M, Singh LN, Wan L, Dreyfuss G. 2010. U1 snRNP protects pre-mRNAs from premature cleavage and polyadenylation. *Nature* **468:** 664–668. doi:10.1038/nature09479

Kamieniarz-Gdula K, Proudfoot NJ. 2019. Transcriptional control by premature termination: a forgotten mechanism. *Trends Genet* **35:** 553–564. doi:10.1016/j.tig.2019.05.005

Kamieniarz-Gdula K, Gdula MR, Panser K, Nojima T, Monks J, Wiśniewski JR, Riepsaame J, Brockdorff N, Pauli A, Proudfoot NJ. 2019. Selective roles of vertebrate PCF11 in premature and full-length transcript termination. *Mol Cell* **74:** 158–172 e159. doi:10.1016/j.molcel.2019.01.027

Kapp LD, Abrams EW, Marlow FL, Mullins MC. 2013. The integrator complex subunit 6 (Ints6) confines the dorsal organizer in vertebrate embryogenesis. *PLoS Genet* **9:** e1003822. doi:10.1371/journal.pgen.1003822

Kapranov P, Cheng J, Dike S, Nix DA, Duttagupta R, Willingham AT, Stadler PF, Hertel J, Hackermuller J, Hofacker IL, et al. 2007. RNA maps reveal new RNA classes and a possible function for pervasive transcription. *Science* **316:** 1484–1488. doi:10.1126/science.1138341

Kearse MG, Goldman DH, Choi J, Nwaezeapu C, Liang D, Green KM, Goldstrohm AC, Todd PK, Green R, Wilusz JE. 2019. Ribosome queuing enables non-AUG translation to be resistant to multiple protein synthesis inhibitors. *Genes Dev* **33:** 871–885. doi:10.1101/gad.324715.119

Kiledjian M. 2018. Eukaryotic RNA 5′-end NAD^+ capping and DeNADding. *Trends Cell Biol* **28:** 454–464. doi:10.1016/j.tcb.2018.02.005

Kramer MC, Liang D, Tatomer DC, Gold B, March ZM, Cherry S, Wilusz JE. 2015. Combinatorial control of *Drosophila* circular RNA expression by intronic repeats, hnRNPs, and SR proteins. *Genes Dev* **29:** 2168–2182. doi:10.1101/gad.270421.115

Krebs AR, Imanci D, Hoerner L, Gaidatzis D, Burger L, Schübeler D. 2017. Genome-wide single-molecule footprinting reveals high RNA polymerase II turnover at paused promoters. *Mol Cell* **67:** 411–422 e414. doi:10.1016/j.molcel.2017.06.027

Kristensen LS, Andersen MS, Stagsted LVW, Ebbesen KK, Hansen TB, Kjems J. 2019. The biogenesis, biology and characterization of circular RNAs. *Nat Rev Genet* **20:** 675–691. doi:10.1038/s41576-019-0158-7

Kuehner JN, Brow DA. 2008. Regulation of a eukaryotic gene by GTP-dependent start site selection and transcription attenuation. *Mol Cell* **31:** 201–211. doi:10.1016/j.molcel.2008.05.018

Lai F, Gardini A, Zhang A, Shiekhattar R. 2015. Integrator mediates the biogenesis of enhancer RNAs. *Nature* **525:** 399–403. doi:10.1038/nature14906

Lee NN, Chalamcharla VR, Reyes-Turcu F, Mehta S, Zofall M, Balachandran V, Dhakshnamoorthy J, Taneja N, Yamanaka S, Zhou M, et al. 2013. Mtr4-like protein coordinates nuclear RNA processing for heterochromatin assembly and for telomere maintenance. *Cell* **155:** 1061–1074. doi:10.1016/j.cell.2013.10.027

Li X, Yang L, Chen LL. 2018. The biogenesis, functions, and challenges of circular RNAs. *Mol Cell* **71:** 428–442. doi:10.1016/j.molcel.2018.06.034

Liang D, Wilusz JE. 2014. Short intronic repeat sequences facilitate circular RNA production. *Genes Dev* **28:** 2233–2247. doi:10.1101/gad.251926.114

Liang D, Tatomer DC, Luo Z, Wu H, Yang L, Chen LL, Cherry S, Wilusz JE. 2017. The output of protein-coding genes shifts to circular RNAs when the pre-mRNA processing machinery is limiting. *Mol Cell* **68:** 940–954 e943. doi:10.1016/j.molcel.2017.10.034

Lubas M, Christensen MS, Kristiansen MS, Domanski M, Falkenby LG, Lykke-Andersen S, Andersen JS, Dziembowski A, Jensen TH. 2011. Interaction profiling identifies the human nuclear exosome targeting complex. *Mol Cell* **43:** 624–637. doi:10.1016/j.molcel.2011.06.028

Lutz CS, Murthy KG, Schek N, O'Connor JP, Manley JL, Alwine JC. 1996. Interaction between the U1 snRNP-A protein and the 160-kD subunit of cleavage-polyadenylation specificity factor increases polyadenylation efficiency in vitro. *Genes Dev* **10:** 325–337. doi:10.1101/gad.10.3.325

Mao YS, Sunwoo H, Zhang B, Spector DL. 2011. Direct visualization of the co-transcriptional assembly of a nuclear body by noncoding RNAs. *Nat Cell Biol* **13:** 95–101. doi:10.1038/ncb2140

Marzluff WF, Wagner EJ, Duronio RJ. 2008. Metabolism and regulation of canonical histone mRNAs: life without a poly(A) tail. *Nat Rev Genet* **9:** 843–854. doi:10.1038/nrg2438

Merino E, Yanofsky C. 2005. Transcription attenuation: a highly conserved regulatory strategy used by bacteria. *Trends Genet* **21:** 260–264. doi:10.1016/j.tig.2005.03.002

Moore MJ, Proudfoot NJ. 2009. Pre-mRNA processing reaches back to transcription and ahead to translation. *Cell* **136:** 688–700. doi:10.1016/j.cell.2009.02.001

Naville M, Gautheret D. 2010. Transcription attenuation in bacteria: theme and variations. *Brief Funct Genomics* **9:** 178–189. doi:10.1093/bfgp/elq008

Nechaev S, Fargo DC, dos Santos G, Liu L, Gao Y, Adelman K. 2010. Global analysis of short RNAs reveals widespread promoter-proximal stalling and arrest of Pol II in *Drosophila*. *Science* **327:** 335–338. doi:10.1126/science.1181421

Ntini E, Järvelin AI, Bornholdt J, Chen Y, Boyd M, Jørgensen M, Andersson R, Hoof I, Schein A, Andersen PR, et al. 2013. Polyadenylation site-induced decay of upstream transcripts enforces promoter directionality. *Nat Struct Mol Biol* **20:** 923–928. doi:10.1038/nsmb.2640

Oegema R, Baillat D, Schot R, van Unen LM, Brooks A, Kia SK, Hoogeboom AJM, Xia Z, Li W, Cesaroni M, et al. 2017. Human mutations in integrator complex subunits link transcriptome integrity to brain development. *PLoS Genet* **13:** e1006809. doi:10.1371/journal.pgen.1006809

Oh JM, Di C, Venters CC, Guo J, Arai C, So BR, Pinto AM, Zhang Z, Wan L, Younis I, et al. 2017. U1 snRNP telescripting regulates a size-function-stratified human genome. *Nat Struct Mol Biol* **24:** 993–999. doi:10.1038/nsmb.3473

Patop IL, Wüst S, Kadener S. 2019. Past, present, and future of circRNAs. *EMBO J* **38:** e100836. doi:10.15252/embj.2018100836

Preker P, Nielsen J, Kammler S, Lykke-Andersen S, Christensen MS, Mapendano CK, Schierup MH, Jensen TH. 2008. RNA exosome depletion reveals transcription upstream of active human promoters. *Science* **322:** 1851–1854. doi:10.1126/science.1164096

Rubtsova MP, Vasilkova DP, Moshareva MA, Malyavko AN, Meerson MB, Zatsepin TS, Naraykina YV, Beletsky AV, Ravin NV, Dontsova OA. 2019. Integrator is a key component of human telomerase RNA biogenesis. *Sci Rep* **9:** 1701. doi:10.1038/s41598-018-38297-6

Rutkowski RJ, Warren WD. 2009. Phenotypic analysis of deflated/Ints7 function in *Drosophila* development. *Dev Dyn* **238:** 1131–1139. doi:10.1002/dvdy.21922

Sasaki YT, Ideue T, Sano M, Mituyama T, Hirose T. 2009. MENε/β noncoding RNAs are essential for structural integrity of nuclear paraspeckles. *Proc Natl Acad Sci* **106:** 2525–2530. doi:10.1073/pnas.0807899106

Shah N, Maqbool MA, Yahia Y, El Aabidine AZ, Esnault C, Forné I, Decker TM, Martin D, Schüller R, Krebs S, et al. 2018. Tyrosine-1 of RNA polymerase II CTD controls global termination of gene transcription in mammals. *Mol Cell* **69:** 48–61 e46. doi:10.1016/j.molcel.2017.12.009

Shi Y. 2017. Mechanistic insights into precursor messenger RNA splicing by the spliceosome. *Nat Rev Mol Cell Biol* **18:** 655–670. doi:10.1038/nrm.2017.86

Shi Y, Manley JL. 2015. The end of the message: multiple protein–RNA interactions define the mRNA polyadenylation site. *Genes Dev* **29:** 889–897. doi:10.1101/gad.261974.115

Skaar JR, Ferris AL, Wu X, Saraf A, Khanna KK, Florens L, Washburn MP, Hughes SH, Pagano M. 2015. The Integrator complex controls the termination of transcription at diverse classes of gene targets. *Cell Res* **25:** 288–305. doi:10.1038/cr.2015.19

So BR, Di C, Cai Z, Venters CC, Guo J, Oh JM, Arai C, Dreyfuss G. 2019. A complex of U1 snRNP with cleavage and polyadenylation factors controls telescripting, regulating mRNA transcription in human cells. *Mol Cell* **76:** 590–599 e594. doi:10.1016/j.molcel.2019.08.007

Stadelmayer B, Micas G, Gamot A, Martin P, Malirat N, Koval S, Raffel R, Sobhian B, Severac D, Rialle S, et al. 2014. Integrator complex regulates NELF-mediated RNA polymerase II pause/release and processivity at coding genes. *Nat Commun* **5:** 5531. doi:10.1038/ncomms6531

Steinmetz EJ, Conrad NK, Brow DA, Corden JL. 2001. RNA-binding protein Nrd1 directs poly(A)-independent 3′-end formation of RNA polymerase II transcripts. *Nature* **413:** 327–331. doi:10.1038/35095090

Steurer B, Janssens RC, Geverts B, Geijer ME, Wienholz F, Theil AF, Chang J, Dealy S, Pothof J, van Cappellen WA, et al. 2018. Live-cell analysis of endogenous GFP-RPB1 uncovers rapid turnover of initiating and promoter-paused RNA polymerase II. *Proc Natl Acad Sci* **115:** E4368–E4376. doi:10.1073/pnas.1717920115

Sullivan KD, Steiniger M, Marzluff WF. 2009. A core complex of CPSF73, CPSF100, and Symplekin may form two different cleavage factors for processing of poly(A) and histone mRNAs. *Mol Cell* **34:** 322–332. doi:10.1016/j.molcel.2009.04.024

Sunwoo H, Dinger ME, Wilusz JE, Amaral PP, Mattick JS, Spector DL. 2009. MEN ε/β nuclear-retained non-coding RNAs are up-regulated upon muscle differentiation and are essential components of paraspeckles. *Genome Res* **19:** 347–359. doi:10.1101/gr.087775.108

Tatomer DC, Elrod ND, Liang D, Xiao MS, Jiang JZ, Jonathan M, Huang KL, Wagner EJ, Cherry S, Wilusz JE. 2019. The Integrator complex cleaves nascent mRNAs to attenuate transcription. *Genes Dev* **33:** 1525–1538. doi:10.1101/gad.330167.119

Thiebaut M, Colin J, Neil H, Jacquier A, Séraphin B, Lacroute F, Libri D. 2008. Futile cycle of transcription initiation and termination modulates the response to nucleotide shortage in *S. cerevisiae*. *Mol Cell* **31:** 671–682. doi:10.1016/j.molcel.2008.08.010

Tian B, Manley JL. 2017. Alternative polyadenylation of mRNA precursors. *Nat Rev Mol Cell Biol* **18:** 18–30. doi:10.1038/nrm.2016.116

van den Berg DLC, Azzarelli R, Oishi K, Martynoga B, Urbán N, Dekkers DHW, Demmers JA, Guillemot F. 2017. Nipbl interacts with Zfp609 and the integrator complex to regulate cortical neuron migration. *Neuron* **93:** 348–361. doi:10.1016/j.neuron.2016.11.047

Vasiljeva L, Kim M, Terzi N, Soares LM, Buratowski S. 2008. Transcription termination and RNA degradation contribute to silencing of RNA polymerase II transcription within heterochromatin. *Mol Cell* **29:** 313–323. doi:10.1016/j.molcel.2008.01.011

Venters CC, Oh JM, Di C, So BR, Dreyfuss G. 2019. U1 snRNP telescripting: suppression of premature transcription termination in introns as a new layer of gene regulation. *Cold Spring Harb Perspect Biol* **11:** a032235. doi:10.1101/cshperspect.a032235

Wagschal A, Rousset E, Basavarajaiah P, Contreras X, Harwig A, Laurent-Chabalier S, Nakamura M, Chen X, Zhang K, Meziane O, et al. 2012. Microprocessor, Setx, Xrn2, and Rrp6 co-operate to induce premature termination of transcription by RNAPII. *Cell* **150:** 1147–1157. doi:10.1016/j.cell.2012.08.004

Wang ZF, Whitfield ML, Ingledue TC, Dominski Z, Marzluff WF. 1996. The protein that binds the 3′ end of histone mRNA: a novel RNA-binding protein required for histone pre-mRNA processing. *Genes Dev* **10:** 3028–3040. doi:10.1101/gad.10.23.3028

Wilusz JE. 2016. Long noncoding RNAs: re-writing dogmas of RNA processing and stability. *Biochim Biophys Acta* **1859:** 128–138. doi:10.1016/j.bbagrm.2015.06.003

Wilusz JE. 2018. A 360 degrees view of circular RNAs: from biogenesis to functions. *Wiley Interdiscip Rev RNA* **9:** e1478. doi:10.1002/wrna.1478

Wilusz JE, Spector DL. 2010. An unexpected ending: noncanonical 3′ end processing mechanisms. *RNA* **16:** 259–266. doi:10.1261/rna.1907510

Wilusz JE, Freier SM, Spector DL. 2008. 3′ end processing of a long nuclear-retained noncoding RNA yields a tRNA-like cytoplasmic RNA. *Cell* **135:** 919–932. doi:10.1016/j.cell.2008.10.012

Wilusz JE, JnBaptiste CK, Lu LY, Kuhn CD, Joshua-Tor L, Sharp PA. 2012. A triple helix stabilizes the 3′ ends of long noncoding RNAs that lack poly(A) tails. *Genes Dev* **26:** 2392–2407. doi:10.1101/gad.204438.112

Wright S. 1993. Regulation of eukaryotic gene expression by transcriptional attenuation. *Mol Biol Cell* **4:** 661–668. doi:10.1091/mbc.4.7.661

Wu Y, Albrecht TR, Baillat D, Wagner EJ, Tong L. 2017. Molecular basis for the interaction between Integrator subunits IntS9 and IntS11 and its functional importance. *Proc Natl Acad Sci* **114:** 4394–4399. doi:10.1073/pnas.1616605114

Xie M, Zhang W, Shu MD, Xu A, Lenis DA, DiMaio D, Steitz JA. 2015. The host Integrator complex acts in transcription-independent maturation of herpesvirus microRNA 3′ ends. *Genes Dev* **29:** 1552–1564. doi:10.1101/gad.266973.115

Zhang Y, Yang L, Chen LL. 2014. Life without A tail: new formats of long noncoding RNAs. *Int J Biochem Cell Biol* **54:** 338–349. doi:10.1016/j.biocel.2013.10.009

Zhang Y, Koe CT, Tan YS, Ho J, Tan P, Yu F, Sung WK, Wang H. 2019. the integrator complex prevents dedifferentiation of intermediate neural progenitors back into neural stem cells. *Cell Rep* **27:** 987–996 e983. doi:10.1016/j.celrep.2019.03.089

3′ UTRs Regulate Protein Functions by Providing a Nurturing Niche during Protein Synthesis

CHRISTINE MAYR

Cancer Biology and Genetics Program, Memorial Sloan Kettering Cancer Center, New York, New York 10065, USA

Correspondence: mayrc@mskcc.org

Messenger RNAs (mRNAs) are the templates for protein synthesis as the coding region is translated into the amino acid sequence. mRNAs also contain 3′ untranslated regions (3′ UTRs) that harbor additional elements for the regulation of protein function. If the amino acid sequence of a protein is necessary and sufficient for its function, we call it 3′ UTR–independent. In contrast, functions that are accomplished by protein complexes whose formation requires the presence of a specific 3′ UTR are 3′ UTR–dependent protein functions. We showed that 3′ UTRs can regulate protein activity without affecting protein abundance, and alternative 3′ UTRs can diversify protein functions. We currently think that the regulation of protein function by 3′ UTRs is facilitated by the local environment at the site of protein synthesis, which we call the nurturing niche for nascent proteins. This niche is composed of the mRNA and the bound proteins that consist of RNA-binding proteins and recruited proteins. It enables the formation of specific protein complexes, as was shown for TIS granules, a recently discovered cytoplasmic membraneless organelle. This finding suggests that changing the niche for nascent proteins will alter protein activity and function, implying that cytoplasmic membraneless organelles can regulate protein function in a manner that is independent of protein abundance.

The influence of nurture on nature has classically been studied with respect to human behavior and is usually used to characterize the influence of external factors, such as parents, social relationships, and the surrounding culture on the development of a child (Collins et al. 2000). Nurture also affects nature during the development of organs, including the brain, in which neuronal activity influences the construction of cortical networks (Ben-Ari 2002). Here, I propose that this principle is also true for protein functions regulated by 3′ untranslated regions (3′ UTRs). We showed that proteins with the same amino acid sequence that are encoded from mRNA isoforms with alternative 3′ UTRs have different functions (Berkovits and Mayr 2015; Ma and Mayr 2018; Lee and Mayr 2019). This suggests that protein functions are influenced by the local environment at the site of protein synthesis, which will be called the nurturing niche for nascent proteins. Proteins are born into a niche provided by the mRNA and its bound factors. In this analogy, 3′ UTRs act as the "social relationships" of nascent proteins as they influence certain protein functions. The influence of 3′ UTRs will not entirely change the nature of a protein, as, for example, a kinase will remain a kinase, but 3′ UTRs are able to recruit protein interactors that influence the function of the protein, thus changing its qualitative properties and fate. Furthermore, when a protein is translated within a larger RNA granule, the surrounding mRNAs can serve as the "surrounding culture," again changing protein interactors and functions. In this review, I will summarize our current understanding of the regulation of protein functions by the nurturing niche provided by 3′ UTRs (Fig. 1).

3′ UTR SHORTENING DOES NOT CORRELATE WITH INCREASED mRNA LEVELS IN MOST TRANSCRIPTOME-WIDE STUDIES

My laboratory studies the functions of 3′ UTRs. When I started my laboratory the boundaries of human 3′ UTRs were not well-annotated. Therefore, we established a sequencing method called 3′-seq that allowed us to identify all 3′ UTRs of the transcriptome (Lianoglou et al. 2013). The obtained reads overlap the junction between the 3′ UTR and the poly(A) tail, thus providing single-nucleotide resolution of 3′ UTR ends. Importantly, we validated the obtained data using an independent method and showed that 3′-seq is quantitative. This means that in addition to identifying 3′ UTR ends, we can quantify the reads and calculate the expression of individual 3′ UTR isoforms.

We applied 3′-seq to a series of cell lines and normal human tissues and found that nearly one-half of human genes use alternative cleavage and polyadenylation to generate mRNA transcripts with alternative 3′ UTRs (Lianoglou et al. 2013). Our validation experiments showed that in the vast majority of cases the coding region was identical, and the only difference in the mRNA transcript was found in the 3′ UTR. We were surprised that such a large number of mRNAs that generate short or long 3′ UTRs encode proteins with identical amino acid sequence. This finding motivated us to study the functions of alternative 3′ UTRs.

At the time, we and others thought that one of the major functions of 3′ UTRs is the regulation of protein abun-

---

© 2019 Mayr. This article is distributed under the terms of the Creative Commons Attribution-NonCommercial License, which permits reuse and redistribution, except for commercial purposes, provided that the original author and source are credited.

Published by Cold Spring Harbor Laboratory Press; doi:10.1101/sqb.2019.84.039206

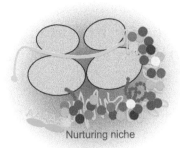

Figure 1. The 3′ UTR with the bound RNA-binding proteins and recruited proteins creates a nurturing niche for nascent proteins. Shown are two ribosomes translating an mRNA (light green). The nascent peptide is depicted in dark green. The RNA-binding proteins bound to the 3′ UTR are in yellow, orange, or red and the recruited proteins are in blue shaded colors.

dance, mostly through regulation of mRNA stability (Barreau et al. 2005; Sandberg et al. 2008; Bartel 2009; Mayr and Bartel 2009). When we compared alternative 3′ UTR isoform abundance and overall mRNA abundance of a gene from two different conditions or cell types, we expected to see a correlation between 3′ UTR shortening and higher mRNA levels. However, in our data set, we did not observe such a relationship (Fig. 2; Lianoglou et al. 2013). Importantly, several other transcriptome-wide studies using different experimental systems agreed with our finding (Spies et al. 2013; Gruber et al. 2014; Zhang et al. 2016; Jia et al. 2017). Overall, from the mRNAs that significantly changed their 3′ UTR isoform levels, <20% also changed their mRNA abundance levels. Instead, across conditions and cell types the majority of mRNAs either changed their mRNA abundance levels or they changed their 3′ UTR isoform expression (Fig. 2; Lianoglou et al. 2013). This suggested that mRNA abundance and 3′ UTR ratios are orthogonal measures of gene expression.

In addition to the lack of correlation between 3′ UTR shortening and mRNA abundance, no association between 3′ UTR isoform length and changes in protein levels was found during differentiation of embryonic stem cells. This study showed that 87% of genes with significant 3′ UTR isoform changes did not show differences in protein abundance (Brumbaugh et al. 2018). However, it is worth pointing out that there are exceptions: Several classes of genes with low mRNA transcript stability, which include cell cycle regulators, cytokines, and oncogenes, indeed use their 3′ UTRs to change their protein levels substantially. This regulation is largely mediated by microRNAs and AU-rich elements (Kontoyiannis et al. 1999; Mayr and Bartel 2009; Herranz et al. 2015). Another exception happens in development during maternal to zygotic transition, in which several groups showed that many mRNAs are cleared using elements found in 3′ UTRs (Giraldez et al. 2006; Benoit et al. 2009).

The overall lack of a relationship between 3′ UTR shortening and mRNA abundance then prompted some researchers to propose that alternative 3′ UTR isoform expression may be random and may not matter (Spies et al. 2013; Neve and Furger 2014; Xu and Zhang

2018). This proposal, however, is inconsistent with the known roles of regulatory elements in 3′ UTRs. First of all, 3′ UTRs, including alternative 3′ UTRs, are known to regulate mRNA localization (Lécuyer et al. 2007; An et al. 2008; Taliaferro et al. 2016; Mayr 2017; Tushev et al. 2018; Ciolli Mattioli et al. 2019). Second, early comparative genomic analyses found conserved 3′ UTR sequences when comparing homologous genes across species, but noticed highly divergent 3′ UTR sequences within an organism when comparing similar proteins. For example, they observed that actin family members that encode similar proteins have very different 3′ UTRs. However, these 3′ UTR sequences were conserved across organisms, suggesting that 3' UTRs contain additional genetic information to distinguish the functions of highly similar proteins (Yaffe et al. 1985). Although 3′ UTRs are usually less conserved than coding regions, high sequence conservation of 3′ UTR regulatory elements was confirmed in subsequent genome-wide studies (Siepel et al. 2005; Xie et al. 2005). Third, when comparing 3′ UTR length of genes across species, we and others found that 3′ UTR length has expanded substantially during evolution of more complex animals (Chen et al. 2012; Mayr 2016). And, last, the use of replicates showed coordinated changes in 3′ UTR isoform expression in different cell types and conditions, which speaks against random 3′ UTR isoform choice (Lianoglou et al. 2013; Lee et al. 2018; Singh et al. 2018).

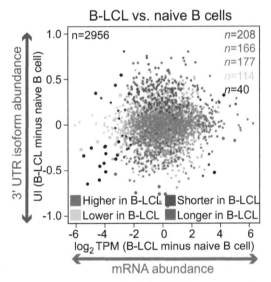

Figure 2. mRNAs either change their abundance or they change their 3′ UTR ratio during B-cell transformation. Shown are changes in mRNA levels (x-axis) versus changes in 3′ UTR isoform abundance (y-axis) in naive B cells before and after immortalization using Epstein–Barr virus transformation (called B lymphoblastic cells [B-LCLs]). Changes in 3′ UTR isoform abundance are given as difference in UTR index (UI). The UI is the fraction of reads that map to the short 3′ UTR isoform out of all the reads mapping to the 3′ UTR. Genes with significant mRNA or 3′ UTR isoform changes are color-coded. Gray indicates no significant change, and black indicates genes with a significant change in both mRNA levels and 3′ UTR isoform abundance. (Adapted from Lianoglou et al. 2013, published by Cold Spring Harbor Laboratory Press.)

DISCOVERY OF 3′ UTR–DEPENDENT PROTEIN FUNCTIONS

To investigate functions of alternative 3′ UTRs, we studied the *CD47* gene, which generates mRNA transcripts with short or long 3′ UTRs and encodes a plasma membrane receptor that acts as a "don't eat me signal" (Jaiswal et al. 2009). We fused green fluorescent protein (GFP) to the coding region of CD47 and then added either its short or long 3′ UTR. When we transfected these constructs into cells, we observed that CD47 protein encoded by the mRNA isoform with the long 3′ UTR (CD47-LU) perfectly localized to the plasma membrane, whereas CD47 protein encoded by the mRNA isoform with the short 3′ UTR (CD47-SU) mostly localized intracellularly (Berkovits and Mayr 2015). Interestingly, mRNA localization of the two constructs was similar and both of them localized to the endoplasmic reticulum (ER). This suggested that the alternative 3′ UTRs of *CD47* contain information for mRNA-independent protein localization.

Furthermore, we noticed that the cells transfected with CD47-LU formed lamellipodia, whereas the cells that were transfected with CD47-SU did not. As lamellipodia formation is a sign of active RAC1, we examined the amount of RAC1-GTP (active RAC1) in the cells after transfection of either CD47-SU or CD47-LU. This revealed that transfection of CD47-LU increased RAC1 activation, whereas transfection of CD47-SU did not. This was a surprising observation as the protein that is generated from the constructs is identical. This finding indicated that the alternative 3′ UTRs of *CD47* can control CD47 protein function as only CD47-LU was able to activate a downstream signaling pathway. This was a remarkable discovery as it shows that 3′ UTRs can change the qualitative property and the function of a protein (Berkovits and Mayr 2015).

We then investigated how the difference in protein function was regulated and found that the RNA-binding protein HuR only bound to the long, but not to the short, 3′ UTR of *CD47*. One of the most abundant interactors of HuR is the protein SET, a small acidic protein that is highly expressed in the nucleus and cytoplasm of nearly all cell types (Brennan and Steitz 2001). Despite the similarity in name, SET should not be confused with a SET domain, a domain found in methyltransferases. Instead, the protein SET has many different functions and was initially discovered as an inhibitor of the phosphatase PP2A (Li et al. 1996). It also acts as a histone chaperone that prevents histone acetylation and is regarded as a "reader" of nonacetylated lysines (Seo et al. 2001; Wang et al. 2016).

We performed coimmunoprecipitation of GFP-tagged CD47-SU or CD47-LU and asked if endogenous SET binds. We found that SET only interacted with CD47-LU, but not with CD47-SU. Therefore, we regard the protein–protein interaction between SET and CD47-LU to be 3′ UTR–dependent. Furthermore, the binding of SET to CD47 also depends on specific amino acids in the cytoplasmic domains of CD47. CD47 protein has five transmembrane domains and is translated on the surface of the ER. The cytoplasmic domains contain positively charged amino acids and mutation of five of 10 amino acids abrogated the binding of SET to CD47. As HuR binds to AU-rich and U-rich motifs in the 3′ UTR of *CD47* and SET binds to positively charged amino acids in CD47 protein, our data indicates that establishment of the 3′ UTR–dependent interaction between SET and CD47 requires motifs in the mRNA as well as in the protein encoded by the mRNA (Berkovits and Mayr 2015).

MASS SPECTROMETRY IDENTIFIES 3′ UTR–INDEPENDENT AND 3′ UTR–DEPENDENT PROTEIN INTERACTORS OF THE E3 UBIQUITIN LIGASE BIRC3

In the case of CD47, we identified the 3′ UTR–dependent protein interactor SET through a literature search. However, this approach is not scalable. Therefore, we established an experimental strategy to identify 3′ UTR–dependent protein interactors for basically any candidate. For our proof-of-principle experiments we chose BIRC3, a nonmembrane bound E3 ligase that can be translated from an mRNA with a short or long 3′ UTR. We generated constructs in which we fused GFP to the coding region of BIRC3 and then added either the short or long *BIRC3* 3′ UTR. Our goal was to identify 3′ UTR–dependent BIRC3 protein interactors. We transfected the constructs into cells grown in SILAC media, followed by GFP coimmunoprecipitation, followed by quantitative mass spectrometry. This revealed that some protein interactors bound equally well to BIRC3 protein regardless if it was translated from the short or long 3′ UTR isoform. We called these 3′ UTR–independent interactors. However, we found many BIRC3 protein interactors that bound better to BIRC3 protein translated from the long 3′ UTR isoform. We validated a number of interactors and were able to confirm 80% of the candidates that we set out to validate (Fig. 3A; Lee and Mayr 2019). This experiment revealed that BIRC3 has 3′ UTR–independent and 3′ UTR–dependent interactors that could be separated into different functional categories. BIRC3, which is also called cIAP2, is known to regulate NF-κB signaling and apoptosis (Srinivasula and Ashwell 2008; Beug et al. 2012). The regulation of cell death is consistent with the identification of DIABLO and XIAP, two 3′ UTR–independent BIRC3 interactors and known regulators of cell death (Srinivasula and Ashwell 2008). In contrast, the 3′ UTR–dependent interaction partners have known roles in the regulation of chromatin, mitochondria, and protein trafficking. These functions have previously not been associated with BIRC3 (Gyrd-Hansen and Meier 2010).

We then set out to find a biologically relevant 3′ UTR–dependent function of BIRC3. This is a function of BIRC3 that can only be accomplished by BIRC3 protein translated from the long 3′ UTR isoform (BIRC3-LU). We identified CXCR4-dependent B-cell migration as such a function because we observed impaired B-cell migration in BIRC3 knockout cells and in cells with knockdown of the long 3′ UTR isoform. As these cells are unable to

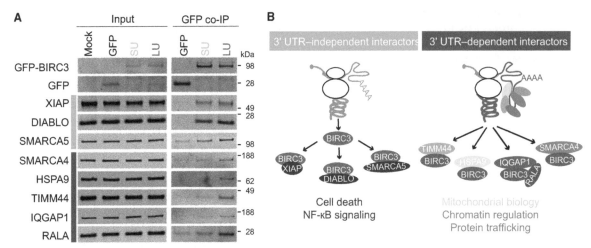

Figure 3. BIRC3 has 3′ UTR–independent as well as 3′ UTR–dependent interaction partners. (*A*) Western blot validation of endogenous BIRC3 interactors. 3′ UTR–independent (orange bar) and 3′ UTR–dependent (red bar) BIRC3 protein interactors identified by GFP coimmunoprecipitation in HEK293T cells after transfection of constructs containing the coding region of BIRC3 fused to GFP and the short 3′ UTR (SU) or the long 3′ UTR (LU). 1% of input was loaded. (*B*) 3′ UTR–independent BIRC3 functions are mediated by 3′ UTR–independent protein interactors (brown), whereas 3′ UTR–dependent BIRC3 functions are mediated by 3′ UTR–dependent interactors (blue and purple colors). 3′ UTR–dependent functions can only be accomplished by BIRC3-LU. (Reprinted from Lee and Mayr 2019, with permission from Elsevier.)

mediate migration, despite the expression of BIRC3 protein translated from the short 3′ UTR isoform (BIRC3-SU), our data indicates that regulation of migration is a 3′ UTR–dependent function of BIRC3. We further showed that two of the validated 3′ UTR–dependent BIRC3 interactors, IQGAP1 and RALA, were both required for migration and were mediators for one of the 3′ UTR–dependent functions of BIRC3 (Lee and Mayr 2019).

In summary, we found that the E3 ligase BIRC3 has 3′ UTR–independent and 3′ UTR–dependent interactors (Fig. 3B). It seems that the 3′ UTR–independent interactors form complexes with BIRC3 mostly in a posttranslational manner as protein abundance of BIRC3 or the interactors was a crucial determinant for interaction. In contrast, protein abundance does not seem to be the most important factor to establish 3′ UTR–dependent interactions. This interpretation is based on the observation that knockdown of the long *BIRC3* 3′ UTR isoform did not change overall BIRC3 protein levels, suggesting that only a minor fraction of total BIRC3 protein consists of BIRC3-LU, which was sufficient to mediate the 3′ UTR–dependent functions. Importantly, BIRC3 uses the 3′ UTR–dependent protein interactors to diversify its functions in a way that is independent of protein localization (Lee and Mayr 2019).

3′ UTRs CONTROL PROTEIN ACTIVITY WITHOUT AFFECTING PROTEIN ABUNDANCE

So far, we had only studied the functions of alternative 3′ UTRs. However, we reasoned that 3′ UTR–mediated protein complex formation is likely to happen at the majority of 3′ UTRs and not only at alternative 3′ UTRs. Therefore, as our next candidate, we chose the 3′ UTR of the *TP53* gene, which generates a single 3′ UTR isoform (Lianoglou et al. 2013). Moreover, we previously had only used GFP-tagged expression constructs and decided to use CRISPR to delete the 3′ UTR at the endogenous locus (Mayr 2019). Our goal was to delete the majority of 3′ UTR regulatory elements while keeping mRNA processing intact in order to not disturb protein production. To do so, we used two guide RNAs and deleted the DNA sequence encoding the 3′ UTR between the stop codon and ~150 bp upstream of the polyadenylation signal. We generated clones with a homozygous deletion of the *TP53* 3′ UTR in a human cell line and found that the 3′ UTR deletion did not influence overall mRNA or protein levels of p53 (S Mitschka and C Mayr, unpubl.). This is in contrast to the widely accepted opinion that one of the major functions of 3′ UTRs is the regulation of mRNA or protein abundance (Barreau et al. 2005; Bartel 2009). Despite no difference in p53 protein levels, in the cells with 3′ UTR deletion, we observed a phenotypic difference that is consistent with premature activation of p53 (S Mitschka and C Mayr, unpubl.). This indicates that the *TP53* 3′ UTR regulates the activity of p53 protein.

We used CRISPR-mediated manipulation to investigate the 3′ UTR–dependent functions of a second candidate. Also in the case of *PTEN*, the 3′ UTR did not affect overall mRNA or protein levels, but instead seemed to alter the enzymatic activity of the PTEN phosphatase (B Kwon, SH Lee, and C Mayr, unpubl.). 3′ UTRs were also manipulated using CRISPR by others. Again, deletion of the long 3′ UTR of Dscam1 did not change Dscam1 protein levels, but impaired axon outgrowth in flies (Zhang et al. 2019). In our hands, all so far tested 3′ UTRs controlled protein activity without affecting protein abundance, indicating that this function of 3′ UTRs is widespread. This has also been suggested by others (Fernandes 2019; Ribero 2019).

HOW IS INFORMATION TRANSFERRED FROM 3′ UTRs TO PROTEINS?

We showed that 3′ UTRs can control protein functions (Berkovits and Mayr 2015; Lee and Mayr 2019), suggesting that 3′ UTRs contain genetic information for the regulation of protein function. We set out to investigate how the information is transferred from 3′ UTRs to proteins. We showed previously that CD47 function is regulated by the 3′ UTR–dependent binding of SET to CD47 protein. SET binds to both the long *CD47* 3′ UTR (as it binds HuR and HuR binds to the long 3′ UTR) as well as to CD47 protein. After the establishment of the SET–CD47 interaction, the *CD47* 3′ UTR is no longer necessary, suggesting that SET is transferred from the 3′ UTR to the protein. As the information contained in the long *CD47* 3′ UTR is transferred to the protein through SET binding, we set out to study how SET transfer is regulated.

The RNA-binding protein HuR is important in this process as knockdown of HuR abrogated binding of SET to the CD47 protein. However, HuR was not sufficient as SET interacted better with CD47 when it was translated from an mRNA with the entire long *CD47* 3′ UTR compared to a 3′ UTR that only consisted of HuR-binding sites (Berkovits and Mayr 2015). Therefore, we hypothesized that a second RNA-binding protein cooperates with HuR to accomplish SET transfer and we set out to identify this protein.

As the 3′ UTR of *CD47* is 4194 nucleotides long, we started with the identification of a region that contains the majority of information for SET transfer. We generated three 3′ UTR pieces and tested their capacity to mediate CD47 cell surface localization, one of the readouts for SET binding (Berkovits and Mayr 2015). We found that all three pieces contained some, but none of them contained the majority, of the information (Fig. 4) (W Ma and C Mayr, unpubl.). This suggested the presence of a repeated motif that is required for SET transfer. To identify it, we generated several short artificial 3′ UTRs that contained different repeated motifs and tested their capacity for CD47 cell surface localization. This led to the discovery of "artificial UTR 5" (aUTR5), derived from the TNF-α AU-rich element that contains six AU-rich elements (Kontoyiannis et al. 1999). This short 3′ UTR fragment was able to fully recapitulate the function of the long *CD47* 3′ UTR with respect to SET binding and CD47 cell surface localization (Ma and Mayr 2018).

With the aUTR5, we performed RNA affinity pull-down experiments and identified the RNA-binding proteins that bind specifically to this 3′ UTR compared with a size-matched control derived from the short 3′ UTR of *CD47*. Using mass spectrometry analysis, we identified HuR and TIS11B as specific interactors of the aUTR5 (Ma and Mayr 2018). TIS11B is an RNA-binding protein also known as ZFP36L1 or BRF1 that binds to the canonical AU-rich element AUUUA (Lai et al. 2002; Stoecklin et al. 2002; Stumpo et al. 2004; Lykke-Andersen and Wagner 2005; Herranz et al. 2015). TIS11B is known to bind to the 3′ UTRs of cytokines and destabilizes their mRNAs (Stoecklin et al. 2002; Lykke-Andersen and Wagner 2005; Herranz et al. 2015). Knockdown of TIS11B abrogated the 3′ UTR–dependent binding of SET to CD47 and decreased CD47 cell surface localization. However, it did not change expression of the proteins involved in 3′ UTR–mediated surface expression of CD47, including CD47, HuR, and SET, that all have several AU-rich elements in their 3′ UTRs, suggesting that not all AU-rich elements have destabilizing capacity. This raised the question of how TIS11B would mediate SET transfer (Ma and Mayr 2018).

We hypothesized that TIS11B might generate a "special" environment for SET transfer and performed imaging of TIS11B. We found that endogenous TIS11B formed granular assemblies in the vicinity of the ER. Live cell confocal imaging with Airyscan mode showed that TIS11B assembles into a tubule-like network that is intertwined with the ER and covers a large portion of the ER (Ma and Mayr 2018). TIS11B assemblies are RNA granules as they contain specific mRNAs and proteins. Therefore, we called them TIS granules. TIS granules were detected under steady state cultivation conditions and in the absence of stress in all cell types investigated, including in primary cells directly isolated from human skin (M Pan, W Ma, C Mayr, and J Young, unpubl.). Not all TIS11B molecules assemble into TIS granules, because we also detected soluble TIS11B protein in the cytoplasm.

To start to investigate the role of TIS11B or TIS granules in SET transfer, we visualized *CD47* mRNAs. Using RNA-FISH (fluorescent in situ hybridization targeting ribonucleic acid molecules), we observed that *CD47-LU* mRNA transcripts localized to TIS granules, whereas *CD47-SU* mRNA transcripts localized to the ER but mostly to regions not covered by TIS granules. As CD47 protein was detected at the site of mRNA localization, we concluded that CD47-LU is translated on the ER in the region of TIS granules, whereas CD47-SU is translated on a different domain of the rough ER that is not associated with TIS granules (Ma and Mayr 2018). This indicates that one function of the alternative 3′ UTRs of *CD47* is the regulation of local translation on different subdomains of the ER.

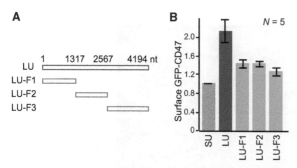

Figure 4. All fragments of the *CD47* 3′ UTR contain information for SET transfer. (*A*) Schematic showing the full-length *CD47* 3′ UTR (LU) together with the three fragments (LU-F1, LU-F2, LU-F3) tested. The nucleotide positions for the boundaries of the LU fragments are shown. (*B*) A readout for SET binding is surface expression of GFP-CD47. Shown are flow cytometry results of the indicated samples. SU is the short *CD47* 3′ UTR and is used as negative control, whereas LU is the positive control.

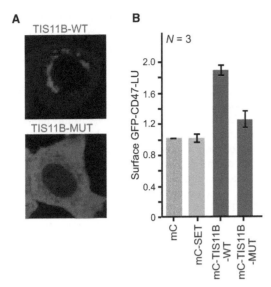

Figure 5. TIS granules are necessary for SET binding to CD47. (*A*) Wild-type TIS11B (TIS11B-WT) forms TIS granules, whereas mutant TIS11B (TIS11B-MUT) does not. (*B*) Presence of TIS11B-WT substantially increases SET binding, but expression of TIS11B that is unable to form TIS granules (TIS11B-MUT) has a small effect. (mC) mCherry. Shown as in Figure 4B. (Adapted from Ma and Mayr 2018, with permission from Elsevier.)

Next, we investigated in more detail how the protein–protein interaction between SET and CD47 is formed. Usually, when two proteins are able to bind to each other, overexpression of the proteins results in increased interaction—a principle that is often used when proteins are tagged with FLAG and HA to test their interaction. However, overexpression of SET and CD47 did not result in increased interaction (Ma and Mayr 2018). Instead, we observed that overexpression of the RNA-binding protein TIS11B resulted in better interaction between SET and CD47-LU (Fig. 5). This raised the question whether overexpression of TIS11B protein or the presence of TIS granules is crucial. To distinguish between the two scenarios we needed a mutant of TIS11B that behaves like the wild-type protein, but is unable to assemble into TIS granules. Through hypothesis-driven mutagenesis of TIS11B we obtained such a mutant. This mutant revealed that the presence of TIS granules is required for the binding of SET to CD47-LU (Fig. 5; Ma and Mayr 2018). This observation allowed us to predict the following: CD47-SU is a protein that is translated on the ER outside of the TIS granule region and does not bind to SET. If translation in the TIS granule region is required for the interaction of SET and CD47, CD47-SU will bind to SET if CD47-SU is translated in the TIS granule region. We set up an experiment in which we forced CD47-SU to be translated in the TIS granule region and were able to recapitulate the prediction. This finding indicates that certain protein–protein interactions can only be established when a protein is translated within a particular subcellular compartment and not outside the compartment (Ma and Mayr 2018). As 3′ UTRs contribute to the local environment within TIS granules (W Ma and C Mayr, unpubl.), our finding indicates that one function of mRNAs is to create a niche during protein synthesis.

Interestingly, in the TIS granule region we found a paradoxical relationship between SET binding and SET concentration: The interaction between SET and CD47 was increased despite lower abundance of SET in the TIS granule region compared to the cytoplasm (Fig. 6; Ma and Mayr 2018). A similar observation was recently obtained with a different experimental system in vitro. Within a CAPRIN1 phase-separated environment the activity of the CNOT7 deadenylation enzyme was seven times higher than in buffer, despite a lower concentration of CNOT7 in the condensate than outside (Kim et al. 2019). The activity increase did not happen in other phase-separated environments and was specific for CAPRIN1. The authors suggested that the protein solvent environment modulates the enzymatic activity of CNOT7 (Kim et al. 2019).

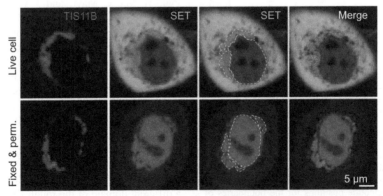

Figure 6. The environment in TIS granules is different from the cytoplasm. Confocal imaging of HeLa cells after transfection of GFP-TIS11B (red) and mCherry-SET (green). (*Top*) SET is relatively depleted from the TIS granule region (demarcated by the white dotted line). The yellow dotted line demarcates the nucleus. (*Bottom*) After fixing and permeabilization of cells, SET disappears from the cytoplasm but is retained in the TIS granule region, suggesting that the biochemical environment in the TIS granule region is different from the cytoplasm. (Adapted from Ma and Mayr 2018, with permission from Elsevier.)

CONCLUSIONS

These findings have several important implications.

1. The most straightforward way how membraneless organelles and condensates may influence reactions is through partitioning and the concentration of molecules for reactions (Banani et al. 2017; Shin and Brangwynne 2017; Kato and McKnight 2018). This has been beautifully illustrated in signaling puncta formed by phosphorylated LAT. Within these LAT clusters, kinases are enriched and phosphatases are excluded, thus facilitating signaling reactions (Su et al. 2016).

2. In addition to regulating partitioning, membraneless organelles were also shown to provide a different biochemical environment that can affect physical and structural properties of biomolecules: In buffer, a short piece of double-stranded DNA remained double-stranded, whereas within a Ddx4 phase-separated environment it was destabilized, leading to sort of unwinding of the double-stranded DNA, possibly through cation–π interactions acting on the backbone of the DNA (Nott et al. 2016).

3. More recently, it was shown that the solvent environment provided by a CAPRIN1 phase-separated condensate increased the enzymatic activity of CNOT7 deadenylase. This was a remarkable result as the solvent was more important for the regulation of the enzymatic activity than the concentration of the enzyme (Kim et al. 2019). This finding is reminiscent of the observation that SET interacts with CD47 better in the TIS granule region than outside, despite a lower concentration of SET in the TIS granule region (Ma and Mayr 2018). These two examples provide a major conceptual advance in our understanding of how cellular reactions are regulated. So far, most of the time, it has been assumed that a high overall or local concentration is a key determinant of enzymatic activity, but the examples show that the solvent or the local environment created by the mRNA together with the bound proteins strongly influence reactions.

4. Specific mRNAs are enriched in TIS granules and one of the determinants for their enrichment are the motifs located in their 3′ UTRs (Ma and Mayr 2018). This implies that additional functions of mRNAs are to compartmentalize the cytoplasm and to contribute to the generation of a local environment. The mRNA, its bound proteins, and the emergent properties acquired by their formation of higher-order assemblies create a local niche. In the future, it will be important to dissect the contribution of the enriched proteins and the enriched mRNAs to the material properties of TIS granules and other membraneless organelles.

5. It is highly likely that in addition to TIS granules other membraneless organelles exist that allow translation and cotranslational protein complex assembly. This would mean that the cytoplasm is much more compartmentalized than previously thought, and it will be interesting to identify new cytoplasmic membraneless organelles and to determine the reactions that are promoted or inhibited. This may have important implications for synthetic and cell biology as we may be able to promote certain reactions through the generation of designer membraneless organelles (Reinkemeier et al. 2019).

6. As shown for TIS granules, the local environment influences processes during protein translation and plays a role in peritranslational protein complex assembly (Natan et al. 2017; Ma and Mayr 2018). The niche formed at the site of protein synthesis may act as a filter to promote certain 3′ UTR–dependent protein interactions over others (Fig. 7A–C). As most 3′ UTRs contain thousands of nucleotides, many RNA-binding proteins are likely to bind to a specific 3′ UTR. It is currently unknown how many RNA-binding proteins bind to a 3′ UTR, but mass spectrometry analyses performed in *Caenorhabditis elegans* found approximately 10 to 30 different proteins bound per 1000 nucleotides of mRNA (Theil et al. 2019). A complementary analysis in HeLa cells using high-resolution imaging estimated how often a single RNA-binding protein was bound to an mRNA and detected between

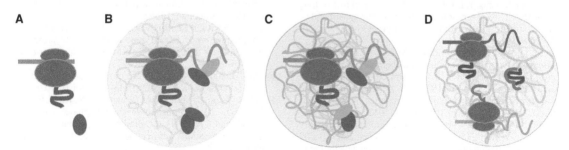

Figure 7. The niche at the site of protein synthesis affects peritranslational protein complex formation. (*A*) Translation of an mRNA (light green) without 3′ UTR may result in the lack of an RNA granule at the site of translation and may prevent cotranslational protein complex assembly. The ribosome is shown in gray and the newly synthesized protein is shown in dark green. (*B*) As in *A*, but the presence of a local RNA granule (yellow) allows a specific protein–protein interaction between a 3′ UTR–recruited protein (dark blue) and the newly synthesized protein (dark green). (*C*) As in *B*, but a different type of RNA granule (purple) allows the formation of a different 3′ UTR–dependent protein complex. (*D*) Translation of two different mRNAs (light red, light blue) within a larger RNA granule (gray) containing a diversity of mRNAs may allow co- or peritranslational protein complex formation.

2 and 34 molecules with a median of 5–8 (Mateu-Regué et al. 2019). Each RNA-binding protein may recruit several effector proteins, thus generating a niche. It is conceivable that the composition of the niche may affect the binding kinetics of proteins with the nascent chain. This can be tested if a protein is forced to be translated in a different niche. One way to change the niche is to swap the 3′ UTR or to recruit the mRNA into a different membraneless organelle (Fig. 7A–C).

7. It has long been thought that "mRNA regulons" exist, in which functionally related groups of mRNAs are coregulated (Keene 2007). However, there is recent evidence that cytoplasmic mRNP (messenger RNA nucleoprotein complex) granules consist of single mRNAs, which would preclude coregulation of mRNAs (Mateu-Regué et al. 2019). However, formation of larger mRNA granules such as TIS granules that contain many mRNAs may enable the mingling of different mRNAs to allow coregulation as well as co- or peritranslational protein complex assembly of transcripts that are translated in vicinity to each other (Fig. 7D; Mayr 2018; Shiber et al. 2018; Schwarz and Beck 2019).

ACKNOWLEDGMENTS

I thank W. Ma for coining the term "nurturing niche" and several members of the Mayr laboratory, including W. Ma, E. Horste, X. Chen, and N. Robertson, for critical comments on the manuscript and for helpful discussions. This work was funded by the National Institutes of Health (NIH) Director's Pioneer Award (DP1-GM123454), the Pershing Square Sohn Cancer Research Alliance, and the National Cancer Institute (NCI) Cancer Center Support Grant (P30 CA008748).

Competing interest statement: The author declares no competing interests.

REFERENCES

An JJ, Gharami K, Liao GY, Woo NH, Lau AG, Vanevski F, Torre ER, Jones KR, Feng Y, Lu B, et al. 2008. Distinct role of long 3′ UTR BDNF mRNA in spine morphology and synaptic plasticity in hippocampal neurons. *Cell* **134:** 175–187. doi:10.1016/j.cell.2008.05.045

Banani SF, Lee HO, Hyman AA, Rosen MK. 2017. Biomolecular condensates: organizers of cellular biochemistry. *Nat Rev Mol Cell Biol* **18:** 285–298. doi:10.1038/nrm.2017.7

Barreau C, Paillard L, Osborne HB. 2005. AU-rich elements and associated factors: are there unifying principles? *Nucleic Acids Res* **33:** 7138–7150. doi:10.1093/nar/gki1012

Bartel DP. 2009. MicroRNAs: target recognition and regulatory functions. *Cell* **136:** 215–233. doi:10.1016/j.cell.2009.01.002

Ben-Ari Y. 2002. Excitatory actions of gaba during development: the nature of the nurture. *Nat Rev Neurosci* **3:** 728–739. doi:10.1038/nrn920

Benoit B, He CH, Zhang F, Votruba SM, Tadros W, Westwood JT, Smibert CA, Lipshitz HD, Theurkauf WE. 2009. An essential role for the RNA-binding protein Smaug during the *Drosophila* maternal-to-zygotic transition. *Development* **136:** 923–932. doi:10.1242/dev.031815

Berkovits BD, Mayr C. 2015. Alternative 3′ UTRs act as scaffolds to regulate membrane protein localization. *Nature* **522:** 363–367. doi:10.1038/nature14321

Beug ST, Cheung HH, LaCasse EC, Korneluk RG. 2012. Modulation of immune signalling by inhibitors of apoptosis. *Trends Immunol* **33:** 535–545. doi:10.1016/j.it.2012.06.004

Brennan CM, Steitz JA. 2001. HuR and mRNA stability. *Cell Mol Life Sci* **58:** 266–277. doi:10.1007/PL00000854

Brumbaugh J, Di Stefano B, Wang X, Borkent M, Forouzmand E, Clowers KJ, Ji F, Schwarz BA, Kalocsay M, Elledge SJ, et al. 2018. Nudt21 controls cell fate by connecting alternative polyadenylation to chromatin signaling. *Cell* **172:** 629–631. doi:10.1016/j.cell.2017.12.035

Chen CY, Chen ST, Juan HF, Huang HC. 2012. Lengthening of 3′UTR increases with morphological complexity in animal evolution. *Bioinformatics* **28:** 3178–3181. doi:10.1093/bioinformatics/bts623

Ciolli Mattioli C, Rom A, Franke V, Imami K, Arrey G, Terne M, Woehler A, Akalin A, Ulitsky I, Chekulaeva M. 2019. Alternative 3′ UTRs direct localization of functionally diverse protein isoforms in neuronal compartments. *Nucleic Acids Res* **47:** 2560–2573. doi:10.1093/nar/gky1270

Collins WA, Maccoby EE, Steinberg L, Hetherington EM, Bornstein MH. 2000. Contemporary research on parenting. The case for nature and nurture. *Am Psychol* **55:** 218–232. doi:10.1037/0003-066X.55.2.218

Fernandes N, Buchan JR. 2019. *RPS28B* mRNA acts as a scaffold promoting cis-translational interaction of proteins driving P-body assembly. bioRxiv doi:10.1101/696823

Giraldez AJ, Mishima Y, Rihel J, Grocock RJ, Van Dongen S, Inoue K, Enright AJ, Schier AF. 2006. Zebrafish MiR-430 promotes deadenylation and clearance of maternal mRNAs. *Science* **312:** 75–79. doi:10.1126/science.1122689

Gruber AR, Martin G, Müller P, Schmidt A, Gruber AJ, Gumienny R, Mittal N, Jayachandran R, Pieters J, Keller W, et al. 2014. Global 3′ UTR shortening has a limited effect on protein abundance in proliferating T cells. *Nat Commun* **5:** 5465. doi:10.1038/ncomms6465

Gyrd-Hansen M, Meier P. 2010. IAPs: from caspase inhibitors to modulators of NF-κB, inflammation and cancer. *Nat Rev Cancer* **10:** 561–574. doi:10.1038/nrc2889

Herranz N, Gallage S, Mellone M, Wuestefeld T, Klotz S, Hanley CJ, Raguz S, Acosta JC, Innes AJ, Banito A, et al. 2015. mTOR regulates MAPKAPK2 translation to control the senescence-associated secretory phenotype. *Nat Cell Biol* **17:** 1205–1217. doi:10.1038/ncb3225

Jaiswal S, Jamieson CH, Pang WW, Park CY, Chao MP, Majeti R, Traver D, van Rooijen N, Weissman IL. 2009. CD47 is upregulated on circulating hematopoietic stem cells and leukemia cells to avoid phagocytosis. *Cell* **138:** 271–285. doi:10.1016/j.cell.2009.05.046

Jia X, Yuan S, Wang Y, Fu Y, Ge Y, Ge Y, Lan X, Feng Y, Qiu F, Li P, et al. 2017. The role of alternative polyadenylation in the antiviral innate immune response. *Nat Commun* **8:** 14605. doi:10.1038/ncomms14605

Kato M, McKnight SL. 2018. A solid-state conceptualization of information transfer from gene to message to protein. *Annu Rev Biochem* **87:** 351–390. doi:10.1146/annurev-biochem-061516-044700

Keene JD. 2007. RNA regulons: coordination of post-transcriptional events. *Nat Rev Genet* **8:** 533–543. doi:10.1038/nrg2111

Kim TH, Tsang B, Vernon RM, Sonenberg N, Kay LE, Forman-Kay JD. 2019. Phospho-dependent phase separation of FMRP and CAPRIN1 recapitulates regulation of translation and deadenylation. *Science* **365:** 825–829. doi:10.1126/science.aax4240

Kontoyiannis D, Pasparakis M, Pizarro TT, Cominelli F, Kollias G. 1999. Impaired on/off regulation of TNF biosynthesis in mice lacking TNF AU-rich elements: implications for joint and gut-associated immunopathologies. *Immunity* **10:** 387–398. doi:10.1016/S1074-7613(00)80038-2

Lai WS, Kennington EA, Blackshear PJ. 2002. Interactions of CCCH zinc finger proteins with mRNA: non-binding tristetraprolin mutants exert an inhibitory effect on degradation of AU-rich element-containing mRNAs. *J Biol Chem* **277:** 9606–9613. doi:10.1074/jbc.M110395200

Lécuyer E, Yoshida H, Parthasarathy N, Alm C, Babak T, Cerovina T, Hughes TR, Tomancak P, Krause HM. 2007. Global analysis of mRNA localization reveals a prominent role in organizing cellular architecture and function. *Cell* **131:** 174–187. doi:10.1016/j.cell.2007.08.003

Lee SH, Mayr C. 2019. Gain of additional BIRC3 protein functions through 3′-UTR-mediated protein complex formation. *Mol Cell* **74:** 701–712 e709. doi:10.1016/j.molcel.2019.03.006

Lee SH, Singh I, Tisdale S, Abdel-Wahab O, Leslie CS, Mayr C. 2018. Widespread intronic polyadenylation inactivates tumour suppressor genes in leukaemia. *Nature* **561:** 127–131. doi:10.1038/s41586-018-0465-8

Li M, Makkinje A, Damuni Z. 1996. The myeloid leukemia-associated protein SET is a potent inhibitor of protein phosphatase 2A. *J Biol Chem* **271:** 11059–11062. doi:10.1074/jbc.271.19.11059

Lianoglou S, Garg V, Yang JL, Leslie CS, Mayr C. 2013. Ubiquitously transcribed genes use alternative polyadenylation to achieve tissue-specific expression. *Genes Dev* **27:** 2380–2396. doi:10.1101/gad.229328.113

Lykke-Andersen J, Wagner E. 2005. Recruitment and activation of mRNA decay enzymes by two ARE-mediated decay activation domains in the proteins TTP and BRF-1. *Genes Dev* **19:** 351–361. doi:10.1101/gad.1282305

Ma W, Mayr C. 2018. A membraneless organelle associated with the endoplasmic reticulum enables 3′UTR-mediated protein–protein interactions. *Cell* **175:** 1492–1506 e1419. doi:10.1016/j.cell.2018.10.007

Mateu-Regué A, Christiansen J, Bagger FO, Winther O, Hellriegel C, Nielsen FC. 2019. Single mRNP analysis reveals that small cytoplasmic mRNP granules represent mRNA singletons. *Cell Rep* **29:** 736–748 e734. doi:10.1016/j.celrep.2019.09.018

Mayr C. 2016. Evolution and biological roles of alternative 3′UTRs. *Trends Cell Biol* **26:** 227–237. doi:10.1016/j.tcb.2015.10.012

Mayr C. 2017. Regulation by 3′-untranslated regions. *Annu Rev Genet* **51:** 171–194. doi:10.1146/annurev-genet-120116-024704

Mayr C. 2018. Protein complexes assemble as they are being made. *Nature* **561:** 186–187. doi:10.1038/d41586-018-05905-4

Mayr C. 2019. What are 3′ UTRs doing? *Cold Spring Harb Perspect Biol* **11:** a034728. doi:10.1101/cshperspect.a034728

Mayr C, Bartel DP. 2009. Widespread shortening of 3′UTRs by alternative cleavage and polyadenylation activates oncogenes in cancer cells. *Cell* **138:** 673–684. doi:10.1016/j.cell.2009.06.016

Natan E, Wells JN, Teichmann SA, Marsh JA. 2017. Regulation, evolution and consequences of cotranslational protein complex assembly. *Curr Opin Struct Biol* **42:** 90–97. doi:10.1016/j.sbi.2016.11.023

Neve J, Furger A. 2014. Alternative polyadenylation: less than meets the eye? *Biochem Soc Trans* **42:** 1190–1195. doi:10.1042/BST20140054

Nott TJ, Craggs TD, Baldwin AJ. 2016. Membraneless organelles can melt nucleic acid duplexes and act as biomolecular filters. *Nat Chem* **8:** 569–575. doi:10.1038/nchem.2519

Reinkemeier CD, Girona GE, Lemke EA. 2019. Designer membraneless organelles enable codon reassignment of selected mRNAs in eukaryotes. *Science* **363:** eaaw2644. doi:10.1126/science.aaw2644

Ribero DM, Prod'homme A, Teixeira A, Zanzoni A, Brun C. 2019. The role of 3′UTR-protein complexes in the regulation of protein multifunctionality and subcellular localization. bioRxiv doi:10.1101/784702

Sandberg R, Neilson JR, Sarma A, Sharp PA, Burge CB. 2008. Proliferating cells express mRNAs with shortened 3′ untranslated regions and fewer microRNA target sites. *Science* **320:** 1643–1647. doi:10.1126/science.1155390

Schwarz A, Beck M. 2019. The benefits of cotranslational assembly: a structural perspective. *Trends Cell Biol* **29:** 791–803. doi:10.1016/j.tcb.2019.07.006

Seo SB, McNamara P, Heo S, Turner A, Lane WS, Chakravarti D. 2001. Regulation of histone acetylation and transcription by INHAT, a human cellular complex containing the set oncoprotein. *Cell* **104:** 119–130. doi:10.1016/S0092-8674(01)00196-9

Shiber A, Döring K, Friedrich U, Klann K, Merker D, Zedan M, Tippmann F, Kramer G, Bukau B. 2018. Cotranslational assembly of protein complexes in eukaryotes revealed by ribosome profiling. *Nature* **561:** 268–272. doi:10.1038/s41586-018-0462-y

Shin Y, Brangwynne CP. 2017. Liquid phase condensation in cell physiology and disease. *Science* **357:** eaaf4382. doi:10.1126/science.aaf4382

Siepel A, Bejerano G, Pedersen JS, Hinrichs AS, Hou M, Rosenbloom K, Clawson H, Spieth J, Hillier LW, Richards S, et al. 2005. Evolutionarily conserved elements in vertebrate, insect, worm, and yeast genomes. *Genome Res* **15:** 1034–1050. doi:10.1101/gr.3715005

Singh I, Lee SH, Sperling AS, Samur MK, Tai YT, Fulciniti M, Munshi NC, Mayr C, Leslie CS. 2018. Widespread intronic polyadenylation diversifies immune cell transcriptomes. *Nat Commun* **9:** 1716. doi:10.1038/s41467-018-04112-z

Spies N, Burge CB, Bartel DP. 2013. 3′ UTR-isoform choice has limited influence on the stability and translational efficiency of most mRNAs in mouse fibroblasts. *Genome Res* **23:** 2078–2090. doi:10.1101/gr.156919.113

Srinivasula SM, Ashwell JD. 2008. IAPs: what's in a name? *Mol Cell* **30:** 123–135. doi:10.1016/j.molcel.2008.03.008

Stoecklin G, Colombi M, Raineri I, Leuenberger S, Mallaun M, Schmidlin M, Gross B, Lu M, Kitamura T, Moroni C. 2002. Functional cloning of BRF1, a regulator of ARE-dependent mRNA turnover. *EMBO J* **21:** 4709–4718. doi:10.1093/emboj/cdf444

Stumpo DJ, Byrd NA, Phillips RS, Ghosh S, Maronpot RR, Castranio T, Meyers EN, Mishina Y, Blackshear PJ. 2004. Chorioallantoic fusion defects and embryonic lethality resulting from disruption of *Zfp36L1*, a gene encoding a CCCH tandem zinc finger protein of the Tristetraprolin family. *Mol Cell Biol* **24:** 6445–6455. doi:10.1128/MCB.24.14.6445-6455.2004

Su X, Ditlev JA, Hui E, Xing W, Banjade S, Okrut J, King DS, Taunton J, Rosen MK, Vale RD. 2016. Phase separation of signaling molecules promotes T cell receptor signal transduction. *Science* **352:** 595–599. doi:10.1126/science.aad9964

Taliaferro JM, Vidaki M, Oliveira R, Olson S, Zhan L, Saxena T, Wang ET, Graveley BR, Gertler FB, Swanson MS, et al. 2016. Distal alternative last exons localize mRNAs to neural projections. *Mol Cell* **61:** 821–833. doi:10.1016/j.molcel.2016.01.020

Theil K, Imami K, Rajewsky N. 2019. Identification of proteins and miRNAs that specifically bind an mRNA in vivo. *Nat Commun* **10:** 4205. doi:10.1038/s41467-019-12050-7

Tushev G, Glock C, Heumüller M, Biever A, Jovanovic M, Schuman EM. 2018. Alternative 3′ UTRs modify the localization, regulatory potential, stability, and plasticity of mRNAs in neuronal compartments. *Neuron* **98:** 495–511 e496. doi:10.1016/j.neuron.2018.03.030

Wang D, Kon N, Lasso G, Jiang L, Leng W, Zhu WG, Qin J, Honig B, Gu W. 2016. Acetylation-regulated interaction between p53 and SET reveals a widespread regulatory mode. *Nature* **538:** 118–122. doi:10.1038/nature19759

Xie X, Lu J, Kulbokas EJ, Golub TR, Mootha V, Lindblad-Toh K, Lander ES, Kellis M. 2005. Systematic discovery of regulatory motifs in human promoters and 3′ UTRs by comparison of several mammals. *Nature* **434:** 338–345. doi:10.1038/nature03441

Xu C, Zhang J. 2018. Alternative polyadenylation of mammalian transcripts is generally deleterious, not adaptive. *Cell Syst* **6:** 734–742 e734. doi:10.1016/j.cels.2018.05.007

Yaffe D, Nudel U, Mayer Y, Neuman S. 1985. Highly conserved sequences in the 3′ untranslated region of mRNAs coding for homologous proteins in distantly related species. *Nucleic Acids Res* **13:** 3723–3737. doi:10.1093/nar/13.10.3723

Zhang KX, Tan L, Pellegrini M, Zipursky SL, McEwen JM. 2016. Rapid changes in the translatome during the conversion of growth cones to synaptic terminals. *Cell Rep* **14:** 1258–1271. doi:10.1016/j.celrep.2015.12.102

Zhang Z, So K, Peterson R, Bauer M, Ng H, Zhang Y, Kim JH, Kidd T, Miura P. 2019. Elav-mediated exon skipping and alternative polyadenylation of the *Dscam1* gene are required for axon outgrowth. *Cell Rep* **27:** 3808–3817 e3807. doi:10.1016/j.celrep.2019.05.083

The THO Complex as a Paradigm for the Prevention of Cotranscriptional R-Loops

Rosa Luna, Ana G. Rondón, Carmen Pérez-Calero, Irene Salas-Armenteros, and Andrés Aguilera

Centro Andaluz de Biología Molecular y Medicina Regenerativa CABIMER, Universidad de Sevilla-CSIC-Universidad Pablo de Olavide, 41092 Seville, Spain

Corresponding author: aguilo@us.es

Different proteins associate with the nascent RNA and the RNA polymerase (RNAP) to catalyze the transcription cycle and RNA export. If these processes are not properly controlled, the nascent RNA can thread back and hybridize to the DNA template forming R-loops capable of stalling replication, leading to DNA breaks. Given the transcriptional promiscuity of the genome, which leads to large amounts of RNAs from mRNAs to different types of ncRNAs, these can become a major threat to genome integrity if they form R-loops. Consequently, cells have evolved nuclear factors to prevent this phenomenon that includes THO, a conserved eukaryotic complex acting in transcription elongation and RNA processing and export that upon inactivation causes genome instability linked to R-loop accumulation. We revise and discuss here the biological relevance of THO and a number of RNA helicases, including the THO partner UAP56/DDX39B, as a paradigm of the cellular mechanisms of cotranscriptional R-loop prevention.

During eukaryotic gene expression, the nascent RNA is cotranscriptionally processed into its mature form through a series of events that will render an exportable messenger ribonucleoparticle (mRNP). Multiple RNA-binding proteins (RBPs) coat the transcripts interacting in a dynamic manner. The carboxy-terminal domain (CTD) of the RNA polymerase II (RNAPII) couples transcription with mRNP biogenesis acting as a scaffold for RNA-processing factors and RBPs. This intimate connection favors a cross talk between transcription and RNA maturation and export that ensures that only properly matured mRNPs exit the nucleus (Müller-McNicoll and Neugebauer 2013; Bentley 2014).

Unexpectedly, we now know that mRNP biogenesis is crucially important to maintain genome integrity. Mutations in a number of mRNP factors confer genome instability detected as DNA damage, recombination, or chromosomal rearrangements in yeast and human cells (Luna et al. 2005; Paulsen et al. 2009; Wahba et al. 2011; Stirling et al. 2012). This instability is associated with DNA–RNA hybrids (Huertas and Aguilera 2003; Li and Manley 2005), which has led to the model that suboptimally assembled mRNPs would favor the formation of R-loops, structures containing a DNA–RNA hybrid and the displaced single-strand DNA (ssDNA) (Aguilera 2005).

R-loops participate in immunoglobulin class-switching recombination, mitochondrial replication, and gene expression regulation, but, in most cases, they are nonscheduled and can compromise genome integrity. The mechanism behind R-loop-mediated genome instability is still unclear. Even though the R-loop ssDNA sequence is more vulnerable to genotoxic agents or DNA-damaging enzymes like cytidine deaminases (Gómez-González and Aguilera 2007; Basu et al. 2011), the prevalent idea is that R-loop main effects are caused by replication impairment (Crossley et al. 2019; García-Muse and Aguilera 2019).

Cells have evolved mechanisms to mitigate the harmful consequences of R-loops, either preventing their formation or removing them via ribonucleases, helicases, and other enzymes (García-Muse and Aguilera 2019). During the last decade, an increasing number of RBPs have been related to R-loop-mediated genome instability (Paulsen et al. 2009; Wahba et al. 2011; Stirling et al. 2012). Whether all these factors have or do not have a direct role preventing R-loop accumulation—and by which mechanisms—is still unclear. Here, we discuss the role of RNA-binding and -processing factors in R-loop dynamics and genome instability, with special emphasis on the THO complex and RNA helicases.

THE THO COMPLEX: A KEY FACTOR IN mRNP BIOGENESIS

THO is a conserved eukaryotic complex required for mRNP biogenesis and the maintenance of genome integrity (Chávez et al. 2000; Luna et al. 2012). In yeast, THO is formed by five subunits: Tho2, Hpr1, Mft1, and Thp2, which form a highly robust complex, and Tex1, which is less tightly associated (Fig.1A; Chávez et al. 2000; Peña et al. 2012). Hpr1/THOC1, Tho2/THOC2, and Tex1/THOC3 (referred to yeast/human genes from now on) are the most evolutionary conserved and together with THOC5, THOC6, and THOC7 form the metazoan complex (Rehwinkel et al. 2004; Masuda et al. 2005; Furumizu et al. 2010; Castellano-Pozo et al. 2013). The

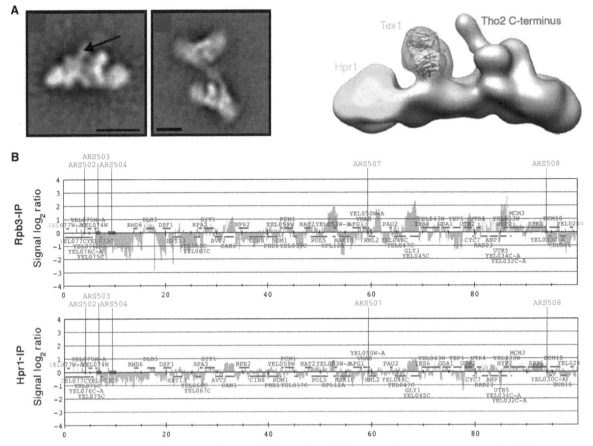

Figure 1. Structure and chromatin distribution of the yeast THO complex. (*A*) Two-dimensional average images of the THO complex as monomer (croissant form, *left*) and dimer (butterfly form, *right*) and a schematic representation mapping some of the components. (Reprinted, with permission, from Peña et al. 2012.) (*B*) Genomic distribution of the THO complex (Hpr1 subunit) and the RNA polymerase subunit Rpb3 in wild-type yeast cells as determined by chromatin immunoprecipitation with DNA microarray (ChIP-chip). (Reprinted, with permission, from Gómez-González et al. 2011.) Peaks showing a statistically significant presence are indicated in yellow.

three-dimensional electron microscopy (EM) structure of yeast THO revealed that the unstructured carboxy-terminal region of Tho2 binds to nucleic acids and is crucial for THO function (Fig. 1A; Peña et al. 2012). Although removal of any of its four main subunits destabilizes THO and confers a strong transcription and genome stability defect in yeast (Huertas et al. 2006; García-Rubio et al. 2008), this is not the case for Tex1 (Luna et al. 2005; Peña et al. 2012). The THO complex interacts with the RNA helicase Sub2/UAP56, as well as with the RBP and export factor Yra1/ALY, to form a larger complex named TREX that links transcription and export (Sträßer et al. 2002). Nevertheless, these interactions seem transient, being detected in substoichiometric amounts if THO is purified under high-salt conditions, and they do not contribute to the integrity of the THO complex (Sträßer et al. 2002; Huertas et al. 2006; Peña et al. 2012; Ren et al. 2017).

THO is cotranscriptionally recruited to chromatin by redundant mechanisms involving the transcription machinery, the nascent mRNA, and other mRNP components (Jimeno et al. 2002; Gómez-González et al. 2011; Meinel et al. 2013). It is found in active open reading frames (ORFs) with a tendency to accumulate at the 3′ end, as observed for yeast Hpr1 (Fig. 1B). THO is found in chromatin in association with RNA-processing and -splicing factors (Masuda et al. 2005; Cheng et al. 2006; Saguez et al. 2008; Chanarat et al. 2011; Katahira et al. 2013) and colocalizes with nuclear speckles (Masuda et al. 2005). As other factors needed for RNA maturation, THO interacts with the phosphorylated CTD of RNAPII probably as an initial step to be loaded into the mRNP (Meinel et al. 2013). Thus, THO is recruited and stably bound to the transcript via a network of contacts with RNAPII and the pre-mRNA or pre-mRNP.

In addition to the previously mentioned Sub2/UAP56 and Yra1/ALY, other proteins contact with THO to form a mature mRNP (Hurt et al. 2004; Dufu et al. 2010; Folco et al. 2012; Chang et al. 2013). By remodeling the RNA, the Sub2/UAP56 DEAD-box helicase facilitates the interaction of several RBPs, including Yra1/ALY, which works as an adaptor for the mRNA export factor Mex67/NXF1 (Dufu et al. 2010). THO interactions with mRNA export factors could explain THO's requirement for bulk mRNA export in yeast and in mammalian cells, including spliced RNAs that are otherwise retained in the speckles (Dias et al. 2010; Viphakone et al. 2012). Indeed, different

THO subunits have been shown to control 3′-end formation and polyadenylation and gene gating at nuclear pores, consistent with a central role of THO in transcription and RNA export via its interaction with factors with a specialized function in those processes (Saguez et al. 2008; Katahira et al. 2013; Mouaikel et al. 2013).

Finally, THO has been shown to work during the biogenesis of all types of RNAPII-driven RNAs, including small interfering RNAs (siRNAs), microRNAs (miRNAs), Piwi-interacting RNAs (piRNAs), viral RNAs, and small nucleolar RNAs (snoRNAs) (Boyne et al. 2008; Jauvion et al. 2010; Larochelle et al. 2012; Francisco-Mangilet et al. 2015; Hur et al. 2016; Zhang et al. 2018). Some evidence has also been provided for a physical and genetic association of THO with the ribosomal DNA (rDNA) transcription (Zhang et al. 2016), but a putative role in RNAPI-driven RNA biogenesis is unclear yet. In any case, THO exerts its main function at RNAPII-driven RNAs all over the genome.

THO AS A PARADIGM FOR R-LOOP PREVENTION

The two main proteins of THO, Tho2 and Hpr1, were initially identified by the increase in genome instability occurring in yeast mutants (Aguilera and Klein 1990; Piruat and Aguilera 1998). The link between RNA export and genome stability was suggested by the observation that THO/TREX was required for mRNA export (Jimeno et al. 2002; Sträßer et al. 2002) and the discovery that DNA–RNA hybrids accumulate in yeast THO mutants (Huertas and Aguilera 2003). The role of THO in transcription and the maintenance of genome integrity are conserved in *Caenorhabditis elegans* and human cells (Domínguez-Sánchez et al. 2011; Castellano-Pozo et al. 2013). As in yeast, genome-instability phenotypes are suppressed by overexpression of RNase H (Fig. 2A) and exacerbated by the AID deaminase that acts on the ssDNA of R-loops (Domínguez-Sánchez et al. 2011). Human THOC1-depleted cells also accumulate R-loops as determined by immunofluorescence using the anti-DNA–RNA antibody S9.6 (Fig. 2B).

DNA–RNA hybrids have been proposed to negatively affect transcription by acting as roadblocks for RNAPs in bacteria (Drolet et al. 1995) and yeast (Huertas and Aguilera 2003). Indeed, hybrids impair RNAP elongation in vitro (Tous and Aguilera 2007; Belotserkovskii et al. 2017). Moreover, RNAPI piles up in rDNA if RNaseH activity is removed, supporting an inhibitory role of DNA–RNA hybrids in transcription (El Hage et al. 2010), as also previously shown in bacterial rDNA transcription (Drolet et al. 1995). Nevertheless, it is not clear whether the inefficient transcription derives from an ability

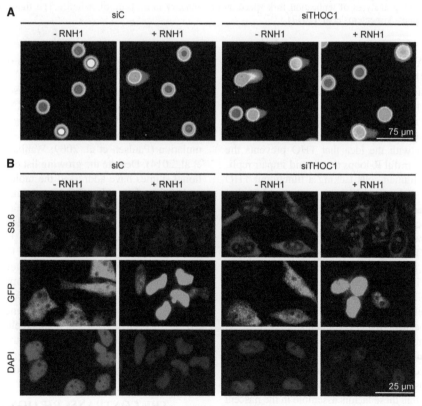

Figure 2. R-loop and DNA damage accumulation in THO-complex-depleted cells. (*A*) Alkaline single-cell electrophoresis showing a higher accumulation of DNA breaks, seen as DNA tails, in siTHOC1 cells, compared to control siC cells. A reduction of DNA breaks can be observed in siTHOC1 cells with RNH1 overexpression (+RNH1). (*B*) Immunofluorescence analysis of R-loop presence using the S9.6 antibody (red) in siC and siTHOC1 HeLa cells with normal (−RNH1) or high levels of RNH1 (+RNH1) obtained by overexpression of nuclear GFP-RNase H1 (green). The figure shows partial data previously used in Salas-Armenteros et al. (2017).

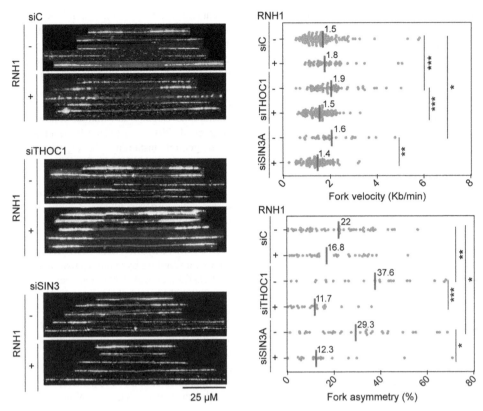

Figure 3. Replication fork progression impairment caused by THOC1 and SIN3 depletion. Representative pictures (*left*) and data plots (*right*) of DNA combing analyses of replication fork speed and asymmetry (a measure of stalling). The figure shows partial data previously used in Salas-Armenteros et al. (2017).

of the DNA–RNA hybrid to either retain or stall the RNAP from which the RNA moiety was born or to act as a roadblock for successive upstream RNAPs. Similarly, R-loops are also an obstacle for the replication machinery either by themselves or by retaining the RNAPII at the DNA (García-Muse and Aguilera 2016; Hamperl and Cimprich 2016). Consistent with the idea that THO prevents the accumulation of harmful R-loops that would impair replication, yeast and human cells lacking a functional THO complex display R-loop-mediated hyper-recombination and DNA break accumulation that are associated with an increase in transcription-replication conflicts (Wellinger et al. 2006; Gómez-González et al. 2011; Salas-Armenteros et al. 2017). This conclusion is supported by the observation that THO depletion in human cells confers DNA replication impairment that is suppressed by RNase H, as seen by DNA combing (Fig. 3).

A distinctive phenotype of yeast THO mutants is the transcription-associated hyper-recombination, which is also shared by mutations in other specific mRNP factors (Huertas and Aguilera 2003; Luna et al. 2005). On the other hand, in human cells, depletion of the RNA processing factor SRSF1 causes higher mutation rates (Li and Manley 2005). It is likely that these factors assemble in the nascent RNA, preventing its hybridization with DNA and the formation of R-loops (Aguilera 2005). Consistent with this interpretation could be the observations that overexpression of RBPs Sub2/UAP56, Tho1/CIP29, or RNPS1 suppressed the phenotypes of yeast THO mutants or SRSF1-depleted cells (Jimeno et al. 2002, 2006; Li et al. 2007).

Nowadays, proteins involved in transcription, chromatin modification, RNA export, splicing, RNA 3'-end cleavage and polyadenylation, rRNA processing, and RNA surveillance have been associated with DNA–RNA hybrid accumulation (Paulsen et al. 2009; Wahba et al. 2011; Chan et al. 2014). Despite the growing list of factors, few functional studies have addressed the molecular mechanisms underlying their role in R-loop homeostasis. Thus, although depletion of several splicing factors causes DNA damage that is suppressed by RNase H overexpression, not all of them protect the genome by preventing R-loops (Paulsen et al. 2009; Chan et al. 2014). For instance, depletion of the spliceosome-associated factor Spp381/MFAP1, which interacts with THO, also compromises genome stability, but in an R-loop-independent manner (Salas-Armenteros et al. 2019). Transcriptome data of Spp381/MFAP1 defective yeast and human cells support the idea that splicing factors may contribute to genome integrity indirectly by regulating the expression of other genes and not necessarily by R-loop accumulation.

THE CONTRANSCRIPTIONAL ROLE OF THO IN CHROMATIN

Transcription and RNA processing are coordinated processes occurring in the context of active chromatin, but how RNA processing relates with chromatin is still poorly understood. In a search for new THO interactors, we recently identified the mammalian histone deacetylase complex mSin3A (Salas-Armenteros et al. 2017). This interaction is paralleled by common cellular effects as Sin3A depletion causes R-loop-mediated genome instability comparable to THO depletion. On the other hand, inhibition of histone deacetylation using siSIN3 or the deacetylase inhibitors TSA and SAHA results in R-loop accumulation. Reciprocally, inhibiting acetylation by anacardic acid suppresses DNA breaks and R-loops in THOC1-depleted cells. In both siSIN3 and siTHOC1 cells, the replication fork progresses faster, consistent with a more open chromatin associated with histone hyperacetylation, but replication stalling is increased (Fig. 3), both phenotypes being suppressed by RNH1 overexpression (Salas-Armenteros et al. 2017). Altogether, these data suggest that the state of chromatin in the absence of SIN3 promotes R-loop and transcription–replication conflicts. In agreement with the idea that an open chromatin favors DNA–RNA hybrids, these colocalize genome-wide with DNase I hypersensitive sites, chromatin decondensation, and regions with reduced nucleosome occupancy in human cell lines (Sanz et al. 2016), and yeast mutations in histone H3 and H4 tails induce R-loop accumulation (García-Pichardo et al. 2017). Thus, a cotranscriptional cross talk between RNA-binding factors like THO and chromatin modifiers like mSin3A could represent a coordinated mechanism to prevent R-loop formation by transient cotranscriptional closing of chromatin, thus reducing DNA accessibility of the nascent RNA (Salas-Armenteros et al. 2017).

UAP56: THE HIDDEN LINK BETWEEN THO AND RNP BIOGENESIS

In addition to factors potentially preventing R-loop accumulation, R-loop-removal activities are necessary to preserve genome integrity. The main activity relies on ribonucleases H that digest the RNA moieties of the DNA–RNA hybrids. These are of two classes: RNase H1, which removes DNA–RNA hybrids like primers in Okazaki fragments, and RNase H2, which apart from acting on DNA–RNA hybrids, initiates ribonucleotide excision repair (Cerritelli and Crouch 2019). However, hybrids can also be eliminated by specific helicases that unwind DNA–RNA.

The RNA-dependent ATPase Sub2/UAP56 (also termed DDX39B) of the DDX family of ATP-dependent RNA helicases physically interacts with THO (Sträßer et al. 2002; Ren et al. 2017). UAP56 was initially identified as a splicing factor (Fleckner et al. 1997), but as for other previously termed "splicing factors," it might have a more general role in mRNP biogenesis. It may act as an RNA chaperone by unwinding RNA secondary structures or adding/removing proteins to render a mature exportable mRNP. Consistently, UAP56 is essential for export of most mRNAs in *Drosophila melanogaster*, *Saccharomyces cerevisiae*, *C. elegans*, and human regardless of whether or not these derive from intron-containing genes (Herold et al. 2003; MacMorris et al. 2003; Kapadia et al. 2006).

UAP56 has a paralog in humans, URH49 (also termed DDX39A), with a 90% of amino acid homology (Pryor et al. 2004). Whereas UAP56 is a ubiquitous protein, URH49 has a more restrictive expression pattern, but both seem to play similar roles according to the RNA export phenotypes of siUAP56 and siURH49 cells (Kapadia et al. 2006). URH49 also interacts with THO (Dufu et al. 2010). Nonetheless, UAP56 and URH49 are not totally interchangeable as depletion of UAP56 or URH49 downregulates a different set of genes in cell lines (Yamazaki et al. 2010).

UAP56 was shown to impact transcription elongation (Domínguez-Sánchez et al. 2011), and to explore further its role we have analyzed the recruitment of UAP56 to chromatin in the human cell line K562, regularly used in genome-wide occupancy of different transcription factors, secondary DNA structures, and chromatin modifications. Notably, UAP56 associates with the majority of the transcribed genes, as it localizes to most RNAPII active sites (Fig.4; Pérez-Calero et al. 2020). This is paralleled by previous data on the yeast counterpart Sub2 (Gómez-González et al. 2011). Thus, as is the case with THO, UAP56 is a general factor of RNAPII-driven transcription with a general role in mRNP biogenesis and export.

Importantly, Sub2/UAP56 is also required to preserve genome integrity. Several lines of evidence support this link. Thus, mutations of the yeast SUB2 gene confer similar defects in transcription and genome instability to those of *THO* mutants suggesting a similar function. Indeed, Sub2 overexpression partially suppresses those phenotypes (Jimeno et al. 2002). In human cells, UAP56 depletion causes DNA damage and double-strand breaks (DSBs) (Domínguez-Sánchez et al. 2011). Notably, we have recently observed that it is dependent on transcription and hybrids (Pérez-Calero et al. 2020). Although the same analysis has not been done for yeast Sub2 so far, these results could explain the ability of overexpressed Sub2 to suppress the growth defect of *hpr1* mutants (Jimeno et al. 2002, 2006) and the mutational characterization of Sub2 helicase domains (Saguez et al. 2013). Because UAP56 is an ATP-dependent RNA helicase that can unwind substrates with 5′ or 3′ overhangs or blunt ends in vitro (Shen et al. 2008), UAP56 could be the main activity required for the THO complex to facilitate proper mRNP assembly and export and, thus, prevent cotranscriptional R-loops.

RNA HELICASES TO UNWIND COTRANSCRIPTIONAL DNA–RNA HYBRIDS

Over the last several years, a number of RNA helicases have been reported to be involved in R-loop dynamics. Senataxin (Sen1/SETX) was the first reported of this class. It is required for efficient transcription termination of RNAPII and RNAPIII (Mischo et al. 2011; Skourti-Stathaki et al. 2011; Han et al. 2017; Rivosecchi et al. 2019).

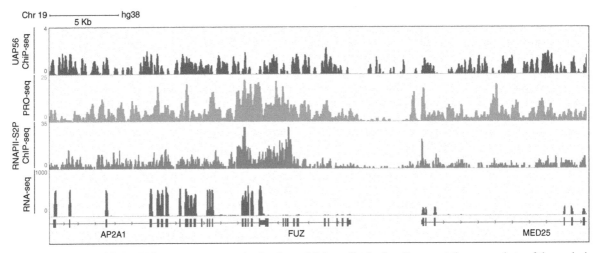

Figure 4. Genome-wide UAP56 occupancy, expression levels, and R-loop distribution. Representative screenshots of the analysis of the distribution in K562 cells of chromatin-bound UAP56 (green) and RNA-seq (purple) (Pérez-Calero et al. 2020) together with PRO-seq (blue) and RNAPII-S2P ChIP-seq (orange) data from ENCODE (SRP045616 and ENCSR000EGF accession numbers, respectively).

By resolving R-loops formed over G-rich pause sites near poly(A) signals, SETX grants Xrn2-mediated RNA degradation triggering transcription termination (Skourti-Stathaki et al. 2011). Data suggest that BRCA1 promotes SETX recruitment at R-loop-forming termination regions and protects them from R-loop-mediated DNA damage (Hatchi et al. 2015). Sen1/SETX influence is not restricted to transcription but extends to other DNA templated processes probably modulated by protein interactions. Thus, Sen1/SETX associates with actively replicating regions and facilitates replication and DNA repair, clearing R-loops from the template (Alzu et al. 2012).

Following Sen1/SETX, the RNA helicases Dbp2/DDX5, DHX9, Dbp5/DDX19, DDX21, DDX23, AQR, Sgs1/BLM, Mph1/FANCM, Pif1, RECQ5, and DDX1 were also implicated in R-loop homeostasis (García-Muse and Aguilera 2019). Depletion of any of these proteins increases R-loops and some cause RNaseH-sensitive DNA damage. Nonetheless, only Dbp2/DDX5, DDX19, DDX21, Sgs1/BLM, DHX9, Pif1, and FANCM have been shown to unwind hybrids together with other nucleic acid duplexes in vitro (Chakraborty and Grosse 2011; Chang et al. 2017; Hodroj et al. 2017; Song et al. 2017; Mohammad et al. 2018; Mersaoui et al. 2019; Silva et al. 2019). This has not been tested in the other cases, and for most, a comparative study of the efficiency removing DNA–RNA hybrids versus RNA–RNA duplexes is lacking. Yet, the mechanism by which these helicases control R-loop dynamics in vivo is uncertain. Their relevance may depend on the affinity to bind and act on hybrids compared to other nucleic acid structures, their different interactions with other mRNP components, or whether they have a preference to act on particular RNA sequences. In addition, RNA helicases are in many circumstances modulated by posttranslational modifications that control their activity, localization, and interacting partners, probably limiting their contribution to R-loop control (Sloan and Bohnsack 2018).

Despite the potential ability of RNA helicases to control R-loops in vivo by unwinding DNA–RNA hybrids, it is difficult to separate this activity from what may be its major activity, RNA–RNA unwinding, that would ensure mRNP assembly and export. This might be the case for Dbp2/DDX5 and DHX9, which in the end facilitate the access of RNA export and splicing factors to the mRNP (Chakraborty et al. 2018; Mersaoui et al. 2019) or RECQ5, an RNAPII-interacting RNA helicase that promotes topoisomerase 1 interaction with the elongating RNAPII (Li et al. 2015). Alternatively, the RNA helicases that facilitate transcription elongation or termination could avoid hybridization of the nascent transcript by reducing the possibility of RNA threading back. This might be the case for DDX21, which promotes transcription elongation through the activation of the pTEFb elongation factor, or DDX5 and SETX, which induce termination recruiting Xrn2 and/or ensuring RNA degradation (Skourti-Stathaki et al. 2011; Song et al. 2017; Mersaoui et al. 2019). Therefore, whether these RNA helicases have a direct role in unwinding DNA–RNA hybrids or whether their ability to reduce R-loop accumulation is a consequence of remodeling or releasing mRNPs is yet unclear.

The uncertainty about the way RNA helicases prevent R-loops is fed by the observations that other RNA helicases could contribute to form R-loops. Thus, DHX9 also opens R-loops in vitro and was also proposed to induce R-loops in the absence of splicing factors (Chakraborty and Grosse 2011; Chakraborty et al. 2018). In addition, DDX1 binds to G-quarters present in ncRNAs from immunoglobulin genes and mediates their conversion into R-loops, promoting class-switching recombination (Ribeiro de Almeida et al. 2018), but also shows an in vitro DNA–RNA unwinding activity that has been proposed to act at DSB sites (Li et al. 2008). Thus, it seems that these RNA helicases could perform opposite roles in R-loop dynamics, arguing whether the R-loop protecting role in vivo rather relies on their RNA–RNA helicase activity. Further analysis is required to clarify this conundrum.

Furthermore, the fact that inactivation of an RNA helicase of those detected causes the same phenotype of DNA–RNA hybrid accumulation argues against the idea that they all act in vivo by unwinding them, because this would imply no redundancy at all. Instead, cells may use a set of nonredundant RNA helicases with a final effect on R-loop homeostasis, if they act on specific RNA sites, structures, or metabolic steps required to generate an export-competent mRNP refractory to form DNA–RNA hybrids. In this sense, DDX21 and DDX23 seem to be specifically recruited to RNAPII stalled at R-loops, thus rescuing transcript elongation (Song et al. 2017; Sridhara et al. 2017), whereas SETX removes hybrids at the 3′-end region of genes and DDX19 has been suggested to target the replisome to overcome the transcription–replication conflicts caused by R-loops (Skourti-Stathaki et al. 2011; Hodroj et al. 2017). In any case, these possibilities need to be further confirmed. The diversity of RNA helicases involved in R-loop control could also be due to a temporal specificity of action during the cell cycle in some cases. Indeed, yeast Sen1 helicase is cell cycle–regulated as it is expressed mainly during the S and G_2 phases (Mischo et al. 2018).

The UAP56 RNA helicase is also recruited to actively transcribed genes (C Pérez-Calero, unpubl.), thus being a potential RNA helicase with a direct role in removing cotranscriptional DNA–RNA hybrids working together with THO. It has been shown that yeast Sub2 can unwind both RNA–RNA and RNA–DNA duplexes (Saguez et al. 2013), but the relationship of this activity with a potential role in R-loop prevention in vivo has not been established. Assaying whether human UAP56 removes DNA–RNA hybrids in vitro and in vivo would respond to this question. To summarize, the recent reports suggesting the potential of RNA helicases to remove DNA–RNA hybrids open the possibility that, apart from contributing to R-loop prevention by promoting mRNP assembly and by talking to histone deacetylases, such as mSIN3A, that would transiently close chromatin upstream of the RNAPII, THO may also facilitate RNA–DNA unwinding by specific helicases as a backup mechanism to resolve accidental R-loops (Fig. 5), a candidate for this being UAP56.

CONCLUSION

The nascent RNAs may suppose a risk to genome dynamics and function if they hybridize back to the DNA template. As a counterpart, specific RNA biogenesis factors assure that hybrids are not accumulated during transcription. A paradigmatic example is the THO complex, which is able to prevent cotranscriptional R-loop formation by warranting an optimal package of the mRNP and by talking to histone deacetylases like Sin3A, likely promoting transient closing of chromatin. On the other hand, although RNase H is believed to be the major activity resolving R-loops in vivo, it would have the negative counterpart of cleaving long pre-mRNAs at the hybrid site. The identification of various DDX/DHX helicases with the potential to unwind DNA–RNA hybrids suggests the possibility that helicases rather than nucleases resolve unscheduled cotranscriptional R-loops. Given the prominent transcriptional role of THO and its interaction with the Sub2/UAP56 RNA helicase, we believe that THO function in R-loop prevention may also rely on an associated DNA–RNA unwinding factor.

ACKNOWLEDGMENTS

Research in A.A.'s laboratory is funded by the European Research Council (Advanced Program), the Spanish Ministry of Economy and Competitiveness, and the Junta de Andalucía and Regional European Funds (FEDER).

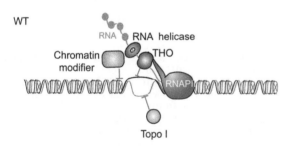
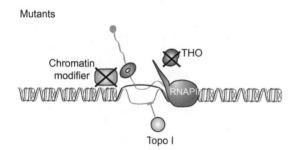

Figure 5. Model for the function of THO in R-loop prevention. THO is a key cotranscriptional factor in R-loop prevention that could act by promoting RNP assembly, by talking to the Sin3A histone deacetylase to promote transient chromatin closing, and by facilitating the recruitment and actions of RNA helicases with a potential to unwind DNA–RNA hybrids.

REFERENCES

Aguilera A. 2005. mRNA processing and genomic instability. *Nat Struct Mol Biol* **12:** 737–738. doi:10.1038/nsmb0905-737

Aguilera A, Klein HL. 1990. *HPR1*, a novel yeast gene that prevents intrachromosomal excision recombination, shows carboxy-terminal homology to the *Saccharomyces cerevisiae TOP1* gene. *Mol Cell Biol* **10:** 1439–1451. doi:10.1128/MCB.10.4.1439

Alzu A, Bermejo R, Begnis M, Lucca C, Piccini D, Carotenuto W, Saponaro M, Brambati A, Cocito A, Foiani M, et al. 2012. Senataxin associates with replication forks to protect fork integrity across RNA-polymerase-II-transcribed genes. *Cell* **151:** 835–846. doi:10.1016/j.cell.2012.09.041

Basu U, Meng FL, Keim C, Grinstein V, Pefanis E, Eccleston J, Zhang T, Myers D, Wasserman CR, Wesemann DR, et al. 2011. The RNA exosome targets the AID cytidine deaminase to both strands of transcribed duplex DNA substrates. *Cell* **144:** 353–363. doi:10.1016/j.cell.2011.01.001

Belotserkovskii BP, Soo Shin JH, Hanawalt PC. 2017. Strong transcription blockage mediated by R-loop formation within a G-rich homopurine-homopyrimidine sequence localized in the vicinity of the promoter. *Nucleic Acids Res* **45:** 6589–6599. doi:10.1093/nar/gkx403

Bentley DL. 2014. Coupling mRNA processing with transcription in time and space. *Nat Rev Genet* **15:** 163–175. doi:10.1038/nrg3662

Boyne JR, Colgan KJ, Whitehouse A. 2008. Recruitment of the complete hTREX complex is required for Kaposi's sarcoma–associated herpesvirus intronless mRNA nuclear export and virus replication. *PLoS Pathog* **4:** e1000194. doi:10.1371/journal.ppat.1000194

Castellano-Pozo M, García-Muse T, Aguilera A. 2013. R-loops cause replication impairment and genome instability during meiosis. *EMBO Rep* **13:** 923–929. doi:10.1038/embor.2012.119

Cerritelli SM, Crouch RJ. 2019. RNases H: multiple roles in maintaining genome integrity. *DNA Repair (Amst)* **84:** 102742. doi:10.1016/j.dnarep.2019.102742

Chakraborty P, Grosse F. 2011. Human DHX9 helicase preferentially unwinds RNA-containing displacement loops (R-loops) and G-quadruplexes. *DNA Repair (Amst)* **10:** 654–665. doi:10.1016/j.dnarep.2011.04.013

Chakraborty P, Huang JTJ, Hiom K. 2018. DHX9 helicase promotes R-loop formation in cells with impaired RNA splicing. *Nat Commun* **9:** 4346. doi:10.1038/s41467-018-06677-1

Chan YA, Aristizabal MJ, Lu PY, Luo Z, Hamza A, Kobor MS, Stirling PC, Hieter P. 2014. Genome-wide profiling of yeast DNA:RNA hybrid prone sites with DRIP-chip. *PLoS Genet* **10:** e1004288. doi:10.1371/journal.pgen.1004288

Chanarat S, Seizl M, Sträßer K. 2011. The Prp19 complex is a novel transcription elongation factor required for TREX occupancy at transcribed genes. *Genes Dev* **25:** 1147–1158. doi:10.1101/gad.623411

Chang CT, Hautbergue GM, Walsh MJ, Viphakone N, van Dijk TB, Philipsen S, Wilson SA. 2013. Chtop is a component of the dynamic TREX mRNA export complex. *EMBO J* **32:** 473–486. doi:10.1038/emboj.2012.342

Chang EY, Novoa CA, Aristizabal MJ, Coulombe Y, Segovia R, Chaturvedi R, Shen Y, Keong C, Tam AS, Jones SJM, et al. 2017. RECQ-like helicases Sgs1 and BLM regulate R-loop-associated genome instability. *J Cell Biol* **216:** 3991–4005. doi:10.1083/jcb.201703168

Chávez S, Beilharz T, Rondón AG, Erdjument-Bromage H, Tempst P, Svejstrup JQ, Lithgow T, Aguilera A. 2000. A protein complex containing Tho2, Hpr1, Mft1 and a novel protein, Thp2, connects transcription elongation with mitotic recombination in *Saccharomyces cerevisiae*. *EMBO J* **19:** 5824–5834. doi:10.1093/emboj/19.21.5824

Cheng H, Dufu K, Lee CS, Hsu JL, Dias A, Reed R. 2006. Human mRNA export machinery recruited to the 5′ end of mRNA. *Cell* **127:** 1389–1400. doi:10.1016/j.cell.2006.10.044

Crossley MP, Bocek M, Cimprich KA. 2019. R-loops as cellular regulators and genomic threats. *Mol Cell* **73:** 398–411. doi:10.1016/j.molcel.2019.01.024

Dias AP, Dufu K, Lei H, Reed R. 2010. A role for TREX components in the release of spliced mRNA from nuclear speckle domains. *Nat Commun* **1:** 97. doi:10.1038/ncomms1103

Domínguez-Sánchez MS, Barroso S, Gómez-González B, Luna R, Aguilera A. 2011. Genome instability and transcription elongation impairment in human cells depleted of THO/TREX. *PLoS Genet* **7:** e1002386. doi:10.1371/journal.pgen.1002386

Drolet M, Phoenix P, Menzel R, Masse E, Liu LF, Crouch RJ. 1995. Overexpression of RNase H partially complements the growth defect of an *Escherichia coli* Δ*topA* mutant: R-loop formation is a major problem in the absence of DNA topoisomerase I. *Proc Natl Acad Sci* **92:** 3526–3530. doi:10.1073/pnas.92.8.3526

Dufu K, Livingstone MJ, Seebacher J, Gygi SP, Wilson SA, Reed R. 2010. ATP is required for interactions between UAP56 and two conserved mRNA export proteins, Aly and CIP29, to assemble the TREX complex. *Genes Dev* **24:** 2043–2053. doi:10.1101/gad.1898610

El Hage A, French SL, Beyer AL, Tollervey D. 2010. Loss of topoisomerase I leads to R-loop-mediated transcriptional blocks during ribosomal RNA synthesis. *Genes Dev* **24:** 1546–1558. doi:10.1101/gad.573310

Fleckner J, Zhang M, Valcárcel J, Green MR. 1997. U2AF65 recruits a novel human DEAD box protein required for the U2 snRNP–branchpoint interaction. *Genes Dev* **11:** 1864–1872. doi:10.1101/gad.11.14.1864

Folco EG L, Dufu K CS, Yamazaki T, Reed R. 2012. The proteins PDIP3 and ZC11A associate with the human TREX complex in an ATP-dependent manner and function in mRNA export. *PLoS ONE* **7:** e43804. doi:10.1371/journal.pone.0043804

Francisco-Mangilet AG, Karlsson P, Kim MH, Eo HJ, Oh SA, Kim JH, Kulcheski FR, Park SK, Manavella PA. 2015. THO2, a core member of the THO/TREX complex, is required for microRNA production in *Arabidopsis*. *Plant J* **82:** 1018–1029. doi:10.1111/tpj.12874

Furumizu C, Tsukaya H, Komeda Y. 2010. Characterization of *EMU*, the *Arabidopsis* homolog of the yeast THO complex member *HPR1*. *RNA* **16:** 1809–1817. doi:10.1261/rna.2265710

García-Muse T, Aguilera A. 2016. Transcription-replication conflicts: how they occur and how they are resolved. *Nat Rev Mol Cell Biol* **17:** 553–563. doi:10.1038/nrm.2016.88

García-Muse T, Aguilera A. 2019. R-loops: from physiological to pathological roles. *Cell* **179:** 604–618. doi:10.1016/j.cell.2019.08.055

García-Pichardo D, Cañas JC, García-Rubio ML, Gómez-González B, Rondón AG, Aguilera A. 2017. Histone mutants separate R-loop formation from genome instability induction. *Mol Cell* **66:** 597–609.e595. doi:10.1016/j.molcel.2017.05.014

García-Rubio M, Chávez S, Huertas P, Tous C, Jimeno S, Luna R, Aguilera A. 2008. Different physiological relevance of yeast THO/TREX subunits in gene expression and genome integrity. *Mol Genet Genomics* **279:** 123–132. doi:10.1007/s00438-007-0301-6

Gómez-González B, Aguilera A. 2007. Activation-induced cytidine deaminase action is strongly stimulated by mutations of the THO complex. *Proc Natl Acad Sci* **104:** 8409–8414. doi:10.1073/pnas.0702836104

Gómez-González B, García-Rubio M, Bermejo R, Gaillard H, Shirahige K, Marín A, Foiani M, Aguilera A. 2011. Genome-wide function of THO/TREX in active genes prevents R-loop-dependent replication obstacles. *EMBO J* **30:** 3106–3119. doi:10.1038/emboj.2011.206

Hamperl S, Cimprich KA. 2016. Conflict resolution in the genome: how transcription and replication make it work. *Cell* **167:** 1455–1467. doi:10.1016/j.cell.2016.09.053

Han Z, Libri D, Porrua O. 2017. Biochemical characterization of the helicase Sen1 provides new insights into the mechanisms of non-coding transcription termination. *Nucleic Acids Res* **45:** 1355–1370. doi:10.1093/nar/gkw1230

Hatchi E, Skourti-Stathaki K, Ventz S, Pinello L, Yen A, Kamieniarz-Gdula K, Dimitrov S, Pathania S, McKinney KM, Eaton ML, et al. 2015. BRCA1 recruitment to transcriptional pause sites is required for R-loop-driven DNA damage repair. *Mol Cell* **57:** 636–647. doi:10.1016/j.molcel.2015.01.011

Herold A, Teixeira L, Izaurralde E. 2003. Genome-wide analysis of nuclear mRNA export pathways in *Drosophila*. *EMBO J* **22:** 2472–2483. doi:10.1093/emboj/cdg233

Hodroj D, Recoli B, Serhal K, Martinez S, Tsanov N, Abou Merhi R, Maiorano D. 2017. An ATR-dependent function for the Ddx19 RNA helicase in nuclear R-loop metabolism. *EMBO J* **36:** 1182–1198. doi:10.15252/embj.201695131

Huertas P, Aguilera A. 2003. Cotranscriptionally formed DNA: RNA hybrids mediate transcription elongation impairment and transcription-associated recombination. *Mol Cell* **12:** 711–721. doi:10.1016/j.molcel.2003.08.010

Huertas P, Garcia-Rubio ML, Wellinger RE, Luna R, Aguilera A. 2006. An *hpr1* point mutation that impairs transcription and mRNP biogenesis without increasing recombination. *Mol Cell Biol* **26:** 7451–7465. doi:10.1128/MCB.00684-06

Hur JK, Luo Y, Moon S, Ninova M, Marinov GK, Chung YD, Aravin AA. 2016. Splicing-independent loading of TREX on nascent RNA is required for efficient expression of dual-strand piRNA clusters in *Drosophila*. *Genes Dev* **30:** 840–855. doi:10.1101/gad.276030.115

Hurt E, Luo MJ, Röther S, Reed R, Sträßer K. 2004. Cotranscriptional recruitment of the serine–arginine-rich (SR)-like proteins Gbp2 and Hrb1 to nascent mRNA via the TREX complex. *Proc Natl Acad Sci* **101:** 1858–1862. doi:10.1073/pnas.0308663100

Jauvion V, Elmayan T, Vaucheret H. 2010. The conserved RNA trafficking proteins HPR1 and TEX1 are involved in the production of endogenous and exogenous small interfering RNA in *Arabidopsis*. *Plant Cell* **22:** 2697–2709. doi:10.1105/tpc.110.076638

Jimeno S, Rondón AG, Luna R, Aguilera A. 2002. The yeast THO complex and mRNA export factors link RNA metabolism with transcription and genome instability. *EMBO J* **21:** 3526–3535. doi:10.1093/emboj/cdf335

Jimeno S, Luna R, Garcia-Rubio M, Aguilera A. 2006. Tho1, a novel hnRNP, and Sub2 provide alternative pathways for mRNP biogenesis in yeast THO mutants. *Mol Cell Biol* **26:** 4387–4398. doi:10.1128/MCB.00234-06

Kapadia F, Pryor A, Chang TH, Johnson LF. 2006. Nuclear localization of poly(A)$^+$ mRNA following siRNA reduction of expression of the mammalian RNA helicases UAP56 and URH49. *Gene* **384:** 37–44. doi:10.1016/j.gene.2006.07.010

Katahira J, Okuzaki D, Inoue H, Yoneda Y, Maehara K, Ohkawa Y. 2013. Human TREX component Thoc5 affects alternative polyadenylation site choice by recruiting mammalian cleavage factor I. *Nucleic Acids Res* **41:** 7060–7072. doi:10.1093/nar/gkt414

Larochelle M, Lemay JF, Bachand F. 2012. The THO complex cooperates with the nuclear RNA surveillance machinery to control small nucleolar RNA expression. *Nucleic Acids Res* **40:** 10240–10253. doi:10.1093/nar/gks838

Li X, Manley JL. 2005. Inactivation of the SR protein splicing factor ASF/SF2 results in genomic instability. *Cell* **122:** 365–378. doi:10.1016/j.cell.2005.06.008

Li X, Niu T, Manley JL. 2007. The RNA binding protein RNPS1 alleviates ASF/SF2 depletion-induced genomic instability. *RNA* **13:** 2108–2115. doi:10.1261/rna.734407

Li L, Monckton EA, Godbout R. 2008. A role for DEAD box 1 at DNA double-strand breaks. *Mol Cell Biol* **28:** 6413–6425. doi:10.1128/MCB.01053-08

Li M, Pokharel S, Wang JT, Xu X, Liu Y. 2015. RECQ5-dependent SUMOylation of DNA topoisomerase I prevents transcription-associated genome instability. *Nat Commun* **6:** 6720. doi:10.1038/ncomms7720

Luna R, Jimeno S, Marín M, Huertas P, García-Rubio M, Aguilera A. 2005. Interdependence between transcription and mRNP processing and export, and its impact on genetic stability. *Mol Cell* **18:** 711–722. doi:10.1016/j.molcel.2005.05.001

Luna R, Rondón AG, Aguilera A. 2012. New clues to understand the role of THO and other functionally related factors in mRNP biogenesis. *Biochim Biophys Acta* **1819:** 514–520. doi:10.1016/j.bbagrm.2011.11.012

MacMorris M, Brocker C, Blumenthal T. 2003. UAP56 levels affect viability and mRNA export in *Caenorhabditis elegans*. *RNA* **9:** 847–857. doi:10.1261/rna.5480803

Masuda S, Das R, Cheng H, Hurt E, Dorman N, Reed R. 2005. Recruitment of the human TREX complex to mRNA during splicing. *Genes Dev* **19:** 1512–1517. doi:10.1101/gad.1302205

Meinel DM, Burkert-Kautzsch C, Kieser A, O'Duibhir E, Siebert M, Mayer A, Cramer P, Söding J, Holstege FC, Sträßer K. 2013. Recruitment of TREX to the transcription machinery by its direct binding to the phospho-CTD of RNA polymerase II. *PLoS Genet* **9:** e1003914. doi:10.1371/journal.pgen.1003914

Mersaoui S, Yu Z, Coulombe Y, Karam M, Busatto FF, Masson JY, Richard S. 2019. Arginine methylation of the DDX5 helicase RGG/RG motif by PRMT5 regulates resolution of RNA: DNA hybrids. *EMBO J* **38:** e100986. doi:10.15252/embj.2018100986

Mischo HE, Gómez-González B, Grzechnik P, Rondón AG, Wei W, Steinmetz L, Aguilera A, Proudfoot NJ. 2011. Yeast Sen1 helicase protects the genome from transcription-associated instability. *Mol Cell* **41:** 21–32. doi:10.1016/j.molcel.2010.12.007

Mischo HE, Chun Y, Harlen KM, Smalec BM, Dhir S, Churchman LS, Buratowski S. 2018. Cell-cycle modulation of transcription termination factor Sen1. *Mol Cell* **70:** 312–326. doi:10.1016/j.molcel.2018.03.010

Mohammad JB, Wallgren M, Sabouri N. 2018. The Pif1 signature motif of Pfh1 is necessary for both protein displacement and helicase unwinding activities, but is dispensable for strand-annealing activity. *Nucleic Acids Res* **46:** 8516–8531. doi:10.1093/nar/gky654

Mouaikel J, Causse SZ, Rougemaille M, Daubenton-Carafa Y, Blugeon C, Lemoine S, Devaux F, Darzacq X, Libri D. 2013. High-frequency promoter firing links THO complex function to heavy chromatin formation. *Cell Rep* **5:** 1082–1094. doi:10.1016/j.celrep.2013.10.013

Müller-McNicoll M, Neugebauer KM. 2013. How cells get the message: dynamic assembly and function of mRNA-protein complexes. *Nat Rev Genet* **14:** 275–287. doi:10.1038/nrg3434

Paulsen RD, Soni DV, Wollman R, Hahn AT, Yee MC, Guan A, Hesley JA, Miller SC, Cromwell EF, Solow-Cordero DE, et al. 2009. A genome-wide siRNA screen reveals diverse cellular processes and pathways that mediate genome stability. *Mol Cell* **35:** 228–239. doi:10.1016/j.molcel.2009.06.021

Peña A, Gewartowski K, Mroczek S, Cuéllar J, Szykowska A, Prokop A, Czarnocki-Cieciura M, Piwowarski J, Tous C, Aguilera A, et al. 2012. Architecture and nucleic acids recognition mechanism of the THO complex, an mRNP assembly factor. *EMBO J* **31:** 1605–1616. doi:10.1038/emboj.2012.10

Pérez-Calero C, Bayona-Feliu A, Xue X, Barroso S, Muñoz S, González-Basallote V, Sung P, Aguilera A. 2020. UAP56/DDX39B is a major co-transcriptional RNA–DNA helicase that unwinds harmful R loops genome-wide. *Genes Dev* doi:10.1101/gad336024.119

Piruat JI, Aguilera A. 1998. A novel yeast gene, *THO2*, is involved in RNA pol II transcription and provides new evidence for transcriptional elongation-associated recombination. *EMBO J* **17:** 4859–4872. doi:10.1093/emboj/17.16.4859

Pryor A, Tung L, Yang Z, Kapadia F, Chang TH, Johnson LF. 2004. Growth-regulated expression and G0-specific turnover of the mRNA that encodes URH49, a mammalian DExH/D box protein that is highly related to the mRNA export protein UAP56. *Nucleic Acids Res* **32:** 1857–1865. doi:10.1093/nar/gkh347

Ren Y, Schmiege P, Blobel G. 2017. Structural and biochemical analyses of the DEAD-box ATPase Sub2 in association with THO or Yra1. *eLife* **6:** e20070. doi:10.7554/eLife 20070

Rehwinkel J, Herold A, Gari K, Köcher T, Rode M, Ciccarelli FL, Wilm M, Izaurralde E. 2004. Genome-wide analysis of mRNAs regulated by the THO complex in *Drosophila melanogaster*. *Nat Struct Mol Biol* **11:** 558–566. doi:10.1038/nsmb759

Ribeiro de Almeida C, Dhir S, Dhir A, Moghaddam AE, Sattentau Q, Meinhart A, Proudfoot NJ. 2018. RNA helicase DDX1 converts RNA G-quadruplex structures into R-loops to promote *IgH* class switch recombination. *Mol Cell* **70:** 650–662.e658. doi:10.1016/j.molcel.2018.04.001

Rivosecchi J, Larochelle M, Teste C, Grenier F, Malapert A, Ricci EP, Bernard P, Bachand F, Vanoosthuyse V. 2019. Senataxin homologue Sen1 is required for efficient termination of RNA polymerase III transcription. *EMBO J* **38:** e101955. doi:10.15252/embj.2019101955

Saguez C, Schmid M, Olesen JR, Ghazy MA, Qu X, Poulsen MB, Nasser T, Moore C, Jensen TH. 2008. Nuclear mRNA surveillance in THO/sub2 mutants is triggered by inefficient polyadenylation. *Mol Cell* **31**: 91–103. doi:10.1016/j.molcel.2008.04.030

Saguez C, Gonzales FA, Schmid M, Boggild A, Latrick CM, Malagon F, Putnam A, Sanderson L, Jankowsky E, Brodersen DE, et al. 2013. Mutational analysis of the yeast RNA helicase Sub2p reveals conserved domains required for growth, mRNA export, and genomic stability. *RNA* **19**: 1363–1371. doi:10.1261/rna.040048.113

Salas-Armenteros I, Pérez-Calero C, Bayona-Feliu A, Tumini E, Luna R, Aguilera A. 2017. Human THO-Sin3A interaction reveals new mechanisms to prevent R-loops that cause genome instability. *EMBO J* **36**: 3532–3547. doi:10.15252/embj.201797208

Salas-Armenteros I, Barroso SI, Rondón AG, Pérez M, Andújar E, Luna R, Aguilera A. 2019. Depletion of the MFAP1/SPP381 splicing factor causes R-loop-independent genome instability. *Cell Rep* **28**: 1551–1563.e1557. doi:10.1016/j.celrep.2019.07.010

Sanz L, Hartono SR, Li YW, Steyaert S, Rajpurkar A, Ginno PA, Xu X, Chédin F. 2016. Prevalent, dynamic, and conserved R-loop structures associate with specific epigenomic signatures in mammals. *Mol Cell* **63**: 167–178. doi:10.1016/j.molcel.2016.05.032

Shen H, Zheng X, Shen J, Zhan L, Zhao R, Green MR. 2008. Distinct activities of the DExD/H-box splicing factor hUAP56 facilitate stepwise assembly of the spliceosome. *Genes Dev* **22**: 1796–1803. doi:10.1101/gad.1657308

Silva B, Pentz R, Figueira AM, Arora R, Lee YW, Hodson C, Wischnewski H, Deans AJ, Azzalin CM. 2019. FANCM limits ALT activity by restricting telomeric replication stress induced by deregulated BLM and R-loops. *Nat Commun* **10**: 2253. doi:10.1038/s41467-019-10179-z

Skourti-Stathaki K, Proudfoot NJ, Gromak N. 2011. Human senataxin resolves RNA/DNA hybrids formed at transcriptional pause sites to promote Xrn2-dependent termination. *Mol Cell* **42**: 794–805. doi:10.1016/j.molcel.2011.04.026

Sloan KE, Bohnsack MT. 2018. Unravelling the mechanisms of RNA helicase regulation. *Trends Biochem Sci* **43**: 237–250. doi:10.1016/j.tibs.2018.02.001

Song C, Hotz-Wagenblatt A, Voit R, Grummt I. 2017. SIRT7 and the DEAD-box helicase DDX21 cooperate to resolve genomic R-loops and safeguard genome stability. *Genes Dev* **31**: 1370–1381. doi:10.1101/gad.300624.117

Sridhara SC, Carvalho S, Grosso AR, Gallego-Paez LM, Carmo-Fonseca M, de Almeida SF. 2017. Transcription dynamics prevent RNA-mediated genomic instability through SRPK2-dependent DDX23 phosphorylation. *Cell Rep* **18**: 334–343. doi:10.1016/j.celrep.2016.12.050

Stirling PC, Chan YA, Minaker SW, Aristizabal MJ, Barrett I, Sipahimalani P, Kobor MS, Hieter P. 2012. R-loop-mediated genome instability in mRNA cleavage and polyadenylation mutants. *Genes Dev* **26**: 163–175. doi:10.1101/gad.179721.111

Sträßer K, Masuda S, Mason P, Pfannstiel J, Oppizzi M, Rodriguez-Navarro S, Rondón AG, Aguilera A, Struhl K, Reed R, et al. 2002. TREX is a conserved complex coupling transcription with messenger RNA export. *Nature* **417**: 304–308. doi:10.1038/nature746

Tous C, Aguilera A. 2007. Impairment of transcription elongation by R-loops in vitro. *Biochem Biophys Res Commun* **360**: 428–432. doi:10.1016/j.bbrc.2007.06.098

Viphakone N, Hautbergue GM, Walsh M, Chang CT, Holland A, Folco EG, Reed R, Wilson SA. 2012. TREX exposes the RNA-binding domain of Nxf1 to enable mRNA export. *Nat Commun* **3**: 1006. doi:10.1038/ncomms2005

Wahba L, Amon JD, Koshland D, Vuica-Ross M. 2011. RNase H and multiple RNA biogenesis factors cooperate to prevent RNA:DNA hybrids from generating genome instability. *Mol Cell* **44**: 978–988. doi:10.1016/j.molcel.2011.10.017

Wellinger RE, Prado F, Aguilera A. 2006. Replication fork progression is impaired by transcription in hyperrecombinant yeast cells lacking a functional THO complex. *Mol Cell Biol* **26**: 3327–3334. doi:10.1128/MCB.26.8.3327-3334.2006

Yamazaki T, Fujiwara N, Yukinaga H, Ebisuya M, Shiki T, Kurihara T, Kioka N, Kambe T, Nagao M, Nishida E, et al. 2010. The closely related RNA helicases, UAP56 and URH49, preferentially form distinct mRNA export machineries and coordinately regulate mitotic progression. *Mol Biol Cell* **21**: 2953–2965. doi:10.1091/mbc.e09-10-0913

Zhang Y, French SL, Beyer AL, Schneider DA. 2016. The transcription factor THO promotes transcription initiation and elongation by RNA polymerase I. *J Biol Chem* **291**: 3010–3018. doi:10.1074/jbc.M115.673442

Zhang G, Tu S, Yu T, Zhang XO, Parhad SS, Weng Z, Theurkauf WE. 2018. Co-dependent assembly of *Drosophila* piRNA precursor complexes and piRNA cluster heterochromatin. *Cell Rep* **24**: 3413–3422.e3414. doi:10.1016/j.celrep.2018.08.081

U1 snRNP Telescripting Roles in Transcription and Its Mechanism

Chao Di,[1,2] Byung Ran So,[1,2] Zhiqiang Cai,[1] Chie Arai,[1] Jingqi Duan,[1] and Gideon Dreyfuss[1]

[1]*Howard Hughes Medical Institute, Department of Biochemistry and Biophysics, University of Pennsylvania School of Medicine, Philadelphia, Pennsylvania 19104-6148, USA*

Correspondence: gdreyfuss@hhmi.upenn.edu

Telescripting is a fundamental cotranscriptional gene regulation process that relies on U1 snRNP (U1) to suppress premature 3′-end cleavage and polyadenylation (PCPA) in RNA polymerase II (Pol II) transcripts, which is necessary for full-length transcription of thousands of protein-coding (pre-mRNAs) and long noncoding (lncRNA) genes. Like U1 role in splicing, telescripting requires U1 snRNA base-pairing with nascent transcripts. Inhibition of U1 base-pairing with U1 snRNA antisense morpholino oligonucleotide (U1 AMO) mimics widespread PCPA from cryptic polyadenylation signals (PASs) in human tissues, including PCPA in introns and last exons' 3′-untranslated regions (3′ UTRs). U1 telescripting–PCPA balance changes generate diverse RNAs depending on where in a gene it occurs. Long genes are highly U1-telescripting-dependent because of PASs in introns compared to short genes. Enrichment of cell cycle control, differentiation, and developmental functions in long genes, compared to housekeeping and acute cell stress response genes in short genes, reveals a gene size–function relationship in mammalian genomes. This polarization increased in metazoan evolution by previously unexplained intron expansion, suggesting that U1 telescripting could shift global gene expression priorities. We show that that modulating U1 availability can profoundly alter cell phenotype, such as cancer cell migration and invasion, underscoring the critical role of U1 homeostasis and suggesting it as a potential target for therapies. We describe a complex of U1 with cleavage and polyadenylation factors that silences PASs in introns and 3′ UTR, which gives insights into U1 telescripting mechanism and transcription elongation regulation.

Telescripting is an activity of U1 snRNP (U1) that suppresses premature transcription of RNA polymerase II (Pol II) transcription at 3′-end processing signals, generally cleavage and polyadenylation (CPA) signals (PASs) in introns. Premature CPA (PCPA) (Kaida et al. 2010) was uncovered from transcriptome profiling of human cells transfected with antisense morpholino oligonucleotides (AMOs), which showed that AMO masking of a U1 snRNA 5′ sequence, whose base pairs with 5′ splice sites (5′ss) is necessary for the first step in splicing (Lerner et al. 1980; Mount et al. 1983; Padgett et al. 1983). These experiments showed PCPA in thousands of nascent transcripts of protein coding genes (pre-mRNAs) and long noncoding RNAs (lncRNAs), indicating that U1 is necessary to suppress PCPA and thereby promote full-length transcription (hence the term telescripting). Telescripting is an additional U1 function, separate from its role in splicing. Here, we review U1 telescripting emergence as a key regulator of gene expression, some of its biological roles, current understanding of its mechanism, and outstanding questions for future research.

U1 snRNP comprises a single noncoding small nuclear RNA (164 nt in human) and 10 proteins (U1-specific U1-70K, U1A, U1C, and seven Sm proteins common to spliceosomal snRNPs). RNA sequencing (RNA-seq) from cells transfected with U1 snRNA antisense morpholino oligonucleotide (U1 AMO) titrated to mask all or nearly all U1 snRNA 5′ sequence ("high U1 AMO"), compared to control nonspecific AMO (cAMO), showed unfamiliar and striking "Z patterns" in thousands of genes, consisting of RNAs extending several kilobases (typically, ~1–3 kb and up to tens of kilobases) from the transcription start site (TSS) that end abruptly in an intron (Fig. 1). At their ends these RNAs had 3′-poly(A)s specified by canonical PAS hexamers, AAUAAA and variants thereof, indistinguishable from classical PASs at the ends of full-length pre-mRNAs and lncRNAs (Proudfoot and Brownlee 1976; Tian and Manley 2017). The widespread PCPA elicited by high U1 AMO revealed that U1 is a PCPA suppressor. Importantly, detection of 3′-poly(A) tags at U1 AMO-induced PCPA positions in normal tissues in human and other organisms (Derti et al. 2012; Oh et al. 2017) demonstrated that PCPA and U1 telescripting are physiological processes, and U1 AMO is a useful tool to study it.

As expected, U1 AMO also caused widespread splicing inhibition, evident in intron retention in many introns that are not PCPAed. However, U1 telescripting appears to be additional and separable from U1's role in splicing. As PCPA frequently occurs in introns well before their 3′ss is transcribed, PCPA is not secondary to splicing inhibition (of the same intron) and U1 telescripting is U1-specific and may not require other spliceosomal snRNPs (U2,

---

[2]These authors contributed equally to this work.

© 2019 Di et al. This article is distributed under the terms of the Creative Commons Attribution-NonCommercial License, which permits reuse and redistribution, except for commercial purposes, provided that the original author and source are credited.

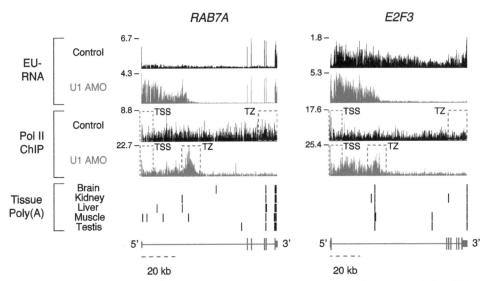

Figure 1. Premature 3′-end cleavage and polyadenylation (PCPA) terminates Pol II elongation cotranscriptionally in gene bodies. Genome browser (version hg19) views of representative PCPAed genes are shown with nascent RNA-seq and Pol II ChIP-seq (Oh et al. 2017) reads aligned to the human genome (e.g., *RAB7A* and *E2F3*). HeLa cells were pulse-labeled for 5 min with 5′-ethynyluridine (EU) at 8 h transfection with a 25-mer control or U1 AMO to U1 snRNA 5′ sequence. Pol II near transcription start sites (TSSs) and Pol II near termination zones (TZs) are indicated as boxes. 3′-poly(A) sites detected in various human tissues (Derti et al. 2012) are shown as bars and indicate natural PCPAs. Tissue differences suggest a role for cell-specific factors in PCPA/telescripting or metabolism of PCPAed RNAs. Gene structure lines show introns (horizontal lines) and exons (vertical bars), respectively.

U4, U5, U6, or the minor snRNPs U11, U12, U4atac, U6atac) (Kaida et al. 2010). Although telescripting and splicing require U1 base-pairing to the nascent transcript and U1 bound at 5′ss can function in both, telescripting can also be supplied by U1 base-paired to sequences that cannot function as 5′ss, giving U1 telescripting more potential sites (Berg et al. 2012).

Pol II chromatin immunoprecipitation-sequencing (ChIP-seq) maps complemented the nascent RNA maps (Fig. 1; Gilmour and Lis 1984; Oh et al. 2017), indicating that U1 telescripting acts cotranscriptionally, as opposed to post-transcriptionally (e.g., by reprocessing from full-length transcripts). Like CPA at normal 3′ ends (the ultimate PASs), Pol II ChIP-seq tapered off within a few kilobases from PCPA locations, a gradual post-3′-end transcription termination zone (TZ) (Fong et al. 2015) consistent with the torpedo termination model (Connelly and Manley 1988; Proudfoot 2016). Pol II occupancy downstream from TZs through the rest of the gene was sharply reduced or eliminated. Notably, U1 AMO did not inhibit transcription initiation (and promoter-proximal pause Pol II release into elongation); rather, transcription continued to stream into genes and frequently increased up to PCPA points, evident by high pol II occupancy from the TSSs to PCPA points in U1 AMO compared to control. Thus, full-length Pol II transcription elongation in most genes in vertebrates depends on averting PCPA termination at PAS checkpoints, which U1 telescripting provides.

BIOLOGICAL ROLES OF U1 TELESCRIPTING

Technically, PAS selection in genes that have more than one can be classified as alternative polyadenylation (APA). Although APA, involving choice among tandem PASs in terminal exons 3′-untranslated regions (3′ UTRs), and a few instances of APA in gene bodies had been known previously (Wang et al. 2008; Hartmann and Valcárcel 2009; Di Giammartino et al. 2011), the PCPA-U1 telescripting phenomenon as a major transcription control was unexpected.

Transcriptome Control and Proteome Diversification

RNA-seq and PCPA maps developed from Z-pattern-detecting algorithms and nongenomic 3′-poly(A) reads at various U1 AMO doses helped identify key telescripting features and its biological roles. PCPA in a gene can be complete, effectively shutting down full-length transcription or partial (in some genes even at high U1 AMO), affecting a fraction of a gene's transcripts (e.g., *RAB7A* and *E2F3*; Fig. 1), and can occur at multiple PASs in the same gene. Any PCPA is at the expense of full-length transcription; however, it can potentially generate diversity of RNAs and proteins. Figure 2 illustrates various outcomes of PCPA at different locations that U1 telescripting can regulate. The relative abundance of the different isoforms can be manipulated with U1 AMO dose (and correspondingly decreases available U1), which increases usage of TSS-proximal PAS with higher U1 AMO, consistent with a cotranscriptional process whereby the first PAS encountered by Pol II that is unprotected by U1 is PCPAed. Pervasive PCPA at multiple cryptic PASs in introns counters the textbook view that PASs function primarily in the 3′ UTR.

PCPA in the 1–3 kb from the first 5′ss (which is <0.5 kb from the TSS) in intron1 makes polyadenylated RNAs that are generally rapidly degraded by nuclear exosomes

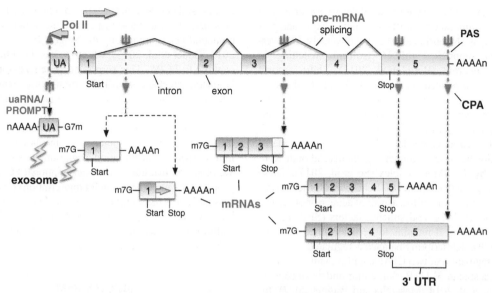

Figure 2. Diverse outcomes of PCPA (indicated by red arrows at polyadenylation signals [PASs]), recapitulated by U1 AMO, reveal U1 telescripting key role in shaping the transcriptome. The wide blue arrows indicate bidirectional transcription of RNA polymerase II (Pol II) in the sense (*right*) and antisense (*left*) directions. Upstream antisense (uaRNAs)/ promoter-upstream transcripts (PROMPTs) are indicated. Start and stop indicate translation start and end positions, respectively. m7G (monomethyl guanosine) indicates capped 5′ ends and AAAAn represent 3′-poly(A) tails. CPA indicates cleavage and polyadenylation at the canonical full-length gene transcript. 3′ UTR shortening removes binding elements for mRNA regulating RBPs and miRNAs.

(Almada et al. 2013; Ntini et al. 2013; Lubas et al. 2015; Meola et al. 2016; Chiu et al. 2018). These RNAs are difficult to detect by standard total RNA-seq but are readily detected if exosomes are inhibited or with high U1 AMO, which results in massive PCPA that overwhelms exosomes. PCPA detection is also enhanced by selective RNA-seq of nascent RNAs. This, and 3′-poly(A) tags in tissues, show TSS-proximal PCPA, indicating that U1 telescripting is a normal source of transcription attrition (Oh et al. 2017; So et al. 2019). Stealthy PCPA likely play a significant role in gene expression down-regulation. Conditions in which such drastic PCPA occurs in nature have not been described, although we envision toxins, viruses, and other pathogens evolved mechanisms to inhibit U1 telescripting as means to shut off host mRNA synthesis.

In a few cases, TSS-proximal PCPAed RNAs are relatively stable and have a translation open reading frame (ORF) extending from exon1 into intron1, where PCPA occurs. These can be functional mRNAs that encode the same amino terminus as the full-length mRNA, but a different carboxyl terminus and function, such as epidermal growth factor receptor (EGFR) and related receptors (Vorlová et al. 2011).

Promoter Directionality

TSS-proximal PCPA occurs naturally in noncoding upstream antisense RNAs (uaRNAs [Almada et al. 2013], also known as PROMPTs [Preker et al. 2008; Ntini et al. 2013]) transcribed from divergent Pol II promoters in many genes (Fig. 2). uaRNAs tend to have high ratios of PASs to U1 binding sites, causing their PCPA and rapid elimination (Almada et al. 2013; Ntini et al. 2013). In contrast, over a similar distance from the TSS (1–3 kb), RNA in the sense direction have high ratios of U1 base-pairing sites to PASs. Preferential uaRNA pruning contributes to promoter directionality in the sense, protein coding transcripts thereby enhancing production of mRNAs. These observations highlight the general role of PCPA and telescripting in shaping the transcriptome.

Short mRNA Isoform Switching, Cell Stimulation, and Oncogenicity

Despite U1's high abundance (∼1,000,000 per cell in HeLa [Baserga and Steitz 1993]), even small changes in available U1 levels have profound effects that are not readily detected in high U1 AMO because of toxicity and drastic PCPA closer to TSS. Low U1 AMO (masking <15% of U1), in addition to low level PCPA, causes widespread 3′ UTR shortening (increased usage of proximal PAS in the last exon) and shifts to shorter mRNA isoforms (Berg et al. 2012), that are hallmarks of and play an important role in stimulated states in immune cells and neurons, cell proliferation, and cancer (Niibori et al. 2007; Flavell et al. 2008; Sandberg et al. 2008; Mayr and Bartel 2009; Lianoglou et al. 2013; Xia et al. 2014; de Morree et al. 2019). Although it maintains mRNA's full-length coding sequence, 3′ UTR shortening removes binding sites for RBPs and miRNAs that regulate translation, stability and localization thereby altering these functions.

Low U1 AMO dose-dependently mimics short mRNA isoform switching because of intronic PCPA induced by transient transcription up-regulation in stimulated neurons (Berg et al. 2012). For example, PCPA in an intron in *homer-1*, which encodes a synaptogenesis scaffold protein, creates a short (carboxy-terminal deleted) isoform with antagonistic function that buffers overstimulation and prevents epilepsy (Niibori et al. 2007). Conversely, U1 overexpression suppressed mRNA isoform switching in stimulated neurons (Berg et al. 2012). Short isoform switching in stimulated neurons tracked with the transient burst in transcription, during which absolute U1 levels did not decrease, which could create transient telescripting deficit because of increased competition for U1 (Berg et al. 2012). U1 levels inevitably lag behind rapidly up-regulated transcription because U1 biogenesis entails elaborate RNP assembly by the SMN complex (So et al. 2017). These findings suggest that transient U1 shortage (for telescripting and splicing) is a built-in aspect of acute response. The CPA factors (CPAFs) and splicing factors (SFs) may be similarly affected. It remains to be determined how the critical PCPA–telescripting balance is regulated.

Down-regulation of two CPAFs, CFIm25 and CFIm68, elicits widespread 3′ UTR shortening and is oncogenic (Masamha et al. 2014; Masamha and Wagner 2017). Interestingly, low U1 AMO also dose-dependently increased cancer cells' migration and invasion in vitro by up to 500% (Oh et al. 2020). U1 overexpression had the opposite effect. U2 AMO and high U1 AMO were toxic in <24 h. In addition to 3′ UTR length changes (generally shortening in low U1 AMO and lengthening in U1 overexpression), numerous transcriptome changes that could contribute to the altered phenotype were observed, including alternative splicing, and mRNA expression levels of proto-oncogenes and tumor suppressors. These findings reveal an unexpected role for U1 homeostasis (available U1 and its activity) in relation to oncogenic and activated cell states and suggest U1 as a potential target for their modulation (Oh et al. 2020). Recent reports of oncogenic mutations in U1 5′ sequence in cancer patients (Shuai et al. 2019; Suzuki et al. 2019) support this notion.

Selective Telescripting Dependence of Long Introns: A New Layer of Gene Regulation

Nearly 1,000 human genes show no evidence of PCPA at any U1 AMO dose (Oh et al. 2017). The PCPAed and non-PCPAed genes have strikingly different lengths (medians of 39 kb vs. 14.2 kb, respectively; overall expressed genes 22.8 kb) derived almost entirely from intron length. Thus, U1 telescripting is selectively required for full-length transcription of long genes, an unprecedented gene regulation mechanism based on gene length. Remarkably, at high U1 AMO, many small PCPA-resistant genes (median 6.8 kb) were up-regulated, spliced robustly, and produced more mRNA and protein (up to fivefold) (Oh et al. 2017). Gene ontology revealed that short PCPA-insensitive genes are enriched in primary response genes that are induced during acute cell stimulation and necessary to enhance cell survival and adapt to adverse environmental changes, including *MYC* (proto-oncogene), *CYR61* (chemo-resistance), and *GADD45B*. These primary response functions are underrepresented in long genes, which are instead more highly expressed in differentiated tissues, and enriched in specialized functions. Tumor suppressors, DNA damage responsive, neuronal, and developmental genes are among the longest (Bertagnolli et al. 2013; Gabel et al. 2015), which makes them highly susceptible to PCPA. This surprising gene size–function relationship creates a stratified system that depends on U1 telescripting and can be actuated to rapidly shift expression priorities by cotranscriptional PCPA.

Tracking the gene groups in divergent metazoans, showed that gene–size function polarization increased by selective intron expansion in evolution, suggesting it was beneficial. We proposed that transient cotranscriptional PCPA in long genes decreases competition for transcription and RNA processing factors that normally limit small genes' expression, boosting their up-regulation in acute phase and facilitating shifts in polarized priorities (e.g., growth vs. differentiation) (Oh et al. 2017; Venters et al. 2019).

MECHANISM

A Complex of U1 with CPAFs Suppresses PASs in Introns

A general U1 telescripting model (Berg et al. 2012; Venters et al. 2019) incorporating the observations described above and in publications cited therein proposed that CPAFs bind nascent Pol II transcripts cotranscriptionally, aided by interactions with Pol II carboxy-terminal domain (CTD) and various RNA-binding proteins (RBPs). This recruitment is initiated early on near the TSS (Dantonel et al. 1997; McCracken et al. 1997; Calvo and Manley 2003) and could leave the first PAS in the RNA vulnerable to PCPA. U1 base-paired at the first 5′ss, which is enhanced by the 5′-cap binding complex (CBC), prevents PCPA from downstream PASs nearby in the intron. However, the effective range of U1 telescripting seems to be limited (<1 kb, possibly 2–3 kb). Thus, U1 base-paired to 5′ss would be insufficient to suppress PASs in longer introns (as in most mammalian genes), requiring additional U1 binding near PASs in the intron, which may explain the need for extra U1. Despite 1:1 stoichiometry with other snRNPs in U2 spliceosomes (U2, U4, U5, U6), U1 is several-fold more abundant than U4 and U6 in human cells (Baserga and Steitz 1993). Short introns have fewer PASs, stochastically, and kinetic competition from splicing lessens PCPA likelihood. Except for the 3′ UTR, exons are short (median ∼145 nt in humans) and may be protected by bound splicing factors.

To understand the molecular mechanism of U1 telescripting, it is required to know where U1 and CPAFs are bound on nascent transcripts relative to PCPA locations and to identify potential interactions between U1 and CPAFs. High degeneracy of U1 base-pairing and numerous PAS variants required their binding sites to be determined experimentally. These objectives were addressed by minimal RNA–protein and protein–protein cross-linking (XL) in cells with formaldehyde, followed by RNase digestion (to <150 nt), and high stringency parallel immunopurifications (IPs) with multiple antibodies to capture U1 and CPAF interactions. We have applied this procedure to the U1-specific proteins, several CPAFs, other snRNPs, and splicing factors in control and U1 AMO–transfected cells and used mass spectrometry (MS) and RNA-seq to define the XLIPs' protein composition and stoichiometry and map RNA fragments bound to them (So et al. 2019). This procedure is broadly applicable for comprehensive RNP definition in cells.

U1 and CPAF XLIPs-RNA-seq mapped to expected positions, including 5′ss and PASs in terminal exons, respectively. Additionally, U1 and CPAF XLIPs coincided with PCPA locations in introns (Fig. 3). The compositions and arrangement of U1 proteins and CPAFs and the sequences they bind in U1 snRNA and PASs, respectively, are shown in Figure 4. XLIP-seq of the U1-specific proteins and CPAFs of the three major CPA subunits (CFIm25, Fip1, and CstF64; CFIm, CPSF, and CstF, respectively; Fig. 4) coincided in the same peaks (in a 100-nt window). In contrast, essential splicing factors (e.g., SF3B1) were absent from the same peaks, but present at splice sites. XLIPs-MS uncovered a novel complex comprising U1 and CPAFs (U1–CPAFs), that binds and suppresses PASs in introns and 3′ UTRs (Figs. 3 and 4). U1–CPAFs are distinct from U1 complexes with spliceosomes. These observations suggested that U1 telescripting is mediated by direct binding to CPAFs at PASs, ruling out several alternative scenarios. U1–CPAFs as the telescripting complex can explain early studies showing that U1 prevents PAS usage in HIV-1 transcripts (Ashe et al. 1995, 1997) by U1's base-pairing at cryptic 5′ss in the 5′ long terminal repeat (5′ LTR), and in bovine papilloma virus pre-mRNA (Gunderson et al. 1998).

Comparison of XLIPs from U1 AMO and cAMO and purification of U1–CPAFs in active/uninhibited state with biotin-labeled U1 AMO transfected into cells (triggering PCPA), showed that U1 AMO remodels, but does not disrupt, U1–CPAFs (So et al. 2019). At least two changes in

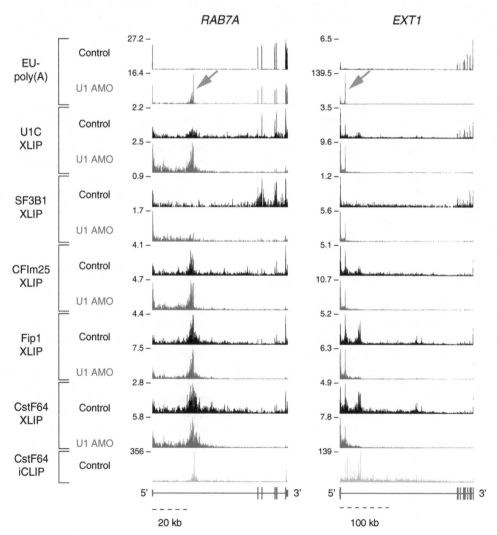

Figure 3. Peaks of U1 and CPAF XLIPs RNA-seq colocalized at PCPA locations in introns. Genome browser views of representative PCPA genes are shown with XLIPs, poly(A)-seq (So et al. 2019), and iCLIP (Yao et al. 2012) (e.g., *RAB7A* and *EXT1*) from HeLa cells transfected with either control or U1 AMO. The major PCPA points, identified from U1 AMO-induced 5-min EU-labeled, oligo(dT)-selected RNA-seq are indicated with arrows. U1C, SF3B1, and CPAF (CFIm25, Fip1, and CstF64) XLIPs and UV iCLIP-seq (CstF64) coincide (100-nt window) with PCPA points.

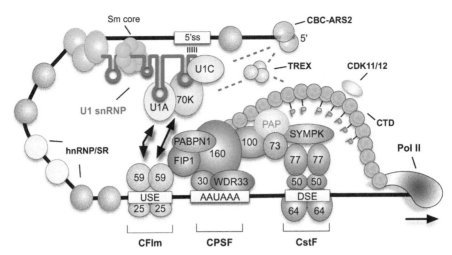

Figure 4. A model of U1–CPAFs complex function in telescripting. Schematic representation of components and interactions (red arrows) at the cryptic PAS in the first introns, based on XLIPs-MS and XLIP-RNA-seq. The U1-specific U1-70K and U1A proteins bind stem loop (SL1) 1 and 2, respectively. U1C binds 5′-end U1 snRNA and associates with U1 through U1-70K. A heptameric Sm ring on U1's Sm site is shown between SL3 and SL4. Known CPAF subunits (CFIm, CPSF, and CstF), bound to upstream element (USE), a PAS hexamer, downstream element (DSE) are indicated, respectively. Interactions with cap binding complex (CBC)/ARS2 complex and transcription and nuclear export complex (TREX) are shown as dashed lines. The blocking arrow from U1 to the CPAFs represents CPA inhibitory activities, including of U1A. The shown interactions are eliminated or altered by U1 AMO binding to the U1 snRNA 5′ end, which remodels U1–CPAFs, allowing the CPA stimulating CFIm68 to associate, likely replacing nonstimulatory CFIm59, and disrupts U1A-CPAFs association. (Modified, from So et al. 2019, with permission from Elsevier.)

U1–CPAFs induced by U1 AMO help us understand U1 telescripting: loss of CPA-inhibitory interactions and gain of CPA-stimulating factor, represented by U1A and CFIm68, respectively. U1-free U1A and U1A bound and in specific contexts in the last exon, can inhibit CPA by binding to poly(A) polymerase (PAP) (Boelens et al. 1993; Gunderson et al. 1994; Klein Gunnewiek et al. 2000; Phillips and Gunderson 2003; Workman et al. 2014). U1A associations with CPAFs were abolished by U1 AMO, which is compatible with a role for U1A in telescripting in the context of U1–CPAFs. In contrast, U1C and U1-70K's binding to CPAFs were not disrupted with U1 AMO and may explain why U1–CPAFs remain largely intact despite loss of U1 base-pairing. U1 AMO replaced CFIm59 binding to CFIm25 with CFIm68. CFIm68/CFIm25 is CPA-stimulating, but CFIm59/CFIm25 is not (Rüegsegger et al. 1998; Kim et al. 2010; Zhu et al. 2018). Thus, U1 is a CPA-regulating subunit of the U1–CPAFs that is held together and suppresses PASs by multiple protein–protein interactions that remain to be fully defined.

Additional U1–CPAFs associations detected in the XLIPs included the CBC/ARS2 (Giacometti et al. 2017), exosomes, transcription and nuclear export complex (TREX) (Silla et al. 2018) and other transcription factors, chromatin remodelers, and CDK12, known to phosphorylate Pol II CTD (Fig. 4). Recent studies have shown that U1 AMO induces Pol II pausing and PCPA at the first nucleosome barrier, generally in the first part of long introns (Chiu et al. 2018). These barriers thus serve U1-controlled transcription checkpoints requiring chromatin remodelers and CDK12 to get through. U1s bound at the first 5′ss and possibly additional sites upstream of the intronic checkpoints suppress PAS clusters at these nucleosome barriers, thereby regulating Pol II flow. Interestingly, CDK12 is frequently mutated in cancer and contributes to BRCAness (loss of DNA damage repair, DDR) (Cancer Genome Atlas Research Network 2011; Abeshouse et al. 2015; Quereda et al. 2019). CDK12 knockdown or its inhibition with THZ531, have also been recently shown to cause PCPA (Dubbury et al. 2018; Krajewska et al. 2019). Thus, U1 and CDK12 are both necessary for telescripting, and THZ531 provides an additional tool for future studies.

CONCLUSION

Studies on U1 telescripting add to U1's established role in splicing, placing it at the center of gene expression regulation, in both transcription and RNA processing. It is a remarkable range of functions for one small RNP, and consequently changes in U1 homeostasis have profound impacts on cell survival and adaptation to stimuli and cell behavior. The ability to modulate these biological processes experimentally, such as with U1 AMO, should have many applications, including potential therapies. Future research on U1 telescripting mechanism and U1 abundance regulation will help realize these opportunities.

ACKNOWLEDGMENTS

We are grateful to the members of our laboratory for helpful discussions. This work was supported by the National Institutes of Health (NIH) (R01 GM112923 to G.D.). G.D. is an Investigator of the Howard Hughes Medical Institute.

REFERENCES

Abeshouse A, Ahn J, Akbani R, Ally A, Amin S, Andry CD, Annala M, Aprikian A, Armenia J, Arora A. 2015. The molecular taxonomy of primary prostate cancer. *Cell* **163:** 1011–1025. doi:10.1016/j.cell.2015.10.025

Almada AE, Wu X, Kriz AJ, Burge CB, Sharp PA. 2013. Promoter directionality is controlled by U1 snRNP and polyadenylation signals. *Nature* **499:** 360–363. doi:10.1038/nature12349

Ashe MP, Griffin P, James W, Proudfoot NJ. 1995. Poly(A) site selection in the HIV-1 provirus: inhibition of promoter-proximal polyadenylation by the downstream major splice donor site. *Genes Dev* **9:** 3008–3025. doi:10.1101/gad.9.23.3008

Ashe MP, Pearson LH, Proudfoot NJ. 1997. The HIV-1 5′ LTR poly(A) site is inactivated by U1 snRNP interaction with the downstream major splice donor site. *EMBO J* **16:** 5752–5763. doi:10.1093/emboj/16.18.5752

Baserga SJ, Steitz JA. 1993. The diverse world of small ribonucleoproteins. In *The RNA world* (ed. Gesteland RF, Atkins JF), pp. 359–381. Spring Harbor Laboratory Press, New York.

Berg MG, Singh LN, Younis I, Liu Q, Pinto AM, Kaida D, Zhang Z, Cho S, Sherrill-Mix S, Wan L, et al. 2012. U1 snRNP determines mRNA length and regulates isoform expression. *Cell* **150:** 53–64. doi:10.1016/j.cell.2012.05.029

Bertagnolli NM, Drake JA, Tennessen JM, Alter O. 2013. SVD identifies transcript length distribution functions from DNA microarray data and reveals evolutionary forces globally affecting GBM metabolism. *PLoS One* **8:** e78913. doi:10.1371/journal.pone.0078913

Boelens WC, Jansen EJ, van Venrooij WJ, Stripecke R, Mattaj IW, Gunderson SI. 1993. The human U1 snRNP-specific U1A protein inhibits polyadenylation of its own pre-mRNA. *Cell* **72:** 881–892. doi:10.1016/0092-8674(93)90577-D

Calvo O, Manley JL. 2003. Strange bedfellows: polyadenylation factors at the promoter. *Genes Dev* **17:** 1321–1327. doi:10.1101/gad.1093603

Cancer Genome Atlas Research Network. 2011. Integrated genomic analyses of ovarian carcinoma. *Nature* **474:** 609. doi:10.1038/nature10166

Chiu AC, Suzuki HI, Wu X, Mahat DB, Kriz AJ, Sharp PA. 2018. Transcriptional pause sites delineate stable nucleosome-associated premature polyadenylation suppressed by U1 snRNP. *Molecular cell* **69:** 648–663. doi:10.1016/j.molcel.2018.01.006

Connelly S, Manley JL. 1988. A functional mRNA polyadenylation signal is required for transcription termination by RNA polymerase II. *Genes Dev* **2:** 440–452. doi:10.1101/gad.2.4.440

Dantonel J-C, Murthy KGK, Manley JL, Tora L. 1997. Transcription factor TFIID recruits factor CPSF for formation of 3′ end of mRNA. *Nature* **389:** 399–402. doi:10.1038/38763

de Morree A, Klein JDD, Gan Q, Farup J, Urtasun A, Kanugovi A, Bilen B, van Velthoven CTJ, Quarta M, Rando TA. 2019. Alternative polyadenylation of Pax3 controls muscle stem cell fate and muscle function. *Science* **366:** 734–738. doi:10.1126/science.aax1694

Derti A, Garrett-Engele P, Macisaac KD, Stevens RC, Sriram S, Chen R, Rohl CA, Johnson JM, Babak T. 2012. A quantitative atlas of polyadenylation in five mammals. *Genome Res* **22:** 1173–1183. doi:10.1101/gr.132563.111

Di Giammartino DC, Nishida K, Manley JL. 2011. Mechanisms and consequences of alternative polyadenylation. *Mol Cell* **43:** 853–866. doi:10.1016/j.molcel.2011.08.017

Dubbury SJ, Boutz PL, Sharp PA. 2018. CDK12 regulates DNA repair genes by suppressing intronic polyadenylation. *Nature* **564:** 141–145. doi:10.1038/s41586-018-0758-y

Flavell SW, Kim TK, Gray JM, Harmin DA, Hemberg M, Hong EJ, Markenscoff-Papadimitriou E, Bear DM, Greenberg ME. 2008. Genome-wide analysis of MEF2 transcriptional program reveals synaptic target genes and neuronal activity-dependent polyadenylation site selection. *Neuron* **60:** 1022–1038. doi:10.1016/j.neuron.2008.11.029

Fong N, Brannan K, Erickson B, Kim H, Cortazar MA, Sheridan RM, Nguyen T, Karp S, Bentley DL. 2015. Effects of transcription elongation rate and Xrn2 exonuclease activity on RNA polymerase II termination suggest widespread kinetic competition. *Mol Cell* **60:** 256–267. doi:10.1016/j.molcel.2015.09.026

Gabel HW, Kinde B, Stroud H, Gilbert CS, Harmin DA, Kastan NR, Hemberg M, Ebert DH, Greenberg ME. 2015. Disruption of DNA-methylation-dependent long gene repression in Rett syndrome. *Nature* **522:** 89–93. doi:10.1038/nature14319

Giacometti S, Benbahouche NEH, Domanski M, Robert MC, Meola N, Lubas M, Bukenborg J, Andersen JS, Schulze WM, Verheggen C, et al. 2017. Mutually exclusive CBC-containing complexes contribute to RNA fate. *Cell Rep* **18:** 2635–2650. doi:10.1016/j.celrep.2017.02.046

Gilmour DS, Lis JT. 1984. Detecting protein–DNA interactions in vivo: distribution of RNA polymerase on specific bacterial genes. *Proc Natl Acad Sci* **81:** 4275–4279. doi:10.1073/pnas.81.14.4275

Gunderson SI, Beyer K, Martin G, Keller W, Boelens WC, Mattaj IW. 1994. The human U1A snRNP protein regulates polyadenylation via a direct interaction with poly(A) polymerase. *Cell* **76:** 531–541. doi:10.1016/0092-8674(94)90116-3

Gunderson SI, Polycarpou-Schwarz M, Mattaj IW. 1998. U1 snRNP inhibits pre-mRNA polyadenylation through a direct interaction between U1 70K and poly(A) polymerase. *Mol Cell* **1:** 255–264. doi:10.1016/S1097-2765(00)80026-X

Hartmann B, Valcárcel J. 2009. Decrypting the genome's alternative messages. *Curr Opin Cell Biol* **21:** 377–386. doi:10.1016/j.ceb.2009.02.006

Kaida D, Berg MG, Younis I, Kasim M, Singh LN, Wan L, Dreyfuss G. 2010. U1 snRNP protects pre-mRNAs from premature cleavage and polyadenylation. *Nature* **468:** 664–668. doi:10.1038/nature09479

Kim S, Yamamoto J, Chen Y, Aida M, Wada T, Handa H, Yamaguchi Y. 2010. Evidence that cleavage factor Im is a heterotetrameric protein complex controlling alternative polyadenylation. *Genes Cells* **15:** 1003–1013. doi:10.1111/j.1365-2443.2010.01436.x

Klein Gunnewiek JM, Hussein RI, van Aarssen Y, Palacios D, de Jong R, van Venrooij WJ, Gunderson SI. 2000. Fourteen residues of the U1 snRNP-specific U1A protein are required for homodimerization, cooperative RNA binding, and inhibition of polyadenylation. *Mol Cell Biol* **20:** 2209–2217. doi:10.1128/MCB.20.6.2209-2217.2000

Krajewska M, Dries R, Grassetti AV, Dust S, Gao Y, Huang H, Sharma B, Day DS, Kwiatkowski N, Pomaville M, et al. 2019. CDK12 loss in cancer cells affects DNA damage response genes through premature cleavage and polyadenylation. *Nat Commun* **10:** 1757. doi:10.1038/s41467-019-09703-y

Lerner MR, Boyle JA, Mount SM, Wolin SL, Steitz JA. 1980. Are snRNPs involved in splicing? *Nature* **283:** 220–224. doi:10.1038/283220a0

Lianoglou S, Garg V, Yang JL, Leslie CS, Mayr C. 2013. Ubiquitously transcribed genes use alternative polyadenylation to achieve tissue-specific expression. *Genes Dev* **27:** 2380–2396. doi:10.1101/gad.229328.113

Lubas M, Andersen PR, Schein A, Dziembowski A, Kudla G, Jensen TH. 2015. The human nuclear exosome targeting complex is loaded onto newly synthesized RNA to direct early ribonucleolysis. *Cell Rep* **10:** 178–192. doi:10.1016/j.celrep.2014.12.026

Masamha CP, Wagner EJ. 2017. The contribution of alternative polyadenylation to the cancer phenotype. *Carcinogenesis* **39:** 2–10. doi:10.1093/carcin/bgx096

Masamha CP, Xia Z, Yang J, Albrecht TR, Li M, Shyu AB, Li W, Wagner EJ. 2014. CFIm25 links alternative polyadenylation to glioblastoma tumour suppression. *Nature* **510:** 412–416. doi:10.1038/nature13261

Mayr C, Bartel DP. 2009. Widespread shortening of 3′UTRs by alternative cleavage and polyadenylation activates oncogenes in cancer cells. *Cell* **138:** 673–684. doi:10.1016/j.cell.2009.06.016

McCracken S, Fong N, Yankulov K, Ballantyne S, Pan G, Greenblatt J, Patterson SD, Wickens M, Bentley DL. 1997. The C-terminal domain of RNA polymerase II couples mRNA processing to transcription. *Nature* **385:** 357–361. doi:10.1038/385357a0

Meola N, Domanski M, Karadoulama E, Chen Y, Gentil C, Pultz D, Vitting-Seerup K, Lykke-Andersen S, Andersen JS, Sandelin A, et al. 2016. Identification of a nuclear exosome decay pathway for processed transcripts. *Mol Cell* **64:** 520–533. doi:10.1016/j.molcel.2016.09.025

Mount SM, Pettersson I, Hinterberger M, Karmas A, Steitz JA. 1983. The U1 small nuclear RNA-protein complex selectively binds a 5′ splice site in vitro. *Cell* **33:** 509–518. doi:10.1016/0092-8674(83)90432-4

Niibori Y, Hayashi F, Hirai K, Matsui M, Inokuchi K. 2007. Alternative poly(A) site-selection regulates the production of alternatively spliced *vesl-1/homer1* isoforms that encode postsynaptic scaffolding proteins. *Neurosci Res* **57:** 399–410. doi:10.1016/j.neures.2006.11.014

Ntini E, Järvelin AI, Bornholdt J, Chen Y, Boyd M, Jørgensen M, Andersson R, Hoof I, Schein A, Andersen PR, et al. 2013. Polyadenylation site-induced decay of upstream transcripts enforces promoter directionality. *Nat Struct Mol Biol* **20:** 923–928. doi:10.1038/nsmb.2640

Oh JM, Di C, Venters CC, Guo J, Arai C, So BR, Pinto AM, Zhang Z, Wan L, Younis I, et al. 2017. U1 snRNP telescripting regulates a size-function-stratified human genome. *Nat Struct Mol Biol* **24:** 993–999. doi:10.1038/nsmb.3473

Oh JM, Venters CC, Di C, Pinto AM, Wan L, Younis I, Cai Z, Arai C, So BR, Duan J, et al. 2020. U1 snRNP regulates cancer cell migration and invasion in vitro. *Nat Commun* **11:** 1. doi:10.1038/s41467-019-13993-7

Padgett RA, Mount SM, Steitz JA, Sharp PA. 1983. Splicing of messenger RNA precursors is inhibited by antisera to small nuclear ribonucleoprotein. *Cell* **35:** 101–107. doi:10.1016/0092-8674(83)90212-X

Phillips C, Gunderson S. 2003. Sequences adjacent to the 5′ splice site control U1A binding upstream of the IgM heavy chain secretory poly(A) site. *J Biol Chem* **278:** 22102–22111. doi:10.1074/jbc.M301349200

Preker P, Nielsen J, Kammler S, Lykke-Andersen S, Christensen MS, Mapendano CK, Schierup MH, Jensen TH. 2008. RNA exosome depletion reveals transcription upstream of active human promoters. *Science* **322:** 1851–1854. doi:10.1126/science.1164096

Proudfoot NJ. 2016. Transcriptional termination in mammals: stopping the RNA polymerase II juggernaut. *Science* **352:** aad9926. doi:10.1126/science.aad9926

Proudfoot NJ, Brownlee GG. 1976. 3′ non-coding region sequences in eukaryotic messenger RNA. *Nature* **263:** 211–214. doi:10.1038/263211a0

Quereda V, Bayle S, Vena F, Frydman SM, Monastyrskyi A, Roush WR, Duckett DR. 2019. Therapeutic targeting of CDK12/CDK13 in triple-negative breast cancer. *Cancer Cell* **36:** 545–558.e7. doi:10.1016/j.ccell.2019.09.004

Rüegsegger U, Blank D, Keller W. 1998. Human pre-mRNA cleavage factor I_m is related to spliceosomal SR proteins and can be reconstituted in vitro from recombinant subunits. *Mol Cell* **1:** 243–253. doi:10.1016/S1097-2765(00)80025-8

Sandberg R, Neilson JR, Sarma A, Sharp PA, Burge CB. 2008. Proliferating cells express mRNAs with shortened 3′ untranslated regions and fewer microRNA target sites. *Science* **320:** 1643–1647. doi:10.1126/science.1155390

Shuai S, Suzuki H, Diaz-Navarro A, Nadeu F, Kumar SA, Gutierrez-Fernandez A, Delgado J, Pinyol M, López-Otín C, Puente XS, et al. 2019. The U1 spliceosomal RNA is recurrently mutated in multiple cancers. *Nature* **574:** 712–716. doi:10.1038/s41586-019-1651-z

Silla T, Karadoulama E, Mąkosa D, Lubas M, Jensen TH. 2018. The RNA exosome adaptor ZFC3H1 functionally competes with nuclear export activity to retain target transcripts. *Cell Rep* **23:** 2199–2210. doi:10.1016/j.celrep.2018.04.061

So BR, Zhang Z, Dreyfuss G. 2017. The function of survival motor neuron complex and its role in spinal muscular atrophy pathogenesis. In *Spinal muscular atrophy* (ed. Sumner CJ, Paushkin S, Ko C-P), pp. 99–111. Elsevier, New York.

So BR, Di C, Cai Z, Venters CC, Guo J, Oh J-M, Arai C, Dreyfuss G. 2019. A complex of U1 snRNP with cleavage and polyadenylation factors controls telescripting, regulating mRNA transcription in human cells. *Mol Cell* **76:** 590–599.e4. doi:10.1016/j.molcel.2019.08.007

Suzuki H, Kumar SA, Shuai S, Diaz-Navarro A, Gutierrez-Fernandez A, De Antonellis P, Cavalli FMG, Juraschka K, Farooq H, Shibahara I. 2019. Recurrent non-coding U1-snRNA mutations drive cryptic splicing in Shh medulloblastoma. *Nature* **574:** 707–711. doi:10.1038/s41586-019-1650-0

Tian B, Manley JL. 2017. Alternative polyadenylation of mRNA precursors. *Nat Rev Mol Cell Biol* **18:** 18–30. doi:10.1038/nrm.2016.116

Venters CC, Oh JM, Di C, So BR, Dreyfuss G. 2019. U1 snRNP telescripting: suppression of premature transcription termination in introns as a new layer of gene regulation. *Cold Spring Harb Perspect Biol* **11:** a032235. doi:10.1101/cshperspect.a032235

Vorlová S, Rocco G, Lefave CV, Jodelka FM, Hess K, Hastings ML, Henke E, Cartegni L. 2011. Induction of antagonistic soluble decoy receptor tyrosine kinases by intronic polyA activation. *Mol Cell* **43:** 927–939. doi:10.1016/j.molcel.2011.08.009

Wang ET, Sandberg R, Luo S, Khrebtukova I, Zhang L, Mayr C, Kingsmore SF, Schroth GP, Burge CB. 2008. Alternative isoform regulation in human tissue transcriptomes. *Nature* **456:** 470–476. doi:10.1038/nature07509

Workman E, Veith A, Battle DJ. 2014. U1A regulates 3′ processing of the survival motor neuron mRNA. *J Biol Chem* **289:** 3703–3712. doi:10.1074/jbc.M113.538264

Xia Z, Donehower LA, Cooper TA, Neilson JR, Wheeler DA, Wagner EJ, Li W. 2014. Dynamic analyses of alternative polyadenylation from RNA-seq reveal a 3′-UTR landscape across seven tumour types. *Nat Commun* **5:** 5274. doi:10.1038/ncomms6274

Yao C, Biesinger J, Wan J, Weng L, Xing Y, Xie X, Shi Y. 2012. Transcriptome-wide analyses of CstF64-RNA interactions in global regulation of mRNA alternative polyadenylation. *Proc Natl Acad Sci* **109:** 18773–18778. doi:10.1073/pnas.1211101109

Zhu Y, Wang X, Forouzmand E, Jeong J, Qiao F, Sowd GA, Engelman AN, Xie X, Hertel KJ, Shi Y. 2018. Molecular mechanisms for CFIm-mediated regulation of mRNA alternative polyadenylation. *Mol Cell* **69:** 62–74. doi:10.1016/j.molcel.2017.11.031

Functional and Mechanistic Interplay of Host and Viral Alternative Splicing Regulation during Influenza Infection

MATTHEW G. THOMPSON AND KRISTEN W. LYNCH

Department of Biochemistry and Biophysics Perelman School of Medicine, University of Pennsylvania, Philadelphia, Pennsylvania 19104, USA

Correspondence: klync@pennmedicine.upenn.edu

Alternative splicing is a pervasive gene regulatory mechanism utilized by both mammalian cells and viruses to expand their genomic coding capacity. The process of splicing and the RNA sequences that guide this process are the same in mammalian and viral transcripts; however, viruses lack the splicing machinery and therefore must usurp both the host spliceosome and many of the associated regulatory proteins in order to correctly process their genes. Here, we use the example of the influenza A virus to both describe how viruses utilize host splicing factors to regulate their own splicing and provide examples of how viral infection can, in turn, alter host splicing. Importantly, we show that at least some of the viral-induced changes in host splicing occur in genes that alter the efficiency of influenza replication. We emphasize the importance of increased understanding of the mechanistic interplay between host and viral splicing, and its functional consequences, in uncovering potential antiviral vulnerabilities.

Influenza A virus (IAV) is a ubiquitous and significant health threat, resulting in 290,000–650,000 deaths per year worldwide (World Health Organization 2019). In the United States alone, IAV is estimated to result 12,000–56,000 deaths annually (Centers for Disease Control 2019), burdening the economy with an estimated $11.2 billion cost (Putri et al. 2018). Although efforts are ongoing to treat the virus, there still is no universal cure or preventative vaccine. This lack of treatment is due, in part, to the virus's ability to rapidly mutate and develop resistance (Hussain et al. 2017). Therefore, it is important to further understand how IAV and host cells interact during infection in order to develop new avenues for antiviral therapies.

The IAV genome is comprised of eight single-stranded, negative-sense RNA segments that are transcribed and replicated in the nucleus. Like many other nuclear-expressed viruses, several of the transcripts expressed by IAV undergo alternative splicing to generate distinct protein-coding open reading frames. Specifically, at least three of the eight RNA segments (M, NS, and PB2) have been reported to express at least two proteins through regulated splicing (Fig. 1; Palese and Shaw 2013; Yamayoshi et al. 2015; Fabozzi et al. 2018).

In the case of both mammalian and influenza genes, splicing occurs through recognition and joining of sequences in the RNA by the spliceosome—a multicomponent enzymatic complex (Fig. 2). Although the sequences specifically bound by the spliceosome are known as the "splice sites," sequences outside of these splice site regions bind regulatory factors to control the efficiency of spliceosomal binding and function (Fig. 2). Hundreds of RNA-binding proteins (RBPs) exist in mammalian cells that have been shown to be able to function as splicing regulatory factors (Fu and Ares 2014). Ultimately it is the activity of such regulatory factors that determine where and when splicing of any given transcript occurs. Such alternative splicing of mammalian genes often takes the form of skipping or inclusion of specific exons in a transcript. In contrast, the alternative splicing most observed among the IAV genes involves the retention or removal of a single intron (Fig. 1).

The IAV NS segment encodes the NS1 protein when the single intron in its transcript is retained, whereas the NS2 protein is encoded upon removal of the intron (Fig. 1). Both the NS1 and NS2 proteins are required for successful viral replication. NS1 has multiple functions in countering the antiviral response of host and promoting IAV gene expression, whereas NS2 promotes vRNA export and packaging and forms part of the viral particle (Palese and Shaw 2013). The first 56 nt of the NS1 and NS2 transcripts are identical, resulting in 13 shared amino-terminal amino acids. However, upon removal of the intron, the reading frame is altered and the downstream amino acids are divergent between NS1 and NS2. Similarly, recent work by Yamayoshi et al. show that the PB2 segment contains a putative intronic sequence from nucleotides 1513 to 1894 of the 2341-nt transcript (Fig. 1; Yamayoshi et al. 2015) that, when removed, encodes a protein termed PB2-S1. Interestingly, not all IAV strains express this splicing pattern and resultant protein, and PB2-S1 null viruses do not show altered viral replication rates. Therefore, the significance of PB2 segment splicing remains to be determined.

M segment splicing is perhaps the best characterized of all IAV segments at both the functional and mechanistic

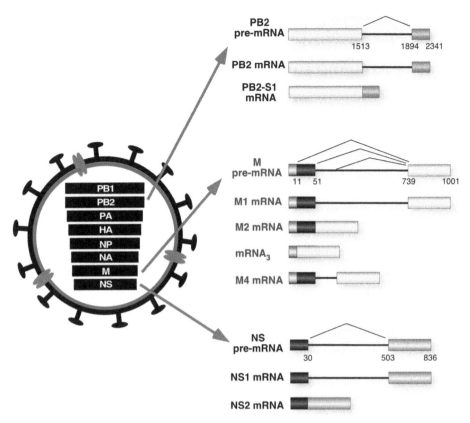

Figure 1. Diagram of influenza virus A showing the eight genome segments and the splicing patterns of the three segments best documented to undergo alternative splicing.

level. Like the other IAV segments described above, M segment splicing involves removal or retention of an intron. To date, the M segment has been shown to produce four alternate RNA isoforms: M1, M2, mRNA$_3$, and M4 (Fig. 1). The M2, mRNA$_3$, and M4 isoforms all share a common 3′ splice site (3′ SS) joined to different 5′ SSs, whereas the M1 isoform is the unspliced form. The M1 and M2 isoforms are the most abundant of the M segment RNAs and are both translated into proteins that are necessary for the viral life cycle (Palese and Shaw 2013). In contrast, the function of mRNA$_3$ and M4 splice isoforms is not understood. Jackson and Lamb showed that deletion of the mRNA$_3$ splicing isoform does not influence IAV replication in cell culture (Jackson and Lamb 2008), whereas the M4 isoform exists in only trace amounts and has not been explored functionally.

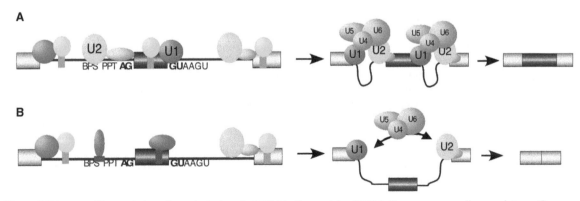

Figure 2. Diagram of the regulation of exon inclusion via RNA-binding proteins (RBPs). Boxes are exons; lines are introns. Conserved sequences at the splice sites are shown. Blue/purple circles labeled "U" are core spliceosome components that assemble in a step-wise fashion on the substrate. Green (*A*) and red (*B*) circles represent RBPs functioning to recruit the spliceosome to enhance exon inclusion (*A*) or repress spliceosome recognition of splice sites to induce exon skipping (*B*).

At least for the NS1/NS2 and M1/M2 proteins, the balance of expression is critical for viral replication. Therefore, the splicing of the NS and M segment RNAs must be carefully regulated. The control of NS1 to NS2 splicing has yet to be fully understood. What is known is that NS1 to NS2 splicing is regulated by a weak 5′ splice site (5′SS) that diverges from the canonical sequence (Dubois et al. 2014). This deviation from the canonical motif results in altered base-pairing of the U1 snRNP to the 5′ SS leading to inefficient splicing. Interestingly, Chua et al. propose that the NS1 5′SS is weak by design (Chua et al. 2013). In their model, they show that inefficient splicing causes NS2 levels to slowly accumulate throughout IAV infection at a rate that is advantageous to the virus. In contrast to NS, regulation of the M segment splicing has been studied more thoroughly and occurs through regulation of both the 5′ and 3′ splice sites by the activity of host RBPs. Here we describe work by ourselves and others in defining the mechanism by which M segment splicing is regulated by host proteins, what this may mean for the splicing of host genes, and how the regulation of host and viral splicing regulation together impact viral replication.

REGULATION OF IAV M SEGMENT SPLICING

As mentioned above, all of the spliced forms of the M segment utilize a common 3′SS (Fig. 1). Previous work has shown that host protein SRSF1 can repress this 3′SS, thus promoting the production of the M1 isoform over the spliced variants (Shih and Krug 1996). In addition, structural analysis of the 3′SS has shown that 3′SS availability, and thus spliceosomal recognition of the intron–exon boundary, may be regulated via the confirmation of the 3′SS (Moss et al. 2012). These 3′SS studies provide an understanding of how high levels of the M1 isoform are produced; however, they do not address why M2 is the predominant spliced form. A previous model for M segment 5′SS selection suggested that mRNA$_3$ splicing was repressed by the binding of viral polymerase components over the mRNA$_3$ 5′SS to promote M1 and M2 splice products (Shih et al. 1995). However, a later study refuted this claim by showing that a IAV protein with RNA-binding activity, NS1, is sufficient to repress mRNA$_3$ production and is likely a key contributor in the production of M1 and M2 splice products over mRNA$_3$ (Robb and Fodor 2012). The mechanism by which NS1 controls M1 to M2 splicing was initially unclear. However, NS1 has a large interactome that includes many host RBPs (Thulasi Raman and Zhou 2016; Kuo et al. 2018), suggesting that host proteins might likewise contribute significantly to the regulation of M1 versus M2 expression.

One of the earliest characterized binding partners of the IAV NS1 protein is the aptly named NS1 binding protein (NS1-BP) (Wolff et al. 1998). NS1-BP localizes to nuclear speckles, a subnuclear compartment that harbors a high concentration of splicing-related proteins (Wolff et al. 1998). An important breakthrough in understanding the regulation of M2 splicing came with the demonstration that knockdown of NS1-BP results in decreased levels of M2 and increased levels of M1, suggesting it had a direct role in M segment splicing regulation (Tsai et al. 2013). Additionally, the same study showed that knockdown of hnRNP K, a known host splicing regulator and binding partner of NS1-BP, also resulted in inhibition of M2 splicing (Tsai et al. 2013). Interestingly, NS1 has been proposed to cause relocalization of NS1-BP and other splicing factors away from nuclear speckles (Wolff et al. 1998; Fortes et al. 1995). To determine if subnuclear localization of proteins or RNA is important to the splicing of the M segment, a series of microscopy experiments were done to follow the localization of M1 and M2 splice isoforms in the context of NS1, NS1-BP, and hnRNP K depletion (Mor et al. 2016). Several key observations were made in this study. First, the authors show that M segment RNA localizes to nuclear speckles and that this localization is dependent on NS1 and NS1-BP. Second, M2 RNA is enriched specifically at speckles, suggesting that M1 to M2 splicing takes place within or in close proximity to nuclear speckles. Finally, although hnRNP K is not required for localization of M1 RNA to speckles, it is required to promote M2 splicing. Putting these observations together, the authors propose a model in which NS1, NS1-BP, and hnRNP K control M1 to M2 splicing through a nuclear speckle–dependent pathway (Fig. 3; Mor et al. 2016).

To dig more deeply into the mechanisms by which NS1-BP and hnRNP K regulate M2 splicing, we recently carried out a biochemical study to localize the binding sites of NS1-BP and hnRNP K on the M1 transcript (Thompson et al. 2018). To identify the site(s) of hnRNP K and NS1-BP binding to the M1 transcript we first carried out in vitro UV cross-linking assays with mammalian nuclear extract and a series of truncations of the M1 mRNA (Fig. 4A). These experiments showed that hnRNP K binds predominantly to a sequence within 30 nt downstream from the M2 5′SS (Thompson et al. 2018). Previous work has shown that hnRNP K binds preferentially to polycytosine (pC) tracts (Thisted et al. 2001). Notably, mutation of two pC tracts at nt 69–71 and 78–84, respectively, abrogates hnRNP K cross-linking to the 1–106 fragment (Fig. 4A). The binding of hnRNP K close to the M2 5′SS, and the

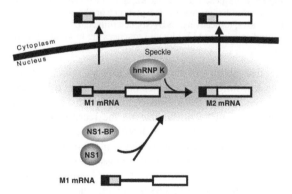

Figure 3. Diagram of the role of NS1, NS1-BP, and hnRNP K in promoting M2 splicing. NS1 and NS1-BP traffic the unspliced message to nuclear speckles. Within speckles hnRNP K promotes M1 to M2 splicing. Whether or not the message is spliced, trafficking to the speckles enhances export to the cytoplasm.

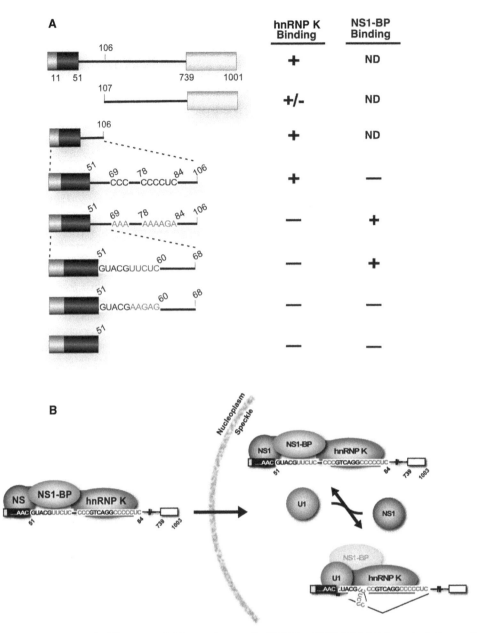

Figure 4. Sequences around the M2 5′SS are essential for recruitment of hnRNP K and NS1-BP. (*A*) Schematics indicate substrates used in UV cross-linking assays. The presence or absence of association of hnRNP K and NS1-BP with each indicated substrate is indicated by + and −, respectively. (ND) Not determined. (*B*) Detailed model for the association of hnRNP K and NS1-BP with the M substrate to control splicing to M2. Mutant sequences are in gray; wild-type sequences are in maroon, blue, and black.

fact that this protein directly promotes M2 splicing, led us to ask if hnRNP K promotes the recruitment of the spliceosome to the 5′SS. Specifically, we investigated the impact of hnRNP K binding on the recruitment of the U1 snRNP—a complex of RNA (the U1 snRNA) and proteins that initially recognizes and binds 5′SS. By both RNA affinity and psoralen cross-linking assays, we indeed find that binding of the U1 snRNA and its associated proteins to the M2 5′SS is promoted in an hnRNP K–dependent manner. Together these data provide a clear model for how hnRNP K enhances the production of the M2 RNA in the speckles through facilitating the recruitment of the U1 snRNP (Fig. 4B).

Interestingly, a cross-linking signal was also observed for NS1-BP upon either mutation or deletion of the pC tracts (Fig. 4A). Further truncations of the M1 transcript revealed this apparent NS1-BP binding to be immediately overlapping the M2 5′SS sequence. However, as NS1-BP does not contain any canonical RNA-binding domains (Adams et al. 2000), we consider this interaction to be dependent on protein–protein interactions. Consistent with this conclusion, the presence of NS1, either in vitro or in infected cells, results in a different association of NS1-BP with the M transcript, in which the association of NS1-BP with RNA is dependent on the proximal binding of hnRNP K but not sequences within the 5′SS

(Thompson et al. 2018). Because NS1-BP can interact in a trimeric complex with NS1 and hnRNP K, these data suggest that NS1 and hnRNP K effectively "hold" NS1-BP over the 5′SS through protein–protein interactions. This model suggests that this trimeric complex thus hinders splicing until the M RNA is trafficked to the nuclear speckles, where we propose that the high level of splicing factors and the activity of hnRNP K favor recruitment of the U1 snRNP and the productive use of the M2 5′SS (Fig. 4B).

REGULATION OF HOST SPLICING BY INFLUENZA INFECTION

Although NS1-BP has been studied with respect to IAV and the viral NS1 protein, its role in uninfected cells has not been well-characterized. The observation that NS1-BP associates with RNA in the absence of virus led to the question of whether NS1-BP might have activity as a regulator of splicing of human genes. We therefore carried out a quantitative analysis of splicing of approximately 5500 known alternative exons in A549 cells in the absence or presence of knockdown of NS1-BP or hnRNP K using the previously described RASL platform (Martinez et al. 2015; Li et al. 2012). Consistent with the previously described role of hnRNP K as a splicing regulator, we observed that splicing of approximately 200 out of the approximately 5500 exons surveyed is dependent on hnRNP K (Fig. 5A). Strikingly, more than one-half of these hnRNP K–dependent exons were also regulated in an NS1-BP-dependent manner, whereas only a handful of exons were identified as potentially NS1-BP-dependent, but not dependent on hnRNP K (Fig. 5A). Moreover, although NS1-BP and hnRNP K both showed enhancer and silencer activities (i.e., increased or decreased exon inclusion), almost all of the NS1-BP and hnRNP K–dependent exons were regulated in the same direction by both of these proteins (Fig. 5B), consistent with a model in which these proteins are working in concert. The limited scope of splicing events interrogated by RASL makes it unfeasible to assess sequence enrichment within the hnRNP K and NS1-BP coregulated genes. However, several of the NS1-BP/hnRNP K–regulated host genes show obvious proximal pY and pC tracts downstream from the enhanced 5′SS, similar to the M2 5′SS (Thompson et al. 2018). Taken together, these data reveal a previously unappreciated function for NS1-BP in host gene regulation and suggest that hnRNP K and NS1-BP form a complex that specifically regulates a subset of host splicing events.

The fact that hnRNP K and NS1-BP coordinately regulate host splicing events also indicates that IAV hijacks a preexisting widespread cellular splicing regulatory relationship between hnRNP K and NS1-BP to carry out its own M1 to M2 splicing. Such a model immediately raises the question of whether IAV hijacking of hnRNP K and NS1-BP also perturbs host splicing through either sequestering these proteins away from host targets or redirecting their activity to other genes beyond the M segment. Indeed, changes in the splicing patterns of human genes upon IAV infection was recently reported by Fabozzi et al. (2018). Using seasonal and laboratory strains of IAV H3N2, these authors showed IAV-induced alternative splicing changes in at least 775 genes (Fabozzi et al. 2018), although this reported impact of IAV on splicing is likely an underestimate given the relatively low sequencing depth in the study.

To specifically ask if hnRNP K and NS1-BP coregulated host splicing events are altered in the context of IAV infection, we first directly assessed the splicing of several of the genes from the RASL study for changes upon infection with the WSN strain of IAV. Notably, we observe IAV-induced changes in >60% of the genes tested (Thompson et al. 2018). Moreover, the impact of IAV infection on splicing implies two distinct mechanisms. For example, in the case of the genes *CASP8* and *INF2*, the impact of IAV infection was the opposite of that observed upon hnRNP K and NS1-BP knockdown, suggesting that IAV enhances the activity of hnRNP K and NS1-BP on some host target genes (Fig. 6). In contrast,

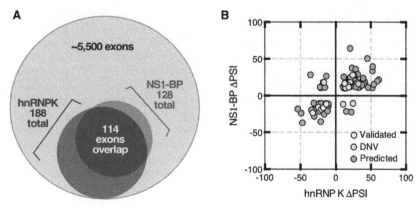

Figure 5. (*A*) Number of total surveyed exons that are differentially spliced upon depletion of hnRNP K (red), NS1-BP (blue), with overlap shown in purple. (*B*) Correlation of the impact on exon inclusion of depletion of hnRNP K or NS1-BP. Purple are events from overlap in *A*. Yellow and gray represent those events validated or not, respectively, by reverse transcription polymerase chain reaction (RT-PCR). Plotted is the difference in percent spliced isoform (ΔPSI) between wild type and NS1-BP (*y*-axis) or hnRNP K (*x*-axis) depletion.

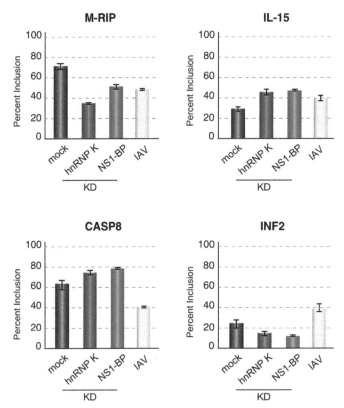

Figure 6. Examples of genes that are differentially spliced upon depletion of hnRNP K and NS1-BP and are also regulated upon IAV infection. The percent of transcripts including the variable exon are plotted for each condition. (KD) The protein has been targeted for knockdown.

splicing of *M-RIP* and *IL-15* in IAV-infected cells phenocopied hnRNP K or NS1-BP depletion (Fig. 6), suggesting that for a different set of target genes IAV infection inhibits hnRNP K/NS1-BP association and/or function.

To expand this analysis, we have recently carried out a high-depth RNA-seq experiment comparing IAV-infected versus uninfected A549 cells. Specifically, we sequenced poly(A) RNA from cells 6 and 12 h after infection with the WSN strain of IAV and compared the splicing to that of poly(A) RNA from uninfected cells using the MAJIQ algorithm (Vaquero-Garcia et al. 2016). The robustness of infection was confirmed by both the presence of reads from IAV RNA as well as up-regulation of several interferon-inducible genes. Strikingly, we observe approximately 900 significant splicing changes in host transcripts at both time points of IAV infection, impacting approximately 600 genes. The vast majority (~90%) of these genes do not show any changes in expression levels in response to IAV, indicating that regulation of these transcripts is solely at the level of splicing. Over the time course of the infection, these splicing events tend to increase their difference relative to wild type, consistent with what would be expected for IAV-induced splicing regulation.

Importantly, several of the IAV-induced splicing changes occur in genes that have previously been described to regulate IAV infection (Fig. 7). For example, *CLK1* is known to be an upstream regulator of splicing of host and viral transcripts through interactions with other host splicing factors and is particularly important for the production of IAV M2 (Duncan et al. 1997; Aubol et al. 2016; Dominguez et al. 2016). IAV-induced splicing of *CLK1* is predicted to increase the production of the full-length protein by promoting inclusion of all exons (Fig. 7). Therefore IAV-regulation of *CLK1* via splicing is likely to promote viral replication through promoting a correct balance of M1 to M2 expression. Similarly, *RAB11FIP3* and *PPIP5K2*, two additional genes harboring IAV-induced splicing events, have been shown to regulate IAV replication (Fig. 7). *RAB11FIP3* competes with IAV vRNPs for RAB11 on endosomes in the particle assembly pathway, thus regulating viral egress (Vale-Costa et al. 2016), whereas *PPIP5K2* is required for efficient IFN-β production (Pulloor et al. 2014), thus inhibiting viral replication via the innate immune response. The IAV-induced splicing changes in both of these genes results in the insertion of additional peptide sequences within the canonical encoded protein. The functional impact of these altered reading frames remains to be explored and will be an interesting area of future study (see Conclusion).

CONCLUSION

The work described above highlights how IAV uses host machinery to expand the coding potential of its genome

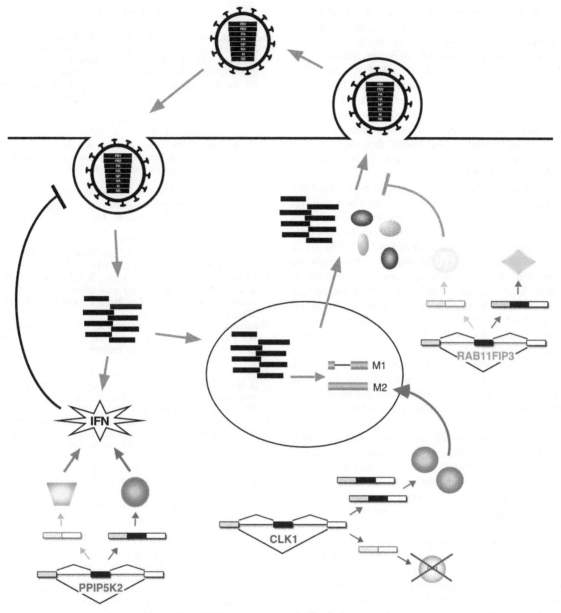

Figure 7. Model of the putative role of CLK1, RAB11FIP3, and PPIP5K2 in IAV replication and the possible impact of alternative splicing of the genes encoding these proteins in promoting or limiting IAV infection. Gray arrows indicate normal viral life cycle; red arrows indicate viral-induced splicing patterns of the three genes and resulting predicted protein and functional impact.

through alternative splicing. In particular the splicing event we focus on, that of M1 to M2 splicing, is essential for viral replication. A major hurdle for the development of antiviral therapies is the ability of the virus to mutate and evolve rapidly. However, virus cannot confer the same adaptive response on host proteins. Therefore, identifying essential interactions between the host splicing regulatory machinery and IAV RNAs and proteins opens the door to therapeutic approaches that target host proteins to prevent viral replication.

Beyond the potential for novel therapeutics, the study of viral splicing provides important insights to general mechanisms of splicing regulation. For example, the requirement for M1 RNA to be trafficked to nuclear speckles for subsequent export or splicing is unusual. Typically host genes are thought to be spliced in a cotranscriptional manner, with the splicing machinery being recruited from the speckles to the sites of transcription (Merkhofer et al. 2014). The fact that hnRNP K and NS1-BP regulate splicing of many host genes in addition to the M transcript suggests that a subset of host RNA may also be spliced in a speckle-dependent manner at least in the context of IAV infection. Indeed, the involvement of NS1 in trafficking and splicing the M segment RNA with hnRNP K and NS1-BP raises the possibility that NS1 could be directing a specific subset of hnRNP K/NS1-BP–regulated RNAs through nuclear speckles, which would otherwise be spliced cotranscriptionally. Recent experiments from the Fontoura laboratory do show that NS1 binds a population of host transcripts during IAV infection (Zhang et al. 2018). Therefore, determining if hnRNP K– and NS1-BP-regulated splicing events are bound by NS1 could reveal host RNAs that might be subject to M segment-like splicing mechanisms. Importantly, if it is determined that nuclear speckle–dependent splicing pathways are exclusively an IAV-driven phenomenon, then targeting this pathway therapeutically may prove to be an additional viable antiviral strategy.

More broadly, a major open question is to define the mechanism(s) by which IAV infection alters host splicing. As described above, one potential mechanism is via IAV-induced changes in hnRNP K/NS1-BP activity. For example, the presence of NS1 could draw hnRNP K/NS1-BP to speckles, as discussed in the preceding paragraph, or could alter the recruitment of these host proteins to specific RNAs through a combination of protein–protein and protein–RNA interactions. Alternatively, the abundance of IAV M RNA could function as a molecular sponge to sequester hnRNP K/NS1-BP away from host transcripts, as has been shown to occur for other viral RNAs and host proteins (Barnhart et al. 2013; Michalski et al. 2019) Consistent with this model, our and other RNA-seq experiments have shown that during infection IAV RNA accounts for >30% of polyadenylated transcripts in the cell.

Finally, the fact that IAV infection induces so many alterations in host splicing begs the question of whether these are advantageous for the virus, part of the host defense, or both. The fact that several genes that harbor IAV-regulated splicing events are known to have pro or antiviral activities strongly suggests that changes in host splicing induced upon viral infection impact viral replication. We predict that splicing changes induced specifically by the virus (e.g., via NS1 or viral RNA load) are likely to be proviral, as the virus would avoid causing such changes if they hinder viral replication. In contrast, it is possible that a subset of the observed splicing changes induced upon infection are actually part of the host-driven innate immune response. Such innate immune triggered splicing has been observed in other systems (Carpenter et al. 2014) and would be expected to be antiviral. Future studies are needed to tease apart viral- versus host-driven splicing regulation and to specifically test the influence of alternative splicing on viral growth by specifically modulating splicing by antisense or CRISPR approaches.

ACKNOWLEDGMENTS

The authors thank our collaborators Beatriz Fontoura, Adolfo Garcia-Sastre, YuhMin Chook, and Sara Cherry for discussions throughout this work. K.W.L. is supported by National Institutes of Health (NIH) grant R35 GM118048, and M.G.T., K.W.L., and the work described herein are supported by NIH grant R01 AI125524.

REFERENCES

Adams J, Kelso R, Cooley L. 2000. The kelch repeat superfamily of proteins: propellers of cell function. *Trends Cell Biol* **10:** 17–24. doi:10.1016/S0962-8924(99)01673-6

Aubol BE, Wu G, Keshwani MM, Movassat M, Fattet L, Hertel KJ, Fu X-D, Adams JA. 2016. Release of SR proteins from CLK1 by SRPK1: a symbiotic kinase system for phosphorylation control of pre-mRNA splicing. *Mol Cell* **63:** 218–228. doi:10.1016/j.molcel.2016.05.034

Barnhart MD, Moon SL, Emch AW, Wilusz CJ, Wilusz J. 2013. Changes in cellular mRNA stability, splicing, and polyadenylation through HuR protein sequestration by a cytoplasmic RNA virus. *Cell Rep* **5:** 909–917. doi:10.1016/j.celrep.2013.10.012

Carpenter S, Ricci EP, Mercier BC, Moore MJ, Fitzgerald KA. 2014. Post-transcriptional regulation of gene expression in innate immunity. *Nat Rev Immunol* **14:** 361–376. doi:10.1038/nri3682

Centers for Disease Control. 2019. *Estimated influenza illnesses, medical visits, hospitalizations, and deaths averted by vaccination in the United States CDC*. CDC, Atlanta.

Chua MA, Schmid S, Perez JT, Langlois RA, Tenoever BR. 2013. Influenza A virus utilizes suboptimal splicing to coordinate the timing of infection. *Cell Rep* **3:** 23–29. doi:10.1016/j.celrep.2012.12.010

Dominguez D, Tsai Y-H, Weatheritt R, Wang Y, Blencowe BJ, Wang Z. 2016. An extensive program of periodic alternative splicing linked to cell cycle progression. *Elife* **5:** e10288. doi:10.7554/eLife.10288

Dubois J, Terrier O, Rosa-Calatrava M, De Virologie L, Virpath H, Vircell G, Claude U, Lyon B, De Lyon U. 2014. Influenza viruses and mRNA splicing : doing more with less. *mBio* **5:** 1–13. doi:10.1128/mBio.00070-14

Duncan PI, Stojdl DF, Marius RM, Bell JC. 1997. In vivo regulation of alternative pre-mRNA splicing by the Clk1 protein kinase. *Mol Cell Biol* **17:** 5996–6001. doi:10.1128/MCB.17.10.5996

Fabozzi G, Oler AJ, Liu P, Chen Y, Mindaye S, Dolan MA, Kenney H, Gucek M, Zhu J, Rabin RL, et al. 2018. Strand-specific dual RNA-seq of bronchial epithelial cells infected with influenza A/H3N2 viruses reveals splicing of gene segment 6 and novel host–virus interactions. *J Virol* **92:** e00518-18. doi:10.1128/JVI.00518-18

Fortes P, Lamond A, Ortín J. 1995. Influenza virus NS1 protein alters the subnuclear localization of cellular splicing components. *J Gen Virol* **76:** 1001–1007. doi:10.1099/0022-1317-76-4-1001

Fu X-D, Ares M. 2014. Context-dependent control of alternative splicing by RNA-binding proteins. *Nat Rev Genet* **15:** 689–701. doi:10.1038/nrg3778

Hussain M, Galvin H, Haw TY, Nutsford A, Husain M. 2017. Drug resistance in influenza A virus: the epidemiology and management. *Infect Drug Resist* **10:** 121–134. doi:10.2147/IDR.S105473

Jackson D, Lamb RA. 2008. The influenza A virus spliced messenger RNA M mRNA3 is not required for viral replication in tissue culture. *J Gen Virol* **89:** 3097–3101. doi:10.1099/vir.0.2008/004739-0

Kuo R-L, Chen C-J, Tam E-H, Huang C-G, Li L-H, Li Z-H, Su P-C, Liu H-P, Wu C-C. 2018. Interactome analysis of NS1 protein encoded by influenza A H7N9 virus reveals an inhibitory role of NS1 in host mRNA maturation. *J Proteome Res* **17:** 1474–1484. doi:10.1021/acs.jproteome.7b00815

Li H, Qiu J, Fu XD. 2012. RASL-seq for massively parallel and quantitative analysis of gene expression. *Curr Protoc Mol Biol* **Chapter 4:** Unit 4.13.1-9. doi:10.1002/0471142727.mb0413s98

Martinez NM, Agosto L, Qiu J, Mallory MJ, Gazzara MR, Barash Y, Fu X-D, Lynch KW. 2015. Widespread JNK-dependent alternative splicing induces a positive feedback loop through CELF2-mediated regulation of MKK7 during T-cell activation. *Genes Dev* **29:** 2054–2066. doi:10.1101/gad.267245.115

Merkhofer EC, Hu P, Johnson TL. 2014. Introduction to cotranscriptional RNA splicing. *Methods Mol Biol* **1126:** 83–96. doi:10.1007/978-1-62703-980-2_6

Michalski D, Ontiveros JG, Russo J, Charley PA, Anderson JR, Heck AM, Geiss BJ, Wilusz J. 2019. Zika virus noncoding sfRNAs sequester multiple host-derived RNA-binding proteins and modulate mRNA decay and splicing during infection. *J Biol Chem* **294:** 16282–16296. doi:10.1074/jbc.RA119.009129

Mor A, White A, Zhang K, Thompson M, Esparza M, Muñoz-Moreno R, Koide K, Lynch KW, García-Sastre A, Fontoura BMA. 2016. Influenza virus mRNA trafficking through host nuclear speckles. *Nat Microbiol* **1:** 16069. doi:10.1038/nmicrobiol.2016.69

Moss WN, Dela-Moss LI, Kierzek E, Kierzek R, Priore SF, Turner DH. 2012. The 3′ splice site of influenza A segment 7 mRNA can exist in two conformations: a pseudoknot and a hairpin. *PLoS ONE* **7:** e38323. doi:10.1371/journal.pone.0038323

Palese P, Shaw M. 2013. Orthomyxoviridae. In *Fields virology*, 6th ed. (ed. Fields BN, Knipe DM, Howley PM), pp. 1151–1185. Lippincott Williams & Wilkins, Philadelphia.

Pulloor NK, Nair S, Kostic AD, Bist P, Weaver JD, Riley AM, Tyagi R, Uchil PD, York JD, Snyder SH, et al. 2014. Human genome-wide RNAi screen identifies an essential role for inositol pyrophosphates in type-I interferon response. *PLoS Pathog* **10:** e1003981. doi:10.1371/journal.ppat.1003981

Putri WCWS, Muscatello DJ, Stockwell MS, Newall AT. 2018. Economic burden of seasonal influenza in the United States. *Vaccine* **36:** 3960–3966. doi:10.1016/j.vaccine.2018.05.057

Robb NC, Fodor E. 2012. The accumulation of influenza A virus segment 7 spliced mRNAs is regulated by the NS1 protein. *J Gen Virol* **93:** 113–118. doi:10.1099/vir.0.035485-0

Shih SR, Krug RM. 1996. Novel exploitation of a nuclear function by influenza virus: the cellular SF2/ASF splicing factor controls the amount of the essential viral M2 ion channel protein in infected cells. *EMBO J* **15:** 5415–5427. doi:10.1002/j.1460-2075.1996.tb00925.x

Shih SR, Nemeroff ME, Krug RM. 1995. The choice of alternative 5′ splice sites in influenza virus M1 mRNA is regulated by the viral polymerase complex. *Proc Natl Acad Sci* **92:** 6324–6328. doi:10.1073/pnas.92.14.6324

Thisted T, Lyakhov DL, Liebhaber SA. 2001. Optimized RNA targets of two closely related triple KH domain proteins, heterogeneous nuclear ribonucleoprotein K and αCP-2KL, suggest distinct modes of RNA recognition. *J Biol Chem* **276:** 17484–17496. doi:10.1074/jbc.M010594200

Thompson MG, Muñoz-Moreno R, Bhat P, Roytenberg R, Lindberg J, Gazzara MR, Mallory MJ, Zhang K, García-Sastre A, Fontoura BMA, et al. 2018. Co-regulatory activity of hnRNP K and NS1-BP in influenza and human mRNA splicing. *Nat Commun* **9:** 2407. doi:10.1038/s41467-018-04779-4

Thulasi Raman SN, Zhou Y. 2016. Networks of host factors that interact with NS1 protein of influenza A virus. *Front Microbiol* **7:** 654. doi:10.3389/fmicb.2016.00654

Tsai P-L, Chiou N-T, Kuss S, García-Sastre A, Lynch KW, Fontoura BMA, Garcia-Sastre A, Lynch KW, Fontoura BMA. 2013. Cellular RNA binding proteins NS1-BP and hnRNP K regulate influenza A virus RNA splicing. *PLoS Pathog* **9:** e1003460. doi:10.1371/journal.ppat.1003460

Vale-Costa S, Alenquer M, Sousa AL, Kellen B, Ramalho J, Tranfield EM, Amorim MJ. 2016. Influenza A virus ribonucleoproteins modulate host recycling by competing with Rab11 effectors. *J Cell Sci* **129:** 1697–1710. doi:10.1242/jcs.188409

Vaquero-Garcia J, Barrera A, Gazzara MR, González-Vallinas J, Lahens NF, Hogenesch JB, Lynch KW, Barash Y. 2016. A new view of transcriptome complexity and regulation through the lens of local splicing variations. *Elife* **5:** e11752. doi:10.7554/eLife.11752

Wolff T, O'Neill RE, Palese P. 1998. NS1-binding protein (NS1-BP): a novel human protein that interacts with the influenza A virus nonstructural NS1 protein is relocalized in the nuclei of infected cells. *J Virol* **72:** 7170–7180.

World Health Organization. 2019. *WHO influenza (seasonal)*. WHO, Geneva.

Yamayoshi S, Watanabe M, Goto H, Kawaoka Y. 2015. Identification of a novel viral protein expressed from the PB2 segment of influenza A virus. *J Virol* **90:** 444–456. doi:10.1128/JVI.02175-15

Zhang L, Wang J, Muñoz-Moreno R, Kim M, Sakthivel R, Mo W, Shao D, Anantharaman A, García-Sastre A, Conrad NK, et al. 2018. Influenza virus NS1 protein–RNA interactome reveals intron targeting. *J Virol* **92:** e01634-18. doi:10.1128/JVI.01634-18

Small RNA Function in Plants: From Chromatin to the Next Generation

Jean-Sébastien Parent,[1,2,3] Filipe Borges,[1,2,4] Atsushi Shimada,[1,2] and Robert A. Martienssen[1,2]

[1]Howard Hughes Medical Institute, Cold Spring Harbor, New York 11724
[2]Cold Spring Harbor Laboratory, Cold Spring Harbor, New York 11724
Correspondence: martiens@cshl.edu

Small RNA molecules can target a particular virus, gene, or transposable element (TE) with a high degree of specificity. Their ability to move from cell to cell and recognize targets in *trans* also allows building networks capable of regulating a large number of related targets at once. In the case of epigenetic silencing, small RNA may use the widespread distribution of TEs in eukaryotic genomes to coordinate many loci across developmental and generational time. Here, we discuss the intriguing role of plant small RNA in targeting transposons and repeats in pollen and seeds. Epigenetic reprogramming in the germline and early seed development provides a mechanism to control genome dosage, imprinted gene expression, and incompatible hybridizations via the "triploid block."

Epigenetic phenomena have long been recognized and even harnessed in plants. Examples include early studies from Barbara McClintock on how transposons act as controlling elements through the production of a "repressor substance" (McClintock 1961), imprinting and the nonreciprocal fate of hyperploid seeds (Blakeslee et al. 1920; Kermicle 1970), transgenerational paramutation (Coe 1966; Chandler 2010), and somaclonal variation (Ong-Abdullah et al. 2015). It is now widely accepted that small RNA molecules (sRNAs) (21- to 24-nt-long in plants) actively participate in all these phenomena, but how these molecules themselves influence cell fate in subsequent generations remains mysterious (Heard and Martienssen 2014). Exciting recent studies in plants and in other eukaryotes now point to new roles for these small RNAs that are demanding further investigation.

From developmental signaling to virus resistance, sRNAs have many important functions in plant cells, yet the majority of them target repeats and transposable elements (TEs) previously considered largely inert (Borges and Martienssen 2015). Active transposons present the cell with an important challenge given their propensity to copy themselves and "move" within the genome, as well as causing chromosomal instability and thereby necessitating tight control. What makes sRNAs particularly well-suited to silence these elements is their ability to recognize sequence homology in *trans*. Indeed, host defense against invasive species appears to be a conserved function of sRNAs and associated factors going all the way back to prokaryotes (Swarts et al. 2014). More recent studies have cast a new light on McClintock's early observations in maize, showing that transposons provide an important source of regulatory elements for endogenous gene expression (Chuong et al. 2017). Indeed, the massive expansion of specific transposon families in different eukaryotic genomes provides abundant regulatory opportunities for sRNA-mediated control of gene expression at a genome-wide scale. Here, we summarize recent findings on how plant sRNAs regulate cell fate and transgenerational inheritance of epigenetic states.

GENERATING SMALL RNA MOLECULES

sRNAs in plants have different sizes owing to a variety of DICER-LIKE (DCL) proteins each specializing in a given size class (Borges and Martienssen 2015). In flowering plants, repeated regions and TEs typically produce small-interfering RNA molecules (siRNAs) that are 23- to 24-nt-long, although abundant 21-nt and 22-nt siRNAs are also produced in some species or specific developmental contexts and are known as epigenetically activated siRNA (easiRNA) (Nuthikattu et al. 2013; Creasey et al. 2014). easiRNA precursors can be transcribed by the RNA polymerase II (Pol II) or the plant-specific RNA polymerase IV, which shares many subunits with Pol II and is generally required for TE siRNA production. Pol IV generates short transcripts (Blevins et al. 2015; Zhai et al. 2015) that are immediately converted into double-stranded RNA by the RNA DEPENDENT RNA POLYMERASE2 (RDR2), which forms a complex with Pol IV and is required for polymerase activity (Singh et al. 2019).

---

[3]Present address: Ottawa Research and Development Center, Agriculture and Agri-Food Canada, Ottawa, Ontario K1A 0C6, Canada
[4]Present address: Institut Jean-Pierre Bourgin, Institut National de la Recherche Agronomique de Versailles-Grignon, 78026 Versailles Cedex, France

© 2019 Parent et al. This article is distributed under the terms of the Creative Commons Attribution License, which permits unrestricted reuse and redistribution provided that the original author and source are credited.

Importantly, the emergence of Pol IV allowed transcription of heterochromatic regions that are refractory to Pol II, thus allowing abundant production of siRNAs from transcriptionally silent transposons and repeats. One important factor required for Pol IV recruitment to a subset of heterochromatic loci is the SAWADEE HOMOEDOMAIN HOMOLOG1 (SHH1) protein, which binds to the dimethylated lysine 9 on histone 3 (H3K9me2) and resembles homeobox transcription factors (Law et al. 2013). Targeting of transcription to heterochromatin also occurs at Piwi-interacting RNA (piRNA) clusters in *Drosophila*, via the methylated H3K9 binding protein Rhino, presumably for much the same reason (Mohn et al. 2014; Yu et al. 2018). More recently, it was shown that the different members of the SWI2/SNF2-type chromatin remodeling protein family CLASSY (CLSY) also assist in Pol IV recruitment at specific targets (Zhou et al. 2018). Although CLSY1 and 2 appear to act together with SHH1, the recruiting partner(s) of CLSY3 and 4 are still unknown.

Paradoxically, RNA polymerase is required for expression, whereas sRNAs are associated with silencing. Perhaps for this reason, there is a constant need for surveillance of Pol II transcripts to intercept potentially problematic transcripts derived from transposons. In *Arabidopsis thaliana*, Pol II-dependent TE transcripts are rapidly targeted by a variety of endogenous microRNAs (miRNAs) that trigger production of 21- to 22-nt siRNAs via RDR6, DCL2, and DCL4 (Creasey et al. 2014). Interestingly, certain miRNA families in plants evolved to target long terminal repeat (LTR) retrotransposons specifically. Most notably the miR845 family targets retrotransposons at the conserved primer-binding site (PBS) where transfer RNAs (tRNAs) bind to initiate reverse transcription (Borges et al. 2018). Targeting the PBS with small RNA is a common mechanism for transposon control in both mammals and plants, via 3′ fragments of mature tRNA in mammals, and via miRNA derived from rearranged tRNA in plants (Šurbanovski et al. 2016; Schorn et al. 2017). At least in mammals, 3′CCA-tRF are potent inhibitors of retrotransposition and might provide a uniquely sensitive means to monitor transposon activity in eukaryotic genomes (Schorn et al. 2017).

In plants, miRNAs are produced by the enzyme DCL1, whereas secondary siRNAs can be the product of either DCL2, DCL3, or DCL4 (or any combination of the three) working in combination with RDR6 (Borges and Martienssen 2015). Although most sRNAs from TEs are 24 nt in length, there are a few repeated loci in *Arabidopsis* that generate 21- and 22-nt secondary siRNAs (Panda et al. 2016). These siRNAs resemble easiRNAs, which arise in specific genetic, cellular, and/or temporal contexts and depend on DCL2 and DCL4. For instance, the massive activation of transposons in *DEFECTIVE IN DNA METHYLATION1* (*DDM1*) mutants causes easiRNA accumulation in *Arabidopsis*, maize, and tomato (Slotkin et al. 2009; Creasey et al. 2014; Corem et al. 2018; Fu et al. 2018). easiRNAs also arise in support cells within gametophytes and in the seed, where TEs are epigenetically reactivated during reprogramming (Slotkin et al. 2009; Calarco et al. 2012; Ibarra et al. 2012). Intriguingly, easiRNAs in pollen depend on a noncanonical pathway involving components of both the siRNA and secondary siRNA pathways in plants, including RNA Pol IV, DCL2, and DCL4 (Borges et al. 2018; Martinez et al. 2018). This is especially interesting as heterochromatin is lost from the vegetative nucleus (VN) (the "nurse cell" in pollen) as well as from the microspore. In seeds the same pathway is required for the biogenesis of DCL4 isoform-dependent siRNAs (disiRNAs). These 21-nt small RNAs depend on a DCL4 isoform found in *Arabidopsis* and potentially in other Brassicaceae like *Capsella rubella*, which includes a nuclear localization signal (NLS) and depends on loss of DNA methylation for expression (Pumplin et al. 2016). Loss of DNA methylation and expression of this isoform is found in pollen and in endosperm, consistent with these observations (Pumplin et al. 2016; our unpublished results). A recent study also showed that sRNA molecules of 21, 22, and 24 nt are also dependent on Pol IV in the microspores of *Arabidopsis* and *C. rubella* (Fig. 1A) (Wang et al. 2020), indicating that biogenesis of gametophytic easiRNA is conserved at least in Brassicaceae.

Importantly, easiRNAs in pollen accumulate in sperm cells, but arise in the microspore and the VN, suggesting that they might move from cell to cell (Slotkin et al. 2009) (Fig. 1B), together with other small RNAs. This was recently showed in *Arabidopsis* pollen as transgene small RNA products made by DCL2 and DCL4 moved from the VN to the sperm cells to silence target transcripts (Martínez et al. 2016). One question that remains is whether these molecules diffuse passively or if they are actively transported by a protein factor, such as a member of the ARGONAUTE (AGO) family. Regardless, there is significant potential for sRNAs loaded in sperm cells to be delivered to the egg and central cells and contribute to early embryo and endosperm development (Fig. 1B,C), as has been shown for at least one miRNA (Zhao et al. 2018). One idea is that easiRNAs mediate the transition from cytoplasmic posttranscriptional gene silencing (PTGS) to nuclear transcriptional gene silencing (TGS) during epigenetic reprogramming, to establish or reinforce transgenerational silencing at imprinted loci (Teixeira and Colot 2010; Borges and Martienssen 2015).

REPROGRAMMING CHROMATIN WITH SMALL RNA

Epigenetic reprogramming refers to the erasure and resetting of epigenetic marks acquired during the life of the parent, and in mammals it occurs in the gametes and in the embryo (Heard and Martienssen 2014). The extent of reprogramming in plants is much less pronounced, and epigenetic inheritance far more common, as the plant germline maintains high levels of DNA methylation during the sporophytic (somatic) to gametophytic (germline) transition (Calarco et al. 2012; Ibarra et al. 2012; Ingouff et al. 2017; Walker et al. 2018). In plants, sRNAs can trigger cytosine methylation (mC) in all sequence contexts (CG, CHG, and CHH, where H = A, C, or T) via RNA-

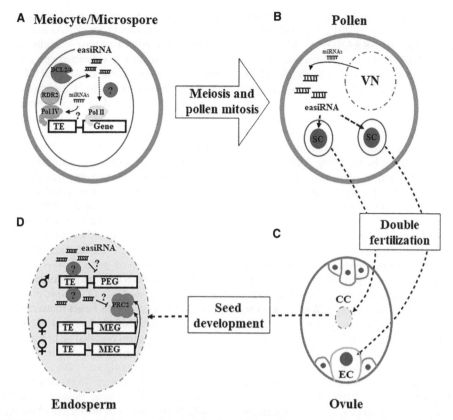

Figure 1. Small RNA movement and influence in plant reproduction. (*A*) In meiocytes and/or microspores, Pol IV produces small RNA from activated transposable elements through the different DCL proteins. These molecules influence the transcriptional state of different genes including some involved in pollen development. Pollen-specific miRNA molecules like miR845b in *Arabidopsis* that are presumably synthesized at meiosis or before and contribute to easiRNA production. (*B*) In mature pollen, the sperm cells (SC) are mostly inactive but receive epigenetic information in the form of easiRNA molecules produced in the vegetative nucleus (VN). These molecules are again dependent on Pol IV but also miRNA targeting transposon RNA molecules. (*C*) At fertilization, the two sperm cells fuse with the egg cell (EC) and central cell (CC) of the ovule delivering the haploid paternal genome and potentially the small RNA complement previously loaded into the sperm cells. (*D*) In the endosperm nuclei of the developing seed, imprinted alleles with specific epigenetic marks are treated differently by the transcriptional machinery. Depicted is a model of a paternally expressed gene where the maternal alleles have been silenced by the PRC2 complex. The maternal silencing is, however, incapable of reaching the paternal that is protected by the epigenetic state conferred by small RNA targeting working with yet unidentified factors.

directed DNA methylation (RdDM) (Matzke and Mosher 2014; Borges and Martienssen 2015; Matzke et al. 2015). CG and CHG methylation are symmetric on both strands and can be maintained during DNA replication by chromatin remodeling and histone modification, but mCHH is asymmetric and requires RdDM (Law and Jacobsen 2010; Borges and Martienssen 2015). Although high levels of mCG are found in male meiocytes, microspores, and differentiated sperm nuclei in the pollen, mCHH is almost completely erased in the male germline (Calarco et al. 2012; Ibarra et al. 2012; Walker et al. 2018), only to be restored to high levels in the mature embryo (An et al. 2017; Bouyer et al. 2017; Kawakatsu et al. 2017). The role of easiRNA during this form of reprogramming is still poorly understood, but might contribute to the restoration of CHH methylation during embryogenesis. In *C. rubella*, it was shown that Pol IV is essential for pollen development (Wang et al. 2020). The lack of severe phenotype in *Arabidopsis Pol IV* mutants may be due to a lower content of active transposons, although fertile *Pol IV* mutants in maize suggest that other mechanisms may be involved. In maize, Pol IV is required for paramutation, a classical epigenetic phenomenon in which epigenetic marks can be acquired from another allele and maintained in following generations (Hollick 2017). How those epialleles are formed and how they acquire the ability to be propagated through cell divisions remains a tantalizing mystery, as DNA methylation seems to play only a minor role. It has, however, been shown in *Arabidopsis* that some components of small RNA pathways are both required and sufficient to initiate de novo silencing of naive alleles (Fultz and Slotkin 2017; Gallego-Bartolomé et al. 2019).

Small RNA-loaded AGO4 (and likely its homologs AGO6 and AGO9 in the case of *A. thaliana*) is recruited to chromatin via noncoding RNA scaffolds produced by another plant-specific DNA-dependent RNA Polymerase, Pol V. Target recognition triggers the slicing of the RNA molecule (Liu et al. 2018) and subsequently, recruitment of factors such as INVOLVED IN DE NOVO METHYLATION2 (IDN2) (Böhmdorfer et al. 2014). IDN2 binds to long (10- to 11-nt) 5′ overhangs, presumably reflecting the cleaved RNA molecule still bound to the small RNA, and in turn recruits a variety of chromatin remodeling proteins responsible for silencing (Zhu et al. 2013; Liu et al. 2016), including DOMAINS REARRANGED METHYL-TRANSFERASE2 (DRM2) that methylates corresponding DNA in all sequence contexts (Böhmdorfer et al. 2014). Such methylation is thought to be recognized by members of the SU(VAR)3-9 HOMOLOG (SUVH) through their SRA methylated DNA-binding domains (Du et al. 2014) that can then establish H3K9 methylation in the case of SUVH4, 5, and 6 (Stroud et al. 2014) or propagate DNA methylation by recruiting additional Pol V, in the case of SUVH2 and 9 (Johnson et al. 2014). In doing so, the RdDM pathway perpetuates an environment that is refractory to Pol II transcription thereby keeping TEs silent.

Pol IV mutations lead to genome-wide depletion of 24-nt sRNAs (Zhou et al. 2018), but CHH methylation is only lost at a subset of those targets. The difference between the small RNA–dependent and –independent effects on asymmetric DNA methylation levels appears to be related to the presence or absence of Pol V at the locus (Wierzbicki et al. 2012). Indeed, the precise extent of the regions silenced by sRNAs appears to be dictated solely by the activity of Pol V (Böhmdorfer et al. 2016; Liu et al. 2018). It has been proposed that Pol V recruitment is guided by epigenetic marks like mC and H3K9me2, but the correlation is far from perfect (Böhmdorfer et al. 2016). Importantly, if it simply relies on preexisting marks, Pol V recruitment could not account for de novo establishment of silencing. Recruitment of Pol V is conditioned by the presence of members of the DDR (DEFECTIVE IN RNA-DIRECTED DNA METHYLATION1 (DRD1), DEFECTIVE IN MERISTEM SILENCING3 (DMS3), and RNA-DIRECTED DNA METHYLATION1 (RDM1)) complex (Zhong et al. 2012) whose partial structure has recently come to light (Wongpalee et al. 2019). Interestingly, the targeting of either DMS3 or RDM1 to a naive locus with a zinc finger is sufficient to trigger strong de novo silencing, but only in the presence of Pol V (Gallego-Bartolomé et al. 2019), consistent with an early recruiting role. Moreover, RDM1 has the capacity to bind to DNA (Gao et al. 2010), and DRD1 is the only protein associated with Pol V that has ATPase activity and chromatin remodeling potential for transcription (Wongpalee et al. 2019) as do other Rad54 homologs (Amitani et al. 2006). All in all, the activity of the DDR complex appears as a key to small RNA–guided epigenetic silencing.

FUNCTIONAL ROLE OF SMALL RNA IN THE GERMLINE

In *Arabidopsis*, the RdDM pathway targets mostly small repeats, the long terminal repeats of retrotransposons and certain families of DNA transposons (Matzke et al. 2015). Interestingly, these targets are distributed not only in pericentromeric regions where these elements are most abundant, but also in intergenic regions along the chromosome arms in *Arabidopsis*, tomato, and maize (Gouil and Baulcombe 2016). The proximity between these targeted regions and neighboring genes, therefore, creates the opportunity for sRNAs to influence their transcriptional activity. This type of control is observed in the seed where small RNA-directed epigenetic silencing activity is more pronounced (An et al. 2017; Bouyer et al. 2017; Kawakatsu et al. 2017). Imprinted genes, or alleles differentially expressed in the endosperm depending on the parent of origin are a striking example of such control. The maternally expressed imprinted genes *FWA*, *SDC*, *MOP9.5*, *DOG1-LIKE 4* (*DOGL4*), and *ALLANTOINASE* (*ALN*) were all shown to be methylated in the endosperm on the paternal side, and this was dependent on sRNA targeting a repeated element found in the promoter (Fig. 1D) (Vu et al. 2013; Zhu et al. 2018; Iwasaki et al. 2019). Interestingly, *DOGL4* and *ALN* are involved in seed dormancy, thereby linking sRNA pathways to this agriculturally important trait. Similar examples of small RNAs regulating imprinted gene expression on the maternal side have also been shown to influence seed development and ultimately seed size (Kirkbride et al. 2019).

It is clear that sRNAs target transcriptionally active regions to silence them, but several questions remain as to how this mechanism is scaled up to hundreds or thousands of genes that must be expressed or silenced in a coordinated fashion (Chuong et al. 2017). This invokes the genetic concept of "balance" to explain phenotypic variation associated with genome dosage (Birchler and Veitia 2007). Early studies of distinctive phenotypes dependent on the dosage of parental chromosomes in plants and flies trace back to the 1920s (Blakeslee et al. 1920; Belling and Blakeslee 1923; Bridges 1925). A few years later in maize, McClintock also reported the spontaneous appearance of "a triploid individual notably more vigorous than its diploid sibs" (McClintock 1929). Albert Blakeslee, working at Cold Spring Harbor Laboratory, first reported the inability to hybridize closely related species or parents with different ploidy, which is often associated with failure in endosperm function and seed collapse. For example, interploidy hybridizations with paternal excess result in endosperm overproliferation and seed abortion, sometimes known as the "triploid block" (Köhler et al. 2010). sRNAs and TEs were recently implicated in the triploid block response in *Arabidopsis*, as the loss of miR845 and Pol IV-dependent easiRNA in diploid pollen restored viability of triploid seeds almost completely (Erdmann et al. 2017; Borges et al. 2018; Martinez et al. 2018). Strikingly, this is highly reminiscent of hybrid dysgenesis in *Drosophila*, although in this case piRNAs protect the hybrid (Malone et al. 2015; Martienssen 2010). Thus, plants and animals use similar small RNA guides to control transposon activity and dosage in hybrid genomes.

The triploid block response led to the endosperm balance number (EBN) hypothesis, which was developed in the early 1980s in potato and then extended to many other crop species, to explain the ratio between maternal and paternal chromosomes via genetic factors required for development of a normal seed (Johnston and Hanneman 1982; Ehlenfeldt and Hanneman 1984; Carputo et al. 1997, 1999). We can now propose that such genetic factors may include small RNA loci, imprinted genes, and TEs that were found up-regulated in abortive interploid and interspecific hybrid seeds (Josefsson et al. 2006; Lu et al. 2012; Stoute et al. 2012; Kirkbride et al. 2015; Florez-Rueda et al. 2016; Roth et al. 2018). This includes maternally expressed genes (MEGs) that encode important components of the Polycomb repressor complex (PRC2), such as *MEDEA* and *FERTILIZATION-INDEPENDENT SEED2* (*FIS2*), that silence the paternal alleles via deposition of H3K27me3 (Lafon-Placette and Köhler 2015; Satyaki and Gehring 2017). Small RNAs and the RdDM pathway are also required for genomic imprinting (Vu et al. 2013; Erdmann et al. 2017; Zhu et al. 2018; Iwasaki et al. 2019; Kirkbride et al. 2019), whereas mutations in PRC2 result in endosperm defects closely resembling the triploid block. These observations lend support to the old idea that endosperm failure in interploidy crosses results from disruption of genomic imprinting (Köhler et al. 2010). Indeed, previous studies in *Arabidopsis* have shown that loss-of-function mutations in PEGs are able to suppress the triploid block, allowing the formation of viable triploid seeds (Kradolfer et al. 2013; Wolff et al. 2015). Taken together, current models propose that paternally inherited easiRNAs mediate "endosperm balance" by targeting TEs flanking imprinted genes, thus regulating their expression this way. For instance, if MEGs are direct targets of easiRNA regulation, this could result in PRC2 depletion and up-regulation of PEGs. However, there is still no clear evidence that PRC2 activity is impaired in abortive triploid seeds. On the other hand, targeting TEs flanking PEGs could result in their direct up-regulation in a dosage-dependent manner (Martinez et al. 2018), though target loci involved in this response have not been identified either.

Although defects in the RdDM pathway typically have a minor impact on somatic phenotypes in *Arabidopsis* (Sasaki et al. 2012), mutants losing both 24-nt siRNA and maintenance of DNA and histone methylation show various developmental phenotypes because of the misregulation of imprinted genes (Zemach et al. 2013). For example, *SDC*, an imprinted gene normally expressed maternally in endosperm cells, is up-regulated during the somatic developmental stage in that context, resulting in curly leaf and short stature phenotypes (Henderson and Jacobsen 2008; Zemach et al. 2013). In addition, reduced fertility and floral defects are also observed in the mutant defective in DDM1 and RDR6-dependent 21-nt and 22-nt sRNAs (Creasey et al. 2014), suggesting an important role of epigenetically activated 21-nt and 22-nt sRNAs. These defects are more severe when 24-nt siRNA is further depleted (Sasaki et al. 2012), and triple mutants suffer from chromosome missegregation (A Shimada and RA Martienssen, unpubl. data), reminiscent of the regulation of chromosome segregation by RNA interference in *Schizosaccharomyces pombe* (Gutbrod and Martienssen 2020). Thus in somatic tissues, sRNA molecules cooperatively work as a backup system for DNA methylation to maintain plant fitness and chromosome integrity.

It is revealing that the RdDM pathway is only found intact in the angiosperms—flowering plants (Lee et al. 2011; Huang et al. 2015). For example, key factors appear to be absent or incomplete in the gymnosperm Norway spruce, *Pices abies* (Ma et al. 2015; Matzke et al. 2015; Ausin et al. 2016; Pei et al. 2019). It is tempting to speculate that RdDM evolved with the appearance of double fertilization and genomic imprinting in the endosperm of flowering plants, a tissue that has no equivalent in gymnosperms. It has been speculated that this pathway evolved as a facilitator of polyploidization, which has had a crucial role in the evolution of angiosperms (Matzke et al. 2015), but is relatively rare in gymnosperms (Ickert-Bond et al. 2020). This is also consistent with a key role for sRNA in mediating genome dosage responses during interploidy hybridization barriers in the endosperm (Erdmann et al. 2017; Borges et al. 2018; Martinez et al. 2018; Moreno-Romero et al. 2019; Satyaki and Gehring 2019). sRNA molecules may indeed target all the different copies of a given TE and mark them with DNA methylation, insuring their transcriptional control and potentially accelerating their decay and elimination. In line with this hypothesis, it appears that concerted changes in sRNA abundance and epigenetic changes occur mostly at transposons in autotetraploid rice (Zhang et al. 2015) and in rapeseed following whole genome duplication (Cheng et al. 2016). There is also evidence of a transient increase in sRNA levels in newly synthesized allopolyploid *Brassica napus* (Martinez Palacios et al. 2019), whereas CHH methylation levels at TEs flanking genes are anticorrelated with their expression and may, therefore, be related to the establishment of subgenome dominance in interspecies crosses (Edger et al. 2017). All this points to an active role for sRNAs in unexpected "genomic shocks" that occur when distantly related genomes meet (McClintock 1984).

CONCLUSION

Since their discovery, sRNAs have been implicated in a plethora of biological processes ranging from gene regulation to defense against invaders. Among these important functions, their ability to be inherited by the next generation and influence it remains one of the most intriguing. Flowering plants present a fantastic opportunity to study such phenomenon because of their propensity to transmit epigenetic information to their offspring. The knowledge gained is already being applied to address emerging agricultural and environmental challenges.

ACKNOWLEDGMENTS

The authors thank Claudia Köhler, Daniel Grimanelli, and members of the Martienssen laboratory for fruitful discussions. This research was supported by the Howard Hughes Medical Institute and by grants from the National Science Foundation Plant Genome Research Program to R.A.M.

REFERENCES

Amitani I, Baskin RJ, Kowalczykowski SC. 2006. Visualization of Rad54, a chromatin remodeling protein, translocating on single DNA molecules. *Mol Cell* **23**: 143–148. doi:10.1016/j.molcel.2006.05.009

An YC, Goettel W, Han Q, Bartels A, Liu Z, Xiao W. 2017. Dynamic changes of genome-wide DNA methylation during soybean seed development. *Sci Rep* **7**: 12263. doi:10.1038/s41598-017-12510-4

Ausin I, Feng S, Yu C, Liu W, Kuo HY, Jacobsen EL, Zhai J, Gallego-Bartolome J, Wang L, Egertsdotter U, et al. 2016. DNA methylome of the 20-gigabase Norway spruce genome. *Proc Natl Acad Sci* **113**: E8106–E8113. doi:10.1073/pnas.1618019113

Belling J, Blakeslee AF. 1923. The reduction division in haploid, diploid, triploid and tetraploid Daturas. *Proc Natl Acad Sci* **9**: 106–111. doi:10.1073/pnas.9.4.106

Birchler JA, Veitia RA. 2007. The gene balance hypothesis: from classical genetics to modern genomics. *Plant Cell* **19**: 395–402. doi:10.1105/tpc.106.049338

Blakeslee AF, Belling J, Farnham ME. 1920. Chromosomal duplication and Mendelian phenomena in Datura mutants. *Science* **52**: 388–390. doi:10.1126/science.52.1347.388

Blevins T, Podicheti R, Mishra V, Marasco M, Wang J, Rusch D, Tang H, Pikaard CS. 2015. Identification of Pol IV and RDR2-dependent precursors of 24 nt siRNAs guiding de novo DNA methylation in *Arabidopsis*. *eLife* **4**: e09591. doi:10.7554/eLife.09591

Böhmdorfer G, Rowley MJ, Kucinski J, Zhu Y, Amies I, Wierzbicki AT. 2014. RNA-directed DNA methylation requires stepwise binding of silencing factors to long non-coding RNA. *Plant J* **79**: 181–191. doi:10.1111/tpj.12563

Böhmdorfer G, Sethuraman S, Rowley MJ, Krzyszton M, Rothi MH, Bouzit L, Wierzbicki AT. 2016. Long non-coding RNA produced by RNA polymerase V determines boundaries of heterochromatin. *eLife* **5**: e19092. doi:10.7554/eLife.19092

Borges F, Martienssen RA. 2015. The expanding world of small RNAs in plants. *Nat Rev Mol Cell Biol* **16**: 727–741. doi:10.1038/nrm4085

Borges F, Parent J-S, van Ex F, Wolff P, Martínez G, Köhler C, Martienssen RA. 2018. Transposon-derived small RNAs triggered by miR845 mediate genome dosage response in *Arabidopsis*. *Nat Genet* **50**: 186–192. doi:10.1038/s41588-017-0032-5

Bouyer D, Kramdi A, Kassam M, Heese M, Schnittger A, Roudier F, Colot V. 2017. DNA methylation dynamics during early plant life. *Genome Biol* **18**: 179. doi:10.1186/s13059-017-1313-0

Bridges CB. 1925. Sex in relation to chromosomes and genes. *Am Soc Nat* **59**: 127–137. doi:10.1086/280023

Calarco JP, Borges F, Donoghue MTA, Van Ex F, Jullien PE, Lopes T, Gardner R, Berger F, Feijó JA, Becker JD, et al. 2012. Reprogramming of DNA methylation in pollen guides epigenetic inheritance via small RNA. *Cell* **151**: 194–205. doi:10.1016/j.cell.2012.09.001

Carputo D, Barone A, Cardi T, Sebastiano A, Frusciante L, Peloquin SJ. 1997. Endosperm balance number manipulation for direct in vivo germplasm introgression to potato from a sexually isolated relative (*Solanum commersonii* Dun). *Proc Natl Acad Sci* **94**: 12013–12017. doi:10.1073/pnas.94.22.12013

Carputo D, Monti L, Werner JE, Frusciante L. 1999. Uses and usefulness of endosperm balance number. *Theor Appl Genet* **98**: 478–484. doi:10.1007/s001220051095

Chandler VL. 2010. Paramutation's properties and puzzles. *Science* **330**: 628–629. doi:10.1126/science.1191044

Cheng F, Sun C, Wu J, Schnable J, Woodhouse MR, Liang J, Cai C, Freeling M, Wang X. 2016. Epigenetic regulation of subgenome dominance following whole genome triplication in *Brassica rapa*. *New Phytol* **211**: 288–299. doi:10.1111/nph.13884

Chuong EB, Elde NC, Feschotte C. 2017. Regulatory activities of transposable elements: from conflicts to benefits. *Nat Rev Genet* **18**: 71–86. doi:10.1038/nrg.2016.139

Coe EH. 1966. The properties, origin, and mechanism of conversion-type inheritance at the B locus in maize. *Genetics* **53**: 1035–1063.

Corem S, Doron-Faigenboim A, Jouffroy O, Maumus F, Arazi T, Bouché N. 2018. Redistribution of CHH methylation and small interfering RNAs across the genome of tomato *ddm1* mutants. *Plant Cell* **30**: 1628–1644. doi:10.1105/tpc.18.00167

Creasey KM, Zhai J, Borges F, Van Ex F, Regulski M, Meyers BC, Martienssen RA. 2014. miRNAs trigger widespread epigenetically activated siRNAs from transposons in *Arabidopsis*. *Nature* **508**: 411–415. doi:10.1038/nature13069

Du J, Johnson LM, Groth M, Feng S, Hale CJ, Li S, Vashisht AA, Gallego-Bartolome J, Wohlschlegel JA, Patel DJ, et al. 2014. Mechanism of DNA methylation-directed histone methylation by KRYPTONITE. *Mol Cell* **55**: 495–504. doi:10.1016/j.molcel.2014.06.009

Edger PP, Smith R, McKain MR, Cooley AM, Vallejo-Marin M, Yuan Y, Bewick AJ, Ji L, Platts AE, Bowman MJ, et al. 2017. Subgenome dominance in an interspecific hybrid, synthetic allopolyploid, and a 140-year-old naturally established neo-allopolyploid monkeyflower. *Plant Cell* **29**: 2150–2167. doi:10.1105/tpc.17.00010

Ehlenfeldt MK, Hanneman RE. 1984. The use of Endosperm Balance Number and 2n gametes to transfer exotic germplasm in potato. *Theor Appl Genet* **68**: 155–161. doi:10.1007/BF00252332

Erdmann RM, Satyaki PR V, Klosinska M, Gehring M. 2017. A small RNA pathway mediates global allelic dosage in endosperm. *Cell Rep* **21**: 3364–3372. doi:10.1016/j.celrep.2017.11.078

Florez-Rueda AM, Paris M, Schmidt A, Widmer A, Grossniklaus U, Städler T. 2016. Genomic imprinting in the endosperm is systematically perturbed in abortive hybrid tomato seeds. *Mol Biol Evol* **33**: 2935–2946. doi:10.1093/molbev/msw175

Fu F-F, Dawe RK, Gent JI. 2018. Loss of RNA-directed DNA methylation in maize chromomethylase and DDM1-type nucleosome remodeler mutants. *Plant Cell* **30**: 1617–1627. doi:10.1105/tpc.18.00053

Fultz D, Slotkin RK. 2017. Exogenous transposable elements circumvent identity-based silencing, permitting the dissection of expression-dependent silencing. *Plant Cell* **29**: 360–376. doi:10.1105/tpc.16.00718

Gallego-Bartolomé J, Liu W, Kuo PH, Feng S, Ghoshal B, Gardiner J, Zhao JMC, Park SY, Chory J, Jacobsen SE. 2019. Co-targeting RNA polymerases IV and V promotes efficient de novo DNA methylation in *Arabidopsis*. *Cell* **176**: 1068–1082.e19. doi:10.1016/j.cell.2019.01.029

Gao Z, Liu H-L, Daxinger L, Pontes O, He X, Qian W, Lin H, Xie M, Lorkovic ZJ, Zhang S, et al. 2010. An RNA polymerase II- and AGO4-associated protein acts in RNA-directed DNA methylation. *Nature* **465**: 106–109. doi:10.1038/nature09025

Gouil Q, Baulcombe DC. 2016. DNA methylation signatures of the plant chromomethyltransferases. *PLOS Genet* **12**: e1006526. doi:10.1371/journal.pgen.1006526

Gutbrod MJ, Martienssen RA. 2020. Conserved chromosomal functions of RNA interference. *Nat Rev Genet* **21**: 311–331. doi:10.1038/s41576-019-0203-6

Heard E, Martienssen RA. 2014. Transgenerational epigenetic inheritance: myths and mechanisms. *Cell* **157**: 95–109. doi:10.1016/j.cell.2014.02.045

Henderson IR, Jacobsen SE. 2008. Tandem repeats upstream of the *Arabidopsis* endogene SDC recruit non-CG DNA methylation and initiate siRNA spreading. *Genes Dev* **22**: 1597–1606. doi:10.1101/gad.1667808

Hollick JB. 2017. Paramutation and related phenomena in diverse species. *Nat Rev Genet* **18**: 5–23. doi:10.1038/nrg.2016.115

Huang Y, Kendall T, Forsythe ES, Dorantes-Acosta A, Li S, Caballero-Pérez J, Chen X, Arteaga-Vázquez M, Beilstein MA, Mosher RA. 2015. Ancient origin and recent innovations of RNA polymerase IV and V. *Mol Biol Evol* **32**: 1788–1799. doi:10.1093/molbev/msv060

Ibarra CA, Feng X, Schoft VK, Hsieh T-F, Uzawa R, Rodrigues JA, Zemach A, Chumak N, Machlicova A, Nishimura T, et al. 2012. Active DNA demethylation in plant companion cells reinforces transposon methylation in gametes. *Science* **337:** 1360–1364. doi:10.1126/science.1224839

Ickert-Bond SM, Sousa A, Min Y, Loera I, Metzgar J, Pellicer J, Hidalgo O, Leitch IJ. 2020. Polyploidy in gymnosperms—insights into the genomic and evolutionary consequences of polyploidy in *Ephedra*. *Mol Phylogenet Evol* **147:** 106786. doi:10.1016/j.ympev.2020.106786

Ingouff M, Selles B, Michaud C, Vu TM, Berger F, Schorn AJ, Autran D, Van Durme M, Nowack MK, Martienssen RA, et al. 2017. Live-cell analysis of DNA methylation during sexual reproduction in *Arabidopsis* reveals context and sex-specific dynamics controlled by noncanonical RdDM. *Genes Dev* **31:** 72–83. doi:10.1101/gad.289397.116

Iwasaki M, Hyvärinen L, Piskurewicz U, Lopez-Molina L. 2019. Non-canonical RNA-directed DNA methylation participates in maternal and environmental control of seed dormancy. *eLife* **8:** e37434. doi:10.7554/eLife.37434

Johnson LM, Du J, Hale CJ, Bischof S, Feng S, Chodavarapu RK, Zhong X, Marson G, Pellegrini M, Segal DJ, et al. 2014. SRA- and SET-domain-containing proteins link RNA polymerase V occupancy to DNA methylation. *Nature* **507:** 124–128. doi:10.1038/nature12931

Johnston SA, Hanneman RE. 1982. Manipulations of endosperm balance number overcome crossing barriers between diploid *Solanum* species. *Science* **217:** 446–448. doi:10.1126/science.217.4558.446

Josefsson C, Dilkes B, Comai L. 2006. Parent-dependent loss of gene silencing during interspecies hybridization. *Curr Biol* **16:** 1322–1328. doi:10.1016/j.cub.2006.05.045

Kawakatsu T, Nery JR, Castanon R, Ecker JR. 2017. Dynamic DNA methylation reconfiguration during seed development and germination. *Genome Biol* **18:** 171. doi:10.1186/s13059-017-1251-x

Kermicle JL. 1970. Dependence of the *R*-mottled aleurone phenotype in maize on mode of sexual transmission. *Genetics* **66:** 69–85.

Kirkbride RC, Yu HH, Nah G, Zhang C, Shi X, Chen ZJ. 2015. An epigenetic role for disrupted paternal gene expression in postzygotic seed abortion in *Arabidopsis* interspecific hybrids. *Mol Plant* **8:** 1766–1775. doi:10.1016/j.molp.2015.09.009

Kirkbride RC, Lu J, Zhang C, Mosher RA, Baulcombe DC. 2019. Maternal small RNAs mediate spatial–temporal regulation of gene expression, imprinting, and seed development in *Arabidopsis*. *Proc Natl Acad Sci* **116:** 2761–2766. doi:10.1073/pnas.1807621116

Köhler C, Mittelsten Scheid O, Erilova A. 2010. The impact of the triploid block on the origin and evolution of polyploid plants. *Trends Genet* **26:** 142–148. doi:10.1016/j.tig.2009.12.006

Kradolfer D, Wolff P, Jiang H, Siretskiy A, Köhler C. 2013. An imprinted gene underlies postzygotic reproductive isolation in *Arabidopsis thaliana*. *Dev Cell* **26:** 525–535. doi:10.1016/j.devcel.2013.08.006

Lafon-Placette C, Köhler C. 2015. Epigenetic mechanisms of postzygotic reproductive isolation in plants. *Curr Opin Plant Biol* **23:** 39–44. doi:10.1016/j.pbi.2014.10.006

Law JA, Jacobsen SE. 2010. Establishing, maintaining and modifying DNA methylation patterns in plants and animals. *Nat Rev Genet* **11:** 204–220. doi:10.1038/nrg2719

Law JA, Du J, Hale CJ, Feng S, Krajewski K, Palanca AMS, Strahl BD, Patel DJ, Jacobsen SE. 2013. Polymerase IV occupancy at RNA-directed DNA methylation sites requires SHH1. *Nature* **498:** 385–389. doi:10.1038/nature12178

Lee EK, Cibrian-Jaramillo A, Kolokotronis SO, Katari MS, Stamatakis A, Ott M, Chiu JC, Little DP, Stevenson DW, McCombie WR, et al. 2011. A functional phylogenomic view of the seed plants. *PLoS Genet* **7:** e1002411. doi:10.1371/journal.pgen.1002411

Liu Z-W, Zhou J-X, Huang H-W, Li Y-Q, Shao C-R, Li L, Cai T, Chen S, He X-J. 2016. Two components of the RNA-directed DNA methylation pathway associate with MORC6 and silence loci targeted by MORC6 in *Arabidopsis*. *PLOS Genet* **12:** e1006026. doi:10.1371/journal.pgen.1006026

Liu W, Duttke SH, Hetzel J, Groth M, Feng S, Gallego-Bartolome J, Zhong Z, Kuo HY, Wang Z, Zhai J, et al. 2018. RNA-directed DNA methylation involves co-transcriptional small-RNA-guided slicing of polymerase V transcripts in *Arabidopsis*. *Nat Plants* **4:** 181–188. doi:10.1038/s41477-017-0100-y

Lu J, Zhang C, Baulcombe DC, Chen ZJ. 2012. Maternal siRNAs as regulators of parental genome imbalance and gene expression in endosperm of *Arabidopsis* seeds. *Proc Natl Acad Sci* **109:** 5529–5534. doi:10.1073/pnas.1203094109

Ma L, Hatlen A, Kelly LJ, Becher H, Wang W, Kovarik A, Leitch IJ, Leitch AR. 2015. Angiosperms are unique among land plant lineages in the occurrence of key genes in the RNA-directed DNA methylation (RdDM) Pathway. *Genome Biol Evol* **7:** 2648–2662. doi:10.1093/gbe/evv171

Malone CD, Lehmann R, Teixeira FK. 2015. The cellular basis of hybrid dysgenesis and Stellate regulation in *Drosophila*. *Curr Opin Genet Dev* **34:** 88–94. doi:10.1016/j.gde.2015.09.003

Martienssen RA. 2010. Heterochromatin, small RNA and postfertilization dysgenesis in allopolyploid and interploid hybrids of *Arabidopsis*. *New Phytol* **186:** 46–53. doi:10.1111/j.1469-8137.2010.03193.x

Martínez G, Panda K, Köhler C, Slotkin RK. 2016. Silencing in sperm cells is directed by RNA movement from the surrounding nurse cell. *Nat Plants* **2:** 16030. doi:10.1038/nplants.2016.30

Martinez G, Wolff P, Wang Z, Moreno-Romero J, Santos-González J, Conze LL, DeFraia C, Slotkin K, Köhler C. 2018. Paternal easiRNAs regulate parental genome dosage in *Arabidopsis*. *Nat Genet* **50:** 193–198. doi:10.1038/s41588-017-0033-4

Martinez Palacios P, Jacquemot MP, Tapie M, Rousselet A, Diop M, Remoue C, Falque M, Lloyd A, Jenczewski E, Lassalle G, et al. 2019. Assessing the response of small RNA populations to allopolyploidy using resynthesized *Brassica napus* allotetraploids. *Mol Biol Evol* **36:** 709–726. doi:10.1093/molbev/msz007

Matzke MA, Mosher RA. 2014. RNA-directed DNA methylation: an epigenetic pathway of increasing complexity. *Nat Rev Genet* **15:** 394–408. doi:10.1038/nrg3683

Matzke MA, Kanno T, Matzke AJM. 2015. RNA-directed DNA methylation: the evolution of a complex epigenetic pathway in flowering plants. *Annu Rev Plant Biol* **66:** 243–267. doi:10.1146/annurev-arplant-043014-114633

McClintock B. 1929. A cytological and genetical study of triploid maize. *Genetics* **14:** 180–222.

McClintock B. 1961. Some parallels between gene control systems in maize and in bacteria. *Am Nat* **95:** 265–277. doi:10.1086/282188

McClintock B. 1984. The significance of responses of the genome to challenge. *Science* **226:** 792–801. doi:10.1126/science.15739260

Mohn F, Sienski G, Handler D, Brennecke J. 2014. The Rhino–Deadlock–Cutoff complex licenses noncanonical transcription of dual-strand piRNA clusters in *Drosophila*. *Cell* **157:** 1364–1379. doi:10.1016/j.cell.2014.04.031

Moreno-Romero J, Del Toro-De León G, Yadav VK, Santos-González J, Köhler C. 2019. Epigenetic signatures associated with imprinted paternally-expressed genes in the *Arabidopsis* endosperm. *Genome Biol* **20:** 41. doi:10.1186/s13059-019-1652-0

Nuthikattu S, McCue AD, Panda K, Fultz D, DeFraia C, Thomas EN, Slotkin RK. 2013. The initiation of epigenetic silencing of active transposable elements is triggered by RDR6 and 21-22 nucleotide small interfering RNAs. *Plant Physiol* **162:** 116–131. doi:10.1104/pp.113.216481

Ong-Abdullah M, Ordway JM, Jiang N, Ooi S-E, Kok S-Y, Sarpan N, Azimi N, Hashim AT, Ishak Z, Rosli SK, et al. 2015. Loss of *Karma* transposon methylation underlies the mantled somaclonal variant of oil palm. *Nature* **525:** 533–537. doi:10.1038/nature15365

Panda K, Ji L, Neumann DA, Daron J, Schmitz RJ, Slotkin RK. 2016. Full-length autonomous transposable elements are preferentially targeted by expression-dependent forms of RNA-directed DNA methylation. *Genome Biol* **17**: 170. doi:10.1186/s13059-016-1032-y

Pei L, Zhang L, Li J, Shen C, Qiu P, Tu L, Zhang X, Wang M. 2019. Tracing the origin and evolution history of methylation-related genes in plants. *BMC Plant Biol* **19**: 1–13. doi:10.1186/s12870-019-1923-7

Pumplin N, Sarazin A, Jullien PE, Bologna NG, Oberlin S, Voinnet O. 2016. DNA methylation influences the expression of *DICER-LIKE4* isoforms, which encode proteins of alternative localization and function. *Plant Cell* **28**: 2786–2804. doi:10.1105/tpc.16.00554

Roth M, Florez-Rueda AM, Paris M, Städler T. 2018. Wild tomato endosperm transcriptomes reveal common roles of genomic imprinting in both nuclear and cellular endosperm. *Plant J* **95**: 1084–1101. doi:10.1111/tpj.14012

Sasaki T, Kobayashi A, Saze H, Kakutani T. 2012. RNAi-independent de novo DNA methylation revealed in *Arabidopsis* mutants of chromatin remodeling gene *DDM1*. *Plant J* **70**: 750–758. doi:10.1111/j.1365-313X.2012.04911.x

Satyaki PRV, Gehring M. 2017. DNA methylation and imprinting in plants: machinery and mechanisms. *Crit Rev Biochem Mol Biol* **52**: 163–175. doi:10.1080/10409238.2017.1279119

Satyaki PRV, Gehring M. 2019. Paternally acting canonical RNA-directed DNA methylation pathway genes sensitize *Arabidopsis* endosperm to paternal genome dosage. *Plant Cell* **31**: 1563–1578. doi:10.1105/tpc.19.00047

Schorn AJ, Gutbrod MJ, LeBlanc C, Martienssen R. 2017. LTR-retrotransposon control by tRNA-derived small RNAs. *Cell* **170**: 61–71.e11. doi:10.1016/j.cell.2017.06.013

Singh J, Mishra V, Wang F, Huang H-Y, Pikaard CS. 2019. Reaction mechanisms of Pol IV, RDR2, and DCL3 drive RNA channeling in the siRNA-directed DNA methylation pathway. *Mol Cell* **75**: 576–589.e5. doi:10.1016/j.molcel.2019.07.008

Slotkin RK, Vaughn M, Borges F, Tanurdžić M, Becker JD, Feijó JA, Martienssen RA. 2009. Epigenetic reprogramming and small RNA silencing of transposable elements in pollen. *Cell* **136**: 461–472. doi:10.1016/j.cell.2008.12.038

Stoute AI, Varenko V, King GJ, Scott RJ, Kurup S. 2012. Parental genome imbalance in *Brassica oleracea* causes asymmetric triploid block. *Plant J* **71**: 503–516. doi:10.1111/j.1365-313X.2012.05015.x

Stroud H, Do T, Du J, Zhong X, Feng S, Johnson L, Patel DJ, Jacobsen SE. 2014. Non-CG methylation patterns shape the epigenetic landscape in *Arabidopsis*. *Nat Struct Mol Biol* **21**: 64–72. doi:10.1038/nsmb.2735

Šurbanovski N, Brilli M, Moser M, Si-Ammour A. 2016. A highly specific microRNA-mediated mechanism silences LTR retrotransposons of strawberry. *Plant J* **85**: 70–82. doi:10.1111/tpj.13090

Swarts DC, Makarova K, Wang Y, Nakanishi K, Ketting RF, Koonin E V, Patel DJ, van der Oost J. 2014. The evolutionary journey of Argonaute proteins. *Nat Struct Mol Biol* **21**: 743–753. doi:10.1038/nsmb.2879

Teixeira FK, Colot V. 2010. Repeat elements and the *Arabidopsis* DNA methylation landscape. *Heredity (Edinb)* **105**: 14–23. doi:10.1038/hdy.2010.52

Vu TM, Nakamura M, Calarco JP, Susaki D, Lim PQ, Kinoshita T, Higashiyama T, Martienssen RA, Berger F. 2013. RNA-directed DNA methylation regulates parental genomic imprinting at several loci in *Arabidopsis*. *Development* **140**: 2953–2960. doi:10.1242/dev.092981

Walker J, Gao H, Zhang J, Aldridge B, Vickers M, Higgins JD, Feng X. 2018. Sexual-lineage-specific DNA methylation regulates meiosis in *Arabidopsis*. *Nat Genet* **50**: 130–137. doi:10.1038/s41588-017-0008-5

Wang Z, Butel N, Santos-González J, Borges F, Yi J, Martienssen R, Martinez G, Köhler C. 2020. Polymerase IV plays a crucial role in pollen development in *Capsella*. *Plant Cell* **32**: 950–966 doi:10.1105/tpc.19.00938

Wierzbicki AT, Cocklin R, Mayampurath A, Lister R, Rowley MJ, Gregory BD, Ecker JR, Tang H, Pikaard CS. 2012. Spatial and functional relationships among Pol V-associated loci, Pol IV-dependent siRNAs, and cytosine methylation in the *Arabidopsis* epigenome. *Genes Dev* **26**: 1825–1836. doi:10.1101/gad.197772.112

Wolff P, Jiang H, Wang G, Santos-González J, Köhler C. 2015. Paternally expressed imprinted genes establish postzygotic hybridization barriers in *Arabidopsis thaliana*. *eLife* **4**: e10074. doi:10.7554/eLife.10074

Wongpalee SP, Liu S, Gallego-Bartolomé J, Leitner A, Aebersold R, Liu W, Yen L, Nohales MA, Kuo PH, Vashisht AA, et al. 2019. CryoEM structures of *Arabidopsis* DDR complexes. *Nat Commun* **10**: 3916. doi:10.1038/s41467-019-11759-9

Yu B, Lin YA, Parhad SS, Jin Z, Ma J, Theurkauf WE, Zhang ZZ, Huang Y. 2018. Structural insights into Rhino–Deadlock complex for germline piRNA cluster specification. *EMBO Rep* **19**: e45418. doi:10.15252/embr.201745418

Zemach A, Kim MY, Hsieh PH, Coleman-Derr D, Eshed-Williams L, Thao K, Harmer SL, Zilberman D. 2013. The *Arabidopsis* nucleosome remodeler DDM1 allows DNA methyltransferases to access H1-containing heterochromatin. *Cell* **153**: 193–205. doi:10.1016/j.cell.2013.02.033

Zhai J, Bischof S, Wang H, Feng S, Lee TF, Teng C, Chen X, Park SY, Liu L, Gallego-Bartolome J, et al. 2015. A one precursor one siRNA model for Pol IV-dependent siRNA biogenesis. *Cell* **163**: 445–455. doi:10.1016/j.cell.2015.09.032

Zhang J, Liu Y, Xia E-H, Yao Q-Y, Liu X-D, Gao L-Z. 2015. Autotetraploid rice methylome analysis reveals methylation variation of transposable elements and their effects on gene expression. *Proc Natl Acad Sci* **112**: E7022–E7029. doi:10.1073/pnas.1515170112

Zhao Y, Wang S, Wu W, Li L, Jiang T, Zheng B. 2018. Clearance of maternal barriers by paternal miR159 to initiate endosperm nuclear division in *Arabidopsis*. *Nat Commun* **9**: 5011. doi:10.1038/s41467-018-07429-x

Zhong X, Hale CJ, Law JA, Johnson LM, Feng S, Tu A, Jacobsen SE. 2012. DDR complex facilitates global association of RNA polymerase V to promoters and evolutionarily young transposons. *Nat Struct Mol Biol* **19**: 870–875. doi:10.1038/nsmb.2354

Zhou M, Palanca AMS, Law JA. 2018. Locus-specific control of the de novo DNA methylation pathway in *Arabidopsis* by the CLASSY family. *Nat Genet* **50**: 865–873. doi:10.1038/s41588-018-0115-y

Zhu Y, Rowley MJ, Böhmdorfer G, Wierzbicki AT. 2013. A SWI/SNF chromatin-remodeling complex acts in noncoding RNA-mediated transcriptional silencing. *Mol Cell* **49**: 298–309. doi:10.1016/j.molcel.2012.11.011

Zhu H, Xie W, Xu D, Miki D, Tang K, Huang C-F, Zhu J-K. 2018. DNA demethylase ROS1 negatively regulates the imprinting of *DOGL4* and seed dormancy in *Arabidopsis thaliana*. *Proc Natl Acad Sci* **115**: E9962–E9970. doi:10.1073/pnas.1812847115

A Nuclear RNA Degradation Pathway Helps Silence Polycomb/H3K27me3-Marked Loci in *Caenorhabditis elegans*

Anna Mattout,[1,3] Dimos Gaidatzis,[1] Véronique Kalck,[1] and Susan M. Gasser[1,2]

[1]*Friedrich Miescher Institute for Biomedical Research, CH-4058 Basel, Switzerland*
[2]*University of Basel, Faculty of Science, CH-4056 Basel, Switzerland*
Correspondence: susan.gasser@fmi.ch

In fission yeast and plants, RNA-processing pathways contribute to heterochromatin silencing, complementing well-characterized pathways of transcriptional repression. However, it was unclear whether this additional level of regulation occurs in metazoans. In a genetic screen, we uncovered a pathway of silencing in *Caenorhabditis elegans* somatic cells, whereby the highly conserved, RNA-binding complex LSM2-8 contributes to the repression of heterochromatic reporters and endogenous genes bearing the Polycomb mark H3K27me3. Importantly, the LSM2-8 complex works cooperatively with a 5′–3′ exoribonuclease, XRN-2, and disruption of the pathway leads to selective mRNA stabilization. LSM2-8 complex–mediated RNA degradation does not target nor depend on H3K9me2/3, unlike previously described pathways of heterochromatic RNA degradation. Up-regulation of *lsm-8*-sensitive loci coincides with a localized drop in H3K27me3 levels in the *lsm-8* mutant. Put into the context of epigenetic control of gene expression, it appears that targeted RNA degradation helps repress a subset of H3K27me3-marked genes, revealing an unappreciated layer of regulation for facultative heterochromatin in animals.

Genomic DNA is organized into highly condensed, dark-staining heterochromatin, which correlates with reduced gene expression (Trojer and Reinberg 2007; Wenzel et al. 2011; Saksouk et al. 2015), and less condensed open chromatin, which is transcribed. Whereas the primary mechanism for both constitutive (enriched for H3K9me2/3) and facultative (enriched for H3K27me3) heterochromatic gene repression is thought to involve the inhibition of RNA polymerases, robust pathways that silence at the co- and posttranscriptional levels have also been documented in fission yeast and plants (Buhler 2009; Wang et al. 2016).

A genome-wide screen in our laboratory that monitored the derepression of a heterochromatic, multicopy reporter in *Caenorhabditis elegans* embryos identified 29 factors essential for silencing. Most of these were chromatin modifiers and transcription-related proteins (Towbin et al. 2012) (Figs. 1A and 2A), but we also recovered three subunits of the evolutionarily conserved like-SM protein (LSM) complex (Fig. 1A). A role for these proteins in the repression of heterochromatin was unexpected given the reported roles for the two LSM complexes, LSM1-7 and LSM2-8, in general RNA metabolism but not in transcriptional control per se (Kufel et al. 2004; Beggs 2005; Tharun 2009; Perea-Resa et al. 2012; Golisz et al. 2013; Cornes et al. 2015). Whereas pathways of RNA processing and degradation were reported to contribute to silencing in fission yeast and plants (Buhler et al. 2007; Vasiljeva et al. 2008; Keller et al. 2012; Zofall et al. 2012; Yamanaka et al. 2013; Egan et al. 2014; Chalamcharla et al. 2015; Tucker et al. 2016), heterochromatic silencing through transcript degradation had not been documented in animals.

The RNA interference (RNAi) screen that identified *gut-2/lsm-2*, *lsm-5*, and *lsm-6* monitored the robust derepression or desilencing of a *C. elegans* heterochromatic reporter, which consisted of an integrated array of several hundred copies of a green fluorescent protein (GFP)-encoding reporter gene driven from a ubiquitously active promoter (Towbin et al. 2012). In a copy-number-dependent manner, these heterochromatic reporters carry the histone modifications H3K9me2/3 and H3K27me3, but lack H3K4 methylation, and are sequestered at the nuclear envelope, thereby mimicking many aspects of endogenous heterochromatin (Meister et al. 2010; Towbin et al. 2010). As expected, the loss of either of the two H3K9 methyltransferases (*met-2* and *set-25*) or the EZH2 homolog *mes-2*, which deposits H3K27me3, led to reporter derepression (Towbin et al. 2012). It was surprising, however, to detect three subunits of an RNA processing complex with the same phenotype. Therefore, we examined in more detail how LSM proteins regulate gene expression within heterochromatic domains in *C. elegans* (Mattout et al. 2020).

LSM PROTEINS REDUCE mRNA FROM HETEROCHROMATIC REPORTERS SELECTIVELY

The initial screen for reporter derepression monitored GFP expression in embryos (Fig. 1B). The elimination of LSM proteins by *lsm-2*, *lsm-5*, or *lsm-6* RNAi, on the other

---

[3]Present address: Université Paul Sabatier-CNRS UMR 5088, 31062 Toulouse, France

© 2019 Mattout et al. This article is distributed under the terms of the Creative Commons Attribution-NonCommercial License, which permits reuse and redistribution, except for commercial purposes, provided that the original author and source are credited.

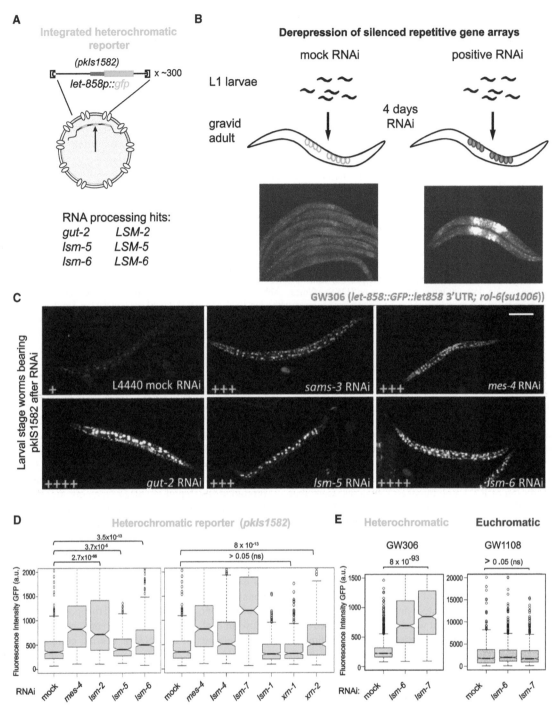

Figure 1. LSM proteins silence heterochromatic reporters integrated as high-copy-number gene arrays, but not euchromatic reporters. (*A*) The integrated, high-copy-number heterochromatic reporter *pkIs1582* used in the genome-wide screen (Towbin et al. 2012). RNA interference (RNAi)-based derepression monitored in progeny of all stages by increased green fluorescent protein (GFP) fluorescence. (*B*) An RNAi-based derepression screen was used to monitor derepression of *pkIs1582* (GFP) in embryos within young adult worms. The initial screen was reported in Towbin et al. (2012). (*C*) Fluorescence microscopy of *pkIs1582* (GFP) derepression in L4 larvae treated with the indicated RNAi clones, versus control (mock/L4440). Bar, 100 μm. A + indicates relative intensity of GFP fluorescence. (*D*) Quantitation of GFP intensity of heterochromatic reporter *pkIs1582* in L1 larvae (GW306) following indicated RNAi by the worm sorter. L1 progeny were scored after RNAi of indicated genes, or control RNAi (mock: L4440). Shown are box plots of fluorescence intensity in arbitrary units (a.u.), with whiskers = 1st and 3rd quartiles; black lines, median; black circles, outliers. The notch around the median represents 95% confidence interval of the median. Quantification and statistical analysis were based on $n = 375$ for each condition. (*E*) Quantitation of derepression of the heterochromatic (GW306) as well as the euchromatic (GW1108) reporters in L1 progeny under *lsm-6*, *lsm-7*, and control RNAi conditions, as in *D*. Numbers of L1 worms scored are $n = 2000$ (GW306-mock), 1068 (GW306-*lsm6*), 613 (GW306-*lsm7*) and 875 (GW1108-mock), 1110 (GW1108-*lsm6*), 1026 (GW1108-*lsm7*), pooled from three independent experiments. Indicated *P*-values by two-tailed unpaired Student's *t*-test (n.s. = nonsignificant or $P > 0.05$). Strains are described in Table 1. (Data in panels *C,D* are found in different formats in Mattout et al. 2020. Reproduced with permission.)

Table 1. Strains and reporters used in this study

Strain	Original name/OC	Genotype	Reference
GW1	N2[a]	Wild-type, Bristol isolate	
GW76		gwIs4 [baf-1p::GFP-lacI::let-858 3′UTR; myo-3p::RFP] X	Meister et al. 2010
GW306	NL2507	pkIs1582[let-858p::GFP:: let-858 3′UTR; rol-6(su1006)]V	Towbin et al. 2012
GW566		gwIs39 [baf-1p::GFP-LacI::let-858 3′UTR; vit-5p::GFP] III; gwIs4 [baf-1p:: GFP-lacI::let-858 3′UTR; myo-3p::RFP] X	Towbin et al. 2012
GW694		gwIs85[his-72p::mcherry-set-25::his-72 3′UTR; unc-119(+)] unc-119(ed3) ttTi5605 II	Towbin et al. 2012
GW299		gwIs25 [tbb-1p::wmCherry-LacI::tbb-2 3′UTR unc-119(+)] unc-119(ed3)	Mattout et al. 2020
GW637		met-2(n4256) set-25(n5021) III; gwIs4 [baf-1p::GFP-lacI::let-858 3′UTR; myo-3p::RFP] X	Towbin et al. 2012
GW214		hpl-2(tm1489) III; gwIs4 [baf-1p::GFP-lacI::let-858 3′UTR; myo-3p::RFP] X	Towbin et al. 2012
GW468		mes-2(bn11) unc-4(e120)/mnC1 dpy-10e128() unc-52(e444)II; gwIs4[myo-3:: RFP baf-1::GFP-lacI let-858] X.	Towbin et al. 2012
GW1004		lsm4:GFP-FLAGx3	Mattout et al. 2020
GW1109	HW1390 ocx4	lsm-8 (xe17 [myo2p::mcherry::unc54 3′UTR])IV	Mattout et al. 2020
GW1125	ocx6	lsm-8 (xe17 [myo2p::mcherry::unc54 3′UTR])IV/nT1[qIs51](IV;V)	Mattout et al. 2020
GW1119	ocx6	lsm-8 (xe17 [myo2p::mcherry::unc54 3′UTR])IV/nT1[qIs51](IV;V); pkIs1582 [let-858::GFP rol-6(su1006)]V	Mattout et al. 2020
GW1120	ocx6	lsm-8 (xe17 [myo2p::mcherry::unc54 3′UTR])IV; pkIs1582[let-858::GFP rol-6 (su1006)]V	Mattout et al. 2020
GW1148	ocx6	met-2(n4256) set-25(n5021) III; lsm-8 (xe17 [myo2p::mcherry::unc54 3′UTR]) IV/nT1[qIs51](IV;V)	Mattout et al. 2020
GW1080	VC2785* ocx5	lsm-4&ada-2(ok3151)/mIn1[mIs14 dpy-10(e128)]II	Mattout et al. 2020

Strains with deletion alleles and reporters obtained from the *Caenorhabditis elegans* knockout consortium or made by the CRISPR-Cas9 system were outcrossed two to six times to the N2 (WT) strain. Worms were grown on OP50 and maintained at 22.5°C.
[a](CGC) *Caenorhabditis* Genetics Center, (OCx4) outcrossed four times.

hand, was detectable throughout worm development—that is, in larvae and in adult worms (L4 larvae shown in Fig. 1C; see also Mattout et al. 2020). This was also true for other RNAi hits from the initial screen—namely, *sams-3* RNAi and *mes-4* RNAi (Fig. 1C). Elevated GFP expression from the heterochromatic array after RNAi treatment was scored at all stages of worm development and in nearly every somatic cell type, except germline cells (see below). Given that Piwi-interacting RNAs (piRNAs) are known to contribute to silencing uniquely in germline cells, we hypothesize that the effect of *lsm* gene down-regulation may be masked by redundant silencing pathways, like piRNAs, that act specifically in the gonad. Alternatively, the LSM-mediated silencing pathway may not be functional in germ cells.

To obtain a quantitative value for reporter expression, we monitored the increase in GFP fluorescence intensity using the COPAS BioSort, which measures fluorescence on a worm-by-worm basis for hundreds to thousands of worms, using flow cytometry. Using the worm sorter, we compared the effects of *lsm-6*, *lsm-2/gut-2*, and *lsm-5* RNAi with the knockdown of further shared subunits of the two LSM complexes: LSM-3, LSM-4, and LSM-7 (Fig. 1D). The RNAi clone we used against *lsm-3* had several predicted targets; therefore, the results were inconclusive (data not shown). However, in case of both *lsm-4* and *lsm-7* RNAi, a significant ($P < 10^{-13}$) increase in average GFP fluorescence of twofold to fourfold was scored (see Mattout et al. 2020). The variable degrees of GFP derepression following RNAi against these LSM subunits likely stems from differences in RNAi efficiency.

Given that the LSM complex has been implicated in RNA processing (Kufel et al. 2004; Beggs 2005; Tharun 2009; Perea-Resa et al. 2012; Golisz et al. 2013; Cornes et al. 2015), it was possible that somatic cell derepression was specific to the reporter construct used in the screen. We ruled this out by monitoring the effects of *lsm-2*, *lsm-5*, or *lsm-6* RNAi on the expression of four different heterochromatic reporters each integrated in the *C. elegans* genome as a large array (200–300 copies) at a single locus (Mattout et al. 2020). Each reporter carried a different promoter, gene, and 3′ UTR, yet all showed reporter up-regulation following *lsm-2*, *lsm-5*, or *lsm-6* RNAi (Mattout et al. 2020). In contrast, two euchromatic reporters, which were single-copy transgenes integrated into a nonheterochromatic region of the genome, showed no change in expression under the same conditions (data shown for GW1108, *eft-3p::gfp cb-unc-119*(+) chr II in Fig. 1E). This was true for euchromatic reporters with both high and low levels of expression, indicating that LSM complex subunits are selectively involved in the repression of loci that have heterochromatic, but not euchromatic features. Derepression levels were comparable or greater than the loss of our positive control, MES-4, a histone H3K36 methyltransferase recovered in the original screen (Towbin et al. 2012), and persisted during somatic cell differentiation and in postmitotic cells.

We confirmed that the up-regulation of GFP in these worms indeed reflects changes on the mRNA level, rather than GFP protein synthesis or turnover. This was achieved by quantifying the reporter mRNA by quantitative polymerase chain reaction using reverse transcription (RT-qPCR) following RNAi against various LSM subunits, notably *lsm-6* or *lsm-7* down-regulation (Mattout et al. 2020).

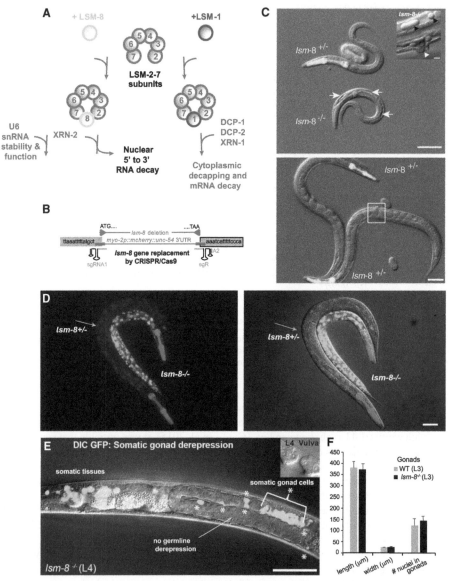

Figure 2. LSM-8 is required for heterochromatic reporter silencing, implicating the LSM2-8 complex. (*A*) The two main LSM complexes and functions (Beggs 2005; Tharun 2009). (*B*) Schematic view of the *lsm-8* deletion/gene replacement created by CRISPR–Cas9. (*C*) Differential interference contrast (DIC) images of L1/L2 larvae (*upper* panel) and young adults (*lower* panel) together with dual-channel fluorescence for *lsm-8*$^{+/-}$ (yellow pharynx in merge) and *lsm-8*$^{-/-}$ (red pharynx only) worms. Red and green channels are merged to distinguish heterozygous from homozygous worms. The *inset* in the *upper* panel shows the protruding vulva of an *lsm-8*$^{-/-}$ adult (marked with white triangle) and the accumulating vesicles (black triangles). The normal vulva is boxed in the *lower* panel. Bars = 100 µm. (*D*) Red and green fluorescence channels are merged and coupled with DIC images of one *lsm-8*$^{-/-}$ worm and one *lsm-8*$^{+/-}$ heterozygote bearing the *pkIs1582* reporter (GW1120). Note that this strain has no nT1 balancer with the green fluorescent pharynx marker. GFP fluorescence in nuclei throughout the *lsm-8*$^{-/-}$ worm is visible. Bar, 100 µm. (*E*) Merged DIC images with fluorescent images as in *D*. *lsm-8*$^{-/-}$ mutant L4 larvae bearing *pkIs1582* reporter is shown with fully developed gonad (encircled in red). The *lsm-8*$^{-/-}$ mutant is 100% sterile. Somatic nuclei and somatic cells of the gonad show GFP expression from the heterochromatic array. Bar, 50 µm. (This image also appeared in Mattout et al. 2020. Reproduced with permission.) (*F*) Quantification of the length, width, and gonad nuclei count from the DAPI staining of L3 WT and *lsm-8*$^{-/-}$ larvae. $n = 8$.

LSM2-8 CONTRIBUTES TO HETEROCHROMATIC REPORTER SILENCING

Studies in yeast and plants have established that there are two LSM complexes (LSM1-7 and LSM2-8) (Fig. 2A) that share six subunits, whereas the seventh subunit in each complex, LSM-1 or LSM-8, participates only in one (Beggs 2005; Tharun 2009). The two complexes also interact with distinct cofactors. RNAi against *lsm-1*, one of the nonshared subunits, did not derepress the heterochromatic reporter nor did the depletion of the 5′ → 3′ exoribonucleases, XRN-1, which is reported to act together with LSM1-7 in the cytoplasm to mediate cytoplasmic RNA decay (Figs. 1D and 2A). On the other hand, RNAi against *xrn-2*, a nuclear RNase that is thought to degrade RNA together with LSM2-8 in budding yeast (Tharun 2009), did show significant heterochromatic reporter derepression (Fig. 1D). Null-allele worms lacking *lsm-1* or the decapping factor *dcap-2* also failed to show array derepression (Mattout et al. 2020). This argued against a role for LSM1-7 in heterochromatic reporter silencing and suggested instead that LSM2-8 and an evolutionarily conserved nuclear 5′ → 3′ exoribonuclease called XRN-2 (HsXRN2, SpDhp1, or ScRat1) are involved (Miki et al. 2016).

Because LSM-8, the only subunit unique for the nuclear LSM2-8 complex, was unresponsive to RNAi in our hands, it was essential to create a genomic deletion allele of the *lsm-8* gene. A full *lsm-8* deletion was generated by CRISPR–Cas9, and simultaneously we inserted a red fluorescent marker gene with pharynx-specific expression. We could track the null allele by fluorescence microscopy, but found that the homozygous $lsm\text{-}8^{-/-}$ worms were 100% sterile. To propagate the *lsm-8* deletion, the mutation was balanced with a green fluorescence–marked balancer *nT1 [qIs51]*, which is also expressed in the pharynx. With dual fluorescence, we could distinguish offspring that are heterozygous for $lsm\text{-}8^{-/+}$ (i.e., that have red and green fluorescence in the pharynx) from those that are homozygous ($lsm8^{-/-}$; red pharynx only) (Fig. 2C). GFP fluorescence from the heterochromatic array was then monitored in red pharynx–marked homozygous, $lsm\text{-}8^{-/-}$ animals with or without (Fig. 1D) the *nT1* balancer. Indeed, deficiency for LSM-8 strongly derepressed the heterochromatic array, yielding GFP fluorescences at levels similar to those observed after *lsm-2*, *lsm-5*, *lsm-6*, or *lsm-7* RNAi (Fig. 2D). Although mutant germline and gonad development up to L3 and L4 larval stages was similar to that in wild-type (N2) worms (Fig. 2E,F), there was no oocyte maturation in the $lsm\text{-}8^{-/-}$ mutant. Nonetheless, we could monitor GFP derepression in all somatic tissues of larvae and adults, including the somatic cells of the gonad, but not in the germline itself (Fig. 2E).

Although the $lsm\text{-}8^{-/-}$ animals can develop to adulthood, the adults have protruding vulva (see inset, Fig. 2C, white arrowhead), show empty cavities or vacuoles in differentiated tissues (black triangle), and die prematurely (Mattout et al. 2020). Other *lsm* mutants, such as *lsm-2* and *lsm-5*, are phenotypically similar to *lsm-8* mutants, with clear protruding vulva, the presence of vacuoles, and 100% sterility. Adult populations of the *lsm-8* homozygous phenotype started dying after 6 days, whereas a similar loss of viability in wild-type worms would not occur before 10–12 days. Worms mutant for *lsm-1* did not present these phenotypes (Cornes et al. 2015), confirming the distinct roles played by the two LSM complexes.

LSM2-8 IS REQUIRED TO MAINTAIN SILENT ENDOGENOUS HETEROCHROMATIN

Although LSM2-8 was clearly implicated in heterochromatic array silencing, it was important to establish whether or not it also controls the expression of endogenous transcripts. The data sets for RNA-seq of WT and homozygous $lsm8^{-/-}$ sorted L3 and L1 larvae are presented in Mattout et al. (2020) and data are deposited with the NCBI Gene Expression Omnibus (GSE92851).

The *lsm-8* data sets were compared with RNA-seq data from similarly sorted $met\text{-}2^{-/-}$ $set\text{-}25^{-/-}$ L3 larvae, which lack detectable H3K9 methylation (Towbin et al. 2012), and from L3 worms of the triple mutant, $met\text{-}2^{-/-}$ $set\text{-}25^{-/-}$; $lsm8^{-/-}$, to determine if the LSM2-8 silencing pathways would be epistatic with the classic H3K9 methylation–mediated repression. In each case, the mutant and wild-type worms were sorted by fluorescence (nongreen pharynx indicating homozygosity) and by size, to generate a uniform population of L3 stage larvae. Mutant and wild-type L1 populations were also examined following sorting for L1 size larvae (Mattout et al. 2020). For the L3 populations, shifts in developmental timing arising during the sorting process were compensated by comparing matched replicates by making use of a characteristic, temporal fluctuation in gene expression characterized by the Großhans laboratory (Hendriks et al. 2014). With developmentally matched mutants and control larvae, we analyzed the RNA-seq data using edgeR, to determine the genes that changed significantly in each of three genetic backgrounds and WT strains (Mattout et al. 2020). Two or three analytical criteria were used, with the most stringent being the combined criteria of a false discovery rate (FDR) < 0.05 and a fold change (Fc) in average RNA level >4 (Mattout et al. 2020). In Figure 3A, we highlight genes with average altered expression rates of Fc > 4 only.

We found that the deletion of *lsm-8* led to the strong and reproducible up-regulation of 122 genes (FDR < 0.05 and Fc > 4), with very few genes being down-regulated (Mattout et al. 2020). With less stringent cutoff values, even higher numbers of loci were selectively derepressed in $lsm\text{-}8^{-/-}$ (Fig. 3A): Only FDR < 0.05 identified 147 genes up-regulated; only Fc > 4 identified 204 genes; and only Fc > 2 identified 1332 genes. Interestingly, a significant number of transcripts sensitive to $lsm\text{-}8^{-/-}$ were involved in the innate immune response and the regulation of cell shape, but they did not reflect a single developmental stage. Genes derepressed in the mutant in L1 larvae overlapped only partially (5%–10% based on the thresholds used) with those derepressed in the L3 stage, and among those overlapping, many are involved in the innate immune response (Mattout et al. 2020). We confirmed the $lsm\ 8^{-/-}$ larvae results by performing RNA-seq experiments on the *lsm-4* mutant (Fig. 4A). The patterns of genes derepressed were overlapping but not identical, in *lsm-4* and *lsm-8* mutants (Fig. 4B). This is to be expected, given that *lsm-4* also eliminates the activity of the LSM1-7 complex as well as the LSM2-8. Indeed, the distribution of LSM-4-GFP is diffuse in both cytoplasm and nucleus, and in the nucleus it appears to be excluded from foci that bind the H3K9me3 histone methyltransferase (HMT), SET-25 (labeled with mCherry, Fig. 4C).

In yeast and plants, RNA-binding complexes feed back to establish H3K9 methylation on centromeric and other heterochromatin (Buhler et al. 2007; Vasiljeva et al. 2008; Keller et al. 2012; Zofall et al. 2012; Yamanaka et al. 2013; Egan et al. 2014; Chalamcharla et al. 2015; Tucker et al. 2016). It was therefore of interest to compare RNA-seq changes provoked by the loss of LSM2-8 with loss of canonical heterochromatin marks, H3K9me2 or me3. We

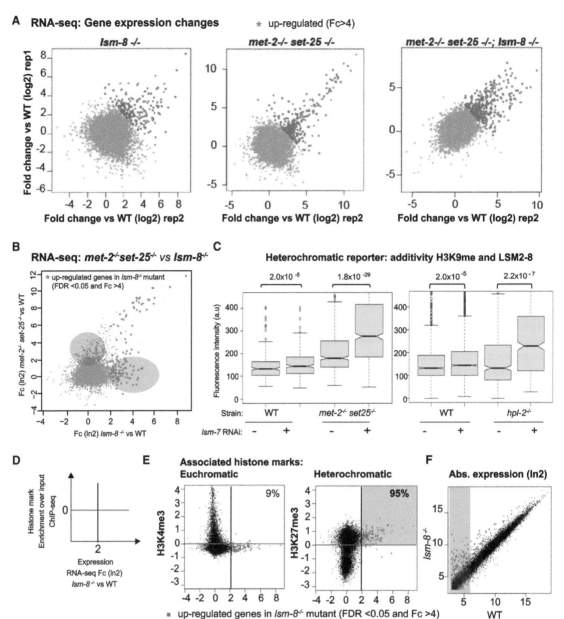

Figure 3. LSM2-8 silences hundreds of genes and 95% of these carry the Polycomb mark H3K27me3. (*A*) Deletion of *lsm-8* (*lsm-8*[−/−]) derepresses significantly more than 200 genes (Fc > 4). Relative gene expression profiles are shown as scatter plots, with fold change (Fc > 4) for two RNA-seq replicas of L3 sorted worms of the indicated genotype versus WT. Each dot corresponds to a gene. The original RNA-seq data are available at NCBI Gene Expression Omnibus number GSE92851. The same data set was analyzed in Mattout et al. (2020), but with (FDR < 0.05 and Fc > 4). (*B*) Scatter plot comparing the relative gene expression between the *lsm-8* (*x*-axis) and the *met-2 set-25* double mutant (*y*-axis). Common up-regulated genes are shaded yellow; 36% of genes up-regulated in the *lsm-8* mutant (FDR < 0.05 and Fc > 4) are also up-regulated (FDR < 0.05 and Fc > 4) in the *met-2 set-25* mutant. *lsm-8*[−/−]-specific up-regulated genes are shaded in pink; *met-2*[−/−] *set-25*[−/−]-specific are in blue. (This panel is reproduced from Figure 3C in Mattout et al. 2020 with permission.) (*C*) LSM2-8-mediated silencing of the heterochromatic reporter is independent of H3K9 methylation. Quantitation of derepression of GFP expressed from the *gwIs4* heterochromatic reporter in L1 progeny in WT and *met-2 set-25* mutant genotypes, after control or *lsm-7* RNAi. Quantitation of GFP derepression expressed from the *gwIs4* heterochromatic reporter in L1 progeny in WT and *met-2 set-25* mutant genotypes, respectively, from strains GW76 and GW637, after control or *lsm-7* RNAi. *P*-values indicated; two-tailed unpaired *t*-test. Quantification and statistical analyses were based on *n* = 1460, 2399, 1593, and 1189 worms for conditions indicated from *left* to *right*, pooled from three independent experiments. *N* = 339, 1004, 426, 673 same after *lsm-7* RNAi in the *hpl-2*[−/−] mutant background. (*D,E*) Scatter plots compare the average gene expression changes in *lsm-8*[−/−] worms (*x*-axis in log$_2$, RNA-seq L3 stage) versus enrichment for the indicated euchromatic or heterochromatic histone modification (*y*-axis in log$_2$, ModEncode data of WT L3 stage). Up-regulated genes (FDR > 0.05 and Fc > 4) in the *lsm-8*[−/−] mutant are in red to the *right* of the black line, and genes enriched for the histone mark are *above* the red line (enriched over input). % indicates genes in upper right zone. (The panels in *E* are reproduced from Fig. 4B of Mattout et al. 2020, with permission.) (*F*) Scatter plot of absolute gene expression (log$_2$) of *lsm-8*[−/−] versus WT. Red dots as in *E* and each dot corresponds to a gene. Values < 6 (log$_2$) are considered as very low expression: 6 (log$_2$) corresponds to about 60 RNA-seq reads per gene, whereas 10 (log$_2$) corresponds to about 1000 reads per gene. Most *lsm-8*[−/−] derepressed genes have low steady-state level expression in WT worms but moderate expression levels in the mutant (pink shading). (This panel is reproduced from Fig. 4C of Mattout et al. 2020, with permission.)

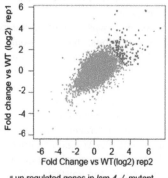
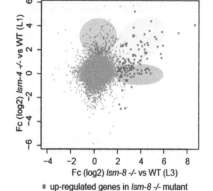
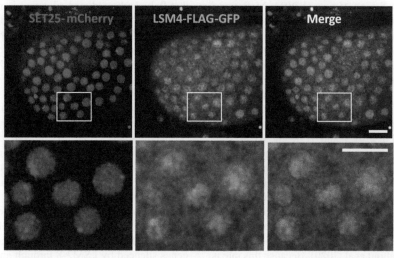

Figure 4. *lsm-8* and *lsm-4* deletions derepress overlapping gene sets in L1 and L3 larvae. (*A*) Deletion of *lsm-4* (*lsm-4*$^{-/-}$) leads to L1 larval arrest and derepresses more than 100 genes significantly. Relative gene expression profiles are shown as scatter plots of fold change (Fc) in log$_2$ of two RNA-seq replicas of L1 stage worms, *lsm-4*$^{-/-}$ versus WT. Each dot corresponds to a gene. The class of genes that were most significantly affected control body morphogenesis (*P* = 0.0072). (*B*) Scatter plot comparing the relative gene expression between the *lsm-8* (*x*-axis) and the *lsm-4* mutant (*y*-axis). Common up-regulated genes are shaded yellow; 53% of genes up-regulated in *lsm-8* (FDR < 0.05 and Fc > 4) are also up-regulated to some extent (50% increase) in *lsm-4*. Genes up-regulated in *lsm-8*$^{-/-}$ only are shaded pink, and those in *lsm-4*$^{-/-}$ only in blue. (*C*) Z-projection of live confocal images of embryonic progeny issued from the cross between the strain (GW1004) expressing LSM-4-Flag-GFP (GFP, green channel) and the strain (GW694) SET-25, the H3K9me3 HMT, fused to mCherry. The merged image shows no significant overlap on the brightest foci of SET-25. The *boxed* area is enlarged to show nuclear foci in *bottom* panels. Bars, 10 μm.

found that 36% of the genes derepressed because of the loss of H3K9 methylation (*met-2*$^{-/-}$ *set-25*$^{-/-}$) overlapped with those depressed in the *lsm-8*$^{-/-}$ mutant at the L3 stage (Fig. 3B; Mattout et al. 2020). To test the additivity of the LSM2-8 and H3K9me-mediated repression pathways, we compared the RNA-seq data of the single mutants with the triple mutant (*lsm 8*$^{-/-}$; *met-2*$^{-/-}$ *set-25*$^{-/-}$) (Fig. 3A) and found that there were clear subsets of up-regulated genes that were unique to either the *lsm-8* or the *met-2 set-25* mutant, and that genes showing mild up-regulation upon loss of either *lsm-8* or *met-2 set-25* (Fc ∼ 2) were more strongly expressed when combined (FDR < 0.05 and Fc > 4) (Mattout et al. 2020). Thus, the LSM2-8 pathway of silencing for endogenous loci is distinct from that mediated by H3K9me2/3, even though some genes are targeted by both pathways. Consistently using the heterochromatic reporter pkIS1582, we could show that *lsm-7* RNAi derepresses in an additive fashion with the loss of either H3K9me3 (*met-2 set-25*) or the loss of the H3K9me3 reader, HPL-2 (Fig. 3C).

LSM2-8 TARGETS POLYCOMB-MARKED GENES

To ask what the *lsm-8*-sensitive genes have in common, we compared our RNA-seq data to the normalized ChIP-seq data of common histone modifications generated by ModEncode (Fig. 3D–E). In worms, as in most organisms, H3K4me2, H3K4me3, and H3K27ac are associated with active genes (Liu et al. 2011; Wenzel et al. 2011; Ho et al. 2014), whereas H3K9me2/3 and H3K27me3 are repressive histone marks that colocalize with heterochromatin (Wenzel et al. 2011). The genes that were up-regulated in the $lsm\text{-}8^{-/-}$ mutant were depleted for active marks (H3K4me3) (Fig. 4E) in WT L3 larvae, and showed no particular enrichment for H3K9me2 nor for H3K9me3 (Mattout et al. 2020). Strikingly, however, 95% of genes that were derepressed by loss of LSM-8 were enriched for the repressive Polycomb mark, H3K27me3 (Fig. 3E). This was true not only for strongly up-regulated genes, but also for genes that were mildly up-regulated upon loss of LSM-8 (less than fourfold) and for windows of 500 bp across the genome. Thus, the overriding feature of genes or domains silenced by LSM2-8 was their enrichment for H3K27me3.

Consistent with the fact that H3K27me3 is the hallmark of Polycomb-repressed genes (Liu et al. 2011; Margueron and Reinberg 2011; Grossniklaus and Paro 2014; Conway et al. 2015), the absolute steady state level of the $lsm\text{-}8$-sensitive mRNAs was very low or almost undetectable in WT larvae (Fig. 3F). Although H3K27me3 is deposited in *C. elegans* by a PRC2-like complex comprising MES-2/E(z)/Ezh2, MES-3, and MES-6/Esc (Ketel et al. 2005; Yuzyuk et al. 2009; Gaydos et al. 2014) and mediates cell type–specific repression of developmentally regulated genes, in worms H3K27me3 and H3K9me3 ChIP-seq signals colocalize at many sites across the *C. elegans* genome (Ho et al. 2014). Interestingly, based on the ModEncode data sets from WT L3 larvae, 83% of the genes that are reproducibly enriched for H3K9me3 are also enriched for H3K27me3, and 41% of the genes enriched for H3K27me3 also carry H3K9me3. Among the genes up-regulated by $lsm\text{-}8$ mutation, we find that 40%–43% also carry H3K9me3 (Mattout et al. 2020). Because this rate is the same as the genome-wide coincidence of paired K9/K27 methylation, it appears that nucleosomes or domains bearing both repressive marks are not significantly enriched among $lsm\text{-}8$ targets. Most likely, H3K9me2 or me3 marks are irrelevant, for LSM2-8 recruitment or silencing.

LSM2-8 SILENCES THROUGH RNA DEGRADATION

Given that the LSM2-8 complex is known to stabilize and bind U6 snRNA in yeast and plants (Beggs 2005; Perea-Resa et al. 2012; Zhou et al. 2014), and that the LSM2-8 complex coprecipitates both with a factor involved in U6 snRNA stability (Ruegger et al. 2015) and with U6 snRNA itself in *C. elegans* (Mattout et al. 2020), we analyzed the RNA-seq data for splicing defects. We compared the exon–exon junction reads in the $lsm\text{-}8$ RNA-seq data with those from WT worms, to see if LSM2-8-mediated silencing coincides with defects in splicing, but we found irregular splicing events at only 18 exon–exon junctions out of 134,836 splice junctions (<0.02%) (Mattout et al. 2020). Thus, impaired splicing is most likely not the cause of $lsm\text{-}8^{-/-}$-triggered derepression.

If $lsm\ 8^{-/-}$ worms reduced the expression of PRC2 or PRC1 homologs, then it would follow that histone H3K27me3 levels would be reduced, and Polycomb-target genes would be derepressed. By checking the relevant mRNA levels, we could rule this out as a mechanism (Mattout et al. 2020). Nonetheless, to explore the relationship of LSM2-8-mediated silencing to H3K27me3, we treated the heterochromatic reporter–bearing $lsm\text{-}8^{-/-}$ worms with RNAi against either *mes-2* (EZH2) or *xrn-2*. As controls, we combined the $lsm\text{-}8$ mutant with RNAi against $lsm\text{-}7$ for epistasis and with *mes-4* (H3K36me HMT) or *set-25* (H3K9me HMT), which both show additivity with $lsm\text{-}8$ (Fig. 5A). Although *xrn-2* RNAi is epistatic with the $lsm\text{-}8$ background, the depletion of *mes-2* increased reporter expression, albeit less than that observed for *set-25* or *mes-4* RNAi (Fig. 5A). This argues that Polycomb-mediated silencing is not exclusively dependent on LSM-8. Again, this can be explained by the fact that H3K27me3 is known to repress on the level of transcription. The strong epistasis observed with XRN-2, on the other hand, argued that LSM2-8 may silence through RNA turnover or degradation (Fig 5A). Comparison of RNA-seq data from worms treated with *xrn-2* RNAi at L4 stage (Miki et al. 2016) and $lsm\text{-}8$ mutant worms at L3 confirms significant overlap of up-regulated endogenous genes (Fig. 5B). Indeed, 71% of the genes up-regulated by $lsm\text{-}8^{-/-}$ were also up-regulated by *xrn-2* RNAi (Fig. 5B, yellow) and 95% of those genes are enriched for H3K27me3. Again, this argues that LSM-8 and XRN-2 function on the same pathway. Nonetheless, a subset (<1/3) of LSM2-8 target genes (pink) were not affected by the down-regulation of XRN-2. It is unclear if those reflect experimental differences or a subset of genes silenced by LSM2-8 through another mechanism.

LSM2-8 ACTS TOGETHER WITH RNA DEGRADATION ENZYMES

To find other genes implicated in this potential nuclear RNA decay function, we tested candidate cofactors for heterochromatic reporter derepression and examined their genetic relationship with $lsm\text{-}8^{-/-}$. Heterochromatic array derepression at the L1 stage was monitored by GFP expression, following a targeted RNAi screen against genes that are co-regulated with LSM2-8 subunits and/or functionally implicated in RNA processing—namely, RNA Pol II subunits (*rpb-12*, *rpb-7*) and the type II poly(A) binding protein (*pabp-2*). Like *xrn-2* RNAi, RNAi against *rpb-12*, *rpb-7*, or *pabp-2* on their own derepress the heterochromatic reporter (Fig. 5A), whereas the depletion of six additional genes that are implicated in aspects of RNA metabolism does not (*F49C12.11*, *T13F2.2*, *cgh-1*—the human ortholog of DDX6, *B0495.8*, *C50D2.8*, and *pab-2*—the human ortholog of PABPC1) (Mattout et al. 2020). Important, nonetheless, is the epistasis between derepression by *xrn-2* RNAi, *rpb-12*, *rpb-7*, and *pabp-2*

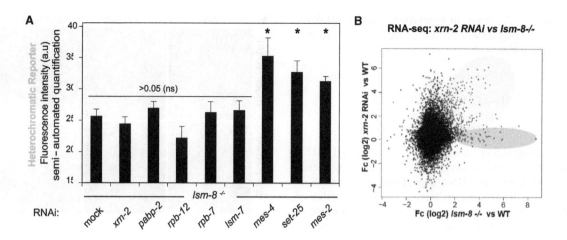
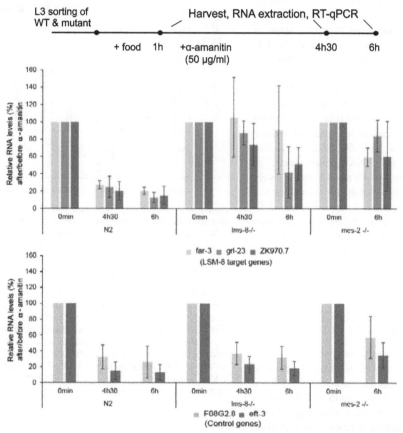

Figure 5. XRN-2 and other RNA factors show epistasis with *lsm-8* deficiency for heterochromatic array derepression. (*A*) RNAi experiments were performed in the *lsm-8*$^{-/-}$ (GW1119) strain. The GFP derepression of the heterochromatic reporter *pkIs1582* was quantified by semiautomated fluorescence microscopy on homozygous *lsm-8*$^{-/-}$ progeny under indicated RNAi conditions. *P*-values (Students *t*-test) are indicated: (*) $P < 0.005$. This indicates additivity between the *lsm-8* derepression and derepression by the RNAi used. Bars = S.E.M., $N=2$, $n=55, 45, 22, 11, 10, 23, 25, 25, 85$. (These data are reproduced from Fig. 6F of Mattout et al. 2020, with permission.) (*B*) Scatter plot comparing relative gene expression changes of *lsm-8*$^{-/-}$ L3 larvae and *xrn-2* RNAi treated L4 (Miki et al. 2016). Common up-regulated genes are shaded yellow; 71% of genes up-regulated in the *lsm-8* mutant (Fc > 4) are also up-regulated to some extent (50% increase) in *xrn-2* depleted worms. *lsm-8*$^{-/-}$-specific up-regulated genes are shaded pink. (This panel is reproduced from Fig. 6A of Mattout et al. 2020, with permission.) (*C*) WT and *lsm-8*$^{-/-}$ worms were sorted, refed with OP50 in liquid culture for 1 h at room temperature, and treated with 50 μg/mL final concentration of α-amanitin, which inhibits Pol II and Pol III transcription. RNA was isolated at time 0, 4.5, and 6 h, as indicated for each independent experiment. RNA levels of three transcripts affected by LSM-8 and two control transcripts (expression not affected by LSM-8) were determined by RT-qPCR and normalized to 18S rRNA levels, which are insensitive to α-amanitin. The value at 0 h is defined as 100%. Error bars represent standard deviation of the mean from at least three independent experiments. (Data are reproduced from Extended Data Fig. 8A in Mattout et al. 2020, with permission.)

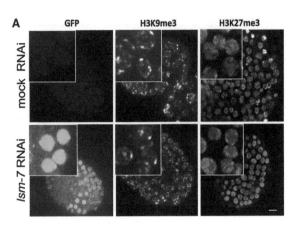

and the *lsm-8* mutant. In conclusion, PABP-2, RPB-12, and RPB-7 may act with XRN-2 on the LSM2-8-mediated silencing pathway.

LSM2-8 CONTRIBUTES TO THE DEGRADATION OF mRNAs FROM POLYCOMB-MARKED GENES

The cooperation between the RNA-binding LSM2-8 complex and XRN-2 suggests that LSM2-8 may silence genes by triggering mRNA degradation. It would follow that *lsm-8* deletion might increase the RNA stability of specific LSM2-8-sensitive transcripts in vivo because of a failure to degrade these RNAs by XRN-2. This was confirmed for the mRNAs of three endogenous *lsm-8*-sensitive genes—*far-3*, *grl-23*, and *ZK970.2*—by monitoring mRNA stability after the addition of α-amanitin, an inhibitor of RNA Pol II and Pol III elongation. The rate of mRNA decay was monitored at the L3 stage of sorted WT and *lsm-8*$^{-/-}$ worms over 6 h using RT-qPCR. RNA levels were normalized to that of the 18S rRNA. Deletion of *lsm-8* delayed the turnover of *far-3*, *grl-23*, and *ZK970.2* mRNAs, whereas the mRNA and pre-mRNA of two control genes (*eft-3* and *F08G2.8*) showed no change in turnover rates (Fig. 5C; Mattout et al. 2020). Consistently, transcript levels from these *lsm-8*-sensitive genes, *far-3* and *ZK970.2*, were also strongly up-regulated in *xrn-2* RNAi treated worms as well, with a fold change in $\log_2$ of 4.9 and 4.0, respectively (Mattout et al. 2020). The decay of the three *lsm-8*-sensitive mRNAs was also attenuated in the Polycomb-deficient strain (*mes-2*) (Mattout et al. 2020). Thus, our data suggest that MES-2-mediated repression of LSM-8 regulated genes acts at least in part through mRNA degradation (Mattout et al. 2020).

LSM2-8 ACTION FEEDS BACK TO MAINTAIN H3K27me3 LEVELS

Although *mes-2* and *lsm-8* depletion appeared to be additive with respect to derepression, one cannot conclude that the LSM2-8 complex acts independently of MES-2 and H3K27me3, especially considering the striking correlation of LSM-8-sensitivity with H3K27me3 (Fig. 3E). We initially tested for global changes in repressive epigenetic marks, H3K9me3 and H3K27me3, by immunostaining of embryos following *lsm-7* RNAi. However, we found no global drop in staining (Fig. 6A). Nonetheless, when we performed quantitative chromatin immunoprecipitation (ChIP) for H3K27me3 on genes that are sensitive to *lsm-8* ablation (i.e., genes silenced by the LSM2-8 complex in WT worms), we observed a significant decrease (>50%) in H3K27me3 levels (Fig. 7B). This was not detected for *lsm-8*-insensitive H3K27me3-marked genes. This suggests that the LSM2-8 complex selectively feeds back to maintain H3K27me3 levels, either directly or indirectly, at a subset of H3K27me3-silenced loci. The fact that it is not detectable by immunostaining argues that H3K27me3 lev-

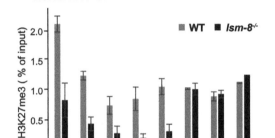

Figure 6. LSM2-8 helps maintain full H3K27me3 levels on silenced loci. (*A*) Z-projection of confocal images of wild-type (GW566) embryos treated with control or *lsm-7* RNAi (GFP, live imaging) or stained with antibodies against H3K9me3 and H3K27me3, as indicated. Nuclei are enlarged in the *insets*. Bar, 5 μm. (*B*) H3K27me3 ChIP-qPCR on target genes in WT and *lsm-8*$^{-/-}$ worms. Three categories of genes were assessed. Five genes were monitored that are up-regulated in *lsm-8*$^{-/-}$ worms and enriched for H3K27me3 mark in WT (*lsm-8* target genes), three genes were monitored that are enriched for the H3K27me3 mark in WT but do not change in expression in the *lsm-8* mutant (nontarget genes), and the control category is represented by *ska-1*, a gene not enriched for H3K27me3 in WT and with no expression change (control). $N=3$, $n=3$, bars (mean ± S.E.M.). (Data appear in Fig. 7A of Mattout et al. 2020 and are reproduced with permission.)

els may not be affected globally, although many Polycomb targets do show significant derepression.

CONCLUSION

We have shown that in nematodes a conserved nuclear RNA-binding complex, LSM2-8, contributes to full repression of heterochromatic reporters and genomic regions that carry the H3K27me3 epigenetic mark. A related, cytoplasmic complex, LSM1-7, appears to have no specific role in heterochromatic silencing. Unlike heterochromatin-linked RNA degradation pathways in plants and fission yeast, LSM-2-8-mediated repression is inde-

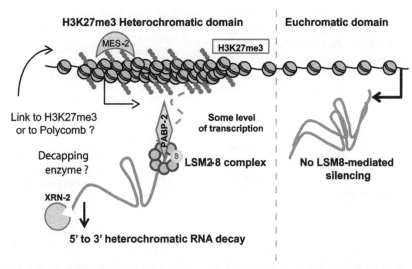

Figure 7. Model for the LSM2-8 pathway of silencing. We showed that the LSM-8-mediated silencing pathway makes use of XRN-2 ribonuclease and involves other factors, such as PABP-2 (*H.s*: PABPN1). We hypothesize that RNA arising from H3K27me3 genomic regions that are controlled by the LSM2-8 complex acquire a specific feature during transcription (e.g., a specific structure, RNA modification, 3′ UTR, poly(A/U) tail, or specific RNA-binding protein(s)), which allows recognition and processing by LSM2-8. LSM2-8-mediated silencing also feeds back to regulate H3K27me3 levels on LSM-8-regulated genes, although it is unclear if the interaction with PRC2 complex is direct (dotted arrow). LSM-2-8-mediated silencing of H3K27me3-bound regions defines a new mechanism for selective post-/cotranscriptional silencing for facultative heterochromatin through RNA decay.

pendent of and additive with H3K9me3 and H3K9me HMTs. Our data suggest that the LSM2-8 complex acts primarily on RNA stability, selectively mediating the decay of transcripts from H3K27me3-tagged genes through the 5′–3′ exoribonuclease XRN-2. The pathway does not, however, involve XRN-1 nor the decapping enzyme DCAP-2. The LSM2-8-mediated silencing occurs at Polycomb H3K27me3-marked genes, with only ~40% of the LSM2-8 targets also carrying H3K9me3. This represents the stochastic frequency of H3K27me3 and H3K9me3 coincidence on genes in L3 stage larvae, suggesting that H3K27me3 and not H3K9me3 is the recognition signal for mRNA degradation. As sketched in our model in Figure 7, we propose that the LSM2-8 complex mediates gene silencing primarily through the exonuclease XRN-2, which would degrade the transcripts from a subset of H3K27me3-marked genes either co- or posttranscriptionally. This represents a novel pathway for Polycomb-mediated gene silencing in animals, which appears to act in addition to transcriptional repression.

The fact that only a subset of transcripts arising from H3K27me3-marked genes is found to be significantly derepressed in *lsm-8* mutant worms suggests that there may be additional criteria that provide specificity for LSM2-8 control. Using a less stringent cutoff for derepression in the *lsm-8* RNA-seq data set (Mattout et al. 2020), we find many more genes up-regulated, and >93% still bear H3K27me3. It is important to note that the derepression of tissue-specific transcripts (i.e., those that are expressed in only a few cells of larvae or adult worms) will not be detectable by our RNA-seq methods that monitor RNA in whole worm extracts. This was confirmed to be the case for one HOX gene called *egl-5*, which is known to be regulated by MES-2 (EZH2) (Mattout et al. 2020). By analyzing the cell type–specific derepression of *egl-5* using fluorescence microscopy, we were able to show similar levels of derepression of this gene in male tail cells, in both *lsm-8* and *mes-2* mutants (Mattout et al. 2020), although its derepression level by RNA-seq was only ~1.5-fold.

The molecular basis of LSM2-8 specificity for H3K27me3 provides a framework for future studies exploring how the epigenetic information can be transferred from the chromatin to an RNA-binding complex. We propose that the RNAs arising from H3K27me3-marked genomic regions are recognized and bound selectively by the LSM2-8 complex, and that this likely requires a feature that is acquired during transcription from that locus—for example, a specific structure, RNA modification, cap, a poly-A/U tail, or specific RNA-binding protein(s) recognized by LSM2-8. It is important to note that we scored a drop in H3K27me3 specifically on LSM2-8-sensitive genes in $lsm\text{-}8^{-/-}$ animals, even though the overall staining level of H3K27me3 appeared unchanged (Fig. 6). This suggests that RNA degradation helps maintain H3K27me3 levels. In other words, transcriptional repression by this epigenetic mark may be transferred from the RNA-binding complex LSM2-8 to chromatin.

An RNA decay-mediated pathway involving the cofactors PABP-2, RPB-12, and RBP-7 appears to function independently of DACP-2 and LSM-1 and additively with HPL-2 and the H3K9me3 HMT, SET-25. The epistasis with type II poly(A) binding protein (PABP-2) (Hurschler et al. 2011) (*Hs*PABPN1 and *Sp*Pab2) is particularly interesting, because it appears to regulate 3′ UTR and poly(A) tail length in the nucleus (Kuhn et al. 2009). Moreover, it binds nascent RNAs early during the elongation step

(Lemieux and Bachand 2009; Beaulieu et al. 2012). Given that the LSM2-8 complex is known to bind to the 3′ oligo(U) tail of the U6 snRNA (Zhou et al. 2014), as well as the 3′ poly(A+) tail of nuclear RNAs (Kufel et al. 2004), PABP-2 may regulate its specificity by modulating the 3′ end of mRNAs at H3K27me3-marked domains. The participation of RNA Pol II subunits RPB-7 and -12 in LSM2-8-mediated gene silencing is also suggestive of an involvement of RNA Pol II in silencing, just as the *Schizosaccharomyces pombe* RBP-7 homolog has been implicated in centromeric repeat RNAi-directed silencing (Djupedal et al. 2005). Interestingly, fission yeast lacks H3K27me3; thus, the link to chromatin status may have evolved at a later stage. In *S. cerevisiae*, the same subunit plays a different, but direct role in Pat1/Lsm1-7-mediated mRNA decay in the cytoplasm (Lotan et al. 2007; Haimovich et al. 2013). One speculative scenario, based on functions described in other organisms, is that *rpb-7* and *rpb-12* RNA Pol II subunits might be loaded on LSM2-8-regulated transcripts to signal to that these should be degraded by XRN-2. An unknown decapping enzyme(s) may also be involved, which would allow RNA degradation by XRN-2 in the nucleus.

We conclude that facultative heterochromatin in metazoans can be silenced through a mechanism of selective transcript degradation and not only by transcriptional repression. LSM2-8 may mediate gene silencing by linking a specific epigenetic state to transcript degradation, adding an additional layer of control over differentiation and development.

ACKNOWLEDGMENTS

The accession number for the RNA-seq data is NCBI Gene Expression Omnibus GSE92851. Some strains were provided by the *Caenorhabditis* Genetics Center (CGC), which is funded by the National Institutes of Health Office of Research Infrastructure Programs (P40 OD010440). We thank I. Katiç, the Friedrich Miescher Institute Genomics and Microscopy facilities, and M. Bühler, H. Großhans, and W. Filipowicz for discussions. The authors acknowledge the support of Marie Curie Intra-European grant (no. PIEF-GA-2010-276589) and the Swiss National Science Foundation (SNF) Marie-Heim Vögtlin grant (nos. PMPDP3_151381 and PMPDP3_168717) to A.M.; SNF grant (no. 310030B_156936) to S.M.G.; and support of the Novartis Research Foundation. This project received funding from the European Research Council under the European Union's Horizon 2020 Research and Innovation programme (Epiherigans—grant agreement no. 743312 to S.M.G.).

REFERENCES

Beaulieu YB, Kleinman CL, Landry-Voyer A-M, Majewski J, Bachand F. 2012. Polyadenylation-dependent control of long noncoding RNA expression by the poly (A)-binding protein nuclear 1. *PLoS Genet* 8: e1003078. doi:10.1371/journal.pgen.1003078

Beggs JD. 2005. Lsm proteins and RNA processing. *Biochem Soc Trans* 33: 433–438. doi:10.1042/BST0330433

Bühler M. 2009. RNA turnover and chromatin-dependent gene silencing. *Chromosoma* 118: 141–151. doi:10.1007/s00412-008-0195-z

Bühler M, Haas W, Gygi SP, Moazed D. 2007. RNAi-dependent and -independent RNA turnover mechanisms contribute to heterochromatic gene silencing. *Cell* 129: 707–721. doi:10.1016/j.cell.2007.03.038

Chalamcharla VR, Folco HD, Dhaksnamoorthy J, Grewal SI. 2015. Conserved factor Dhp1/Rat1/Xrn2 triggers premature transcription termination and nucleates heterochromatin to promote gene silencing. *Proc Natl Acad Sci* 112: 15548–15555. doi:10.1073/pnas.1522127112

Conway E, Healy E, Bracken AP. 2015. PRC2 mediated H3K27 methylations in cellular identity and cancer. *Curr Opin Cell Biol* 37: 42–48. doi:10.1016/j.ceb.2015.10.003

Cornes E, Porta-De-La-Riva M, Aristizábal-Corrales D, Brokate-Llanos AM, García-Rodríguez FJ, Ertl I, Díaz M, Fontrodona L, Reis K, Johnsen R, et al. 2015. Cytoplasmic LSM-1 protein regulates stress responses through the insulin/IGF-1 signaling pathway in *Caenorhabditis elegans*. *RNA* 21: 1544–1553. doi:10.1261/rna.052324.115

Djupedal I, Portoso M, Spåhr H, Bonilla C, Gustafsson CM, Allshire RC, Ekwall K. 2005. RNA Pol II subunit Rpb7 promotes centromeric transcription and RNAi-directed chromatin silencing. *Genes Dev* 19: 2301–2306. doi:10.1101/gad.344205

Egan ED, Braun CR, Gygi SP, Moazed D. 2014. Post-transcriptional regulation of meiotic genes by a nuclear RNA silencing complex. *RNA* 20: 867–881. doi:10.1261/rna.044479.114

Gaydos LJ, Wang W, Strome S. 2014. Gene repression. H3K27me and PRC2 transmit a memory of repression across generations and during development. *Science* 345: 1515–1518. doi:10.1126/science.1255023

Golisz A, Sikorski PJ, Kruszka K, Kufel J. 2013. Arabidopsis thaliana LSM proteins function in mRNA splicing and degradation. *Nucleic Acids Res* 41: 6232–6249. doi:10.1093/nar/gkt296

Gonzalez-Sandoval A, Towbin Benjamin D, Kalck V, Cabianca DS, Gaidatzis D, Hauer MH, Geng L, Wang L, Yang T, Wang X, et al. 2015. Perinuclear anchoring of H3K9-methylated chromatin stabilizes induced cell fate in *C. elegans* embryos. *Cell* 163: 1333–1347. doi:10.1016/j.cell.2015.10.066

Grossniklaus U, Paro R. 2014. Transcriptional silencing by Polycomb-group proteins. *Cold Spring Harb Perspect Biol* 6: a019331. doi:10.1101/cshperspect.a019331

Haimovich G, Choder M, Singer RH, Trcek T. 2013. The fate of the messenger is pre-determined: a new model for regulation of gene expression. *Biochim Biophys Acta* 1829: 643–653. doi:10.1016/j.bbagrm.2013.01.004

Hendriks G-J, Gaidatzis D, Aeschimann F, Großhans H. 2014. Extensive oscillatory gene expression during *C. elegans* larval development. *Mol Cell* 53: 380–392. doi:10.1016/j.molcel.2013.12.013

Ho JW, Jung YL, Liu T, Alver BH, Lee S, Ikegami K, Sohn KA, Minoda A, Tolstorukov MY, Appert A, et al. 2014. Comparative analysis of metazoan chromatin organization. *Nature* 512: 449–452. doi:10.1038/nature13415

Hurschler BA, Harris DT, Grosshans H. 2011. The type II poly(A)-binding protein PABP-2 genetically interacts with the let-7 miRNA and elicits heterochronic phenotypes in *Caenorhabditis elegans*. *Nucleic Acids Res* 39: 5647–5657. doi:10.1093/nar/gkr145

Keller C, Adaixo R, Stunnenberg R, Woolcock KJ, Hiller S, Bühler M. 2012. HP1[Swi6] mediates the recognition and destruction of heterochromatic RNA transcripts. *Mol Cell* 47: 215–227. doi:10.1016/j.molcel.2012.05.009

Ketel CS, Andersen EF, Vargas ML, Suh J, Strome S, Simon JA. 2005. Subunit contributions to histone methyltransferase activities of fly and worm Polycomb group complexes. *Mol Cell Biol* 25: 6857–6868. doi:10.1128/MCB.25.16.6857-6868.2005

Kufel J, Bousquet-Antonelli C, Beggs JD, Tollervey D. 2004. Nuclear pre-mRNA decapping and 5′ degradation in yeast require the Lsm2-8p complex. *Mol Cell Biol* **24:** 9646–9657. doi:10.1128/MCB.24.21.9646-9657.2004

Kühn U, Gündel M, Knoth A, Kerwitz Y, Rüdel S, Wahle E. 2009. Poly (A) tail length is controlled by the nuclear poly(A)-binding protein regulating the interaction between poly(A) polymerase and the cleavage and polyadenylation specificity factor. *J Biol Chem* **284:** 22803–22814. doi:10.1074/jbc.M109.018226

Lemieux C, Bachand F. 2009. Cotranscriptional recruitment of the nuclear poly (A)-binding protein Pab2 to nascent transcripts and association with translating mRNPs. *Nucleic Acids Res* **37:** 3418–3430. doi:10.1093/nar/gkp207

Liu T, Rechtsteiner A, Egelhofer TA, Vielle A, Latorre I, Cheung M-S, Ercan S, Ikegami K, Jensen M, Kolasinska-Zwierz P, et al. 2011. Broad chromosomal domains of histone modification patterns in *C. elegans*. *Genome Res* **21:** 227–236. doi:10.1101/gr.115519.110

Lotan R, Goler-Baron V, Duek L, Haimovich G, Choder M. 2007. The Rpb7p subunit of yeast RNA polymerase II plays roles in the two major cytoplasmic mRNA decay mechanisms. *J Cell Biol* **178:** 1133–1143. doi:10.1083/jcb.200701165

Margueron R, Reinberg D. 2011. The Polycomb complex PRC2 and its mark in life. *Nature* **469:** 343–349. doi:10.1038/nature09784

Mattout A, Gaidatzis D, Padeken J, Schmid C, Aeschlimann F, Kalck V, Gasser SM. 2020. LSM2-8 and XRN-2 contribute to the silencing of H3K27me3-marked genes through targeted RNA decay. *Nature Cell Biol* doi:10.1038/s41556-020-0504-1

Meister P, Towbin BD, Pike BL, Ponti A, Gasser SM. 2010. The spatial dynamics of tissue-specific promoters during *C. elegans* development. *Genes Dev* **24:** 766–782. doi:10.1101/gad.559610

Miki TS, Carl SH, Stadler MB, Großhans H. 2016. XRN2 autoregulation and control of polycistronic gene expression in *Caenorhabditis elegans*. *PLoS Gen* **12:** e1006313. doi:10.1371/journal.pgen.1006313

Perea-Resa C, Hernández-Verdeja T, López-Cobollo R, del Mar Castellano M, Salinas J. 2012. LSM proteins provide accurate splicing and decay of selected transcripts to ensure normal *Arabidopsis* development. *Plant Cell* **24:** 4930–4947. doi:10.1105/tpc.112.103697

Rüegger S, Miki TS, Hess D, Großhans H. 2015. The ribonucleotidyl transferase USIP-1 acts with SART3 to promote U6 snRNA recycling. *Nucleic Acids Res* **43:** 3344–3357. doi:10.1093/nar/gkv196

Saksouk N, Simboeck E, Déjardin J. 2015. Constitutive heterochromatin formation and transcription in mammals. *Epigenetics Chromatin* **8:** 3. doi:10.1186/1756-8935-8-3

Tharun S. 2009. Roles of eukaryotic Lsm proteins in the regulation of mRNA function. *Int Rev Cell Mol Biol* **272:** 149–189. doi:10.1016/S1937-6448(08)01604-3

Towbin BD, Meister P, Pike BL, Gasser SM. 2010. Repetitive transgenes in *C. elegans* accumulate heterochromatic marks and are sequestered at the nuclear envelope in a copy-number- and lamin-dependent manner. *Cold Spring Harb Symp Quant Biol* **75:** 555–565. doi:10.1101/sqb.2010.75.041

Towbin BD, González-Aguilera C, Sack R, Gaidatzis D, Kalck V, Meister P, Askjaer P, Gasser SM. 2012. Step-wise methylation of histone H3K9 positions heterochromatin at the nuclear periphery. *Cell* **150:** 934–947. doi:10.1016/j.cell.2012.06.051

Trojer P, Reinberg D. 2007. Facultative heterochromatin: is there a distinctive molecular signature? *Mol Cell* **28:** 1–13. doi:10.1016/j.molcel.2007.09.011

Tucker JF, Ohle C, Schermann G, Bendrin K, Zhang W, Fischer T, Zhang K. 2016. A novel epigenetic silencing pathway involving the highly conserved 5′-3′ exoribonuclease Dhp1/Rat1/Xrn2 in *Schizosaccharomyces pombe*. *PLoS Gen* **12:** e1005873. doi:10.1371/journal.pgen.1005873

Vasiljeva L, Kim M, Terzi N, Soares LM, Buratowski S. 2008. Transcription termination and RNA degradation contribute to silencing of RNA polymerase II transcription within heterochromatin. *Mol Cell* **29:** 313–323. doi:10.1016/j.molcel.2008.01.011

Wang J, Jia ST, Jia S. 2016. New insights into the regulation of heterochromatin. *Trends Genetics* **32:** 284–294. doi:10.1016/j.tig.2016.02.005

Wenzel D, Palladino F, Jedrusik-Bode M. 2011. Epigenetics in *C. elegans*: facts and challenges. *Genesis* **49:** 647–661. doi:10.1002/dvg.20762

Yamanaka S, Mehta S, Reyes-Turcu FE, Zhuang F, Fuchs RT, Rong Y, Robb GB, Grewal SI. 2013. RNAi triggered by specialized machinery silences developmental genes and retrotransposons. *Nature* **493:** 557–560. doi:10.1038/nature11716

Yuzyuk T, Fakhouri TH, Kiefer J, Mango SE. 2009. The polycomb complex protein *mes-2/E (z)* promotes the transition from developmental plasticity to differentiation in *C. elegans* embryos. *Dev Cell* **16:** 699–710. doi:10.1016/j.devcel.2009.03.008

Zhou L, Hang J, Zhou Y, Wan R, Lu G, Yin P, Yan C, Shi Y. 2014. Crystal structures of the Lsm complex bound to the 3′ end sequence of U6 small nuclear RNA. *Nature* **506:** 116–120. doi:10.1038/nature12803

Zofall M, Yamanaka S, Reyes-Turcu FE, Zhang K, Rubin C, Grewal SI. 2012. RNA elimination machinery targeting meiotic mRNAs promotes facultative heterochromatin formation. *Science* **335:** 96–100. doi:10.1126/science.1211651

To Process or to Decay: A Mechanistic View of the Nuclear RNA Exosome

Mahesh Lingaraju,[1] Jan M. Schuller,[1] Sebastian Falk,[2] Piotr Gerlach,[1]
Fabien Bonneau,[1] Jérôme Basquin,[1] Christian Benda,[1] and Elena Conti[1]

[1]Max-Planck-Institute of Biochemistry, Department of Structural Cell Biology, D-82152 Martinsried/Munich, Germany

[2]Max Perutz Labs, Department of Structural and Computational Biology, University of Vienna, 1030, Vienna, Austria

Correspondence: conti@biochem.mpg.de

The RNA exosome was originally discovered in yeast as an RNA-processing complex required for the maturation of 5.8S ribosomal RNA (rRNA), one of the constituents of the large ribosomal subunit. The exosome is now known in eukaryotes as the major 3′–5′ RNA degradation machine involved in numerous processing, turnover, and surveillance pathways, both in the nucleus and the cytoplasm. Yet its role in maturing the 5.8S rRNA in the pre-60S ribosomal particle remains probably the most intricate and emblematic among its functions, as it involves all the RNA unwinding, degradation, and trimming activities embedded in this macromolecular complex. Here, we propose a comprehensive mechanistic model, based on current biochemical and structural data, explaining the dual functions of the nuclear exosome—the constructive versus the destructive mode.

The RNA exosome is a major 3′–5′ exoribonuclease that targets a wide range of substrates in different cellular compartments in all eukaryotes studied to date (for reviews, see Chlebowski et al. 2013; Zinder and Lima 2017; Schmid and Jensen 2019). In RNA metabolism, the exosome may be considered as a "Dr. Jekyll and Mr. Hyde" complex: it can either trim or destroy (Fig. 1). For example, exosome-mediated degradation eliminates transfer RNAs (tRNAs) in the nucleus that are in excess or defective (Kadaba et al. 2004; Gudipati et al. 2012) and a myriad of transcripts generated from pervasive transcription such as cryptic unstable transcripts (CUTs) in yeast (LaCava et al. 2005; Wyers et al. 2005) or promoter upstream transcripts in human cells (PROMPTs) (Preker et al. 2008). Likewise, exosome-mediated degradation eliminates messenger RNAs (mRNAs) in the cytoplasm (for review, see Schaeffer et al. 2011). In contrast, in nuclear RNA biogenesis pathways the exosome partially and specifically trims ribosomal RNA (rRNA) and small nuclear/nucleolar RNA precursors (Mitchell et al. 1996, 1997; Allmang et al. 1999). Taken together, it is not surprising that mutations in the RNA exosome are associated with human pathologies, including autoimmune and neurodegenerative disorders (for reviews, see Staals and Pruijn 2011; Fasken et al. 2020).

Tremendous progress has been made in the past decade to decipher the properties, architecture, and interplay of the exosome and its cofactors, first using *Saccharomyces cerevisiae* as a model organism (Liu et al. 2006, 2016; Bonneau et al. 2009; Makino et al. 2013a, 2015; Wasmuth et al. 2014, 2017; Kowalinski et al. 2016; Zinder et al. 2016; Falk et al. 2017a) and lately the human orthologs (Gerlach et al. 2018; Weick et al. 2018). Strategies developed from biochemical and genetic studies to "immobilize" the exosome in defined functional states allowed visualization of different snapshots of this complex at near-atomic resolution. However, because of the technical challenges posed by the sheer size and complexity of the factors involved at both the protein and RNA levels, the molecular basis explaining how the exosome can function constructively in RNA processing pathways and destructively in RNA decay pathways has evaded the field. In the past few years, the "resolution revolution" in cryo-electron microscopy has contributed to important advancements, including visualization of not only the entire nuclear exosome holocomplex (Gerlach et al. 2018; Weick et al. 2018) but also the complex processing of rRNA from a precursor of the large ribosomal subunit (Schuller et al. 2018). In this review, we focus on individual structural snapshots and extrapolate possible mechanisms by which the nuclear exosome may execute its decay and processing functions.

ARCHITECTURE OF THE RNA EXOSOME: THE CORE COMPLEX

The RNA exosome is an ancient machine with prokaryotic ancestry that has been likened to a proteasome for RNA (for reviews, see van Hoof and Parker 1999; Lorentzen and Conti 2006; Makino et al. 2013b). In eukaryotes, the core complex is formed by 10 different subunits (Exo10) (Fig. 2). Studies in yeast and human revealed that

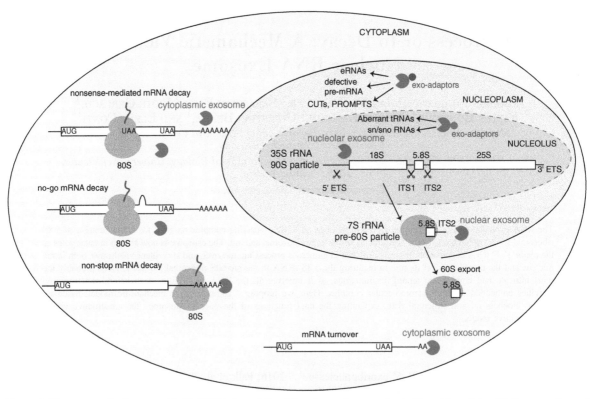

Figure 1. The exosome is a juggernaut in 3′–5′ RNA degradation. The scheme shows a subset of the wide variety of RNA processing and degradation activities of the eukaryotic exosome (shown in red, with nuclear adaptor proteins represented in violet) in the nucleolus, the nucleoplasm, and the cytoplasm. In the nucleoplasm, the exosome is involved in the maturation of 5.8S rRNA and also interacts with several exosome adaptors to form complexes such as NEXT and PAXT to target various substrates (see also Puno et al. 2019). Endoribonucleolytic cleavages involved in ribosome processing (not involving the exosome) are indicated with scissors. Exosome-mediated cytoplasmic pathways include mRNA turnover and translation-dependent surveillance pathways.

nine exosome subunits assemble into a barrel-like structure that lacks catalytic activity (Exo9) (Liu et al. 2006; Dziembowski et al. 2007). Exo9 is organized into an upper ring of three "cap" subunits (Rrp4, Rrp40, and Csl4) with S1/KH domains similar to those found in RNA-binding proteins and a lower ring of six subunits with the fold typical of a bacterial 3′–5′ ribonuclease, RNase PH, but lacking functional active sites. The Exo9 barrel is traversed by a prominent central channel that spans from the narrow entry pore at the top of upper ring to the side of the lower ring adopting an L-shaped structure that has been conserved from the archaeal ancestor complex (Lorentzen et al. 2007; Bonneau et al. 2009). In eukaryotes, this internal channel feeds RNA substrates to the 10th subunit, the Rrp44 ribonuclease (also known as Dis3) (Makino et al. 2013a). The Rrp44 amino-terminal PIN domain provides the high-affinity binding that anchors Rrp44 to Exo9 (Bonneau et al. 2009) and in addition contains an accessible endonuclease active site (Lebreton et al. 2008; Schaeffer et al. 2009; Schneider et al. 2009; Han and van Hoof 2016). The carboxy-terminal domain of Rrp44 shares similar structural features to RNase II of *Escherichia coli* and degrades RNAs in the 3′–5′ direction, cleaving one nucleotide at a time in a processive manner—that is, performing sequential rounds of cleavages without disso-

ciating from the RNA substrate (Frazao et al. 2006; Lorentzen et al. 2008). This domain harbors the principal exoribonuclease activity of the exosome complex (Liu et al. 2006; Dziembowski et al. 2007). In *S. cerevisiae*, the RNase II-like domain of Rrp44 can swing from an open conformation (Bonneau et al. 2009; Makino et al. 2015) to a closed conformation oriented toward Exo9 to receive an RNA substrate coming from the central channel (Fig. 2A,B; Makino et al. 2013a). Biochemical and structural studies using a catalytically inactive Exo10 mutant have shown that the footprint of an RNA bound to the complex spans ∼30 nt (Bonneau et al. 2009; Makino et al. 2013a). Data in yeast suggest that this internal channel serves as the major path used by exosome substrates in vivo (Schneider et al. 2012). Yeast Exo10 is largely similar in structure and function to its human counterpart, but with two important differences. First, in human cells there are two homologs of Rrp44—namely, DIS3 in the nucleus and DIS3L in the cytoplasm—resulting in compartment-specific variants of the exosome core (Staals et al. 2010; Tomecki et al. 2010). Second, current data suggest that the DIS3 ribonuclease may retain an open conformation both in the unbound state and in the RNA-bound state, resulting in a slightly longer RNA channel than denoted in yeast (Gerlach et al. 2018; Weick et al. 2018).

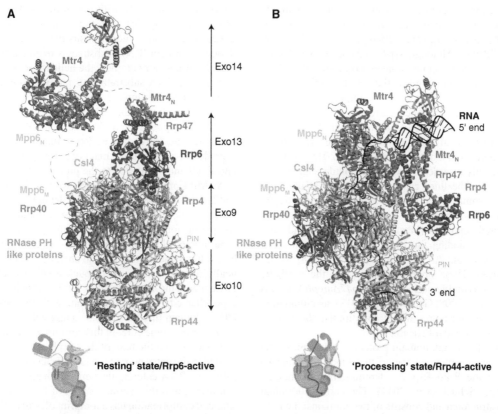

Figure 2. Architecture of the nuclear exosome holocomplex: Exo14n. (*A*) Composite structure of eukaryotic Exo13n (cartoon representation) in resting mode and in distributive degradation mode (e.g., Rrp6-mediated). The model is a superposition of several crystal structures: Exo10–Rrp6–Rrp47 (Makino et al. 2015), Exo9-Mpp6$_M$ (Falk et al. 2017a; Wasmuth et al. 2017), Rrp6–Rrp47–Mtr4$_N$ (Schuch et al. 2014), and the Rrp6-RNA exosome complex (Wasmuth et al. 2014). The helicase core of Mtr4 (Jackson et al. 2010; Weir et al. 2010) is shown in a hypothetical arrangement, linked to the Exo13n core by the interaction of the DExH core with Mpp6$_N$ (Schuller et al. 2018; Weick et al. 2018) and by the interaction of Mtr4$_N$ with the Rrp6$_N$–Rrp47$_N$ heterodimerization module (Schuch et al. 2014). N stands for amino-terminal domain; M for middle domain. The low-complexity carboxy-terminal sequences of Mpp6 and Rrp6 are flexible and not shown. Exo9: the upper ring of the S1/KH proteins is shown in warm colors (Csl4 in yellow, Rrp4 in orange and Rrp40 in salmon) and the lower ring of six RNase-PH like proteins in gray. The processive ribonuclease Rrp44 (light pink) is positioned at the lower ring of Exo9 (here in the open conformation without RNA). The nuclear cofactors Mpp6 (cyan), Rrp6 (red), Rrp47 (purple), and Mtr4 (blue) are at the top of the upper ring. A cartoon model of the nuclear exosome in the resting state/Rrp6 active state, corresponding to the cartoon model in Figure 5, is shown below. (*B*) Composite structure of eukaryotic Exo14n (cartoon representation) in processive mode (i.e., Rrp44-mediated). The model is based on the cryo-EM structure of a pre-60S-bound complex (Schuller et al. 2018), the crystal structure of RNA-bound Exo10-Rrp6$_C$ (Makino et al. 2013a), and cryo-EM maps of the human Exo14 (Gerlach et al. 2018; Weick et al. 2018). The RNA is threaded through Mtr4 into the exosome central channel and to the Rrp44 ribonuclease (in closed conformation). A cartoon model of the nuclear exosome in the Rrp44 active state, corresponding to the cartoon model in Figure 5, is shown *below*.

ARCHITECTURE OF THE RNA EXOSOME: THE NUCLEAR COFACTORS

In the yeast nucleus, Exo10 associates with four conserved cofactors with physical interactions (Schuch et al. 2014) that are functionally important (for review, see Butler and Mitchell 2011) to form the nuclear exosome holocomplex (Exo14n). Purifications of the endogenous complex from yeast indicate that in vivo three of them (the nonprocessive ribonuclease Rrp6 and the small proteins Rrp47 and Mpp6) form a stable assembly with Exo10 (nuclear Exo13 or Exo13n), whereas the fourth cofactor, the Mtr4 helicase, appears to be more weakly or transiently incorporated into the nuclear exosome holocomplex (Falk et al. 2017a). All nuclear cofactors bind Exo9 at the top of the S1/KH ring, essentially on the opposite side with respect to Rrp44 (Fig. 2; Wasmuth et al. 2014; Makino et al. 2015; Falk et al. 2017a; Wasmuth et al. 2017). Rrp6 is organized into several distinct domains, each with specific functions. The amino-terminal domain of Rrp6 heterodimerizes with Rrp47 (Schuch et al. 2014). The carboxy-terminal region wraps around the cap subunit Csl4, providing the primary anchor to Exo9 (Makino et al. 2013a) and also contains a nuclear localization signal (Callahan and Butler 2008). The central region binds and degrades RNAs via its HRDC and DEDD domains, respectively (Midtgaard et al. 2006; Schuch et al. 2014; Zinder et al. 2016). The Rrp6 exoribonuclease activity cleaves single nucleotides in the 3′–5′ direction in a distributive manner (i.e., dissociating from the RNA substrate after each round of cleavage). The difference between the distributive properties of Rrp6 and the processive properties of Rrp44 can be reconciled with the geometry of their active site: shallow and exposed to solvent in the case of Rrp6 and buried in an internal channel in case of Rrp44. Although Rrp44 is the general and ubiquitous degradation engine of the exosome complex, Rrp6 appears to have trimming functions in the nucleus.

When the nuclear exosome holocomplex is in a resting state, the central exoribonuclease region of Rrp6 binds flat on top of cap subunit Rrp4 (Wasmuth et al. 2014; Zinder et al. 2016) and is in turn surmounted by the Rrp6–Rrp47 heterodimerization module (Makino et al. 2015). Mpp6 is an intrinsically disordered protein that binds with its central region to the third cap subunit, Rrp40 (Falk et al. 2017a; Wasmuth et al. 2017). Mpp6 and the Rrp6–Rrp47 unit both contribute to recruiting RNA via their unstructured segments (Wasmuth et al. 2017) and both contribute to binding the Mtr4 helicase (Schuch et al. 2014; Falk et al. 2017a; Schuller et al. 2018; Weick et al. 2018). These observations explain the synthetic lethality of $rrp6\Delta mpp6\Delta$ and of $rrp47\Delta mpp6\Delta$ strains (Milligan et al. 2008; Garland et al. 2013), as eliminating both Mtr4-anchor points would compromise the recruitment of this essential helicase.

Mtr4 is also a multidomain protein. At the amino terminus is an unstructured region that binds the concave surface of the Rrp6–Rrp47 heterodimerization module (Fig. 2A,B; Schuch et al. 2014). The carboxy-terminal region is structured and consists of two domains: the helicase DExH domain and the arch (also known as SK) domain (Jackson et al. 2010; Weir et al. 2010). The DExH domain is catalytically active in unwinding RNA substrates in an ATP-dependent manner; RNA enters at the top of the DExH core and the unwound 3′ end exits at the base (Weir et al. 2010). The arch domain faces the entrance of the helicase channel at the top of the DExH core and contains a KOW domain capable of binding both single-stranded and structured RNAs (Jackson et al. 2010; Weir et al. 2010). In addition to RNA binding, both domains of Mtr4 serve as protein binding platforms: the DExH core, for example, binds the amino-terminal region of Mpp6 (Gerlach et al. 2018; Schuller et al. 2018; Weick et al. 2018), and the KOW domain binds so-called arch-interacting motifs (AIMs) of adaptor proteins, such as in the ribosomal biogenesis factor Nop53 (Thoms et al. 2015; Falk et al. 2017b; Lingaraju et al. 2019). The essential unwinding activity of Mtr4 likely serves to prepare substrates for threading into the narrow entry pore of the exosome core, which is only wide enough to accommodate single-stranded RNA substrates (Makino et al. 2013a).

MECHANISTIC MODEL OF THE EXOSOME IN ITS RNA DECAY FUNCTION

Based on structures determined through X-ray crystallography and cryo-electron microscopy (cryo-EM) as well as biochemical data (Makino et al. 2013a, 2015; Wasmuth et al. 2014, 2017; Zinder et al. 2016; Falk et al. 2017a; Gerlach et al. 2018; Weick et al. 2018), we can propose the molecular mechanisms with which the nuclear exosome processively degrades and thus eliminates RNA substrates. The working model posits that when an RNA or ribonucleoprotein (RNP) substrate is recruited to Mtr4 or to an Mtr4-adaptor complex, the helicase activity of Mtr4 unwinds the RNA substrate, thus progressively extruding the unwound 3′ end toward the top of Exo13n, where Mtr4 is loosely positioned through its interactions with Rrp6–Rrp47 and Mpp6. In Exo13n, the RNA 3′ end will first encounter the active site of Rrp6 and stochastically be degraded in a distributive manner before getting threaded into Exo9 and channeled to the processive site of Rrp44.

As Mtr4 productively engages with an RNA substrate and Exo14n reaches its processive degradation mode, the helicase core displaces the nuclease domain of Rrp6 by competing for the same binding site on Exo10, the cap subunit Rrp4 (Fig. 2B). The amino-terminal Rrp6–Rrp47 heterodimerization module also changes its position drastically, as it detaches from the nuclease domain of Rrp6 to interact with the arch domain of Mtr4. The helicase adopts a peculiar edge-on conformation, with the base of the DExH domain tilted at an ~45° angle with respect to the top of the exosome core. In this conformation, the RNA 3′ end exiting from the base of the DExH domain of Mtr4 is guided to the entry pore of the Exo10 channel. The overall path of an RNA entering the helicase channel of Mtr4 and continuing into the central channel of Exo10 in a single-stranded conformation has a footprint of ~40 nt (Falk et al. 2017a). The processive mode of Exo14n results in a final product of 4 nt (i.e., in the degradation of almost the entire transcript).

Interestingly, the Mtr4 conformation on top of the exosome core and the threading RNA observed not only in structures of the yeast complex (Schuller et al. 2018) but also in the human complex (Gerlach et al. 2018; Weick et al. 2018) suggest the mechanism and conformational regulation are evolutionarily conserved aspects of the nuclear exosome. The proposed mechanism is expected to lead to complete decay of any substrate unwound by Mtr4. Effectively, this appears to be the case for most RNA substrates, including structured transcripts such as tRNAs, with the notable exception of the 5.8S rRNA during pre-60S biogenesis.

CONSTRUCTIVE FUNCTION OF THE RNA EXOSOME IN RIBOSOMAL RNA PROCESSING

During the biogenesis and maturation of the pre-60S subunit, the nuclear exosome recognizes a massive ribonucleoprotein particle at a specific maturation stage, remodels it, and dissociates when the rRNA has been trimmed to a precise point. The exosome specifically recognizes the pre-60S particles at the 7S rRNA maturation stage. The 7S rRNA is a 5.8S rRNA precursor with a 3′-end extension of ~140 nt that is generated by an upstream endonucleolytic cleavage in the internal transcribed spacer 2 (ITS2) (Thomson et al. 2013; Woolford and Baserga 2013; Gasse et al. 2015; Turowski and Tollervey 2015). The 7S rRNA is trimmed by the sequential action of the two nuclear exosome ribonucleases, Rrp44 and Rrp6 (Fig. 3). In the first step, Rrp44 shortens the 7S rRNA to form a 5.8S precursor carrying a 30-nt 3′ extension (5.8S+30) (Briggs et al. 1998). This intermediate is then shortened by Rrp6 to form the 6S rRNA (a 5.8S with a 6- to 8-nt 3′ extension), which is then exported to the cytoplasm for the final trimming (Thomson and Tollervey 2010). Notably, the +30-nt RNA processing defect of 5.8S rRNA in Rrp6Δ strains (Briggs et al. 1998) matches the footprint of the 30-nt path of Exo10 in vitro (Bonneau et al. 2009; Makino et al. 2013a). Human cells have a similar 5.8S+40 intermediate (Tafforeau et al. 2013), again consistent with the longer footprint of the human exosome core in vitro (Gerlach et al. 2018).

The specificity with which the exosome ribonucleases recognize different pre-rRNA substrates and the accuracy with which they stop degrading at specific positions in the transcripts have long remained a mystery, particularly because the end points do not correlate with the presence of structured elements in the rRNA. An important aspect of this process is that the rRNA precursors are not trimmed in isolation, but in the context of RNPs that also contain ribosomal proteins and ribosome assembly factors. Cryo-EM reconstructions of pre-60S biogenesis intermediate at the 7S maturation stage have shown how the ITS2 RNP forms the so-called "foot structure" of the pre-60S. The "foot structure" contains ∼60 nt of ITS2 folded into an intertwined RNA structure bound by several ribosomal assembly factors, including Nop53 and Nop7 (Fig. 4A; Wu et al. 2016). The remaining ∼80 nt of ITS2 is presumably extended into solvent as is the amino-terminal region of Nop53, which has been shown to recruit the Mtr4 helicase (Thoms et al. 2015; Falk et al. 2017b; Fromm et al. 2017). Thus, the two elements of the pre-60S particle that are recognized by the RNA exosome (the 7S 3′ end and amino terminus of Nop53) are exposed to solvent and accessible to the RNA processing machinery.

MECHANISTIC MODEL OF THE NUCLEAR EXOSOME IN ITS RNA PROCESSING FUNCTION

In light of the published data, we propose the following model. We envision an initial recognition step in the precise processing of 5.8S rRNA precursor where the Mtr4 arch domain binds the flexible amino-terminal region of Nop53 (Thoms et al. 2015; Wu et al. 2016) and threads the accessible 3′ RNA extension of ITS2 into the helicase channel. Once Mtr4 is loaded with its physiological RNA substrate, it can "hand over" the RNA to the Rrp44 activity in the exosome (Figs. 4B and 5-II), presumably with a similar sequence of steps described for the processive degradative activity above. The main difference is that at this stage the 7S rRNA is likely to bypass the Rrp6 active site altogether: the endonucleolytic cleavage that generates the 7S rRNA leaves a 2′,3′-cyclic phosphate at the 3′ end of the exosome substrate (Gasse et al. 2015), and such a modified 3′ end is unlikely to fit the geometry of the Rrp6 active site (our unpubl. observations). In contrast, the active site of Rrp44 is larger and can, in principle, accommodate a 2′,3′-cyclic phosphate (Meaux and van Hoof 2006; Zinder et al. 2016). As the 3′ end of the 7S rRNA reaches the active site of Rrp44 and

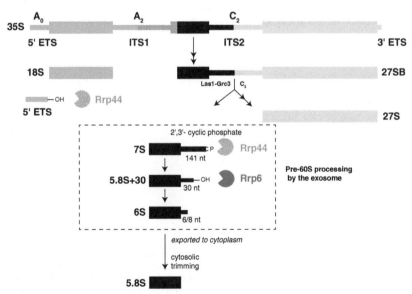

Figure 3. Scheme of yeast rRNA processing. The ribosomal pre-rRNA is transcribed as a single long polycistronic transcript, the 35S rRNA precursor that contains the 5′-ETS region (light orange-thin rectangle), the 18S rRNA (light orange-thick rectangle), the ITS1 (internal transcribed spacer region), 5.8S rRNA (black), ITS2, 25S rRNA (light blue), and the 3′ ETS. The 5′-ETS sequence after cleavage at the A_0 site is degraded in an Mtr4-dependent manner by the nuclear RNA exosome. For 5.8S maturation, the 27SB pre-rRNA is cleaved at site C2 by the Las1 endonuclease complex, yielding a 3′ RNA with a 2′,3′-cyclic phosphate. The resulting product is first processed by the processive action of Rrp44 (pink) (degrades about 110 nt) to yield the 5.8S+30 intermediate, followed by the distributive action of the Rrp6 (red) ribonuclease. (The scheme is adapted from Schuller et al. 2018.)

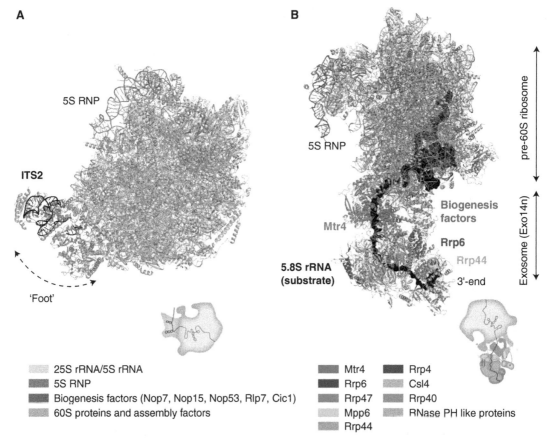

Figure 4. Nuclear exosome in preribosome processing. (*A*) Cartoon representation of the cryo-EM structure of a 7S pre-60S substrate (Wu et al. 2016) showing the foot structure that is remodeled by the exosome. (*B*) Cartoon representation of the cryo-EM structure of Exo14n determined by "immobilizing" the complex on a 7S pre-60S substrate at the 5.8S + 30 intermediate state using an Rrp6 active site mutant (Schuller et al. 2018). The pre-60S is shown with ribosomal proteins in wheat, the 5.8S rRNA in black, the 25S rRNA and 5S rRNAs in gray, ribosomal biogenesis factors in green, and the exosome components are colored similarly as in Figure 2. *Below* each panel are the corresponding cartoon models as they appear in Figure 5.

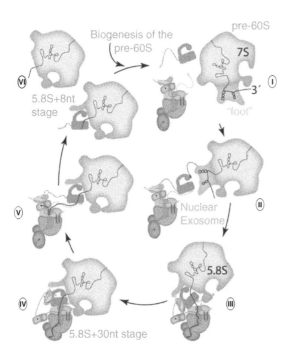

Figure 5. Model of exosome-mediated pre-60S particle processing. Schematic representation depicting the compositional and conformational rearrangements during 5.8S rRNA maturation. The steps show the recruitment to the 7S particle via the Nop53–Mtr4 interaction (I–II); the processive degradation step, resulting in remodeling of the ITS2 and associated factors (the "foot" structure, II–III); the 5.8S +30 particle as a transient state where Mtr4 adopts a strained high-energy position on Exo9 (IV); the distributive trimming step after Mtr4 releases the RNA from the exosome channel (V). In this model, Rrp6 can degrade the 5.8S extension up to 6–8 nt (6S rRNA), which is then exported to the cytoplasm.

starts to be processively degraded, Mtr4 can unwind the rest of ITS2.

Almost all the ribosome biogenesis factors in the foot structure are removed at this stage (Fromm et al. 2017; Schuller et al. 2018), including Nop53, the factor that initially recruited Mtr4 (Thoms et al. 2015). Mtr4, however, forms other contacts with the pre-60S as the foot structure is remodeled: the KOW domain and the DExH core bind the 25s rRNA at domain I and domain V, respectively (Schuller et al. 2018). The only biogenesis factor to remain in the foot structure at this stage is Nop7, which engages in an interaction with the carboxy-terminal helix of Rrp47 (Schuller et al. 2018) (Fig. 4B). Exo14n is expected to degrade the ITS2 extension from 110 to ∼40 nt in the processive conformation, generating a 3′-OH end that is compatible with both Rrp6 and Rrp44 (Fig. 5-III). However, further degradation by Rrp44 is predicted to result in substantial deformation. The sheer mass of the pre-60S would likely compress Exo14n and, in particular, Mtr4 from its resting edge-on conformation to a strained flat-on conformation on the exosome core (Fig. 5-IV). Based on in vivo data (Briggs et al. 1998), the strain on Exo14n will likely reach its limit at 30 nt, when further compaction is physically not possible. At this point, we speculate that the strained arch domain may flip the helicase from the top of Exo13n, simultaneously ejecting the remaining 30 nt of the 5.8S rRNA extension from the exosome channel and rendering them accessible to Rrp6 (Fig. 5-V).

We envision that in the distributive step, Rrp6 is repositioned on top of the exosome core (Zinder et al. 2016). Degradation by Rrp6 at the last 6–8 nt will likely stop upon encountering a physical block, possibly by clashing against the helicase Mtr4 that could still be docked onto the ribosome and bound to the 5.8S rRNA extension at this step. The model implicates that Mtr4 might have to be actively removed from the large ribosomal subunit and the 3′ end before it is exported to the cytoplasm (Fig. 5-VI). How this could be achieved remains unclear, although it is possible that it involves the action of one of the AAA$^+$-ATPases that are involved in remodeling steps of the pre-ribosomal particles.

CONCLUSION

The biochemical and structural studies to date support the notion proposed a decade ago (Bonneau et al. 2009) that the exosome core functions as a macromolecular cage to channel RNA substrates for degradation. The entrance to the cage is gated by cofactors that either limit or grant access to the central degradation channel. In its resting state in the nucleus, Rrp6 regulates the access to the processive ribonuclease of the complex. In the processive degradation mode, when Mtr4 is loaded with an RNA, the ATPase displaces the negative regulator with a large conformational change and injects the substrate into the processive degradation channel. This central channel of the exosome is used for both RNA decay and for 5.8S rRNA processing. The difference between complete and partial degradation resides in the physical constraints imposed by the substrate, a very large and complex RNP particle in the case of 5.8S rRNA processing. Looking back, the exosome was thus discovered (Mitchell et al. 1997) by studying the exception (partial degradation and processing) to the rule (complete degradation and decay). Looking forward, we expect that the exosome may integrate additional layers of negative regulation to avoid unleashing uncontrolled degradation. Given the crucial roles of Mtr4 in providing the substrate remodeling activity and in providing a hub for interacting adaptor proteins, it is reasonable to expect regulation at this pivotal intersection to target RNA substrate to the RNA exosome. Unraveling exosome helicase regulation will be an important avenue of future research.

ACKNOWLEDGMENTS

We are grateful to Courtney Long and members of the group for input and discussion on the manuscript. We also thank Sandra Schuller for the models and the drawings in Figure 5. This study was supported by funding from the Max-Planck-Gesellschaft, the European Commission (ERC Advanced Investigator Grant EXORICO), and the German Research Foundation (DFG SFB1035, GRK1721, SFB/TRR 237) to E.C. and a Boehringer Ingelheim Fonds fellowship to M.L.

REFERENCES

Allmang C, Kufel J, Chanfreau G, Mitchell P, Petfalski E, Tollervey D. 1999. Functions of the exosome in rRNA, snoRNA and snRNA synthesis. *EMBO J* **18:** 5399–5410. doi:10.1093/emboj/18.19.5399

Bonneau F, Basquin J, Ebert J, Lorentzen E, Conti E. 2009. The yeast exosome functions as a macromolecular cage to channel RNA substrates for degradation. *Cell* **139:** 547–559. doi:10.1016/j.cell.2009.08.042

Briggs MW, Burkard KT, Butler JS. 1998. Rrp6p, the yeast homologue of the human PM-Scl 100-kDa autoantigen, is essential for efficient 5.8 S rRNA 3′ end formation. *J Biol Chem* **273:** 13255–13263. doi:10.1074/jbc.273.21.13255

Butler JS, Mitchell P. 2011. Rrp6, rrp47 and cofactors of the nuclear exosome. *Adv Exp Med Biol* **702:** 91–104. doi:10.1007/978-1-4419-7841-7_8

Callahan KP, Butler JS. 2008. Evidence for core exosome independent function of the nuclear exoribonuclease Rrp6p. *Nucleic Acids Res* **36:** 6645–6655. doi:10.1093/nar/gkn743

Chlebowski A, Lubas M, Jensen TH, Dziembowski A. 2013. RNA decay machines: the exosome. *Biochim Biophys Acta* **1829:** 552–560. doi:10.1016/j.bbagrm.2013.01.006

Dziembowski A, Lorentzen E, Conti E, Séraphin B. 2007. A single subunit, Dis3, is essentially responsible for yeast exosome core activity. *Nat Struct Mol Biol* **14:** 15–22. doi:10.1038/nsmb1184

Falk S, Bonneau F, Ebert J, Kogel A, Conti E. 2017a. Mpp6 incorporation in the nuclear exosome contributes to RNA channeling through the Mtr4 helicase. *Cell Rep* **20:** 2279–2286. doi:10.1016/j.celrep.2017.08.033

Falk S, Tants JN, Basquin J, Thoms M, Hurt E, Sattler M, Conti E. 2017b. Structural insights into the interaction of the nuclear exosome helicase Mtr4 with the preribosomal protein Nop53. *RNA* **23:** 1780–1787. doi:10.1261/rna.062901.117

Fasken MB, Morton DJ, Kuiper EG, Jones SK, Leung SW, Corbett AH. 2020. The RNA exosome and human disease. *Methods Mol Biol* **2062:** 3–33. doi:10.1007/978-1-4939-9822-7_1

Frazao C, McVey CE, Amblar M, Barbas A, Vonrhein C, Arraiano CM, Carrondo MA. 2006. Unravelling the dynamics of RNA degradation by ribonuclease II and its RNA-bound complex. *Nature* **443**: 110–114. doi:10.1038/nature05080

Fromm L, Falk S, Flemming D, Schuller JM, Thoms M, Conti E, Hurt E. 2017. Reconstitution of the complete pathway of ITS2 processing at the pre-ribosome. *Nat Commun* **8**: 1787. doi:10.1038/s41467-017-01786-9

Garland W, Feigenbutz M, Turner M, Mitchell P. 2013. Rrp47 functions in RNA surveillance and stable RNA processing when divorced from the exoribonuclease and exosome-binding domains of Rrp6. *RNA* **19**: 1659–1668. doi:10.1261/rna.039388.113

Gasse L, Flemming D, Hurt E. 2015. Coordinated ribosomal ITS2 RNA processing by the Las1 complex integrating endonuclease, polynucleotide kinase, and exonuclease activities. *Mol Cell* **60**: 808–815. doi:10.1016/j.molcel.2015.10.021

Gerlach P, Schuller JM, Bonneau F, Basquin J, Reichelt P, Falk S, Conti E. 2018. Distinct and evolutionary conserved structural features of the human nuclear exosome complex. *eLife* **7**: e38686. doi:10.7554/eLife.38686

Gudipati RK, Xu Z, Lebreton A, Séraphin B, Steinmetz LM, Jacquier A, Libri D. 2012. Extensive degradation of RNA precursors by the exosome in wild-type cells. *Mol Cell* **48**: 409–421. doi:10.1016/j.molcel.2012.08.018

Han J, van Hoof A. 2016. The RNA exosome channeling and direct access conformations have distinct in vivo functions. *Cell Rep* **16**: 3348–3358. doi:10.1016/j.celrep.2016.08.059

Jackson RN, Klauer AA, Hintze BJ, Robinson H, van Hoof A, Johnson SJ. 2010. The crystal structure of Mtr4 reveals a novel arch domain required for rRNA processing. *EMBO J* **29**: 2205–2216. doi:10.1038/emboj.2010.107

Kadaba S, Krueger A, Trice T, Krecic AM, Hinnebusch AG, Anderson J. 2004. Nuclear surveillance and degradation of hypomodified initiator tRNA^{Met} in *S. cerevisiae*. *Genes Dev* **18**: 1227–1240. doi:10.1101/gad.1183804

Kowalinski E, Kogel A, Ebert J, Reichelt P, Stegmann E, Habermann B, Conti E. 2016. Structure of a cytoplasmic 11-subunit RNA exosome complex. *Mol Cell* **63**: 125–134. doi:10.1016/j.molcel.2016.05.028

LaCava J, Houseley J, Saveanu C, Petfalski E, Thompson E, Jacquier A, Tollervey D. 2005. RNA degradation by the exosome is promoted by a nuclear polyadenylation complex. *Cell* **121**: 713–724. doi:10.1016/j.cell.2005.04.029

Lebreton A, Tomecki R, Dziembowski A, Séraphin B. 2008. Endonucleolytic RNA cleavage by a eukaryotic exosome. *Nature* **456**: 993–996. doi:10.1038/nature07480

Lingaraju M, Johnsen D, Schlundt A, Langer LM, Basquin J, Sattler M, Heick Jensen T, Falk S, Conti E. 2019. The MTR4 helicase recruits nuclear adaptors of the human RNA exosome using distinct arch-interacting motifs. *Nat Commun* **10**: 3393. doi:10.1038/s41467-019-11339-x

Liu Q, Greimann JC, Lima CD. 2006. Reconstitution, activities, and structure of the eukaryotic RNA exosome. *Cell* **127**: 1223–1237. doi:10.1016/j.cell.2006.10.037

Liu JJ, Niu CY, Wu Y, Tan D, Wang Y, Ye MD, Liu Y, Zhao W, Zhou K, Liu QS, et al. 2016. CryoEM structure of yeast cytoplasmic exosome complex. *Cell Res* **26**: 822–837. doi:10.1038/cr.2016.56

Lorentzen E, Conti E. 2006. The exosome and the proteasome: nano-compartments for degradation. *Cell* **125**: 651–654. doi:10.1016/j.cell.2006.05.002

Lorentzen E, Dziembowski A, Lindner D, Séraphin B, Conti E. 2007. RNA channelling by the archaeal exosome. *EMBO Rep* **8**: 470–476. doi:10.1038/sj.embor.7400945

Lorentzen E, Basquin J, Tomecki R, Dziembowski A, Conti E. 2008. Structure of the active subunit of the yeast exosome core, Rrp44: diverse modes of substrate recruitment in the RNase II nuclease family. *Mol Cell* **29**: 717–728. doi:10.1016/j.molcel.2008.02.018

Makino DL, Baumgartner M, Conti E. 2013a. Crystal structure of an RNA-bound 11-subunit eukaryotic exosome complex. *Nature* **495**: 70–75. doi:10.1038/nature11870

Makino DL, Halbach F, Conti E. 2013b. The RNA exosome and proteasome: common principles of degradation control. *Nat Rev Mol Cell Biol* **14**: 654–660. doi:10.1038/nrm3657

Makino DL, Schuch B, Stegmann E, Baumgartner M, Basquin C, Conti E. 2015. RNA degradation paths in a 12-subunit nuclear exosome complex. *Nature* **524**: 54–58. doi:10.1038/nature14865

Meaux S, van Hoof A. 2006. Yeast transcripts cleaved by an internal ribozyme provide new insight into the role of the cap and poly(A) tail in translation and mRNA decay. *RNA* **12**: 1323–1337. doi:10.1261/rna.46306

Midtgaard SF, Assenholt J, Jonstrup AT, Van LB, Jensen TH, Brodersen DE. 2006. Structure of the nuclear exosome component Rrp6p reveals an interplay between the active site and the HRDC domain. *Proc Natl Acad Sci* **103**: 11898–11903. doi:10.1073/pnas.0604731103

Milligan L, Decourty L, Saveanu C, Rappsilber J, Ceulemans H, Jacquier A, Tollervey D. 2008. A yeast exosome cofactor, Mpp6, functions in RNA surveillance and in the degradation of noncoding RNA transcripts. *Mol Cell Biol* **28**: 5446–5457. doi:10.1128/MCB.00463-08

Mitchell P, Petfalski E, Tollervey D. 1996. The 3′ end of yeast 5.8S rRNA is generated by an exonuclease processing mechanism. *Genes Dev* **10**: 502–513. doi:10.1101/gad.10.4.502

Mitchell P, Petfalski E, Shevchenko A, Mann M, Tollervey D. 1997. The exosome: a conserved eukaryotic RNA processing complex containing multiple 3′→5′ exoribonucleases. *Cell* **91**: 457–466. doi:10.1016/S0092-8674(00)80432-8

Preker P, Nielsen J, Kammler S, Lykke-Andersen S, Christensen MS, Mapendano CK, Schierup MH, Jensen TH. 2008. RNA exosome depletion reveals transcription upstream of active human promoters. *Science* **322**: 1851–1854. doi:10.1126/science.1164096

Puno MR, Weick EM, Das M, Lima CD. 2019. SnapShot: the RNA exosome. *Cell* **179**: 282.e1. doi:10.1016/j.cell.2019.09.005

Schaeffer D, Tsanova B, Barbas A, Reis FP, Dastidar EG, Sanchez-Rotunno M, Arraiano CM, van Hoof A. 2009. The exosome contains domains with specific endoribonuclease, exoribonuclease and cytoplasmic mRNA decay activities. *Nat Struct Mol Biol* **16**: 56–62. doi:10.1038/nsmb.1528

Schaeffer D, Clark A, Klauer AA, Tsanova B, van Hoof A. 2011. Functions of the cytoplasmic exosome. *Adv Exp Med Biol* **702**: 79–90. doi:10.1007/978-1-4419-7841-7_7

Schmid M, Jensen TH. 2019. The nuclear RNA exosome and its cofactors. *Adv Exp Med Biol* **1203**: 113–132. doi:10.1007/978-3-030-31434-7_4

Schneider C, Leung E, Brown J, Tollervey D. 2009. The N-terminal PIN domain of the exosome subunit Rrp44 harbors endonuclease activity and tethers Rrp44 to the yeast core exosome. *Nucleic Acids Res* **37**: 1127–1140. doi:10.1093/nar/gkn1020

Schneider C, Kudla G, Wlotzka W, Tuck A, Tollervey D. 2012. Transcriptome-wide analysis of exosome targets. *Mol Cell* **48**: 422–433. doi:10.1016/j.molcel.2012.08.013

Schuch B, Feigenbutz M, Makino DL, Falk S, Basquin C, Mitchell P, Conti E. 2014. The exosome-binding factors Rrp6 and Rrp47 form a composite surface for recruiting the Mtr4 helicase. *EMBO J* **33**: 2829–2846. doi:10.15252/embj.201488757

Schuller JM, Falk S, Fromm L, Hurt E, Conti E. 2018. Structure of the nuclear exosome captured on a maturing preribosome. *Science* **360**: 219–222. doi:10.1126/science.aar5428

Staals RH, Pruijn GJ. 2011. The human exosome and disease. *Adv Exp Med Biol* **702**: 132–142. doi:10.1007/978-1-4419-7841-7_11

Staals RH, Bronkhorst AW, Schilders G, Slomovic S, Schuster G, Heck AJ, Raijmakers R, Pruijn GJ. 2010. Dis3-like 1: a novel exoribonuclease associated with the human exosome. *EMBO J* **29**: 2358–2367. doi:10.1038/emboj.2010.122

Tafforeau L, Zorbas C, Langhendries JL, Mullineux ST, Stamatopoulou V, Mullier R, Wacheul L, Lafontaine DL. 2013. The complexity of human ribosome biogenesis revealed by systematic nucleolar screening of pre-rRNA processing factors. *Mol Cell* **51:** 539–551. doi:10.1016/j.molcel.2013.08.011

Thoms M, Thomson E, Bassler J, Gnadig M, Griesel S, Hurt E. 2015. The exosome is recruited to RNA substrates through specific adaptor proteins. *Cell* **162:** 1029–1038. doi:10.1016/j.cell.2015.07.060

Thomson E, Tollervey D. 2010. The final step in 5.8S rRNA processing is cytoplasmic in *Saccharomyces cerevisiae*. *Mol Cell Biol* **30:** 976–984. doi:10.1128/MCB.01359-09

Thomson E, Ferreira-Cerca S, Hurt E. 2013. Eukaryotic ribosome biogenesis at a glance. *J Cell Sci* **126:** 4815–4821. doi:10.1242/jcs.111948

Tomecki R, Kristiansen MS, Lykke-Andersen S, Chlebowski A, Larsen KM, Szczesny RJ, Drazkowska K, Pastula A, Andersen JS, Stepien PP, et al. 2010. The human core exosome interacts with differentially localized processive RNases: hDIS3 and hDIS3L. *EMBO J* **29:** 2342–2357. doi:10.1038/emboj.2010.121

Turowski TW, Tollervey D. 2015. Cotranscriptional events in eukaryotic ribosome synthesis. *Wiley Interdiscip Rev RNA* **6:** 129–139. doi:10.1002/wrna.1263

van Hoof A, Parker R. 1999. The exosome: a proteasome for RNA? *Cell* **99:** 347–350. doi:10.1016/S0092-8674(00)81520-2

Wasmuth EV, Januszyk K, Lima CD. 2014. Structure of an Rrp6-RNA exosome complex bound to poly(A) RNA. *Nature* **511:** 435–439. doi:10.1038/nature13406

Wasmuth EV, Zinder JC, Zattas D, Das M, Lima CD. 2017. Structure and reconstitution of yeast Mpp6-nuclear exosome complexes reveals that Mpp6 stimulates RNA decay and recruits the Mtr4 helicase. *eLife* **6:** e29062. doi:10.7554/eLife.29062

Weick EM, Puno MR, Januszyk K, Zinder JC, DiMattia MA, Lima CD. 2018. Helicase-dependent RNA decay illuminated by a cryo-EM structure of a human nuclear RNA exosome-MTR4 complex. *Cell* **173:** 1663–1677 e1621. doi:10.1016/j.cell.2018.05.041

Weir JR, Bonneau F, Hentschel J, Conti E. 2010. Structural analysis reveals the characteristic features of Mtr4, a DExH helicase involved in nuclear RNA processing and surveillance. *Proc Natl Acad Sci* **107:** 12139–12144. doi:10.1073/pnas.1004953107

Woolford JL Jr, Baserga SJ. 2013. Ribosome biogenesis in the yeast *Saccharomyces cerevisiae*. *Genetics* **195:** 643–681. doi:10.1534/genetics.113.153197

Wu S, Tutuncuoglu B, Yan K, Brown H, Zhang Y, Tan D, Gamalinda M, Yuan Y, Li Z, Jakovljevic J, et al. 2016. Diverse roles of assembly factors revealed by structures of late nuclear pre-60S ribosomes. *Nature* **534:** 133–137. doi:10.1038/nature17942

Wyers F, Rougemaille M, Badis G, Rousselle JC, Dufour ME, Boulay J, Regnault B, Devaux F, Namane A, Séraphin B, et al. 2005. Cryptic pol II transcripts are degraded by a nuclear quality control pathway involving a new poly(A) polymerase. *Cell* **121:** 725–737. doi:10.1016/j.cell.2005.04.030

Zinder JC, Lima CD. 2017. Targeting RNA for processing or destruction by the eukaryotic RNA exosome and its cofactors. *Genes Dev* **31:** 88–100. doi:10.1101/gad.294769.116

Zinder JC, Wasmuth EV, Lima CD. 2016. Nuclear RNA exosome at 3.1 Å reveals substrate specificities, RNA paths, and allosteric inhibition of Rrp44/Dis3. *Mol Cell* **64:** 734–745. doi:10.1016/j.molcel.2016.09.038

Long Noncoding RNAs in Development and Regeneration of the Neural Lineage

HADAS HEZRONI, ROTEM BEN TOV PERRY, AND IGOR ULITSKY

Department of Biological Regulation, Weizmann Institute of Science, Rehovot 76100, Israel

Correspondence: igor.ulitsky@weizmann.ac.il

Long noncoding RNAs (lncRNAs) are gathering increasing attention toward their roles in different biological systems. In mammals, the richest repertoires of lncRNAs are expressed in the brain and in the testis, and the diversity of lncRNAs in the nervous system is thought to be related to the diversity and the complexity of its cell types. Supporting this notion, many lncRNAs are differentially expressed between different regions of the brain or in particular cell types, and many lncRNAs are dynamically expressed during embryonic or postnatal neurogenesis. Less is known about the functions of these genes, if any, but they are increasingly implicated in diverse processes in health and disease. Here, we review the current knowledge about the roles and importance of lncRNAs in the central and peripheral nervous systems and discuss the specific niches within gene regulatory networks that might be preferentially occupied by lncRNAs.

Tens of thousands of loci in mammalian genomes produce long noncoding RNAs (lncRNAs), which are generally less abundant (by roughly an order of magnitude) and much more tissue-specific than protein-coding genes. A typical mammalian cell type expresses (at one transcript per cell, on average) at least ~50% of all protein-coding genes, but only approximately 1000 distinct lncRNAs (<5% of the currently annotated lncRNAs). Most lncRNAs are produced from regions distal to other genes, but some, enriched with the more highly and broadly expressed ones, are produced in close proximity to promoters of other genes (most commonly in a divergent orientation) or from other "genic" regions. Another subset of lncRNAs, which is also rather abundant, serves as precursors for small RNAs, such as microRNAs and small nucleolar RNAs (snoRNAs), and the interplay between the two classes has been reviewed elsewhere (Ulitsky 2018).

Most lncRNAs are spliced, and produce "linear" isoforms that start with a cap and end with a poly(A) tail, which help stabilize the RNA, but there is an increasing appreciation of the diversity of circular RNAs (circRNAs), which are lncRNAs produced by backsplicing (Ebbesen et al. 2016). We will focus here only on linear RNA forms, as functions of the circular forms in the neural system were recently reviewed (Sekar and Liang 2019).

Studies describing where lncRNAs are expressed, utilizing techniques used to measure mRNA expression, have overall progressed rapidly. There is also an increasing number of reports on what lncRNAs do in the nervous system, described below. It is overall much less clear how lncRNAs function, but various mechanisms have been proposed for both *cis* and *trans* actions. Overall, there are no reasons to think that the mechanisms utilized by lncRNAs in the nervous system are different than those used in other tissues, but there are so far few similarities between mechanisms proposed for different lncRNAs (for a recent review, see Kopp and Mendell 2018), and so transfer of mechanistic understanding from one gene to others remains difficult.

There are several reasons to be interested in what lncRNAs do specifically in the nervous system. One is that the brain expresses a particularly rich repertoire of lncRNAs, rivaled only by that expressed in germ cells. Together with the rich cell type diversity of the nervous system, it is possible that lncRNAs often regulate cell fate decisions in the brain, and so neuronal systems can be particularly appealing for dissecting such functions. Another reason is the common diseases that affect the brain, and in which lncRNAs are commonly dysregulated (see below). From this perspective, lncRNAs can act as interesting biomarkers, potential therapeutic targets, or even potential therapeutic agents (although delivery of large RNAs remains a formidable challenge). As biomarkers, lncRNAs can be particularly appealing because of their tissue specificity, although the low abundance of the vast majority of lncRNAs will likely make it difficult to robustly detect them in distal tissues (e.g., detect in the blood lncRNAs arising from neural cells in the brain).

DIFFERENTIAL EXPRESSION OF lncRNAs IN THE BRAIN AND DURING NEURAL DIFFERENTIATION

Several large-scale resources of gene expression are now available and describe where lncRNAs are expressed at various resolutions (Table 1). Both human and mouse brains express a relatively rich repertoire of lncRNAs (Ravasi et al. 2006; Cabili et al. 2011; Derrien et al. 2012), many of which display unique temporal and spatial expression patterns within the central nervous system

Table 1. Large-scale resources of long noncoding RNA (lncRNA) expression in the nervous system

Name	URL	Tissues	Methodology	lncRNA annotations
GTEx	https://gtexportal.org/home/index.html	54 human tissues, including 13 brain regions	RNA-seq	GENCODE/Ensebml
Allen brain atlas	https://portal.brain-map.org/	Various	RNA-seq, single-cell RNA-seq, and FISH in human and mouse fetal and adult brain	RefSeq
EBI expression atlas	https://www.ebi.ac.uk/gxa/home	Various studies, including atlases of tissues and cell lines and differential expression	RNA-seq	GENCODE/Ensembl
ENCODE	https://www.encodeproject.org/	Mouse tissues (mostly adult) and human and mouse cell lines	RNA-seq	GENCODE
Eurexpress	http://www.eurexpress.org/ee	Mouse in situ images	FISH	Various (mapped to FISH probes)
Kaessmann laboratory Evo-Devo resource	https://apps.kaessmannlab.org/lncRNA_app/	Developmental time course in 7 tissues in 7 species	RNA-seq	De novo reconstruction based on RNA-seq data

(CNS) and during neuronal and glial cell differentiation (Mercer et al. 2008, 2010; Ramos et al. 2013; Goff et al. 2015). The genome-wide expression patterns of mouse lncRNAs have been studied in cells isolated from different brain tissues, in neural stem cells (NSCs) isolated from the subventricular zone (SVZ) and differentiated in vitro for four days, and in embryonic stem cells (ESCs) differentiated into neural progenitor cells (NPCs) (Ramos et al. 2013). Cells in different regions of the brain, as well as cells at different stages of differentiation, had distinct lncRNA expression patterns. Mean expression specificity score in these data sets was 0.57 for lncRNAs compared to 0.45 for mRNAs, demonstrating that lncRNAs exhibit greater brain region and temporal specificity than mRNAs (Ramos et al. 2013). Another study generated a map of expression dynamics and regulatory effects of 13 lncRNA loci in mouse brain, using mouse models where the lncRNA locus was replaced with a LacZ cassette, and found dynamic and broad expression profiles in the brain, with some lncRNAs presenting highly restricted expression in few brain regions (Goff et al. 2015). These data suggest that some lncRNAs have the potential to carry out specific functions in different types of neural cells and during different stages of acquisition of a neuronal cell identity.

SUBCELLULAR EXPRESSION OF lncRNAs

The complex morphology of cells in the nervous system opens the question of *where* within the neuronal cells lncRNAs preferentially reside and act. For individual lncRNAs this is typically addressed using fluorescence in situ hybridization (FISH), recently with single-molecule approaches, which found highly variable localization patterns for lncRNAs, ranging from a single focal region in the nucleus to mRNA-like distribution in the cytoplasm (Cabili et al. 2015). More systematically, physical fractionation of cells followed by high-throughput sequencing can be used to quantify subcellular localization of all transcripts (Sterne-Weiler et al. 2013). In neurons, it is of particular interest to ask whether some lncRNAs transit to the neurites, axons, or dendrites, as there they can have unique and specific functions in neuronal pathways. Using mouse ESC (mESC)-derived neurons and RNA-seq on neurite and soma fractions, the Chekulaeva laboratory recently found that 12 of the 550 tested lncRNAs were at least twofold enriched in neurites (Zappulo et al. 2017) (~2% compared to ~7% of mRNAs using similar criteria), suggesting that at least in this cell type, a modest fraction of lncRNAs travels to neurites. Additional studies identified specific lncRNAs in axonal fractions (Briese et al. 2016) or sequenced RNA from axons or dendrites but did not specifically describe lncRNAs (Minis et al. 2014; Nijssen et al. 2018; Farris et al. 2019). lncRNA-centric analysis of these and new data will likely shed light on how prevalent are lncRNAs in neuronal processes.

METHODS FOR STUDYING lncRNA FUNCTIONS IN THE NERVOUS SYSTEM

Whereas the approaches described above are useful for describing *where* lncRNAs may act, understanding the functions of lncRNAs requires perturbations, which are generally difficult in the nervous system. Challenges include cell accessibility, the postmitotic state of most relevant cells, and the related transfection difficulties. When considering lncRNAs, these constraints meet additional ones, derived from low and sometimes nuclear-enriched expression of most lncRNAs, and their common overlap with other functional elements, such as other genes or enhancers. These difficulties can be addressed by an increasingly diverse toolbox that now contains many modalities that are used in lncRNA research (Fig. 1). These can be broadly divided into genome-editing approaches, which nowadays rely almost exclusively on CRISPR-Cas9 and which are typically applied in either mouse zygotes (for generating transgenic animals) or mESCs (for either generating transgenic animals or deriving cellular models that can then be differentiated to neuronal cells). In principle, it is also possible to edit the genomes of neuronal cell lines, but the typically large and variable number of chromosomes in these cells make it difficult to obtain homozygously edited cells. Genome editing has obvious advantages in its capability to completely ablate

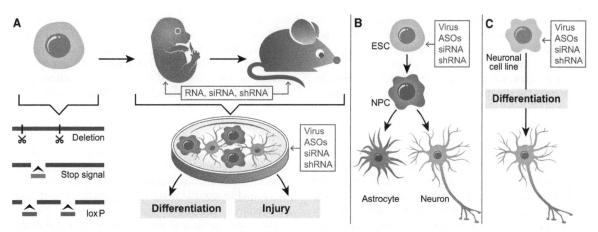

Figure 1. Methods for the study of long noncoding RNAs (lncRNAs). (*A*) Transgenic animals can be generated using introduction of genome engineering reagents into pluripotent stem cells, mouse zygotes, or early embryos, to induce deletions of lncRNA loci or parts thereof, and insertion of various "stop" signals or insertion of loxP sites that can be used for conditional manipulations. The resulting cells can then be used to derive and breed animal models. The indicated reagents can then be introduced into wild type (WT) or transgenic animals in the embryonic (e.g., using electroporation) or adult stages. Primary cells can be derived from embryonic or adult neural tissues and used in culture, when the indicated methods of manipulation are available and can be applied prior to differentiation of the cells into more mature populations or in vitro injury models. (*B*) Embryonic stem cells (ESCs) can be manipulated using the engineering methods described in *A* or by introducing the indicated reagents, and then differentiated into neural progenitor cells (NPCs) and glial or neuronal cells. (*C*) Neuronal cell lines, such as Neuro2a/N2a, can be transfected with various reagents and then differentiated into postmitotic neuronal cells. (siRNA) Small interfering RNA, (shRNA) short hairpin RNA, (ASOs) antisense oligonucleotides.

loci and the resulting uniform cellular population. However, several caveats have to be considered. First, as lncRNA loci may overlap other elements, pinpointing phenotypes resulting specifically from lncRNA loss has to typically rely on complementary evidence from other methods (Bassett et al. 2014; Kopp and Mendell 2018). Some editing approaches, like introduction of polyadenylation sequences, are considered relatively "safe" as they do not delete any DNA sequences, but recent discrepancies between the results obtained after editing the *Hand2as* locus hint that even this approach should be interpreted with caution (Anderson et al. 2016; Han et al. 2019). Promoter deletions are commonly used for lncRNA inactivation and provide a useful compromise between deletion of a relatively short stretch of DNA (compared to whole locus deletion) while having reasonable likelihood to eliminate transcription, although activation of cryptic promoters is often observed (Lavalou et al. 2019).

Another caveat is that successful elimination of lncRNA production still makes it difficult to distinguish between the function of the RNA product and the act of its transcription (Kopp and Mendell 2018). The recent description of introduction loci of self-cleaving ribozymes into lncRNA loci is an exciting progress in this direction (Tuck et al. 2018), but the efficiency of these elements in reducing gene expression is variable, and it is not yet clear how fast the cleavage occurs. Because of that, it is not yet clear if self-cleaving ribozymes may also affect transcription elongation.

The last major caveat is mostly relevant to editing done in cultured cells and, in particular, in pluripotent stem cells. The act of genome editing and, in particular, the double-strand breaks that it usually involves can trigger a stress response that may have lasting effects on the cells and on their ability to later differentiate into different lineages. The general process of establishing stable clones, which typically requires disruptive actions such as cell dissociation and several passages, can also have lasting effects. As a result of these, any edited clone will differ from the original cells and from other clones in ways that may or may not relate to the genetic changes that were acquired. Analysis of multiple independently established clones, as well as of control clones that experienced a very similar procedure (ideally, editing in a bystander locus), can help alleviate these concerns, but this remains very laborious and is still not routinely performed.

Transient introduction of perturbation reagents, such as siRNAs or antisense oligos, typically does not suffer from these caveats but is associated with many other concerns. These perturbations may not lower expression levels below those required for function, may lead to off-target effects (which are well-understood for RNAi, but almost entirely unexplored for antisense oligonucleotides [ASOs] or GapmeRs), and the introduction of the reagents via transfection or infection may activate a stress response. Because of these concerns, it is essential to deploy and compare different combinations of perturbation techniques, which often include genome editing and transient perturbations, ideally using the same system and the same phenotypic assays, although it is not always possible, in particular given the challenges involved in working with neuronal cells.

FUNCTIONS OF lncRNAs IN DEVELOPMENT OF THE NERVOUS SYSTEM

Different in vivo and in vitro models have been used to study the functions of lncRNAs during neuronal differentiation and during nervous system development (Table 2). The roles of *Cyrano* and *Megamind*, two deeply conserved lncRNAs, were studied during the development of zebrafish embryos by injection of morpholino (MO) antisense oligos, which led to severe defects in brain development (Ulitsky et al. 2011), which were not observed in genetic models later generated for these lncRNAs (Goudarzi et al. 2019; Kok et al. 2015). The function and mode of function of *Cyrano* was also studied by the generation of knockout (KO) mice (Kleaveland et al. 2018). Loss of *Cyrano* did not result in any detectable effects on survival, appearance or behavior in mice, but it led to the accumulation of miR-7 and down-regulation of *Cdr1as*, a circRNA known to regulate neuronal activity (Piwecka et al. 2017; Kleaveland et al. 2018). The function of *Megamind* (also known as *Tunar*) was also studied during neuronal differentiation of mESCs. In that system, *Megamind* knockdown (KD) with shRNAs reduced the differentiation efficiency, demonstrating its potentially conserved role in the acquisition of neuronal cell fate (Lin et al. 2014).

Several studies found roles for lncRNAs during mouse brain development. KO of *Evf2*, which is transcribed from the *Dlx-5/6* ultraconserved region (Feng et al. 2006; Bond et al. 2009), led to reduced numbers of GABAergic interneurons in early postnatal hippocampus and dentate gyrus (Bond et al. 2009). *Pantr2 (linc-Brn1b)* KO mice have defects in the proliferation of cortical progenitors in the subventricular zone (SVZ) during development (Sauvageau et al. 2013). In addition to the generation of mouse models, NSCs can be isolated from embryonic, postnatal, or adult mice and used for targeting different lncRNAs, optionally followed by in vitro differentiation. *Pnky* was studied in the mouse cortex using different methods: electroporation of shRNAs to ventricular zone stem cells during development, infection of shRNAs to cultured NSCs from postnatal mouse brain (Ramos et al. 2015), and generation of conditional KO mice (Andersen et al. 2019), which showed that *Pnky* interacts with PTBP1 splicing factor and acts in *trans* to regulate proliferation of NSCs and neurogenesis. In vitro differentiation of mouse embryonic NSCs was used to study the function of *lnc-OPC*. KD of this lncRNA using shRNAs during differentiation of NSCs to oligodendrocyte progenitor cells (OPCs) significantly repressed OPC markers (Dong et al. 2015). Overall these studies have shown that lncRNAs can modulate progenitor proliferation and differentiation and have profound effects on brain development. lncRNAs are also involved in developmental processes in the postnatal nervous system, such as myelination, retinal development, and neuronal outgrowth. The function of four lncRNAs highly expressed in oligodendrocyte lineage cells, *lncOL1-4*, was studied by siRNA transfection during oligodendrocyte differentiation of primary postnatal OPCs. KD of each of these lncRNAs repressed myelin genes (He et al. 2017). *lncOL1* was further studied in KO mice, in which severe defects in myelination were observed at P9 and postnatal week 3 (He et al. 2017). During retinal development, *Six3OS* is transcribed from an independent promoter separated by ~4 kb from the promoter of *Six3*. *Six3OS* was studied by overexpression or KD with shRNAs in P0.5 mouse retina and was found to be involved in retinal cell specification by binding factors known to co-regulate target genes with SIX3 and potentially recruiting histone-modifying enzymes to SIX3 targets (Rapicavoli et al. 2011). KD of *Six3OS* in adult SVZ NSCs during differentiation reduced neurogenesis, suggesting that, in addition to its role in retinal development, this lncRNA has roles also in neuronal differentiation (Ramos et al. 2013). Another lncRNA, *BDNF-AS*, which was studied both by transfection of siRNAs or ASOs to neurospheres and by intracerebroventricular delivery of ASOs to adult mice was found to regulate neuronal outgrowth and differentiation both in vitro and in vivo (Modarresi et al. 2012).

Other studies focused on in vitro neuronal differentiation of ESCs or neuroblastoma cell lines, which allow the interrogation of several lncRNAs in parallel. ESCs can be differentiated into NPCs and further into different types of neurons and glial cells, and in recent years, they have been used to identify and characterize the functions of lncRNAs in this process. *RMST* and three additional lncRNAs have been identified in a study based on neuronal differentiation of human ESCs (hESCs) (Ng et al. 2012), which also found that their KD with siRNAs blocked neurogenesis. A later study found that *RMST* interacts with SOX2 and promotes SOX2 binding to its target sites in the genome during neuronal differentiation (Ng et al. 2013). In haploid mESCs, the overexpression and genetic perturbation of *lnc-Nr2f1* led to dysregulation of hundreds of genes as the cells were differentiated into neurons, with down-regulated genes related to neuronal pathfinding and axon guidance (Ang et al. 2019). Conversely, overexpression of *lnc-Nr2f1* during ESC differentiation and during reprogramming of mouse embryonic fibroblasts (MEFs) to induced neurons (iNs) led to increased neurite length and increased neural conversion (Ang et al. 2019). Another study focused on the differentiation of mESCs to motor neurons (MNs) (Yen et al. 2018). Differentiated MNs were enriched with *Dlk-Dio3* locus-derived lncRNAs including *Meg3*. The depletion of *Meg3* resulted in dysregulation of progenitor and caudal Hox genes (Yen et al. 2018).

Neuroblastoma cell lines are another in vitro model of neuronal cells that are generally more accessible to perturbations compared to primary cells. These cells, including the commonly used mouse Neuro2a/N2a and human SH-SY5Y cell lines, express many neuronal markers, and when exposed to specific neurotrophic factors, they further differentiate, exit the cell cycle, and extend neurites. *Paupar*, which is transcribed from a locus upstream of *Pax6*, was studied in N2a cells by transfection of shRNAs and was found to act both in *cis*, by regulating *Pax6* expression, and in *trans*, by binding and regulating Pax6 activity, affecting cell cycle progression and neural differentiation (Vance et al. 2014). A later study found that *Paupar* binds the KAP1 chromatin regulator and promotes KAP1 occupancy to Pax6 targets (Pavlaki et al. 2018). Interestingly, in pancreatic α cells *Paupar* appears to regulate splicing and not expression of *Pax6* (Singer et al. 2019). *Dali*, also studied with shRNAs in N2a cells, was reported to locally regulate the transcription of the *Pou3f3*

Table 2. Roles of long noncoding RNAs (lncRNAs) in neural development

Name	Systems studied	Methods of perturbation	Phenotype	Suggested mode of action	Reference(s)
Cyrano/ OIP5-AS1	Zebrafish embryos, adult mice	MO antisense oligos, deletion of conserved region and whole gene deletion	Zebrafish: defects in neural tube opening, loss of neurons in the retina and tectum, enlarged nasal placodes. These phenotypes were not observed when the whole locus was deleted.	Mouse: Cyrano pairing to miR-7 promotes miR-7 degradation, leading to increased expression of Cdr1as circRNA in neuronal cell bodies and processes.	Ulitsky et al. 2011; Kleaveland et al. 2018; Goudarzi et al. 2019
Megamind/ Tunar	Zebrafish embryo, mESCs	MO antisense oligos, shRNAs	Zebrafish: smaller heads and eyes, enlarged brain ventricles, loss of neurons in the retina and tectum. These phenotypes were not observed when the whole locus was deleted. mESCs: reduced differentiation efficiency	Tunar interacts with PTBP1, hnRNP-K, and Nucleolin and mediates recruitment of PTBP1, hnRNP-K, and NCL to the neural gene promoters during neuronal differentiation of mESCs.	Ulitsky et al. 2011; Lin et al. 2014; Kok et al. 2015
Evf2	Mouse embryos and adults, C17 and MN9D neural cell lines	Insertion of a triple polyadenylation signal into exon 1, overexpression	Reduced numbers of GABAergic interneurons, reduced synaptic inhibition	Evf2 recruits DLX and MECP2 TFs to DNA regulatory elements in the Dlx5/6 intergenic region, regulating expression levels of Dlx5 and Dlx6; Evf2 inhibits BRG1 ATPase and chromatin remodeling activities.	Bond et al. 2009; Feng et al. 2006; Cajigas et al. 2015, 2018
Pantr2 / linc-Brn1b	Mouse embryos	Replacement of entire gene locus with a lacZ reporter cassette	Reduction in the number of intermediate progenitor cells in the cerebral cortex	N/A	Sauvageau et al. 2013
Pnky	Mouse embryos, cultured postnatal NSCs	shRNAs, conditional KO by flanking the entire gene with loxP sites	Cultured NSCs: decreased population of proliferating cells in the ventricular zone, increased neurogenesis. In vivo cortical development: malformation of the postnatal neocortex, increasing the number of deep-layer neurons while decreasing the population of upper-layer neurons	Pnky interacts with PTBP1 splicing factor, regulating a common set of transcripts related to neuronal differentiation.	Ramos et al. 2015; Andersen et al. 2019
lnc-OPC	Cultures of mouse embryonic NSCs	shRNAs	Decreased expression of OPC markers	N/A	Dong et al. 2015
BDNF-AS	Mouse neurospheres, adult mice	siRNAs, ASOs	Increased neurite outgrowth and neuronal differentiation	BDNF-AS recruits EZH2 to the promoter of BDNF, inhibiting its transcription.	Modarresi et al. 2012
lncOL1	Postnatal mice, cultured OPCs	siRNAs, deletion of exon 1	Absence of myelinated axons at P9, reduced numbers of myelinated axons at postnatal week 3	lncOL1 interacts with Suz1 to promote oligodendrocyte maturation through Suz12-mediated repression of an inhibitory network that maintains the OPC state.	He et al. 2017
Six3os	Postnatal mice, cultured adult NSCs	Overexpression, shRNAs	Postnatal mice: following KD, decrease in the fraction of bipolar cells, increase in Muller glia; following overexpression, reduction in syntaxin staining. Cultured NSCs: reduced numbers of Tuj1- and Olig2-expressing cells, increased numbers of GFAP-expressing cells	Six3os binds Ezh2 and Eya1/3/4, modulating the expression of Six3 target genes by acting as a transcriptional scaffold.	Rapicavoli et al. 2011; Ramos et al. 2013
RMST	hESCs	siRNAs	Arrested neurogenesis	RMST interacts with SOX2 and promotes SOX2 binding to its target sites in the genome during neuronal differentiation.	Ng et al. 2013, 2012

Continued

Table 2. Continued

Name	Systems studied	Methods of perturbation	Phenotype	Suggested mode of action	Reference(s)
lnc-Nr2f1	mESCs, reprogramming of MEFs to iNs	KO by insertion of a poly (A) transcriptional termination signal, overexpression	Overexpression: increased neurite length, induction of genes with functions in axon guidance, increased conversion of MEFs to iNs KO: down-regulation of neuronal pathfinding and axon guidance genes	lnc-Nr2f1 binds to distinct genomic loci regulating neuronal genes.	Ang et al. 2019
Meg3, Rian, Mirg	mESCs, mouse embryos	Deletion of maternal IG-DMR, shRNAs	Dysregulation of progenitor and caudal Hox genes, motor axon innervation defects	Meg3 facilitates the binding of PRC2 and Jarid2, leading to global down-regulation of H3K27me3 and aberrant expression of progenitor and caudal Hox genes in postmitotic MNs.	Yen et al. 2018
Paupar	N2a cells, cultured neurospheres, postnatal mice	shRNAs	Reduced proliferation, increased neurite outgrowth, reduced number and altered morphology of olfactory bulb neurons	Paupar binds KAP1 and promotes KAP1 occupancy and H3K9me3 deposition at PAX6 bound loci.	Vance et al. 2014; Pavlaki et al. 2018
Dali	N2a cells	shRNAs	Reduced neurite outgrowth following differentiation	Dali interacts with the POU3F TF and the DNA methyltransferase DNMT1 and directly binds genomic loci to regulate gene expression.	Chalei et al. 2014
lncND	hESCs, SH-SY5Y cells, mouse embryo	siRNAs, in utero electroporation of human transcript to mouse radial glia cells	SH-SY5Y cells: reduced cell proliferation and increased neuronal differentiation Mouse embryo: expansion of radial glia population	lncND sequesters miR-143-3p in NPCs, leading to increased production of NOTCH proteins.	Rani et al. 2016

(MO) Morpholino, (circRNA) circular RNAs, (mESCs) mouse embryonic stem cells, (shRNAs) short hairpin RNAs, (TFs) transcription factors, (N/A) not applicable, (NSCs) neural stem cells, (KO) knockout, (siRNAs) small interfering RNAs, (ASO) antisense oligonucleotide, (OPCs) oligodendrocyte progenitor cells, (KD) knockdown, (hESCs) human embryonic stem cells, (MEFs) mouse embryonic fibroblasts, (iNs) induced neurons, (MNs) motor neurons, (NPCs) neural progenitor cells.

transcription factor (TF) and distally bind active promoters and regulate the expression of genes related to neuronal differentiation through physical association with the POU3F3 protein (Chalei et al. 2014).

The combination of different in vitro and in vivo systems can be used to efficiently characterize the functions of lncRNAs in neuronal differentiation. For example, *lncND*, a primate-specific lncRNA, was found to regulate Notch signaling during neuronal differentiation by sequestering miR-143-3p (Rani et al. 2016). hESCs were used to characterize the expression dynamics of *lncND* during neuronal differentiation, siRNA transfection in SH-SY5Y cells was then used to study the functions of *lncND* during proliferation and differentiation, and, finally, injection of the human *lncND* transcript to the developing mouse cortex, which resulted in expansion of a radial glia population, supported the hypothesis that *lncND* contributed to the expansion of radial glia in higher primates (Rani et al. 2016).

THE ROLE OF lncRNAs IN NEUROREGENERATION

Neurons within the peripheral nervous system (PNS) can undergo axon outgrowth, which may lead to substantial functional recovery, while this process is limited within the CNS. Following axotomy, PNS neurons activate a unique regenerative transcriptional program and up-regulate numerous regeneration-associated genes (RAGs) (Curcio and Bradke 2018). The differences in the ability of CNS neurons to activate RAG expression, along with extracellular environmental factors, can explain why CNS neurons do not regenerate successfully (Ma and Willis 2015). Modulating critical hubs of RAG transcription can therefore have important therapeutic implications (Gao et al. 2016).

The next section describes the role of lncRNAs in nerve regeneration following different types of injuries, as demonstrated by in vitro and in vivo methods (Fig. 2 and Table 3).

lncRNAs IN PERIPHERAL NERVE INJURY (PNI)

Glial cells and molecules in the extracellular matrix (ECM) orchestrate axonal regeneration upon injury. Successful PNS axon regeneration is largely attributed to Schwann cell response via proliferation, migration, and remyelination (Toy and Namgung 2013). In parallel, in PNS neurons, multiple signaling pathways monitoring gene expression in the soma, together with pathways in the growth cone, control the balance of microtubule assembly, dynamics, and stabilization to achieve optimal axon growth (Glenn and Talbot 2013; Saijilafu et al. 2013a,b). lncRNAs have been suggested to play important roles in these well-coordinated and complicated processes, in particular in stress responses, plasticity, and axonal outgrowth (Yu et al. 2013; Yao and Yu 2019). A common model to study mammalian axon regeneration in PNS neurons is sciatic nerve injury (SNI). The sensory neurons extending into the sciatic nerve are located in the L4–L6 dorsal root ganglia (DRGs) (Angius et al. 2012). The transcriptome of DRG neurons was studied using microarrays (Yu et al. 2013) and RNA-seq (Perry et al. 2018), and hundreds of genes were differentially expressed following SNI. In vitro, silencing of *BC089918* and *uc.217* using siRNAs in rat cultured neurons increased neurite outgrowth through an unknown mechanism (Yu et al. 2013; Yao et al. 2015), and *uc.217* repression led to increased levels of *Gal* and *Vip* RAGs (Yao et al. 2015). We found that siRNA-mediated KD of *Silc1* and *Norris1* lncRNAs led to a reduction in total axonal outgrowth without any apparent effect on cell viability. *Silc1* is transcribed from an intergenic region ~200 kb upstream of the *Sox11* TF, a known master regulator of neurogenesis that has been shown to be required for both embryonic and adult neurogenesis in mice (Jankowski et al. 2006, 2009, 2018). To study *Silc1* function in vivo, we generated *Silc1* KO mice by deletion of *Silc1* promoter and exon 1 using CRISPR–Cas9. *Silc1* KO mice have reduced levels of *Sox11* in neuronal cells and exhibit delayed regeneration following

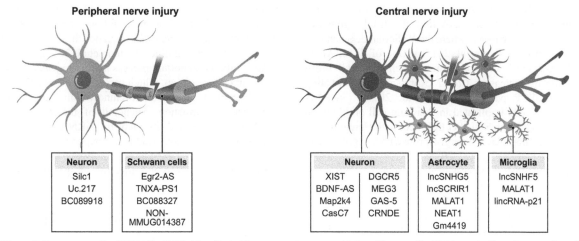

Figure 2. Long noncoding RNAs (lncRNAs) implicated in response to neuronal injury. Names of lncRNAs are indicated next to the cell types and injury types in which they have been studied.

Table 3. Roles of long noncoding RNAs (lncRNAs) in neuroregeneration

Name	Systems studied	Methods of perturbation	Phenotype	Suggested mode of action	Reference(s)
BC089918	SNI in rat model and DRG culture cells	siRNAs	Negative effect on neurite outgrowth of DRG culture cells	N/A	Yu et al. 2013
Uc.217	SNI in rat model and DRG culture cells	siRNAs	Negative effect on neurite outgrowth of DRG culture cells	N/A	Yao et al. 2015
Silc1	SNI in mouse and DRG culture cells	siRNAs, KO by deletion of promoter and first exon	Reduction in total axonal outgrowth; delayed regeneration following sciatic injury	*cis*-acting activation of *Sox11*	Perry et al. 2018
NONMMUG014387	SNI in mouse and Schwann cells	Overexpression	Increased proliferation of Schwann cells	Activation of Wnt/PCP pathway	Pan et al. 2017a
BC088327	Rat SNI and mouse Schwann cells	siRNAs	Increased Schwann cells proliferation	N/A	Wang et al. 2018
Egr2-AS	Mouse SNI and DRG explant culture	Overexpression, GapmeRs	Increased demyelination	Inhibition of Egr2 expression	Martinez-Moreno et al. 2017
TNXA-PS1	Rat SNI and Schwann cells culture	siRNAs	Reduced Schwann cell migration	Sponging miR-24-3p/miR-152-3p and regulation of Dusp1	Yao et al. 2018
BDNF-AS	Rat acute SCI and hypoxia cellular model, neuronal cell lines	siRNAs	Increased neuronal cell apoptosis	Sponging miR-130b-5p to regulate PRDM5	Zhang et al. 2018a
DGCR5	Rat acute SCI and hypoxia cellular model, neuronal cell lines	siRNAs	Reduced neuronal apoptosis	Binding and negative regulation of PRDM5	Zhang et al. 2018b
SNHG5	Rat SCI and primary cultured astrocytes	siRNAs, overexpression	Enhanced viability of astrocytes and microglia	N/A	Jiang and Zhang 2018
GAS5	Mouse TBI and primary culture neuronal cells	siRNAs	Increased neuronal apoptosis	Sponging miR-335 to activate Rasa1	Wang et al. 2017; Dai et al. 2019
CRNDE	Rat TBI	siRNAs	Increased nerve repair	N/A	Yi et al. 2019

(SNI) Spinal nerve injury, (DRG) dorsal root ganglion, (siRNAs) small interfering RNAs, (N/A) not applicable, (KO) knockout, (SCI) spinal cord injury, (TBI) traumatic brain injury.

SNI. Therefore, *Silc1* regulates neuroregeneration in cultured cells and in vivo, through *cis*-acting activation of the *Sox11* TF, through a currently unknown mechanism (Perry et al. 2018).

lncRNAs are also important for Schwann cells regulation after PNI. lncRNAs in Schwann cells were profiled using microarrays with RNA from the distal segment of the mouse sciatic nerve, where the Wallerian degeneration process occurs (Pan et al. 2017b). One of these, *NONMMUG014387*, promoted mouse Schwann cell proliferation by increasing *Cthrc1* expression and activating the Wnt/PCP pathway (Pan et al. 2017a). lncRNA *BC088327* was implicated in Schwann cell proliferation in a rat model with SNI after treatment with Heregulin-1β. *BC088327* may play a synergistic role with heregulin-1β in repairing PNI (Wang et al. 2018).

The myelination process is also crucial for PNS nerve regeneration. Recently it was found that the EGR2 TF is an important modulator in this process in mice. *Egr2* is down-regulated during SNI, whereas *Egr2-AS* lncRNA, transcribed antisense to the proximal promoter of *Egr2*, is up-regulated. Overexpressing *Egr2-AS* in mouse DRG explant cultures results in inhibition of *Egr2* mRNA expression and induces demyelination, suggesting that this lncRNA acts in *trans*. Inhibition of *Egr2-AS* in vivo using GapmeRs at the time of SNI, rescues the inhibition of *Egr2* transcript expression and affects EGR2-regulated genes to delay demyelination. Mechanistically, it was proposed that *Egr2-AS* recruits H3K27me3, AGO1, AGO2, and EZH2 to the *Egr2* promoter following SNI. Furthermore, expression of *Egr2-AS* is regulated through ERK1/2 signaling to YY1 (Martinez-Moreno et al. 2017).

An additional important process in PNS regeneration is Schwann cell migration, which can precede and enhance axonal repair by guiding axon reinnervation and controlling synaptic formation. Silencing of the *TNAX-PS1* lncRNA in vitro and in vivo by siRNAs promoted Schwann cell migration. Additional experiments showed that *TNXA-PS1* might exert its regulatory role by sponging miR-24-3p/miR-152-3p and affecting Dusp1 expression (Yao et al. 2018).

lncRNAs IN SPINAL CORD INJURY

Primary spinal cord injury (SCI) is commonly caused by direct trauma or pathological alterations; it is much more complicated than PNI because of extensive cell loss, axonal disruption, glial scar, and shortage of growth-permissive factors (Estrada and Müller 2014). This is followed by secondary injury mechanisms, including glutaminergic excitotoxicity, oxidative stress, increased adaptive immune responses, Wallerian degeneration, and scar tissue formation, leading to further structural and functional disturbances (Ahuja et al. 2017). Various types of ncRNAs were implicated in these processes (Ning et al. 2014; Qin et al. 2018; Li et al. 2019b; Pinchi et al. 2019; Yao and Yu 2019; Zhou et al. 2019). The roles of lncRNAs after SCI have been recently thoroughly reviewed (Li et al. 2019b), and so we will present only a few examples here. *BDNF-AS* lncRNA is up-regulated in a rat acute SCI model. *BDNF-AS* KD by siRNAs in neuronal cell lines reduced neuronal cell apoptosis in hypoxic conditions. It was suggested that *BDNF-AS* associates with miR-130b-5p, which is repressed during SNI. In vivo, *BDNF-AS* KD with siRNAs inhibited the expression of PRDM5, supposedly through competitive binding with miR-130b-5p, resulting in decreased apoptosis (Zhang et al. 2018a). In contrast, *DGCR5* lncRNA was down-regulated in this model and in neurons treated with hypoxia (Zhang et al. 2018b). Using in vitro and in vivo methods it was shown that *DGCR5* suppresses neuronal apoptosis through directly binding and negatively regulating PRDM5, thereby ameliorating SCI (Zhang et al. 2018b).

lncRNA *SNHG5* is an example of a lncRNA that is induced following SCI and enhances the viability of astrocytes and microglia. *SNHG5* overexpression promoted cell viability of both astrocytes or microglia, while its down-regulation by siRNAs led to cell death (Jiang and Zhang 2018). *SNHG5* overexpression in vivo results in reduced motor recovery post-SCI in rats. Additional experiments demonstrated that *SNHG5* enhances the expression of both KLF4 and eNOS (Jiang and Zhang 2018). Additional lncRNAs that have a proposed role in SCI are *lncSCRIR1*, *Casc7*, *MALAT1*, *XIST*, and *Map2K4* (Li et al. 2019b).

lncRNAs IN TRAUMATIC BRAIN INJURY

Traumatic brain injury (TBI) is a significant source of morbidity and mortality in the adult population. TBI causes secondary biochemical changes that contribute to neurological dysfunction, delayed neuroinflammation, and nerve cell death (Kabadi and Faden 2014). lncRNA dysregulation after TBI was observed in the cerebral cortex and hippocampus (Zhong et al. 2016; Wang et al. 2017). The current progress of studies on lncRNAs in TBI was recently reviewed (Zhang and Wang 2019; Li et al. 2019a), so we will mention only few examples that were recently published. *Gas5* was found using microarrays to be induced in the rat hippocampus after TBI (Wang et al. 2017). In mouse neuronal cells *GAS5* could up-regulate *Rasa1* expression and promote neuronal apoptosis following TBI. In vivo experiments indicated that Changqin NO. 1, a traditional Chinese medicine, had neuroprotective effects by inhibiting neuronal apoptosis via the *GAS5*/miR-335/*Rasa1* axis (Dai et al. 2019). Furthermore, *GAS5* silencing protected against hypoxic/ischemic-induced brain injury in vivo and primary hippocampal neuron injury in vitro, suggesting a potential therapeutic approach of *GAS5* inhibition in the treatment of neonatal brain damage (Wang et al. 2017; Zhao et al. 2018).

CRNDE was up-regulated in serum of TBI patients compared to healthy controls. In vivo studies using TBI rat model showed that silencing of *CRNDE* improves neurobehavioral function, inhibits the expression of neuroinflammatory factors, and inhibits neuronal apoptosis and autophagy in TBI rats. Repression of *CRNDE* also promoted the expression of differentiation markers in neurons and the directional growth and regeneration of nerve fibers (Yi et al. 2019). Other lncRNAs that have a reported function in TBI are *NEAT1*, *MEG3*, *MALAT1*, *Gm4419*, and *lincRNA-p21* (Li et al. 2019a; Zhang and Wang 2019).

lncRNAs IN NEUROLOGICAL DISEASES

Supporting the importance of lncRNAs in the development and functions of the CNS, dysregulation of lncRNAs in the nervous system has been linked with changes occurring in human neurological diseases, including neurodegenerative diseases and psychiatric disorders.

Miat (also known as *Gomafu*) is a lncRNA that is localized to a subnuclear domain in a distinct subset of differentiating neurons in the mouse nervous system (Sone et al. 2007; Mercer et al. 2010), and its expression is regulated by neuronal activity (Barry et al. 2014). *Miat* is down-regulated in postmortem cortex of subjects with and its KD in neuronal cultures led to changes in alternative splicing that resembled those observed in schizophrenia patients (Barry et al. 2014). Another study identified 125 lncRNAs with aberrant expression in schizophrenia patients (Chen et al. 2016), and a co-expression network analysis suggested that specific lncRNAs are associated with early-onset schizophrenia (Ren et al. 2015).

BACE1-AS was shown to regulate the expression level of *BACE1*, a crucial enzyme in Alzheimer's disease (AD) pathophysiology. *BACE1-AS* levels were elevated in subjects with AD and in amyloid precursor protein transgenic mice (Faghihi et al. 2008). Several other lncRNAs have been suggested to have roles in AD (Luo and Chen 2016), and microarray analysis of postmortem AD tissues (Zhou and Xu 2015) and a rat model of AD (Zhou and Xu 2015; Yang et al. 2017) identified hundreds of dysregulated lncRNAs.

The expression level of *Megamind/TUNA* lncRNA, discussed above, was shown to be associated with Huntington's disease (HD) neuropathological grade in patients' brains (Lin et al. 2014), and a microarray-based study found that several lncRNAs, including *TUG1*, *NEAT1*, *MEG3*, and *DGCR5*, are dysregulated in the brains of HD patients (Johnson 2012).

Different lncRNAs, such as *naPINK1*, *NEAT1 PINK1-AS*, *BC200*, and *Sox2OT*, were found to be dysregulated in Parkinson's disease (PD) patients' brains (Wan et al. 2017). Recently it was found that lncRNA *Neat1* is significantly up-regulated in the midbrain of PD mice, and it was suggested that *Neat1* promotes MPTP-induced autophagy in PD by stabilizing PINK1 protein (Yan et al. 2018).

An increasing number of studies report on lncRNAs as being implicated in additional neurodegenerative diseases including amyotrophic lateral sclerosis (ALS), multiple system atrophy (MSA), frontotemporal lobar degeneration (FTLD), and glaucoma (Riva et al. 2016; Quan et al. 2017; Wan et al. 2017; Wang et al. 2017).

CONCEPTUAL ROLES OF lncRNAs IN THE NERVOUS SYSTEM

We summarized here a large number of recent reports on various functional outcomes of lncRNA perturbations in the nervous system. What ties together these observations? One trend is that a large number of the observed phenotypes were related to changes in proliferation of progenitors or their ability to give rise to particular populations. Importantly, these are also likely the easiest phenotypes to score in the nervous system, which provides a possible explanation for their prevalence in the literature.

Beyond the need to regulate the balance between proliferation and differentiation and to specify and then "lock" different fates, cells in the nervous system have several features that may be particularly prone to lncRNA regulation. First, the vast majority of the cells in the adult nervous system are postmitotic and so cannot rely on replication-dependent histone exchange for chromatin regulation. Transcription- or lncRNA-mediated effects thus may replace some of the regulatory processes that rely more heavily on replication-associated mechanisms in more proliferative cells (e.g., in colon or blood). Such mechanisms likely play an outsized role in the ability of neurons to regenerate in the absence of cell proliferation. Another prominent feature of the nervous system is the requirement for rapid changes in gene expression upon specific cues, such as neuronal activity, which requires maintenance of chromatin in a particularly poised state, as well as extensive post-transcriptional regulation, that acts to limit the expression window of genes induced in the early stages of the response or allows fast maturation and export of particularly long genes (Mauger et al. 2016). Last, the size and complex morphology of neural cells dictate the need for complex post-transcriptional regulation, which may benefit from RNA-based control. Future studies will elucidate which of these or other features of the neural environment preferentially rely on lncRNA function and may explain why neural cells express a particularly vibrant complement of lncRNA genes.

ACKNOWLEDGMENTS

Research in the Ulitsky laboratory is supported by the Israel Science Foundation (2406/18 and 852/19); European Research Council project lincSAFARI; MOST-PRC program; German–Israel foundation for Scientific Research and Development (GIF, I-1455-417.13/2018); The EU Joint Programme—Neurodegenerative Disease Research (JPND ERA-Net localMND); BIRAX—Britain Israel Research and Academic Exchange Partnership; United States–Israel Binational Science Foundation; Lapon Raymond; and the Abramson Family Center for Young Scientists. I.U. is incumbent of the Sygnet Career Development Chair for Bioinformatics.

REFERENCES

Ahuja CS, Wilson JR, Nori S, Kotter MRN, Druschel C, Curt A, Fehlings MG. 2017. Traumatic spinal cord injury. *Nat Rev Dis Primers* **3:** 17018. doi:10.1038/nrdp.2017.18

Andersen RE, Hong SJ, Lim JJ, Cui M, Harpur BA, Hwang E, Delgado RN, Ramos AD, Liu SJ, Blencowe BJ, et al. 2019. The long noncoding RNA *Pnky* is a *trans*-acting regulator of cortical development in vivo. *Dev Cell* **49:** 632–642.e7. doi:10.1016/j.devcel.2019.04.032

Anderson KM, Anderson DM, McAnally JR, Shelton JM, Bassel-Duby R, Olson EN. 2016. Transcription of the non-coding RNA upperhand controls *Hand2* expression and heart development. *Nature* **539:** 433–436. doi:10.1038/nature20128

Ang CE, Ma Q, Wapinski OL, Fan S, Flynn RA, Lee QY, Coe B, Onoguchi M, Olmos VH, Do BT, et al. 2019. The novel lncRNA is pro-neurogenic and mutated in human neurodevelopmental disorders. *Elife* **8:** e41770. doi:10.7554/eLife.41770

Angius D, Wang H, Spinner RJ, Gutierrez-Cotto Y, Yaszemski MJ, Windebank AJ. 2012. A systematic review of animal models used to study nerve regeneration in tissue-engineered scaffolds. *Biomaterials* **33:** 8034–8039.

Barry G, Briggs JA, Vanichkina DP, Poth EM, Beveridge NJ, Ratnu VS, Nayler SP, Nones K, Hu J, Bredy TW, et al. 2014. The long non-coding RNA Gomafu is acutely regulated in response to neuronal activation and involved in schizophrenia-associated alternative splicing. *Mol Psychiatry* **19:** 486–494. doi:10.1038/mp.2013.45

Bassett AR, Akhtar A, Barlow DP, Bird AP, Brockdorff N, Duboule D, Ephrussi A, Ferguson-Smith AC, Gingeras TR, Haerty W, et al. 2014. Considerations when investigating lncRNA function in vivo. *Elife* **3:** e03058. doi:10.7554/eLife.03058

Bond AM, Vangompel MJW, Sametsky EA, Clark MF, Savage JC, Disterhoft JF, Kohtz JD. 2009. Balanced gene regulation by an embryonic brain ncRNA is critical for adult hippocampal GABA circuitry. *Nat Neurosci* **12:** 1020–1027. doi:10.1038/nn.2371

Briese M, Saal L, Appenzeller S, Moradi M, Baluapuri A, Sendtner M. 2016. Whole transcriptome profiling reveals the RNA content of motor axons. *Nucleic Acids Res* **44:** e33. doi:10.1093/nar/gkv1027

Cabili MN, Trapnell C, Goff L, Koziol M, Tazon-Vega B, Regev A, Rinn JL. 2011. Integrative annotation of human large intergenic noncoding RNAs reveals global properties and specific subclasses. *Genes Dev* **25:** 1915–1927. doi:10.1101/gad.17446611

Cabili MN, Dunagin MC, McClanahan PD, Biaesch A, Padovan-Merhar O, Regev A, Rinn JL, Raj A. 2015. Localization and abundance analysis of human lncRNAs at single-cell and single-molecule resolution. *Genome Biol* **16:** 20. doi:10.1186/s13059-015-0586-4

Cajigas I, Leib DE, Cochrane J, Luo H, Swyter KR, Chen S, Clark BS, Thompson J, Yates JR, Kingston RE, et al. 2015. *Evf2* lncRNA/BRG1/DLX1 interactions reveal RNA-dependent inhibition of chromatin remodeling. *Development* **142:** 2641–2652. doi:10.1242/dev.126318

Cajigas I, Chakraborty A, Swyter KR, Luo H, Bastidas M, Nigro M, Morris ER, Chen S, VanGompel MJW, Leib D, et al. 2018. The *Evf2* ultraconserved enhancer lncRNA functionally and spatially organizes megabase distant genes in the developing forebrain. *Mol Cell* **71:** 956–972.e9. doi:10.1016/j.molcel.2018.07.024

Chalei V, Sansom SN, Kong L, Lee S, Montiel JF, Vance KW, Ponting CP. 2014. The long non-coding RNA *Dali* is an epigenetic regulator of neural differentiation. *Elife* **3:** e04530. doi:10.7554/eLife.04530

Chen S, Sun X, Niu W, Kong L, He M, Li W, Zhong A, Lu J, Zhang L. 2016. Aberrant expression of long non-coding RNAs in schizophrenia patients. *Med Sci Monit* **22:** 3340–3351. doi:10.12659/MSM.896927

Curcio M, Bradke F. 2018. Axon regeneration in the central nervous system: facing the challenges from the inside. *Annu Rev Cell Dev Biol* **34:** 495–521. doi:10.1146/annurev-cellbio-100617-062508

Dai X, Yi M, Wang D, Chen Y, Xu X. 2019. Changqin NO. 1 inhibits neuronal apoptosis via suppressing GAS5 expression in a traumatic brain injury mice model. *Biol Chem* **400:** 753–763. doi:10.1515/hsz-2018-0340

Derrien T, Johnson R, Bussotti G, Tanzer A, Djebali S, Tilgner H, Guernec G, Martin D, Merkel A, Knowles DG, et al. 2012. The GENCODE v7 catalog of human long noncoding RNAs: analysis of their gene structure, evolution, and expression. *Genome Res* **22:** 1775–1789. doi:10.1101/gr.132159.111

Dong X, Chen K, Cuevas-Diaz Duran R, You Y, Sloan SA, Zhang Y, Zong S, Cao Q, Barres BA, Wu JQ. 2015. Comprehensive identification of long non-coding RNAs in purified cell types from the brain reveals functional lncRNA in OPC fate determination. *PLoS Genet* **11:** e1005669. doi:10.1371/journal.pgen.1005669

Ebbesen KK, Kjems J, Hansen TB. 2016. Circular RNAs: identification, biogenesis and function. *Biochim Biophys Acta* **1859:** 163–168. doi:10.1016/j.bbagrm.2015.07.007

Estrada V, Müller HW. 2014. Spinal cord injury—there is not just one way of treating it. *F1000Prime Rep* **6:** 84. doi:10.12703/P6-84

Faghihi MA, Modarresi F, Khalil AM, Wood DE, Sahagan BG, Morgan TE, Finch CE, St Laurent G III, Kenny PJ, Wahlestedt C. 2008. Expression of a noncoding RNA is elevated in Alzheimer's disease and drives rapid feed-forward regulation of β-secretase. *Nat Med* **14:** 723–730. doi:10.1038/nm1784

Farris S, Ward JM, Carstens KE, Samadi M, Wang Y, Dudek SM. 2019. Hippocampal subregions express distinct dendritic transcriptomes that reveal differences in mitochondrial function in CA2. *Cell Rep* **29:** 522–539.e6. doi:10.1016/j.celrep.2019.08.093

Feng J, Bi C, Clark BS, Mady R, Shah P, Kohtz JD. 2006. The *Evf-2* noncoding RNA is transcribed from the Dlx-5/6 ultraconserved region and functions as a Dlx-2 transcriptional coactivator. *Genes Dev* **20:** 1470–1484. doi:10.1101/gad.1416106

Gao Y, Yang Z, Li X. 2016. Regeneration strategies after the adult mammalian central nervous system injury—biomaterials. *Regen Biomater* **3:** 115–122. doi:10.1093/rb/rbw004

Glenn TD, Talbot WS. 2013. Signals regulating myelination in peripheral nerves and the Schwann cell response to injury. *Curr Opin Neurobiol* **23:** 1041–1048. doi:10.1016/j.conb.2013.06.010

Goff LA, Groff AF, Sauvageau M, Trayes-Gibson Z, Sanchez-Gomez DB, Morse M, Martin RD, Elcavage LE, Liapis SC, Gonzalez-Celeiro M, et al. 2015. Spatiotemporal expression and transcriptional perturbations by long noncoding RNAs in the mouse brain. *Proc Natl Acad Sci* **112:** 6855–6862. doi:10.1073/pnas.1411263112

Goudarzi M, Berg K, Pieper LM, Schier AF. 2019. Individual long non-coding RNAs have no overt functions in zebrafish embryogenesis, viability and fertility. *Elife* **8:** e40815. doi:10.7554/eLife.40815

Han X, Zhang J, Liu Y, Fan X, Ai S, Luo Y, Li X, Jin H, Luo S, Zheng H, et al. 2019. The lnc RNA *Hand2os1/Uph* orchestrates heart development through regulation of precise expression of *Hand2*. *Development* **146:** dev176198. doi:10.1242/dev.176198

He D, Wang J, Lu Y, Deng Y, Zhao C, Xu L, Chen Y, Hu Y-C, Zhou W, Lu QR. 2017. lncRNA functional networks in oligodendrocytes reveal stage-specific myelination control by an *lncOL1*/Suz12 complex in the CNS. *Neuron* **93:** 362–378. doi:10.1016/j.neuron.2016.11.044

Jankowski MP, Cornuet PK, McIlwrath S, Koerber HR, Albers KM. 2006. SRY-box containing gene 11 (Sox11) transcription factor is required for neuron survival and neurite growth. *Neuroscience* **143:** 501–514. doi:10.1016/j.neuroscience.2006.09.010

Jankowski MP, McIlwrath SL, Jing X, Cornuet PK, Salerno KM, Koerber HR, Albers KM. 2009. Sox11 transcription factor modulates peripheral nerve regeneration in adult mice. *Brain Res* **1256:** 43–54. doi:10.1016/j.brainres.2008.12.032

Jankowski MP, Miller L, Richard Koerber H. 2018. Increased expression of transcription factor SRY-box-containing gene 11 (Sox11) enhances neurite growth by regulating neurotrophic factor responsiveness. *Neuroscience* **382:** 93–104. doi:10.1016/j.neuroscience.2018.04.037

Jiang Z-S, Zhang J-R. 2018. lncRNA SNHG5 enhances astrocytes and microglia viability via upregulating KLF4 in spinal cord injury. *Int J Biol Macromol* **120:** 66–72. doi:10.1016/j.ijbiomac.2018.08.002

Johnson R. 2012. Long non-coding RNAs in Huntington's disease neurodegeneration. *Neurobiol Dis* **46:** 245–254. doi:10.1016/j.nbd.2011.12.006

Kabadi SV, Faden AI. 2014. Neuroprotective strategies for traumatic brain injury: improving clinical translation. *Int J Mol Sci* **15:** 1216–1236.

Kleaveland B, Shi CY, Stefano J, Bartel DP. 2018. A network of noncoding regulatory RNAs acts in the mammalian brain. *Cell* **174:** 350–362.e17. doi:10.1016/j.cell.2018.05.022

Kok FO, Shin M, Ni C-W, Gupta A, Grosse AS, van Impel A, Kirchmaier BC, Peterson-Maduro J, Kourkoulis G, Male I, et al. 2015. Reverse genetic screening reveals poor correlation between morpholino-induced and mutant phenotypes in zebrafish. *Dev Cell* **32:** 97–108. doi:10.1016/j.devcel.2014.11.018

Kopp F, Mendell JT. 2018. Functional classification and experimental dissection of long noncoding RNAs. *Cell* **172:** 393–407. doi:10.1016/j.cell.2018.01.011

Lavalou P, Eckert H, Damy L, Constanty F, Majello S, Bitetti A, Graindorge A, Shkumatava A. 2019. Strategies for genetic inactivation of long noncoding RNAs in zebrafish. *RNA* **25:** 897–904. doi:10.1261/rna.069484.118

Li Z, Han K, Zhang D, Chen J, Xu Z, Hou L. 2019a. The role of long noncoding RNA in traumatic brain injury. *Neuropsychiatr Dis Treat* **15:** 1671–1677. doi:10.2147/NDT.S206624

Li Z, Ho IHT, Li X, Xu D, Wu WKK, Chan MTV, Li S, Liu X. 2019b. Long non-coding RNAs in the spinal cord injury: novel spotlight. *J Cell Mol Med* **23:** 4883–4890. doi:10.1111/jcmm.14422

Lin N, Chang K-Y, Li Z, Gates K, Rana ZA, Dang J, Zhang D, Han T, Yang C-S, Cunningham TJ, et al. 2014. An evolutionarily conserved long noncoding RNA TUNA controls pluripotency and neural lineage commitment. *Mol Cell* **53:** 1005–1019. doi:10.1016/j.molcel.2014.01.021

Luo Q, Chen Y. 2016. Long noncoding RNAs and Alzheimer's disease. *Clin Interv Aging* **11:** 867–872. doi:10.2147/cia.s107037

Ma TC, Willis DE. 2015. What makes a RAG regeneration associated? *Front Mol Neurosci* **8:** 43. doi:10.3389/fnmol.2015.00043

Martinez-Moreno M, O'Shea TM, Zepecki JP, Olaru A, Ness JK, Langer R, Tapinos N. 2017. Regulation of peripheral myelination through transcriptional buffering of Egr2 by an antisense long non-coding RNA. *Cell Rep* **20:** 1950–1963. doi:10.1016/j.celrep.2017.07.068

Mauger O, Lemoine F, Scheiffele P. 2016. Targeted intron retention and excision for rapid gene regulation in response to neuronal activity. *Neuron* **92:** 1266–1278. doi:10.1016/j.neuron.2016.11.032

Mercer TR, Dinger ME, Sunkin SM, Mehler MF, Mattick JS. 2008. Specific expression of long noncoding RNAs in the mouse brain. *Proc Natl Acad Sci* **105:** 716–721. doi:10.1073/pnas.0706729105

Mercer TR, Qureshi IA, Gokhan S, Dinger ME, Li G, Mattick JS, Mehler MF. 2010. Long noncoding RNAs in neuronal–glial fate specification and oligodendrocyte lineage maturation. *BMC Neurosci* **11:** 14. doi:10.1186/1471-2202-11-14

Minis A, Dahary D, Manor O, Leshkowitz D, Pilpel Y, Yaron A. 2014. Subcellular transcriptomics—dissection of the mRNA composition in the axonal compartment of sensory neurons. *Dev Neurobiol* **74:** 365–381. doi:10.1002/dneu.22140.

Modarresi F, Faghihi MA, Lopez-Toledano MA, Fatemi RP, Magistri M, Brothers SP, van der Brug MP, Wahlestedt C. 2012. Inhibition of natural antisense transcripts in vivo results in gene-specific transcriptional upregulation. *Nat Biotechnol* **30:** 453–459. doi:10.1038/nbt.2158

Ng S-Y, Johnson R, Stanton LW. 2012. Human long non-coding RNAs promote pluripotency and neuronal differentiation by association with chromatin modifiers and transcription factors. *EMBO J* **31:** 522–533. doi:10.1038/emboj.2011.459

Ng S-Y, Bogu GK, Soh BS, Stanton LW. 2013. The long noncoding RNA *RMST* interacts with SOX2 to regulate neurogenesis. *Mol Cell* **51:** 349–359. doi:10.1016/j.molcel.2013.07.017

Nijssen J, Aguila J, Hoogstraaten R, Kee N, Hedlund E. 2018. Axon-seq decodes the motor axon transcriptome and its modulation in response to ALS. *Stem Cell Rep* **11:** 1565–1578. doi:10.1016/j.stemcr.2018.11.005

Ning B, Gao L, Liu R-H, Liu Y, Zhang N-S, Chen Z-Y. 2014. microRNAs in spinal cord injury: potential roles and therapeutic implications. *Int J Biol Sci* **10:** 997–1006. doi:10.7150/ijbs.9058

Pan B, Shi Z-J, Yan J-Y, Li J-H, Feng S-Q. 2017a. Long noncoding RNA NONMMUG014387 promotes Schwann cell proliferation after peripheral nerve injury. *Neural Regen Res* **12:** 2084–2091. doi:10.4103/1673-5374.221168

Pan B, Zhou H-X, Liu Y, Yan J-Y, Wang Y, Yao X, Deng Y-Q, Chen S-Y, Lu L, Wei Z-J, et al. 2017b. Time-dependent differential expression of long non-coding RNAs following peripheral nerve injury. *Int J Mol Med* **39:** 1381–1392. doi:10.3892/ijmm.2017.2963

Pavlaki I, Alammari F, Sun B, Clark N, Sirey T, Lee S, Woodcock DJ, Ponting CP, Szele FG, Vance KW. 2018. The long noncoding RNA *Paupar* promotes KAP1-dependent chromatin changes and regulates olfactory bulb neurogenesis. *EMBO J* **37:** e98219. doi:10.15252/embj.201798219.

Perry RB-T, Hezroni H, Goldrich MJ, Ulitsky I. 2018. Regulation of neuroregeneration by long noncoding RNAs. *Mol Cell* **72:** 553–567.e5. doi:10.1016/j.molcel.2018.09.021

Pinchi E, Frati A, Cantatore S, D'Errico S, Russa RL, Maiese A, Palmieri M, Pesce A, Viola RV, Frati P, et al. 2019. Acute spinal cord injury: a systematic review investigating miRNA families involved. *Int J Mol Sci* **20:** E1841. doi:10.3390/ijms20081841.

Piwecka M, Glažar P, Hernandez-Miranda LR, Memczak S, Wolf SA, Rybak-Wolf A, Filipchyk A, Klironomos F, Cerda Jara CA, Fenske P, et al. 2017. Loss of a mammalian circular RNA locus causes miRNA deregulation and affects brain function. *Science* **357:** eaam8526. doi:10.1126/science.aam8526.

Qin C, Liu C-B, Yang D-G, Gao F, Zhang X, Zhang C, Du L-J, Yang M-L, Li J-J. 2018. Circular RNA expression alteration and bioinformatics analysis in rats after traumatic spinal cord injury. *Front Mol Neurosci* **11:** 497. doi:10.3389/fnmol.2018.00497

Quan Z, Zheng D, Qing H. 2017. Regulatory roles of long noncoding RNAs in the central nervous system and associated neurodegenerative diseases. *Front Cell Neurosci* **11:** 175. doi:10.3389/fncel.2017.00175

Ramos AD, Diaz A, Nellore A, Delgado RN, Park K-Y, Gonzales-Roybal G, Oldham MC, Song JS, Lim DA. 2013. Integration of genome-wide approaches identifies lncRNAs of adult neural stem cells and their progeny in vivo. *Cell Stem Cell* **12:** 616–628. doi:10.1016/j.stem.2013.03.003

Ramos AD, Andersen RE, Liu SJ, Nowakowski TJ, Hong SJ, Gertz C, Salinas RD, Zarabi H, Kriegstein AR, Lim DA. 2015. The long noncoding RNA *Pnky* regulates neuronal differentiation of embryonic and postnatal neural stem cells. *Cell Stem Cell* **16:** 439–447. doi:10.1016/j.stem.2015.02.007

Rani N, Nowakowski TJ, Zhou H, Godshalk SE, Lisi V, Kriegstein AR, Kosik KS. 2016. A primate lncRNA mediates notch signaling during neuronal development by sequestering miRNA. *Neuron* **90:** 1174–1188. doi:10.1016/j.neuron.2016.05.005

Rapicavoli NA, Poth EM, Zhu H, Blackshaw S. 2011. The long noncoding RNA *Six3OS* acts in trans to regulate retinal development by modulating Six3 activity. *Neural Dev* **6:** 32. doi:10.1186/1749-8104-6-32

Ravasi T, Suzuki H, Pang KC, Katayama S, Furuno M, Okunishi R, Fukuda S, Ru K, Frith MC, Gongora MM, et al. 2006. Experimental validation of the regulated expression of large numbers of non-coding RNAs from the mouse genome. *Genome Res* **16:** 11–19. doi:10.1101/gr.4200206

Ren Y, Cui Y, Li X, Wang B, Na L, Shi J, Wang L, Qiu L, Zhang K, Liu G, et al. 2015. A co-expression network analysis reveals lncRNA abnormalities in peripheral blood in early-onset schizophrenia. *Prog Neuropsychopharmacol Biol Psychiatry* **63:** 1–5. doi:10.1016/j.pnpbp.2015.05.002

Riva P, Ratti A, Venturin M. 2016. The long non-coding RNAs in neurodegenerative diseases: novel mechanisms of pathogenesis. *Curr Alzheimer Res* **13:** 1219–1231. doi:10.2174/1567205013666160622112234

Saijilafu, Hur E-M, Liu C-M, Jiao Z, Xu W-L, Zhou F-Q. 2013a. PI3K-GSK3 signalling regulates mammalian axon regeneration by inducing the expression of Smad1. *Nat Commun* **4:** 2690. doi:10.1038/ncomms3690

Saijilafu, Zhang B-Y, Zhou F-Q. 2013b. Signaling pathways that regulate axon regeneration. *Neurosci Bull* **29:** 411–420. doi:10.1007/s12264-013-1357-4

Sauvageau M, Goff LA, Lodato S, Bonev B, Groff AF, Gerhardinger C, Sanchez-Gomez DB, Hacisuleyman E, Li E, Spence M, et al. 2013. Multiple knockout mouse models reveal lincRNAs are required for life and brain development. *Elife* **2:** e01749. doi:10.7554/eLife.01749

Sekar S, Liang WS. 2019. Circular RNA expression and function in the brain. *Noncoding RNA Res* **4:** 23–29. doi:10.1016/j.ncrna.2019.01.001

Singer RA, Arnes L, Cui Y, Wang J, Gao Y, Guney MA, Burnum-Johnson KE, Rabadan R, Ansong C, Orr G, et al. 2019. The long noncoding RNA *Paupar* modulates PAX6 regulatory activities to promote α cell development and function. *Cell Metab* doi:10.1016/j.cmet.2019.09.013

Sone M, Hayashi T, Tarui H, Agata K, Takeichi M, Nakagawa S. 2007. The mRNA-like noncoding RNA Gomafu constitutes a novel nuclear domain in a subset of neurons. *J Cell Sci* **120:** 2498–2506. doi:10.1242/jcs.009357

Sterne-Weiler T, Martinez-Nunez RT, Howard JM, Cvitovik I, Katzman S, Tariq MA, Pourmand N, Sanford JR. 2013. Frac-seq reveals isoform-specific recruitment to polyribosomes. *Genome Res* **23:** 1615–1623. doi:10.1101/gr.148585.112

Toy D, Namgung U. 2013. Role of glial cells in axonal regeneration. *Exp Neurobiol* **22:** 68–76. doi:10.5607/en.2013.22.2.68

Tuck AC, Natarajan KN, Rice GM, Borawski J, Mohn F, Rankova A, Flemr M, Wenger A, Nutiu R, Teichmann S, et al. 2018. Distinctive features of lincRNA gene expression suggest widespread RNA-independent functions. *Life Sci Alliance* **1:** e201800124. doi:10.26508/lsa.201800124

Ulitsky I. 2018. Interactions between short and long noncoding RNAs. *FEBS Lett* **592:** 2874–2883. doi:10.1002/1873-3468.13085

Ulitsky I, Shkumatava A, Jan CH, Sive H, Bartel DP. 2011. Conserved function of lincRNAs in vertebrate embryonic development despite rapid sequence evolution. *Cell* **147:** 1537–1550. doi:10.1016/j.cell.2011.11.055

Vance KW, Sansom SN, Lee S, Chalei V, Kong L, Cooper SE, Oliver PL, Ponting CP. 2014. The long non-coding RNA *Pau-*

par regulates the expression of both local and distal genes. *EMBO J* **33:** 296–311. doi:10.1002/embj.201386225

Wan P, Su W, Zhuo Y. 2017. The role of long noncoding RNAs in neurodegenerative diseases. *Mol Neurobiol* **54:** 2012–2021. doi:10.1007/s12035-016-9793-6

Wang C-F, Zhao C-C, Weng W-J, Lei J, Lin Y, Mao Q, Gao G-Y, Feng J-F, Jiang J-Y. 2017. Alteration in long non-coding RNA expression after traumatic brain injury in rats. *J Neurotrauma* **34:** 2100–2108. doi:10.1089/neu.2016.4642.

Wang H, Wu J, Zhang X, Ding L, Zeng Q. 2018. Microarray analysis of the expression profile of lncRNAs reveals the key role of lncRNA BC088327 as an agonist to heregulin-1β-induced cell proliferation in peripheral nerve injury. *Int J Mol Med* **41:** 3477–3484. doi:10.3892/ijmm.2018.3571.

Yan W, Chen Z-Y, Chen J-Q, Chen H-M. 2018. LncRNA NEAT1 promotes autophagy in MPTP-induced Parkinson's disease through stabilizing PINK1 protein. *Biochem Biophys Res Commun* **496:** 1019–1024. doi:10.1016/j.bbrc.2017.12.149

Yang B, Xia Z-A, Zhong B, Xiong X, Sheng C, Wang Y, Gong W, Cao Y, Wang Z, Peng W. 2017. Distinct hippocampal expression profiles of long non-coding RNAs in an Alzheimer's disease model. *Mol Neurobiol* **54:** 4833–4846. doi:10.1007/s12035-016-0038-5.

Yao C, Yu B. 2019. Role of long noncoding RNAs and circular RNAs in nerve regeneration. *Front Mol Neurosci* **12:** 165. doi:10.3389/fnmol.2019.00165

Yao C, Wang J, Zhang H, Zhou S, Qian T, Ding F, Gu X, Yu B. 2015. Long non-coding RNA uc.217 regulates neurite outgrowth in dorsal root ganglion neurons following peripheral nerve injury. *Eur J Neurosci* **42:** 1718–1725. doi:10.1111/ejn.12966

Yao C, Wang Y, Zhang H, Feng W, Wang Q, Shen D, Qian T, Liu F, Mao S, Gu X, et al. 2018. lncRNA TNXA-PS1 modulates Schwann cells by functioning as a competing endogenous RNA following nerve injury. *J Neurosci* **38:** 6574–6585. doi:10.1523/JNEUROSCI.3790-16.2018

Yen Y-P, Hsieh W-F, Tsai Y-Y, Lu Y-L, Liau ES, Hsu H-C, Chen Y-C, Liu T-C, Chang M, Li J, et al. 2018. *Dlk1-Dio3* locus-derived lncRNAs perpetuate postmitotic motor neuron cell fate and subtype identity. *Elife* **7:** e38080. doi:10.7554/eLife.38080.

Yi M, Dai X, Li Q, Xu X, Chen Y, Wang D. 2019. Downregulated lncRNA CRNDE contributes to the enhancement of nerve repair after traumatic brain injury in rats. *Cell Cycle* **18:** 2332–2343. doi:10.1080/15384101.2019.1647024

Yu B, Zhou S, Hu W, Qian T, Gao R, Ding G, Ding F, Gu X. 2013. Altered long noncoding RNA expressions in dorsal root ganglion after rat sciatic nerve injury. *Neurosci Lett* **534:** 117–122. doi:10.1016/j.neulet.2012.12.014

Zappulo A, van den Bruck D, Ciolli Mattioli C, Franke V, Imami K, McShane E, Moreno-Estelles M, Calviello L, Filipchyk A, Peguero-Sanchez E, et al. 2017. RNA localization is a key determinant of neurite-enriched proteome. *Nat Commun* **8:** 583. doi:10.1038/s41467-017-00690-6

Zhang L, Wang H. 2019. Long non-coding RNA in CNS injuries: a new target for therapeutic intervention. *Mol Ther Nucleic Acids* **17:** 754–766. doi:10.1016/j.omtn.2019.07.013

Zhang H, Li D, Zhang Y, Li J, Ma S, Zhang J, Xiong Y, Wang W, Li N, Xia L. 2018a. Knockdown of lncRNA BDNF-AS suppresses neuronal cell apoptosis via downregulating miR-130b-5p target gene PRDM5 in acute spinal cord injury. *RNA Biol* **15:** 1071–1080. doi:10.1080/15476286.2017.1408764

Zhang H, Wang W, Li N, Li P, Liu M, Pan J, Wang D, Li J, Xiong Y, Xia L. 2018b. LncRNA DGCR5 suppresses neuronal apoptosis to improve acute spinal cord injury through targeting PRDM5. *Cell Cycle* **17:** 1992–2000. doi:10.1080/15384101.2018.1509622

Zhao R-B, Zhu L-H, Shu J-P, Qiao L-X, Xia Z-K. 2018. GAS5 silencing protects against hypoxia/ischemia-induced neonatal brain injury. *Biochem Biophys Res Commun* **497:** 285–291. doi:10.1016/j.bbrc.2018.02.070

Zhong J, Jiang L, Cheng C, Huang Z, Zhang H, Liu H, He J, Cao F, Peng J, Jiang Y, et al. 2016. Altered expression of long noncoding RNA and mRNA in mouse cortex after traumatic brain injury. *Brain Res* **1646:** 589–600. doi:10.1016/j.brainres.2016.07.002

Zhou X, Xu J. 2015. Identification of Alzheimer's disease-associated long noncoding RNAs. *Neurobiol Aging* **36:** 2925–2931. doi:10.1016/j.neurobiolaging.2015.07.015

Zhou Z-B, Du D, Chen K-Z, Deng L-F, Niu Y-L, Zhu L. 2019. Differential expression profiles and functional predication of circular ribonucleic acid in traumatic spinal cord injury of rats. *J Neurotrauma* **36:** 2287–2297. doi:10.1089/neu.2018.6366

How Complementary Targets Expose the microRNA 3′ End for Tailing and Trimming during Target-Directed microRNA Degradation

Paulina Pawlica,[1] Jessica Sheu-Gruttadauria,[2,3] Ian J. MacRae,[2] and Joan A. Steitz[1]

[1]Department of Molecular Biophysics and Biochemistry, Howard Hughes Medical Institute, Yale University School of Medicine, New Haven, Connecticut 06536, USA

[2]Department of Integrative Structural and Computational Biology, The Scripps Research Institute, La Jolla, California 92037, USA

Correspondence: joan.steitz@yale.edu

microRNAs (miRNAs) are crucial for posttranscriptional regulation of messenger RNAs. "Classical" miRNA targets predominantly interact with the miRNA seed sequence located near the miRNA 5′ end. Interestingly, certain transcripts that exhibit extensive complementarity to the miRNAs 3′ region, instead of being subjected to regulation, induce miRNA decay in a process termed target-directed miRNA degradation (TDMD). Here, we review recent advances in understanding the molecular mechanisms of TDMD. Specifically, we discuss how extensive miRNA complementarity to TDMD-inducing targets results in displacement of the miRNA 3′ end from its protective pocket in the Argonaute protein. Unprotected miRNA 3′ ends are then available for enzymatic attack by still-unidentified cellular enzymes. Identification of these cellular enzymes and discovery of additional TDMD-inducing transcripts are subjects for future research.

microRNAs AND TARGET-DIRECTED microRNA DEGRADATION

microRNAs (miRNAs) are small noncoding RNAs (ncRNAs) that are essential for posttranscriptional regulation of more than one-half of messenger RNAs (mRNAs) in human cells (Bartel 2018). Aberrant levels of cellular miRNAs are linked to disease. miRNAs associate with and act through Argonaute (Ago) proteins, which are the primary components of the RNA-induced silencing complex. In addition, Ago protects miRNAs from degradation (Winter and Diederichs 2011) by selective association with the 5′ and 3′ ends with binding pockets in the Ago MID and PAZ domains, respectively (Wang et al. 2008). Despite this protection, miRNA stability varies greatly (Duffy et al. 2015; Marzi et al. 2016), and miRNA levels are tightly regulated (e.g., during development or the cell cycle [Monticelli et al. 2005; Rissland et al. 2011]). But the mechanisms of miRNA regulation are not well-understood and only two miRNA decay pathways have been described: Tudor SN-mediated miRNA decay (TumiD) (Elbarbary et al. 2017) and target-directed miRNA degradation (TDMD) (Ameres et al. 2010; Baccarini et al. 2011; Libri et al. 2012; Marcinowski et al. 2012; Lee et al. 2013; Bitetti et al. 2018; Cazalla 2018; Ghini et al. 2018; Kleaveland et al. 2018). TDMD is the subject of this review.

"Classical" miRNA targets usually interact with miRNA via base-pairing with the miRNA seed sequence (nucleotides [nt] 2–8 from the miRNA 5′ end); this interaction is sometimes accompanied by supplementary base-pairing to miRNA nt 13–16 (Fig. 1; Bartel 2018). Such interactions result in either mRNA degradation via deadenylation and decapping and/or translational repression. Interestingly, when a miRNA and its target are perfectly complementary, Ago2—one of four Ago proteins found in humans—is able to cleave the target (Liu et al. 2004), which is then quickly degraded by cellular exonucleases. However, when a target exhibits extensive base-pairing to the 3′ region of the miRNA combined with central mismatches that likely prevent Ago2-mediated cleavage, the miRNA, instead of the target, is subjected to degradation in a process known as TDMD (reviewed recently in de la Mata and Grosshans 2018; Fuchs Wightman et al. 2018).

Targets that are able to selectively induce miRNA decay, known as TDMD targets, represent various RNA classes, including ncRNAs—HSUR1 (small nuclear RNA from herpesvirus saimiri), UL144–145 (intergenic region from human cytomegalovirus), and Cyrano (cellular long ncRNA)—as well as mRNAs: m169 (from murine cytomegalovirus), NREP, and Serpine1 (Table 1; Libri et al. 2012; Marcinowski et al. 2012; Lee et al. 2013; Bitetti et al. 2018; Cazalla 2018; Ghini et al. 2018; Kleaveland et al. 2018). These RNAs vary not only in class and origin, but also seem to have flexible requirements for base-pairing with miRNAs to induce TDMD. In addition, HSUR1 mutagenesis further expanded the range of possible miRNA–TDMD target interactions (Table 2; Sheu-Gruttadauria and Pawlica et al. 2019).

---

[3]Present address: Department of Cellular and Molecular Pharmacology, Howard Hughes Medical Institute, University of California San Francisco, San Francisco, California 94143, USA

© 2019 Pawlica et al. This article is distributed under the terms of the Creative Commons Attribution-NonCommercial License, which permits reuse and redistribution, except for commercial purposes, provided that the original author and source are credited.

Published by Cold Spring Harbor Laboratory Press; doi:10.1101/sqb.2019.84.039321

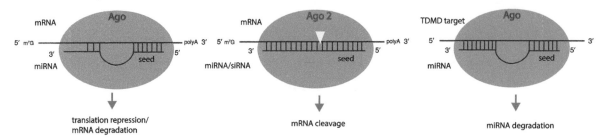

Figure 1. Three known modes of miRNA–target interaction on Ago proteins result in different outcomes.

Thus, bioinformatic attempts to identify additional TDMD targets have proved challenging (Fuchs Wightman et al. 2018; Ghini et al. 2018), and only a handful of TDMD targets have been discovered. Moreover, the molecular mechanism of TDMD remained elusive for nearly 10 years. Our recent paper provides critical structural insights into how TDMD targets trap Ago2 in an extended conformation with the miRNA 3′ end displayed for cellular tailing (nontemplated nucleotide additions) and trimming while remaining bound to Ago proteins (Sheu-Gruttadauria and Pawlica et al. 2019).

RECENT INSIGHTS INTO TDMD

Prior to target association, Ago displays only the miRNA seed sequence for base-pairing. Upon binding to "classical" targets, Ago and the miRNA undergo a conformational shift to reveal the supplementary nucleotides in preordered A-form, which then can bind supplementary target sequences (Schirle et al. 2014; Sheu-Gruttadauria et al. 2019). These interactions do not destabilize the miRNA 3′ region, which remains sequestered within a narrow RNA binding cleft with the 3′ end securely bound in the Ago PAZ domain. Thus, when binding canonical miRNA target RNAs, Ago protects the miRNA from degradation. However, when a TDMD target binds, the central RNA binding cleft of Ago opens and the PAZ domain shifts away from the protein body. Importantly, the miRNA 3′ end is dislodged from its binding pocket within the PAZ domain (Fig. 2; Sheu-Gruttadauria and Pawlica et al. 2019).

Confirmation that the function of a TDMD target is to expose the miRNA 3′ end comes from the analysis of miRNAs associated with Ago1 and 2 proteins each containing in the PAZ domain a Phe → Ala substitution predicted to impair miRNA 3′ end binding. Even in the absence of a TDMD target, mutation of the Ago PAZ domain pocket causes the fraction of mature miRNAs bound to Ago to decrease, whereas the fraction of tailed and trimmed miRNA isoforms (isomiRs) increases (Sheu-Gruttadauria and Pawlica et al. 2019). In the presence of a TDMD target these isomiRs disappear, suggesting that the TDMD target further dislodges the miRNA 3′ end or that the Ago2–miRNA complex transitions into a conformation that enables recognition by TDMD-mediating enzymes. Indeed, several cellular enzymes implicated in TDMD can be modeled onto the Ago TDMD conforma-

Table 1. Known miRNA–TDMD target interactions

miR-27a HSUR1 ncRNA, HSV	5′UUCACAGUGGCUAAGUUCCG 3′UAGUGUCUAA–AUUCAAGGUCA
miR-27a m169 3′UTR, MCMV	5′UUCACAGUGGCUAAGUUCCGC 3′AAGUGUCGAAUAA–UAAGGCGUC
miR-17 UL144-145 ncRNA, HCMV	5′CAAAGUGCUU- - -AC - - -AGUGCAGGUAG 3′AUUUCACGAAGAAAAAAAUCACGUCCU
miR-20a UL144-145 ncRNA, HCMV	5′CAAAGUGCUU- - - A - - -UAGUGCAGGUAG 3′AUUUCACGAAGAAAAAAAUCACGUCCU
miR-29b NREP 3′ UTR	5′UAGCACCAUUUGAAAUCAGUGUU 3′AUCGUGGUAAG-U- -AGUCACAGA
miR-7 Cyrano ncRNA	5′UGGAAGAC-UA-GUGAUUUUGUUGUU 3′ACCUUCUGUAACCACUAAAACAACAA
miR-30b Serpine1 3′ UTR	5′UGUAAACA- - U- -CCUACACUCAGCU 3′ACAUUUGUUCAGUGGAUGUGAGACUUU
miR-30c Serpine1 3′ UTR	5′UGUAAACA- - U - -CCUACACUCUCAGC 3′ACAUUUGUUCAGUGGAUGUGAGACUUU

Adapted from Sheu-Gruttadauria and Pawlica et al. 2019, with permission from Elsevier.
Complementary regions are colored green, nucleotides in the central bulge are colored red, and noncanonical base pairs are colored blue.
(TDMD) Targeted directed microRNA degradation, (ncRNA) noncoding RNA, (HSV) herpesvirus saimiri, (UTR) untranslated region, (MCMV) murine cytomegalovirus, (HCMV) human cytomegalovirus, (NREP) neuronal regeneration-related protein.

Table 2. Observed base-pairing interactions of miRNAs and their respective TDMD targets

	Seed interaction			Central mismatches			Pairing to miRNA 3′ region		
	Start	Length	ΔG	Start	Length	ΔG	Start	Length	ΔG
In miRNA	1–2	6–10	−9.6–−21.9	8–12	1–6	0–6.7	10–15	6–14	−11.2–−20.9
In target		6–10	−9.6–−21.9		0–8	0–6.7		6–14	−11.2–−20.9

Calculations of the start, length, and hybridization energy (kcal/mol) of the two paired and one unpaired regions in miRNA–TDMD target interactions. Start denotes nucleotides from the miRNA 5′ end (Ameres et al. 2010; Baccarini et al. 2011; Libri et al. 2012; Marcinowski et al. 2012; Lee et al. 2013; Bitetti et al. 2018; Cazalla 2018; Ghini et al. 2018; Kleaveland et al. 2018; Sheu-Gruttadauria and Pawlica et al. 2019).
(miRNA) microRNA, (TDMD) targeted-directed microRNA degradation.

tion poised to access the exposed miRNA 3′ end. These results, together with the evidence that tailing and trimming of miRNAs take place on Ago proteins (Marcinowski et al. 2012; de la Mata et al. 2015; Haas et al. 2016; Sheu-Gruttadauria and Pawlica et al. 2019), strongly suggest that the enzymes involved in TDMD either are already associated with Ago or are promptly recruited to Ago when the miRNA 3′ end becomes available.

OUTSTANDING QUESTIONS

An important question is which cellular enzymes are responsible for miRNA decay during TDMD. TDMD is associated with miRNA tailing and trimming (Ameres et al. 2010; Marcinowski et al. 2012; de la Mata et al. 2015; Ghini et al. 2018; Kleaveland et al. 2018; Sheu-Gruttadauria and Pawlica et al. 2019). Yet evidence for tailing as an essential step in miRNA decay is lacking. In fact, evidence that tailing is uncoupled from TDMD is accumulating. In mammalian cells there are 11 terminal nucleotidyl transferases (TENTs) with differing activities (mainly either adenylation or uridylation) (Warkocki et al. 2018). Many of these have been implicated in altering miRNA stability via tailing (Jones et al. 2009; Katoh et al. 2009; Boele et al. 2014; Gutiérrez-Vázquez et al. 2017). In addition, deep sequencing of miRNAs in the presence of TDMD targets revealed adenylation (Ghini et al. 2018; Kleaveland et al. 2018), uridylation (Baccarini et al. 2011; Haas et al. 2016; Sheu-Gruttadauria and Pawlica et al. 2019), and mixed A/U tails (Baccarini et al. 2011; de la Mata et al. 2015; Ghini et al. 2018; Sheu-Gruttadauria and Pawlica et al. 2019). However, TUT2 (also known as TENT2, GLD-2) knockout has no effect on Cyrano-mediated TDMD (Kleaveland et al. 2018), and TUT1 (also known as TENT1) knockdown likewise does not alter m169-mediated TDMD (Haas et al. 2016). Similarly, single and simultaneous knockouts of two of the main suspects that could catalyze miR-27a U-tailing in the presence of HSUR1 (Sheu-Gruttadauria and Pawlica et al. 2019)—TUT4 (also known as TENT3A or ZCCHC11)

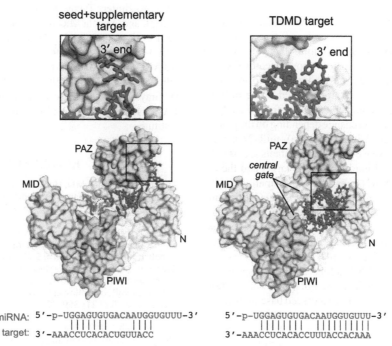

Figure 2. Binding of a TDMD target exposes the miRNA 3′ end. Crystal structures of the Ago2–miRNA complex bound to a "classical" miRNA target with seed + supplementary pairing (*left*, PDB 6N4O) or a TDMD-inducing target RNA (*right*, PDB 6MDZ). *Insets* show close-up views of the miRNA 3′ end bound to the PAZ domain in the seed + supplementary structure or exposed to enzymatic attack in the TDMD conformation. (*Below*) Observed miRNA–target base pairs are shown schematically.

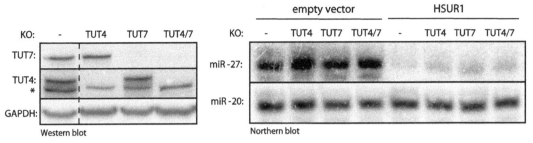

Figure 3. TUT4 and TUT7 are not required for HSUR1-mediated TDMD. *(Left)* Western blot showing the levels of TUT4, TUT7, and GAPDH in single clones of BJAB cells generated after CRISPR–Cas9-mediated knockout of TUT4, TUT7, or both. *(Right)* Northern blot showing the impact of single or simultaneous knockouts of TUT4 and TUT7 on HSUR1's ability to induce TDMD of miR-27. (KO) Knockout, (*) nonspecific band.

and TUT7 (also known as TENT3B or ZCCHC6)—do not appear to reverse HSUR1-mediated TDMD (Fig. 3). It therefore seems likely that the exposed miRNA 3′ end can be subjected either to tailing or to trimming, but tailing is not essential for ensuing decay. In addition, the trimming 3′-to-5′ exonucleases responsible for miRNA degradation remain to be identified. Those that have been examined and do not appear to function in TDMD include PARN (Kleaveland et al. 2018) and Dis3L2 (Haas et al. 2016; Kleaveland et al. 2018). Our HSUR1 and Ago mutagenesis studies suggest that TDMD enzymes do not require a lengthy tail of adenylates/uridylates as is needed for Dis3L2 (Faehnle et al. 2014) and PARN (Astrom et al. 1992; Wu et al. 2005), but do require association with Ago protein (Sheu-Gruttadauria and Pawlica et al. 2019). It is also possible that more than one exonuclease acts on an exposed miRNA 3′ end, just as multiple tailing activities clearly participate in both general miRNA decay and TDMD.

An interesting observation from our HSUR1 mutagenesis analyses underscores the role of target complementarity in the generation of isomiRs that cannot result from earlier miRNA processing steps. Specifically, the miRNA-target architecture on Ago appears to modulate the activity of enzymes that are recruited to the miRNA 3′ end. These results complement those from a study of TUT7, whose activity can change depending on the pairing status of its substrate (Kim et al. 2015). miR-27 isomiR generation in the presence of HSUR1 mutants seems to be enhanced by base-pairing between the target and the extreme 3′ nt of the miRNA, as well as by small and symmetrical central mismatches (Sheu-Gruttadauria and Pawlica et al. 2019). Also, interestingly, certain elongated miR-27 isomiRs may acquire altered ability to repress target mRNAs depending on the mode of miRNA-target interaction (miRNA-like vs. siRNA-like). Moreover, it is becoming clear that miRNA tailing can alter the specificity of miRNA repression, as in the case of miR-27 uridylation, which can compensate for imperfect seed base-pairing and repress noncanonical targets (Yang et al. 2019).

Another outstanding question is what defines a TDMD target. We have learned that base-pairing to the miRNA seed sequence and the miRNA 3′ region, as well as the central mismatches, are necessary, but not sufficient for miRNA decay (Lee et al. 2013; P Pawlica, unpubl. results). It is critical that a miRNA binding site be exposed in an unstructured region of the TDMD target RNA (Pawlica et al. 2016). However, the great majority of transcripts fulfilling these criteria does not induce TDMD. It is possible that additional protein-binding motifs are located near miRNA binding sites (Lee et al. 2013; Kleaveland et al. 2018), whose identification would greatly enhance the pace of discovery of novel TDMD targets. Finally, it remains unclear why TDMD targets appear to have different cell type–specific potency (de la Mata et al. 2015; Kleaveland et al. 2018; P Pawlicka, unpubl. results).

In summary, our study brings us closer to understanding the molecular mechanism of TDMD, but many challenging questions still remain.

ACKNOWLEDGMENTS

We are grateful to Therese Yario for technical assistance, Kazimierz Tycowski for critical reading of the manuscript, and Angela Miccinello for editorial support. This work was supported by National Institutes of Health (NIH) grants GM115649, GM104475, and GM127090 (I.J.M.), and CA016038 (J.A.S.). P.P. is funded by NIH fellowship K99/R00 (K99GM129412). J.S.-G. is a Hanna H. Gray Fellow of the Howard Hughes Medical Institute. J.A.S. is an investigator of the Howard Hughes Medical Institute.

REFERENCES

Ameres SL, Horwich MD, Hung JH, Xu J, Ghildiyal M, Weng Z, Zamore PD. 2010. Target RNA-directed trimming and tailing of small silencing RNAs. *Science* **328:** 1534–1539. doi:10.1126/science.1187058

Astrom J, Astrom A, Virtanen A. 1992. Properties of a HeLa cell 3′ exonuclease specific for degrading poly(A) tails of mammalian mRNA. *J Biol Chem* **267:** 18154–18159.

Baccarini A, Chauhan H, Gardner TJ, Jayaprakash AD, Sachidanandam R, Brown BD. 2011. Kinetic analysis reveals the fate of a microRNA following target regulation in mammalian cells. *Curr Biol* **21:** 369–376. doi:10.1016/j.cub.2011.01.067

Bartel DP. 2018. Metazoan microRNAs. *Cell* **173:** 20–51. doi:10.1016/j.cell.2018.03.006

Bitetti A, Mallory AC, Golini E, Carrieri C, Carreño Gutiérrez H, Perlas E, Pérez-Rico YA, Tocchini-Valentini GP, Enright AJ, Norton WHJ, et al. 2018. MicroRNA degradation by a conserved target RNA regulates animal behavior. *Nat Struct Mol Biol* **25:** 244–251. doi:10.1038/s41594-018-0032-x

Boele J, Persson H, Shin JW, Ishizu Y, Newie IS, Sokilde R, Hawkins SM, Coarfa C, Ikeda K, Takayama K, et al. 2014. PAPD5-mediated 3′ adenylation and subsequent degradation of miR-21 is disrupted in proliferative disease. *Proc Natl Acad Sci* **111**: 11467–11472. doi:10.1073/pnas.1317751111

Cazalla D. 2018. Novel roles for Sm-class RNAs in the regulation of gene expression. *RNA Biol* **15**: 856–862. doi:10.1080/15476286.2018.1467176

de la Mata M, Gaidatzis D, Vitanescu M, Stadler MB, Wentzel E, Scheiffele P, Filipowicz W, Grosshans H. 2015. Potent degradation of neuronal miRNAs induced by highly complementary targets. *EMBO Rep* **16**: 500–511. doi:10.15252/embr.201540078

de la Mata M, Grosshans H. 2018. Turning the table on miRNAs. *Nat Struct Mol Biol* **25**: 195–197. doi: 10.1038/s41594-018-0040-x

Duffy EE, Rutenberg-Schoenberg M, Stark CD, Kitchen RR, Gerstein MB, Simon MD. 2015. Tracking distinct RNA populations using efficient and reversible covalent chemistry. *Mol Cell* **59**: 858–866. doi:10.1016/j.molcel.2015.07.023

Elbarbary RA, Miyoshi K, Myers JR, Du P, Ashton JM, Tian B, Maquat LE. 2017. Tudor-SN-mediated endonucleolytic decay of human cell microRNAs promotes G_1/S phase transition. *Science* **356**: 859–862. doi:10.1126/science.aai9372

Faehnle CR, Walleshauser J, Joshua-Tor L. 2014. Mechanism of Dis3L2 substrate recognition in the Lin28-let-7 pathway. *Nature* **514**: 252–256. doi:10.1038/nature13553

Fuchs Wightman F, Giono LE, Fededa JP, de la Mata M. 2018. Target RNAs strike back on microRNAs. *Front Genet* **9**: 435. doi:10.3389/fgene.2018.00435

Ghini F, Rubolino C, Climent M, Simeone I, Marzi MJ, Nicassio F. 2018. Endogenous transcripts control miRNA levels and activity in mammalian cells by target-directed miRNA degradation. *Nat Commun* **9**: 3119. doi:10.1038/s41467-018-05182-9

Gutiérrez-Vázquez C, Enright AJ, Rodríguez-Galán A, Pérez-García A, Collier P, Jones MR, Benes V, Mizgerd JP, Mittelbrunn M, Ramiro AR, et al. 2017. 3′ uridylation controls mature microRNA turnover during CD4 T-cell activation. *RNA* **23**: 882–891. doi:10.1261/rna.060095.116

Haas G, Cetin S, Messmer M, Chane-Woon-Ming B, Terenzi O, Chicher J, Kuhn L, Hammann P, Pfeffer S. 2016. Identification of factors involved in target RNA-directed microRNA degradation. *Nucleic Acids Res* **44**: 2873–2887. doi:10.1093/nar/gkw040

Jones MR, Quinton LJ, Blahna MT, Neilson JR, Fu S, Ivanov AR, Wolf DA, Mizgerd JP. 2009. Zcchc11-dependent uridylation of microRNA directs cytokine expression. *Nat Cell Biol* **11**: 1157–1163. doi:10.1038/ncb1931

Katoh T, Sakaguchi Y, Miyauchi K, Suzuki T, Kashiwabara S, Baba T, Suzuki T. 2009. Selective stabilization of mammalian microRNAs by 3′ adenylation mediated by the cytoplasmic poly(A) polymerase GLD-2. *Genes Dev* **23**: 433–438. doi: 10.1101/gad.1761509

Kim B, Ha M, Loeff L, Chang H, Simanshu DK, Li S, Fareh M, Patel DJ, Joo C, Kim VN. 2015. TUT7 controls the fate of precursor microRNAs by using three different uridylation mechanisms. *EMBO J* **34**: 1801–1815. doi:10.15252/embj.201590931

Kleaveland B, Shi CY, Stefano J, Bartel DP. 2018. A network of noncoding regulatory RNAs acts in the mammalian brain. *Cell* **174**: 350–362.e17. doi:10.1016/j.cell.2018.05.022

Lee S, Song J, Kim S, Kim J, Hong Y, Kim Y, Kim D, Baek D, Ahn K. 2013. Selective degradation of host microRNAs by an intergenic HCMV noncoding RNA accelerates virus production. *Cell Host Microbe* **13**: 678–690. doi:10.1016/j.chom.2013.05.007

Libri V, Helwak A, Miesen P, Santhakumar D, Borger JG, Kudla G, Grey F, Tollervey D, Buck AH. 2012. Murine cytomegalovirus encodes a miR-27 inhibitor disguised as a target. *Proc Natl Acad Sci* **109**: 279–284. doi:10.1073/pnas.1114204109

Liu J, Carmell MA, Rivas FV, Marsden CG, Thomson JM, Song JJ, Hammond SM, Joshua-Tor L, Hannon GJ. 2004. Argonaute2 is the catalytic engine of mammalian RNAi. *Science* **305**: 1437–1441. doi:10.1126/science.1102513

Marcinowski L, Tanguy M, Krmpotic A, Rädle B, Lisnić VJ, Tuddenham L, Chane-Woon-Ming B, Ruzsics Z, Erhard F, Benkartek C, et al. 2012. Degradation of cellular mir-27 by a novel, highly abundant viral transcript is important for efficient virus replication in vivo. *PLoS Pathog* **8**: e1002510. doi:10.1371/journal.ppat.1002510

Marzi MJ, Ghini F, Cerruti B, de Pretis S, Bonetti P, Giacomelli C, Gorski MM, Kress T, Pelizzola M, Muller H, et al. 2016. Degradation dynamics of microRNAs revealed by a novel pulse-chase approach. *Genome Res* **26**: 554–565. doi:10.1101/gr.198788.115

Monticelli S, Ansel KM, Xiao C, Socci ND, Krichevsky AM, Thai TH, Rajewsky N, Marks DS, Sander C, Rajewsky K, et al. 2005. MicroRNA profiling of the murine hematopoietic system. *Genome Biol* **6**: R71. doi:10.1186/gb-2005-6-8-r71

Pawlica P, Moss WN, Steitz JA. 2016. Host miRNA degradation by *Herpesvirus saimiri* small nuclear RNA requires an unstructured interacting region. *RNA* **22**: 1181–1189. doi:10.1261/rna.054817.115

Rissland OS, Hong SJ, Bartel DP. 2011. MicroRNA destabilization enables dynamic regulation of the miR-16 family in response to cell-cycle changes. *Mol Cell* **43**: 993–1004. doi:10.1016/j.molcel.2011.08.021

Schirle NT, Sheu-Gruttadauria J, MacRae IJ. 2014. Structural basis for microRNA targeting. *Science* **346**: 608–613. doi:10.1126/science.1258040

Sheu-Gruttadauria J, Pawlica P, Klum SM, Wang S, Yario TA, Schirle Oakdale NT, Steitz JA, MacRae IJ. 2019. Structural basis for target-directed microRNA degradation. *Molecular Cell* **75**: 1243–1255 e1247. doi:10.1016/j.molcel.2019.06.019

Sheu-Gruttadauria J, Xiao Y, Gebert LF, MacRae IJ. 2019. Beyond the seed: structural basis for supplementary microRNA targeting by human Argonaute2. *EMBO J* **38**: e101153. doi:10.15252/embj.2018101153

Wang Y, Sheng G, Juranek S, Tuschl T, Patel DJ. 2008. Structure of the guide-strand-containing Argonaute silencing complex. *Nature* **456**: 209–213. doi:10.1038/nature07315

Warkocki Z, Liudkovska V, Gewartowska O, Mroczek S, Dziembowski A. 2018. Terminal nucleotidyl transferases (TENTs) in mammalian RNA metabolism. *Philos Trans R Soc Lond B Biol Sci* **373**: 20180162. doi:10.1098/rstb.2018.0162

Winter J, Diederichs S. 2011. Argonaute proteins regulate microRNA stability: increased microRNA abundance by Argonaute proteins is due to microRNA stabilization. *RNA Biol* **8**: 1149–1157. doi:10.4161/rna.8.6.17665

Wu M, Reuter M, Lilie H, Liu Y, Wahle E, Song H. 2005. Structural insight into poly(A) binding and catalytic mechanism of human PARN. *EMBO J* **24**: 4082–4093. doi:10.1038/sj.emboj.7600869

Yang A, Bofill-De Ros X, Shao TJ, Jiang M, Li K, Villanueva P, Dai L, Gu S. 2019. 3′ uridylation confers miRNAs with noncanonical target repertoires. *Mol Cell* **75**: 511–522 e514. doi:10.1016/j.molcel.2019.05.014

Dicer's Helicase Domain: A Meeting Place for Regulatory Proteins

SARAH R. HANSEN,[1] ADEDEJI M. ADEROUNMU,[1,2] HELEN M. DONELICK,[1,2] AND BRENDA L. BASS[1]

[1]*Department of Biochemistry, University of Utah, Salt Lake City, Utah 84112-5650, USA*

Correspondence: bbass@biochem.utah.edu

The function of Dicer's helicase domain has been enigmatic since its discovery. Why do only some Dicers require ATP, despite a high degree of sequence conservation in their helicase domains? We discuss evolutionary considerations based on differences between vertebrate and invertebrate antiviral defense, and how the helicase domain has been co-opted in extant organisms as the binding site for accessory proteins. Many accessory proteins are double-stranded RNA binding proteins, and we propose models for how they modulate Dicer function and catalysis.

The gene we now know as Dicer was first revealed in a screen for double-stranded RNA (dsRNA) binding proteins (dsRBPs) (Bass et al. 1994) before the discovery of the surprising phenomena of RNA interference (RNAi) (Fire et al. 1998). Once RNAi was discovered, the unique features of the mysterious gene, which included an amino-terminal helicase domain and tandem ribonuclease III (RNase III) domains (Fig. 1A), led to the proposal that it was key to the process (Bass 2000). Indeed, subsequent studies in *Drosophila melanogaster* S2 cells (Bernstein et al. 2001) and *Caenorhabditis elegans* (Knight and Bass 2001) confirmed that Dicer was key to RNAi.

The observed RNase III domains placed Dicer in a well-characterized family of endoribonucleases, first discovered in *Escherichia coli* (Robertson et al. 1968) and now known to exist in Bacteria and Eukarya, but only infrequently in Archaea (Court et al. 2013; Nicholson 2014). Because RNAi had been correlated with the cleavage of dsRNA into smaller pieces (Zamore et al. 2000), the function of the RNase III domains was immediately predictable, and we now know that they catalyze cleavage of dsRNA into microRNAs (miRNAs) and short interfering RNAs (siRNAs) (Wilson and Doudna 2013; Ha and Kim 2014). In contrast to the RNase III domains, the function of the amino-terminal helicase domain was enigmatic when Dicer was discovered and, in many respects, remains so today. However, in recent years progress has been made in understanding this domain, and one goal of this review is to summarize recent insights.

Soon after the discovery of RNAi, it was recognized that, in plants, the mysterious phenomena of gene silencing in response to transgene expression and of cosuppression involved dsRNA and were based in an antiviral defense pathway (Waterhouse et al. 2001; Ding and Voinnet 2007). In addition, studies in invertebrates showed that RNAi targeted transposons (Ketting et al. 1999), remnants of ancient viruses. These observations paved the way for the realization that Dicer's helicase domain is most similar to helicases that act as viral sensors in mammalian innate immunity, the RIG-I-like receptors (RLRs; e.g., RIG-I, MDA5, LGP2) (Fairman-Williams et al. 2010; Ahmad and Hur 2015), and the demonstration for invertebrate Dicers that mutations in the helicase domain increase susceptibility to viral infection (Deddouche et al. 2008; Marques et al. 2013).

Here we take the view that the ancestral function of the helicase domain was in antiviral defense, and in certain extant organisms this domain has been co-opted for other cellular functions, such as regulation of miRNA processing, and production of endogenous siRNAs (endo-siRNAs). These modern-day functions often require cellular dsRNA binding proteins (dsRBPs) that directly interact with the helicase domain to modulate its functions. So far, these accessory dsRBPs all contain multiple dsRNA binding motifs (dsRBMs; Fig. 1B): an ∼65-amino acid motif that folds into an αβββα topology (Masliah et al. 2013; Gleghorn and Maquat 2014). Although there are many enzymes that include dsRBMs in conjunction with catalytic domains, such as adenosine deaminases that act on RNA (ADARs), RNA-dependent protein kinase (PKR), Dicer, and most RNase III enzymes, others simply consist of multiple dsRBMs separated by linkers. For the most part, it is the latter that have been found to interact with Dicer, and these will be our main focus. Each of these dsRBPs contains two amino-terminal "Type A" dsRBMs, which are more conserved and competent to bind dsRNA (Gleghorn and Maquat 2014). As indicated (Fig. 1B), some of these proteins also contain a carboxy-terminal "Type B" dsRBM, divergent motifs that frequently lack residues important for binding dsRNA, display little or no affinity for dsRNA, and typically mediate protein–protein interactions (Gleghorn and Maquat 2014).

---

[2]These authors contributed equally to this work.

© 2019 Hansen et al. This article is distributed under the terms of the Creative Commons Attribution-NonCommercial License, which permits reuse and redistribution, except for commercial purposes, provided that the original author and source are credited.

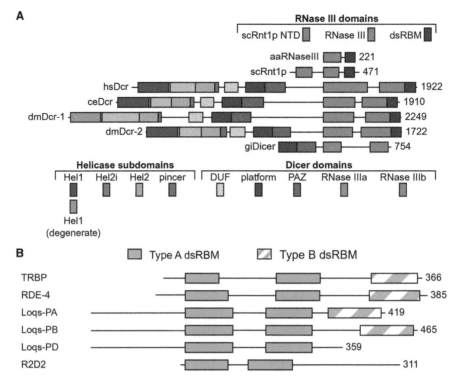

Figure 1. (*A*) Domains of Dicer and RNase III open-reading frames including *Aquifex aeolicus* RNase III (aaRNaseIII), *Saccharomyces cerevisiae* Rnt1p (scRnt1p), *Homo sapiens* Dicer (hsDcr), *Caenorhabditis elegans* Dicer (ceDcr), *Drosophila melanogaster* Dicers (dmDcr-1 and dmDcr-2), and *Giardia intestinalis* Dicer (giDicer). Domains are to scale and anchored at the amino terminus of aaRNase III, with the carboxy-terminal amino acid numbered on the right. (*Top*) color coding of RNase III domains, including amino-terminal domain (NTD). (*Bottom*) Color coding of helicase subdomains and Dicer domains. (*B*) Domain organization of Dicer-interacting dsRBPs noting Type A (solid) and Type B (striped) double-stranded RNA binding motifs (dsRBMs). Open-reading frames are to scale and anchored by the amino terminus of the first dsRBM. Numbers at the far right indicate the carboxy-terminal amino acid.

THE HELICASE DOMAINS OF VERTEBRATE AND INVERTEBRATE DICERS HAVE DISTINCT ACTIVITIES

Like other RLRs, Dicer's helicase domain belongs to the Superfamily 2 group of helicases and is comprised of three subdomains: Hel1, Hel2i, and Hel2 (Fig. 1A; Fairman-Williams et al. 2010; Luo et al. 2013; Ahmad and Hur 2015). Helicase domains have diverse functions, including the unwinding of nucleic acid double helices, promoting conformational changes, and translocation along nucleic acids (Singleton et al. 2007; Jarmoskaite and Russell 2014; Ahmad and Hur 2015). These functions are all fueled by ATP hydrolysis, and indeed, initial studies in *D. melanogaster* and *C. elegans* extracts showed a clear requirement for ATP (Zamore et al. 2000; Ketting et al. 2001). Thus, it came as a surprise when *Homo sapiens* Dicer (hsDcr) was purified to homogeneity and found to cleave dsRNA in the complete absence of ATP (Provost et al. 2002; Zhang et al. 2002).

Indeed, so far, a vertebrate Dicer has never been observed to require ATP. In contrast, there are multiple examples of invertebrate Dicers, those of plants, and that of *Schizosaccharomyces pombe* that show an ATP requirement in vitro (Liu et al. 2003; Colmenares et al. 2007; Welker et al. 2011; Fukudome and Fukuhara 2017). Further, a mutation in the helicase Walker A motif of *D. melanogaster* Dicer-2 (dmDcr-2), a motif critical for ATP binding and hydrolysis, eliminates processing of siRNAs, establishing an ATP dependence in vivo (Lee et al. 2004). Similarly, point mutations in any of three different helicase motifs in another invertebrate Dicer, *C. elegans* DCR-1 (ceDCR-1), produced animals that were defective for processing endo-siRNAs (Welker et al. 2010).

An important clue to understanding the function of Dicer's helicase domain came with the observation that, although *C. elegans* with mutations in the helicase domain of ceDCR-1 were defective for processing endo-siRNAs, these animals contained normal levels of miRNAs (Pavelec et al. 2009; Welker et al. 2010). Although termini of endo-siRNA precursors are not well characterized, miRNA precursors have a distinct 3′ overhang, and, thus, the *C. elegans* studies led to in vitro studies with purified dmDcr-2 and *C. elegans* extracts that showed these invertebrate Dicers cleave dsRNA differently depending on its termini (Welker et al. 2011; Sinha et al. 2015). When dmDcr-2 encounters a dsRNA with a blunt terminus, an optimal reaction ensues, whereby the helicase domain enables ATP-dependent, processive cleavage of the dsRNA. In contrast, dsRNA with 3′-overhanging termini promote a suboptimal, distributive cleavage that does not require ATP. The different termini-dependent reaction modes were confusing until a cryo-electron microscopy (cryo-EM) study of dmDcr-2 was reported (Sinha

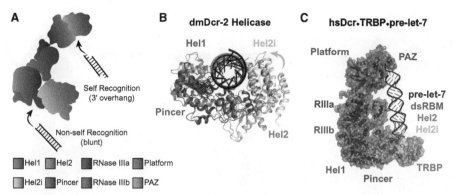

Figure 2. (*A*) A cartoon illustrating that dmDcr-2 uses its Platform•PAZ domains to recognize substrates with 3′ overhangs (self) and the helicase domain to recognize dsRNA with blunt termini (non-self). (*Bottom*) Color coding of domains, used throughout the figure. (*B*) Superimposition of dmDcr-2 helicase conformations shows an open, C-shaped, conformation in the apo structure (light, PDB: 6BUA) and a closed conformation in the presence of blunt dsRNA and an ATP analog (dark, PDB: 6BU9). In the closed conformation, Hel2 and Hel2i swivel toward Hel1 to clamp on the dsRNA terminus. (*C*) Structure of the hsDcr•TRBP•pre-let-7 complex with domains labeled (PDB: 5ZAL). Only the carboxy-terminal Type B motif of TRBP was resolved in the structure (Liu et al. 2018).

et al. 2018). Although numerous studies had implicated the Platform•PAZ domains in mediating Dicer's recognition of dsRNA termini (Fig. 1A), the cryo-EM study showed that dsRNA with blunt termini binds to the helicase domain (Fig. 2A). Because viral dsRNA in some cases has a blunt terminus (Schlee 2013), this led to the simple model that the helicase domain recognizes non-self dsRNA, such as viral dsRNA, threading it through the helicase domain to processively cleave the dsRNA, whereas the Platform•PAZ domains recognizes termini of cellular, or "self," dsRNA, such as miRNAs, acting distributively to make a single dsRNA cleavage with each binding event.

Based on phylogenetic studies (Zou et al. 2009; Mukherjee et al. 2013, 2014), it seems possible that ancient Dicers, which predated the split of invertebrate and vertebrate lineages, had helicase domains that discriminated blunt dsRNA as "non-self." The emergence of RLRs allowed for loss of "non-self" recognition by the helicase domain of certain Dicers, and this occurred in different ways. In *H. sapiens*, recognition of blunt dsRNA was taken over by RIG-I, and as interferon signaling evolved, the helicase domain of hsDcr lost the ability to recognize blunt dsRNA and hydrolyze ATP. Instead, hsDcr was coopted for miRNA processing, possibly enabled by interacting with dsRBPs such as TRBP. (However, we note that the helicase domains of mammalian Dicers are highly conserved, and it remains possible that ATP-dependent functions of mammalian Dicers will be discovered.)

At the other extreme, in arthropods like *D. melanogaster*, gene duplication allowed dmDcr-2 to exclusively provide antiviral functions and optimize these by coupling recognition of blunt viral dsRNA to processivity (Welker et al. 2011; Sinha et al. 2015). A second Dicer in *D. melanogaster*, dmDcr-1, became dedicated to the miRNA pathway, and its Hel1 domain became degenerate (Fig. 1A). Like mammals, *C. elegans* encode only a single Dicer, but unlike mammals, *C. elegans* do not have an interferon pathway, and a single Dicer participates in processing miRNA and endo-siRNA precursors, and the antiviral response. ceDCR-1 has not been characterized as a pure protein, and it is unclear whether it is capable of discriminating non-self or, alternatively, is assisted by other proteins, such as the Dicer-related helicase, DRH-1, a RIG-I ortholog required for the *C. elegans* antiviral response (Thivierge et al. 2011; Ashe et al. 2013; Guo et al. 2013; Sowa et al. 2020).

HOW DO dsRBPs MODULATE DICER FUNCTION?

All of the Dicers that have been purified to homogeneity are able to process dsRNA in vitro, in the complete absence of dsRBP accessory factors. Our thesis is that dsRBPs evolved to lend new activities to an enzyme that was specialized for processing viral dsRNA, and indeed, Dicer-interacting dsRBPs confer additional activities to Dicer, often by using their carboxy-terminal Type B dsRBM to interact with the helicase domain (Fig. 1B). Below we review some of the most well-studied dsRBPs known to interact with Dicer and speculate on the mechanisms by which they modulate Dicer activity. We consider two general mechanisms. First, for the PD isoform of Loquacious, we discuss a mechanism whereby "tethering" of the carboxy-terminal region to the helicase domain promotes a conformational change in the helicase domain itself (Fig 2B). Second, and possibly applicable to all Dicer-interacting dsRBPs, we consider a mechanism whereby the carboxy-terminal region of the dsRBP (e.g., TRBP in Fig. 2C) acts as a tether to bring in amino-terminal dsRBMs to engage with the dsRNA substrate and directly modulate catalysis by the RNase III domains.

Mechanistic Insights from RNase III Enzymes

In the proposed catalytic mechanism of RNase III (Fig. 3A), the dsRNA substrate is initially recognized by a single Type A dsRBM. The dsRBM moves the dsRNA into the "catalytic valley" formed between the two RNase III

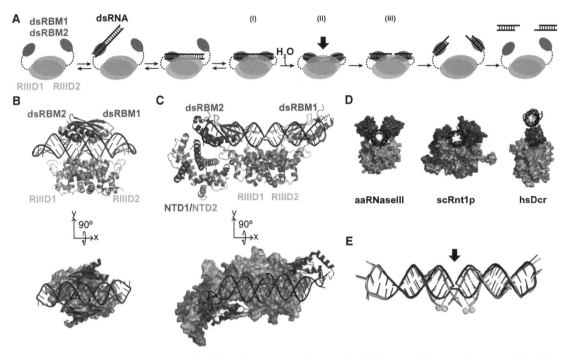

Figure 3. (*A*) The proposed catalytic pathway for RNase III adapted for Dicer models (Nicholson 2014). Double-stranded RNA binding motifs (dsRBMs) and RNase III domains are numbered and colored with similar shades for each monomer. Dashed lines represent covalent linkage of dsRBMs in RNase III or Dicer or noncovalent linkage for those provided *in trans* by accessory dsRBPs. The dsRBMs engage the dsRNA (i), then "squeeze" the helix into the dimeric active sites, creating a bend in the dsRNA helix (ii) and promoting hydrolysis of the phosphodiester bonds on opposite sides of the same minor groove (iii). The arrow in ii signifies the bent helix state at the dimeric active site. For Dicer, the pathway proceeds with its carboxy-terminal dsRBM or, possibly, with dsRBMs bent by an accessory dsRBP. (*B,C*) Crystal structures of aaRNaseIII (*B*; PDB: 2NUF) and scRnt1p (*C*; PDB: 5T16) with dsRNA positioned in the catalytic valley (Gan et al. 2008; Song et al. 2017). In the side view the proteins are shown as cartoons, and in the top-down view (*below*) the domains beneath the dsRNA are shown as surfaces. (*D*) Looking down the helical axis of dsRNA for aaRNaseIII (PDB: 2NUF), scRnt1p (PDB: 5T16), and the hsDcr•TRBP•pre-let-7 cryo-EM structure (PDB: 5ZAL). For hsDcr, only let-7 pre-miRNA and the RNase IIIa, RNase IIIb, and dsRBM domains are shown (residues 1293–1913), with the structured linker between RNase IIIa and RNase IIIb shown in gray (Gan et al. 2008; Song et al. 2017; Liu et al. 2018). (*E*) scRnt1p crystal structures were aligned by their RNase III domains to illustrate the straight (black) and bent (orange) conformations of the dsRNA helix (PDB: 5T16 and 4OOG). Active site Mg^{2+} ions that facilitate cleavage are represented by green spheres (two per monomer, observed only in PDB: 4OOG).

monomers, and the second dsRBM associates to form the precatalytic complex, where two dsRBMs secure the dsRNA at the active site (Fig 3A, i). Substrate-loaded and postcatalytic structures of *Aquifex aeolicus* RNase III (aaRNaseIII) and *Saccharomyces cerevisiae* RNase III (scRnt1p) reveal two conformations of dsRNA in the catalytic valley: a bent and a typical (unbent) A-form RNA helix (Fig. 3B–E; Gan et al. 2005; Liang et al. 2014; Song et al. 2017). Presumably, the bent helix observed postcatalytically also represents the catalytic site arrangement immediately before cleavage by hydrolysis (Fig. 3A, ii), whereas the straightened helix represents the conformation when dsRNA is first bound and then as the first step toward product release (Fig. 3A, iii; Gan et al. 2005; Song et al. 2017). Analysis of these structures suggests that protein and RNA components undergo conformational changes during catalysis causing the scissile bond to be "pulled" into the active sites. Although the dsRBMs do not undergo a significant conformational change between these steps, there is a change in their relative positions (Gan et al. 2008), raising the possibility that the dsRBMs "slide" along the bound dsRNA to "squeeze" the scissile bonds into the active site to enable catalysis.

Although recent cryo-EM structures of dmDcr-2 and hsDcr represent a significant advance in resolution (Liu et al. 2018; Sinha et al. 2018), neither structure shows dsRNA engaged at the RNase III active sites and, thus, does not offer insight into the cleavage-competent state. In fact, in the cryo-EM structure of the hsDcr•TRBP•pre-miRNA complex (Fig. 2C), Dicer's dsRBM occupies the catalytic valley where the dsRNA substrate would be predicted to bind for cleavage, and instead, the pre-let-7 substrate interacts with the opposite face of hDcr's carboxy-terminal dsRBM (Fig. 3D, hsDcr; Liu et al. 2018). The absence of dsRNA in the catalytic valley of hsDcr is not surprising as the complex was intentionally reconstituted in the absence of Mg^{2+}, and based on studies of aaRNase III, magnesium is necessary for a catalytically competent state (Blaszczyk et al. 2001). In fact, the arrangement of pre-let-7 in the hsDcr•TRBP•pre-miRNA complex is similar to that in the crystal structure of dsRNA bound by aaRNaseIII outside of the catalytic valley in the absence of Mg^{2+} (Gan et al. 2008).

Based on existing RNase III structures, we propose a mechanism for how some accessory dsRBPs might contribute to Dicer activity. To form a structure analogous to aaRNaseIII and scRnt1p, the dsRNA substrate would need to be engaged by two RNase III domains and two dsRBMs. Metazoan Dicers contain two RNase III domains, one dsRBM at the carboxyl terminus, and the DUF283 domain, which adopts a dsRBM-like αββα fold (Fig. 1; Qin et al. 2010). Although the DUF283 domain may function as one of the dsRBMs that contribute to an RNase III–type mechanism (Liu et al. 2018), it has been shown to interact with ssRNA, not dsRNA (Kurzynska-Kokorniak et al. 2016). However, certain observations are consistent with the carboxy-terminal dsRBM of hsDcr playing a role. In vitro, hsDcr lacking the carboxy-terminal dsRBM shows a fourfold reduction in initial rates of cleavage of miRNA and siRNA precursors (Ma et al. 2008), but this truncation has no effect on binding affinity for the miRNA precursor and only causes a slight increase in the affinity for the siRNA precursor. This is consistent with the idea that hsDcr's carboxy-terminal dsRBM functions similarly to one of the RNase III dsRBMs, playing a direct role in catalysis by promoting the transition state (Fig. 3A).

It seems feasible that, at least in some cases, Dicer accessory dsRBPs also contribute to catalysis by lending one or more dsRBMs to the transition state. This role is not absolutely required, as purified Dicer is catalytically active in vitro in the absence of binding partners and its carboxy-terminal dsRBM (Ma et al. 2008). Additionally, *Giardia intestinalis* Dicer lacks the helicase, DUF283, and dsRBM domains, yet is active in vitro in the absence of accessory proteins (Fig. 1; MacRae et al. 2006). Instead, one role of dsRBMs in Dicer and its accessory proteins would be to enhance the rate of cleavage, consistent with hsDcr truncation experiments (Ma et al. 2008). This contribution may be particularly required for certain suboptimal dsRNA substrates.

TRBP

In humans, the most well-studied cofactor of Dicer is the HIV-transactivating response RNA-binding protein (TRBP), first discovered as a binding partner for the TAR RNA structure of HIV (Gatignol et al. 1991). Later, immunoprecipitation experiments showed that TRBP interacts with hsDcr and Argonaute 2 (Ago2) as part of the RNA-induced silencing complex (RISC), which escorts the small RNA to its complementary mRNA (Chendrimada et al. 2005). TRBP is important for normal development, and ablation of the *tarbp2* gene in mice causes increased mortality before weaning, and surviving mice are generally smaller in size with spermatogenesis defects (Zhong et al. 1999; Ding et al. 2015). Conditional knockout of TRBP in cardiac and skeletal muscle leads to defective development of these tissues (Ding et al. 2015, 2016). It seems likely that most of these defects derive from aberrant miRNA processing, and this is supported by studies in both mice and human cells (Kim et al. 2014; Ding et al. 2016). In vitro studies using synthetic miRNA and siRNA precursors show that TRBP modulates Dicer function by promoting optimal substrate selection, affecting cleavage rate and fidelity, and strand selection, and loading the strand into RISC (Noland et al. 2011; Lee and Doudna 2012; Noland and Doudna 2013; Kim et al. 2014; Wilson et al. 2015; Fareh et al. 2016). However, so far, there are no mechanistic explanations for how TRBP coordinates with Dicer to perform these reported functions.

As illustrated (Fig. 1B), the first two dsRBMs of TRBP are Type A motifs, and the third is a Type B motif, consistent with observations that TRBP binds dsRNA with the first two motifs (Parker et al. 2008) and interacts with Dicer's helicase domain with its third dsRBM (Wilson et al. 2015; Liu et al. 2018). According to our model (Fig. 3A), and consistent with structural information (Liu et al. 2018), the carboxy-terminal Type B dsRBM of TRBP tethers it to the helicase domain (Fig. 2C), whereas the amino-terminal Type A dsRBMs engage the dsRNA substrate to bring it to the RNase III catalytic sites. Although only dsRBM3 was resolved in the Dicer-TRBP cryo-EM structure, a structure of dsRBM1 and dsRBM2 in complex with siRNA was solved by nuclear magnetic resonance (NMR), electron paramagnetic resonance (EPR), and single-molecule spectroscopy (Masliah et al. 2018). This structure shows that dsRBM1 and dsRBM2 bind on the same side of the dsRNA helix, leaving a large accessible surface area of the dsRNA for interaction with the RNase III active site.

One or both of TRBP's amino-terminal dsRBMs, possibly in collaboration with Dicer's carboxy-terminal dsRBM, could assist in docking the dsRNA substrate into the catalytic valley of the RNase III domains, stimulating cleavage activity by "squeezing" the dsRNA into the active site (Fig. 3A, ii). Indeed, in vitro studies show that TRBP•hsDcr complexes have a higher affinity for both miRNA and siRNA precursors compared to hsDcr alone, and similarly, TRBP increases the rate of dicing for both substrates in multiple-turnover experiments (Chakravarthy et al. 2010). However, single-turnover experiments in the same study show that the addition of TRBP increases the cleavage rate for an siRNA precursor, without a significant effect on cleavage of the miRNA precursor. The latter may suggest that dsRBMs of accessory proteins participate in catalysis for some substrates, but not all. Although these in vitro observations are consistent with our model (Fig. 3A), without more detailed kinetic analyses, as well as information from additional structural and mutagenesis studies, it is difficult to deconvolute effects on substrate binding from those affecting the chemical step of catalysis.

The model of Figure 3A is also consistent with the observation that TRBP enhances cleavage site fidelity in vitro and in vivo (Kim et al. 2014; Wilson et al. 2015). Bending the dsRNA helix into the catalytic valley and locking the substrate in place using dsRBMs of TRBP could reduce substrate movement that leads to cleavage at the wrong sites. Further, Dicer's increased affinity for siRNA-sized products in the presence of TRBP (Chakravarthy et al. 2010) might not be so surprising given the stable interactions observed with the cleaved dsRNA in postcatalytic RNase III structures (Fig. 3B; Gan et al. 2005; Liang et al. 2014). This scenario might also explain the observation that TRBP participates in siRNA strand selection and loading into Ago2 after Dicer cleavage (Noland et al. 2011; Noland and Doudna 2013). At the extreme, we envision that TRBP could have a role in the precleavage, cleavage, and postcleavage phases of Dicer-mediated RNA processing in vivo (Fig. 3A).

Loquacious

The Loquacious (Loqs) proteins were discovered as interactors of dmDcr-1 and dmDcr-2 and are required for miRNA and siRNA production, respectively (Lee et al. 2004; Jiang et al. 2005; Liu et al. 2007). Because of alternative splicing, there are four Loqs isoforms: Loqs-PA and Loqs-PB, which interact with dmDcr-1 and affect miRNA levels; Loqs-PC, whose function remains elusive; and Loqs-PD, which interacts with dmDcr-2 and affects endo-siRNA biogenesis (Hartig et al. 2009; Zhou et al. 2009). All four isoforms contain two amino-terminal Type A dsRBMs, whereas only Loqs-PA and Loqs-PB contain an additional Type B dsRBM3 that interacts with dmDcr-1's helicase domain (Fig. 1B; Ye et al. 2007).

Our thesis is that dsRBPs evolved to lend new activities to a Dicer enzyme that was specialized for processing viral dsRNA, and this idea is exemplified by Loqs-PD. At least for the viruses tested, Loqs-PD is not necessary for dmDcr-2 to target viral dsRNA (Marques et al. 2013), and although not definitively proven, we presume this is because viral dsRNA has termini that can be recognized by dmDcr-2 alone (Fig. 2A). However, Loqs-PD is required for processing a subset of endo-siRNA precursors (Hartig et al. 2009; Zhou et al. 2009). In *D. melanogaster*, endo-siRNAs map to dsRNA arising from convergent transcription, inverted repeats, and transposons (Chung et al. 2008; Czech et al. 2008; Ghildiyal et al. 2008; Kawamura et al. 2008; Okamura et al. 2008; Watanabe et al. 2008), and the precursors for these siRNAs are predicted to have ill-defined termini, such as long, frayed ends. Consistent with the requirement of Loqs-PD for production of endo-siRNAs, in vitro studies show that addition of Loqs-PD to dmDcr-2 allows it to function in a termini-independent manner, and this property is dependent on ATP (Sinha et al. 2015). Based on in vitro studies, the simple model is that Loqs-PD allows dmDcr-2 to bind internal regions of a dsRNA in a manner that promotes the closed, cleavage-competent conformation of the helicase domain (Fig. 2B). The latter is supported by limited proteolysis experiments (Sinha et al. 2015) in which Loqs-PD is observed to promote a protease-resistant dmDcr-2 fragment predicted to correlate with the closed conformation of the helicase domain observed in a cryo-EM structure of a complex containing dmDcr-2, blunt dsRNA, and an ATP analog (Sinha et al. 2018).

Interestingly, Loqs-PD is one of the few dsRBPs that interact with the helicase domain that does not have a Type-B domain at its carboxyl terminus (Fig. 1B). In fact, the region carboxy-terminal to the second dsRBM of Loqs-PD is too short to form the $\alpha\beta\beta\beta\alpha$ fold of other dsRBMs, although it is possible that some elements of the dsRBM secondary structure may remain. At the carboxyl terminus of Loqs-PD are 22 amino acids specific to this isoform that are necessary for efficient endo-siRNA biogenesis in vivo (Hartig and Förstemann 2011) and that have been shown to cross-link to the Hel2 domain of dmDcr-2 (Trettin et al. 2017).

The "tethering" interaction with Dicer occurs at the carboxyl terminus of all Loqs isoforms, and it is also possible that the Type A motifs, dsRBM1 and dsRBM2, facilitate binding of the dsRNA substrate and promote catalysis by analogy to RNase III enzymes (Fig. 3A). Indeed, Loqs-PD greatly increases dmDcr-2 cleavage rates in vitro (Trettin et al. 2017). Interestingly, efficient cleavage of an optimal dmDcr-2 substrate, blunt dsRNA, does not require dsRBM1, but cleavage of a suboptimal substrate, dsRNA with a 3′ overhang, requires both amino-terminal motifs (Trettin et al. 2017). Again, this raises the possibility that the requirement of dsRBMs in Dicer's accessory dsRBPs is substrate-dependent.

R2D2

All of the dsRBPs discussed so far act in facilitating cleavage by Dicer. At least for metazoan Dicers, after cleavage, the miRNA or siRNA product must be passed to an Argonaute protein to form the RISC, which escorts the small RNA to its complementary mRNA. We know little about the mechanism of this process, although studies of *D. melanogaster* R2D2 offer the most insight and reiterate the key role played by dsRBPs in facilitating Dicer functions. In early RNAi studies, siRNA-generating activity was monitored in fractionated extracts from S2 cells, and dmDcr-2 and R2D2 copurified through six chromatographic steps (Liu et al. 2003); the stability of R2D2 relied on the formation of the dmDcr-2•R2D2 complex. Embryos from *r2d2* deletion mutant flies were defective in silencing, but in vitro assays using recombinant proteins showed similar cleavage activity by dmDcr-2 alone and the dmDcr-2•R2D2 complex (Liu et al. 2003, 2006). These observations suggested R2D2 was important downstream from the cleavage step, and indeed, although R2D2 was not required for siRNA biogenesis in whole fly lysates, without R2D2, only minimal mRNA could be properly cleaved by RISC (Liu et al. 2006). Further, efficient transfer of a biotinylated siRNA to Argonaute 2 (dmAgo2) was observed with a dmDcr-2•R2D2 complex, but not with dmDcr-2 alone (Liu et al. 2003). Together these results indicate that R2D2 functions with dmDcr-2 downstream from siRNA production to facilitate siRNA loading into RISC.

R2D2 has an extended carboxy-terminal region (Fig. 1B) that interacts with dmDcr-2's helicase domain (Nishida et al. 2013), but the helicase subdomain involved in the interaction is unknown. Although there is no annotated dsRBM in this region, Type B dsRBMs are sometimes difficult to detect by sequence alignment and current motif search algorithms, because of their lack of conservation (Gleghorn and Maquat 2014). Individually, dmDcr-2 and R2D2 bind poorly to siRNA, but as a complex bind strongly (Liu et al. 2006). For the guide strand to be properly loaded onto dmAgo2, the siRNA needs to unwind.

However, the dmDcr-2•R2D2 complex lacks siRNA unwinding activity, suggesting the pre-RISC complex, including dmAgo2, must be assembled for unwinding and transfer to occur (Tomari et al. 2004). Although the dmDcr-2•R2D2 complex cannot unwind siRNA on its own, R2D2 appears to aid in strand selection for eventual loading, like TRBP with hsDcr (Noland et al. 2011). As both TRBP and R2D2 have been reported to be involved in guide strand selection and loading into RISC (Tomari et al. 2004), perhaps R2D2 binds dmDcr-2 similarly via the Hel2i subdomain (Noland et al. 2011; Noland and Doudna 2013).

Although R2D2 has been detected in immunoprecipitated complexes of dmDcr-2•Loqs-PD (Miyoshi et al. 2010), it is unclear if both proteins can interact with dmDcr-2 simultaneously. If R2D2 binds the Hel2i subdomain of dmDcr-2, like TRBP for hsDcr, could Loqs-PD bind Hel2 at the same time or would R2D2 and Loqs-PD compete for binding? If Loqs-PD and R2D2 could bind at the same time, the dmDcr-2•R2D2•Loqs-PD complex could act as a well-oiled machine, efficiently cleaving dsRNA into siRNA products, with the help of Loqs-PD, and then allowing transfer to dmAgo2, possibly in one fluid motion. Alternatively, if these two cofactors compete for binding the helicase domain, they might act as a switch for dmDcr-2's functions, as either an siRNA biogenesis factory or a stable siRNA transporter.

OUTSTANDING QUESTIONS

We focused on a few dsRBPs that interact with Dicer's helicase domain to modulate its functions. We expect that other dsRBPs not covered here will emphasize paradigms we have discussed, as well as point to currently unimagined methods of regulation. Indeed, it is exciting to imagine the future biochemical and structural studies that undoubtedly will offer additional insight.

As mentioned for R2D2 and Loqs-PD, of interest will be the question of whether accessory dsRBPs interact simultaneously with Dicer or have mutually exclusive interactions. Although R2D2 and Loqs-PD may bind to different helicase subdomains, TRBP and PACT both bind to the same interface of the Hel2i subdomain of hsDcr and thus interact in a mutually exclusive manner (Wilson et al. 2015). Although reported associations of dsRBPs are sometimes indirect, mediated by their common ligand, dsRNA, in many cases interactions are direct. Further, it seems possible that even dsRNA-mediated interactions could have biological relevance. The RNA editing enzyme ADAR1 interacts directly with hsDcr to modulate processing of pre-miRNA (Ota et al. 2013) and has been reported to have direct as well as indirect interactions with other dsRBPs (Herbert 2019). As a further layer of complexity, for mammals, a subset of the dsRBPs that interact with the helicase domain of mammalian Dicer have also been shown to interact with other RLRs (Takahashi et al. 2018; van der Veen et al. 2018). Understanding this complexity is an important goal for future studies.

ACKNOWLEDGMENTS

We thank Dr. Nels Elde for helpful discussions on the evolution of Dicer. This work was supported by funding from the National Institute of General Medical Sciences (R01GM121706) to B.L.B. H.M.D was supported by a National Institutes of Health Chemical Biology training grant (T32GM122740).

REFERENCES

Ahmad S, Hur S. 2015. Helicases in antiviral immunity: dual properties as sensors and effectors. *Trends Biochem Sci* **40:** 576–585. doi:10.1016/j.tibs.2015.08.001

Ashe A, Bélicard T, Le Pen J, Sarkies P, Frézal L, Lehrbach NJ, Félix M-A, Miska EA. 2013. A deletion polymorphism in the *Caenorhabditis elegans* RIG-I homolog disables viral RNA dicing and antiviral immunity. *eLife* **2:** e00994. doi:10.7554/eLife.00994

Bass BL. 2000. Double-stranded RNA as a template for gene silencing. *Cell* **101:** 235–238. doi:10.1016/S0092-8674(02)71133-1

Bass BL, Hurst SR, Singer JD. 1994. Binding properties of newly identified *Xenopus* proteins containing dsRNA-binding motifs. *Curr Biol* **4:** 301–314. doi:10.1016/S0960-9822(00)00069-5

Bernstein E, Caudy AA, Hammond SM, Hannon GJ. 2001. Role for a bidentate ribonuclease in the initiation step of RNA interference. *Nature* **409:** 363–366. doi:10.1038/35053110

Blaszczyk J, Tropea JE, Bubunenko M, Routzahn KM, Waugh DS, Court DL, Ji X. 2001. Crystallographic and modeling studies of RNase III suggest a mechanism for double-stranded RNA cleavage. *Structure* **9:** 1225–1236. doi:10.1016/S0969-2126(01)00685-2

Chakravarthy S, Sternberg SH, Kellenberger CA, Doudna JA. 2010. Substrate-specific kinetics of Dicer-catalyzed RNA processing. *J Mol Biol* **404:** 392–402. doi:10.1016/j.jmb.2010.09.030

Chendrimada TP, Gregory RI, Kumaraswamy E, Norman J, Cooch N, Nishikura K, Shiekhattar R. 2005. TRBP recruits the Dicer complex to Ago2 for microRNA processing and gene silencing. *Nature* **436:** 740–744. doi:10.1038/nature03868

Chung W-J, Okamura K, Martin R, Lai EC. 2008. Endogenous RNA interference provides a somatic defense against *Drosophila* transposons. *Curr Biol* **18:** 795–802. doi:10.1016/j.cub.2008.05.006

Colmenares SU, Buker SM, Buhler M, Dlakić M, Moazed D. 2007. Coupling of double-stranded RNA synthesis and siRNA generation in fission yeast RNAi. *Mol Cell* **27:** 449–461. doi:10.1016/j.molcel.2007.07.007

Court DL, Gan J, Liang Y-H, Shaw GX, Tropea JE, Costantino N, Waugh DS, Ji X. 2013. RNase III: genetics and function; structure and mechanism. *Annu Rev Genet* **47:** 405–431. doi:10.1146/annurev-genet-110711-155618

Czech B, Malone CD, Zhou R, Stark A, Schlingeheyde C, Dus M, Perrimon N, Kellis M, Wohlschlegel JA, Sachidanandam R, et al. 2008. An endogenous small interfering RNA pathway in *Drosophila*. *Nature* **453:** 798–802. doi:10.1038/nature07007

Deddouche S, Matt N, Budd A, Mueller S, Kemp C, Galiana-Arnoux D, Dostert C, Antoniewski C, Hoffmann JA, Imler J-L. 2008. The DExD/H-box helicase Dicer-2 mediates the induction of antiviral activity in *Drosophila*. *Nat Immunol* **9:** 1425–1432. doi:10.1038/ni.1664

Ding S-W, Voinnet O. 2007. Antiviral immunity directed by small RNAs. *Cell* **130:** 413–426. doi:10.1016/j.cell.2007.07.039

Ding J, Chen J, Wang Y, Kataoka M, Ma L, Zhou P, Hu X, Lin Z, Nie M, Deng Z-L, et al. 2015. Trbp regulates heart function through microRNA-mediated *Sox6* repression. *Nat Genet* **47:** 776–783. doi:10.1038/ng.3324

Ding J, Nie M, Liu J, Hu X, Ma L, Deng Z-L, Wang D-Z. 2016. Trbp is required for differentiation of myoblasts and normal regeneration of skeletal muscle. *PLoS One* **11:** e0155349. doi:10.1371/journal.pone.0155349

Fairman-Williams ME, Guenther U-P, Jankowsky E. 2010. SF1 and SF2 helicases: family matters. *Curr Opin Struct Biol* **20:** 313–324. doi:10.1016/j.sbi.2010.03.011

Fareh M, Yeom K-H, Haagsma AC, Chauhan S, Heo I, Joo C. 2016. TRBP ensures efficient Dicer processing of precursor microRNA in RNA-crowded environments. *Nat Commun* **7:** 13694. doi:10.1038/ncomms13694

Fire A, Xu S, Montgomery MK, Kostas SA, Driver SE, Mello CC. 1998. Potent and specific genetic interference by double-stranded RNA in *Caenorhabditis elegans*. *Nature* **391:** 806–811. doi:10.1038/35888

Fukudome A, Fukuhara T. 2017. Plant dicer-like proteins: double-stranded RNA-cleaving enzymes for small RNA biogenesis. *J Plant Res* **130:** 33–44. doi:10.1007/s10265-016-0877-1

Gan J, Tropea JE, Austin BP, Court DL, Waugh DS, Ji X. 2005. Intermediate states of ribonuclease III in complex with double-stranded RNA. *Structure* **13:** 1435–1442. doi:10.1016/j.str.2005.06.014

Gan J, Shaw G, Tropea JE, Waugh DS, Court DL, Ji X. 2008. A stepwise model for double-stranded RNA processing by ribonuclease III. *Mol Microbiol* **67:** 143–154. doi:10.1111/j.1365-2958.2007.06032.x

Gatignol A, Buckler-White A, Berkhout B, Jeang KT. 1991. Characterization of a human TAR RNA-binding protein that activates the HIV-1 LTR. *Science (New York, NY)* **251:** 1597–1600. doi:10.1126/science.2011739

Ghildiyal M, Seitz H, Horwich MD, Li C, Du T, Lee S, Xu J, Kittler ELW, Zapp ML, Weng Z, et al. 2008. Endogenous siRNAs derived from transposons and mRNAs in *Drosophila* somatic cells. *Science (New York, NY)* **320:** 1077–1081. doi:10.1126/science.1157396

Gleghorn ML, Maquat LE. 2014. "Black sheep" that don't leave the double-stranded RNA-binding domain fold. *Trends Biochem Sci* **39:** 328–340. doi:10.1016/j.tibs.2014.05.003

Guo X, Zhang R, Wang J, Ding S-W, Lu R. 2013. Homologous RIG-I-like helicase proteins direct RNAi-mediated antiviral immunity in *C. elegans* by distinct mechanisms. *Proc Natl Acad Sci* **110:** 16085–16090. doi:10.1073/pnas.1307453110

Ha M, Kim VN. 2014. Regulation of microRNA biogenesis. *Nat Rev Mol Cell Biol* **15:** 509–524. doi:10.1038/nrm3838

Hartig JV, Förstemann K. 2011. Loqs-PD and R2D2 define independent pathways for RISC generation in *Drosophila*. *Nucleic Acids Res* **39:** 3836–3851. doi:10.1093/nar/gkq1324

Hartig JV, Esslinger S, Böttcher R, Saito K, Förstemann K. 2009. Endo-siRNAs depend on a new isoform of loquacious and target artificially introduced, high-copy sequences. *EMBO J* **28:** 2932–2944. doi:10.1038/emboj.2009.220

Herbert A. 2019. ADAR and immune silencing in cancer. *Trends Cancer* **5:** 272–282. doi:10.1016/j.trecan.2019.03.004

Jarmoskaite I, Russell R. 2014. RNA helicase proteins as chaperones and remodelers. *Annu Rev Biochem* **83:** 697–725. doi:10.1146/annurev-biochem-060713-035546

Jiang F, Ye X, Liu X, Fincher L, McKearin D, Liu Q. 2005. Dicer-1 and R3D1-L catalyze microRNA maturation in *Drosophila*. *Genes Dev* **19:** 1674–1679. doi:10.1101/gad.1334005

Kawamura Y, Saito K, Kin T, Ono Y, Asai K, Sunohara T, Okada TN, Siomi MC, Siomi H. 2008. *Drosophila* endogenous small RNAs bind to Argonaute 2 in somatic cells. *Nature* **453:** 793–797. doi:10.1038/nature06938

Ketting RF, Haverkamp TH, van Luenen HG, Plasterk RH. 1999. *mut-7* of *C. elegans*, required for transposon silencing and RNA interference, is a homolog of Werner syndrome helicase and RNaseD. *Cell* **99:** 133–141. doi:10.1016/S0092-8674(00)81645-1

Ketting RF, Fischer SE, Bernstein E, Sijen T, Hannon GJ, Plasterk RH. 2001. Dicer functions in RNA interference and in synthesis of small RNA involved in developmental timing in *C. elegans*. *Genes Dev* **15:** 2654–2659. doi:10.1101/gad.927801

Kim Y, Yeo J, Lee JH, Cho J, Seo D, Kim J-S, Kim VN. 2014. Deletion of human *tarbp2* reveals cellular microRNA targets and cell-cycle function of TRBP. *Cell Rep* **9:** 1061–1074. doi:10.1016/j.celrep.2014.09.039

Knight SW, Bass BL. 2001. A role for the RNase III enzyme DCR-1 in RNA interference and germ line development in *Caenorhabditis elegans*. *Science (New York, NY)* **293:** 2269–2271. doi:10.1126/science.1062039

Kurzynska-Kokorniak A, Pokornowska M, Koralewska N, Hoffmann W, Bienkowska-Szewczyk K, Figlerowicz M. 2016. Revealing a new activity of the human Dicer DUF283 domain in vitro. *Sci Rep* **6:** 23989-13. doi:10.1038/srep23989

Lee HY, Doudna JA. 2012. TRBP alters human precursor microRNA processing in vitro. *RNA* **18:** 2012–2019. doi:10.1261/rna.035501.112

Lee YS, Nakahara K, Pham JW, Kim K, He Z, Sontheimer EJ, Carthew RW. 2004. Distinct roles for *Drosophila* Dicer-1 and Dicer-2 in the siRNA/miRNA silencing pathways. *Cell* **117:** 69–81. doi:10.1016/S0092-8674(04)00261-2

Liang Y-H, Lavoie M, Comeau M-A, Abou Elela S, Ji X. 2014. Structure of a eukaryotic RNase III postcleavage complex reveals a double-ruler mechanism for substrate selection. *Mol Cell* **54:** 431–444. doi:10.1016/j.molcel.2014.03.006

Liu Q, Rand TA, Kalidas S, Du F, Kim H-E, Smith DP, Wang X. 2003. R2D2, a bridge between the initiation and effector steps of the *Drosophila* RNAi pathway. *Science (New York, NY)* **301:** 1921–1925. doi:10.1126/science.1088710

Liu X, Jiang F, Kalidas S, Smith D, Liu Q. 2006. Dicer-2 and R2D2 coordinately bind siRNA to promote assembly of the siRISC complexes. *RNA* **12:** 1514–1520. doi:10.1261/rna.101606

Liu X, Park JK, Jiang F, Liu Y, McKearin D, Liu Q. 2007. Dicer-1, but not Loquacious, is critical for assembly of miRNA-induced silencing complexes. *RNA* **13:** 2324–2329. doi:10.1261/rna.723707

Liu Z, Wang J, Cheng H, Ke X, Sun L, Zhang QC, Wang H-W. 2018. Cryo-EM structure of human dicer and its complexes with a pre-miRNA substrate. *Cell* **173:** 1191–1203.e12. doi:10.1016/j.cell.2018.03.080

Luo D, Kohlway A, Pyle AM. 2013. Duplex RNA activated ATPases (DRAs): platforms for RNA sensing, signaling and processing. *RNA Biol* **10:** 111–120. doi:10.4161/rna.22706

Ma E, MacRae IJ, Kirsch JF, Doudna JA. 2008. Autoinhibition of human dicer by its internal helicase domain. *J Mol Biol* **380:** 237–243. doi:10.1016/j.jmb.2008.05.005

MacRae IJ, Zhou K, Li F, Repic A, Brooks AN, Cande WZ, Adams PD, Doudna JA. 2006. Structural basis for double-stranded RNA processing by Dicer. *Science (New York, NY)* **311:** 195–198. doi:10.1126/science.1121638

Marques JT, Wang J-P, Wang X, de Oliveira KPV, Gao C, Aguiar ERGR, Jafari N, Carthew RW. 2013. Functional specialization of the small interfering RNA pathway in response to virus infection. *PLoS Pathog* **9:** e1003579. doi:10.1371/journal.ppat.1003579

Masliah G, Barraud P, Allain FH-T. 2013. RNA recognition by double-stranded RNA binding domains: a matter of shape and sequence. *Cell Mol Life Sci* **70:** 1875–1895. doi:10.1007/s00018-012-1119-x

Masliah G, Maris C, König SL, Yulikov M, Aeschimann F, Malinowska AL, Mabille J, Weiler J, Holla A, Hunziker J, et al. 2018. Structural basis of siRNA recognition by TRBP double-stranded RNA binding domains. *EMBO J* **37:** e97089. doi:10.15252/embj.201797089

Miyoshi K, Miyoshi T, Hartig JV, Siomi H, Siomi MC. 2010. Molecular mechanisms that funnel RNA precursors into endogenous small-interfering RNA and microRNA biogenesis pathways in *Drosophila*. *RNA* **16:** 506–515. doi:10.1261/rna.1952110

Mukherjee K, Campos H, Kolaczkowski B. 2013. Evolution of animal and plant dicers: early parallel duplications and recurrent adaptation of antiviral RNA binding in plants. *Mol Biol Evol* **30:** 627–641. doi:10.1093/molbev/mss263

Mukherjee K, Korithoski B, Kolaczkowski B. 2014. Ancient origins of vertebrate-specific innate antiviral immunity. *Mol Biol Evol* **31:** 140–153. doi:10.1093/molbev/mst184

Nicholson AW. 2014. Ribonuclease III mechanisms of double-stranded RNA cleavage. *WIREs RNA* **5:** 31–48. doi:10.1002/wrna.1195

Nishida KM, Miyoshi K, Ogino A, Miyoshi T, Siomi H, Siomi MC. 2013. Roles of R2D2, a cytoplasmic D2 body component, in the endogenous siRNA pathway in *Drosophila*. *Mol Cell* **49:** 680–691. doi:10.1016/j.molcel.2012.12.024

Noland CL, Doudna JA. 2013. Multiple sensors ensure guide strand selection in human RNAi pathways. *RNA* **19:** 639–648. doi:10.1261/rna.037424.112

Noland CL, Ma E, Doudna JA. 2011. siRNA repositioning for guide strand selection by human Dicer complexes. *Mol Cell* **43:** 110–121. doi:10.1016/j.molcel.2011.05.028

Okamura K, Chung W-J, Ruby JG, Guo H, Bartel DP, Lai EC. 2008. The *Drosophila* hairpin RNA pathway generates endogenous short interfering RNAs. *Nature* **453:** 803–806. doi:10.1038/nature07015

Ota H, Sakurai M, Gupta R, Valente L, Wulff B-E, Ariyoshi K, Iizasa H, Davuluri RV, Nishikura K. 2013. ADAR1 forms a complex with Dicer to promote microRNA processing and RNA-induced gene silencing. *Cell* **153:** 575–589. doi:10.1016/j.cell.2013.03.024

Parker GS, Maity TS, Bass BL. 2008. dsRNA binding properties of RDE-4 and TRBP reflect their distinct roles in RNAi. *J Mol Biol* **384:** 967–979. doi:10.1016/j.jmb.2008.10.002

Pavelec DM, Lachowiec J, Duchaine TF, Smith HE, Kennedy S. 2009. Requirement for the ERI/DICER complex in endogenous RNA interference and sperm development in *Caenorhabditis elegans*. *Genetics* **183:** 1283–1295. doi:10.1534/genetics.109.108134

Provost P, Dishart D, Doucet J, Frendewey D, Samuelsson B, Rådmark O. 2002. Ribonuclease activity and RNA binding of recombinant human Dicer. *EMBO J* **21:** 5864–5874. doi:10.1093/emboj/cdf578

Qin H, Chen F, Huan X, Machida S, Song J, Yuan YA. 2010. Structure of the *Arabidopsis thaliana* DCL4 DUF283 domain reveals a noncanonical double-stranded RNA-binding fold for protein–protein interaction. *RNA* **16:** 474–481. doi:10.1261/rna.1965310

Robertson HD, Webster RE, Zinder ND. 1968. Purification and properties of ribonuclease III from *Escherichia coli*. *J Biol Chem* **243:** 82–91.

Schlee M. 2013. Master sensors of pathogenic RNA—RIG-I like receptors. *Immunobiology* **218:** 1322–1335. doi:10.1016/j.imbio.2013.06.007

Singleton MR, Dillingham MS, Wigley DB. 2007. Structure and mechanism of helicases and nucleic acid translocases. *Annu Rev Biochem* **76:** 23–50. doi:10.1146/annurev.biochem.76.052305.115300

Sinha NK, Trettin KD, Aruscavage PJ, Bass BL. 2015. *Drosophila* Dicer-2 cleavage is mediated by helicase- and dsRNA termini-dependent states that are modulated by Loquacious-PD. *Mol Cell* **58:** 406–417. doi:10.1016/j.molcel.2015.03.012

Sinha NK, Iwasa J, Shen PS, Bass BL. 2018. Dicer uses distinct modules for recognizing dsRNA termini. *Science (New York, NY)* **359:** 329–334. doi:10.1126/science.aaq0921

Song H, Fang X, Jin L, Shaw GX, Wang Y-X, Ji X. 2017. The functional cycle of Rnt1p: five consecutive steps of double-stranded RNA processing by a eukaryotic RNase III. *Structure* **25:** 353–363. doi:10.1016/j.str.2016.12.013

Sowa JN, Jiang H, Somasundaram L, Tecle E, Xu G, Wang D, Troemel ER. 2020. The *Caenorhabditis elegans* RIG-I homolog DRH-1 mediates the intracellular pathogen response upon viral infection. *J Virol* **94:** e01173-19. doi:10.1128/JVI.01173-19

Takahashi T, Nakano Y, Onomoto K, Murakami F, Komori C, Suzuki Y, Yoneyama M, Ui-Tei K. 2018. LGP2 virus sensor regulates gene expression network mediated by TRBP-bound microRNAs. *Nucleic Acids Res* **46:** 9134–9147. doi:10.1093/nar/gky575

Thivierge C, Makil N, Flamand M, Vasale JJ, Mello CC, Wohlschlegel J, Conte D, Duchaine TF. 2011. Tudor domain ERI-5 tethers an RNA-dependent RNA polymerase to DCR-1 to potentiate endo-RNAi. *Nat Struct Mol Biol* **19:** 90–97. doi:10.1038/nsmb.2186

Tomari Y, Matranga C, Haley B, Martinez N, Zamore PD. 2004. A protein sensor for siRNA asymmetry. *Science (New York, NY)* **306:** 1377–1380. doi:10.1126/science.1102755

Trettin KD, Sinha NK, Eckert DM, Apple SE, Bass BL. 2017. Loquacious-PD facilitates *Drosophila* Dicer-2 cleavage through interactions with the helicase domain and dsRNA. *Proc Natl Acad Sci* **114:** E7939–E7948. doi:10.1073/pnas.1707063114

van der Veen AG, Maillard PV, Schmidt JM, Lee SA, Deddouche Grass S, Borg A, Kjær S, Snijders AP, Reis e Sousa C. 2018. The RIG-I-like receptor LGP2 inhibits Dicer-dependent processing of long double-stranded RNA and blocks RNA interference in mammalian cells. *EMBO J* **37:** e97479.

Watanabe T, Totoki Y, Toyoda A, Kaneda M, Kuramochi-Miyagawa S, Obata Y, Chiba H, Kohara Y, Kono T, Nakano T, et al. 2008. Endogenous siRNAs from naturally formed dsRNAs regulate transcripts in mouse oocytes. *Nature* **453:** 539–543. doi:10.1038/nature06908

Waterhouse PM, Wang MB, Lough T. 2001. Gene silencing as an adaptive defence against viruses. *Nature* **411:** 834–842. doi:10.1038/35081168

Welker NC, Pavelec DM, Nix DA, Duchaine TF, Kennedy S, Bass BL. 2010. Dicer's helicase domain is required for accumulation of some, but not all, *C. elegans* endogenous siRNAs. *RNA* **16:** 893–903. doi:10.1261/rna.2122010

Welker NC, Maity TS, Ye X, Aruscavage PJ, Krauchuk AA, Liu Q, Bass BL. 2011. Dicer's helicase domain discriminates dsRNA termini to promote an altered reaction mode. *Mol Cell* **41:** 589–599. doi:10.1016/j.molcel.2011.02.005

Wilson RC, Doudna JA. 2013. Molecular mechanisms of RNA interference. *Annu Rev Biophys* **42:** 217–239. doi:10.1146/annurev-biophys-083012-130404

Wilson RC, Tambe A, Kidwell MA, Noland CL, Schneider CP, Doudna JA. 2015. Dicer-TRBP complex formation ensures accurate mammalian microRNA biogenesis. *Mol Cell* **57:** 397–407. doi:10.1016/j.molcel.2014.11.030

Ye X, Paroo Z, Liu Q. 2007. Functional anatomy of the *Drosophila* microRNA-generating enzyme. *J Biol Chem* **282:** 28373–28378. doi:10.1074/jbc.M705208200

Zamore PD, Tuschl T, Sharp PA, Bartel DP. 2000. RNAi: double-stranded RNA directs the ATP-dependent cleavage of mRNA at 21 to 23 nucleotide intervals. *Cell* **101:** 25–33. doi:10.1016/S0092-8674(00)80620-0

Zhang H, Kolb FA, Brondani V, Billy E, Filipowicz W. 2002. Human Dicer preferentially cleaves dsRNAs at their termini without a requirement for ATP. *EMBO J* **21:** 5875–5885. doi:10.1093/emboj/cdf582

Zhong J, Peters AH, Lee K, Braun RE. 1999. A double-stranded RNA binding protein required for activation of repressed messages in mammalian germ cells. *Nat Genet* **22:** 171–174. doi:10.1038/9684

Zhou R, Czech B, Brennecke J, Sachidanandam R, Wohlschlegel JA, Perrimon N, Hannon GJ. 2009. Processing of *Drosophila* endo-siRNAs depends on a specific Loquacious isoform. *RNA* **15:** 1886–1895. doi:10.1261/rna.1611309

Zou J, Chang M, Nie P, Secombes CJ. 2009. Origin and evolution of the RIG-I like RNA helicase gene family. *BMC Evol Biol* **9:** 85-14. doi:10.1186/1471-2148-9-85

Reconstitution of siRNA Biogenesis In Vitro: Novel Reaction Mechanisms and RNA Channeling in the RNA-Directed DNA Methylation Pathway

JASLEEN SINGH[1] AND CRAIG S. PIKAARD[1,2]

[1]*Department of Molecular and Cellular Biochemistry and Department of Biology,*
[2]*Howard Hughes Medical Institute, Indiana University, Bloomington, Indiana 47405, USA*
Correspondence: cpikaard@indiana.edu

Eukaryotes deploy RNA-mediated gene silencing pathways to guard their genomes against selfish genetic elements, such as transposable elements and invading viruses. In plants, RNA-directed DNA methylation (RdDM) is used to silence selfish elements at the level of transcription. This process involves 24-nt short interfering RNAs (siRNAs) and longer noncoding RNAs to which the siRNAs base-pair. Recently, we showed that 24-nt siRNA biogenesis could be recapitulated in the test tube using purified enzymes, yielding biochemical answers to numerous questions left unresolved by prior genetic and genomic studies. Interestingly, each enzyme has activities that program what happens in the next step, thus channeling the RNAs within the RdDM pathway and restricting their diversion into alternative pathways. However, a similar mechanistic understanding is lacking for other important steps of the RdDM pathway. We discuss some of the steps most in need of biochemical investigation and important questions still in need of answers.

Small noncoding RNAs (RNAs that do not encode proteins) play important roles in restricting the proliferation of selfish genetic elements, either by base-pairing with messenger RNAs to interfere with their stability or function, thereby inhibiting protein synthesis, or by interfering with transcription to prevent RNA synthesis (Borges and Martienssen 2015; Holoch and Moazed 2015; Martienssen and Moazed 2015; Maillard et al. 2019). In the latter strategy, the small RNAs guide chemical modifications at matching chromosomal DNA sequences. These modifications include addition or removal of chemical groups on the histone proteins that wrap the DNA and, in many eukaryotes, including humans and plants, methylation of the DNA (Law and Jacobsen 2010; Martienssen and Moazed 2015). Collectively, these chemical modifications contribute to chromatin environments that are refractory to promoter-dependent transcription by DNA-dependent RNA polymerases I, II, or III.

In plants, the major transcriptional gene silencing pathway is RNA-directed DNA methylation (RdDM) (Zhou and Law 2015; Wendte and Pikaard 2017). The process has been elucidated, primarily in *Arabidopsis thaliana*, using tools of genetics, genomics, cell biology, and molecular biology. Collectively, these studies have provided a working understanding of what the pathway accomplishes, but how the enzymes and RNAs of the pathway function at a biochemical level remains unclear. Progress in illuminating these biochemical details is the subject of this perspective.

KEY ENZYMES AND RNAs OF THE RdDM PATHWAY

The central aspects of RdDM can be grasped by focusing on the RNA synthesizing and processing enzymes that are key to the pathway. Two are nuclear multisubunit RNA polymerases, abbreviated as Pol IV and Pol V (Haag and Pikaard 2011), that evolved in plants as specialized forms of DNA-dependent RNA polymerase II (Pol II) (Ream et al. 2009). The third is RNA-DEPENDENT RNA POLYMERASE 2 (RDR2) (Xie et al. 2004). Figure 1 provides a simplified view of the RdDM pathway (for a more complete view, see Wendte and Pikaard 2017), with Pol IV and RDR2 collaborating to produce double-stranded RNAs (dsRNAs) that are then cut (diced) by DICER-LIKE 3 (DCL3) (Xie et al. 2004) into short interfering RNA (siRNA) duplexes, single strands of which somehow become associated with ARGONAUTE4 (AGO4) (Zilberman et al. 2003) or related Argonaute proteins. AGO4–siRNA complexes are then recruited to sites of Pol V transcription, apparently via base-pairing between the siRNAs and nascent Pol V transcripts (Wierzbicki et al. 2008, 2009). Protein–protein interactions can also occur between AGO4 and the largest subunit of Pol V (NRPE1) (El-Shami et al. 2007) and between AGO4 and the Pol V–associated protein, SUPPRESSOR of TY INSERTION 5-LIKE (SPT5L) (Lahmy et al. 2016). AGO–siRNA–Pol V complexes then facilitate recruitment of chromatin-modifying enzymes, including the de novo DNA methyltrans-

© 2019 Singh and Pikaard. This article is distributed under the terms of the Creative Commons Attribution-NonCommercial License, which permits reuse and redistribution, except for commercial purposes, provided that the original author and source are credited.

Published by Cold Spring Harbor Laboratory Press; doi:10.1101/sqb.2019.84.039842

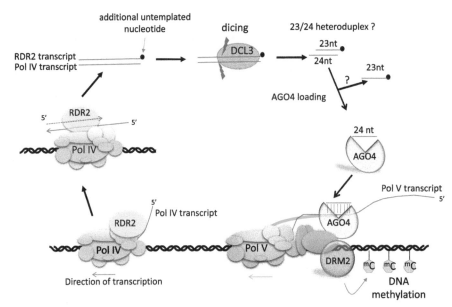

Figure 1. Key events of the RNA-directed DNA methylation pathway. Pol IV and RDR2 physically associate and their reactions are coupled to produce double-stranded RNAs (dsRNAs). These dsRNAs include a nontemplated nucleotide at the 3′ end of the RDR2 strands attributable to RDR2's terminal transferase activity. Upon dicing by DICER-LIKE 3 (DCL3), 24- and 23-nt siRNAs are generated, possibly as a heteroduplex (hence the question mark), with 23-nt RNAs enriched for the untemplated 3′ nucleotides. We hypothesize that the 23-nt siRNAs serve as the passenger strands for the associated 24-nt guide RNAs that are loaded into ARGO-NAUTE 4 (AGO4). Resulting siRNA–AGO4 complexes find their target sites by base-pairing to Pol V transcripts and interacting with the carboxy-terminal domain (CTD) of the Pol V largest subunit. The cytosine methyltransferase DRM2 is ultimately recruited and carries out regional de novo methylation of cytosines in any sequence context.

ferase, DOMAINS REARRANGED METHYLTRANSFERASE 2 (DRM2) (Cao and Jacobsen 2002), the plant ortholog of the mammalian enzymes, DNMT3a and DNMT3b (Law and Jacobsen 2010). Extensive methylation of local deoxycytosines ensues, in all sequence contexts (CG, CHG, and CHH, in which H represents any nucleotide other than G). Repressive histone modifications also occur, with genetic evidence implicating histone deacetylation, catalyzed by HISTONE DEACETYLASE 6 (HDA6) (Aufsatz et al. 2002; He et al. 2009), histone H3 lysine 9 dimethylation (H3K9me2) catalyzed by suppressor of variegation homologs (SUVH), SUVH4, SUVH5, and SUVH6 (Jackson et al. 2004; Ebbs and Bender 2006; Johnson et al. 2007; Blevins et al. 2014), and histone H3 lysine 4 (H3K4) demethylation catalyzed by the Jumonji C (JmjC) domain–containing protein JMJ14 (Deleris et al. 2010; Searle et al. 2010) and nucleosome repositioning by the SWI/SNF complex (Zhu et al. 2013). Collectively, the DNA and histone modifications contribute to a chromatin environment refractory to promoter-dependent transcription of transposon or viral genes.

IN VIVO STUDIES AND UNANSWERED QUESTIONS

Genetic studies showed that mutants defective for Pol IV, RDR2, or DCL3 lack the 24-nt class of siRNAs associated with DNA methylation and gene silencing (Xie et al. 2004; Herr et al. 2005; Onodera et al. 2005; Henderson et al. 2006; Kasschau et al. 2007). Based on the presumptive functions of these enzymes, deduced from similarities to related proteins, Pol IV and RDR2 were hypothesized to generate RNA precursors that DCL3 then diced into 24-nt siRNAs. It seemed logical that Pol IV, as a presumptive DNA-dependent RNA polymerase, might act first, with RDR2 then using the Pol IV transcripts as templates to generate dsRNAs. Cell biological studies provided indirect support for this hypothesis by showing that RDR2 becomes mislocalized in nuclei of cells lacking Pol IV (Pontes et al. 2006). In contrast, Pol IV immunolocalization was largely unchanged in *rdr2* mutants. However, Pol IV was mislocalized in nuclei treated with ribonuclease A, suggesting that Pol IV might transcribe an RNA template (Pontes et al. 2006); thus, the possibility that RDR2 might act prior to Pol IV could not be ruled out.

Pol IV and RDR2 transcripts proved difficult to identify, remaining elusive until 2015, when several laboratories independently examined RNAs that accumulate in Dicer mutants, as expected for siRNA precursors (Blevins et al. 2015; Li et al. 2015; Zhai et al. 2015). The RNAs were shown to be dependent on both Pol IV and RDR2 for their synthesis, to be sensitive to double-strand specific ribonucleases, to accumulate in *dcl3* mutants but be diced into 24-nt siRNAs by DCL3 in vitro (Blevins et al. 2015), and to map to 24-nt siRNA producing loci. Surprisingly, the siRNA precursors were mostly 25–35 nt in length, so short that they could only be diced once (Blevins et al. 2015; Zhai et al. 2015).

A priori, one might predict Pol IV transcripts to accumulate in a *rdr2* mutant if Pol IV acts first, and RDR2 then converts Pol IV transcripts into dsRNAs. Conversely, if

RDR2 acts first, RDR2 transcripts would be expected to accumulate in a *pol iv* mutant. Surprisingly, in single mutants defective for either Pol IV or RDR2, no precursor transcripts were detected (Blevins et al. 2015; Zhai et al. 2015). Could it be that single-stranded transcripts are simply turned over rapidly, thus evading detection? Or might the physical association of Pol IV and RDR2, shown by previous coimmunoprecipitation experiments (Law et al. 2011; Haag et al. 2012), be needed for each enzyme to be active? Initial in vitro transcription assays of Haag et al. (2012) suggested that this might be true for RDR2, but not Pol IV. Or could it be that Pol IV and RDR2 are both needed to synthesize each strand of a dsRNA precursor, perhaps by RDR2 carrying out a priming step and Pol IV a subsequent elongation step, reminiscent of Okazaki fragment synthesis during DNA replication (Bergsch et al. 2019)?

Another puzzling characteristic of Pol IV and RDR2-dependent precursor RNAs, which we refer to as P4R2 RNAs (Blevins et al. 2015), is that they frequently have a 3′-terminal nucleotide that does not match the corresponding genomic DNA (Blevins et al. 2015; Zhai et al. 2015). Whether these mismatches occur at the ends of Pol IV or RDR2 transcripts or both was unclear because there was no definitive way to distinguish Pol IV transcripts from RDR2 transcripts in vivo. One hypothesis suggested that mismatched nucleotides result from misincorporation errors when Pol IV encounters methylated cytosines in the DNA template (Zhai et al. 2015). It was further hypothesized that if the mistake could not be repaired, transcription termination might result. The latter idea, combined with evidence that Pol IV is recruited to regions displaying CG maintenance methylation (Blevins et al. 2014), provided a possible explanation for why Pol IV transcripts are so short.

The Pol IV-misincorporation hypothesis was tested in subsequent biochemical experiments in vitro. Pol IV transcription is relatively error-prone compared to Pol II and Pol V, but there was no difference in nucleotide misincorporation at methylcytosines versus cytosines (Marasco et al. 2017). Cytosine methylation also does not induce Pol IV termination in vitro (Marasco et al. 2017; Singh et al. 2019).

An alternative explanation for template-mismatched nucleotides in P4R2 RNAs stems from the fact that RDR2 displays terminal nucleotidyl transferase activity that can add one or more untemplated nucleotides to the 3′ end of an RNA in vitro (Blevins et al. 2015). But can RDR2 add an untemplated nucleotide to the 3′ ends of Pol IV transcripts, its own transcripts, or both? Once again, because there was no way to distinguish Pol IV from RDR2 transcripts among P4R2 RNAs in vivo, an answer to this question awaited assays able to distinguish their transcripts in vitro.

NEW BIOCHEMICAL ASSAYS PROVIDE IN VITRO ANSWERS TO THE IN VIVO QUESTIONS

As introduced above, numerous questions concerning Pol IV and RDR2 transcription remained unanswered from in vivo studies. Among these were the following: (1) Which enzyme acts first in the biogenesis of siRNA precursors? (2) Does Pol IV make one strand of each dsRNA precursor and RDR2 the other strand, or are both enzymes needed to synthesize each strand? (3) Is there a way to distinguish Pol IV transcripts from RDR2 transcripts? (4) Are 3′-untemplated nucleotides present at the ends of both Pol IV and RDR2 transcripts? (5) Is there any functional significance to these untemplated nucleotides? (6) Why are P4R2 transcripts so short?

Answers to the first four questions came from a biochemical assay we devised using single-stranded bacteriophage M13 genomic DNA as a source of template for dsRNA synthesis by Pol IV and RDR2 (Singh et al. 2019). Pol IV and RDR2 associate and thus copurify from wild-type plants (Law et al. 2011; Haag et al. 2012). However, epitope-tagged Pol IV can be isolated from a *rdr2* null mutant, and epitope-tagged RDR2 can be purified from a *nrpd1* null mutant, lacking the Pol IV largest subunit, so as to be isolated individually (Haag et al. 2012). We found that dsRNAs were synthesized by Pol IV–RDR2 complexes but not by either enzyme individually (Singh et al. 2019). RNA sequencing showed that Pol IV transcripts are exclusively reverse complements of the DNA. Only if RDR2 is also present are RNA transcripts generated in the same 5′→3′ polarity as the template DNA; these RNAs can only be generated by using Pol IV transcripts as templates. The results showed definitively that each enzyme is wholly responsible for synthesizing individual strands of the dsRNA (Singh et al. 2019). Moreover, the phenomenon of template-mismatched nucleotides present at the 3′ ends of P4R2 RNAs in vivo was recapitulated (Blevins et al. 2015; Singh et al. 2019). Importantly, these untemplated nucleotides occur at the 3′ termini of RDR2 transcripts, not Pol IV transcripts. This indicates that RDR2 uses its terminal transferase activity to add an extra nucleotide to the end of its own transcripts but does not act on Pol IV transcripts.

Terminal transferase activity has been shown for a number of RNA-dependent RNA polymerases, including *Arabidopsis* RDR6, *Neurospora crassa* QDE1, and viral polymerases (Ranjith-Kumar et al. 2001; Curaba and Chen 2008; Poranen et al. 2008; Aalto et al. 2010). But what is the significance of this activity, if any? Our findings using the M13 system suggest that RDR2's terminal transferase activity serves a purpose—in specifying the fates of the two RNA strands following dicing. Reactions that included DCL3, in addition to Pol IV and RDR2, generated both 24- and 23-nt siRNAs in vitro (Singh et al. 2019), as is also the case in vivo (Kasschau et al. 2007). The significance of the 23-nt class of siRNAs has never been clear. Interestingly, we found that the template-mismatched 3′ nucleotides attributable to RDR2 terminal transferase activity are primarily found at the ends of 23-nt siRNAs (Singh et al. 2019). This in vitro result fits with analyses that showed that template-mismatched 3′-terminal nucleotides are also enriched among 23-nt siRNAs in vivo (Wang et al. 2016). Based on these observations, we hypothesize that DCL3 interacts with dsRNAs, measures 24 nt from the 5′ end of the Pol IV strand, and makes a staggered cut that leaves the 3′ end of the Pol IV strand overhanging the RDR2 strand by 2 nt, consistent with cutting by other Dicers (Park et al. 2011). Because the RDR2 strand has a 1-nt extension at its 3′ end as a result of RDR2's terminal transferase activity, DCL3 cutting generates a 23-nt RNA from the RDR2 strand, with the template-mismatched nucleotide at its 3′ end (Fig. 1). Because 23-nt siRNAs are not abundant among the RNAs that copurify with immunoprecipitated AGO4 (Qi et al. 2006; Havecker et al. 2010; Wang et al. 2016), we hypothesize that 23-nt RNAs may serve as so-called "passenger strands" for the 24-nt guide strands that become stably associated with AGO4 (Singh et al. 2019). If so, 23-nt siRNAs have a function, and so does the terminal transferase activity of RDR2 that contributes to their biogenesis.

An answer to the question of why P4R2 RNAs in vivo are so short came from a different set of in vitro experiments using synthetic oligo- and polynucleotide DNAs. In these experiments, we found that Pol IV engaged in transcription of a single-stranded DNA template DNA strand terminates ∼12–16 nt after encountering the base-paired nontemplate DNA strand (Singh et al. 2019), in a sequence-independent manner. This suggests a model in which Pol IV initiates on one DNA strand following the melting of duplex DNA to form a transcription bubble, like other DNA-dependent RNA polymerases (Holstege et al. 1997; Bae et al. 2015; Barnes et al. 2015). After transcribing single-stranded DNA within the bubble, Pol IV encounters double-stranded DNA at the edge of the bubble and is only able to transcribe an additional 12–16 nt before terminating. Exactly why Pol IV cannot transcribe further than 12–16 nt into duplex DNA is unclear. However, amino acid changes in several key regions of Pol IV's catalytic center offer some clues (Haag et al. 2009, 2012). In prokaryotic and eukaryotic DNA-dependent RNA polymerases, structural features in their largest subunits known as the "trigger," "rudder," and "zipper" loops affect catalytic activity, melting of the DNA strands ahead of the advancing polymerase, and reannealing of the DNA strands in the wake of the passing polymerase, respectively (Toulokhonov and Landick 2006). The largest subunit of Pol IV (NRPD1) contains amino acid deletions in each of these elements (Haag et al. 2009, 2012; Singh et al. 2019). Collectively, these deletions may hamper Pol IV's ability to plow ahead into duplex DNA and propagate a transcription bubble that translocates along with the polymerase, allowing for processivity.

Does RDR2 Interact with Pol IV Transcripts Prior to Pol IV Termination?

Arabidopsis has numerous small RNA-mediated pathways, involving six RNA-dependent RNA polymerases, four Dicers, and 10 Argonaute proteins (Volpe and Martienssen 2011; Borges and Martienssen 2015). With so many alternative enzymes in play, how can RNAs stay channeled within a specific pathway? Our findings suggest that protein–protein interactions and RNA signals, imparted by the enzymes that synthesize or process the RNAs, account for RNA channeling. Pol IV physically associates with RDR2 (Law et al. 2011; Haag et al. 2012) and the interaction is insensitive to RNase A treatment, suggesting that RNA does not link them (Pontes et al. 2006). Therefore, protein–protein interactions are likely, involving one or more Pol IV subunits that interact directly with RDR2 or indirectly via one or more bridging proteins. In *Arabidopsis*, Pol IV's largest subunit, NRPD1, and seventh-largest subunit, NRPD7, are the only subunits not shared by Pols I, II, III, or V (Ream et al. 2009), making them potential candidates for RDR2 interactors. In maize, however, in which Pol IV and RDR2 (MOP1 in maize) also copurify, the only subunit unique to Pol IV is NRPD1 (Haag et al. 2014). Thus, NRPD1 is the prime suspect for mediating the interaction with RDR2 interaction.

RDR2 initiation is coupled to Pol IV termination and results in RDR2 transcripts that mirror the length of Pol IV transcripts in the M13 system (Singh et al. 2019), suggesting that RDR2 transcribes the Pol IV transcripts end-to-end. However, it remains unclear when RDR2 first engages the Pol IV transcript. One possibility is that RDR2 interacts with the nascent Pol IV transcript as soon as the 5′ end of the transcript emerges from Pol IV's RNA exit channel such that the RNA is already engaged by RDR2 prior to Pol IV termination and transcript release. Alternatively, RDR2 may only capture Pol IV transcripts after their release. Additional biochemical tests, such as protein cross-linking experiments and/or structural studies of Pol IV–RDR2 complexes, will be needed to determine how Pol IV and RDR2 interact and how RNAs are channeled from Pol IV to RDR2.

How Is Pol IV Transcription Initiated?

Genetic evidence indicates that sites of Pol IV transcription are not specified by conventional promoters to which transcription factors bind. Instead, the evidence suggests that chromatin modifications are the signals that facilitate Pol IV recruitment (Law et al. 2013; Blevins et al. 2014). These include cytosine methylation marks that can be maintained following every round of DNA replication by DNA METHYLTRANSFERASE 1 (MET1), the plant ortholog of mammalian DNMT1. MET1 maintenance methylation at CG motifs somehow requires the activity of HDA6 (To et al. 2011; Liu et al. 2012; Blevins et al. 2014), as does maintenance methylation at CHG motifs by CHROMOMETHYLTRANSFERASE 3 (CMT3) (Earley et al. 2010). Nucleosomes assembled using DNA methylated by these enzymes include histone H3 proteins that are dimethylated on lysine 9 (H3K9me2). SAWADEE HOMEODOMAIN HOMOLOG 1 (SHH1; also known as DTF1), a protein with TUDOR-like domains that can associate with *Arabidopsis* Pol IV, can bind these H3K9me2 marks (Law et al. 2013; Zhang et al. 2013). Interestingly, in maize, the paralogous proteins, SHH2a and SHH2b, associate with both Pol IV and Pol V (Haag et al. 2014). Like Pol IV, Pol V depends on MET1 to specify its sites of action (Johnson et al. 2014), somehow facilitated by methylcytosine binding by SUVH2 and SUVH9 (Liu et al. 2014; Jing et al. 2016); thus, both polymerases may have similar epigenetically inherited recruitment signals. But how recruitment leads to Pol IV and Pol V being able to engage and transcribe one strand of DNA wrapped around a histone octamer is unknown. Pol IV–dependent synthesis of 24-nt siRNAs requires one or more members of a family of putative ATP-dependent DNA translocases, CLSY1–CLSY4 (Smith et al. 2007). CLSY1 has been shown to physically interact with SHH1/DTF1 (Zhang et al. 2013) and has been proposed to act as a chromatin remodeler that might alter DNA–histone contacts. However, no biochemical assays of CLSY activity have been reported. Importantly, ATP-dependent DNA translocases can have diverse functions, including helicase activity. This raises the intriguing possibility that CLSY proteins could potentially be involved in the transcription bubble opening to allow Pol IV transcription initiation. Likewise, the putative ATP-dependent DNA translocase, DRD1 (Kanno et al. 2005), is required for Pol V transcription (Wierzbicki et al. 2008; Law et al. 2010; Zhong et al. 2012) and may play a similar role (Pikaard et al. 2012). Clearly, biochemical assays need to be developed to understand how Pol IV and Pol V can initiate transcription in the context of chromatin.

How Does Nontemplate DNA Induce Pol IV Termination?

Pol IV terminates shortly after encountering double-stranded DNA and termination in this manner is somehow required to enable RDR2 to synthesize the second RNA strand (Singh et al. 2019). The biochemical basis for these molecular events is unclear. As discussed previously, amino acid deletions in the Pol IV "trigger," "rudder," and "zipper" loops in the vicinity of the catalytic center may contribute to transcription stalling and termination (for sequence comparisons, see Toulokhonov and Landick 2006; Landick 2009; and Singh et al. 2019). But how termination results in a nucleic acid conformation that enables RDR2 to engage the 3′ end of the released Pol IV transcript and use it as a template is difficult to envision. Structural studies may be necessary to come to a mechanistic understanding of this critical step.

Do Patterns of DCL3 Dicing Program AGO4–siRNA Complex Formation?

AGO4–siRNA complexes immunoprecipitated from *Arabidopsis* contain 24-nt siRNAs, almost exclusively, despite the fact that DCL3 generates both 23- and 24-nt siRNAs (Qi et al. 2006; Kasschau et al. 2007; Havecker et al. 2010; Wang et al. 2016; Singh et al. 2019). As discussed previously, we hypothesize that 23-nt RNAs may serve as passenger strands that help specify the stable association of AGO4 with the 24-nt RNAs base-paired with the 23-nt RNAs (Singh et al. 2019). However, it is not yet clear if 24/23-nt heteroduplexes exist in cells. Likewise, we know little about how AGO4 becomes loaded with siRNAs. In vitro assays that could recapitulate AGO4 loading and siRNA strand choice would provide a means to address these questions.

Are There DCL-Independent Routes of siRNA Biogenesis?

Recent studies suggested that there are DCL-independent routes of siRNA biogenesis, explaining residual levels of DNA methylation in Dicer mutants (Yang et al. 2016; Ye et al. 2016). Briefly, the idea is that Pol IV nascent transcripts might be loaded into AGO4 as single strands that are then trimmed down to 24 nt by the putative 3′ to 5′ exonuclease activity of RRP6L1 or an uncharacterized "trimmer" enzyme. However, using the M13 system in vitro, a significant fraction of Pol IV and RDR2 transcripts happen to be 24 nt in size (Singh et al. 2019). Might these transcripts be capable of being loaded into AGO4? If so, this would be a Dicer- and Trimmer-independent way of producing 24-nt RNAs. Moreover, it is noteworthy that the proposed role for RRP6L1 as an siRNA Trimmer is hypothetical, stemming from genetic evidence that it is involved for RdDM. But there is an alternative explanation for RRP6L1 involvement in RdDM. RRP6L1 interacts with the carboxy-terminal domain of Pol V in yeast two-hybrid assays, and in *rrp6l1* mutants, Pol V transcripts are longer than in wild type (Wendte et al. 2017). Pol V transcripts are similarly longer in *ago4* mutants, suggesting that what RRP6L1 trims are Pol V transcripts sliced by AGO4 (Wendte et al. 2017).

CONCLUSION

Two decades of in vivo studies have identified the players in the RdDM pathway and what happens when they are mutated. However, we are still in the early stages of understanding the biochemical functions of many of these proteins. Having reconstituted the siRNA biogenesis steps of the RdDM pathway in vitro, we can envision a future in which we can load AGO4 with siRNAs of our choosing and have these AGO4–siRNA complexes associate with Pol V engaged in transcription. By considering RdDM pathway as a series of reactions that can be understood in biochemical detail, we are optimistic that RdDM can ultimately be reconstituted in vitro. If so, we may finally understand how the siRNAs, long noncoding RNAs, DNA strands, and proteins are configured to enable the chemical modification of genomic target sites, an understanding not yet in hand for RNA-directed chromatin modification pathways of any organism.

ACKNOWLEDGMENTS

Our work is supported by the National Institutes of Health grant GM077590 and funds to C.S.P. as an Investigator of the Howard Hughes Medical Institute. J.S. was supported, in part, by a Carlos O. Miller fellowship (Indiana University).

REFERENCES

Aalto AP, Poranen MM, Grimes JM, Stuart DI, Bamford DH. 2010. *In vitro* activities of the multifunctional RNA silencing polymerase QDE-1 of *Neurospora crassa*. *J Biol Chem* **285**: 29367–29374. doi:10.1074/jbc.M110.139121

Aufsatz W, Mette MF, Van Der Winden J, Matzke M, Matzke AJ. 2002. HDA6, a putative histone deacetylase needed to enhance DNA methylation induced by double-stranded RNA. *EMBO J* **21:** 6832–6841. doi:10.1093/emboj/cdf663

Bae B, Feklistov A, Lass-Napiorkowska A, Landick R, Darst SA. 2015. Structure of a bacterial RNA polymerase holoenzyme open promoter complex. *eLife* **4:** e08504. doi:10.7554/eLife.08504

Barnes CO, Calero M, Malik I, Graham BW, Spahr H, Lin G, Cohen AE, Brown IS, Zhang Q, Pullara F, et al. 2015. Crystal structure of a transcribing RNA polymerase II complex reveals a complete transcription bubble. *Mol Cell* **59:** 258–269. doi:10.1016/j.molcel.2015.06.034

Bergsch J, Allain FH-T, Lipps G. 2019. Recent advances in understanding bacterial and archaeoeukaryotic primases. *Curr Opin Struct Biol* **59:** 159–167. doi:10.1016/j.sbi.2019.08.004

Blevins T, Pontvianne F, Cocklin R, Podicheti R, Chandrasekhara C, Yerneni S, Braun C, Lee B, Rusch D, Mockaitis K, et al. 2014. A two-step process for epigenetic inheritance in *Arabidopsis*. *Mol Cell* **54:** 30–42. doi:10.1016/j.molcel.2014.02.019

Blevins T, Podicheti R, Mishra V, Marasco M, Tang H, Pikaard CS. 2015. Identification of Pol IV and RDR2-dependent precursors of 24 nt siRNAs guiding de novo DNA methylation in *Arabidopsis*. *eLife* **4:** e09591. doi:10.7554/eLife.09591

Borges F, Martienssen RA. 2015. The expanding world of small RNAs in plants. *Nat Rev Mol Cell Biol* **16:** 727–741. doi:10.1038/nrm4085

Cao X, Jacobsen SE. 2002. Role of the *Arabidopsis* DRM methyltransferases in de novo DNA methylation and gene silencing. *Curr Biol* **12:** 1138–1144. doi:10.1016/S0960-9822(02)00925-9

Curaba J, Chen X. 2008. Biochemical activities of *Arabidopsis* RNA-dependent RNA polymerase 6. *J Biol Chem* **283:** 3059–3066. doi:10.1074/jbc.M708983200

Deleris A, Greenberg MVC, Ausin I, Law RWY, Moissiard G, Schubert D, Jacobsen SE. 2010. Involvement of a Jumonji-C domain–containing histone demethylase in DRM2-mediated maintenance of DNA methylation. *EMBO Rep* **11:** 950–955. doi:10.1038/embor.2010.158

Earley KW, Pontvianne F, Wierzbicki AT, Blevins T, Tucker S, Costa-Nunes P, Pontes O, Pikaard CS. 2010. Mechanisms of HDA6-mediated rRNA gene silencing: suppression of intergenic Pol II transcription and differential effects on maintenance versus siRNA-directed cytosine methylation. *Genes Dev* **24:** 1119–1132. doi:10.1101/gad.1914110

Ebbs ML, Bender J. 2006. Locus-specific control of DNA methylation by the *Arabidopsis* SUVH5 histone methyltransferase. *Plant Cell* **18:** 1166–1176. doi:10.1105/tpc.106.041400

El-Shami M, Pontier D, Lahmy S, Braun L, Picart C, Vega D, Hakimi M-A, Jacobsen SE, Cooke R, Lagrange T. 2007. Reiterated WG/GW motifs form functionally and evolutionarily conserved ARGONAUTE-binding platforms in RNAi-related components. *Genes Dev* **21:** 2539–2544. doi:10.1101/gad.451207

Haag JR, Pikaard CS. 2011. Multisubunit RNA polymerases IV and V: purveyors of non-coding RNA for plant gene silencing. *Nat Rev Mol Cell Biol* **12:** 483–492. doi:10.1038/nrm3152

Haag JR, Pontes O, Pikaard CS. 2009. Metal A and metal B sites of nuclear RNA polymerases Pol IV and Pol V are required for siRNA-dependent DNA methylation and gene silencing. *PLoS One* **4:** e4110. doi:10.1371/journal.pone.0004110

Haag JR, Ream TS, Marasco M, Nicora CD, Norbeck AD, Pasa-Tolic L, Pikaard CS. 2012. In vitro transcription activities of Pol IV, Pol V, and RDR2 reveal coupling of Pol IV and RDR2 for dsRNA synthesis in plant RNA silencing. *Mol Cell* **48:** 811–818. doi:10.1016/j.molcel.2012.09.027

Haag JR, Brower-Toland B, Krieger EK, Sidorenko L, Nicora CD, Norbeck AD, Irsigler A, LaRue H, Brzeski J, McGinnis K, et al. 2014. Functional diversification of maize RNA polymerase IV and V subtypes via alternative catalytic subunits. *Cell Rep* **9:** 378–390. doi:10.1016/j.celrep.2014.08.067

Havecker ER, Wallbridge LM, Hardcastle TJ, Bush MS, Kelly KA, Dunn RM, Schwach F, Doonan JH, Baulcombe DC. 2010. The *Arabidopsis* RNA-directed DNA methylation Argonautes functionally diverge based on their expression and interaction with target loci. *Plant Cell* **22:** 321–334. doi:10.1105/tpc.109.072199

He X-J, Hsu Y-F, Pontes O, Zhu J, Lu J, Bressan RA, Pikaard C, Wang C-S, Zhu J-K. 2009. NRPD4, a protein related to the RPB4 subunit of RNA polymerase II, is a component of RNA polymerases IV and V and is required for RNA-directed DNA methylation. *Genes Dev* **23:** 318–330. doi:10.1101/gad.1765209

Henderson IR, Zhang X, Lu C, Johnson L, Meyers BC, Green PJ, Jacobsen SE. 2006. Dissecting *Arabidopsis thaliana* DICER function in small RNA processing, gene silencing and DNA methylation patterning. *Nat Genet* **38:** 721–725. doi:10.1038/ng1804

Herr AJ, Jensen MB, Dalmay T, Baulcombe DC. 2005. RNA polymerase IV directs silencing of endogenous DNA. *Science* **308:** 118–120. doi:10.1126/science.1106910

Holoch D, Moazed D. 2015. RNA-mediated epigenetic regulation of gene expression. *Nat Rev Genet* **16:** 71–84. doi:10.1038/nrg3863

Holstege FCP, Fiedler U, Timmers HT. 1997. Three transitions in the RNA polymerase II transcription complex during initiation. *EMBO J* **16:** 7468–7480. doi:10.1093/emboj/16.24.7468

Jackson JP, Johnson L, Jasencakova Z, Zhang X, PerezBurgos L, Singh PB, Cheng X, Schubert I, Jenuwein T, Jacobsen SE. 2004. Dimethylation of histone H3 lysine 9 is a critical mark for DNA methylation and gene silencing in *Arabidopsis thaliana*. *Chromosoma* **112:** 308–315. doi:10.1007/s00412-004-0275-7

Jing Y, Sun H, Yuan W, Wang Y, Li Q, Liu Y, Li Y, Qian W. 2016. SUVH2 and SUVH9 couple two essential steps for transcriptional gene silencing in *Arabidopsis*. *Mol Plant* **9:** 1156–1167. doi:10.1016/j.molp.2016.05.006

Johnson LM, Bostick M, Zhang X, Kraft E, Henderson I, Callis J, Jacobsen SE. 2007. The SRA methyl-cytosine-binding domain links DNA and histone methylation. *Curr Biol* **17:** 379–384. doi:10.1016/j.cub.2007.01.009

Johnson LM, Du J, Hale CJ, Bischof S, Feng S, Chodavarapu RK, Zhong X, Marson G, Pellegrini M, Segal DJ, et al. 2014. SRA- and SET-domain-containing proteins link RNA polymerase V occupancy to DNA methylation. *Nature* **507:** 124–128. doi:10.1038/nature12931

Kanno T, Aufsatz W, Jaligot E, Mette MF, Matzke M, Matzke AJ. 2005. A SNF2-like protein facilitates dynamic control of DNA methylation. *EMBO Rep* **6:** 649–655. doi:10.1038/sj.embor.7400446

Kasschau KD, Fahlgren N, Chapman EJ, Sullivan CM, Cumbie JS, Givan SA, Carrington JC. 2007. Genome-wide profiling and analysis of *Arabidopsis* siRNAs. *PLoS Biol* **5:** e57. doi:10.1371/journal.pbio.0050057

Lahmy S, Pontier D, Bies-Etheve N, Laudie M, Feng S, Jobet E, Hale CJ, Cooke R, Hakimi MA, Angelov D, et al. 2016. Evidence for ARGONAUTE4–DNA interactions in RNA-directed DNA methylation in plants. *Genes Dev* **30:** 2565–2570. doi:10.1101/gad.289553.116

Landick R. 2009. Functional divergence in the growing family of RNA polymerases. *Structure* **17:** 323–325. doi:10.1016/j.str.2009.02.006

Law JA, Jacobsen SE. 2010. Establishing, maintaining and modifying DNA methylation patterns in plants and animals. *Nat Rev Genet* **11:** 204–220. doi:10.1038/nrg2719

Law JA, Ausin I, Johnson LM, Vashisht AA, Zhu J-K, Wohlschlegel JA, Jacobsen SE. 2010. A protein complex required for polymerase V transcripts and RNA- directed DNA methylation in *Arabidopsis*. *Curr Biol* **20:** 951–956. doi:10.1016/j.cub.2010.03.062

Law JA, Vashisht AA, Wohlschlegel JA, Jacobsen SE. 2011. SHH1, a homeodomain protein required for DNA methylation, as well as RDR2, RDM4, and chromatin remodeling factors, associate with RNA polymerase IV. *PLoS Genet* **7:** e1002195. doi:10.1371/journal.pgen.1002195

Law JA, Du J, Hale CJ, Feng S, Krajewski K, Palanca AM, Strahl BD, Patel DJ, Jacobsen SE. 2013. Polymerase IV occupancy at RNA-directed DNA methylation sites requires SHH1. *Nature* **498:** 385–389. doi:10.1038/nature12178

Li S, Vandivier LE, Tu B, Gao L, Won SY, Li S, Zheng B, Gregory BD, Chen X. 2015. Detection of Pol IV/RDR2-dependent transcripts at the genomic scale in *Arabidopsis* reveals features and regulation of siRNA biogenesis. *Genome Res* **25:** 235–245. doi:10.1101/gr.182238.114

Liu X, Yu C-W, Duan J, Luo M, Wang K, Tian G, Cui Y, Wu K. 2012. HDA6 directly interacts with DNA methyltransferase MET1 and maintains transposable element silencing in *Arabidopsis*. *Plant Physiol* **158:** 119–129. doi:10.1104/pp.111.184275

Liu Z-W, Shao C-R, Zhang C-J, Zhou J-X, Zhang S-W, Li L, Chen S, Huang H-W, Cai T, He X-J. 2014. The SET domain proteins SUVH2 and SUVH9 are required for Pol V occupancy at RNA-directed DNA methylation loci. *PLoS Genet* **10:** e1003948. doi:10.1371/journal.pgen.1003948

Maillard PV, van der Veen AG, Poirier EZ, Reis ESC. 2019. Slicing and dicing viruses: antiviral RNA interference in mammals. *EMBO J* **38,** e100941. doi:10.15252/embj.2018100941

Marasco M, Li W, Lynch M, Pikaard CS. 2017. Catalytic properties of RNA polymerases IV and V: accuracy, nucleotide incorporation and rNTP/dNTP discrimination. *Nucleic Acids Res* **45:** 11315–11326. doi:10.1093/nar/gkx794

Martienssen R, Moazed D. 2015. RNAi and heterochromatin assembly. *Cold Spring Harb Perspect Biol* **7:** a019323. doi:10.1101/cshperspect.a019323

Onodera Y, Haag JR, Ream T, Costa Nunes P, Pontes O, Pikaard CS. 2005. Plant nuclear RNA polymerase IV mediates siRNA and DNA methylation-dependent heterochromatin formation. *Cell* **120:** 613–622. doi:10.1016/j.cell.2005.02.007

Park J-E, Heo I, Tian Y, Simanshu DK, Chang H, Jee D, Patel DJ, Kim VN. 2011. Dicer recognizes the 5′ end of RNA for efficient and accurate processing. *Nature* **475:** 201–205. doi:10.1038/nature10198

Pikaard CS, Haag JR, Pontes OM, Blevins T, Cocklin R. 2012. A transcription fork model for Pol IV and Pol V–dependent RNA-directed DNA methylation. *Cold Spring Harb Symp Quant Biol* **77:** 205–212. doi:10.1101/sqb.2013.77.014803

Pontes O, Li CF, Costa Nunes P, Haag J, Ream T, Vitins A, Jacobsen SE, Pikaard CS. 2006. The *Arabidopsis* chromatin-modifying nuclear siRNA pathway involves a nucleolar RNA processing center. *Cell* **126:** 79–92. doi:10.1016/j.cell.2006.05.031

Poranen MM, Koivunen MR, Bamford DH. 2008. Nontemplated terminal nucleotidyltransferase activity of double-stranded RNA bacteriophage φ6 RNA-dependent RNA polymerase. *J Virol* **82:** 9254–9264. doi:10.1128/JVI.01044-08

Qi Y, He X, Wang X-J, Kohany O, Jurka J, Hannon GJ. 2006. Distinct catalytic and non-catalytic roles of ARGONAUTE4 in RNA-directed DNA methylation. *Nature* **443:** 1008–1012. doi:10.1038/nature05198

Ranjith-Kumar CT, Gajewski J, Gutshall L, Maley D, Sarisky RT, Kao CC. 2001. Terminal nucleotidyl transferase activity of recombinant Flaviviridae RNA-dependent RNA polymerases: implication for viral RNA synthesis. *J Virol* **75:** 8615–8623. doi:10.1128/JVI.75.18.8615-8623.2001

Ream TS, Haag JR, Wierzbicki AT, Nicora CD, Norbeck AD, Zhu J-K, Hagen G, Guilfoyle TJ, Paša-Tolić L, Pikaard CS. 2009. Subunit compositions of the RNA-silencing enzymes Pol IV and Pol V reveal their origins as specialized forms of RNA polymerase II. *Mol Cell* **33:** 192–203. doi:10.1016/j.molcel.2008.12.015

Searle IR, Pontes O, Melnyk CW, Smith LM, Baulcombe DC. 2010. JMJ14, a JmjC domain protein, is required for RNA silencing and cell-to-cell movement of an RNA silencing signal in *Arabidopsis*. *Genes Dev* **24:** 986–991. doi:10.1101/gad.579910

Singh J, Mishra V, Wang F, Huang HY, Pikaard CS. 2019. Reaction mechanisms of Pol IV, RDR2, and DCL3 drive RNA channeling in the siRNA-directed DNA methylation pathway. *Mol Cell* **75:** 576–589.e5. doi:10.1016/j.molcel.2019.07.008

Smith LM, Pontes O, Searle I, Yelina N, Yousafzai FK, Herr AJ, Pikaard CS, Baulcombe DC. 2007. An SNF2 protein associated with nuclear RNA silencing and the spread of a silencing signal between cells in *Arabidopsis*. *Plant Cell* **19:** 1507–1521. doi:10.1105/tpc.107.051540

To TK, Kim J-M, Matsui A, Kurihara Y, Morosawa T, Ishida J, Tanaka M, Endo T, Kakutani T, Toyoda T, et al. 2011. *Arabidopsis* HDA6 regulates locus-directed heterochromatin silencing in cooperation with MET1. *PLoS Genet* **7:** e1002055. doi:10.1371/journal.pgen.1002055

Toulokhonov I, Landick R. 2006. The role of the lid element in transcription by *E. coli* RNA polymerase. *J Mol Biol* **361:** 644–658. doi:10.1016/j.jmb.2006.06.071

Volpe T, Martienssen RA. 2011. RNA interference and heterochromatin assembly. *Cold Spring Harb Perspect Biol* **3:** a003731. doi:10.1101/cshperspect.a003731

Wang F, Johnson NR, Coruh C, Axtell MJ. 2016. Genome-wide analysis of single non-templated nucleotides in plant endogenous siRNAs and miRNAs. *Nucleic Acids Res* **44:** 7395–7405. doi:10.1093/nar/gkw457

Wendte JM, Pikaard CS. 2017. The RNAs of RNA-directed DNA methylation. *Biochim Biophys Acta* **1860:** 140–148. doi:10.1016/j.bbagrm.2016.08.004

Wendte JM, Haag JR, Singh J, McKinlay A, Pontes OM, Pikaard CS. 2017. Functional dissection of the Pol V largest subunit CTD in RNA-directed DNA methylation. *Cell Rep* **19:** 2796–2808. doi:10.1016/j.celrep.2017.05.091

Wierzbicki AT, Haag JR, Pikaard CS. 2008. Noncoding transcription by RNA polymerase Pol IVb/Pol V mediates transcriptional silencing of overlapping and adjacent genes. *Cell* **135:** 635–648. doi:10.1016/j.cell.2008.09.035

Wierzbicki AT, Ream TS, Haag JR, Pikaard CS. 2009. RNA polymerase V transcription guides ARGONAUTE4 to chromatin. *Nat Genet* **41:** 630–634. doi:10.1038/ng.365

Xie Z, Johansen LK, Gustafson AM, Kasschau KD, Lellis AD, Zilberman D, Jacobsen SE, Carrington JC. 2004. Genetic and functional diversification of small RNA pathways in plants. *PLoS Biol* **2:** e104. doi:10.1371/journal.pbio.0020104

Yang D-L, Zhang G, Tang K, Li J, Yang L, Huang H, Zhang H, Zhu JK. 2016. Dicer-independent RNA-directed DNA methylation in *Arabidopsis*. *Cell Res* **26:** 66–82. doi:10.1038/cr.2015.145

Ye R, Chen Z, Lian B, Rowley MJ, Xia N, Chai J, Li Y, He X-J, Wierzbicki AT, Qi Y. 2016. A Dicer-independent route for biogenesis of siRNAs that direct DNA methylation in *Arabidopsis*. *Mol Cell* **61:** 222–235. doi:10.1016/j.molcel.2015.11.015

Zhai J, Bischof S, Wang H, Feng S, Lee T-F, Teng C, Chen X, Park SY, Liu L, Gallego-Bartolome J, et al. 2015. A one precursor one siRNA model for Pol IV-dependent siRNA biogenesis. *Cell* **163:** 445–455. doi:10.1016/j.cell.2015.09.032

Zhang H, Ma Z-Y, Zeng L, Tanaka K, Zhang C-J, Ma J, Bai G, Wang P, Zhang S-W, Liu Z-W, et al. 2013. DTF1 is a core component of RNA-directed DNA methylation and may assist in the recruitment of Pol IV. *Proc Natl Acad Sci* **110:** 8290–8295. doi:10.1073/pnas.1300585110

Zhong X, Hale CJ, Law JA, Johnson LM, Feng S, Tu A, Jacobsen SE. 2012. DDR complex facilitates global association of RNA polymerase V to promoters and evolutionarily young transposons. *Nat Struct Mol Biol* **19:** 870–875. doi:10.1038/nsmb.2354

Zhou M, Law JA. 2015. RNA Pol IV and V in gene silencing: rebel polymerases evolving away from Pol II's rules. *Curr Opin Plant Biol* **27:** 154–164. doi:10.1016/j.pbi.2015.07.005

Zhu Y, Rowley MJ, Bohmdorfer G, Wierzbicki AT. 2013. A SWI/SNF chromatin-remodeling complex acts in noncoding RNA-mediated transcriptional silencing. *Mol Cell* **49:** 298–309. doi:10.1016/j.molcel.2012.11.011

Zilberman D, Cao X, Jacobsen SE. 2003. ARGONAUTE4 control of locus-specific siRNA accumulation and DNA and histone methylation. *Science* **299:** 716–719. doi:10.1126/science.1079695

RNP Granule Formation: Lessons from P-Bodies and Stress Granules

GIULIA ADA CORBET[1] AND ROY PARKER[1,2]

[1]Department of Biochemistry,
[2]Howard Hughes Medical Institute, University of Colorado at Boulder, Boulder, Colorado 80309, USA

Correspondence: roy.parker@colorado.edu

It is now clear that cells form a wide collection of large RNA–protein assemblies, referred to as RNP granules. RNP granules exist in bacterial cells and can be found in both the cytosol and nucleus of eukaryotic cells. Recent approaches have begun to define the RNA and protein composition of a number of RNP granules. Herein, we review the composition and assembly of RNP granules, as well as how RNPs are targeted to RNP granules using stress granules and P-bodies as model systems. Taken together, these reveal that RNP granules form through the summative effects of a combination of protein–protein, protein–RNA, and RNA–RNA interactions. Similarly, the partitioning of individual RNPs into stress granules is determined by the combinatorial effects of multiple elements. Thus, RNP granules are assemblies generally dominated by combinatorial effects, thereby providing rich opportunities for biological regulation.

Ribonucleoprotein (RNP) granules are a diverse set of biomolecular assemblies of RNA and protein that form within cells. RNP granules in the nucleus, such as Cajal bodies, the nucleolus, and paraspeckles, regulate various steps of RNA and RNP processing (Table 1; Trinkle-Mulcahy and Sleeman 2017; Fox et al. 2018). Cytoplasmic granules, such as P-bodies and stress granules, may regulate RNA translation, RNA degradation, signaling pathways, and stress responses (Sheth and Parker 2003; Kedersha et al. 2013; Protter and Parker 2016; Samir et al. 2019). RNP transport granules in neurons and myogranules in skeletal muscle are thought to facilitate RNA transport and localized translation (Kiebler and Bassell 2006; Vogler et al. 2018). Furthermore, germ granules, such as *Caenorhabditis elegans* P-granules and *Drosophila* polar bodies, form in the germline during development, contain maternally encoded mRNAs, and help specify germ cell fate (Voronina et al. 2011; Wang and Seydoux 2014; Trcek and Lehmann 2019).

RNP granules are also found in bacteria. In *Caulobacter*, bacterial RNP bodies (BR-bodies) form under stress and can be considered similar to eukaryotic stress granules (Al-Husini et al. 2018). It is likely only a matter of time before more RNP granules are discovered in bacteria.

Aberrant RNP granules are also pathological hallmarks of numerous neurodegenerative and neuromuscular diseases, including amyotrophic lateral sclerosis (ALS), frontotemporal lobar degeneration (FTLD), and inclusion body myopathy (Arai et al. 2006; Neumann et al. 2006; Weihl et al. 2008; Dickson et al. 2011). These aggregates often contain many of the same components that are found in canonical RNP granules, such as stress granules and splicing speckles, suggesting that aberrant granules may arise from canonical granules and thereby contribute to neurodegenerative diseases (Taylor et al. 2016; Hsieh et al. 2019; Wolozin and Ivanov 2019).

An unanswered question is how RNP granules assemble, and what interactions allow for multiple distinct RNP granules to exist in cells at any given time. Herein, we review the composition, assembly, and how RNPs are targeted to RNP granules using stress granules (SGs) and P-bodies (PBs) as model systems to understand these issues.

RNP GRANULE COMPOSITION

Proteins in RNP Granules

Three types of approaches have been used to identify components of RNP granules (Fig. 1). First, and sometimes the initial identification of the granule, has been the identification of locally concentrated proteins by immunofluorescence or by the tagging of proteins with green fluorescent protein (GFP). More complete descriptions of the proteins in RNP granules has come from mass spectroscopy of RNP granules purified by differential centrifugation, with or without immunopurification, or by using particle sorting to fractionate granules (Jain et al. 2016; Hubstenberger et al. 2017). A third approach has been to use proximity labeling approaches such as Bio-ID or APEX labeling (Markmiller et al. 2018; Youn et al. 2018), although such proximity labeling approaches are unable to distinguish between interacting proteins within the granule or outside the granule, which can be an issue because even proteins that are highly concentrated in gran-

© 2019 Corbet and Parker. This article is distributed under the terms of the Creative Commons Attribution-NonCommercial License, which permits reuse and redistribution, except for commercial purposes, provided that the original author and source are credited.

Published by Cold Spring Harbor Laboratory Press; doi:10.1101/sqb.2019.84.040329

Table 1. Overview of select ribonucleoprotein (RNP) granules function and composition

RNP granule	Cell type/ modulators	Function	Protein composition	RNA composition	References
Nucleolus	Found in all cell types during interphase	Site of rRNA transcription, pre-rRNA processing, ribosome assembly	snRNPs, ribosomal proteins, and ribosome biogenesis factors: RNA polymerase I, DNA topoisomerase, fibrillarin, nucleolin, p52; RNA-modifying enzymes	snoRNAs, rRNAs	Yasuzumi et al. 1958; Chamberland and Lafontaine 1993; Scheer et al. 1993; Mélèse and Xue 1995; Maden and Hughes 1997; Dundr and Olson 1998; Olson et al. 2000; Fatica and Tollervey 2002; Desterro et al. 2003
Cajal body	All cell types	snRNP assembly, processing	coilin, Sm proteins, SMN, NIOPP140, snRNPs	snRNAs, scaRNAs, snoRNAs	Gall et al. 1999; Narayanan et al. 1999; Sleeman and Lamond 1999; Hebert and Matera 2000; Carmo-Fonseca 2002; Darzacq et al. 2002; Verheggen et al. 2002; Schaffert et al. 2004; Cioce and Lamond 2005; Morris 2008; Staněk et al. 2008; Shaw et al. 2014; Hebert and Poole 2016; Love et al. 2016; Trinkle-Mulcahy and Sleeman 2017
Paraspeckle	All cell types, increase in stress	Unknown	RNA-binding proteins: PSPC1, p54nrb, SFPQ, RBM14, CPSF6	*NEAT1* lncRNA, hyperedited RNAs	Prasanth et al. 2005; Bond and Fox 2009; Clemson et al. 2009; Fox and Lamond 2010; Nakagawa et al. 2011; Naganuma et al. 2012; Naganuma and Hirose 2013; Imamura et al. 2014; Fox et al. 2018
P-bodies	Constitutive but increase in stress	Translation regulation, RNA storage, degradation	mRNA decapping enzymes, exonucleases, RNA-binding proteins, NMD factors, translation repressors, RNA helicases	Enriched for nontranslating mRNAs, biased for AU-rich RNAs, RNA composition similar to stress granules under stress, enriched for RNAs with short poly(A) tails	Schwartz and Parker 1999; Sheth and Parker 2003, 2006; Decker et al. 2007; Eulalio et al. 2007; Teixeira and Parker 2007; Buchan et al. 2008; Pilkington and Parker 2008; Hubstenberger et al. 2017; Courel et al. 2019; Matheny et al. 2019
Stress granules	Induced by stress such as viral infection, heat/cold shock, oxidative stress, starvation	Unknown, hypothesized to regulate translation, RNA stability, cell survival, cell signaling	RNA-binding proteins: G3BP1, UBAP2L, PABPC1, TIA-1, IDR-containing proteins, RNA helicases, translation factors	Enriched for longer and poorly translated mRNAs	Kedersha et al. 1999, 2000, 2005, 2013, 2016; Gilks et al. 2004; Buchan and Parker 2009; Hofmann et al. 2012; Buchan et al. 2013; Jain et al. 2016; Wheeler et al. 2016; Khong et al. 2017; Namkoong et al. 2018; Protter et al. 2018; Matheny et al. 2019; Moon et al. 2019, 2020; Cirillo et al. 2020

Continued

Table 1. *Continued*

RNP granule	Cell type/ modulators	Function	Protein composition	RNA composition	References
P-granules	*Caenorhabditis elegans* germline	Germ cell fate specification, RNA silencing	RNA-binding proteins: PGL-1, PGL-3, MEG-3, LAF-1, IDR-containing proteins, RNA helicases	Maternal mRNAs, poorly translated mRNAs	Voronina et al. 2012; Seydoux 2018; Ouyang et al. 2019; Putnam et al. 2019; Lee et al. 2020
Polar granules	*Drosophila*	Germ cell fate, RNA segregation	RNA-binding proteins: Vasa, Tudor, eIF4A, Aubergine, ME31B, TER94, Oskar	*Nanos*, *cyclin B*, *polar granule component*, *oskar* mRNAs	Mahowald 1971; Hay et al. 1988; Ephrussi and Lehmann 1992; Nakamura et al. 1996; Ferrandon et al. 1997; Thomson and Lasko 2004; Arkov et al. 2006; Jones and Macdonald 2007; Vanzo et al. 2007; Thomson et al. 2008; Little et al. 2015
Myogranules	Regenerating skeletal muscle	Unknown, hypothesized to facilitate localized translation	RNA-binding proteins: TDP-43, FMRP; IDR-containing proteins	Unknown, hypothesized to be long, sarcomeric mRNAs	Vogler et al. 2018
C9ORF72 RNA foci	Neurons, glia, fibroblasts	Aberrant	Hypothesized to contain RNA-binding proteins	*C9ORF72* G_4C_2 repeat, potentially entire transcript and/or other RNA species	DeJesus-Hernandez et al. 2011; Fratta et al. 2012; Almeida et al. 2013; Gendron et al. 2013; Mori et al. 2013; Kumar et al. 2017; Česnik et al. 2019
Myotonic dystrophy type 1 (DM1) granules	Skeletal muscle, neurons	Aberrant	MBNL1, MBNL2	*DMPK* transcript	Taneja et al. 1995; Wang et al. 1995; Timchenko et al. 1996; Davis et al. 1997; Jiang et al. 2004
Bacterial RNP bodies (BR-bodies)	*Caulobacter*, other α-proteobacteria form in stress	RNA degradation, rRNA processing	RNA degradeosome proteins: RNase E, Aconitase, RhlB	Nontranslating RNAs	Al-Husini et al. 2018

(rRNA) Ribosomal RNA, (snRNP) small nuclear ribonucleoprotein, (snoRNA) small nucleolar RNA, (snRNA) small nuclear RNA, (scaRNA) small Cajal body–specific RNA, (lncRNA) long noncoding RNA, (mRNA) messenger RNA, (NMD) nonsense-mediated decay.

ules can have a majority of molecules outside the granule in the cell (Wheeler et al. 2017). Taken together, these approaches have revealed several features of RNP granule proteomes.

RNP granules contain three types of RNA-binding proteins (RBPs). First, there are RBPs that contribute to the assembly of the granule by providing protein-based cross-links between different RNAs or by forming a protein-based assembly. Second, there are RBPs that bind to RNAs resident to a granule and as such are simply carried into the granule by the RNAs. Third, there can be RBPs, such as YB-1 (Bounedjah et al. 2014) and eIF4A for stress granules (Tauber et al. 2020), that bind to RNA and compete for RNA–RNA interactions that would other promote RNP granule formation.

Some RBPs are found in more than one RNP granule. This can be understood for RBPs that bind RNAs and are carried into granules. However, RBPs that play a role in the formation of a specific granule are generally only found in that granule. For example, G3BP1 or Edc3, which are important for SG or PB formation (Tourrière et al. 2003; Decker et al. 2007), respectively, are only found in those granules (Kedersha et al. 2005; Decker et al. 2007). Furthermore, coilin, which is essential for Cajal body (CB) formation, is found specifically within CBs (Raška et al. 1991; Sleeman et al. 2003).

RNP granules can also contain other types of proteins. For example, stress granules are known to contain a number of proteins involved in cellular signaling pathways and SG formation can influence cell signaling (Kedersha et al. 2013). Similarly, SGs can contain proteins involved in metabolism (Jain et al. 2016; Bergovich et al. 2020). Moreover, the nucleolus contains a wide variety of proteins involved in a number of physiological process, such as dealing with misfolded proteins (Boisvert et al. 2007; Frottin et al. 2019). Taken together, the compositional studies suggest that cells have taken advantage of the formation of RNP granules to create localized concentrations of proteins involved in various processes, thereby increasing their effectiveness.

Although multiple RNP granules have been described as having very complex proteomes, the composition of RNP granules may often be much simpler. For example, yeast and mammalian PBs have been described to have more

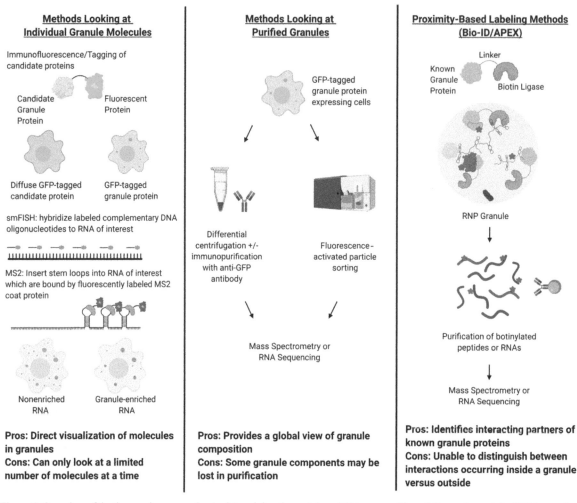

Figure 1. Overview of the three major approaches to determining the protein or RNA composition of ribonucleoprotein (RNP) granules. One approach involves probing for individual proteins or RNAs by fluorescence imaging methods (*left*). A second approach involves purifying RNP granules, followed by RNA sequencing or mass spectrometry to determine the RNA or protein compositions, respectively (*middle*). The third approach relies on tagging RNAs or proteins that interact with known granule proteins, followed by purification of tagged molecules and RNA sequencing or mass spectrometry (*right*).

than 50 or 100 different proteins resident to them, respectively (Hubstenberger et al. 2017). However, a quantitative analysis revealed that only 14 proteins are highly concentrated within yeast PBs, and the other approximately 36 proteins are present at low amounts (Xing et al. 2020). This is significant because 300–500 different proteins have been identified in SGs (Jain et al. 2016) and 200 proteins have been identified in the nucleolus (Thul et al. 2017). An important area of future research will be quantitative analyses of the abundance of these components to determine the primary constituents of any given RNP granule.

RNAs in RNP Granules

The RNAs present in RNP granules have been identified by three general approaches (Fig. 1). In one case, individual RNAs are examined by single-molecule fluorescence in situ hybridization (smFISH) or by tagging the RNAs with a fluorescent system such as MS2. A second approach has been to purify RNP granules and then sequence the copurifying RNAs (Hubstenberger et al. 2017; Khong et al. 2017; Namkoong et al. 2018). A third approach is to identify RNAs that interact with a key component of an RNP granule, inferring that this will identify RNA present in that assembly. This latter approach, followed by smFISH to validate the inferred partitioning, has worked well for the identification of mRNAs in *C. elegans* P-granules (Lee et al. 2020). These experiments have revealed the following features of RNP granules.

First, as inferred from their definition, every RNP granule contains a set of RNAs, although they can vary between different RNP granules. For example, both SG and PBs in the cytosol are made up of a wide distribution of nontranslating messenger RNAs (mRNAs) and long noncoding RNAs (lncRNAs) (Hubstenberger et al. 2017; Khong et al. 2017). Cajal bodies contain a population of small Cajal body–specific RNAs (scaRNAs), small nucleolar RNAs (snoRNAs), and small nuclear RNAs (snRNAs) (Darzacq et al. 2002; Schaffert et al. 2004; Staněk and

Neugebauer 2004; Trinkle-Mulcahy and Sleeman 2017). Some RNP granules appear more specific to particular RNAs. For example, paraspeckles are thought to essentially contain multiple copies of the *NEAT1* RNA (Clemson et al. 2009). Similarly, RNP foci observed from repeat expansion RNAs are inferred to be made up solely of those RNAs with repeat expansions (Česnik et al. 2019). However, given the tendency of RNA to interact with other RNA molecules (Van Treeck et al. 2018; Tauber et al. 2020), one anticipates that additional RNAs, perhaps in relatively nonspecific manners, will be localized to these structures.

Second, some RNP granules form locally when a new population of RNA is produced. This is observed for the nucleolus, wherein the production of nascent ribosomal RNA (rRNA) transcripts is required for the formation of the nucleolus (Mélèse and Xue 1995; Falahati et al. 2016). Similarly, the formation of the histone locus body (HLB) occurs when histone mRNAs are produced at high rates (Shevtsov and Dundr 2011). Paraspeckles also require *NEAT1* transcription for formation and maintenance (Mao et al. 2011; Naganuma and Hirose 2013). One anticipates that anytime a transcript is produced at high rates, and from repeated genes, it would generate a transient RNP granule as a result of the interactions between RNAs and RBPs.

Third, because the assembly of RNPs into granules may be a relatively promiscuous process, one should expect that even highly specialized RNP granules will contain other RNAs. For example, the formation of P-granules in *C. elegans* is critical for the delivery of a few specific mRNAs to the germline precursor cells, yet P-granules contain a large number of mRNA molecules that are simply carried along for the ride and may contribute to the formation of P-granules by acting as scaffolds for RBPs or through RNA–RNA interactions (Lee et al. 2020). In other words, because RNP granule formation requires a critical concentration to form above the activity of RNA decondensers (Tauber et al. 2020), one way of achieving the critical concentration is to include additional nontranslating RNA and lncRNAs that are not necessary to be transported per se but allow the formation of the granule.

HOW DO RNP GRANULES FORM?

It is now clear that RNP granules form from a summation of multivalent interactions between individual RNPs (Fig. 2) (Mittag and Parker 2018; Van Treeck and Parker 2018). The multivalent nature of these interactions allows for the formation of large assemblies containing up to hundreds of RNPs, as in the case of stress granules (Khong

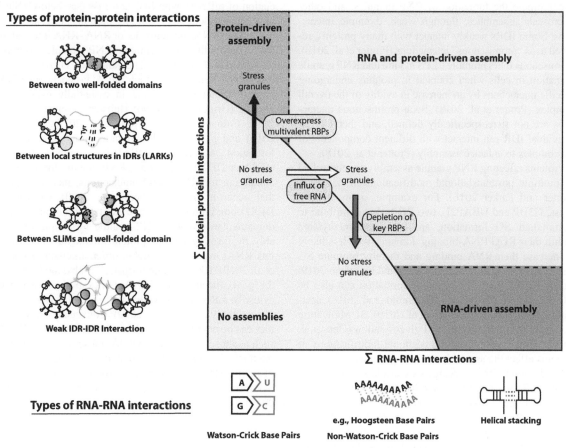

Figure 2. Summation model for RNP granule formation. Protein–protein interactions such as those between intrinsically disordered regions (IDRs), LARKs, short linear motifs (SLiMs), or well-folded domains contribute to granule formation, as do RNA–RNA interactions such as Watson–Crick base pairs and non-Watson–Crick base pairs, and possibly helical stacking between RNAs. Increasing or decreasing the availability of any one type of interaction can induce or prevent RNP granule formation.

et al. 2017). RNPs can interact either through protein–protein, protein–RNA, and/or RNA–RNA interactions.

Protein–protein interactions promoting RNP granule formation occur between RBPs bound to different RNAs and can involve well-folded protein domains (for review, see Mittag and Parker 2018). For example, the dimerization of G3BP1 or EDC3 promotes SG formation in mammals (Tourrière et al. 2003) or PBs in yeast (Ling et al. 2008), respectively. Similarly, depletion of coilin prevents Cajal body formation (Kaiser et al. 2008; Walker et al. 2009; Nizami et al. 2010). However, many RNP granule proteins contain intrinsically disordered regions (IDRs) that can also contribute to RNP granule assembly (Gilks et al. 2004; Decker et al. 2007; Lee et al. 2020). The IDRs of RBPs can promote RNP granule formation in multiple ways, including forming short linear motif (SLiM) interactions with well-folded domains (Jonas and Izaurralde 2013) or by forming local regions of structure that stabilize interactions (Hughes et al. 2018).

Purified RNP granule proteins, or their IDR domains, can also self-assemble in vitro through liquid–liquid phase separation (LLPS), wherein weak multivalent interactions of the IDRs partition those proteins into a concentrated phase in solution (Elbaum-Garfinkle et al. 2015; Lin et al. 2015; Molliex et al. 2015; Nott et al. 2015; Patel et al. 2015; Lee et al. 2020; Martin et al. 2020). This suggested that IDRs also promote the formation of RNP granules, and other macroscale assemblies, through weak, dynamic interactions. Some IDRs weakly interact with many proteins, referred to as "promiscuous" interactions (Protter et al. 2018). Promiscuous IDRs promote LLPS in vitro and RNP granule formation in cells when coupled to proteins undergoing specific interactions by an increase in avidity of the overall complex (Protter et al. 2018). Such promiscuous interactions are not stereospecifically defined, and therefore an individual IDR can interact with different components of the complex to enhance assembly (Protter et al. 2018).

Proteins affecting RNP granule assembly are controlled by multiple posttranslational modifications (reviewed in Protter and Parker 2016). For example, upon arsenite stress, G3BP and UBAP2L, two RBPs that contribute to mammalian SG formation, are arginine-demethylated within their RGG RNA-binding domains, which is likely to increase their RNA binding and thereby promote SG assembly (Tsai et al. 2016; Hofweber and Dormann 2019; Huang et al. 2020). RNP granule formation can also be affected by phosphorylation (Berchtold et al. 2018), acetylation (Gal et al. 2019; Saito et al. 2019), SUMOylation (Jongjitwimol et al. 2016), and glycosylation (Ohn et al. 2008). Thus, multiple posttranslational modifications of proteins affect the assembly of RNP granules.

In addition to RNA serving as a scaffold for the recruitment of RBPs, intermolecular RNA–RNA interactions also contribute to RNP granule assembly. A role for intermolecular RNA–RNA interactions in RNP granules was suggested by the observations that RNA molecules can self-assemble in vitro (Aumiller et al. 2016). For example, base-pairing allows oligonucleotides of repeat expansion RNAs to self-assemble in vitro, suggesting that intermolecular RNA–RNA interactions contribute to the formation of nuclear RNP foci of repeat expansions RNAs (Jain and Vale 2017). Moreover, when purified cellular RNA is incubated under approximately physiological concentrations and salts, RNA assemblies form whose composition is biased toward longer RNAs and largely matches the stress granule transcriptome (Van Treeck et al. 2018). Because longer RNAs are more prone to self-assembly through RNA–RNA interactions, this observation provides a possible biophysical explanation for why SGs and PBs are generally enriched in longer RNAs (Hubstenberger et al. 2017; Khong et al. 2017; Namkoong et al. 2018; Matheny et al. 2019). However, based on the ability of proteins to target RNAs to RNP granules (see below), one anticipates that, on average, longer RNAs will also have more binding sites for proteins that contribute to RNP recruitment. In the *Drosophila* embryo, homotypic intermolecular base-pairing between *oskar* or *bicoid* mRNAs determines whether those mRNAs localize to RNP granules in the anterior or the posterior of the embryo (Ferrandon et al. 1997; Jambor et al. 2011). This latter observation demonstrates that cells can use specific intermolecular RNA–RNA interactions for RNP granule assembly.

Cells have developed multiple mechanisms to modulate intermolecular RNA–RNA interactions and thereby control the process of RNP granule formation. For mRNAs, this includes elongating ribosomes, which limit the association of mRNAs with SGs and PBs (see below). Similarly, monovalent RNA binding proteins, such as YB-1, can bind to RNA and compete for RNA–RNA interactions that would otherwise promote RNP granule assembly (Bounedjah et al. 2014). The family of DEAD/DEVH-box "RNA helicases," which are ATP-dependent RNA-binding proteins, are found in essentially every RNP granule (Gruidl et al. 1996; Hubstenberger et al. 2013, 2017; Calo et al. 2015; Nott et al. 2015; Jain et al. 2016) and play two roles in modulating RNP granule formation. Monovalent DEAD-box proteins such as eIF4A or DDX19A can limit RNP granule formation by competing for RNA–RNA interactions, including those that occur in *trans* (Tauber et al. 2020). Conversely, DEAD-box proteins that have additional multimerization domains can promote RNP granule formation, presumably by forming protein–protein interactions that cross-link RNAs together into higher-order structures (Hondele et al. 2019). RNP granule formation is also modulated by the local concentration of RNA. For example, depletion of cytosolic mRNAs by RNase L induction alters stress granule assembly (Burke et al. 2019). Conversely, RNP granules can be nucleated by high local concentration of RNA, such as those produced by rRNA transcription leading to the nucleolus, or translational shutoff increasing the pool of nontranslating RNA in the cytoplasm, which induces SG formation (Falahati et al. 2016; Protter and Parker 2016).

RECRUITMENT OF RNPs TO RNP GRANULES

A key step in partitioning RNPs into RNP granules is the interaction of RNPs with the surface of an RNP gran-

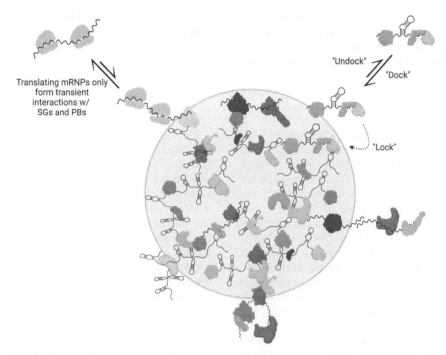

Figure 3. The surface of RNP granules are dynamic and recruit mRNPs. Formation of a transient surface interaction ("docking") can lead to more stable interactions ("locking"). Translating mRNPs only interact transiently with P-bodies (PBs) and stress granules (SGs), whereas nontranslating mRNPs may form transient or stable interactions.

ule and/or the subsequent stable interaction of the RNP with the RNP granule. Although limited information is available about this process, we consider what is known about mRNP partitioning into stress granules and/or P-bodies to begin to address the mechanisms that modulate RNPs interacting with RNP granules.

The imaging of single mRNAs has revealed that mRNPs interact with SGs and PBs in mammalian cells in two manners (Fig. 3). Both stable and dynamic interactions of RNPs have been observed for PBs (Pitchiaya et al. 2019). For both PBs and SGs, one population of mRNAs is relatively stable in these RNP granules and can be rigidly positioned within the granule (Moon et al. 2019). A second population of RNPs transiently interacts with the surface of these granules, showing rapid exchange with the bulk cytosol with an average dwell time on the granule surface of ~10 sec (Moon et al. 2019). Stable associations are increased with granule size and mRNA length and decreased by ribosome association with the mRNA (Moon et al. 2019). These observations suggested a "dock and lock" model wherein RNPs can form transient protein or RNA-based interactions with the surface of an RNP granule, and that surface association can lead to formation of more interactions, leading to a stable association of the RNP with the granule (Moon et al. 2019). Interestingly, surface binding leading to more stable associations through the creation of a high local concentration can also occur on the surface of RNA condensates created from an RNA homopolymer (Tauber et al. 2020), suggesting that surface interactions of RNP granules will be critical in controlling the entry and egress of RNPs.

FACTORS AFFECT RNPs PARTITIONING INTO STRESS GRANULES AND P-BODIES

Examination of the composition of both SGs and PBs has revealed that there can be substantial differences in the partitioning of individual mRNAs into these assemblies. On a correlative level, both assemblies are biased toward longer, poorly translated mRNAs (Hubstenberger et al. 2017; Khong et al. 2017; Namkoong et al. 2018; Matheny et al. 2019). Similar results are seen for the collection of mRNAs enriched in *C. elegans* P-granules (Lee et al. 2020). In addition to these correlations, further experiments have revealed both positive and negative components of RNPs that affects their assembly into these complexes.

Three observations argue that the engagement of an 80S ribosome limits the ability of an mRNP to stably associate with a SG or PB. First, trapping mRNAs in polysomes with translation elongation inhibitors prevents SG, PB, and P-granule formation (Kedersha et al. 2000; Brengues et al. 2005; Lee et al. 2020), demonstrating that these assemblies require mRNAs dissociated from ribosomes for their stable formation. Second, mRNAs with long coding regions are delayed in their accumulation in stress granules after translation inhibition in a ribosome-dependent manner (Khong and Parker 2018). Finally, simultaneous monitoring of translation status and interaction with the RNP granules of single mRNPs revealed that mRNAs engaged with ribosomes can only interact with SGs or PBs in the dynamic mode and are unable to enter the stable state (Moon et al. 2019). Thus, mRNAs engaged with ribosomes can interact with SGs and PBs but are unable to form a stable interaction. This suggests that the energetic cost of partitioning a ribosome into an RNP granule is significant, perhaps because of continued ribosome elongation, an absence of stabilizing interactions between the ribosome and other RNP granule components, or an altered geometry to the mRNA owing to elongating ribosomes (Adivarahan et al. 2018; Khong and Parker 2018).

Individual RNPs can be targeted to RNP granules by the combined action of multiple positive elements. For example, insertion of the *NORAD* lncRNA, which is highly enriched in SGs (Khong et al. 2017), into a reporter efficiently targets that reporter RNA into SGs (Matheny et al. 2020). Moreover, deletion analysis of *NORAD* demonstrates that multiple regions of *NORAD* act in an additive manner to promote SG accumulation of the RNA (Matheny et al. 2020). The number of binding sites for known SG RBPs within the region of *NORAD* inserted into the reporter correlates with increased SG enrichment, whereas control RNAs of the same length, but without known sites for RBPs, showed lesser accumulation in SGs (Matheny et al. 2020). This suggests that the SG accumulation of *NORAD* may be largely dictated by the presence of multiple binding sites for SG proteins and raises the possibility that nonspecific RNA–RNA interactions contribute less to the SG accumulation of *NORAD* RNA.

More direct evidence that some RNA-binding proteins can target RNPs into SGs has come from examining how artificial tethering of proteins to reporter RNAs affects their accumulation in SGs. For example, the tethering of G3BP1 or TIA-1 to a luciferase reporter mRNA increases its accumulation in stress granules in a dose-dependent manner (Matheny et al. 2020). This provides direct evidence that proteins bound to an RNA can dictate its accumulation in an RNP granule. This is consistent with the observations that in *C. elegans*, the P granule protein MEG-3 is required for proper mRNA localization and P-granule formation in germ cells (Lee et al. 2020), and that in neurons, trafficking of specific mRNAs in neuronal transport granules is dependent on sequence motifs for the binding of heterogeneous nuclear ribonucleoprotein (hnRNP) A2 and A3 (Shan et al. 2000). Similarly, sequence elements in the 3′ UTR of *Drosophila* germ granule RNAs are necessary and sufficient for localization in maternal mRNA granules (Rangan et al. 2009).

The observation that *NORAD* contains many elements promoting SG accumulation and the observation that efficient SG targeting by tethered proteins requires 25 sites of tethering suggest a model wherein the accumulation of a given RNP into a SG is due to the summation of many interactions (Matheny et al. 2020). This view is consistent with the strong bias toward longer RNAs in SGs, which by their length would simply have more sites for positive interactions. Interestingly, this summative model predicts that any given protein would have small effects on the set of RNAs that accumulate in a SG. This provides an explanation for the surprising observation that U-2 OS cells lacking G3BP1 and G3BP2, two abundant and important proteins for SG formation, form SGs with essentially the same transcriptome as wild-type cells (Matheny et al. 2020).

HOW DO CELLS SORT RNPs BETWEEN MULTIPLE GRANULES?

The combination of translation status and sets of specific interactions allows cells to form multiple types of RNP granules with distinct RNA and protein compositions in the same compartment. For example, P-bodies in unstressed cells are biased toward poorly translated mRNAs and have a distinctly different mRNA composition than stress granules (Fig. 4). (Hubstenberger et al. 2017; Matheny et al. 2019). However, during stress conditions, when most mRNAs cease translation, the transcriptome of P-bodies and stress granules is very similar (Matheny et al. 2019). During such stress conditions, individual mRNPs then can be understood as partitioning into stress granules or P-bodies based on their RNA binding proteins. Because proteins found in P-bodies or stress granules preferentially interact with themselves (Jonas and Izaurralde 2013; Youn et al. 2018), this would allow two distinct forms of mRNPs that partition into stress granules or P-bodies, respectively. Such distinct mRNPs could be dictated by the status of the poly(A) tail because deadenylated mRNAs are preferentially bound by specific components of P-bodies (Bouveret et al. 2000; Tharun et al. 2000), which would then target deadenylated molecules to P-bodies, consistent with P-bodies failing to stain for poly(A)$^+$ mRNA by oligo(dT) FISH, whereas SGs do (Kedersha et al. 2000; Cougot et al. 2004). Similarly, specific RNA–RNA interactions can target the *bicoid* or *oskar* RNAs to anterior or posterior RNP granules during *Drosophila* oogenesis (Ferrandon et al. 1997; Jambor et al. 2011).

Partitioning of RNPs into different RNP granules may also be controlled by RNA structures that form to limit any RNA–RNA interactions between components. For example, it is suggested in the filamentous fungi, *Ashbya gossypii*, that the tertiary structure of the *CLN3* mRNA, which associates with the Whi3 protein and forms RNP granules, limits the *CLN3* mRNA's possible interactions with other mRNAs that also interact with Whi3, thereby giving rise to distinct RNP granules (Langdon et al. 2018).

CONCLUSION

The combination of available protein and RNA-based interactions determines whether an RNP granule will form and whether an individual RNP will partition into the granule. The diversity of interactions allows for redundancy in RNP granule formation as well as many layers of potential regulation. Future work to understand how the availability of these interactions is regulated will be important for understanding how aberrant granules form in neurological disease.

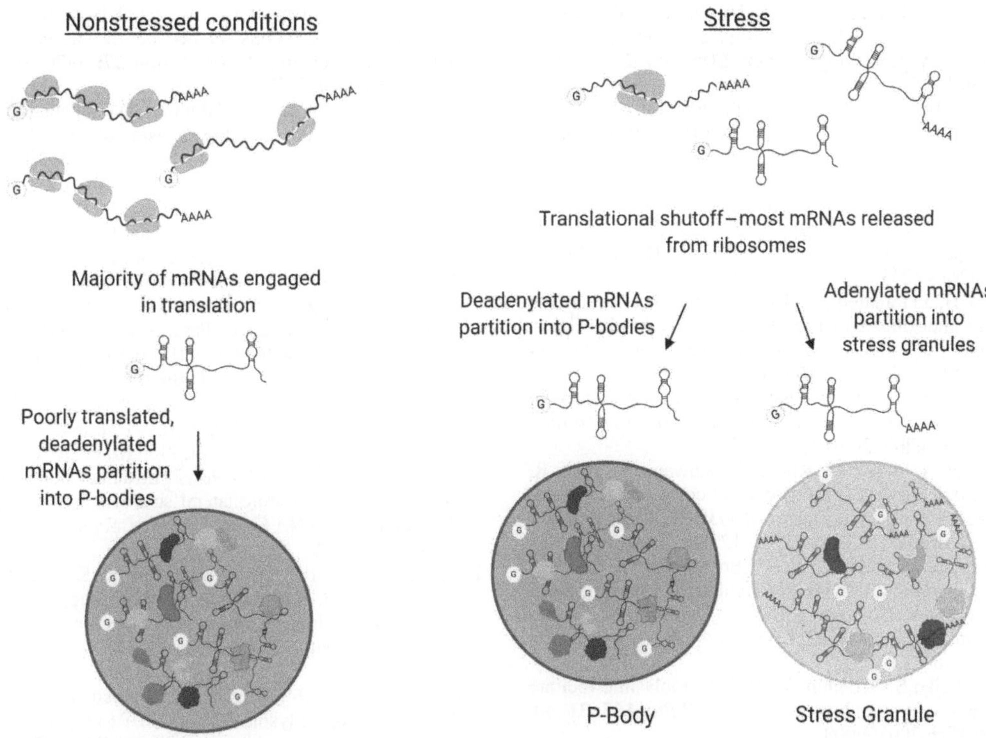

Figure 4. P-bodies and stress granules are composed of nontranslating mRNPs. During nonstressed conditions (*left*), only P-bodies are present, and nontranslating, deadenylated mRNAs partition into P-bodies. During stress (*right*), both P-bodies and stress granules are present in cytosol, and mRNAs may partition into either stress granules or P-bodies, likely dictated by their poly(A) status.

REFERENCES

Adivarahan S, Livingston N, Nicholson B, Rahman S, Wu B, Rissland OS, Zenklusen D. 2018. Spatial organization of single mRNPs at different stages of the gene expression pathway. *Mol Cell* **72:** 727–738.e5. doi:10.1016/j.molcel.2018.10.010

Al-Husini N, Tomares DT, Bitar O, Childers WS, Schrader JM. 2018. α-Proteobacterial RNA degradosomes assemble liquid–liquid phase-separated RNP bodies. *Mol Cell* **71:** 1027–1039. e14. doi:10.1016/j.molcel.2018.08.003

Almeida S, Gascon E, Tran H, Chou HJ, Gendron TF, DeGroot S, Tapper AR, Sellier C, Charlet-Berguerand N, Karydas A, et al. 2013. Modeling key pathological features of frontotemporal dementia with *C9ORF72* repeat expansion in iPSC-derived human neurons. *Acta Neuropathol (Berl)* **126:** 385–399. doi:10.1007/s00401-013-1149-y

Arai T, Hasegawa M, Akiyama H, Ikeda K, Nonaka T, Mori H, Mann D, Tsuchiya K, Yoshida M, Hashizume Y, et al. 2006. TDP-43 is a component of ubiquitin-positive tau-negative inclusions in frontotemporal lobar degeneration and amyotrophic lateral sclerosis. *Biochem Biophys Res Commun* **351:** 602–611. doi:10.1016/j.bbrc.2006.10.093

Arkov AL, Wang J-YS, Ramos A, Lehmann R. 2006. The role of Tudor domains in germline development and polar granule architecture. *Development* **133:** 4053–4062. doi:10.1242/dev .02572

Aumiller WM, Pir Cakmak F, Davis BW, Keating CD. 2016. RNA-based coacervates as a model for membraneless organelles: formation, properties, and interfacial liposome assembly. *Langmuir* **32:** 10042–10053. doi:10.1021/acs.langmuir .6b02499

Berchtold D, Battich N, Pelkmans L. 2018. A systems-level study reveals regulators of membrane-less organelles in human cells. *Mol Cell* **72:** 1035–1049.e5. doi:10.1016/j.molcel.2018.10 .036

Bergovich K, Vu AQ, Yeo GW, Wilhelm JE. 2020. Conserved metabolite regulation of stress granule assembly via AdoMet. *J Cell Biol* doi:10.1083/jcb.2019.04141

Boisvert F-M, van Koningsbruggen S, Navascués J, Lamond AI. 2007. The multifunctional nucleolus. *Nat Rev Mol Cell Biol* **8:** 574–585. doi:10.1038/nrm2184

Bond CS, Fox AH. 2009. Paraspeckles: nuclear bodies built on long noncoding RNA. *J Cell Biol* **186:** 637–644. doi:10.1083/ jcb.200906113

Bounedjah O, Desforges B, Wu T-D, Pioche-Durieu C, Marco S, Hamon L, Curmi PA, Guerquin-Kern J-L, Piétrement O, Pastré D. 2014. Free mRNA in excess upon polysome dissociation is a scaffold for protein multimerization to form stress granules. *Nucleic Acids Res* **42:** 8678–8691. doi:10.1093/nar/gku582

Bouveret E, Rigaut G, Shevchenko A, Wilm M, Séraphin B. 2000. A Sm-like protein complex that participates in mRNA degradation. *EMBO J* **19:** 1661–1671. doi:10.1093/emboj/19 .7.1661

Brengues M, Teixeira D, Parker R. 2005. Movement of eukaryotic mRNAs between polysomes and cytoplasmic processing bodies. *Science* **310:** 486–489. doi:10.1126/science.1115791

Buchan JR, Parker R. 2009. Eukaryotic stress granules: the ins and outs of translation. *Mol Cell* **36:** 932–941. doi:10.1016/j .molcel.2009.11.020

Buchan JR, Muhlrad D, Parker R. 2008. P bodies promote stress granule assembly in *Saccharomyces cerevisiae*. *J Cell Biol* **183:** 441–455. doi:10.1083/jcb.200807043

Buchan JR, Kolaitis R-M, Taylor JP, Parker R. 2013. Eukaryotic stress granules are cleared by granulophagy and Cdc48/VCP function. *Cell* **153:** 1461–1474. doi:10.1016/j.cell.2013.05 .037

Burke JM, Moon SL, Matheny T, Parker R. 2019. RNase L reprograms translation by widespread mRNA turnover escaped by antiviral mRNAs. *Mol Cell* **75:** 1203–1217.e5. doi:10 .1016/j.molcel.2019.07.029

Calo E, Flynn RA, Martin L, Spitale RC, Chang HY, Wysocka J. 2015. RNA helicase DDX21 coordinates transcription and ribosomal RNA processing. *Nature* **518:** 249–253. doi:10.1038/nature13923

Carmo-Fonseca M. 2002. New clues to the function of the Cajal body. *EMBO Rep* **3:** 726–727. doi:10.1093/embo-reports/kvf154

Česnik AB, Darovic S, Mihevc SP, Štalekar M, Malnar M, Motaln H, Lee Y-B, Mazej J, Pohleven J, Grosch M, et al. 2019. Nuclear RNA foci from *C9ORF72* expansion mutation form paraspeckle-like bodies. *J Cell Sci* **132:** jcs224303. doi:10.1242/jcs.224303

Chamberland H, Lafontaine JG. 1993. Localization of snRNP antigens in nucleolus-associated bodies: study of plant interphase nuclei by confocal and electron microscopy. *Chromosoma* **102:** 220–226. doi:10.1007/BF00352395

Cioce M, Lamond AI. 2005. Cajal bodies: a long history of discovery. *Annu Rev Cell Dev Biol* **21:** 105–131. doi:10.1146/annurev.cellbio.20.010403.103738

Cirillo L, Cieren A, Barbieri S, Khong A, Schwager F, Parker R, Gotta M. 2020. UBAP2L forms distinct cores that act in nucleating stress granules upstream of G3BP1. *Curr Biol* **30:** 698–707.e6. doi:10.1016/j.cub.2019.12.020

Clemson CM, Hutchinson JN, Sara SA, Ensminger AW, Fox AH, Chess A, Lawrence JB. 2009. An architectural role for a nuclear noncoding RNA: NEAT1 RNA is essential for the structure of paraspeckles. *Mol Cell* **33:** 717–726. doi:10.1016/j.molcel.2009.01.026

Cougot N, Babajko S, Séraphin B. 2004. Cytoplasmic foci are sites of mRNA decay in human cells. *J Cell Biol* **165:** 31–40. doi:10.1083/jcb.200309008

Courel M, Clément Y, Bossevain C, Foretek D, Vidal Cruchez O, Yi Z, Bénard M, Benassy M-N, Kress M, Vindry C, et al. 2019. GC content shapes mRNA storage and decay in human cells. *ELife* **8:** e49708. doi:10.7554/eLife.49708

Darzacq X, Jády BE, Verheggen C, Kiss AM, Bertrand E, Kiss T. 2002. Cajal body-specific small nuclear RNAs: a novel class of 2′-O-methylation and pseudouridylation guide RNAs. *EMBO J* **21:** 2746–2756. doi:10.1093/emboj/21.11.2746

Davis BM, McCurrach ME, Taneja KL, Singer RH, Housman DE. 1997. Expansion of a CUG trinucleotide repeat in the 3′ untranslated region of myotonic dystrophy protein kinase transcripts results in nuclear retention of transcripts. *Proc Natl Acad Sci* **94:** 7388–7393. doi:10.1073/pnas.94.14.7388

Decker CJ, Teixeira D, Parker R. 2007. Edc3p and a glutamine/asparagine-rich domain of Lsm4p function in processing body assembly in *Saccharomyces cerevisiae*. *J Cell Biol* **179:** 437–449. doi:10.1083/jcb.200704147

DeJesus-Hernandez M, Mackenzie IR, Boeve BF, Boxer AL, Baker M, Rutherford NJ, Nicholson AM, Finch NA, Flynn H, Adamson J, et al. 2011. Expanded GGGGCC hexanucleotide repeat in noncoding region of *C9ORF72* causes chromosome 9p-Linked FTD and ALS. *Neuron* **72:** 245–256. doi:10.1016/j.neuron.2011.09.011

Desterro JMP, Keegan LP, Lafarga M, Berciano MT, O'Connell M, Carmo-Fonseca M. 2003. Dynamic association of RNA-editing enzymes with the nucleolus. *J Cell Sci* **116:** 1805–1818. doi:10.1242/jcs.00371

Dickson DW, Kouri N, Murray ME, Josephs KA. 2011. Neuropathology of frontotemporal lobar degeneration-tau (FTLD-Tau). *J Mol Neurosci* **45:** 384–389. doi:10.1007/s12031-011-9589-0

Dundr M, Olson MOJ. 1998. Partially processed pre-rRNA is preserved in association with processing components in nucleolus-derived foci during mitosis. *Mol Biol Cell* **9:** 2407–2422. doi:10.1091/mbc.9.9.2407

Elbaum-Garfinkle S, Kim Y, Szczepaniak K, Chen CC-H, Eckmann CR, Myong S, Brangwynne CP. 2015. The disordered P granule protein LAF-1 drives phase separation into droplets with tunable viscosity and dynamics. *Proc Natl Acad Sci* **112:** 7189–7194. doi:10.1073/pnas.1504822112

Ephrussi A, Lehmann R. 1992. Induction of germ cell formation by oskar. *Nature* **358:** 387–392. doi:10.1038/358387a0

Eulalio A, Behm-Ansmant I, Schweizer D, Izaurralde E. 2007. P-body formation is a consequence, not the cause, of RNA-mediated gene silencing. *Mol Cell Biol* **27:** 3970–3981. doi:10.1128/MCB.00128-07

Falahati H, Pelham-Webb B, Blythe S, Wieschaus E. 2016. Nucleation by rRNA dictates the precision of nucleolus assembly. *Curr Biol* **26:** 277–285. doi:10.1016/j.cub.2015.11.065

Fatica A, Tollervey D. 2002. Making ribosomes. *Curr Opin Cell Biol* **14:** 313–318. doi:10.1016/S0955-0674(02)00336-8

Ferrandon D, Koch I, Westhof E, Nüsslein-Volhard C. 1997. RNA–RNA interaction is required for the formation of specific *bicoid* mRNA 3′ UTR–STAUFEN ribonucleoprotein particles. *EMBO J* **16:** 1751–1758. doi:10.1093/emboj/16.7.1751

Fox AH, Lamond AI. 2010. Paraspeckles. *Cold Spring Harb Perspect Biol* **2:** a000687. doi:10.1101/cshperspect.a000687

Fox AH, Nakagawa S, Hirose T, Bond CS. 2018. Paraspeckles: where long noncoding RNA meets phase separation. *Trends Biochem Sci* **43:** 124–135. doi:10.1016/j.tibs.2017.12.001

Fratta P, Mizielinska S, Nicoll AJ, Zloh M, Fisher EMC, Parkinson G, Isaacs AM. 2012. *C9orf72* hexanucleotide repeat associated with amyotrophic lateral sclerosis and frontotemporal dementia forms RNA G-quadruplexes. *Sci Rep* **2:** 1016. doi:10.1038/srep01016

Frottin F, Schueder F, Tiwary S, Gupta R, Körner R, Schlichthaerle T, Cox J, Jungmann R, Hartl FU, Hipp MS. 2019. The nucleolus functions as a phase-separated protein quality control compartment. *Science* **365:** 342–347. doi:10.1126/science.aaw9157

Gal J, Chen J, Na D-Y, Tichacek L, Barnett KR, Zhu H. 2019. The acetylation of lysine-376 of G3BP1 regulates RNA binding and stress granule dynamics. *Mol Cell Biol* **39:** e00052-19. doi:10.1128/MCB.00052-19

Gall JG, Bellini M, Wu Z, Murphy C. 1999. Assembly of the nuclear transcription and processing machinery: Cajal bodies (coiled bodies) and transcriptosomes. *Mol Biol Cell* **10:** 4385–4402. doi:10.1091/mbc.10.12.4385

Gendron TF, Bieniek KF, Zhang Y-J, Jansen-West K, Ash PEA, Caulfield T, Daughrity L, Dunmore JH, Castanedes-Casey M, Chew J, et al. 2013. Antisense transcripts of the expanded *C9ORF72* hexanucleotide repeat form nuclear RNA foci and undergo repeat-associated non-ATG translation in c9FTD/ALS. *Acta Neuropathol (Berl)* **126:** 829–844. doi:10.1007/s00401-013-1192-8

Gilks N, Kedersha N, Ayodele M, Shen L, Stoecklin G, Dember LM, Anderson P. 2004. Stress granule assembly is mediated by prion-like aggregation of TIA-1. *Mol Biol Cell* **15:** 5383–5398. doi:10.1091/mbc.e04-08-0715

Gruidl ME, Smith PA, Kuznicki KA, McCrone JS, Kirchner J, Roussell DL, Strome S, Bennett KL. 1996. Multiple potential germ-line helicases are components of the germ-line-specific P granules of *Caenorhabditis elegans*. *Proc Natl Acad Sci* **93:** 13837–13842. doi:10.1073/pnas.93.24.13837

Hay B, Jan LY, Jan YN. 1988. A protein component of *Drosophila* polar granules is encoded by vasa and has extensive sequence similarity to ATP-dependent helicases. *Cell* **55:** 577–587. doi:10.1016/0092-8674(88)90216-4

Hebert MD, Matera AG. 2000. Self-association of coilin reveals a common theme in nuclear body localization. *Mol Biol Cell* **11:** 4159–4171. doi:10.1091/mbc.11.12.4159

Hebert MD, Poole AR. 2016. Towards an understanding of regulating Cajal body activity by protein modification. *RNA Biol* **14:** 761–778. doi:10.1080/15476286.2016.1243649

Hofmann S, Cherkasova V, Bankhead P, Bukau B, Stoecklin G. 2012. Translation suppression promotes stress granule formation and cell survival in response to cold shock. *Mol Biol Cell* **23:** 3786–3800. doi:10.1091/mbc.e12-04-0296

Hofweber M, Dormann D. 2019. Friend or foe—post-translational modifications as regulators of phase separation and RNP granule dynamics. *J Biol Chem* **294:** 7137–7150. doi:10.1074/jbc.TM118.001189

Hondele M, Sachdev R, Heinrich S, Wang J, Vallotton P, Fontoura BMA, Weis K. 2019. DEAD-box ATPases are global regulators of phase-separated organelles. *Nature* **573:** 144–148. doi:10.1038/s41586-019-1502-y

Hsieh Y-C, Guo C, Yalamanchili HK, Abreha M, Al-Ouran R, Li Y, Dammer EB, Lah JJ, Levey AI, Bennett DA, et al. 2019. Tau-mediated disruption of the spliceosome triggers cryptic RNA splicing and neurodegeneration in Alzheimer's disease. *Cell Rep* **29:** 301–316.e10. doi:10.1016/j.celrep.2019.08.104

Huang C, Chen Y, Dai H, Zhang H, Xie M, Zhang H, Chen F, Kang X, Bai X, Chen Z. 2020. UBAP2L arginine methylation by PRMT1 modulates stress granule assembly. *Cell Death Differ* **27:** 227–241. doi:10.1038/s41418-019-0350-5

Hubstenberger A, Noble SL, Cameron C, Evans TC. 2013. Translation repressors, an RNA helicase, and developmental cues control RNP phase transitions during early development. *Dev Cell* **27:** 161–173. doi:10.1016/j.devcel.2013.09.024

Hubstenberger A, Courel M, Bénard M, Souquere S, Ernoult-Lange M, Chouaib R, Yi Z, Morlot J-B, Munier A, Fradet M, et al. 2017. P-body purification reveals the condensation of repressed mRNA regulons. *Mol Cell* **68:** 144–157.e5. doi:10.1016/j.molcel.2017.09.003

Hughes M, Sawaya MR, Boyer DR, Goldschmidt L, Rodriguez JA, Cascio D, Chong L, Gonen T, Eisenberg DS. 2018. Atomic structures of low-complexity protein segments reveal kinked β sheets that assemble networks. *Science* **359:** 698–701. doi:10.1126/science.aan6398

Imamura K, Imamachi N, Akizuki G, Kumakura M, Kawaguchi A, Nagata K, Kato A, Kawaguchi Y, Sato H, Yoneda M, et al. 2014. Long noncoding RNA NEAT1-dependent SFPQ relocation from promoter region to paraspeckle mediates IL8 expression upon immune stimuli. *Mol Cell* **53:** 393–406. doi:10.1016/j.molcel.2014.01.009

Jain A, Vale RD. 2017. RNA phase transitions in repeat expansion disorders. *Nature* **546:** 243–247. doi:10.1038/nature22386

Jain S, Wheeler JR, Walters RW, Agrawal A, Barsic A, Parker R. 2016. ATPase-modulated stress granules contain a diverse proteome and substructure. *Cell* **164:** 487–498. doi:10.1016/j.cell.2015.12.038

Jambor H, Brunel C, Ephrussi A. 2011. Dimerization of *oskar* 3′ UTRs promotes hitchhiking for RNA localization in the *Drosophila* oocyte. *RNA* **17:** 2049–2057. doi:10.1261/rna.2686411

Jiang H, Mankodi A, Swanson MS, Moxley RT, Thornton CA. 2004. Myotonic dystrophy type 1 is associated with nuclear foci of mutant RNA, sequestration of muscleblind proteins and deregulated alternative splicing in neurons. *Hum Mol Genet* **13:** 3079–3088. doi:10.1093/hmg/ddh327

Jonas S, Izaurralde E. 2013. The role of disordered protein regions in the assembly of decapping complexes and RNP granules. *Genes Dev* **27:** 2628–2641. doi:10.1101/gad.227843.113

Jones JR, Macdonald PM. 2007. Oskar controls morphology of polar granules and nuclear bodies in *Drosophila*. *Development* **134:** 233–236. doi:10.1242/dev.02729

Jongjitwimol J, Baldock RA, Morley SJ, Watts FZ. 2016. Sumoylation of eIF4A2 affects stress granule formation. *J Cell Sci* **129:** 2407–2415. doi:10.1242/jcs.184614

Kaiser TE, Intine RV, Dundr M. 2008. De novo formation of a subnuclear body. *Science* **322:** 1713–1717. doi:10.1126/science.1165216

Kedersha NL, Gupta M, Li W, Miller I, Anderson P. 1999. RNA-binding proteins Tia-1 and Tiar link the phosphorylation of Eif-2α to the assembly of mammalian stress granules. *J Cell Biol* **147:** 1431–1442. doi:10.1083/jcb.147.7.1431

Kedersha N, Cho MR, Li W, Yacono PW, Chen S, Gilks N, Golan DE, Anderson P. 2000. Dynamic shuttling of Tia-1 accompanies the recruitment of mRNA to mammalian stress granules. *J Cell Biol* **151:** 1257–1268. doi:10.1083/jcb.151.6.1257

Kedersha N, Stoecklin G, Ayodele M, Yacono P, Lykke-Andersen J, Fritzler MJ, Scheuner D, Kaufman RJ, Golan DE, Anderson P. 2005. Stress granules and processing bodies are dynamically linked sites of mRNP remodeling. *J Cell Biol* **169:** 871–884. doi:10.1083/jcb.200502088

Kedersha N, Ivanov P, Anderson P. 2013. Stress granules and cell signaling: more than just a passing phase? *Trends Biochem Sci* **38:** 494–506. doi:10.1016/j.tibs.2013.07.004

Kedersha N, Panas MD, Achorn CA, Lyons S, Tisdale S, Hickman T, Thomas M, Lieberman J, McInerney GM, Ivanov P, et al. 2016. G3BP–Caprin1–USP10 complexes mediate stress granule condensation and associate with 40S subunits. *J Cell Biol* **212:** 845–860. doi:10.1083/jcb.201508028

Khong A, Parker R. 2018. mRNP architecture in translating and stress conditions reveals an ordered pathway of mRNP compaction. *J Cell Biol* **217:** 4124–4140. doi:10.1083/jcb.201806183

Khong A, Matheny T, Jain S, Mitchell SF, Wheeler JR, Parker R. 2017. The stress granule transcriptome reveals principles of mRNA accumulation in stress granules. *Mol Cell* **68:** 808–820.e5. doi:10.1016/j.molcel.2017.10.015

Kiebler MA, Bassell GJ. 2006. Neuronal RNA granules: movers and makers. *Neuron* **51:** 685–690. doi:10.1016/j.neuron.2006.08.021

Kumar V, Hasan GM, Hassan M. 2017. Unraveling the role of RNA mediated toxicity of C9orf72 repeats in C9-FTD/ALS. *Front Neurosci* **11:** 711. doi:10.3389/fnins.2017.00711

Langdon EM, Qiu Y, Niaki AG, McLaughlin GA, Weidmann CA, Gerbich TM, Smith JA, Crutchley JM, Termini CM, Weeks KM, et al. 2018. mRNA structure determines specificity of a polyQ-driven phase separation. *Science* **360:** 922–927. doi:10.1126/science.aar7432

Lee C-YS, Putnam A, Lu T, He S, Ouyang JPT, Seydoux G. 2020. Recruitment of mRNAs to P granules by condensation with intrinsically-disordered proteins. *Elife* **9:** e52896. doi:10.7554/eLife.52896

Lin Y, Protter DSW, Rosen MK, Parker R. 2015. Formation and maturation of phase-separated liquid droplets by RNA-binding proteins. *Mol Cell* **60:** 208–219. doi:10.1016/j.molcel.2015.08.018

Ling SHM, Decker CJ, Walsh MA, She M, Parker R, Song H. 2008. Crystal structure of human Edc3 and its functional implications. *Mol Cell Biol* **28:** 5965–5976. doi:10.1128/MCB.00761-08

Little SC, Sinsimer KS, Lee JJ, Wieschaus EF, Gavis ER. 2015. Independent and coordinate trafficking of single *Drosophila* germ plasm mRNAs. *Nat Cell Biol* **17:** 558–568. doi:10.1038/ncb3143

Love AJ, Yu C, Petukhova NV, Kalinina NO, Chen J, Taliansky ME. 2016. Cajal bodies and their role in plant stress and disease responses. *RNA Biol* **14:** 779–790. doi:10.1080/15476286.2016.1243650

Maden BEH, Hughes JMX. 1997. Eukaryotic ribosomal RNA: the recent excitement in the nucleotide modification problem. *Chromosoma* **105:** 391–400. doi:10.1007/BF02510475

Mahowald AP. 1971. Polar granules of *Drosophila*. III. The continuity of polar granules during the life cycle of *Drosophila*. *J Exp Zool* **176:** 329–343. doi:10.1002/jez.1401760308

Mao YS, Sunwoo H, Zhang B, Spector DL. 2011. Direct visualization of the co-transcriptional assembly of a nuclear body by noncoding RNAs. *Nat Cell Biol* **13:** 95–101. doi:10.1038/ncb2140

Markmiller S, Soltanieh S, Server KL, Mak R, Jin W, Fang MY, Luo E-C, Krach F, Yang D, Sen A, et al. 2018. Context-dependent and disease-specific diversity in protein interactions within stress granules. *Cell* **172:** 590–604.e13. doi:10.1016/j.cell.2017.12.032

Martin EW, Holehouse AS, Peran I, Farag M, Incicco JJ, Bremer A, Grace CR, Soranno A, Pappu RV, Mittag T. 2020. Valence and patterning of aromatic residues determine the phase behavior of prion-like domains. *Science* **367:** 694–699. doi:10.1126/science.aaw8653

Matheny T, Rao BS, Parker R. 2019. Transcriptome-wide comparison of stress granules and P-bodies reveals that translation plays a major role in RNA partitioning. *Mol Cell Biol* **39:** e00313-19. doi:10.1128/MCB.00313-19

Matheny T, Van Treeck B, Huynh TN, Parker R. 2020. RNA partitioning into stress granules is based on the summa-

tion of multiple interactions. BioRxiv doi:10.1101/2020.04.15
.043646
Mélèse T, Xue Z. 1995. The nucleolus: an organelle formed by the act of building a ribosome. *Curr Opin Cell Biol* **7:** 319–324. doi:10.1016/0955-0674(95)80085-9
Mittag T, Parker R. 2018. Multiple modes of protein–protein interactions promote RNP granule assembly. *J Mol Biol* **430:** 4636–4649. doi:10.1016/j.jmb.2018.08.005
Molliex A, Temirov J, Lee J, Coughlin M, Kanagaraj AP, Kim HJ, Mittag T, Taylor JP. 2015. Phase separation by low complexity domains promotes stress granule assembly and drives pathological fibrillization. *Cell* **163:** 123–133. doi:10.1016/j.cell.2015.09.015
Moon SL, Morisaki T, Khong A, Lyon K, Parker R, Stasevich TJ. 2019. Multicolor single-molecule tracking of mRNA interactions with RNP granules. *Nat Cell Biol* **21:** 162–168. doi:10.1038/s41556-018-0263-4
Moon SL, Morisaki T, Stasevich TJ, Parker R. 2020. Coupling of translation quality control and mRNA targeting to stress granules. BioRxiv doi:10.1101/2020.01.05.895342
Mori K, Lammich S, Mackenzie IRA, Forné I, Zilow S, Kretzschmar H, Edbauer D, Janssens J, Kleinberger G, Cruts M, et al. 2013. hnRNP A3 binds to GGGGCC repeats and is a constituent of p62-positive/TDP43-negative inclusions in the hippocampus of patients with *C9orf72* mutations. *Acta Neuropathol (Berl)* **125:** 413–423. doi:10.1007/s00401-013-1088-7
Morris GE. 2008. The Cajal body. *Biochim Biophys Acta* **1783:** 2108–2115. doi:10.1016/j.bbamcr.2008.07.016
Naganuma T, Hirose T. 2013. Paraspeckle formation during the biogenesis of long non-coding RNAs. *RNA Biol* **10:** 456–461. doi:10.4161/rna.23547
Naganuma T, Nakagawa S, Tanigawa A, Sasaki YF, Goshima N, Hirose T. 2012. Alternative 3′-end processing of long noncoding RNA initiates construction of nuclear paraspeckles. *EMBO J* **31:** 4020–4034. doi:10.1038/emboj.2012.251
Nakagawa S, Naganuma T, Shioi G, Hirose T. 2011. Paraspeckles are subpopulation-specific nuclear bodies that are not essential in mice. *J Cell Biol* **193:** 31–39. doi:10.1083/jcb.201011110
Nakamura A, Amikura R, Mukai M, Kobayashi S, Lasko PF. 1996. Requirement for a noncoding RNA in *Drosophila* polar granules for germ cell establishment. *Science* **274:** 2075–2079. doi:10.1126/science.274.5295.2075
Namkoong S, Ho A, Woo YM, Kwak H, Lee JH. 2018. Systematic characterization of stress-induced RNA granulation. *Mol Cell* **70:** 175–187.e8. doi:10.1016/j.molcel.2018.02.025
Narayanan A, Speckmann W, Terns R, Terns MP. 1999. Role of the box C/D motif in localization of small nucleolar RNAs to coiled bodies and nucleoli. *Mol Biol Cell* **10:** 2131–2147. doi:10.1091/mbc.10.7.2131
Neumann M, Sampathu DM, Kwong LK, Truax AC, Micsenyi MC, Chou TT, Bruce J, Schuck T, Grossman M, Clark CM, et al. 2006. Ubiquitinated TDP-43 in frontotemporal lobar degeneration and amyotrophic lateral sclerosis. *Science* **314:** 130–133. doi:10.1126/science.1134108
Nizami Z, Deryusheva S, Gall JG. 2010. The Cajal body and histone locus body. *Cold Spring Harb Perspect Biol* **2:** a000653. doi:10.1101/cshperspect.a000653
Nott TJ, Petsalaki E, Farber P, Jervis D, Fussner E, Plochowietz A, Craggs TD, Bazett-Jones DP, Pawson T, Forman-Kay JD, et al. 2015. Phase transition of a disordered nuage protein generates environmentally responsive membraneless organelles. *Mol Cell* **57:** 936–947. doi:10.1016/j.molcel.2015.01.013
Ohn T, Kedersha N, Hickman T, Tisdale S, Anderson P. 2008. A functional RNAi screen links O-GlcNAc modification of ribosomal proteins to stress granule and processing body assembly. *Nat Cell Biol* **10:** 1224–1231. doi:10.1038/ncb1783
Olson MOJ, Dundr M, Szebeni A. 2000. The nucleolus: an old factory with unexpected capabilities. *Trends Cell Biol* **10:** 189–196. doi:10.1016/S0962-8924(00)01738-4
Ouyang JPT, Folkmann A, Bernard L, Lee C-Y, Seroussi U, Charlesworth AG, Claycomb JM, Seydoux G. 2019. P granules protect RNA interference genes from silencing by piRNAs. *Dev Cell* **50:** 716–728.e6. doi:10.1016/j.devcel.2019.07.026
Patel A, Lee HO, Jawerth L, Maharana S, Jahnel M, Hein MY, Stoynov S, Mahamid J, Saha S, Franzmann TM, et al. 2015. A liquid-to-solid phase transition of the ALS protein FUS accelerated by disease mutation. *Cell* **162:** 1066–1077. doi:10.1016/j.cell.2015.07.047
Pilkington GR, Parker R. 2008. Pat1 contains distinct functional domains that promote P-body assembly and activation of decapping. *Mol Cell Biol* **28:** 1298–1312. doi:10.1128/MCB.00936-07
Pitchiaya S, Mourao MDA, Jalihal AP, Xiao L, Jiang X, Chinnaiyan AM, Schnell S, Walter NG. 2019. Dynamic recruitment of single RNAs to processing bodies depends on RNA functionality. *Mol Cell* **74:** 521–533.e6. doi:10.1016/j.molcel.2019.03.001
Prasanth KV, Prasanth SG, Xuan Z, Hearn S, Freier SM, Bennett CF, Zhang MQ, Spector DL. 2005. Regulating gene expression through RNA nuclear retention. *Cell* **123:** 249–263. doi:10.1016/j.cell.2005.08.033
Protter DSW, Parker R. 2016. Principles and properties of stress granules. *Trends Cell Biol* **26:** 668–679. doi:10.1016/j.tcb.2016.05.004
Protter DSW, Rao BS, Van Treeck B, Lin Y, Mizoue L, Rosen MK, Parker R. 2018. Intrinsically disordered regions can contribute promiscuous interactions to RNP granule assembly. *Cell Rep* **22:** 1401–1412. doi:10.1016/j.celrep.2018.01.036
Putnam A, Cassani M, Smith J, Seydoux G. 2019. A gel phase promotes condensation of liquid P granules in *Caenorhabditis elegans* embryos. *Nat Struct Mol Biol* **26:** 220–226. doi:10.1038/s41594-019-0193-2
Rangan P, DeGennaro M, Jaime-Bustamante K, Coux R-X, Martinho R, Lehmann R. 2009. Temporal and spatial control of germ plasm RNAs. *Curr Biol* **19:** 72–77. doi:10.1016/j.cub.2008.11.066
Raška I, Andrade LEC, Ochs RL, Chan EKL, Chang C-M, Roos G, Tan EM. 1991. Immunological and ultrastructural studies of the nuclear coiled body with autoimmune antibodies. *Exp Cell Res* **195:** 27–37. doi:10.1016/0014-4827(91)90496-H
Saito M, Hess D, Eglinger J, Fritsch AW, Kreysing M, Weinert BT, Choudhary C, Matthias P. 2019. Acetylation of intrinsically disordered regions regulates phase separation. *Nat Chem Biol* **15:** 51–61. doi:10.1038/s41589-018-0180-7
Samir P, Kesavardhana S, Patmore DM, Gingras S, Malireddi RKS, Karki R, Guy CS, Briard B, Place DE, Bhattacharya A, et al. 2019. DDX3X acts as a live-or-die checkpoint in stressed cells by regulating NLRP3 inflammasome. *Nature* **573:** 590–594. doi:10.1038/s41586-019-1551-2
Schaffert N, Hossbach M, Heintzmann R, Achsel T, Lührmann R. 2004. RNAi knockdown of hPrp31 leads to an accumulation of U4/U6 di-snRNPs in Cajal bodies. *EMBO J* **23:** 3000–3009. doi:10.1038/sj.emboj.7600296
Scheer U, Thiry M, Goessens G. 1993. Structure, function and assembly of the nucleolus. *Trends Cell Biol* **3:** 236–241. doi:10.1016/0962-8924(93)90123-I
Schwartz DC, Parker R. 1999. Mutations in translation initiation factors lead to increased rates of deadenylation and decapping of mRNAs in *Saccharomyces cerevisiae*. *Mol Cell Biol* **19:** 5247–5256. doi:10.1128/MCB.19.8.5247
Seydoux G. 2018. The P granules of *C. elegans*: a genetic model for the study of RNA–protein condensates. *J Mol Biol* **430:** 4702–4710. doi:10.1016/j.jmb.2018.08.007
Shan J, Moran-Jones K, Munro TP, Kidd GJ, Winzor DJ, Hoek KS, Smith R. 2000. Binding of an RNA trafficking response element to heterogeneous nuclear ribonucleoproteins A1 and A2. *J Biol Chem* **275:** 38286–38295. doi:10.1074/jbc.M007642200
Shaw J, Love AJ, Makarova SS, Kalinina NO, Harrison BD, Taliansky ME. 2014. Coilin, the signature protein of Cajal bodies, differentially modulates the interactions of plants with viruses in widely different taxa. *Nucleus* **5:** 85–94. doi:10.4161/nucl.28315

Sheth U, Parker R. 2003. Decapping and decay of messenger RNA occur in cytoplasmic processing bodies. *Science* **300:** 805–808. doi:10.1126/science.1082320

Sheth U, Parker R. 2006. Targeting of aberrant mRNAs to cytoplasmic processing bodies. *Cell* **125:** 1095–1109. doi:10.1016/j.cell.2006.04.037

Shevtsov SP, Dundr M. 2011. Nucleation of nuclear bodies by RNA. *Nat Cell Biol* **13:** 167–173. doi:10.1038/ncb2157

Sleeman JE, Lamond AI. 1999. Newly assembled snRNPs associate with coiled bodies before speckles, suggesting a nuclear snRNP maturation pathway. *Curr Biol* **9:** 1065–1074. doi:10.1016/S0960-9822(99)80475-8

Sleeman JE, Trinkle-Mulcahy L, Prescott AR, Ogg SC, Lamond AI. 2003. Cajal body proteins SMN and Coilin show differential dynamic behaviour in vivo. *J Cell Sci* **116:** 2039–2050. doi:10.1242/jcs.00400

Staněk D, Neugebauer KM. 2004. Detection of snRNP assembly intermediates in Cajal bodies by fluorescence resonance energy transfer. *J Cell Biol* **166:** 1015–1025. doi:10.1083/jcb.200405160

Staněk D, Přidalová-Hnilicová J, Novotný I, Huranová M, Blažíková M, Wen X, Sapra AK, Neugebauer KM. 2008. Spliceosomal small nuclear ribonucleoprotein particles repeatedly cycle through Cajal bodies. *Mol Biol Cell* **19:** 2534–2543. doi:10.1091/mbc.e07-12-1259

Taneja KL, McCurrach M, Schalling M, Housman D, Singer RH. 1995. Foci of trinucleotide repeat transcripts in nuclei of myotonic dystrophy cells and tissues. *J Cell Biol* **128:** 995–1002. doi:10.1083/jcb.128.6.995

Tauber D, Tauber G, Khong A, Van Treeck B, Pelletier J, Parker R. 2020. Modulation of RNA condensation by the DEAD-box protein eIF4A. *Cell* **180:** 411–426.e16. doi:10.1016/j.cell.2019.12.031

Taylor JP, Brown RHB Jr, Cleveland DW. 2016. Decoding ALS: from genes to mechanism. *Nature* **539:** 197–206. doi:10.1038/nature20413

Teixeira D, Parker R. 2007. Analysis of P-body assembly in *Saccharomyces cerevisiae*. *Mol Biol Cell* **18:** 2274–2287. doi:10.1091/mbc.e07-03-0199

Tharun S, He W, Mayes AE, Lennertz P, Beggs JD, Parker R. 2000. Yeast Sm-like proteins function in mRNA decapping and decay. *Nature* **404:** 515–518. doi:10.1038/35006676

Thomson T, Lasko P. 2004. *Drosophila* tudor is essential for polar granule assembly and pole cell specification, but not for posterior patterning. *Genesis* **40:** 164–170. doi:10.1002/gene.20079

Thomson T, Liu N, Arkov A, Lehmann R, Lasko P. 2008. Isolation of new polar granule components in *Drosophila* reveals P body and ER associated proteins. *Mech Dev* **125:** 865–873. doi:10.1016/j.mod.2008.06.005

Thul PJ, Åkesson L, Wiking M, Mahdessian D, Geladaki A, Blal HA, Alm T, Asplund A, Björk L, Breckels LM, et al. 2017. A subcellular map of the human proteome. *Science* **356:** eaal3321. doi:10.1126/science.aal3321

Timchenko LT, Miller JW, Timchenko NA, DeVore DR, Datar KV, Lin L, Roberts R, Caskey CT, Swanson MS. 1996. Identification of a (CUG)n triplet repeat RNA-binding protein and its expression in myotonic dystrophy. *Nucleic Acids Res* **24:** 4407–4414. doi:10.1093/nar/24.22.4407

Tourrière H, Chebli K, Zekri L, Courselaud B, Blanchard JM, Bertrand E, Tazi J. 2003. The RasGAP-associated endoribonuclease G3BP assembles stress granules. *J Cell Biol* **160:** 823–831. doi:10.1083/jcb.200212128

Trcek T, Lehmann R. 2019. Germ granules in *Drosophila*. *Traffic* **20:** 650–660. doi:10.1111/tra.12674

Trinkle-Mulcahy L, Sleeman JE. 2017. The Cajal body and the nucleolus: "In a relationship" or "It's complicated"? *RNA Biol* **14:** 739–751. doi:10.1080/15476286.2016.1236169

Tsai W-C, Gayatri S, Reineke LC, Sbardella G, Bedford MT, Lloyd RE. 2016. Arginine demethylation of G3BP1 promotes stress granule assembly. *J Biol Chem* **291:** 22671–22685. doi:10.1074/jbc.M116.739573

Van Treeck B, Parker R. 2018. Emerging roles for intermolecular RNA–RNA interactions in RNP assemblies. *Cell* **174:** 791–802. doi:10.1016/j.cell.2018.07.023

Van Treeck B, Protter DSW, Matheny T, Khong A, Link CD, Parker R. 2018. RNA self-assembly contributes to stress granule formation and defining the stress granule transcriptome. *Proc Natl Acad Sci* **115:** 2734–2739. doi:10.1073/pnas.1800038115

Vanzo N, Oprins A, Xanthakis D, Ephrussi A, Rabouille C. 2007. Stimulation of endocytosis and actin dynamics by oskar polarizes the *Drosophila* oocyte. *Dev Cell* **12:** 543–555. doi:10.1016/j.devcel.2007.03.002

Verheggen C, Lafontaine DLJ, Samarsky D, Mouaikel J, Blanchard J-M, Bordonné R, Bertrand E. 2002. Mammalian and yeast U3 snoRNPs are matured in specific and related nuclear compartments. *EMBO J* **21:** 2736–2745. doi:10.1093/emboj/21.11.2736

Vogler TO, Wheeler JR, Nguyen ED, Hughes MP, Britson KA, Lester E, Rao B, Betta ND, Whitney ON, Ewachiw TE, et al. 2018. TDP-43 and RNA form amyloid-like myo-granules in regenerating muscle. *Nature* **563:** 508–513. doi:10.1038/s41586-018-0665-2

Voronina E, Seydoux G, Sassone-Corsi P, Nagamori I. 2011. RNA granules in germ cells. *Cold Spring Harb Perspect Biol* **3:** a00274. doi:10.1101/cshperspect.a002774

Voronina E, Paix A, Seydoux G. 2012. The P granule component PGL-1 promotes the localization and silencing activity of the PUF protein FBF-2 in germline stem cells. *Development* **139:** 3732–3740. doi:10.1242/dev.083980

Walker MP, Tian L, Matera AG. 2009. Reduced viability, fertility and fecundity in mice lacking the Cajal body marker protein, coilin. *PLoS One* **4:** e6171. doi:10.1371/journal.pone.0006171

Wang JT, Seydoux G. 2014. P granules. *Curr Biol* **24:** R637–R638. doi:10.1016/j.cub.2014.06.018

Wang J, Pegoraro E, Menegazzo E, Gennarelli M, Hoop RC, Angelini C, Hoffman EP. 1995. Myotonic dystrophy: evidence for a possible dominant-negative RNA mutation. *Hum Mol Genet* **4:** 599–606. doi:10.1093/hmg/4.4.599

Weihl CC, Temiz P, Miller SE, Watts G, Smith C, Forman M, Hanson PI, Kimonis V, Pestronk A. 2008. TDP-43 accumulation in IBM muscle suggests a common pathogenic mechanism with frontotemporal dementia. *J Neurol Neurosurg Psychiatry* **79:** 1186–1189. doi:10.1136/jnnp.2007.131334

Wheeler JR, Matheny T, Jain S, Abrisch R, Parker R. 2016. Distinct stages in stress granule assembly and disassembly. *Elife* **5:** e18413. doi:10.7554/eLife.18413

Wheeler JR, Jain S, Khong A, Parker R. 2017. Isolation of yeast and mammalian stress granule cores. *Methods San Diego Calif* **126:** 12–17. doi:10.1016/j.ymeth.2017.04.020

Wolozin B, Ivanov P. 2019. Stress granules and neurodegeneration. *Nat Rev Neurosci* **20:** 649–666. doi:10.1038/s41583-019-0222-5

Xing W, Muhlrad D, Parker R, Rosen MK. 2020. A quantitative inventory of yeast P body proteins reveals principles of composition and specificity. BioRxiv doi:10.1101/489658

Yasuzumi G, Sawada T, Sugihara R, Kiriyama M, Sugioka M. 1958. Electron microscope researches on the ultrastructure of nucleoli in animal tissues. *Z Zellforsch Mikrosk Anat Vienna Austria* **1948:** 10–23. doi:10.1007/BF00496710

Youn J-Y, Dunham WH, Hong SJ, Knight JDR, Bashkurov M, Chen GI, Bagci H, Rathod B, MacLeod G, Eng SWM, et al. 2018. High-density proximity mapping reveals the subcellular organization of mRNA-associated granules and bodies. *Mol Cell* **69:** 517–532.e11. doi:10.1016/j.molcel.2017.12.020

Biophysical Properties of HP1-Mediated Heterochromatin

SERENA SANULLI,[1] JOHN D. GROSS,[1] AND GEETA J. NARLIKAR[2]

[1]*Department of Pharmaceutical Chemistry,*
[2]*Department of Biochemistry and Biophysics, University of California,*
San Francisco, California 94158, USA

Correspondence: geeta.narlikar@ucsf.edu

Heterochromatin is a classic context for studying the mechanisms of chromatin organization. At the core of a highly conserved type of heterochromatin is the complex formed between chromatin methylated on histone H3 lysine 9 and HP1 proteins. This type of heterochromatin plays central roles in gene repression, genome stability, and nuclear mechanics. Systematic studies over the last several decades have provided insight into the biophysical mechanisms by which the HP1-chromatin complex is formed. Here, we discuss these studies together with recent findings indicating a role for phase separation in heterochromatin organization and function. We suggest that the different functions of HP1-mediated heterochromatin may rely on the increasing diversity being uncovered in the biophysical properties of HP1-chromatin complexes.

Packaging of the eukaryotic genome into chromatin provides the first level of gene regulation by occlusion of specific DNA regions from the transcription machinery (Luger et al. 1997; Jost et al. 2012). The second level of regulation is achieved by the organization of chromatin into transcriptionally active regions called euchromatin and transcriptionally silent regions called heterochromatin (Grewal and Jia 2007; Allshire and Madhani 2018). These two levels of regulation play central roles in the determination of cellular identity and function. The term heterochromatin was originally used by Heitz (1928) to describe regions of chromosomes that appeared to remain condensed throughout the cell cycle (Heitz 1928; Jost et al. 2012). Pioneering genetic and biochemical work by Elgin and colleagues showed that a central component of heterochromatin was the protein HP1, which could bind chromatin that was methylated on histone H3 lysine 9 (James and Elgin 1986; James et al. 1989; Eissenberg et al. 1990). These findings were followed by the discovery of the enzyme responsible for H3K9 methylation by Jenuwein and colleagues (Rea et al. 2000). HP1-mediated heterochromatin is found in several organisms including yeast and humans. Another type of heterochromatin that is also found in multiple organisms is mediated by the Polycomb group of proteins (Simon and Kingston 2009; Schuettengruber et al. 2017; Yu et al. 2019). These proteins act in large complexes and are commonly associated with chromatin that is methylated on histone H3 lysine 27. In this perspective, we will focus on HP1-mediated heterochromatin.

HP1-mediated heterochromatin is often functionally classified into two types: constitutive heterochromatin, which is found near centromeres and telomeres and appears largely invariant with cell type; and facultative heterochromatin, which enables repression of developmentally controlled genes and is cell type–specific (Allshire and Madhani 2018; Nicetto and Zaret 2019). Gene repression is thought to arise from at least two mutually compatible mechanisms: steric occlusion by HP1 binding and chromatin compaction, and recruitment of gene silencing machineries via noncoding RNA based mechanisms (Grewal and Jia 2007; Allshire and Madhani 2018). In this review, we do not discuss the noncoding RNA based pathways and instead refer the reader to some excellent reviews on this topic (Reyes-Turcu and Grewal 2012; Martienssen and Moazed 2015; Allshire and Madhani 2018).

Work from several groups has shown that heterochromatin mediated by HP1 plays other roles in addition to gene repression. One of these roles is enabling correct chromosome segregation (Allshire and Madhani 2018; Janssen et al. 2018). It is proposed that centromeric heterochromatin provides a critical platform for regulating the deposition of the centromeric histone H3 and for regulating cohesion action (Allshire and Madhani 2018). Another role involves providing mechanical stability to cells during interphase (Stephens et al. 2019). Such a role for HP1-mediated heterochromatin is proposed to arise in part from the ability of HP1 to interact with nuclear lamin proteins that provide much of the structural integrity to the nuclear envelope (Stephens et al. 2019). Thus HP1-heterochromatin functions across multiple scales: from the scale of individual genes to the scale of the whole nucleus.

A central and common feature in almost all of these functions is the HP1-chromatin complex. We and others have been focused on understanding the biophysical capabilities of this complex. Below we first summarize prior advances in this area and then describe recent data from our laboratories on the linkage between phase separation by HP1 proteins and chromatin compaction. Finally, we conclude with some speculations on the biophysical basis for the diverse functions performed by heterochromatin.

© 2019 Sanulli et al. This article is distributed under the terms of the Creative Commons Attribution-NonCommercial License, which permits reuse and redistribution, except for commercial purposes, provided that the original author and source are credited.

Published by Cold Spring Harbor Laboratory Press; doi:10.1101/sqb.2019.84.040360

OLIGOMERIZATION BEHAVIOR OF HP1 PROTEINS

Early studies in *Drosophila* found that transposition of a gene from euchromatin into heterochromatin resulted in the silencing of the gene in some cells but not others (Elgin and Reuter 2013). This phenomenon termed position effect variegation was attributed to the spreading of heterochromatic material over the transposed gene. Later studies found that such gene silencing was accompanied by the presence of the H3K9me3 mark and HP1 over the silenced gene (Elgin and Reuter 2013). Therefore, one property attributed to heterochromatin has been the ability to spread across large regions of the genome. From a biophysical perspective, the phenomenon of "spreading" suggests some manner of cooperative action by heterochromatin molecules on chromatin. In this context, as summarized below HP1 molecules can both dimerize and form high-order oligomers.

The HP1 molecule contains two structured domains, the chromodomain (CD) that binds the H3K9me3 mark and the chromoshadow domain (CSD) that forms a homodimer (Canzio et al. 2014; Eissenberg and Elgin 2014) (Fig. 1A). In addition, there are three unstructured regions, the hinge (H), which binds both DNA and RNA, an amino-terminal extension (NTE), and a carboxy-terminal extension (CTE) (Sugimoto et al. 1996; Muchardt et al. 2002; Meehan et al. 2003; Mishima et al. 2013). Cross-linking studies with isolated CD and CSD domains of the human HP1α protein helped uncover the potential of these domains to form higher-order structures (Yamada et al. 1999). High-resolution crystal structures of CSD–CSD dimers from various HP1 paralogs provided a detailed view into the dimerization interface and further indicated how the CSD–CSD dimer could interact with other protein ligands (Brasher et al. 2000; Cowieson et al. 2000; Smothers and Henikoff 2000; Thiru et al. 2004; Kang et al. 2011; Mendez et al. 2013). Overall, most HP1 proteins studied were found to be able to form dimers, although the dissociation constants ranged from low nM to µM (Canzio et al. 2014).

Gel-filtration studies with the major *Schizosaccharomyces pombe* HP1 protein Swi6 first raised the possibility

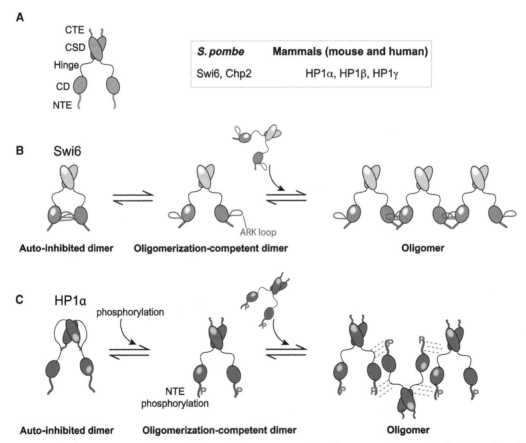

Figure 1. Comparison of HP1 molecules between *Schizosaccharomyces pombe* and mouse and human. (*A*) (*Left*) Domain architecture found in yeast and mammalian HP1 proteins. (*Right*) Nomeclature of HP1 proteins studied biochemically. *S. pombe* has two HP1 proteins, Swi6 and Chp2, whereas mouse and humans have three HP1 proteins that have been studied biochemically, HP1α, Hp1β, and HP1γ. (*B*) Model showing auto-inhibition in Swi6 mediated by the interaction between the ARK loop in red and the chromodomain binding pocket in green. The ARK loop also participates in stabilizing higher-order oligomers by interacting with the CD of a different dimer. (*C*) Model showing auto-inhibition in HP1α mediated by the interaction between the CTE and the hinge. Phosphorylation of the NTE relieves the auto-inhibition and promotes oligomerization via NTE-hinge interactions mediated by the phosphate groups with positively charged hinge residues. The phosphate groups are shown by the red "P" letter.

that the full-length proteins may form oligomers beyond dimers (Wang et al. 2000). Later studies indicated that the gel-filtration results largely reflected a dimeric state of Swi6 (Sadaie et al. 2008). The possibility of higher-order oligomerization was conclusively tested by analytical ultracentrifugation, which enabled quantification of the true masses of the Swi6 oligomers (Canzio et al. 2011; Canzio et al. 2013). These experiments showed that Swi6 could form higher-order oligomers by sequential addition of dimers. Interestingly, Swi6 dimers were found to exist in an auto-inhibited state, and switching out of the auto-inhibited state was required to form higher-order oligomers (Canzio et al. 2013). Auto-inhibition arises because the CD of one monomer within a Swi6 dimer is bound by a loop present in the CD of the other monomer (Fig. 1B). The loop sequence (ARK) mimics the H3 tail sequence that is bound by the CD. Further, the interactions formed in the oligomeric state also appear to involve the CD–ARK loop interaction, but between two dimers rather than within a dimer. Studies performed with the human HP1α protein found that phosphorylation of the NTE (nPhos_HP1α) was essential to switch the protein from an auto-inhibited state to an oligomerization competent state (Fig. 1C) (Larson et al. 2017). In this case, auto-inhibition is thought to be enabled by the interaction of the CTE of one monomer with the hinge of the second monomer. This model for auto-inhibition is consistent with earlier studies suggesting that the CSD region plays an auto-inhibitory role to regulate the ability of the hinge to bind DNA (Mishima et al. 2013). The oligomeric state of nPhos_HP1α was suggested to involve interactions between two of its unstructured regions, the hinge and the NTE (Larson et al. 2017). Thus, analogous to the case with Swi6, common interaction interfaces are used in the auto-inhibited and oligomeric states (Fig. 1B,C).

Overall the studies above provided a key conceptual building block for understanding the biophysical basis of heterochromatin function: the ability of HP1 proteins to form higher-order oligomers in an autoregulated manner.

At the same time, these studies also uncovered interesting differences between HP1 paralogs. For example, although human HP1α formed higher-order oligomers upon phosphorylation, its paralog HP1β did not (Larson et al. 2017). How the sequence differences between HP1α and HP1β contribute to these differences in biophysical behavior is an active area of research, and some additional biophysical differences are mentioned in the sections below.

HP1 INTERACTION WITH CHROMATIN

Several studies have characterized the biophysical behavior of HP1 proteins on chromatin. Early studies have shown that the CD of HP1 proteins can bind the H3K9me3 mark with high specificity (Bannister et al. 2001; Jacobs et al. 2001; Lachner et al. 2001; Jacobs and Khorasanizadeh 2002; Nielsen et al. 2002). Further, amino-terminal phosphorylation of mammalian HP1α has been shown to increase the affinity of its CD for the H3K9me3 mark (Hiragami-Hamada et al. 2011; Nishibuchi et al. 2014). Work with full-length Swi6 has found that four molecules of Swi6 bind one nucleosome (Canzio et al. 2011). Within the Swi6–nucleosome complex, the CSD–CSD dimer was implicated in binding the nucleosome core (Canzio et al. 2013). Later studies, discussed in detail further below, indicated that the CSD–CSD dimer contacts the α1-helix of H2B (Sanulli et al. 2019) (Fig. 2C). Interestingly, the ARK loop within the CD that participates in auto-inhibition and oligomerization was also implicated in binding nucleosomal DNA (Fig. 2A) (Canzio et al. 2013). Together, these findings led to a model where the two unbound CDs act as "sticky ends" that interact with either nearby CDs or nearby nucleosomes (Fig. 2D). In the former possibility, bridging requires both the dimerization and a tetramerization interfaces. This possibility is consistent with and builds on previous models invoking a tetrameric Swi6 complex (Wang et al. 2000). The latter possibility relies mainly on the dimerization interface. Consistent with a nucleosome bridging role, Swi6 was shown to bind with higher specificity for the H3K9me3 mark on nucleosomal arrays with 15 bp versus 47 bp of internucleosomal spacing (Canzio et al. 2011).

Studies with mammalian HP1β have suggested a nucleosome bridging model that largely relies on the HP1 dimer interface (Hiragami-Hamada et al. 2016) (Fig. 2D). Studies with mammalian HP1α have shown that HP1α can bind nucleosome arrays and compact them (Fig. 2B) (Azzaz et al. 2014; Kilic et al. 2018). Further, EM studies have suggested that one HP1α dimer can bridge two nucleosomes (Machida et al. 2018). However, whether HP1β and HP1α can adopt other stoichiometries on nucleosomes is not known. This is because unlike with Swi6 the stoichiometry of these proteins on mononucleosomes in solution has not been directly measured.

In terms of contacting the histone core, the CSD–CSD interface of mammalian HP1 proteins has been shown to interact with the H3αN helix region of the nucleosome core (Fig. 2C) (Dawson et al. 2009; Lavigne et al. 2009; Richart et al. 2012). The different regions of the octamer contacted by mammalian HP1 proteins versus Swi6 are consistent with the different binding specificities identified for their respective CSD–CSD dimers (Fig. 2C) (Smothers and Henikoff 2000; Thiru et al. 2004; Lechner et al. 2005; Mendez et al. 2011; Mendez et al. 2013; Isaac et al. 2017).

Interestingly, phosphorylation of the NTE of mammalian HP1α has been suggested to enhance its specificity for the H3K9me3 mark within nucleosomes while decreasing its affinity for nucleosomes and DNA (Hiragami-Hamada et al. 2011). These findings raise the possibility that phosphorylation dissociates HP1α from the nucleosome core such that it is now bound mainly to the H3 tail.

Overall the studies in this and the prior section have showcased the diverse mechanisms by which HP1 proteins could assemble on chromatin and enable chromatin compaction.

PHASE SEPARATION OF HP1 PROTEINS TO EXPLAIN COMPARTMENTALIZATION

In 2017, two studies reported the ability of HP1 proteins to form phase-separated droplets (Larson et al. 2017;

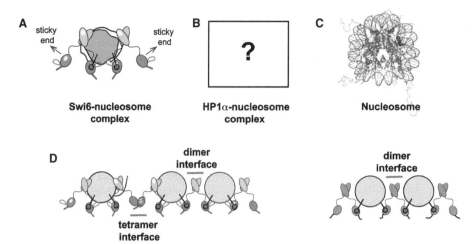

Figure 2. Nucleosomal interactions made by HP1 proteins. (*A*) Model for Swi6 interaction with a mononucleosome. Binding of Swi6 deforms the nucleosome core. The ARK loop on the CD is in red, deformed histone octamer is in orange, loosened DNA is in black, and the H3K9 methylation mark is shown as a red circle on the H3 tail, which is in black. (*B*) The stoichiometry of mammalian HP1α on nucleosomes in solution has not been measured. (*C*) The crystal structure of the nucleosome is shown with the DNA, histone H4, and histone H2A in gray. One copy each of histone H3 and H2B are shown in orange and blue, respectively. The histone helices that are suggested to be contacted by the CSD dimers of Swi6 and the mammalian HP1 proteins, respectively, are shown in dark orange (in H3) and dark blue (in H2B), respectively. (*D*) Two modes of nucleosome bridging. (*Left*) Model for two modes by which Swi6 can bridge across nucleosomes. The tetramerization interface between two dimers facilitates bridging. The tetramerization interface is stabilized by the CD-ARK loop interaction. (*Right*) Model for how HP1β may bridge nucleosomes. The dimer interface facilitates bridging. The two nucleosomes shown could be adjacent nucleosomes on the same chromatin fiber, nonadjacent nucleosomes on the same chromatin fiber, or nucleosomes on two different chromatin fibers.

Strom et al. 2017). One study showed that the *Drosophila* HP1 protein, HP1a, could form droplets in vitro and that heterochromatin foci containing HP1a in early *Drosophila* embryos displayed properties consistent with liquid-like phase-separated states (Strom et al. 2017). In parallel, biochemical studies with human HP1 proteins found that phosphorylation of the NTE of human HP1α promoted higher-order oligomerization and the formation of phase-separated droplets (Larson et al. 2017). Binding to HP1α by DNA was also shown to promote the formation of phase-separated droplets. The phase-separation behavior of HP1α was compatible with at least two proposed roles of heterochromatin in cells. First, the assembly of HP1α on DNA drove rapid compaction of the DNA. Second, chromatin and enzymes known to interact with HP1α such as aurora B kinase could be enriched from solution in the phase-separated nPhos-HP1α droplets. These observations raised the possibility that phase separation could be coupled to the chromatin compaction typically associated with heterochromatin formation. More generally, these results suggested that phase-separation processes could help to compartmentalize chromatin into active and repressed states.

Although HP1α could form phase-separated droplets upon binding DNA, HP1β was shown to be deficient for phase-separation in the presence of DNA (Larson et al. 2017). Instead, recent work has suggested that mammalian HP1β requires the presence of H3K9-methylated chromatin and the histone methylase Suvar39h1 to participate in forming phase-separated droplets (Wang et al. 2019). From a biological perspective, such biophysical differences provide a means to diversify the functions of heterochromatin.

To better understand how chromatin compaction and phase separation by HP1 are linked, we investigated the biophysical consequences of HP1 binding on chromatin. These studies were performed with Swi6 and are described below (Sanulli et al. 2019).

DISORGANIZATION OF HISTONE CORE PROMOTES CHROMATIN COMPACTION INTO DROPLETS

To understand the effect of Swi6 binding on a nucleosome, we used three complementary approaches: hydrogen-deuterium exchange-mass spectrometry (HDX-MS), nuclear magnetic resonance (NMR), and cross-linking mass spectrometry (XL-MS). The HDX-MS approach allows the determination of changes in solvent accessibility, such that increased solvent accessibility of a protein amide backbone region results in increased uptake of deuterium from solution. Given that heterochromatin is thought to be repressive and occlude access to chromatin, we hypothesized that Swi6 binding to a nucleosome would reduce uptake of deuterium by the histone proteins. Instead, we observed widespread increase in the uptake of deuterium by regions of the histone octamer core that are normally buried in the canonical nucleosome conformation. This was a surprising and counterintuitive result. We, therefore, used Methyl-TROSY NMR to further test this finding. Most commonly, NMR is used to study dynamics in proteins that are less than ∼30 kDa. However, methyl-TROSY NMR spectroscopy has made it possible to study dynamics in complexes as large as the 26S proteasome and the nucleosome (Religa et al. 2010; Kato et al. 2011). Applying this approach indicated that binding by Swi6 results in intramolecular conformational dynamics on the msec to μsec timescale in the buried core of the nucleosome. Importantly, the changes in side-chain dynamics detected by this NMR method were consistent with the changes in backbone solvent accessibility detected by HDX-MS. In addition, using XL-MS we found changes in intrahistone cross-links upon Swi6 binding that were consistent with the changes observed by HDX-MS and NMR. Together these results indicated that binding by Swi6 to a nucleosome results in conformational reorganization of the histone core.

Additional XL-MS and NMR studies suggested that the CSD–CSD dimer contacts the nucleosome by interacting with the α1-helix of H2B and further implied that binding to this region of H2B may transiently unfold the helix. Based on these and previous studies described in earlier sections of this review, we proposed a model for how Swi6 could alter nucleosome conformation. In this model, Swi6 makes extensive interactions with the whole nucleosome, including the H3 tail, the H2B α1 helix, and nucleosomal DNA. These interactions are then proposed to loosen intranucleosomal contacts, resulting in greater dynamics and exposure within the histone core (Fig. 2A,C).

To determine the significance of these conformational changes, we turned to a classic method used to study chromatin compaction by HP1 proteins (Azzaz et al. 2014; Hiragami-Hamada et al. 2016). This method uses arrays of nucleosomes, and compaction is measured by the ability to pellet the arrays in the presence of divalent ions such as Mg^{2+} or proteins such as HP1. We found that similar to observations with mammalian HP1 proteins, Swi6 could promote pelleting of nucleosome arrays in a concentration-dependent manner (Azzaz et al. 2014; Hiragami-Hamada et al. 2016). Interestingly, when visualized under a light microscope, the pelleted Swi6-array complexes resembled liquid droplets. The Swi6 concentration dependence for droplet formation with the arrays largely mirrored the concentration dependence for pelleting. These results provided some of the first evidence that chromatin compaction by Swi6 is coupled to phase separation. The droplets could fuse and reform spherical structures, consistent with a liquid-like nature. To assess the role of histone octamer deformation, we used site-specific disulfide bonds to lock a region of the buried H3–H4 interface that showed major perturbations by NMR, HDX-MS, and XL-MS. Specifically, cysteine cross-links were generated between residues H3-I62C and H4-A33C within the histone octamer, and nucleosome arrays were assembled using cross-linked and un-cross-linked octamers. We found that Swi6-mediated pelleting of nucleosome arrays, and corresponding droplet formation were impaired upon the introduction of these cross-links. These results indicated that the increased histone core dynamics resulting from Swi6 binding facilitate the phase separation and compaction of chromatin.

The coupling between Swi6 assembly on chromatin arrays and their compaction into phase-separated droplets gave us an opportunity to revisit the role of Swi6 oligomerization. Here, we describe our results with a mutant in which the ARK loop within the CD is mutated to AAA (LoopX mutant). Mutating the ARK loop reduces the affinity constant for higher-order oligomerization and reduces nucleosome binding (Canzio et al. 2013). Further, the LoopX mutant results in significant defects in heterochromatin assembly, spreading, and function in S. pombe (Canzio et al. 2013). Surprisingly, when used under saturating conditions, we found that the LoopX mutant does not inhibit chromatin phase separation but forms larger droplets than the Swi6 wild-type (WT) protein over time. Interestingly, however, these droplets showed greater wetting behavior than the droplets formed with Swi6 WT, indicating a lower surface tension. A lower surface tension suggests that the inter-molecular interactions that enable droplet formation are weakened consistent with the reduction in the oligomerization affinity of the LoopX mutant (Alberti et al. 2019). We therefore concluded that the ability of Swi6 to compact chromatin templates into phase-separated liquid droplets in vitro is coupled to both its ability to oligomerize and its ability to promote octamer distortion.

In S. pombe, distinct heterochromatin foci can be observed and Swi6 is found to accumulate at these foci (Haldar et al. 2011). To assess the correlation between the properties of Swi6-chromatin complexes in vitro and heterochromatin foci in vivo, we asked how the LoopX mutant affects heterochromatin foci in S. pombe. We found that replacing the wild-type Swi6 protein with the LoopX mutant reduced the number and intensity of Swi6 foci and resulted in a more diffused nuclear localization of the protein. The endogenous expression level of the LoopX protein is comparable to the WT protein, indicating that the observed defects arise from the biophysical properties of the LoopX mutant. Importantly, these results showed that the defects in heterochromatin foci observed in S. pombe with LoopX correlate with the instability of the LoopX-chromatin droplets in vitro.

Although the results above uncovered a functional role for the increased octamer dynamics, the question of why these dynamics are important still remained. Indeed, from a first glance, the increased accessibility and dynamics of buried histone residues appear paradoxical, given the repressive role of heterochromatin. However, these results can be explained if we consider that chromatin compaction and phase separation are linked. Substantial previous work has suggested that the formation of phase-separated assemblies relies on weak multivalent interactions (Alberti et al. 2019). We therefore propose that the transiently exposed histone residues participate in weak and multivalent interactions between nucleosomes, and that these interactions promote the compaction of chromatin into phase-separated states.

Within the above model, we propose that an HP1 protein such as Swi6 would promote chromatin compaction in two ways: (i) by using its nucleosome deformation activity to substantially increase intrinsic histone core dynamics and accessibility, thereby increasing opportunities for weak internucleosomal interactions, and (ii) by using its oligomerization activity to bridge multiple nucleo-

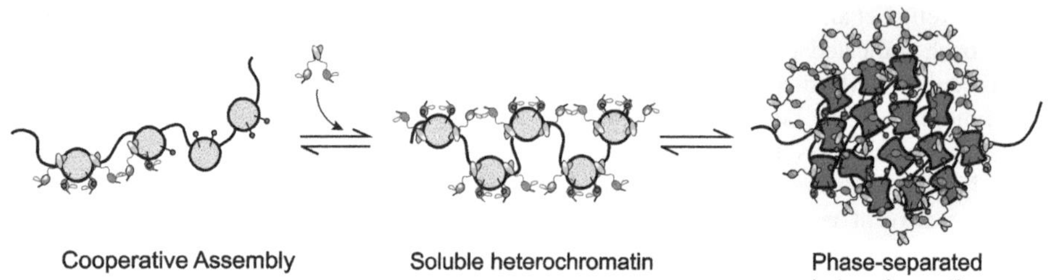

Figure 3. Model for coupling phase separation to chromatin compaction by HP1. Oligomerization by Swi6 and the transiently exposed histone core residues result in multiple weak interactions between the component molecules, which promote phase separation. The deformed nucleosomes are shown in blue.

somes and provide essential modes of multivalency (Fig. 3). Ultimately, although Swi6 increases accessibility at the level of individual nucleosomes, the net effect at the level of multiple nucleosomes is their transformation into a compact phase-separated state where the exposed histone residues become inaccessible because of participation in multivalent interactions. Additional levels of inaccessibility could arise from the specific physico-chemical properties of the phase, which may inhibit the entry of some factors and not others. Overall, such a model is consistent with previous observations of lower histone turnover in *S. pombe* heterochromatin and lower DNA accessibility in *Drosophila* heterochromatin (Aygun et al. 2013; Elgin and Reuter 2013).

A BIOPHYSICAL MECHANISMS TOOL KIT: THE BENEFITS OF DIVERSITY

It is tempting to propose a unified biophysical model for explaining the assembly of the HP1-chromatin complex. However, as described above the studies to date paint a different and more interesting picture. What is emerging is a tool kit of biophysical strategies that are used in different combinations by HP1 paralogs based on the specific chromatin context.

Oligomers versus Dimers

The higher-order oligomerization beyond dimerization observed with Swi6 suggests that oligomerization provides one mechanism to bridge nucleosomes and thereby compact the underlying chromatin. However, bridging could also be accomplished by one HP1 dimer without invoking oligomerization as has been suggested for HP1β. It is possible that bridging via oligomerization makes the assembly process more cooperative than bridging via dimers. In contrast, bridging via dimers would make the process require less HP1 molecules per nucleosome.

Mode of Nucleosome Engagement

The CSD dimers of different HP1 proteins appear to engage different regions of the folded histone core. We speculate that these differences arise from differences in binding specificities of the respective CSD–CSD dimers.

Such differences would allow certain histone posttranslational modifications to regulate the binding of some HP1 molecules but not others. Further, the binding of nonhistone ligands to the CSD dimer could differentially affect the engagement of HP1 molecules with a nucleosome based on the relative affinities for histone versus nonhistone ligands. Finally, it is also possible that in some cases, the CSD dimer does not significantly bind the nucleosome core and the main interactions made by the HP1 dimer are with the histone H3 tail and the nucleosomal DNA.

Octamer Deformation

Our recent findings indicate that Swi6 binding is coupled to the deformation of the nucleosome core. We propose that such deformation is enabled in part due to the unfolding of the H2B α1 helix upon binding by the CSD–CSD dimer. Analogous nucleosome deformation has not been studied for mammalian HP1 proteins. However, we speculate that the previously identified contact between the CSD–CSD dimer of mammalian HP1 proteins and the H3αN helix region of the nucleosome core may contribute to octamer deformation. In this context, it has been suggested that ATP-dependent chromatin remodeling by SWI/SNF family enzymes makes buried histone sites more accessible for binding by mammalian and *Drosophila* HP1 proteins (Lavigne et al. 2009). Interestingly SWI/SNF family remodelers have been suggested to alter nucleosome conformation (Fan et al. 2004; Sinha et al. 2017). Building on these studies we speculate that in some contexts, HP1 proteins may collaborate with ATP-dependent remodelers to drive octamer deformation.

Phase Separation

We propose that the different abilities of HP1 proteins to phase-separate in the presence of chromatin adds to the diversity of HP1-mediated heterochromatin. Thus, we can imagine that in contrast to Swi6, some HP1 paralogs may form phases with chromatin by relying solely on the multivalency provided by their oligomerization and not via octamer deformation. Other HP1 proteins may compact chromatin via nucleosome bridging but not cause phase separation. Overall, we imagine a large range of different types of interaction networks between the HP1 proteins and chromatin within and outside phase-separated heterochromatin states.

The biophysical strategies described above provide fertile ground for conceptualizing how the core HP1-chromatin complex can participate in diverse nuclear functions. For example, it is possible that HP1-heterochromatin at the nuclear lamina exists in a phase-separated state. Because of the nature of the multivalent interactions, such a phase-separated state could possess the viscoelasticity required to withstand mechanical forces exerted on the nucleus. At another extreme, it is possible that heterochromatin that contains a non-phase-separated state of the HP1-chromatin complex is more responsive to developmental cues. We look forward to future studies that will shed light on additional biophysical strategies that are used in the biological regulation of heterochromatin function.

ACKNOWLEDGMENTS

We thank members of the Narlikar laboratory for the many stimulating discussions over the past decade that have helped in formulating this review. This work was supported by the National Institutes of Health (NIH) grant R35 GM127020 to G.J.N.

REFERENCES

Alberti S, Gladfelter A, Mittag T. 2019. Considerations and challenges in studying liquid–liquid phase separation and biomolecular condensates. *Cell* **176**: 419–434. doi:10.1016/j.cell.2018.12.035

Allshire RC, Madhani HD. 2018. Ten principles of heterochromatin formation and function. *Nat Rev Mol Cell Biol* **19**: 229–244. doi:10.1038/nrm.2017.119

Aygün O, Mehta S, Grewal SI. 2013. HDAC-mediated suppression of histone turnover promotes epigenetic stability of heterochromatin. *Nat Struct Mol Biol* **20**: 547–554. doi:10.1038/nsmb.2565

Azzaz AM, Vitalini MW, Thomas AS, Price JP, Blacketer MJ, Cryderman DE, Zirbel LN, Woodcock CL, Elcock AH, Wallrath LL, et al. 2014. Human heterochromatin protein 1α promotes nucleosome associations that drive chromatin condensation. *J Biol Chem* **289**: 6850–6861. doi:10.1074/jbc.M113.512137

Bannister AJ, Zegerman P, Partridge JF, Miska EA, Thomas JO, Allshire RC, Kouzarides T. 2001. Selective recognition of methylated lysine 9 on histone H3 by the HP1 chromo domain. *Nature* **410**: 120–124. doi:10.1038/35065138

Brasher SV, Smith BO, Fogh RH, Nietlispach D, Thiru A, Nielsen PR, Broadhurst RW, Ball LJ, Murzina NV, Laue ED. 2000. The structure of mouse HP1 suggests a unique mode of single peptide recognition by the shadow chromo domain dimer. *EMBO J* **19**: 1587–1597. doi:10.1093/emboj/19.7.1587

Canzio D, Chang EY, Shankar S, Kuchenbecker KM, Simon MD, Madhani HD, Narlikar GJ, Al-Sady B. 2011. Chromodomain-mediated oligomerization of HP1 suggests a nucleosome-bridging mechanism for heterochromatin assembly. *Mol Cell* **41**: 67–81. doi:10.1016/j.molcel.2010.12.016

Canzio D, Liao M, Naber N, Pate E, Larson A, Wu S, Marina DB, Garcia JF, Madhani HD, Cooke R, et al. 2013. A conformational switch in HP1 releases auto-inhibition to drive heterochromatin assembly. *Nature* **496**: 377–381. doi:10.1038/nature12032

Canzio D, Larson A, Narlikar GJ. 2014. Mechanisms of functional promiscuity by HP1 proteins. *Trends Cell Biol* **24**: 377–386. doi:10.1016/j.tcb.2014.01.002

Cowieson NP, Partridge JF, Allshire RC, McLaughlin PJ. 2000. Dimerisation of a chromo shadow domain and distinctions from the chromodomain as revealed by structural analysis. *Curr Biol* **10**: 517–525. doi:10.1016/S0960-9822(00)00467-X

Dawson MA, Bannister AJ, Gottgens B, Foster SD, Bartke T, Green AR, Kouzarides T. 2009. JAK2 phosphorylates histone H3Y41 and excludes HP1α from chromatin. *Nature* **461**: 819–822. doi:10.1038/nature08448

Eissenberg JC, Elgin SC. 2014. HP1α: a structural chromosomal protein regulating transcription. *Trends Genet* **30**: 103–110. doi:10.1016/j.tig.2014.01.002

Eissenberg JC, James TC, Foster-Hartnett DM, Hartnett T, Ngan V, Elgin SC. 1990. Mutation in a heterochromatin-specific chromosomal protein is associated with suppression of position-effect variegation in *Drosophila melanogaster*. *Proc Natl Acad Sci* **87**: 9923–9927. doi:10.1073/pnas.87.24.9923

Elgin SCR, Reuter G. 2013. Position-effect variegation, heterochromatin formation, and gene silencing in *Drosophila*. *Cold Spring Harb Perspect Biol* **5**: a017780. doi:10.1101/cshperspect.a017780

Fan HY, Narlikar GJ, Kingston RE. 2004. Noncovalent modification of chromatin: different remodeled products with different ATPase domains. *Cold Spring Harb Symp Quant Biol* **69**: 183–192. doi:10.1101/sqb.2004.69.183

Grewal SI, Jia S. 2007. Heterochromatin revisited. *Nat Rev Genet* **8**: 35–46. doi:10.1038/nrg2008

Haldar S, Saini A, Nanda JS, Saini S, Singh J. 2011. Role of Swi6/HP1 self-association-mediated recruitment of Clr4/Suv39 in establishment and maintenance of heterochromatin in fission yeast. *J Biol Chem* **286**: 9308–9320. doi:10.1074/jbc.M110.143198

Heitz E. 1928. Das Heterochromatin der Moose. *Jahrb Wiss Bot* **69**: 762–818.

Hiragami-Hamada K, Shinmyozu K, Hamada D, Tatsu Y, Uegaki K, Fujiwara S, Nakayama J. 2011. N-terminal phosphorylation of HP1α promotes its chromatin binding. *Mol Cell Biol* **31**: 1186–1200. doi:10.1128/MCB.01012-10

Hiragami-Hamada K, Soeroes S, Nikolov M, Wilkins B, Kreuz S, Chen C, De La Rosa-Velázquez IA, Zenn HM, Kost N, Pohl W, et al. 2016. Dynamic and flexible H3K9me3 bridging via HP1β dimerization establishes a plastic state of condensed chromatin. *Nat Commun* **7**: 11310. doi:10.1038/ncomms11310

Isaac RS, Sanulli S, Tibble R, Hornsby M, Ravalin M, Craik CS, Gross JD, Narlikar GJ. 2017. Biochemical basis for distinct roles of the heterochromatin proteins Swi6 and Chp2. *J Mol Biol* **429**: 3666–3677. doi:10.1016/j.jmb.2017.09.012

Jacobs SA, Khorasanizadeh S. 2002. Structure of HP1 chromodomain bound to a lysine 9-methylated histone H3 tail. *Science* **295**: 2080–2083. doi:10.1126/science.1069473

Jacobs SA, Taverna SD, Zhang Y, Briggs SD, Li J, Eissenberg JC, Allis CD, Khorasanizadeh S. 2001. Specificity of the HP1 chromo domain for the methylated N-terminus of histone H3. *EMBO J* **20**: 5232–5241. doi:10.1093/emboj/20.18.5232

James TC, Elgin SC. 1986. Identification of a nonhistone chromosomal protein associated with heterochromatin in *Drosophila melanogaster* and its gene. *Mol Cell Biol* **6**: 3862–3872. doi:10.1128/MCB.6.11.3862

James TC, Eissenberg JC, Craig C, Dietrich V, Hobson A, Elgin SC. 1989. Distribution patterns of HP1, a heterochromatin-associated nonhistone chromosomal protein of *Drosophila*. *Eur J Cell Biol* **50**: 170–180.

Janssen A, Colmenares SU, Karpen GH. 2018. Heterochromatin: guardian of the genome. *Annu Rev Cell Dev Biol* **34**: 265–288. doi:10.1146/annurev-cellbio-100617-062653

Jost KL, Bertulat B, Cardoso MC. 2012. Heterochromatin and gene positioning: inside, outside, any side? *Chromosoma* **121:** 555–563. doi:10.1007/s00412-012-0389-2

Kang J, Chaudhary J, Dong H, Kim S, Brautigam CA, Yu H. 2011. Mitotic centromeric targeting of HP1 and its binding to Sgo1 are dispensable for sister-chromatid cohesion in human cells. *Mol Biol Cell* **22:** 1181–1190. doi:10.1091/mbc.e11-01-0009

Kato H, van Ingen H, Zhou B-R, Feng H, Bustin M, Kay LE, Bai Y. 2011. Architecture of the high mobility group nucleosomal protein 2-nucleosome complex as revealed by methyl-based NMR. *Proc Natl Acad Sci* **108:** 12283–12288. doi:10.1073/pnas.1105848108

Kilic S, Felekyan S, Doroshenko O, Boichenko I, Dimura M, Vardanyan H, Bryan LC, Arya G, Seidel CAM, Fierz B. 2018. Single-molecule FRET reveals multiscale chromatin dynamics modulated by HP1α. *Nat Comm* **9:** 235. doi:10.1038/s41467-017-02619-5

Lachner M, O'Carroll D, Rea S, Mechtler K, Jenuwein T. 2001. Methylation of histone H3 lysine 9 creates a binding site for HP1 proteins. *Nature* **410:** 116–120. doi:10.1038/35065132

Larson AG, Elnatan D, Keenen MM, Trnka MJ, Johnston JB, Burlingame AL, Agard DA, Redding S, Narlikar GJ. 2017. Liquid droplet formation by HP1α suggests a role for phase separation in heterochromatin. *Nature* **547:** 236–240. doi:10.1038/nature22822

Lavigne M, Eskeland R, Azebi S, Saint-Andre V, Jang SM, Batsche E, Fan HY, Kingston RE, Imhof A, Muchardt C. 2009. Interaction of HP1 and Brg1/Brm with the globular domain of histone H3 is required for HP1-mediated repression. *PLoS Genet* **5:** e1000769. doi:10.1371/journal.pgen.1000769

Lechner MS, Schultz DC, Negorev D, Maul GG, Rauscher FJ III. 2005. The mammalian heterochromatin protein 1 binds diverse nuclear proteins through a common motif that targets the chromoshadow domain. *Biochem Biophys Res Commun* **331:** 929–937. doi:10.1016/j.bbrc.2005.04.016

Luger K, Mäder AW, Richmond RK, Sargent DF, Richmond TJ. 1997. Crystal structure of the nucleosome core particle at 2.8 Å resolution. *Nature* **389:** 251–260. doi:10.1038/38444

Machida S, Takizawa Y, Ishimaru M, Sugita Y, Sekine S, Nakayama JI, Wolf M, Kurumizaka H. 2018. Structural basis of heterochromatin formation by human HP1. *Mol Cell* **69:** 385–397.e388. doi:10.1016/j.molcel.2017.12.011

Martienssen R, Moazed D. 2015. RNAi and heterochromatin assembly. *Cold Spring Harb Perspect Biol* **7:** a019323. doi:10.1101/cshperspect.a019323

Meehan RR, Kao CF, Pennings S. 2003. HP1 binding to native chromatin in vitro is determined by the hinge region and not by the chromodomain. *EMBO J* **22:** 3164–3174. doi:10.1093/emboj/cdg306

Mendez DL, Kim D, Chruszcz M, Stephens GE, Minor W, Khorasanizadeh S, Elgin SC. 2011. The HP1a disordered C terminus and chromo shadow domain cooperate to select target peptide partners. *Chembiochem* **12:** 1084–1096. doi:10.1002/cbic.201000598

Mendez DL, Mandt RE, Elgin SC. 2013. Heterochromatin Protein 1a (HP1a) partner specificity is determined by critical amino acids in the chromo shadow domain and C-terminal extension. *J Biol Chem* **288:** 22315–22323. doi:10.1074/jbc.M113.468413

Mishima Y, Watanabe M, Kawakami T, Jayasinghe CD, Otani J, Kikugawa Y, Shirakawa M, Kimura H, Nishimura O, Aimoto S, et al. 2013. Hinge and chromoshadow of HP1α participate in recognition of K9 methylated histone H3 in nucleosomes. *J Mol Biol* **425:** 54–70. doi:10.1016/j.jmb.2012.10.018

Muchardt C, Guilleme M, Seeler JS, Trouche D, Dejean A, Yaniv M. 2002. Coordinated methyl and RNA binding is required for heterochromatin localization of mammalian HP1α. *EMBO Rep* **3:** 975–981. doi:10.1093/embo-reports/kvf194

Nicetto D, Zaret KS. 2019. Role of H3K9me3 heterochromatin in cell identity establishment and maintenance. *Curr Opin Genet Dev* **55:** 1–10. doi:10.1016/j.gde.2019.04.013

Nielsen PR, Nietlispach D, Mott HR, Callaghan J, Bannister A, Kouzarides T, Murzin AG, Murzina NV, Laue ED. 2002. Structure of the HP1 chromodomain bound to histone H3 methylated at lysine 9. *Nature* **416:** 103–107. doi:10.1038/nature722

Nishibuchi G, Machida S, Osakabe A, Murakoshi H, Hiragami-Hamada K, Nakagawa R, Fischle W, Nishimura Y, Kurumizaka H, Tagami H, et al. 2014. N-terminal phosphorylation of HP1α increases its nucleosome-binding specificity. *Nucleic Acids Res* **42:** 12498–12511. doi:10.1093/nar/gku995

Rea S, Eisenhaber F, O'Carroll D, Strahl BD, Sun Z-W, Schmid M, Opravil S, Mechtler K, Ponting CP, Allis CD, et al. 2000. Regulation of chromatin structure by site-specific histone H3 methyltransferases. *Nature* **406:** 593–599. doi:10.1038/35020506

Religa TL, Sprangers R, Kay LE. 2010. Dynamic regulation of archaeal proteasome gate opening as studied by TROSY NMR. *Science* **328:** 98–102. doi:10.1126/science.1184991

Reyes-Turcu FE, Grewal SI. 2012. Different means, same end-heterochromatin formation by RNAi and RNAi-independent RNA processing factors in fission yeast. *Curr Opin Genet Dev* **22:** 156–163. doi:10.1016/j.gde.2011.12.004

Richart AN, Brunner CI, Stott K, Murzina NV, Thomas JO. 2012. Characterization of chromoshadow domain-mediated binding of heterochromatin protein 1α (HP1α) to histone H3. *J Biol Chem* **287:** 18730–18737. doi:10.1074/jbc.M111.337204

Sadaie M, Kawaguchi R, Ohtani Y, Arisaka F, Tanaka K, Shirahige K, Nakayama J. 2008. Balance between distinct HP1 family proteins controls heterochromatin assembly in fission yeast. *Mol Cell Biol* **28:** 6973–6988. doi:10.1128/MCB.00791-08

Sanulli S, Trnka MJ, Dharmarajan V, Tibble RW, Pascal BD, Burlingame AL, Griffin PR, Gross JD, Narlikar GJ. 2019. HP1 reshapes nucleosome core to promote phase separation of heterochromatin. *Nature* **575:** 390–394. doi:10.1038/s41586-019-1669-2

Schuettengruber B, Bourbon HM, Di Croce L, Cavalli G. 2017. Genome regulation by polycomb and trithorax: 70 years and counting. *Cell* **171:** 34–57. doi:10.1016/j.cell.2017.08.002

Simon JA, Kingston RE. 2009. Mechanisms of polycomb gene silencing: knowns and unknowns. *Nat Rev Mol Cell Biol* **10:** 697–708. doi:10.1038/nrm2763

Sinha KK, Gross JD, Narlikar GJ. 2017. Distortion of histone octamer core promotes nucleosome mobilization by a chromatin remodeler. *Science* **355:** eaaa3761. doi:10.1126/science.aaa3761

Smothers JF, Henikoff S. 2000. The HP1 chromo shadow domain binds a consensus peptide pentamer. *Curr Biol* **10:** 27–30. doi:10.1016/S0960-9822(99)00260-2

Stephens AD, Banigan EJ, Marko JF. 2019. Chromatin's physical properties shape the nucleus and its functions. *Curr Opin Cell Biol* **58:** 76–84. doi:10.1016/j.ceb.2019.02.006

Strom AR, Emelyanov AV, Mir M, Fyodorov DV, Darzacq X, Karpen GH. 2017. Phase separation drives heterochromatin domain formation. *Nature* **547:** 241–245. doi:10.1038/nature22989

Sugimoto K, Yamada T, Muro Y, Himeno M. 1996. Human homolog of *Drosophila* heterochromatin-associated protein 1 (HP1) is a DNA-binding protein which possesses a DNA-binding motif with weak similarity to that of human centromere protein C (CENP-C). *J Biochem* **120:** 153–159. doi:10.1093/oxfordjournals.jbchem.a021378

Thiru A, Nietlispach D, Mott HR, Okuwaki M, Lyon D, Nielsen PR, Hirshberg M, Verreault A, Murzina NV, Laue ED. 2004. Structural basis of HP1/PXVXL motif peptide interactions and HP1 localisation to heterochromatin. *EMBO J* **23:** 489–499. doi:10.1038/sj.emboj.7600088

Wang G, Ma A, Chow CM, Horsley D, Brown NR, Cowell IG, Singh PB. 2000. Conservation of heterochromatin protein 1 function. *Mol Cell Biol* **20:** 6970–6983. doi:10.1128/MCB.20.18.6970-6983.2000

Wang L, Gao Y, Zheng X, Liu C, Dong S, Li R, Zhang G, Wei Y, Qu H, Li Y, et al. 2019. Histone modifications regulate chromatin compartmentalization by contributing to a phase separation mechanism. *Mol Cell* **76:** 646–659.e6. doi:10.1016/j.molcel.2019.08.019

Yamada T, Fukuda R, Himeno M, Sugimoto K. 1999. Functional domain structure of human heterochromatin protein HP1 (Hsα): involvement of internal DNA-binding and C-terminal self-association domains in the formation of discrete dots in interphase nuclei. *J Biochem* **125:** 832–837. doi:10.1093/oxfordjournals.jbchem.a022356

Yu J-R, Lee CH, Oksuz O, Stafford JM, Reinberg D. 2019. PRC2 is high maintenance. *Genes Dev* **33:** 903–935. doi:10.1101/gad.325050.119

Architectural RNAs for Membraneless Nuclear Body Formation

Tomohiro Yamazaki,[1] Shinichi Nakagawa,[2] and Tetsuro Hirose[1]

[1]Institute for Genetic Medicine, Hokkaido University, Sapporo, 060-0815 Japan
[2]Faculty of Pharmaceutical Sciences, Hokkaido University, Sapporo, 060-0812 Japan
Correspondence: hirose@igm.hokudai.ac.jp

Long noncoding RNAs (lncRNAs) are fundamental regulators of various cellular processes. A subset of lncRNAs, termed architectural RNAs (arcRNAs), function in the formation and maintenance of phase-separated membraneless organelles in multiple eukaryotic species. These membraneless organelles represent an important type of compartmentalization in the crowded cellular environment and have several distinct features. The NEAT1_2 lncRNA is a well-characterized arcRNA that functions as an essential scaffold of paraspeckle nuclear bodies. Here, we describe the biogenesis of paraspeckles on arcRNAs through phase separation, focusing on the specific functions of multiple NEAT1_2 RNA domains and their partner RNA-binding proteins. Finally, we present an updated model of paraspeckle formation and discuss future perspectives of research into arcRNA-instructed architectures of phase-separated nuclear bodies.

ARCHITECTURAL RNAs ARE ESSENTIAL SCAFFOLDS OF PHASE-SEPARATED MEMBRANELESS BODIES

Recently, membraneless organelles and biomolecular condensates, which form through phase separation, have attracted increasing interest (Banani et al. 2017; Alberti et al. 2019). These subcellular compartments play key roles in a number of processes, including the sequestration of biomolecules and the formation of organization hubs to coordinate biochemical reactions and macromolecular assembly (Shin and Brangwynne 2017). A subset of phase-separated cellular bodies are constructed by specific long noncoding RNAs (lncRNAs) (Yamazaki and Hirose 2015; Chujo et al. 2016; Chujo and Hirose 2017), which play diverse regulatory roles in various biological processes, including gene expression (Guttman and Rinn 2012; Geisler and Coller 2013; Quinn and Chang 2016; Schmitt and Chang 2017). Multiple lncRNAs can act as essential architectural scaffolds of nuclear bodies in a variety of eukaryotes, including humans, mice, *Drosophila*, and yeast; hence, we termed them architectural RNAs (arcRNAs) (Yamazaki and Hirose 2015; Chujo et al. 2016; Chujo and Hirose 2017).

A number of arcRNAs are involved in the formation of various nuclear bodies. For example, NEAT1_2, the long isoform of the nuclear paraspeckle assembly transcript 1 (NEAT1) lncRNA, is an essential scaffold of paraspeckle nuclear bodies (Chen and Carmichael 2009; Clemson et al. 2009; Sasaki et al. 2009; Sunwoo et al. 2009); highly repetitive satellite III (HSATIII) lncRNAs, which are induced by conditions such as thermal stress, promote the formation of nuclear stress bodies (nSBs) in primates (Jolly et al. 1999; Aly et al. 2019; Ninomiya et al. 2019); HSATII RNAs are core components of cancer-associated satellite transcript bodies (Hall et al. 2017); intergenic spacer lncRNAs induce the formation of static but reversible amyloid-like solid structures termed amyloid bodies (Audas et al. 2016); histone pre-mRNAs are essential for histone locus body formation (Shevtsov and Dundr 2011); tumor-associated NBL2 transcript (TNBL) aggregates are induced by TNBL RNAs (Dumbović et al. 2018); the perinucleolar compartment is induced by the PNCTR RNA (Yap et al. 2018); Hsrω RNAs are essential cores of ω-speckles in *Drosophila* (Mallik and Lakhotia 2009); and the Mei2 dot is a nuclear body induced by meiRNA in fission yeast (Yamashita et al. 1998). The Sam68 nuclear body and DBC1 bodies are also RNA-dependent bodies, but their arcRNAs have not yet been identified (Mannen et al. 2016). In addition, disease-associated repeat RNAs can also be regarded as a type of arcRNA (Zhang and Ashizawa 2017).

The existence of arcRNAs in a large number of organisms suggests that they are suitable for nucleating subnuclear bodies through phase separation. RNA-binding proteins (RBPs) are enriched with prion-like domains (PLDs), low-complexity domains (LCDs), and intrinsically disordered regions (IDRs), which form weak multivalent molecular interactions that promote biological phase separation (Hennig et al. 2015; Yamazaki and Hirose 2015). Thus, arcRNAs can induce phase separation by increasing the local concentration of these RBPs. More than 1000 human RBPs have been reported to date; these proteins possess a wide variety of biological functions, such as transcriptional regulation and epigenetic chromatin regulation (Baltz et al. 2012; Castello et al. 2012, 2016). RNAs can assemble a specific set of RBPs to combine and integrate their functions, even when the RBPs

© 2019 Yamazaki et al. This article is distributed under the terms of the Creative Commons Attribution-NonCommercial License, which permits reuse and redistribution, except for commercial purposes, provided that the original author and source are credited.

Published by Cold Spring Harbor Laboratory Press; doi:10.1101/sqb.2019.84.039404

would not directly interact (Chujo et al. 2016; Engreitz et al. 2016). This property allows arcRNA-induced bodies to perform multiple tasks in the cell, including acting as molecular sponges to sequestrate proteins and/or RNAs, forming reaction crucible to enhance biochemical reactions, and organizational hubs of chromatin (Chujo et al. 2016; Chujo and Hirose 2017; Hirose et al. 2019). In addition, noncoding RNAs can act as functional molecules themselves, with roles that depend on their specific sequences and do not require protein coding capacity. Architectural noncoding RNAs are able to alter their specific sequences to adapt to circumstantial changes. Furthermore, multiple copies of lncRNAs are synthesized from their gene loci; this process is important for phase separation as it promotes multivalent interactions between high local concentrations of associated IDR proteins. Therefore, lncRNAs act as markers of specific genomic loci. This property of lncRNAs is beneficial to the amplification of signals from specific genomic loci, such as super-enhancers, and phase separation would enhance this effect (Engreitz et al. 2016). Transcription of arcRNAs is essential for the formation of phase-separated nuclear bodies, and inhibition of their transcription leads to rapid disintegration of the bodies. As most arcRNAs are stress-inducible or specifically expressed in certain developmental and pathological conditions, phase-separated nuclear bodies are transiently formed upon the transcription of arcRNAs. Therefore, it is conceivable that arcRNAs spatiotemporally control cellular signals and act as integral hubs for gene regulation.

In the sections below, we focus on NEAT1_2 as a model arcRNA. We describe how NEAT1_2 arcRNAs construct paraspeckle nuclear bodies and determine their biophysical properties, with a special focus on the functional RNA domains of NEAT1 and their partner proteins.

THE CHARACTERISTICS OF PHASE-SEPARATED PARASPECKLE NUCLEAR BODIES CONSTRUCTED BY THE NEAT1_2 arcRNA

Paraspeckles were originally identified as distinct nuclear bodies called interchromatin granule-associated zones (IGAZs) (Visa et al. 1993), which were located adjacent to interchromatin granule clusters (also known as nuclear speckles) (Fig. 1A). In 2002, paraspeckles were identified

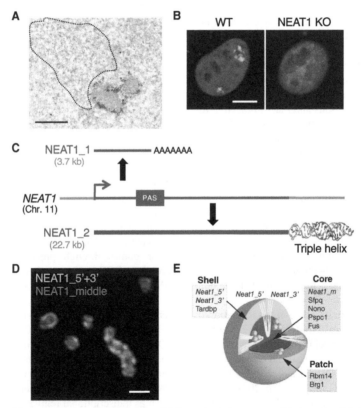

Figure 1. The NEAT1_2 architectural RNA (arcRNA) is essential for maintaining paraspeckle integrity. (*A*) An electron micrograph of a paraspeckle (electron dense structures were detected using gold particles [probes against NEAT1 5′ regions]) and nuclear speckles (black dashed circle) in human HAP1 cells treated with MG132 (5 μM) for 17 h. Scale bar, 500 nm. The image was kindly provided by Dr. Gerard Pierron (Institut Gustave Roussy, CNRS). (*B*) Visualization of paraspeckles in wild-type (WT) and NEAT1 knockout human cells by NONO immunofluorescence (green foci). The diffuse green signals seen in the knockout cells indicate disruption of proper paraspeckle formation. DNA was stained with DAPI (blue). Scale bar, 5 μm. (*C*) Schematic illustration of the NEAT1_1 and NEAT1_2 isoforms of the *NEAT1* gene. (*D*) A super-resolution microscopy image (structured illumination microscopy) of paraspeckles stained with a NEAT1 FISH probe (green, the 5′ and 3′ regions; magenta, the middle region). Scale bar, 500 nm. (*E*) A schematic illustration of NEAT1_2 configuration within a paraspeckle and the core-shell-patch localizations of paraspeckle-localizing proteins (PSPs).

as nuclear bodies enriched in PSPC1, a member of the *Drosophila* behavior human splicing (DBHS) family of proteins that includes SFPQ and NONO (Fig. 1B, left; Fox et al. 2002). Subsequently, paraspeckles were shown to be identical to IGAZs and were found to be RNase-sensitive structures, suggesting a requirement of RNAs for their maintenance (Fox et al. 2005). In 2009, four groups independently reported that the NEAT1 lncRNA, also called Menε/β, is an essential architectural scaffold of paraspeckles, as shown by a lack of discrete paraspeckles in NEAT1 knockdown cells (Fig. 1B, right; Chen and Carmichael 2009; Clemson et al. 2009; Sasaki et al. 2009; Sunwoo et al. 2009). NEAT1 has two isoforms: the longer NEAT1_2/Menβ isoform (22.7 kb in human) is essential for paraspeckle formation, whereas the short NEAT1_1/Menε isoform (3.7 kb in human) is dispensable (Fig. 1C; Naganuma et al. 2012; Yamazaki et al. 2018a,b).

Several subsequent studies identified more than 60 paraspeckle-localizing proteins (PSPs) in humans (Naganuma et al. 2012; Fong et al. 2013; Kawaguchi et al. 2015; Yamazaki and Hirose 2015; Mannen et al. 2016). Most PSPs are RBPs that contain PLDs, LCDs, or IDRs that show significant overlaps with a number of genes associated with amyotrophic lateral sclerosis (Yamazaki and Hirose 2015). Electron microscopy studies revealed that paraspeckles are roundish or oblong structures with constant short axes (∼360 nm in human HeLa cells) (Fig. 1A) and become elongated when NEAT1_2 is up-regulated (Souquere et al. 2010; Hirose et al. 2014b). Notably, the NEAT1_2 lncRNA is spatially organized within paraspeckles; the 5′ and 3′ regions of NEAT1_2 are located in the shells of paraspeckles, whereas the middle region is located in the core region, suggesting that NEAT1_2 forms a looped structure within paraspeckles (Fig. 1A, D; Souquere et al. 2010; West et al. 2016). In addition, PSPs are also spatially organized within paraspeckles, with distinct localizations at the core, shell, and patch regions (Fig. 1E; Kawaguchi et al. 2015; West et al. 2016).

Multiple lines of evidence support the concept that paraspeckles are phase-separated nuclear bodies. First, fluorescence recovery after photobleaching (FRAP) experiments revealed that PSPs are dynamically exchanged between paraspeckles and the surrounding nucleoplasm (Mao et al. 2011; Audas et al. 2016; Wang et al. 2018). Second, paraspeckles are sensitive to 1,6-hexanediol, which mainly disrupts hydrophobic interactions and contributes to the formation of phase-separated subcellular bodies (Yamazaki et al. 2018b). Third, the abilities of the PSPs FUS and RBM14, which are essential for paraspeckle formation, to form hydrogels in vitro are essential for paraspeckle assembly (Hennig et al. 2015; Fox et al. 2018). Fourth, paraspeckles undergo a fusion and fission process termed "kiss-and-run fusion" (Mao et al. 2011; Yang et al. 2019). A recent review article reported that paraspeckles are induced by polymer–polymer phase separation rather than the typical liquid–liquid phase separation (Peng and Weber 2019). Taken together, these findings suggest that paraspeckles are massive phase-separated subnuclear structures with a distinct shape and ordered interior configuration.

BIOGENESIS OF THE NEAT1_2 arcRNA AND PARASPECKLES

Synthesis of the NEAT1 isoforms is regulated by alternative 3′-end processing governed by HNRNPK and TDP-43 (Naganuma et al. 2012; Modic et al. 2019). The 3′ end of NEAT1_2 contains an unusual triple helix structure (3′TH) that is essential for stabilization of the arcRNA (Fig. 1C; Brown et al. 2012; Wilusz et al. 2012; Yamazaki et al. 2018b). Seven PSPs and the SWI/SNF complex are essential for paraspeckle formation (Fig. 2; Naganuma et al. 2012; Kawaguchi et al. 2015; West et al. 2016). Specifically, SFPQ, NONO, and RBM14 are essential for NEAT1_2 stability (Naganuma et al. 2012), and HNRNPK controls isoform switching to the NEAT1_2 isoform by inhibiting polyadenylation of the NEAT1_1

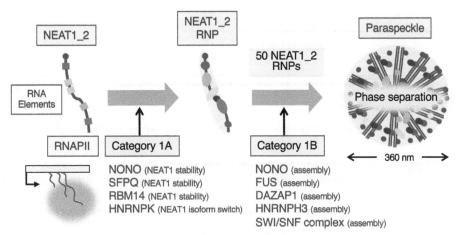

Figure 2. Two distinct processes of paraspeckle formation. NEAT1_2 ribonucleoproteins (RNPs) are transcribed by RNA polymerase II (RNAPII) and then approximately 50 NEAT1_2 RNPs assemble into paraspeckles through phase separation. PSPs that are essential for paraspeckle integrity can be divided into two categories: category 1A includes PSPs required for NEAT1_2 expression and category 1B includes PSPs required for paraspeckle assembly. Putative RNA elements are shown on the NEAT1_2 structure (*left*).

isoform (Fig. 2; Naganuma et al. 2012). NONO, FUS, DAZAP1, HNRNPH3, and the SWI/SNF complex are essential for paraspeckle assembly (Naganuma et al. 2012; Kawaguchi et al. 2015; Yamazaki et al. 2018b), and RNA polymerase II transcription is also essential for paraspeckle formation (Fig. 2). Upon transcriptional inhibition, paraspeckles are rapidly disassembled alongside NEAT1_2 degradation and relocation of PSPs to nucleolar cap structures (Fox et al. 2005; Chujo et al. 2017). Paraspeckles are formed in close proximity to the *NEAT1* gene loci, likely because NEAT1_2 lncRNAs are highly concentrated at these regions during transcription, and thus the environment favors phase separation by facilitating molecular interactions among RNAs and proteins. This evidence suggests cotranscriptional formation of paraspeckles via phase separation (Mao et al. 2011; Yamazaki et al. 2018b). During this process, approximately 50 NEAT1_2 lncRNAs are incorporated into a single spherical paraspeckle (Fig. 2; Chujo et al. 2017). The paraspeckle is a highly dynamic structure; FRAP studies revealed that 60%–70% of PSPs within paraspeckles are recovered within several minutes (Mao et al. 2011; Audas et al. 2016; Wang et al. 2018). In addition, a recent study showed that elongated, oblong paraspeckles show slower mobility than spherical paraspeckles (Wang et al. 2018). These data indicate the existence of slow or immobile fractions of paraspeckles and that paraspeckle dynamics are related to their shape.

FUNCTIONAL RNA DOMAINS OF THE NEAT1_2 arcRNA

Although the biological importance of various lncRNAs has been shown, the specific sequences and structural elements required for their biogenesis and functions remain poorly understood (Hirose et al. 2014a). Nonetheless, some lncRNAs have been studied intensively. A previous study examining the XIST lncRNA revealed the existence of functional modular repeat domains involved in X-chromosome inactivation (Wutz et al. 2002). Subsequent studies identified the subdomains of the XIST lncRNA and their partner proteins required for specific functions (Chu et al. 2015; Monfort and Wutz 2017; Colognori et al. 2019). We recently identified multiple modular domains of NEAT1 involved in its biogenesis and the formation of paraspeckles (Yamazaki et al. 2018b). Similar to that of many other lncRNAs, the cross-species homology of mammalian NEAT1_2 sequences is markedly lower than those of protein-coding mRNAs. In general, mammalian NEAT1_2 sequences only display homology at the 5′ region (~1.6 kb), short stretches in the internal regions, and the 3′TH (Fig. 3A, B). Nevertheless, mouse Neat1_2 can form paraspeckles in human cells, suggesting that the functions of NEAT1_2, at least in paraspeckle formation, are conserved between mouse and human (Fig. 3C; Naganuma et al. 2012).

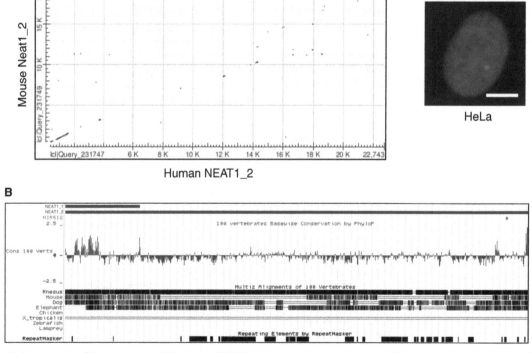

Figure 3. Conservation of the sequences and functions of NEAT1_2 between human and mouse. (*A*) A dot blot showing the homologous regions of the human and mouse NEAT1_2 sequences. The BLAST search parameter was "somewhat similar." (*B*) A screenshot of the UCSC genome browser record for the human *NEAT1* gene locus with modifications. Sequence conservation (PhastCons) and positions of repetitive elements identified by a RepeatMasker algorithm are shown. (*C*) Transient expression of mouse Neat1_2 (mNeat1_2) induces paraspeckles in human HeLa cells (magenta signals). Scale bar, 5 μm.

To dissect the roles of specific sequences within the human NEAT1_2 lncRNA, we performed extensive deletion analyses using CRISPR–Cas9 and the human haploid HAP1 cell line (Yamazaki et al. 2018b). These analyses revealed that multiple functional RNA domains of NEAT1_2 dictate its own biogenesis and the coupled formation of paraspeckles. The 5′ end (first 1 kb) and 3′TH of NEAT1_2 (termed the A domains) are necessary for stability of the lncRNA (Yamazaki et al. 2018b). The regions upstream (2.1–2.8 kb) and downstream (4–5 kb) of the polyadenylation site of NEAT1_1 (termed the B domains) promote the expression of NEAT1_2 by repressing NEAT1_1 polyadenylation, indicating a role in isoform switching from NEAT1_1 to NEAT1_2 (Yamazaki et al. 2018b). The middle domain of NEAT1_2 (termed the C domain) is necessary and sufficient for paraspeckle assembly. Notably, our CRISPR dissection analysis also revealed that the C domain contains several subdomains (C1: 9.8–12 kb; C2: 12–13 kb; C3: 15.4–16.6 kb) that have redundant functions in paraspeckle assembly (Yamazaki et al. 2018b). Such redundancy is also found among the C domain and the unidentified outside domain, suggesting that the presence of multiple redundant domains confers a robust competency of NEAT1_2 to build paraspeckles (Fig. 4).

In our recent study, artificial tethering of NONO, SFPQ, or FUS, but not RBM14, rescued paraspeckle assembly in cells expressing a NEAT1 mutant lacking the C1 and C2 subdomains (Yamazaki et al. 2018b). Moreover, the rescue activity of NONO required dimerization with DBHS family proteins, including SFPQ, PSPC1, and NONO itself, which is also required for the oligomerization of these proteins along the RNA molecule. CLIP-seq data showed that the C domain of NEAT1_2 contains multiple binding sites for NONO/SFPQ, which are likely to be essential for paraspeckle formation (Naganuma et al. 2012; Passon et al. 2012; Lee et al. 2015; Huang et al. 2018; Yamazaki et al. 2018b). The in vitro–transcribed RNA derived from the C2 domain preferentially binds to NONO/SFPQ and induces the formation of higher-order structures that are sensitive to 1,6-hexanediol and depend on NONO/SFPQ (Yamazaki et al. 2018b). In addition, the NONO/SFPQ oligomer seems to recruit additional essential PSPs such as FUS and RBM14, which in turn likely induce phase separation (Hennig et al. 2015; Yamazaki et al. 2018b). We recently found that the NEAT1_2 arcRNA is poorly extracted by conventional RNA purification methods using acid guanidinium thiocyanate-phenol-chloroform reagents such as TRIzol (Thermo Fisher Scientific). This unusual feature, which we termed "semiextractability," depends on NONO, FUS, or the NEAT1_2 C domain, suggesting that the tenacious interactions between these proteins and the C domain, which can probably tolerate guanidium thiocyanate and phenol-containing harsh conditions, contribute to arcRNA-induced cellular body formation (Chujo et al. 2017; T Yamazaki, unpubl. observ.). SFPQ forms higher-order structures by cooperative spreading and coating the nucleic acids, and a similar spreading mechanism is also proposed to play a role in splicing regulation by HNRNPA1 (Zhu et al. 2001; Lee et al. 2015). Hence, this spreading mechanism might play an important role in the initial process of paraspeckle formation.

In addition to the A, B, and C domains, NEAT1_2 contains UG repeat stretches that are evolutionally conserved among multiple species. These repeat sequences are thought to be the major binding sites of TDP-43 (Tollervey et al. 2011). When we deleted the UG repeats from NEAT1_2 using CRISPR–Cas9, recruitment of TDP-43 to the paraspeckles was reduced dramatically (Modic et al. 2019). Furthermore, knock-in of a long UG repeat stretches into the UG-deleted NEAT1_2 mutant recovered the recruitment of TDP-43, suggesting that the UG repeat stretches are necessary and sufficient for TDP-43 recruitment to paraspeckles (Fig. 4; Modic et al. 2019).

As described above, NEAT1_2 is looped and spatially organized within paraspeckles (Fig. 4). Our super-resolution microscopy analyses revealed that deletion of the 3′ region of NEAT1_2 (Δ16.6–22.6-kb mutant) alters its configuration (Yamazaki et al. 2018b). In wild-type cells, the 3′ end of NEAT1_2 is located in the shell of the paraspeckle, but the 3′ end of the Δ16.6–22.6-kb mutant localizes to the paraspeckle core. This finding suggests that the configuration of NEAT1_2 is determined by one or more of its domains. By extension, we speculate that other NEAT1_2 domains will also contribute to the organization of the lncRNA within paraspeckles.

NEAT1_1, which can be regarded as one of the functional domains of the *NEAT1* gene, forms "microspeckles," suggesting a role outside of paraspeckles (Li et al. 2017a). A recent study showed that NEAT1_1-specific knockout mice do not show an aberrant phenotype, although a function of NEAT1_1 under specific conditions cannot be excluded (Adriaens et al. 2019; Isobe et al. 2019). Further investigations may reveal specialized roles of NEAT1_1 under certain cellular conditions.

Several recent studies have suggested the presence of additional functional NEAT1 domains. Genome-wide

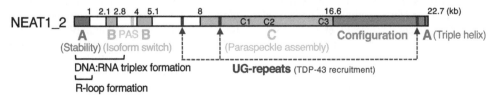

Figure 4. The functional modular domain structure of the human NEAT1_2 arcRNA. The functional domains of NEAT1_2 and their positions (kb, kilobases from the 5′ end of NEAT1_2) are shown. (PAS) Polyadenylation signal.

mapping of R-loop-forming regions revealed that the 5′-terminal region of NEAT1 (~1400 nt) forms a long R-loop (DNA–RNA hybrid), although its biological importance remains unknown (Fig. 4; Dumelie and Jaffrey 2017). Another recent study showed that the 5′-terminal regions of NEAT1, including the GA- and CU-rich regions, form DNA:RNA triplex structures with many human genomic regions (Fig. 4; Senturk Cetin et al. 2019). Consistent with these two reports, the 5′-terminal sequence of NEAT1 is evolutionarily conserved (Fig. 3A), suggesting a sequence-based mechanism for the biological functions of NEAT1, including target gene recognition. In addition, the 3′-terminal region of human NEAT1_2 harbors a pseudo-microRNA (miR-612) that is not processed into mature microRNA (miRNA) but attracts microprocessors to process pri-miRNAs to pre-miRNAs in the nucleus (Jiang et al. 2017). Overall, as described in this section, the multiple modular domains of NEAT1 determine its biogenesis and the composition, biophysical properties, and functions of paraspeckles.

CURRENT MODEL OF PARASPECKLE FORMATION

Based on the studies described above, we propose a model of paraspeckle formation (Fig. 5). In this model, NEAT1_2 lncRNAs are transcribed by RNA polymerase II and PSPs associate with nascent NEAT1_2 transcripts at the *NEAT1* gene loci (Fig. 5, step (1); Mao et al. 2011). PSPs such as NONO and SFPQ, which are essential for paraspeckle assembly, are loaded onto the multiple functionally redundant high-affinity binding sites of NEAT1_2 and then spread and coat NEAT1_2 domains to form base ribonucleoprotein (RNP) complexes through the NOPS and coiled coil domains of NONO and SFPQ (Fig. 5, steps (2) and (3); Lee et al. 2015; Yamazaki et al. 2018b). Subsequently, the NEAT1_2 RNPs further assemble into paraspeckles (50 NEAT1_2 RNPs per paraspeckle sphere), likely through phase separation. Concurrently, FUS and RBM14 are recruited to the RNPs, likely by NONO and SFPQ, and promote paraspeckle assembly via multivalent interactions between their PLDs (Fig. 5, steps (4) and (5); Hennig et al. 2015; West et al. 2016; Chujo et al. 2017; Yamazaki et al. 2018b). During this process, NEAT1_2 RNPs are bundled and folded (Fig. 5, step (5); West et al. 2016).

It is possible that NEAT1_2 lncRNAs interact with each other via direct intermolecular RNA–RNA interactions to contribute to the phase separation process (Fig. 5, step (2); Van Treeck and Parker 2018). This possibility is supported by the fact that weak NEAT1_2 focal signals remain detectable when PSPs are almost completely obliterated by 1,6-hexanediol treatment (Yamazaki et al. 2018b). Several studies have highlighted the importance of RNA–RNA interactions in the formation of cellular bodies, including stress granules, and disease-associated RNA foci caused by aberrant repeat expansions. It is proposed that the sum of the RNA–RNA, RNA–protein, and protein–protein interactions determines whether arcRNA–protein complexes undergo phase separation that eventually leads to cellular body formation (Van Treeck and Parker 2018).

FUTURE PERSPECTIVES

Despite recent advances in our understanding of how the NEAT1_2 domains and their partner proteins dictate paraspeckle formation and functions, many questions still remain. First, although NONO and SFPQ have been identified as partner proteins of the NEAT1_2 C domain, it is important to identify the partner RBPs and specific RNA motifs (e.g., sequences and/or secondary structures) required for proper functioning of the other NEAT1_2 functional domains, especially those involved in inducing phase separation. Second, as NEAT1_2 is involved in various physiological and pathological conditions, it is important to investigate how the molecular functions driven by specific RNA domains/motifs link to physiological functions.

We identified NONO and SFPQ as partner proteins of the NEAT1_2 C domain based on prior knowledge of the protein composition of paraspeckles and the results of experiments using immunofluorescence analyses of PSPs on NEAT1 mutants, MS2 tethering, CLIP-seq, and in vitro RNA pulldown (Yamazaki et al. 2018b). These approaches should also enable the identification of partner proteins of the other functional NEAT1_2 domains. To narrow down the functional domains of NEAT1_2 to precise RNA motifs, a large CLIP-seq data set, including eCLIP data, would be a great resource to differentiate between direct and indirect binding proteins and determine their binding sites (Van Nostrand et al. 2016). Analyses of the secondary structure of NEAT1_2 would also be important (Lu et al. 2016; Lin et al. 2018). For example, it would be useful to identify the specific RNA motifs in the NEAT1_2 C domain required for paraspeckle assembly via phase separation. It would also be interesting to understand how specific NEAT1_2 motifs are conserved between species. Furthermore, it would be helpful to understand how the specific domains and motifs within NEAT1_2 can initiate cellular body formation and determine their properties and functions.

Various molecular and physiological or pathological functions of NEAT1_2 have been reported. One of the molecular functions of paraspeckles is to regulate gene expression by acting as molecular sponges that incorporate specific RBPs and RNAs such as IRAlu-containing RNAs, CTN RNA, AG-rich RNAs, and miRNAs (Chen and Carmichael 2009; Hirose et al. 2014b; Imamura et al. 2014; West et al. 2016; Jiang et al. 2017). In addition, the global association of NEAT1/paraspeckles with chromatin has been shown by multiple methods, including CHART-seq, ChIA-PET, GRID-seq, and RADICL-seq (West et al. 2014; Cai et al. 2016; Li et al. 2017b; Bonetti et al. 2019). Under physiological conditions, NEAT1_2 is required for the establishment of pregnancy, mammary gland development, mitochondrial functions, and efficient regulation of gene expression for pluripotency and

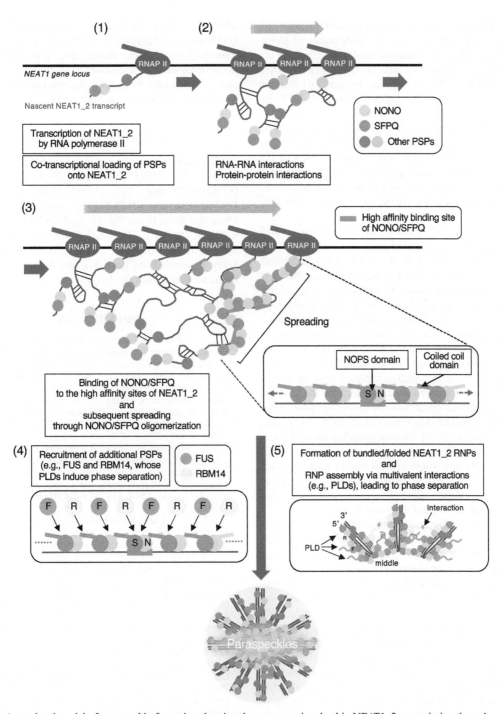

Figure 5. An updated model of paraspeckle formation showing the processes involved in NEAT1_2 transcription through paraspeckle formation via phase separation. See the main text for more details.

cellular differentiation (Nakagawa et al. 2014; Standaert et al. 2014; Wang et al. 2018; Modic et al. 2019). A number of studies have highlighted the importance of NEAT1 as a direct target of the p53 central tumor suppressor protein in cancer (Dong et al. 2018; Nakagawa et al. 2018). NEAT1 is also involved in various other pathological conditions, including neurodegenerative diseases and viral/bacterial infections (Nishimoto et al. 2013; Imamura et al. 2018). Additional research into the relationship between the RNA motifs of NEAT1 and its physiological functions is required to clarify the mechanisms of actions of arcRNAs and the biological significance of nuclear body formation on these molecules.

Several recent studies have shown that arcRNAs are still hidden within genomes. Using our recently developed method that relies on the semiextractable feature of arcRNAs, we performed a genome-wide screen of HeLa cells and identified 45 candidate arcRNAs (Chujo et al. 2017). These candidates include lncRNAs, pre-mRNAs, and repeat-derived RNAs and form nuclear foci that are distinct from known nuclear bodies (Chujo et al. 2017). In addition, another recent study showed that short tandem repeat-containing RNAs (strRNAs) might be a rich source of arcRNAs (Yap et al. 2018). One of the strRNAs is a core of the perinucleolar compartment. As most arcRNAs are induced by various stimuli and/or specific physiological or pathological conditions, searches performed under various conditions will likely expand the repertoire of known arcRNAs.

Because known arcRNAs have partner proteins, several approaches can be used to identify the components of arcRNA-induced cellular bodies. Most PSPs identified to date were found by colocalization screening using a cDNA library and fluorescent fusion proteins (Naganuma et al. 2012; Fong et al. 2013; Mannen et al. 2016). Our recent comprehensive ChIRP-MS (chromatin isolation by RNA purification followed by mass spectrometry) analysis revealed the compositions of nSBs, among which most of the proteins (141) were newly identified (Ninomiya et al. 2019). Proximity labeling methods such as APEX and BioID have also been used to determine the compositions and dynamic compositional changes of phase-separated cellular bodies (Markmiller et al. 2018; Youn et al. 2018; Padron et al. 2019). These approaches could be used to obtain a comprehensive list of the components and their dynamic exchanges within cellular bodies under various cellular conditions. Future studies should also reveal the mechanisms by which cellular body compositions are determined by arcRNAs and their motifs.

It is well known that arcRNAs can induce phase transitions between several material states (e.g., liquid, hydrogel, and solid) of cellular bodies. Exactly how these material properties are determined remains unanswered. In particular, it is unclear which protein components contribute to each state. The binding sites of the proteins on arcRNAs, including their affinities and densities, are still unknown, and the motifs and properties (e.g., amino acid compositions in PLDs, LCDs, or IDRs) of the proteins that determine the material states also require clarification. In vitro reconstitution of arcRNA-induced bodies using purified components or cellular/nuclear extracts should elucidate how they are constructed and how their biophysical properties are determined (Maharana et al. 2018). In addition to paraspeckles, other arcRNA-induced bodies also show various distinct morphologies. For example, nSBs display island structures that are composed of dense cores and shells (Kawaguchi et al. 2015), and amyloid bodies have filamentous structures (Audas et al. 2016). Therefore, it would be interesting to delineate how these morphologies are determined by arcRNAs and their partner proteins.

Identification of both the partner proteins and the precise binding sites as RNP modules should enable artificial control and reprogramming of arcRNAs. Motif-specific functional inhibition (or activation) might be achievable using antisense oligos, small molecules, or CRISPR-based strategies. In future analyses of arcRNAs, theoretical models and simulations, in addition to new methodologies and quantitative analyses, would offer novel frameworks to explain the mechanisms involved in complex biological phase separation.

ACKNOWLEDGMENTS

This work was supported by MEXT KAKENHI grants (to T.Y. [17K15058, 19K06479, 19H05250], S.N. [17H03604], and T.H. [JP26113002, JP16H06279, JP17H03630, and JP17K19335]), and by The Mochida Memorial Foundation for Medical and Pharmaceutical Research (to T.Y.), The Naito Foundation (to T.Y. and S.N.), The Akiyama Life Science Foundation (to T.Y.), Torey Research Fund (to S.N.), and Tokyo Biochemical Research Foundation (to T.H.).

REFERENCES

Adriaens C, Rambow F, Bervoets G, Silla T, Mito M, Chiba T, Hiroshi A, Hirose T, Nakagawa S, Jensen TH, et al. 2019. The lncRNA *NEAT1_1* is seemingly dispensable for normal tissue homeostasis and cancer cell growth. *RNA* **25:** 1681–1695. doi:10.1261/rna.071456.119

Alberti S, Gladfelter A, Mittag T. 2019. Considerations and challenges in studying liquid-liquid phase separation and biomolecular condensates. *Cell* **176:** 419–434. doi:10.1016/j.cell.2018.12.035

Aly MK, Ninomiya K, Adachi S, Natsume T, Hirose T. 2019. Two distinct nuclear stress bodies containing different sets of RNA-binding proteins are formed with HSATIII architectural noncoding RNAs upon thermal stress exposure. *Biochem Biophys Res Commun* **516:** 419–423. doi:10.1016/j.bbrc.2019.06.061

Audas TE, Audas DE, Jacob MD, Ho JJ, Khacho M, Wang M, Perera JK, Gardiner C, Bennett CA, Head T, et al. 2016. Adaptation to stressors by systemic protein amyloidogenesis. *Dev Cell* **39:** 155–168. doi:10.1016/j.devcel.2016.09.002

Baltz AG, Munschauer M, Schwanhausser B, Vasile A, Murakawa Y, Schueler M, Youngs N, Penfold-Brown D, Drew K, Milek M, et al. 2012. The mRNA-bound proteome and its global occupancy profile on protein-coding transcripts. *Mol Cell* **46:** 674–690. doi:10.1016/j.molcel.2012.05.021

Banani SF, Lee HO, Hyman AA, Rosen MK. 2017. Biomolecular condensates: organizers of cellular biochemistry. *Nat Rev Mol Cell Biol* **18:** 285–298. doi:10.1038/nrm.2017.7

Bonetti A, Agostini F, Suzuki AM, Hashimoto K, Pascarella G, Gimenez J, Roos L, Nash AJ, Ghilotti M, Cameron CJF, et al. 2019. RADICL-seq identifies general and cell type-specific principles of genome-wide RNA-chromatin interactions. bioRxiv doi:10.1101/681924

Brown JA, Valenstein ML, Yario TA, Tycowski KT, Steitz JA. 2012. Formation of triple-helical structures by the 3′-end sequences of MALAT1 and MENβ noncoding RNAs. *Proc Natl Acad Sci* **109:** 19202–19207. doi:10.1073/pnas.1217338109

Cai L, Chang H, Fang Y, Li G. 2016. A comprehensive characterization of the function of LincRNAs in transcriptional regulation through long-range chromatin interactions. *Sci Rep* **6:** 36572. doi:10.1038/srep36572

Castello A, Fischer B, Eichelbaum K, Horos R, Beckmann BM, Strein C, Davey NE, Humphreys DT, Preiss T, Steinmetz LM, et al. 2012. Insights into RNA biology from an atlas of mammalian mRNA-binding proteins. *Cell* **149:** 1393–1406. doi:10.1016/j.cell.2012.04.031

Castello A, Fischer B, Frese CK, Horos R, Alleaume AM, Foehr S, Curk T, Krijgsveld J, Hentze MW. 2016. Comprehensive identification of RNA-binding domains in human cells. *Mol Cell* **63:** 696–710. doi:10.1016/j.molcel.2016.06.029

Chen LL, Carmichael GG. 2009. Altered nuclear retention of mRNAs containing inverted repeats in human embryonic stem cells: functional role of a nuclear noncoding RNA. *Mol Cell* **35:** 467–478. doi:10.1016/j.molcel.2009.06.027

Chu C, Zhang QC, da Rocha ST, Flynn RA, Bharadwaj M, Calabrese JM, Magnuson T, Heard E, Chang HY. 2015. Systematic discovery of Xist RNA binding proteins. *Cell* **161:** 404–416. doi:10.1016/j.cell.2015.03.025

Chujo T, Hirose T. 2017. Nuclear bodies built on architectural long noncoding RNAs: unifying principles of their construction and function. *Mol Cells* **40:** 889–896. doi:10.14348/molcells.2017.0263

Chujo T, Yamazaki T, Hirose T. 2016. Architectural RNAs (arcRNAs): a class of long noncoding RNAs that function as the scaffold of nuclear bodies. *Biochim Biophys Acta* **1859:** 139–146. doi:10.1016/j.bbagrm.2015.05.007

Chujo T, Yamazaki T, Kawaguchi T, Kurosaka S, Takumi T, Nakagawa S, Hirose T. 2017. Unusual semi-extractability as a hallmark of nuclear body-associated architectural noncoding RNAs. *EMBO J* **36:** 1447–1462. doi:10.15252/embj.201695848

Clemson CM, Hutchinson JN, Sara SA, Ensminger AW, Fox AH, Chess A, Lawrence JB. 2009. An architectural role for a nuclear noncoding RNA: *NEAT1* RNA is essential for the structure of paraspeckles. *Mol Cell* **33:** 717–726. doi:10.1016/j.molcel.2009.01.026

Colognori D, Sunwoo H, Kriz AJ, Wang CY, Lee JT. 2019. Xist deletional analysis reveals an interdependency between Xist RNA and Polycomb complexes for spreading along the inactive X. *Mol Cell* **74:** 101–117 e110. doi:10.1016/j.molcel.2019.01.015

Dong P, Xiong Y, Yue J, Hanley SJB, Kobayashi N, Todo Y, Watari H. 2018. Long non-coding RNA *NEAT1*: a novel target for diagnosis and therapy in human tumors. *Front Genet* **9:** 471. doi:10.3389/fgene.2018.00471

Dumbović G, Biayna J, Banús J, Samuelsson J, Roth A, Diederichs S, Alonso S, Buschbeck M, Perucho M, Forcales SV. 2018. A novel long non-coding RNA from NBL2 pericentromeric macrosatellite forms a perinucleolar aggregate structure in colon cancer. *Nucleic Acids Res* **46:** 55045–5524. doi:10.1093/nar/gky263

Dumelie JG, Jaffrey SR. 2017. Defining the location of promoter-associated R-loops at near-nucleotide resolution using bis-DRIP-seq. *Elife* **6:** e28306. doi:10.7554/eLife.28306

Engreitz JM, Ollikainen N, Guttman M. 2016. Long non-coding RNAs: spatial amplifiers that control nuclear structure and gene expression. *Nat Rev Mol Cell Biol* **17:** 756–770. doi:10.1038/nrm.2016.126

Fong KW, Li Y, Wang W, Ma W, Li K, Qi RZ, Liu D, Songyang Z, Chen J. 2013. Whole-genome screening identifies proteins localized to distinct nuclear bodies. *J Cell Biol* **203:** 149–164. doi:10.1083/jcb.201303145

Fox AH, Lam YW, Leung AK, Lyon CE, Andersen J, Mann M, Lamond AI. 2002. Paraspeckles: a novel nuclear domain. *Curr Biol* **12:** 13–25. doi:10.1016/S0960-9822(01)00632-7

Fox AH, Bond CS, Lamond AI. 2005. P54nrb forms a heterodimer with PSP1 that localizes to paraspeckles in an RNA-dependent manner. *Mol Biol Cell* **16:** 5304–5315. doi:10.1091/mbc.e05-06-0587

Fox AH, Nakagawa S, Hirose T, Bond CS. 2018. Paraspeckles: where long noncoding RNA meets phase separation. *Trends Biochem Sci* **43:** 124–135. doi:10.1016/j.tibs.2017.12.001

Geisler S, Coller J. 2013. RNA in unexpected places: long non-coding RNA functions in diverse cellular contexts. *Nat Rev Mol Cell Biol* **14:** 699–712. doi:10.1038/nrm3679

Guttman M, Rinn JL. 2012. Modular regulatory principles of large non-coding RNAs. *Nature* **482:** 339–346. doi:10.1038/nature10887

Hall LL, Byron M, Carone DM, Whitfield TW, Pouliot GP, Fischer A, Jones P, Lawrence JB. 2017. Demethylated HSATII DNA and HSATII RNA foci sequester PRC1 and MeCP2 into cancer-specific nuclear bodies. *Cell Rep* **18:** 2943–2956. doi:10.1016/j.celrep.2017.02.072

Hennig S, Kong G, Mannen T, Sadowska A, Kobelke S, Blythe A, Knott GJ, Iyer KS, Ho D, Newcombe EA, et al. 2015. Prion-like domains in RNA binding proteins are essential for building subnuclear paraspeckles. *J Cell Biol* **210:** 529–539. doi:10.1083/jcb.201504117

Hirose T, Mishima Y, Tomari Y. 2014a. Elements and machinery of non-coding RNAs: toward their taxonomy. *EMBO Rep* **15:** 489–507. doi:10.1002/embr.201338390

Hirose T, Virnicchi G, Tanigawa A, Naganuma T, Li R, Kimura H, Yokoi T, Nakagawa S, Bénard M, Fox AH, et al. 2014b. NEAT1 long noncoding RNA regulates transcription via protein sequestration within subnuclear bodies. *Mol Biol Cell* **25:** 169–183. doi:10.1091/mbc.e13-09-0558

Hirose T, Yamazaki T, Nakagawa S. 2019. Molecular anatomy of the architectural NEAT1 noncoding RNA: the domains, interactors, and biogenesis pathway required to build phase-separated nuclear paraspeckles. *Wiley Interdiscip Rev RNA* **10:** e1545. doi:10.1002/wrna.1545

Huang J, Casas Garcia GP, Perugini MA, Fox AH, Bond CS, Lee M. 2018. Crystal structure of a SFPQ/PSPC1 heterodimer provides insights into preferential heterodimerization of human DBHS family proteins. *J Biol Chem* **293:** 6593–6602. doi:10.1074/jbc.RA117.001451

Imamura K, Imamachi N, Akizuki G, Kumakura M, Kawaguchi A, Nagata K, Kato A, Kawaguchi Y, Sato H, Yoneda M, et al. 2014. Long noncoding RNA NEAT1-dependent SFPQ relocation from promoter region to paraspeckle mediates IL8 expression upon immune stimuli. *Mol Cell* **53:** 393–406. doi:10.1016/j.molcel.2014.01.009

Imamura K, Takaya A, Ishida YI, Fukuoka Y, Taya T, Nakaki R, Kakeda M, Imamachi N, Sato A, Yamada T, et al. 2018. Diminished nuclear RNA decay upon *Salmonella* infection upregulates antibacterial noncoding RNAs. *EMBO J* **37:** e97723. doi:10.15252/embj.201797723

Isobe M, Toya H, Mito M, Chiba T, Asahara H, Hirose T, Nakagawa S. 2019. Forced isoform switching of Neat1_1 to Neat1_2 leads to the loss of Neat1_1 and the hyperformation of paraspeckles but does not affect the development and growth of mice. *RNA* doi:10.1261/rna.072587.119

Jiang L, Shao C, Wu QJ, Chen G, Zhou J, Yang B, Li H, Gou LT, Zhang Y, Wang Y, et al. 2017. NEAT1 scaffolds RNA-binding proteins and the Microprocessor to globally enhance pri-miRNA processing. *Nat Struct Mol Biol* **24:** 816–824. doi:10.1038/nsmb.3455

Jolly C, Usson Y, Morimoto RI. 1999. Rapid and reversible relocalization of heat shock factor 1 within seconds to nuclear stress granules. *Proc Natl Acad Sci* **96:** 6769–6774. doi:10.1073/pnas.96.12.6769

Kawaguchi T, Tanigawa A, Naganuma T, Ohkawa Y, Souquere S, Pierron G, Hirose T. 2015. SWI/SNF chromatin-remodeling complexes function in noncoding RNA-dependent assembly of nuclear bodies. *Proc Natl Acad Sci* **112:** 4304–4309. doi:10.1073/pnas.1423819112

Lee M, Sadowska A, Bekere I, Ho D, Gully BS, Lu Y, Iyer KS, Trewhella J, Fox AH, Bond CS. 2015. The structure of human SFPQ reveals a coiled-coil mediated polymer essential for functional aggregation in gene regulation. *Nucleic Acids Res* **43:** 3826–3840. doi:10.1093/nar/gkv156

Li R, Harvey AR, Hodgetts SI, Fox AH. 2017a. Functional dissection of NEAT1 using genome editing reveals substantial localization of the NEAT1_1 isoform outside paraspeckles. *RNA* **23:** 872–881. doi:10.1261/rna.059477.116

Li X, Zhou B, Chen L, Gou LT, Li H, Fu XD. 2017b. GRID-seq reveals the global RNA-chromatin interactome. *Nat Biotechnol* **35:** 940–950. doi:10.1038/nbt.3968

Lin Y, Schmidt BF, Bruchez MP, McManus CJ. 2018. Structural analyses of NEAT1 lncRNAs suggest long-range RNA interactions that may contribute to paraspeckle architecture. *Nucleic Acids Res* **46:** 3742–3752. doi:10.1093/nar/gky046

Lu Z, Zhang QC, Lee B, Flynn RA, Smith MA, Robinson JT, Davidovich C, Gooding AR, Goodrich KJ, Mattick JS, et al. 2016. RNA duplex map in living cells reveals higher-order transcriptome structure. *Cell* **165:** 1267–1279. doi:10.1016/j.cell.2016.04.028

Maharana S, Wang J, Papadopoulos DK, Richter D, Pozniakovsky A, Poser I, Bickle M, Rizk S, Guillen-Boixet J, Franzmann TM, et al. 2018. RNA buffers the phase separation behavior of prion-like RNA binding proteins. *Science* **360:** 918–921. doi:10.1126/science.aar7366

Mallik M, Lakhotia SC. 2009. RNAi for the large non-coding hsrω transcripts suppresses polyglutamine pathogenesis in *Drosophila* models. *RNA Biol* **6:** 464–478. doi:10.4161/rna.6.4.9268

Mannen T, Yamashita S, Tomita K, Goshima N, Hirose T. 2016. The Sam68 nuclear body is composed of two RNase-sensitive substructures joined by the adaptor HNRNPL. *J Cell Biol* **214:** 45–59. doi:10.1083/jcb.201601024

Mao YS, Sunwoo H, Zhang B, Spector DL. 2011. Direct visualization of the co-transcriptional assembly of a nuclear body by noncoding RNAs. *Nat Cell Biol* **13:** 95–101. doi:10.1038/ncb2140

Markmiller S, Soltanieh S, Server KL, Mak R, Jin W, Fang MY, Luo EC, Krach F, Yang D, Sen A, et al. 2018. Context-dependent and disease-specific diversity in protein interactions within stress granules. *Cell* **172:** 590–604 e513. doi:10.1016/j.cell.2017.12.032

Modic M, Grosch M, Rot G, Schirge S, Lepko T, Yamazaki T, Lee FCY, Rusha E, Shaposhnikov D, Palo M, et al. 2019. Cross-regulation between TDP-43 and paraspeckles promotes pluripotency-differentiation transition. *Mol Cell* **74:** 951–965. doi:10.1016/j.molcel.2019.03.041

Monfort A, Wutz A. 2017. Progress in understanding the molecular mechanism of Xist RNA function through genetics. *Philos Trans R Soc Lond B Biol Sci* **372:** 20160368. doi:10.1098/rstb.2016.0368

Naganuma T, Nakagawa S, Tanigawa A, Sasaki YF, Goshima N, Hirose T. 2012. Alternative 3′-end processing of long noncoding RNA initiates construction of nuclear paraspeckles. *EMBO J* **31:** 4020–4034. doi:10.1038/emboj.2012.251

Nakagawa S, Shimada M, Yanaka K, Mito M, Arai T, Takahashi E, Fujita Y, Fujimori T, Standaert L, Marine JC, et al. 2014. The lncRNA Neat1 is required for corpus luteum formation and the establishment of pregnancy in a subpopulation of mice. *Development* **141:** 4618–4627. doi:10.1242/dev.110544

Nakagawa S, Yamazaki T, Hirose T. 2018. Molecular dissection of nuclear paraspeckles: towards understanding the emerging world of the RNP milieu. *Open Biol* **8:** 180150. doi:10.1098/rsob.180150

Ninomiya K, Adachi S, Natsume T, Iwakiri J, Terai G, Asai K, Hirose T. 2019. LncRNA-dependent nuclear stress bodies promote intron retention through SR protein phosphorylation. *EMBO J* e102729. doi:10.15252/embj.2019102729

Nishimoto Y, Nakagawa S, Hirose T, Okano HJ, Takao M, Shibata S, Suyama S, Kuwako K, Imai T, Murayama S, et al. 2013. The long non-coding RNA nuclear-enriched abundant transcript 1_2 induces paraspeckle formation in the motor neuron during the early phase of amyotrophic lateral sclerosis. *Mol Brain* **6:** 31. doi:10.1186/1756-6606-6-31

Padron A, Iwasaki S, Ingolia NT. 2019. Proximity RNA labeling by APEX-seq reveals the organization of translation initiation complexes and repressive RNA granules. *Mol Cell* **75:** 875–887.e875. doi:10.1016/j.molcel.2019.07.030

Passon DM, Lee M, Rackham O, Stanley WA, Sadowska A, Filipovska A, Fox AH, Bond CS. 2012. Structure of the heterodimer of human NONO and paraspeckle protein component 1 and analysis of its role in subnuclear body formation. *Proc Natl Acad Sci* **109:** 4846–4850. doi:10.1073/pnas.1120792109

Peng A, Weber SC. 2019. Evidence for and against liquid–liquid phase separation in the nucleus. *Noncoding RNA* **5:** E50. doi:10.3390/ncrna5040050

Quinn JJ, Chang HY. 2016. Unique features of long non-coding RNA biogenesis and function. *Nat Rev Genet* **17:** 47–62. doi:10.1038/nrg.2015.10

Sasaki YT, Ideue T, Sano M, Mituyama T, Hirose T. 2009. MENε/β noncoding RNAs are essential for structural integrity of nuclear paraspeckles. *Proc Natl Acad Sci* **106:** 2525–2530. doi:10.1073/pnas.0807899106

Schmitt AM, Chang HY. 2017. Long noncoding RNAs: at the intersection of cancer and chromatin biology. *Cold Spring Harb Perspect Med* **7:** a026492. doi:10.1101/cshperspect.a026492

Senturk Cetin N, Kuo CC, Ribarska T, Li R, Costa IG, Grummt I. 2019. Isolation and genome-wide characterization of cellular DNA:RNA triplex structures. *Nucleic Acids Res* **47:** 2306–2321. doi:10.1093/nar/gky1305

Shevtsov SP, Dundr M. 2011. Nucleation of nuclear bodies by RNA. *Nat Cell Biol* **13:** 167–173. doi:10.1038/ncb2157

Shin Y, Brangwynne CP. 2017. Liquid phase condensation in cell physiology and disease. *Science* **357:** 1253. doi:10.1126/science.aaf4382

Souquere S, Beauclair G, Harper F, Fox A, Pierron G. 2010. Highly ordered spatial organization of the structural long noncoding NEAT1 RNAs within paraspeckle nuclear bodies. *Mol Biol Cell* **21:** 4020–4027. doi:10.1091/mbc.e10-08-0690

Standaert L, Adriaens C, Radaelli E, Van Keymeulen A, Blanpain C, Hirose T, Nakagawa S, Marine JC. 2014. The long noncoding RNA Neat1 is required for mammary gland development and lactation. *RNA* **20:** 1844–1849. doi:10.1261/rna.047332.114

Sunwoo H, Dinger ME, Wilusz JE, Amaral PP, Mattick JS, Spector DL. 2009. *MEN* ε/β nuclear-retained non-coding RNAs are up-regulated upon muscle differentiation and are essential components of paraspeckles. *Genome Res* **19:** 347–359. doi:10.1101/gr.087775.108

Tollervey JR, Curk T, Rogelj B, Briese M, Cereda M, Kayikci M, Konig J, Hortobagyi T, Nishimura AL, Zupunski V, et al. 2011. Characterizing the RNA targets and position-dependent splicing regulation by TDP-43. *Nat Neurosci* **14:** 452–458. doi:10.1038/nn.2778

Van Nostrand EL, Pratt GA, Shishkin AA, Gelboin-Burkhart C, Fang MY, Sundararaman B, Blue SM, Nguyen TB, Surka C, Elkins K, et al. 2016. Robust transcriptome-wide discovery of RNA-binding protein binding sites with enhanced CLIP (eCLIP). *Nat Methods* **13:** 508–514. doi:10.1038/nmeth.3810

Van Treeck B, Parker R. 2018. Emerging roles for intermolecular RNA–RNA interactions in RNP assemblies. *Cell* **174:** 791–802. doi:10.1016/j.cell.2018.07.023

Visa N, Puvion-Dutilleul F, Bachellerie JP, Puvion E. 1993. Intranuclear distribution of U1 and U2 snRNAs visualized by high resolution in situ hybridization: revelation of a novel compartment containing U1 but not U2 snRNA in HeLa cells. *Eur J Cell Biol* **60:** 308–321.

Wang Y, Hu SB, Wang MR, Yao RW, Wu D, Yang L, Chen LL. 2018. Genome-wide screening of *NEAT1* regulators reveals cross-regulation between paraspeckles and mitochondria. *Nat Cell Biol* **20:** 1145–1158. doi:10.1038/s41556-018-0204-2

West JA, Davis CP, Sunwoo H, Simon MD, Sadreyev RI, Wang PI, Tolstorukov MY, Kingston RE. 2014. The long noncoding RNAs NEAT1 and MALAT1 bind active chromatin sites. *Mol Cell* **55:** 791–802. doi:10.1016/j.molcel.2014.07.012

West JA, Mito M, Kurosaka S, Takumi T, Tanegashima C, Chujo T, Yanaka K, Kingston RE, Hirose T, Bond C, et al. 2016. Structural, super-resolution microscopy analysis of paraspeckle nuclear body organization. *J Cell Biol* **214:** 817–830. doi:10.1083/jcb.201601071

Wilusz JE, JnBaptiste CK, Lu LY, Kuhn CD, Joshua-Tor L, Sharp PA. 2012. A triple helix stabilizes the 3′ ends of long noncoding RNAs that lack poly(A) tails. *Genes Dev* **26:** 2392–2407. doi:10.1101/gad.204438.112

Wutz A, Rasmussen TP, Jaenisch R. 2002. Chromosomal silencing and localization are mediated by different domains of *Xist* RNA. *Nat Genet* **30:** 167–174. doi:10.1038/ng820

Yamashita A, Watanabe Y, Nukina N, Yamamoto M. 1998. RNA-assisted nuclear transport of the meiotic regulator Mei2p in fission yeast. *Cell* **95:** 115–123. doi:10.1016/S0092-8674(00)81787-0

Yamazaki T, Hirose T. 2015. The building process of the functional paraspeckle with long non-coding RNAs. *Front Biosci (Elite Ed)* **7:** 1–41. doi:10.2741/s420

Yamazaki T, Fujikawa C, Kubota A, Takahashi A, Hirose T. 2018a. CRISPRa-mediated NEAT1 lncRNA upregulation induces formation of intact paraspeckles. *Biochem Biophys Res Commun* **504:** 218–224. doi:10.1016/j.bbrc.2018.08.158

Yamazaki T, Souquere S, Chujo T, Kobelke S, Chong YS, Fox AH, Bond CS, Nakagawa S, Pierron G, Hirose T. 2018b. Functional domains of NEAT1 architectural lncRNA induce paraspeckle assembly through phase separation. *Mol Cell* **70:** 1038–1053.e1037. doi:10.1016/j.molcel.2018.05.019

Yang L-Z, Wang Y, Li S-Q, Yao R-W, Luan P-F, Wu H, Carmichael GG, Chen L-L. 2019. Dynamic imaging of RNA in living cells by CRISPR-Cas13 systems. *Mol Cell* doi:10.1016/j.molcel.2019.10.024

Yap K, Mukhina S, Zhang G, Tan JSC, Ong HS, Makeyev EV. 2018. A short tandem repeat-enriched RNA assembles a nuclear compartment to control alternative splicing and promote cell survival. *Mol Cell* **72:** 525–540.e513. doi:10.1016/j.molcel.2018.08.041

Youn JY, Dunham WH, Hong SJ, Knight JDR, Bashkurov M, Chen GI, Bagci H, Rathod B, MacLeod G, Eng SWM, et al. 2018. High-density proximity mapping reveals the subcellular organization of mRNA-associated granules and bodies. *Mol Cell* **69:** 517–532.e511. doi:10.1016/j.molcel.2017.12.020

Zhang N, Ashizawa T. 2017. RNA toxicity and foci formation in microsatellite expansion diseases. *Curr Opin Genet Dev* **44:** 17–29. doi:10.1016/j.gde.2017.01.005

Zhu J, Mayeda A, Krainer AR. 2001. Exon identity established through differential antagonism between exonic splicing silencer-bound hnRNP A1 and enhancer-bound SR proteins. *Mol Cell* **8:** 1351–1361. doi:10.1016/S1097-2765(01)00409-9

Myriad RNAs and RNA-Binding Proteins Control Cell Functions, Explain Diseases, and Guide New Therapies

BYUNG RAN SO AND GIDEON DREYFUSS

Department of Biochemistry and Biophysics, Howard Hughes Medical Institute, University of Pennsylvania School of Medicine, Philadelphia, Pennsylvania 19104, USA

Correspondence: gdreyfuss@hhmi.upenn.edu

This summary of the 84th *Cold Spring Harbor Laboratory* (CSHL) *Symposium on Quantitative Biology: RNA Control and Regulation*, held in May 2019, highlights key emerging themes in this field, which now impacts nearly every aspect of biology and medicine. Recent discoveries accelerated by technological developments reveal enormous diversity of RNAs and RNA-binding proteins (RBPs) with ever-increasing roles in eukaryotes. Atomic structures and live-cell imaging of transcription, RNA splicing, 3′-end processing, modifications, and degradation machineries provide mechanistic insights, explaining hundreds of diseases caused by their perturbations. This great progress uncovered numerous targets for therapies, some of which have already been successfully exploited, and many opportunities for pharmacological intervention and RNA-guided genome engineering. Myriad unexplained RNAs and RBPs leave the RNA field open for many more exciting discoveries.

Foundational concepts of molecular biology established from studies of prokaryotes—one gene, one mRNA, one protein; the regulation of mRNA synthesis almost entirely by transcription; and cotranscriptional mRNA translation—could be captured in a single electron micrograph and summed up in a simple formula: DNA → mRNA → protein. The now-historical proceedings of CSHL Symposia from the 1950s and 1960s chronicle these insights. However, they could not explain or predict how mRNAs are made and function in eukaryotes, where primary protein-coding gene transcripts (pre-mRNAs, historically called hnRNAs) are not translation-ready mRNAs due to open reading frame (ORF)-disrupting introns, and the translation machinery (mRNA-decoding ribosomes and tRNAs) is in the cytoplasm, segregated from transcription sites on chromatin in the nucleus. Seminal discoveries in the 1970s and 1980s on transcription, splicing of protein-coding pre-mRNAs (to excise introns and join exons), self-splicing RNA catalysis, noncoding (nc) small nuclear RNPs (snRNPs), RNA-binding proteins (RBPs), 3′-end cleavage and polyadenylation, and mRNA translation regulation, transport, and surveillance uncovered an astonishing complexity of post-transcriptional gene regulation in eukaryotes.

The 84th CSHL Symposium showcased remarkable progress in understanding these processes, additional layers of mRNA regulation, numerous ncRNAs, and new RNA functions. Video recordings of 56 30-min oral presentations and discussions, chapters in this treatise, and hundreds of poster presentations provide a complete account of the five-day symposium. A traditional all-inclusive comprehensive summary would be superfluous. Instead, here we highlight some key themes, outstanding questions, and prospects. Presenters' names are noted in parentheses in relevant contexts as starting points for deeper exploration of specific topics. The discussion does not necessarily reflect the presenters' opinions or the main points they made, nor does their mention imply more significant contributions to the topic compared to other researchers in that area.

PRE-mRNA PROCESSING

The main pre-mRNA processing reactions are splicing and 3′-end processing by cleavage and polyadenylation (CPA). Decades of intensive research identified the RNA motifs (sequences and structural elements) and their cognate factors, the inventory of components that mediate these reactions, the chemistry of RNA intermediates, and the stepwise assembly of catalytic complexes. Splicing of each intron occurs in a large complex (spliceosome) that assembles through an elaborate series of binding and rearrangements of five snRNPs (snRNA–protein complexes, generally U1, U2, U4, U5, and U6; JA Steitz) and more than 50 proteins. The key signals, 5′ and 3′ splice sites (ss), are recognized by U1 base pairing and U2 associated proteins, respectively, which commence the assembly process. Various enzymes, particularly RNA helicases, use ATP hydrolysis to effect conformational and compositional transitions necessary for splicing self-catalysis. CPA is specified by polyadenylation signals (PASs), consisting of a hexanucleotide sequence (AAUAAA and variants thereof) flanked by upstream and downstream motifs that recruit three main subcomplexes, comprising 20 proteins. The 3′-end cleavage and poly(A) addition are carried out by protein enzymes.

Progress in understanding these pre-mRNA processing reactions—and other aspects of RNA biology—has been greatly facilitated by technological developments. Extensive biochemical and genetic studies, including reconstitutions with RNA substrates in vitro, purifications from cells, and the use of mutations to arrest reaction intermediates, exemplify discovery schemes used in the field. In addition, crystal structures of many individual constituents and subunits have been determined by X-ray crystallography. Yet, with few exceptions (e.g., ribosomes), the inability to obtain high-quality diffracting crystals of entire complexes left essential information obscure. Single-particle cryo-electron microscopy (cryo-EM), which does not require crystallization, produced in rapid succession a profusion of atomic or near-atomic resolution three-dimensional views of molecular machines that transcribe, process, degrade, and regulate RNAs. This includes spliceosomes at various assembly stages (K Nagai, presented by M Wilkinson), CPA complexes (LA Passmore), exosomes (E Conti), and miRNA and other complexes (JA Steitz, BL Bass).

COTRANSCRIPTIONAL RNA PROCESSING

Despite their intricacies, spliceosomes and CPA complexes are generic machines that are highly conserved in eukaryotes. They attach cotranscriptionally, aided by interactions with the carboxy-terminal domain (CTD) of RNA polymerase II (Pol II) (XD Fu, KM Neugebauer, A Aguilera). This enhances processing efficiency and shortens the lag time (compared to uncoupled in vitro), confers 5′ to 3′ polarity with respect to signal usage (first come, first served), and introduces kinetic effects through the interplay between Pol II speed and signal strength (A Kornblihtt). This coupling has many important consequences. For example, by default, cotranscriptional processing makes CPA most likely at the TSS-proximal PAS. In thousands of genes, cryptic PASs in the first or second intron (which are typically the longest in humans) can elicit transcription-terminating premature CPA (PCPA). This is generally suppressed by U1 snRNP bound upstream nearby (e.g., at the first 5′ss [which also initiates splicing]). This U1 activity (telescripting) is necessary for full-length transcription and used for regulating PAS usage throughout pre-mRNAs or other nascent transcripts that can be modulated by U1 snRNP availability (G Dreyfuss). A lower ratio of U1 snRNP binding sites to PASs in upstream antisense transcripts in many protein-coding genes (which are typically bidirectional) makes them susceptible to PCPA, which enhances sense transcription directionality (PA Sharp).

Without complicating circumstances such as long introns, generic machineries would be sufficient to execute constitutive splicing and CPA. Splicing has some benefits —for example, for eliminating nonsense mutations (LE Maquat)—although it is debatable if the benefits outweigh the costs (intronless genes are perfectly functional). Still, constitutive processing limits the number of mRNAs and proteins a gene can theoretically make to one, regardless of how many introns and PASs it has. However, this is not the case in complex organisms. Their existence depends on making multiple mRNAs and protein isoforms from the same gene. This mRNA diversity is generated primarily by alternative splicing and alternative CPA of pre-mRNAs. It is a compelling rationale for introns and splicing because it obviates the need to increase the number of genes. For example, humans have approximately 20,000 genes, but produce many-fold greater number of protein isoforms, which is necessary for making many different cell types.

RNA-BINDING PROTEINS (RBPs)

Alternative pre-mRNA processing is controlled by a very large assortment of RBPs. There are around 1000 genes in humans. RBPs are part of and play a role in all steps in the life of mRNAs and in every aspect of RNA biology; all RNAs in cells exist as complexes with RBPs (RNPs). Numerous RBPs associate with pre-mRNAs cotranscriptionally (historically called hnRNP proteins), which helps prevent long pre-mRNAs from misfolding and sculpt them for processing. RBPs regulate alternative RNA processing by binding to splicing enhancers and suppressor motifs and thereby promote or hinder the assembly of spliceosomes and CPA factors (CPAFs) at nearby splice sites and PASs. Changes in RBP and snRNP repertoires and activities regulate cell- and developmental-stage-specific alternative processing and cells' responses to environmental changes (K Lynch).

Many of the most abundant prototypical RBPs, the hnRNP proteins, and numerous RBPs with the same domain structure and features that go by other nomenclature are removed with introns, as are snRNPs. However, many others remain associated with mRNAs after splicing, during export to the cytoplasm, where they have additional roles in translation, transport to specific locations, and mRNA stability. They are generally dislodged during the first round of translation and shuttle back to the nucleus (except those on 3′ UTRs). Additional RBPs are added during splicing, thereby marking spliced junctions (exon junction complex), providing positional information that helps detect nonsense codons (LE Maquat). Thus, RBPs link the steps from transcription on chromatin sites to translation on ribosomes and the cell extremities. In another technological feat, single-molecule imaging vividly traces the journey of actin mRNA guided by the RBP ZBP1 to fibroblasts' leading edges, where they translate (RH Singer). The mRNP complexes that guide other long-distance mRNA transport and localized translation, which are crucial in neurons and early development (RB Darnell), are also assembled during processing in the nucleus. Additional RBPs associate with mRNAs in the cytoplasm, such as the poly(A) binding protein, which has roles in translation and mRNA turnover.

GRANULES AND CONDENSATES: THE POWER OF RNP SELF-ASSEMBLY

Hallmark features of RBPs explain their wide range of functions and their roles organizing granular micro-compartments that function in RNA metabolism. Early sequencing described RBPs modular structures, multi-RNA-binding domains (typically two or more RBD/RRMs, KH domains, etc.), and "auxiliary domains" noted for their simple amino acid composition (now referred to as low complexity [LCDs]). Auxiliary domains harbor transport signals and LCD repeats (such as RGGs, proline-rich regions) that can enhance RNA binding and interactions with other RBPs. Similar LCDs are found in hundreds of RBPs and transcription factors (many RBPs have dual functions in transcription). LCDs are relatively disordered and high IDR index and have a high propensity to engage in homotypic, albeit weak, interactions, which makes them form gels at high concentration (S McKnight). RNA lowers the RBP concentration of gelling (also known as phase separation), which high concentrations of RNAs contribute to by RNA–RNA interactions (R Parker, A Gladfelter). Almost any RNA over a few hundred nucleotides that has sufficient single-stranded regions can do this, because RBPs have a certain generic "nonspecific" RNA binding (counter to the misconception that specificity is binary). RNA binding makes LCDs multivalent, which potentiates weak interactions (e.g., NEAT1 RNA [T Hirose]). The result is a massive coalescence of RNAs with many RBPs bound to them. They have a granular appearance that has fascinated microscopists for some time and now have biophysical underpinning, and their significance in many molecular processes is becoming better understood. The forces of self-assembly can give rise to large, readily visible granules in both the nucleus (e.g., Cajal bodies, gems, speckles/interchromatin granules) and cytoplasm (e.g., stress granules). Granules that serve as storage depots of maternal RNP components play important roles in early development (R Lehmann, G Seydoux). The tendency to describe granules as membraneless organelles is somewhat questionable as they can be highly dynamic, and there is a lack of evidence that specific functions can only occur inside of them (as opposed to the idea that they represent a high local concentration of components that can function inside or outside of them).

The same RNP-driven self-assembly principles operate on smaller scales to concentrate transcription factors at super-enhancers (PA Sharp, II Cisse, RA Young), chromatin silencer HP1α to H3K9 methylated chromatin (GJ Narlikar), and mRNAs encoding secretory proteins to the cytoplasmic face of the endoplasmic reticulum (C Mayr).

NONCODING RNAS

Classes of RNAs that do not encode proteins nor are directly involved in translation and instead have defined functions in other processes have been well established. For example, snRNAs (discussed above), snoRNAs (guide RNA modifications/rRNA processing), miRNAs (CS Pikaard), tRNAs (O Rando), transposable elements (M-E Torres-Padilla), and piRNAs (transposable element suppression; J Brennecke, L Joshua-Tor) illustrate the range of roles RNAs have in cell regulation. miRNAs have roles in regulating mRNA translation and stability (D Bartell, JA Steitz, RA Martienssen). A more nebulous grouping, called long noncoding RNAs (lncRNAs), is based on a somewhat arbitrary size cutoff (>~250 nt) and absence of conventional translation ORF. By this definition, there are tens of thousands of lncRNAs. A small fraction of those have specific functions. For example, Xist is essential for X-chromosome silencing (J Lee, N Brockdorff, HY Chang, A Akhtar). Some other lncRNAs have clear and important biological effects (C Dean, J Mendell, J Wysocka)—such as MALAT1, which is highly expressed in and contributes to cancers—and are valuable prognostic markers (DL Spector). However, presently most lncRNAs have no specific functions ascribed to them. The expectation of specific functions, based on proximity to genes of interest, significant expression level correlation with certain conditions, and other considerations, motivates extensive research on individual lncRNAs. On the other hand, the lack of significant sequence conservation among lncRNAs in the same genomic locations in closely related mammals and their very short half-lives (I Ulitsky, NJ Proudfoot) suggest that they may not each have specific or unique functions.

PERVASIVE TRANSCRIPTION, RNA PROCESSING, AND MODIFICATIONS

In our view, the enormous complexity of RNAs that mammalian cells make is puzzling. Deep RNA-sequencing (RNA-seq) shows a much greater diversity of RNAs generated at every step than had been anticipated or can be explained in the foundational framework of molecular biology. Many more will undoubtedly be discovered from single-cell RNA-seq. Pol II transcription is highly pervasive, initiating bidirectional transcription from numerous promoters throughout the genome, unless access to them is blocked. Many alternative TSSs are used in the same genes. Transcription can continue for enormous distances, requiring barriers, checkpoints, and other impediments (as mentioned above and SM Gasser, S Grewal) to slow it down and RNA cleavage to effect termination (K Adelman, JE Wilusz, NJ Proudfoot). There are multiple alternative PASs and other 3′-end processing signals, resulting in numerous RNAs from the same gene and a variety of 3′ ends and tails. Annotation-independent splice junction calling identified many more splicing events than had been anticipated based on what is represented in steady state mRNAs, now in the hundreds of thousands. There is splicing in lncRNAs, recursive splicing in long introns, resplicing in mRNAs, and back splicing, which makes circular RNAs (circRNAs) (L Chen). Enzymatic editing (B Bass) and modifications (C He, S Jaffrey) alter the chemistry of nucleotides. Many of these regulate some aspect of RNA function, but many others may just be bystanders. One can speculate why such complexity arises and what purpose it might have (as G Dreyfuss had at this symposium's conclusion), but it is exciting that so much remains to be deciphered.

MANY DISEASES AND NEW THERAPIES

The large number of components and processes that make and regulate RNAs increase opportunities for something to go wrong. Mutations in RNA processing signals, mRNA untranslated regions, miRNAs, snRNA, RBPs, and RNA processing factors cause minimally hundreds of human diseases. For example, a synonymous mutation in a splicing regulating sequence in *SMN* causes frequent exon 7 skipping, resulting in a SMN deficiency that causes spinal muscular atrophy (SMA). Mutations in splicing factors, such as U2 associated SF3B1, cause hematologic malignancies and other cancers (O Abdel-Wahab, R Bradley), and expanded G-rich repeats in a C9orf72 intron, which sequesters hnRNPF thereby decreasing its availability and impairing splicing, are linked to amyotrophic lateral sclerosis (ALS) (J Manley). Mutations in hnRNPA1 and similar RBPs underlie many neurodegenerative diseases.

Such advances in understanding pathogenic mechanisms have helped develop diagnostic tools and have revealed many targets and approaches for repairing, replacing, or removing them as potential therapies. Remarkable successes have already been achieved—for example, using antisense oligonucleotides to correct SMN splicing (to enhance exon 7 inclusion) for SMA (A Krainer). Improvements in RNA-guided CRISPR–Cas technologies, originally discovered from studies on adaptive immunity in bacteria, enable genome and transcriptome engineering with numerous research and clinical applications (J Doudna and F Zhang). Having atomic 3D maps of RNP complexes reveal numerous potential targets for small molecules and the ability to monitor RNA structure changes (K Weeks) holds promise for pharmacological intervention.

CONCLUSION

Great advances in the RNA field now impact nearly every aspect of biology and medicine. Technological developments, including in RNA-seq and data analysis, mass spectrometry, structure determination by cryo-EM, high-resolution imaging in cells, and genome editing, have greatly accelerated the pace of discovery. The wealth of new information has provided insights, inspired useful applications, and uncovered puzzling complexities for future research.

DORCAS CUMMINGS LECTURE

Video recording of the Dorcas Cummings Lecture is available on Cold Spring Harbor Laboratory's Leading Strand channel on YouTube (playlist 84th CSHL Symposium Interview Series RNA CONTROL AND REGULATION).

Dorcas Cummings Lecture

JENNIFER DOUDNA

Dr. Jennifer Doudna presented the Dorcas Cummings lecture entitled "Editing the Code of Life: The Future of Genome Editing" to friends and neighbors of Cold Spring Harbor Laboratory and Symposium participants on Saturday, June 1, 2019. Dr. Doudna holds the Li Ka Shing Chancellor's Chair in Biomedical and Health Sciences and is a Professor in the Departments of Molecular & Cell Biology and of Chemistry at the University of California, Berkeley, and the Executive Director of the Innovative Genomics Institute.

Thank you. It's a great pleasure to be here to give the Dorcas Cummings Lecture. I've heard this lecture a number of times when I've come to this symposium and I'm pinching myself that I'm actually up here doing this. I do have to tell you something important about Cold Spring Harbor. I was having dinner last night with a couple of colleagues and we were chitchatting about different experiences we'd had here at the lab. We realized that each of us at different times had given our very first scientific lecture here at Cold Spring Harbor, and for each of us it had been a transformative experience. So, I just want to point out that the lab plays a critical role not only in all of the incredible science that you've heard, but also for the next generation and helping people from all around the world come together and think and do science in a way that is really quite unique.

In that spirit, I want to tell you about an area of science that grew out of very small science from just a few labs around the world that were investigating an aspect of biology that initially seemed quite obscure: namely, how bacteria fight viral infection. When a bacterial cell is infected by viruses, these viruses inject their genetic material—their DNA—into the cell. Bacteria are rapidly dividing, rapidly growing organisms, so they don't have much time to defend themselves from this type of infection. A lot of the work in molecular biology over the last several decades has centered around understanding this warfare that goes on between microbes and their invaders.

One system for defense had been missed all those years, and it really came to light only when a few bioinformaticians and scientists in the food industry were studying microbes that are cultivated in making yogurt and cheese and things like that. They figured out that about 40% of bacteria and about 90% of other single-celled organisms called archaea have an adaptive immune system encoded in their DNA. These organisms can capture bits of viral DNA during an infection and integrate a little sequence of that DNA into their own genome at a place called the CRISPR [clustered regularly interspaced short palindromic repeats]. The CRISPR sequence has a distinctive property: It has a series of short repeated elements that flank unique sequences, and each of these unique sequences is an inserted sequence that comes from a virus. This is, effectively, a genetic vaccination card: a way that cells can record over time the viruses they've been exposed to by storing little snippets of the viral DNA in their genome. Importantly, next door to this locus are CRISPR-associated, or *cas*, genes. These genes encode proteins that, together with the CRISPR locus, constitute this adaptive immune system.

Here's how it works: The cell will make a transcript, an RNA copy of the entire CRISPR locus, a piece of RNA that is an exact replica of all the encoded sequences [pre-crRNA]. Those RNA molecules are processed into individual units that each include a sequence that comes from a virus as crRNAs [CRISPR RNAs]. These then combine with Cas proteins to form RNA-guided surveillance complexes that can search the cell looking for matching sequences of DNA. When those matches are found, they recruit the Cas proteins to cleave those targeted sequences and destroy them. If this is a viral DNA, it quickly gets made into mincemeat by this system. It provides a wonderful way for microbes to adapt environmentally to the infectious agents that they encounter.

This video illustrates this in a little more dynamic detail. This shows some bacterial cells that are being invaded by viruses; here's the infection occurring as the virus injects its DNA. If this cell has a CRISPR sequence in the genome, it can integrate a piece of that viral DNA into the CRISPR locus in this very distinctive pattern, flanked by these repeated elements; this is where the viral DNA sequences are stored. Then the cell is able to make an RNA copy, a transient molecule that is chopped into units, each including one of the viral sequences; these become the "zip codes" for the system. These RNAs then combine with a second RNA called "tracr" [*trans*-activating small RNA] and a protein called Cas9. That is the unit that goes searching the cell, looking for DNA that might have a match to the guide RNA from the viral insert. If a match is found, then the DNA unwinds and the Cas protein is

© 2019 Doudna. This article is distributed under the terms of the Creative Commons Attribution-NonCommercial License, which permits reuse and redistribution, except for commercial purposes, provided that the original author and source are credited.

Published by Cold Spring Harbor Laboratory Press; doi:10.1101/sqb.2019.84.040139

able to cut the DNA at a precise position that, in bacteria, triggers destruction of those pieces of DNA. It's a great way to protect the cell from viral infection.

Initially, scientists were studying these and doing a lot of sequencing of bacterial genomes. What became clear was that there is not just one type of CRISPR system but, in fact, quite a few. This figure from a review from a few years ago [Fig. 2A in Mohanraju et al. (2016), *Science* **353:** aad5147] shows that we can divide these CRISPR systems into two different classes called Class 1 and Class 2 that are distinguished by the numbers of Cas proteins that are part of these systems. All of the Class 1 CRISPR systems have several proteins that are required for these systems to function, whereas the Class 2 systems in each case have one single large protein that is the only protein necessary for RNA-guided protection of the cell, and that was something shown genetically.

How did I get involved in all of this? I've been interested in RNA since I was a graduate student and I've been fascinated by how RNA molecules help cells to control the flow of genetic information, whether it's a cellular gene or a viral gene. When I heard from my colleague Jill Banfield about 12 years ago about the existence of these possible adaptive immune systems that might be RNA-based in bacteria, I was intrigued. We started investigating these and that eventually led me to a conference in 2011 where I met Emmanuelle Charpentier. Emmanuelle was studying a bacterial system that had this type of CRISPR element— the Type II CRISPR in Class 2—in the genome that included this gene called *cas9*. When we met up at this conference, we decided to team up to figure out the function of this protein.

It seemed like it must be an extremely interesting protein that would be RNA-guided and somehow targeting DNA, but at the time nobody knew how it worked so that was the question that we asked. What emerged from experiments done by our lab members, Martin Jinek and Kris Chylinski [Jinek et al. (2012), *Science* **337:** 816] is that Cas9 is an RNA-guided protein that uses the CRISPR RNA together with the tracrRNA to recognize double-stranded DNA at a 20-letter sequence that matches the sequence of the guide RNA. When that match occurs, it cuts *both* strands of the DNA double helix and makes a clean break, just like cutting a rope.

Martin Jinek in my lab was investigating this, and being a very good biochemist he was trimming away at these RNAs trying to figure out what was essential for this protein to function. Eventually, he realized that he could link together the CRISPR and tracrRNAs into what we called a "single-guide RNA" that would have both the targeting sequence and the Cas9 binding information in the very same RNA molecule. When Martin did this experiment and showed that you could easily program Cas9 with this single-guide RNA and direct it to cut a desired DNA sequence, we realized that this project had gone from being a curiosity-driven investigation of a bacterial immune system in our lab, which is where it started, to opening the door to something that could be a very important and exciting technology.

The reason for this is because of all of the other beautiful work that had gone on in many labs in the preceding two decades, showing that in eukaryotic cells—like ourselves, or plants, or animals, etc.—when there is a double-stranded DNA break that occurs in the genome, cells can repair those breaks. Rather than degrading DNA like what happens in rapidly growing microbes, in ourselves, those breaks are detected and fixed. Because of that, if you introduce a double-stranded break into a eukaryotic cell genome, you can trigger DNA repair that results either in a gene disruption or the insertion of new genetic information—of donor DNA—at exactly the site of the break.

Scientists had realized that if you could figure out how to introduce a double-stranded DNA break at a desired position you could trigger targeted genome editing, or engineering, as we called it at the time. There were a number of technologies that had been developed to do this—to introduce double-stranded breaks—that were exciting enough that a lot of people were talking about the possibilities of genome editing, but the technologies were too cumbersome for most labs to be able to adopt. CRISPR–Cas9, being a molecule that can be reprogrammed fairly trivially by changing its RNA guide sequence, emerged as a tool that could achieve this goal of cutting DNA easily in different kinds of cells to trigger genome editing.

This illustrates how we imagine that this works in a eukaryotic cell, where the DNA is inside the nucleus. Here's Cas9 with its guide RNA searching through the cell. We imagine that it is quickly binding and releasing DNA, looking for a match to its guide RNA. When that match is eventually found it latches onto the DNA and, by a mechanism we don't fully understand, it triggers melting and opening of that DNA helix to allow an RNA–DNA hybrid to form in the protein. Cutting of the two strands occurs and then repair enzymes in the cell come around, detect the broken DNA, and are able to fix the break by introducing a change to the sequence: in this example, by actually introducing a very small change in the DNA letters at that exact position. This tool has become a very powerful way for scientists working on any area of biology to make modifications to a desired genome to study fundamental questions, as well as to do things that are applied.

What happened next was truly remarkable. This figure from a journal publisher's website [https://www.elsevier.com/research-intelligence/campaigns/crispr] shows that in these years shown here, there were several technologies that had come along for genome editing—meganuclease, ZFN [zinc-finger nuclease], TALEN [transcription activator-like effector nuclease]—and there were some papers published using these technologies that were starting to populate the scientific literature. When CRISPR emerged, the publications using it just took off, and it's still going up exponentially. The reason is that this technology came along at a perfect moment when there was a lot of understanding about genomes, many sequenced genomes, and lots of exciting questions that scientists wanted to be able to ask about gene function. What was missing was an easy way to manipulate those genomes, and that's really what CRISPR offers.

I want to turn now to the opportunities as well as some of the important challenges that are coming forward, given that we now have a tool for genome editing that is widely available, fairly simple and inexpensive to use, and quite effective for the kinds of applications that people are using it for. To start, as a scientist it's been incredibly exciting to see all of the ways that genome editing with CRISPR is being used to do fundamental research. There are many examples, but I want to point to one that was highlighted in the *New York Times*. If you were reading the science section last week, you might've seen a weirdly titled article about CRISPR snail babies. I looked up this paper and it's actually quite fun.

This is the picture that was on the cover of the journal *Development* [May 2019, **146** (9)], and it's all about these CRISPR baby snails. If you look carefully, these two snails look like mirror images of each other. Virtually all snails, for reasons that were unclear, develop with this right-handed coil to their shell and it was extremely rare to have animals occur in the wild that had a left-handed coil to their shell. Developmental biologists have been very interested in this question, partly because there are all sorts of examples across biology—including in humans—where there is a handedness to an organism and the genetics of that have been really unclear. These scientists, [Masanori] Abe and [Reiko] Kuroda, were able to use CRISPR first to look through the genome of the snail to try to figure out some of the possible genes that might be involved in this pathway. Then, using CRISPR, they could actually knock out or disrupt individual genes and ask what happens. They ended up finding a single gene that controls the direction of coiling of the snail shell. These are some images that they took of these snail embryos that have been either treated with CRISPR or not [Abe and Kuroda (2019) *Development* **146**: dev175976]. As the wild-type embryos develop, you can already see at an early stage that they start to coil in the right-handed direction. When they use CRISPR to knock out this single gene, they found that now these embryos were coiling in the left-handed direction, even at a very early stage of development.

This gives you a sense of the kind of question that scientists can now address using a tool like this, but I want to talk about three specific areas where I think we will likely see the impacts of CRISPR in our lives. These also raise important questions that we have to think about, both from an ethical and societal perspective as well as from a regulatory perspective. Of course, there are lots of commercial opportunities that have come along with genome editing as well that are opening the door for companies, but also raise challenges.

The first one is in the area of public health. Ever since I was a graduate student, people have talked about the possibility of "gene drives" in organisms. This is a nice graphic I found on the *Science News* website [https://www.sciencenews.org/blog/science-ticker/gene-drives-arent-ready-wild-report-concludes] that shows how a gene drive works. The left side shows you how normal Mendelian inheritance of genetic traits works, where you have a trait that's passed along to some progeny in a population, but not to all. If you have a gene drive present in this animal, however, this means that you can take a trait and hook it up to a genome editing tool such as CRISPR that's very efficient at introducing a gene into a genome. If you do that in the right way, this trait can be passed along and can propagate through all of the animals in the population very quickly, much faster than you would ever spread a trait with Mendelian genetics.

That sounds like a cool thing, but why do we care? Well, there's a reason that this is shown with mosquitoes as the example because there's an exciting possibility that you could use a gene drive catalyzed by CRISPR to spread a trait that would prevent it from being able to spread mosquito-borne disease in human populations. That would have an incredibly powerful positive impact on human health globally if it could be made to work well and safely. I Googled recently how much money has been invested in research in gene drives using CRISPR and just with three organizations—the Wellcome Trust, the Gates Foundation, and the Tata Group in India—$250 million has gone into this, and that's just what's been publicly announced. There's tremendous excitement about this, but also concern about the potential for this to get out of control if released into the environment. There's a lot of work and a lot of debate right now about how we encourage the science as well as ensure that it's deployed safely.

The second area—and this, in my opinion, is the area where CRISPR will have its biggest global impact in the near term—is in agriculture. Here's an example from a wonderful scientist right here at Cold Spring Harbor labs, Zach Lippmann, who has been using CRISPR–Cas in plants such as tomatoes to do very interesting things. This was a paper that he published a couple of years ago [Soyk et al. (2017), *Cell* **169**: P1142] in which he showed that you could use the CRISPR–Cas9 system in tomato plants to adjust one gene that controls the number of flowers and also the strength of the stems that hold the fruit on the plant. By doing this, you could actually control the yield of tomatoes without affecting anything else about them. Unlike traditional plant breeding, where random changes are introduced into seeds and then traits are selected—typically over months or years—with random changes that come along so we end up with roses that don't have thorns but also don't smell nice, here you have the potential to introduce a very desirable trait such as increasing crop yields but without affecting the quality of the fruit. This is certainly very interesting, and the genetics that are responsible for this in tomatoes apply to lots of other crops. You can imagine all sorts of other traits that you might want to be able to tweak or change in various plants to make them drought-resistant or pest-resistant or to make them more nutritious and things like that.

This is another example of work done at Penn State University. They were able to make a change to a single gene in mushrooms that prevents them from turning brown when you cut them open, very desirable potentially for commercial use. This work was the trigger for the U.S. Department of Agriculture to debate how they might regulate these sorts of products that might come soon to a supermarket or a farmer's market where we could all get access to them. Their conclusion was that if you made a knockout of a gene it would not be considered a genetically modified organism [GMO] because it does not include foreign DNA, and therefore would not be regulated in the U.S. [Waltz (2016) *Nature* **532**: 293]. That was interesting, but not all countries have come to that same conclusion. In Europe, the regulatory agencies decided that any plant that is manipulated by genome editing or any kind of genetic manipulation technology, even if you ended up with a genetically identical product at the end, would now be considered a GMO [https://www.nytimes.com/2018/07/27/science/gmo-europe-crops.html]. This has raised a lot of questions about just thinking fundamentally: How do we regulate this sort of thing? How do we explain to consumers or the public the difference between using a targeted approach to making a single genetic change versus introducing random changes as traditional plant breeding does? And how do we ensure that regulators understand enough themselves about the science so that they get it right and they don't limit the possibilities of this kind of technology in a way that I think will potentially be harmful to people? These questions are ongoing.

I want to turn now to things that are happening in biomedicine with gene editing. One of the interesting and really fun things about the CRISPR field is that as fundamental research on these enzymes has advanced it's been possible to understand enough about the way that these Cas proteins work to take advantage of properties that we didn't understand initially. Here's one example of research on a protein known as Cas12 that, like Cas9, is an RNA-guided double-stranded DNA-targeting enzyme. Cas12 has an additional property where, once it recognizes a target that's defined by its RNA guide, it turns on a *single*-stranded DNA cutting activity that can be useful for detection [East-Seletsky et al. (2016), *Nature* **538**: 270; Chen et al. (2018), *Science* **360**: 436]. This type of protein now can be used to detect the presence of DNA molecules that have a specific sequence by hooking it up with a fluorescently labeled molecule that gets cleaved only when the protein detects its target sequence. That could be extremely useful for what we call "point-of-care diagnostics": being able to tell somebody—maybe from a blood sample or other specimen in a doctor's office—whether they have a viral versus a bacterial infection, the details of what kind of infection they have, or even maybe someday things like if they have DNA circulating in their blood that contains a mutation associated with cancer. I think we'll have more of these coming up as people continue to investigate the fundamental biology of these systems.

I want to turn now to the primary way that people are imagining being able to use gene editing for clinical purposes. There are two kinds of genome editing that can be done. One is called somatic cell editing and the other is called germ cell editing. Somatic editing means making changes to the DNA in an individual that are not heritable. They cannot be passed on to future generations and they affect just that one individual. That's very different from what happens when we edit a germ cell, which would be a sperm or an egg or an embryo where the change to the DNA is heritable and it affects not only the individual, but also their offspring. This is very important. If you think about how genome editing will be used in the future—and even how it's being used right now—it gives humans the possibility of controlling the evolution of organisms or the way that they are studied in laboratories, including ourselves. As this has been contemplated, initially people focused on the opportunities in somatic cell gene editing, the nonheritable types. Going forward, at least for the foreseeable future, the vast majority of opportunities in clinical medicine with genome editing are going to fall in this category.

One example of how this type of technology is likely to be implemented in the not-too-distant future is sickle cell disease, a blood disorder that results from a single genetic change to the DNA. In fact, it's a single A–T base pair that is mutated to T–A in the hemoglobin gene in a blood cell of an affected individual. It leads to the production of a mutated protein with a valine amino acid instead of a glutamic acid that produces cells that have this very characteristic sickled shape. Because of this, these cells tend to occlude blood vessels and cause great pain to patients. We know how to diagnose sickle cell disease; we understand a lot of the biology of the disease, but clinicians don't have anything to offer to patients except for palliative care. Now imagine that you could use CRISPR genome editing to fix the disease-causing mutation. There are different ways that are already being used in laboratories to do this, and there are now several groups that are moving quickly toward applying to the Food and Drug Administration in the U.S. to do a clinical trial to test this in human patients.

There's a documentary called *Human Nature* that was made by Adam Bolt and Elliot Kirschner and kindly sponsored by the Simons Foundation [https://wondercollaborative.org/human-nature-documentary-film/]. The film begins with this boy, David, who is affected with sickle cell disease. He's a normal teenager except that every few weeks when he is affected by these crises, he has to be in the hospital. It's very moving and interesting and they use this as a way to tell the story of gene editing. I highly encourage you to see it when it becomes publicly available [now publicly available].

I think somatic cell editing is likely to be the most prevalent way that genome editing will be used in patients in the future, but there is also this potential to do heritable genome editing, meaning making changes that can be passed on to future generations. One of the very first papers that was published using CRISPR–Cas9 in animals was done in Rudy Jaenisch's lab showing that you could make modifications to the germline of mice very easily using this technology. This has now been done with lots of other kinds of animals—like pigs that are being engineered to be able to use their organs for organ donation, and lots of other examples. It was early in 2014 that a paper was published using CRISPR–Cas9 to edit monkey embryos that, for me, brought to the fore this idea of human germline editing. I wondered even at that time if people were already moving ahead to do this because it seemed like it wouldn't be that difficult.

That motivated me to take a leaf out of the work that had been done by David Baltimore, Paul Berg, and their colleagues back in the 1970s to think about the ethics of molecular cloning. We organized a small meeting in California that resulted in this publication [Baltimore et al. (2015), *Science* **348**: 36] that argued for what we called a "prudent path forward" with genome editing. We did not use the word "moratorium" for a reason, but really this was effectively a call for a moratorium on any clinical use of human germline editing until the technology could be vetted and, more importantly, for the societal discussions to happen around the use of this technology in a way that would affect all of us, potentially, very profoundly. After this publication, there was at least one paper that was published that year in which CRISPR–Cas9 was used in a research setting in human embryos, initially in nonviable embryos but then eventually in viable human embryos, so it certainly showed that one could do this kind of thing. It motivated the National Academies as well as the Royal Society to get together and hold a summit in Washington and then eventually publish this report that came out in February of 2017 [National Academies of Sciences, Engineering, and Medicine. 2017. *Human Genome Editing: Science, Ethics, and Governance.* The National Academies Press, Washington, DC] that also argued for a moratorium on any clinical use of human genome editing.

In November of last year, I received an e-mail the day after Thanksgiving from this gentleman, He Jiankui, who informed me that he had in fact used CRISPR–Cas9 in human embryos, and those had been implanted and resulted in the birth of twin girls in China that had edited genomes. He presented his work at the international summit in Hong Kong on this topic, the second summit that was sponsored by the National Academies to discuss this. As you can imagine, this caused quite an uproar. One of the things that was very clear from his presentation was that it was relatively easy to use CRISPR–Cas9 in human embryos, as we had envisioned, but very difficult to do it well.

To illustrate this, I want to show you this figure [https://twitter.com/RyderLab/status/1068128997656207361/photo/1] from Sean Ryder at the University of Massachusetts. The top panel illustrates a gene called CCR5 [C-C chemokine receptor type 5] that encodes a protein essential for human immune cells to be infected by the HIV [human immunodeficiency virus]. That was the gene that Dr. He decided to target with CRISPR–Cas9 in these human embryos from parents in which the father was HIV-positive. It's been established that there's a few rare individuals in the human population that naturally have a deletion of 32 bp in this CCR5 gene that protects them from HIV infection [second panel]. The stated purpose of Dr. He's study was to introduce this mutation into these girls' genomes to protect them from future infection by HIV, except he showed these data right here [panels 3–5]; these were actual sequences from these girls' genomes. You can see that none of the changes that were introduced look like the naturally occurring deletion. So, yes, the genomes were apparently edited, but none of them received the change that occurred naturally in human populations. Instead, what was done here was to introduce changes that had never been seen in humans and never even been tested in animals, which really is quite shocking. It also emerged that the way that consent was obtained and provisions for following the health care of these girls after birth had really not been done properly either. On many levels, this came across to most people as deeply flawed and really wrong. This highlights why it's very dangerous to do this kind of genetic manipulation before we really understand our own genetics well enough to know how to predict the outcomes of these kinds of manipulations.

So, yes, there's a lot of talk about "CRISPR babies" in the media—babies engineered for things like high IQ, or 20/20 vision, or no baldness, or lower risk for a variety of diseases—but the reality is that we don't understand enough about the human genome and our own genetics to be able to do any of this. Most of these kinds of traits have not just one genetic cause; they have many genes that contribute to those traits. While CRISPR will help uncover those genetics over time, we're not there today. We also know that the technology is, frankly, not appropriate right now to be using in human embryos either. Fortunately, the World Health Organization and the National Academies of Science, Engineering, and Medicine have announced international forums that are going to be looking into this and making what I hope will be very restrictive guidelines that will be in place for anyone that might contemplate using CRISPR–Cas9 or any other genome editing technology in human embryos in the future. I think it's very important at this point not to call for yet another moratorium, but rather to keep a very open dialogue. There's a lot of interest in this and we can't put it back in the box; we really have to be open about it. People need to be discussing it actively, and I'm very glad that these international groups are going to look into it carefully. I hope they will proceed apace to do this in a timely fashion.

It's been incredibly interesting and exciting for me over the last few years to be involved in this field. I came to this from something very different. I'm an RNA biochemist and structural biologist. I was not doing anything with human genetics or any other kind of genetics. We were pulled in this direction because of the power of RNA-guided gene regulation in microbes, leading to the deployment of this technology across many other organisms. Both delivery and control of the DNA repair processes are going to be critical for this technology to go to the next level; we're working hard at Berkeley and with various collaborators on both of these aspects. Of course, fundamental research continues; we continue to understand more about the fundamental biology of CRISPR and other types of pathways and microbes. I personally feel very excited about uncovering what's out there in nature. It's an amazing time to be a biologist. In biology, we have fundamental questions that get asked in the laboratory and we uncover new information. Sometimes, that leads to new technologies, and those new technologies, in turn, allow yet more and new experiments to be performed. Technologies and fundamental science really do go hand in hand.

I'm going to close by thanking a fantastic group in my laboratory: Yavuz Dagdas, Ahmet Yildiz, Eva Nogales, Jill Banfield, and Jamie Cate at the University of California, Berkeley; Emmanuelle Charpentier at the Max Planck Institute in Berlin; and Maria Da Costa and Joel Palefsky at the University of California, San Francisco. We're extremely grateful to the National Institutes of Health, the Paul G. Allen Frontiers Group, the Howard Hughes Medical Institute, the Defense Advanced Research Projects Agency, and the National Science Foundation for their financial support. I also want to call your attention to our Innovative Genomics Institute in the Bay Area. It's a partnership between Berkeley, UCSF, and the Gladstone Institutes that is fostering academic research, but we also encourage partnerships with companies to advance genome editing to solve practical problems, and you can check us out on our website [https://innovativegenomics.org/]. Thank you.

CONVERSATIONS AT THE SYMPOSIUM

Video recording of the Symposium Conversations is available on Cold Spring Harbor Laboratory's Leading Strand channel on YouTube (playlist 84th CSHL Symposium Interview Series RNA CONTROL AND REGULATION).

A Conversation with Karen Adelman[1]

INTERVIEWER: JAN WITKOWSKI

Cold Spring Harbor Laboratory

Karen Adelman is a Professor of Biological Chemistry and Molecular Pharmacology and Co-Director of the Epigenetics and Gene Dynamics Initiative at Harvard Medical School.

Jan Witkowski: The subject of your work [The Integrator Complex] sounds like a wonderful title for a film. What is this 'Integrator'?

Dr. Adelman: It's a very large complex of many proteins—about 14, at least—and it's actually quite understudied. It was first discovered by Ramin Shiekhattar's lab as something that causes the termination at noncoding RNAs like the snRNAs [small nuclear RNAs]: these are RNA components of the spliceosome. It had been thought to be specific for the snRNAs but as people started to peek into the world of Integrator, it was realized that it in fact does bind other loci. What we've recently started to appreciate is that it binds at protein-coding genes and does much of what it does at these snRNAs. Except at snRNAs, Integrator causes the cleavage of the RNA and termination after just a few hundred nucleotides. That's okay for an snRNA because it's just a few hundred nucleotides long; that's the correct 3' end. But when it does this at protein-coding genes, that's obviously not the entirety of the protein-coding gene. When it does this, it makes a short transcript that's prematurely terminated and that's nonfunctional, so it's rapidly degraded. What we have is a situation that by studying noncoding RNAs we've learned something new about protein-coding machinery.

Jan Witkowski: Why does it do this? Why does Integrator truncate these—what presumably are going to be quite normal—full-length messages?

Dr. Adelman: This is a fantastic question. This is why we're just at the beginning of understanding what Integrator does. The canonical models for protein-coding gene expression have largely focused on understanding how the polymerase gets brought to the DNA. The process of elongation has been presumed to some extent that—once the polymerase gets started, it goes to the end and then the cleavage and polyadenylation machinery is what takes it off the DNA finally, and that this process of transcription between start and stop is more or less a nonregulated or default pathway.

The question now is: If that's not true, why? Why would the cell evolve such a thing? The answer, of course, is: We don't know. But, you can speculate that when the polymerase initiates transcription and generates a short RNA and then undergoes this step that my lab has studied for many years, this promoter-proximal pausing, it stops in the promoter region associated with this little 20- to 60-nt RNA. It sits there, and it waits for a signal to be released into the gene.

There are many reasons you can think of that you wouldn't want that polymerase to stay there forever. One is, because this is a leaky process, you may not want the polymerase to eventually escape and transcribe that gene at a low level. The other is that maybe you just want to get the polymerase off of that gene: if the replication fork is coming through, if you have some kind of stress, or if you have something else where you really want to clear out that region. Having a regulated termination factor that clears the genome of this promoter-proximal polymerase, you could envision many reasons why this would be helpful. We envision it mostly as a way of regulating transcription at a level where, at any moment, you can change your mind. You can tell the polymerase instead of terminating to continue on into the gene, so you can very rapidly upregulate the expression of these genes.

Jan Witkowski: They must be quite carefully controlled, because otherwise Integrator could be chopping things from every transcribed gene.

Dr. Adelman: Absolutely. I think these are some of the exciting questions for the future. In the snRNA example, Integrator is selectively recruited to the snRNAs through interactions—presumably with the SNAPc [snRNA-activating protein complex] protein—with a transcription factor that binds to the start site. That's one level of regulation, which is probably selective recruitment. We're trying to understand what might selectively recruit Integrator to specific protein-coding genes.

At the other level in the snRNAs, there's a sequence in the RNA that's recognized by Integrator. It's called a 3' box and, much like the poly(A) site and the canonical cleavage and polyadenylation pathway, it's an AU-rich sequence. When Integrator recognizes it, it seems to stim-

---

[1] This Conversation covers collaborative work done with Eric Wagner (UTMB Galveston).

© 2019 Adelman. This article is distributed under the terms of the Creative Commons Attribution-NonCommercial License, which permits reuse and redistribution, except for commercial purposes, provided that the original author and source are credited.

ulate the cleavage activity. What we're looking for now are the transcription factors that bring Integrator to certain protein-coding genes, the sequences within coding genes that may serve the role of this 3' box, and then any other levels of regulation that may be figuring into this.

We do know that under different conditions, Integrator can be deployed to different sets of genes. We've looked at the metallothionein genes; this is in collaboration with Jeremy Wilusz. He sees that, under normal conditions when cells are happy, Integrator's not at the metallothionein genes. But when you stimulate metallothionein, Integrator suddenly goes to metallothionein genes and helps to attenuate that response so that it doesn't get out of control. In that situation, it's not meant to keep genes "off" under basal conditions. Rather, it's meant to dampen the expression so that an activation process doesn't cascade too far.

Jan Witkowski: Sort of a fail-safe mechanism.

Dr. Adelman: Exactly.

Jan Witkowski: Does Integrator work at all genes, or is it a particular subset or type of genes?

Dr. Adelman: We see it recruited somewhat selectively to a particular subset of maybe 15%–20% of all active genes in the different conditions and cell types that we've looked at. The work that we've done in *Drosophila* cells and a limited number of mammalian cells—mouse cells—is consistent with what's been shown in human cells by the Shiekhattar and Benkirane and Pagano labs. It seems to be not at every gene, but it seems to somehow have a subset that it selects. What we find interesting is that that subset can be different in different cell types we've looked at, and under different conditions. I do think that there's a very selective targeting so that you don't have this crazy endonuclease chopping up all the wrong things, but actually going to the places where gene expression should be attenuated.

Jan Witkowski: It could be a monster.

Dr. Adelman: It could be. But the phenomenon, the concept, has been around for decades in the termination processes that we know and love in bacteria.

Jan Witkowski: I was just going to say, is there a similar transcriptional control?

Dr. Adelman: There are terminators and antiterminators. In particular, in the lambda bacteriophage there's a terminator/antiterminator switch in the promoter-proximal region of some of the master regulatory genes. We also know that this process exists in other organisms. In the budding yeast *Saccharomyces cerevisiae* there's a complex called NNS [Nrd1–Nab3–Sen1] and there too there's an RNA recognition, there's selective recruitment, and there's activity toward both noncoding RNAs and a subset of coding genes. The basic idea that cells use a directed termination is actually out there. It's not something that we are the first to propose. In the mammalian system, this idea that you have regulated promoter-proximal termination has been out there for many years, but to nail down the complex and the conditions where it happens, that's what's exciting and new at this point.

Jan Witkowski: Does that require the development of techniques that were able to do those sorts of experiments?

Dr. Adelman: Absolutely. To be able to separate premature transcription termination from all the subsequent processes of RNA degradation and decay, you need to be looking directly at the nascent transcriptome, to look directly at the nucleotides that are inside the RNA polymerase at any point in time. My lab has developed some techniques, and we've also optimized and used techniques developed by others—like John Lis' lab—to really investigate the nascent transcriptome at single-nucleotide-level resolution.

The data that convinced us we had a handle on what Integrator was doing was the data where we could see at these Integrator-attenuated genes very high levels of polymerase that had gotten onto the promoter, begun transcribing, and then gotten as far as +50 into the gene but didn't make it any farther. Only when we knocked down the Integrator subunits could we then see elongation of the polymerase past that point. That's what really told us that we weren't looking at an RNA stability effect. We weren't looking at something that was posttranscriptional, but we were really looking at a transcriptional cleavage and termination event. Those nascent RNA assays were really essential for being able to understand that.

Jan Witkowski: What's known about how Integrator does it? It's obviously got RNase activity. Did you say how many subunits it is?

Dr. Adelman: The Integrator complex has at least 14 subunits. The field is still trying to understand how these separate into subcomplexes, but clearly there's a subcomplex that contains the Integrator 4, 9, and 11 subunits. Nine and 11 are really interesting, because they're paralogs of two subunits in the normal cleavage and polyadenylation machinery, CPSF [cleavage and polyadenylation specificity factor] 73 and 100. Just like CPSF73 and 100, both proteins have a β-lactamase domain and a β-caspase domain, and this gives them the ability to cleave RNA. But, in both Integrator and the normal cleavage and polyadenylation machinery, one of the two subunits is catalytically active and in the other the active site is sort of degraded so that it still can bind to RNA, but it doesn't cleave RNA. You have an auxiliary subunit and the active subunit, and this arrangement is paralleled in the normal cleavage and polyadenylation machinery and in Integrator. A lot of the insights that we have into how Integrator is cleaving really come from the vast literature on these β-lactamase domains and on the cleavage and polyadenylation machinery.

Jan Witkowski: If I could wave a magic wand for you, what would you really like to know about Integrator?

Dr. Adelman: What I would really like to know about Integrator is very specifically about its targeted recruitment to certain genes. What we have found is that you can take a gene that has accessible chromatin, all the right histone modifications, transcription factor binding, it brings in the polymerase, and yet if you bring in Integrator too, the gene is not on. The gene's expression is very, very low. If you could imagine a situation—a disease context or a developmental defect—where one gene is aberrantly upregulated, if you could bring Integrator selectively to that gene's promoter or find some key part of Integrator to bring to that gene, you could turn it off in a way that doesn't need to supersede all of the important things that the cell does to activate a gene, but—just at that final step—say, "No, we're going to turn this gene off." I think that would be incredibly powerful, both as a therapeutic approach, but also as a basic biology tool for understanding gene activity.

Jan Witkowski: What happens if you knock Integrator out?

Dr. Adelman: It's lethal. We've been working a lot on really rapid degradation techniques to be able to investigate the effects of knocking out Integrator as soon after getting rid of it as possible. We're moving from RNA interference to degrons to be able to get rid of Integrator within an hour of flipping the switch, because, as I mentioned, Integrator also is responsible for the cleavage of snRNAs. If Integrator is nonfunctional for too long, splicing goes awry and then you end up having lots of issues. We're really trying to find the most fast-acting strategies to, in a targeted way, block Integrator's activity. In particular, we're really interested in blocking the activity of the catalytic subunit and then replacing with mutants that are either lacking the ability to interact with the other subunits or lacking the ability to undergo this catalysis.

When you knock Integrator down, there are a lot of defects. A lot of genes get turned on that shouldn't be on. You can rescue with wild-type Integrator 11, but if you try to rescue with a catalytic-dead Integrator 11 subunit, not only do those genes' expression not get turned back down, but they actually are further enhanced. It's more like a dominant negative than a rescue. That suggests that you can also form these subcomplexes that co-opt parts of the machinery that presumably—with a 14-subunit complex—does more than just cut the RNA. We're trying to get into fast-acting strategies to pick apart what the other subunits do as well.

A Conversation with Andrés Aguilera

INTERVIEWER: LARA SZEWCZAK

Senior Scientific Editor, Cell

Andrés Aguilera is Professor of Genetics at the University of Seville and Director of the Andalusian Centre for Molecular Biology and Regenerative Medicine (CABIMER).

Lara Szewczak: When I think about the work that you've done, RNA control and regulation doesn't exactly come to mind, because it's more focused on genome instability. Where do you think the worlds of genome instability and RNA-based control mechanisms come together?

Dr. Aguilera: Our main interest is to understand the mechanisms of genome instability, but more than 20 years ago we entered into trying to understand a number of proteins and genes whose function was to keep the genome stable. Little by little, we began to realize that this genome maintenance function was very much dependent on transcription. Then we entered into the discovery of R-loops as a main source of the instability that was produced in the cells when these proteins disappeared. This turned out to be an important source of genome instability: The RNA that is being produced during transcription can go back to the DNA from where it came from and form a DNA–RNA hybrid. If the formation of this hybrid is not regulated in a natural way, it would lead to a block of the replication fork on the one hand, or it might create a loop in the DNA that may be subject to attack by different genotoxics. That leads to DNA damage, DNA breaks, which is what leads to genome instability. We realized that the main goal of transcription is to produce the RNA, but it deals with many dangers and cells have to be careful that the RNA that is being produced doesn't have the opportunity to go back and form these DNA–RNA hybrids.

Lara Szewczak: That's the R-loop, right?

Dr. Aguilera: Yes, and this is where the two fields join together. This is why I'm trying to understand this aspect of transcription that, so far, had not been studied that much. Many groups are trying to understand whether there is a special machinery in the cell that tries to control these nascent RNAs to keep them from forming these DNA–RNA hybrids that are going to be harmful for the cell.

Lara Szewczak: If you go back to the cells, what's the evidence that these actually do form? Clearly they can, but what's the evidence that they do and that they contribute to genome instability?

Dr. Aguilera: The DNA–RNA hybrids are present in a number of processes that have natural relevance—for example, in class switch region recombination and in mitochondrial DNA replication. We are talking about unscheduled DNA–RNA hybrids that theoretically shouldn't be formed. It can be tested by different manners, by purifying the DNA from the cell and then doing different treatments with RNase A or RNase H so that you are able to distinguish the kind of structures that it forms. But the most popular way of detecting these hybrids now is via a monoclonal antibody that was developed in the '80s. This antibody is used now worldwide to detect the DNA–RNA hybrids by DNA–RNA immunoprecipitation and by immunofluorescence. This antibody can also recognize double-stranded RNA. Therefore, it is very important that in doing this analysis, we always remove the signal that is recovered by the antibody by using RNase H, because RNase H is known to specifically degrade the RNA moiety of the DNA–RNA hybrid. This is the most popular way of detecting it, but now people are developing other ways. For example, we have fused the DNA–RNA hybrid domain of RNase H to GFP [green fluorescent protein], and this is a way to detect the hybrid directly. Some other people are using mutated RNase H so that when it goes to the DNA–RNA hybrids to eliminate the RNA, it is not able to do that. Then it is used as a marker that the DNA–RNA hybrid exists there.

Lara Szewczak: When you use those kinds of approaches, how frequently do you see R-loops happening?

Dr. Aguilera: That is a very important question: frequency. We really don't know. It's very low. Most analyses are done by genome-wide immunoprecipitation and usually this means that we are dealing with a population of cells. That allows us to see where the hybrids are forming, but if we now try to identify the frequency at a particular site in the genome, it's very low. We have tried to see that in single cells and single colonies. We have gone through 500 different colonies to check in a particular site, and we didn't find evidence for the hybrid being formed. That means that the hybrid is not something occurring very frequently, but it is very important. Why? Because once it is formed it can lead to damage. It's the same as a

---

© 2019 Aguilera. This article is distributed under the terms of the Creative Commons Attribution-NonCommercial License, which permits reuse and redistribution, except for commercial purposes, provided that the original author and source are credited.

double-strand break. A spontaneous break doesn't occur very frequently, but when it happens the consequences are very important. The key question here is that if you remove the machinery in the cell that prevents the formation of the hybrids or that eliminates the hybrids, then the frequency goes up causing high level of breaks; this is going to be harmful for the cell.

Lara Szewczak: What do we know about the machinery?

Dr. Aguilera: I would divide the machinery according to three main functions. The first one covers the proteins that prevent the formation of the R-loops during transcription. These usually are proteins that are involved in RNA processing. Their main functions are probably RNA export and transcription. They have to do with RNA metabolism itself, but collaterally they are preventing the formation of the loop, because if the nascent RNA is packed by proteins into an mRNP [messenger ribonucleoprotein] particle, then that particle is not free to hybridize into DNA. That would be the first body of proteins that prevent the formation.

The second mechanism is: If those hybrids form spontaneously, now you need a machinery to remove them. The best-known one is RNase H, but if you think in terms of transcription you don't want to eliminate the RNA that is just being produced. In human cells, genes can be 600 kb and then spontaneously a short region can form the DNA–RNA hybrid. If this is removed by RNase H, then it will cleave the RNA. You destroy the RNA, and you have to start back again. Therefore, you want to eliminate the hybrid but allow the RNA to stay there.

Third, if hybrids still remain there, the problem comes after cells get into S phase and the replication fork comes and finds the hybrid. There are a number of proteins that are involved in DNA repair: Fanconi anemia, BRCA. When you remove them you see accumulation, but not because you form more hybrids, but because the fork gets stalled there and you are not able to remove them and therefore, the hybrids remain there.

Lara Szewczak: You've outlined these three types of machinery. What are you most excited about right now?

Dr. Aguilera: I'm most excited about the idea that during transcription, we never thought that cells had collaterally developed extra functions—for proteins that are already there—to prevent the formation of R-loops. One of the things that we have found is that these RNA-binding proteins talk directly to chromatin remodelers in a way that, once the RNA is coming out of the RNA polymerase and it binds to these proteins—the ones that we are focusing on, for example, are the THO complex and some others—they talk to the chromatin histone deacetylases as a way to, what we believe might be, transiently close chromatin to prevent the RNA of getting back to the DNA. Because when you have hypoacetylated chromatin you have more compacted chromatin and therefore you avoid hybrids. But at the same time these proteins talk to a known RNA helicase—UAP56 in human cells —that is very abundant and is known to be an RNA chaperone that works in splicing and other RNA metabolic steps.

Then, we see in these proteins DNA–RNA unwinding ability. When we overexpress these proteins in a number of conditions we are able to eliminate hybrids. We believe that during transcription there is an associated machinery that is working as a kind of checkpoint on the one side, but at the same time as a kind of machinery that prevents the formation of the R-loops. And if these are formed, the helicase will open the DNA–RNA hybrid so that it will allow the transcription to move on. This is a concept that we never thought before, but I think this explains why we started trying to understand the origin of genome instability and we ended trying to understand what happened with the RNA that is coming out during transcription so that the cells have developed this machinery to eliminate the hybrids.

In bacteria, this is probably not a major problem at this level because in bacteria there is cotranscriptional translation; the ribosomes get onto the RNA at the same time this is being produced. Somehow, that also prevents the formation of the hybrid, because this RNA cannot go back to the DNA. In nature, the danger of this nascent RNA going back to the DNA and forming hybrids that would become harmful is there. Evolution has developed this ancillary machinery just to prevent and eliminate them.

Lara Szewczak: I think there have been recent examples in bacteria where there is uncoupling between transcription and translation, so it seems like a place where maybe some new discovery can go back and happen, if you're thinking about this mechanism.

Dr. Aguilera: Absolutely. Actually, in bacteria there are examples of RNA polymerase backtracking, in which you are really dealing with all this and then you get a higher chance of hybrids remaining there. Even then, the mechanics may be different because the RNA polymerase is moving back. There might be different ways of affecting hybrids.

The field is opening and expanding and probably we have to move cautiously now. One of the bigger issues that I'm finding is that there are many RNA helicases that, when you go to in vitro and check activity, most of them open DNA–RNA in vitro but this doesn't mean they act like that in vivo. We have data that suggests that only a few of the proteins are really involved in this. Many others, we have to be cautious. At the moment, we make conclusions trying to transfer what we see in vitro to what we see in vivo, because many of these proteins are RNA chaperones. It's hard to explain that the cells are going to have so many DNA–RNA helicases so that by just removing one you will get the same phenotype as when you remove the other. That means that there is no redundancy, and that is hard to believe. In that case, I believe that all these proteins are working in the making of the mRNP particle and if you don't make them properly you increase the chance of this RNA going back to the DNA.

Lara Szewczak: Roy Parker's talked about potential different ways to think about all the helicases.

Dr. Aguilera: Absolutely. I believe that the major function of those helicases is this one, because RNA helicases have to live with the standard RNAs. But collaterally some of them—UAP56, for example—are so abundant in the cell that, partially, one of the things that they do is to remove the hybrids. You have to think that this protein works cotranscriptionally. You can see them in the DNA. They are recruited to chromatin at the same time there is transcription; they are there during transcription. It makes sense that they might do that job.

Lara Szewczak: You've got the polymerase and all its accessory factors and now you've got all the factors that are binding the RNA and now these helicases that are there sort of on surveillance. It starts to feel like a very crowded environment.

Dr. Aguilera: Absolutely.

Lara Szewczak: You said that you've got an example where something is talking to chromatin modeler. What do you mean by "talking to" and how direct are these interactions likely to be in this very complex soup?

Dr. Aguilera: I believe that it is a complex structure. I don't think it's a direct interaction. We can see that in vivo they happen in the nuclei but when we check other proteins that we know they also interact with—for example, the THO complex—we can see also that they interact. We believe they are in a large complex that is able to interact. We don't know physically who is making the real interaction. We need to explore this farther, but this is—as you mentioned—a crowded structure. Probably, there are different things going on at different times.

For example, one interesting thing we are approaching now is what happens between G_1 and S phase. We are observing that not all proteins are protecting from hybrids in the different stages of the cell cycle. When you remove some of them you see an increase in G_1 and also in S phase, but some others when you remove them you see an increase in hybrids in S phase, but not in G_1. Therefore, some of these proteins may have a specialized function, probably not only in this specific structure but during the cell cycle. Therefore, we have to deal with many things. When we go into human cells, there is very strong evidence that in neurons hybrids are formed at high levels. Probably what is happening in noncycling cells versus cycling cells is something that we have to explore farther, because the factors that protect one versus another may be different.

Lara Szewczak: When you think about different cell types where you have to worry about R-loops and their biology and the physiology downstream, what are going to be the implications of understanding these mechanisms?

Dr. Aguilera: This is what brings our research to the side of biomedicine. Genome instability is a hallmark of cancer cells. We get many R-loops at the same time leading to genome instability: This is something that we expect to be increased in cancer cells. One thing that we didn't expect was when some years ago we found out that when you remove *BRCA2*, then you increase the R-loops. I think this is a connection that might be somehow present at least in a number of cancers. If that is the case, then certainly the physiology is affected. At the same time, we can consider that this structure might be used as a diagnostic or in some cases even as a target. If the R-loops are enhanced in a number of cancer cells and you get into some molecule that gets into the R-loops and blocks them, you might specifically kill the cells that contain high levels of R-loops versus others. I think there is a good reason to move farther, not only to understand the conceptual part, but the implications.

Lara Szewczak: Your lab focuses a lot on basic research but the idea of a molecule being able to get in either as a diagnostic or as a therapy: Are you dabbling in that space too?

Dr. Aguilera: Absolutely. We are indeed working with companies on a number of tumoral genotoxic agents to see whether they affect better the cells that are accumulating R-loops and whether they target better to regions where the R-loops are accumulating.

A Conversation with David Bartel

INTERVIEWER: RICHARD SEVER

Assistant Director, Cold Spring Harbor Laboratory Press

David Bartel is a Professor of Biology, a Member of the Whitehead Institute, and an Investigator at the Howard Hughes Medical Institute at the Massachusetts Institute of Technology.

Richard Sever: You spoke yesterday about microRNAs, the small RNAs that regulate the activity of genes post-transcriptionally. You made the point that the effects of knocking down these microRNAs could be really debilitating. That was kind of a shock to me because people are always saying that it's a fine-tuning of gene expression. Can you say a little bit more about that?

Dr. Bartel: I like the term "tuning" rather than "fine-tuning" because it actually is more of a tuning. It turns out that these microRNAs do have very striking phenotypes when you knock them out, but even before we knew the knockout phenotypes, we knew that they recognized many targets. We know that there are 90 families of microRNAs highly conserved from humans to fish. Each of those 90 have on average about 400 preferentially conserved targets. That adds up to more than half of the human genes that are conserved targets of microRNAs. On average, they're targeted by four or five different microRNAs. They [the microRNAs] have these very widespread effects.

It is true that for each microRNA–target interaction, that [regulation] can be a rather subtle—maybe 20%—down-regulation, sometimes 30%, sometimes more. But because they have so many targets, when you knock out the microRNA in mice—which is what many labs have been doing; they've knocked out one or several members of the same family of microRNAs—and when they look at the phenotypes, for these 90 conserved families nearly every one of them where knockouts have been reported on, they certainly do see a phenotype. They've looked at over half of them and seen phenotypes. We already know that 15 of these microRNA families, when you knock them out, you have embryonic lethality or perinatal lethality, so that's pretty severe. And there are others that also have [other] very severe phenotypes like blindness, deafness, infertility, seizures, epilepsy, cancer, etc. There's a huge range of phenotypes—and many of them very severe—from the microRNAs. So, we know that they're playing very important roles.

Richard Sever: Are you saying that the severity of the phenotype is not because of the quantitative effect on any given gene, but just because you're hitting a whole bunch of them?

Dr. Bartel: I think that's the easiest way to think about it. In some cases, people have found single individual targets where that 50% down-regulation has a dramatic effect. But even there, they have hundreds of other targets that are conserved more than they would expect by chance. Certainly, over the course of evolution we know that biology cares about many more than single targets, and, again, each of them is regulated by a relatively small amount of tuning. So, that's our current view of what's occurring.

Richard Sever: For those of us stuck in the '90s, how does that compare with the way we think of transcription factors? Do you think that we shouldn't be singling these out as different in the way they work?

Dr. Bartel: Depending upon the transcription factor, you might also see these minor effects on many targets; it may not be that much different. There are some transcription factors that will have a 40- or 50-fold effect on the transcription of a certain mRNA, but I think, more generally, transcription factors might also have widespread effects—on the level of microRNAs—and may not be a lot different for a lot of those interactions.

Richard Sever: The way the microRNAs act is by exerting their effects on messenger RNA [mRNA]. There's two possibilities there: stopping its translation, and degrading it. You've made the point that actually it's much more of one than the other.

Dr. Bartel: What we see is that—with one exception—every place where we've looked, over two-thirds of the repression could be explained by the microRNA recruiting factors that deadenylate the RNA and then cause the RNA to get destabilized. Often, over 90% can be explained by this degradation mechanism, leaving somewhere between 10% (sometimes less) up to a third with this additional mode of repression that's through translational repression —inhibiting, presumably, translation initiation. So, in general, the majority of the effects that we see are through this mRNA destabilization.

The exception, which is really interesting, is that in early zebrafish [embryos], you completely miss what the microRNAs are doing by looking at the changes in mRNA levels. Whereas everywhere else—it's kind of nice—you can perturb the microRNA and be able to know what the targets and the effects of the microRNAs [are] just by looking at changes in mRNA, which is much easier to do than looking at changes in protein or ribosome-protected fragments. But, in the zebrafish [embryos], that would not work at all. That was shown by Antonio Giraldez's lab —that in the early zebrafish embryo, the effects of the microRNAs could only be seen when you look at the changes in ribosome-protected fragments.

Richard Sever: That's this early phase of development in zebrafish. Is it early phases of development in other organisms or is it something unique to a particular lineage?

Dr. Bartel: It's a good question. I think that the same will probably also hold in frogs and in flies, but I'm not sure that people have done those experiments. We do understand what's going on in zebrafish, and we think—based on what we see in these other organisms—that it'll also hold there.

The effects of the microRNAs is that they recruit a protein called TNRC6 [trinucleotide-repeat–containing 6] that in turn recruits the deadenylation complexes, the Pan2–Pan3 or Ccr4–Not complexes, and that causes the poly(A) tail to get shorter. Once the tail gets to be a certain length, you get decapping and decay of the mRNA.

What we see in early zebrafish is that the microRNAs cause the tail to get shorter, and in that developmental context—in that regulatory regime—what we find is that mRNAs with short tails are translated much less efficiently than mRNAs with long tails. There's this very strong coupling in the early zebrafish, before gastrulation, where short-tailed mRNAs are translated much less efficiently than long-tailed [mRNAs]. What's interesting is that that goes away at 6 h postfertilization, which is the [time of] gastrulation of the fish. At that point, the microRNA also causes the tail to get shorter, but in that context, a short tail leads to [mRNA] degradation.

Richard Sever: Sticking with the pregastrulation scenarios, is there an a priori reason for that? Going back to this kind of tunability, if you're stopping a translation rather than degrading the RNA, does that give you more ability to tune? Is it reversible?

Dr. Bartel: I think you're onto it there. I'll just add that we see this same sort of transition in translational control in frogs and in flies, and this coupling between the length of the tail and the efficiency of translation—which is very strong in the early embryo of frogs, flies, fish—is also very strong in oocytes, and then goes away around gastrulation. The reason that we think that there is that coupling between the length of the tail and the efficiency of translation is that this is a time in development—at least up until about 3 h postfertilization—in which there is no transcription. The mRNA is all maternally inherited, and yet the embryo cells still need a way to regulate genes. They can't do it by transcription, and so the way they do it is by changing the tail length. They have this phenomenon called cytoplasmic polyadenylation, which will extend the length of the tail, and that will cause a massive increase in translation. They also have other ways of shortening the tail, so they can adjust the amount of protein output [in either direction].

Richard Sever: Do you get an oscillation there?

Dr. Bartel: Yes, [but more generally,] depending on the gene, you can just get different output at different points in early development through this polyadenylation mechanism.

So, then the question is, "Why does it [this coupling between tail length and translation efficiency] go away?" And the reason we think it goes away is that later in development, transcription is already started up, and that's a great way to regulate genes. So, you don't need this tail-length control to regulate translational output; instead, what the cells do use is transcription, and they also use mRNA stability. That's what the microRNAs are doing; they're changing the stability of the mRNAs—and that's what we see in all these postembryonic contexts.

When the cell is using mRNA stability to regulate genes, then it's not such a great strategy to only translate the mRNAs with long tails, because those are the mRNAs that just came into the cytoplasm, whereas the RNAs that had been there a long time have shorter tails, and the cell would not want to discriminate against them—it wouldn't want to do this "age discrimination" against the older mRNAs. If it's had those RNAs stable, it wants to use them; it wants to use the RNAs with short tails just as much as it wants to use those with long tails—if it's using mRNA stability as a mechanism for gene regulation. So, the idea is that when the cell switches over to this mechanism of using mRNA stability to regulate genes—which they use a lot; you have these massive differences in mRNA stability—then the cells switch away from controlling translation based on the length of the tail.

Richard Sever: When you were looking at the length of the poly(A) tails, you were discriminating between steady state levels and a better experimental approach to make sure you know what the length is and when. Can you tell us a little bit about that?

Dr. Bartel: Sure. What I described so far was in the early zebrafish embryo, and there it's not really a steady state situation. You have development happening, and you can see these tail-length differences very readily for [the targets of] the microRNAs. But interestingly, if you just look in mouse 3T3 cells—the fibroblast cells of mouse—and you're just growing cells at steady state, and you have a microRNA that you've induced that is there at very high levels, and then you just look at the poly(A)-tail lengths of the mRNA targets of that microRNA, surprisingly, at steady state, you see no difference. The distribution of tail lengths is the same whether or not the mRNA is a target of the microRNA or not. That caused us to realize that we shouldn't really be looking at steady state measurements. We need to look at pre-steady state. We need to look at what's happening to the newly transcribed mRNAs and beyond and get measurements of the dynamics that you can't get at steady state.

Richard Sever: What's the timeline that you're thinking about them in discriminating the point at which steady state becomes a reasonable measure?

Dr. Bartel: At about 8 h things have pretty much approached steady state in these cells. There aren't that many mRNAs that have half-lives much longer than that. There's some, but you've pretty much approached steady state by 8 h. So, we'll do a metabolic labeling time course that will take a very early time point, and some intermediate ones, and then 8 h, and then just isolate the RNA that was transcribed over those time intervals and measure the tails. When we do that, we can see there are very clear differences in the microRNA effect; we can see that the microRNAs are clearly shortening the tails [of their targets], which we can see at these intermediate time periods.

Richard Sever: When the tails shorten, is there a threshold at which you're basically done: You're going to get degraded? What's the correlation?

Dr. Bartel: Yes. So we set out to do these experiments to look at this microRNA effect, but we can look at these more general principles of tail-length shortening, deadenylation rates, and then the rate of decay once the mRNA gets to be a short tail length. We can get these measurements, which we knew for a very small set of genes previously [from] people in the '90s, like Ann-Bin Shyu's lab and others in mammalian cells, and Roy Parker's lab in yeast. In mammalian cells, we knew from those early experiments the tail-length dynamics and mRNA-decay dynamics that are associated with tail length for [only] four mRNAs. When we realized that we had the data sets now to be able to look at this for thousands of mRNAs, that was very exciting to us. Now we can use these metabolic-labeling data sets to model—for mRNAs from close to 3000 different genes—we can model the initial tail length when the mRNA goes into the cytoplasm; what is the initial tail length at that early point? And then, how rapidly did the tails get shorter? And then, once they get shorter, how rapidly are they degraded? And we get that information now for close to 3000 mRNAs from 3000 genes.

Richard Sever: Is there a broad spectrum of behavior? Do they all look the same? Does this all happen at 30 nt?

Dr. Bartel: What we see is that the deadenylation rates for mRNAs from different genes can vary widely—a 1000-fold. There are some mRNAs that are deadenylating at 30 nt per minute, and there are others where the tail is shortening at 0.03 nt a minute: 1–2 nt per hour, on average. That's one thing we find.

What's also interesting is that once the tails reach a short length, the rate of decay can also vary a 1000-fold. The short-tailed mRNAs from genes where their mRNAs' tails get shorter more rapidly—once they reach a short tail length, the mRNA decays more rapidly. But, if the tail is deadenylating more slowly, once they've reached a short tail, they get decayed much more slowly. This is very interesting to us. What it does is it prevents the mRNAs that have very rapid deadenylation from accumulating in the cell with very short tails. Otherwise, you could have this big buildup of very short-tailed mRNAs. We just don't see that, and this explains why: It's because for the mRNAs where their tails get shorter more rapidly, they also then decay more rapidly once they reach a short tail length. That allows these huge differences in deadenylation rate—a 1000-fold difference—to impart a similarly large difference in decay rate. And so you could have a really broad spectrum of posttranscriptional behavior for mRNAs from different genes, and of course that, together with transcription and other things, is what gives us the regulation that we have.

A Conversation with Ling-Ling Chen

INTERVIEWER: STEVE MAO

Senior Editor, Science

Ling-Ling Chen is a Professor of Shanghai Institute of Biochemistry and Cell Biology at the Chinese Academy of Sciences.

Steve Mao: Although you set up your independent laboratory less than 10 years ago, you've been studying RNA for a long time. In the last two decades, what do you think of the progress made in the RNA field?

Dr. Chen: We've had the wave of the noncoding RNA being discovered by so many different labs, from small RNAs to long noncoding RNAs [lncRNAs] to circular RNAs [circRNAs]. Those molecules are so fascinating. The discovery of so many of the noncoding RNAs has changed our view of how genes are regulated, how those noncoding RNAs can impact cells in many ways. We're lucky to work in RNA.

Steve Mao: People used to speculate those are junk portions because they do not code any proteins, but now it turns out that they actually code very important RNA species, and many of them play various roles. You started with lncRNA, and later you moved toward a precursor RNA that generates new lncRNAs ended with snoRNA [small nucleolar RNA]. Can you tell us what made you switch?

Dr. Chen: I wouldn't call it a switch, but really a smooth transition. Ten years ago, one of the most exciting discoveries in molecular biology was pervasive transcription of the genome, leading to the discovery of intergenic-region-transcribed long noncoding RNAs [lincRNAs]. We know that the transcription and processing of long noncoding RNAs is different from those of mRNAs [messenger RNAs], but at the ends these mature lncRNAs look similar to mRNAs: They have 5′ m^7G-capping or 3′ poly(A) tails with only a few exceptions, like NEAT1 [Nuclear Enriched Abundant Transcript 1] or MALAT1 [Metastasis-Associated Lung Adenocarcinoma Transcript 1], originally discovered by Dave Spector's lab right here at Cold Spring Harbor. His lab found that the 3′-end processing of these two basically used an RNase P cleavage related to tRNA [transfer RNA] biogenesis, and those RNAs look so different at the 3′ ends from mRNAs.

Luckily, I worked on NEAT1 with Gordon Carmichael at the time. Starting from those Alu elements, I found that mRNAs containing inverted Alu repeats are preferentially retained in the nuclei in bodies called paraspeckles. I found that NEAT1 is a major organizer of paraspeckles. So, triggered by the very interesting, different appearance of NEAT1, I asked the question: Do all RNAs look the same as mRNAs? From there, I began to explore the non-poly(A) transcriptome, which had been ignored by so many people as just junk. We discovered different classes of previously unknown species. These RNAs do not have their own promoters, but rather are processed from the primary Pol II [RNA Polymerase II] transcripts and then stabilized by distinct mechanisms like forming RNA circles, or by the protection of snoRNP [small nucleolar RNA–protein] complexes at one or both ends. More importantly, now we've figured out that some of them can impact important functions of gene regulation in cells that also relate to human diseases including Prader–Willi syndrome and autoimmune diseases like lupus. We're happy to see that these previously thought-of-as-junk things can do something in cells.

Steve Mao: I know that you studied immunology in a lupus setting, and you also studied the differentiation potential of human embryonic stem cells and the function of those RNAs. What's it like to transition from molecular biologist to all these different fields?

Dr. Chen: I was trained as an RNA biologist, so all the RNA stuff from my lab was rooted in molecular biology, but I just take every approach necessary to address the fundamental question I want to ask. For example, my lab routinely uses imaging as one way to look at how exactly a particular lncRNA localized in the cell. That can provide lots of information to predict what they do there. It happens that we found RNA circles because they are slowly generated and rapidly degraded. When they are there, they form these very interesting imperfect intramolecular double-stranded RNA [dsRNA] structures that act as inhibitors of PKR [protein kinase R] activation, which is a protein involved in innate immunity that is related to the autoimmune disease, lupus. For autoimmune diseases, we do collaborate. For example, we get our precious SLE [systemic lupus erythematosus] patient samples from a local collaborator, Dr. Nan Shen at Renji Hospital. For something like this, because I really don't know much of immunology, I tend to do collaborations with col-

© 2019 Chen. This article is distributed under the terms of the Creative Commons Attribution-NonCommercial License, which permits reuse and redistribution, except for commercial purposes, provided that the original author and source are credited.

leagues. For other things, we try to do the most we can in the lab. As I said in the beginning, we try every approach necessary. The ultimate goal is to address what those mystery RNAs do in cells.

Steve Mao: This is a fascinating story: Linking a fundamental question from molecular biology—the role of the circular RNAs—with a very important disease-relevant role in innate immunity. A lot of these RNAs have been moving into the clinics. For example, SPINRAZA®—to treat SMA [spinal muscular atrophy]—is an antisense oligo to block splicing. You've mentioned that your work might provide a direction for therapy in the future to treat lupus patients. Can you elaborate how we can use circular RNA to advance into the clinics?

Dr. Chen: We were actually very excited to see this. Normally, these lupus patients have global reduction of circular RNA suppression accompanied with a burst of PKR activation. If we introduce RNA circles into blood cells that came from an SLE donor, we see reduced PKR activation together with this downstream cascade including interferon responses and interferon-induced genes, and many of those genes have been used as diagnostic markers of SLE. This really suggests that in the future we can introduce the in-vitro-synthesized circles without immunogenicity—in the beginning, with animal models—to see if the circles can reduce this unwanted PKR activation that could benefit patients in the future.

Steve Mao: You've described a very promising technology to try to study circular RNA in this meeting. Can you tell us something about that?

Dr. Chen: The biggest challenge in the circRNA field was that RNA circles have the structural conformation and they have almost the same sequences with the linear RNA that's basically from the same parental genes. The only difference is the backsplice at the junction site. So, because they came from the middle exons of the coding genes, it makes it very hard to interrupt the circular RNAs only without affecting the linear RNA expression with current existing CRISPR–Cas genome editing types of technology, as well as with RNAi because any half of RNAi is targeting the linear RNA.

A couple years ago, Feng Zhang's lab described the exciting CRISPR–Cas13 system. This system is proven to target RNA with high specificity. One of the very unique features of this system is that it is intolerant to mismatches that hybridize to the targeted RNA. I wanted to ask whether this technology could be used to only target and degrade circRNA without affecting linear expression. We did a panel screening and we found one of them worked well, and I'm sure maybe other Cas13 proteins can also do the same thing: discriminate circRNAs from mRNAs.

Importantly, we also can manage the system to apply larger-scale functional circRNA screening. That means that with this new tool added to our existing tool kit we should be able to study circRNA function, not only at the individual level, but also at larger-scale levels in different biological settings. I'm really very excited. Hopefully, in the future we're going to see more individual circRNA functions discovered by different labs.

Steve Mao: A lot of your major discoveries come from this kind of non-hypothesis-driven genome-wide kind of discovery; you mentioned the poly(A^-) portion. The challenge of doing this kind of data-driven research is that not every biologist is trained as a computational biologist, so you really need to have the kind of expertise to analyze the data. Can you say something about your experience collaborating with computer biologists?

Dr. Chen: I do both: genome-wide kind of discovery and hypothesis-driven discovery. Indeed, even for the genome-wide kind of discovery, I first had a hypothesis prior to performing the genome-wide studies. For example, we explored the poly(A^-) portion with the hypothesis that some unknown but unconventionally processed RNAs might be missing from the classical poly(A^+) portion. In terms of computational collaboration, I'm super lucky because my collaborator, Dr. [Li] Yang is the principal investigator at the CAS [Chinese Academy of Sciences] at the PICB [CAS-Max Planck Gesellschaft (MPG) Partner Institute for Computational Biology]. He also is my husband, so we have lots of discussions even when we're taking care of our daughter. My suggestion to do this kind of collaboration is to discuss with your collaborator as much as possible, because this way you can freely deliberate the key questions you want to address.

Steve Mao: There are really not a lot of female scientists, especially at the principal investigator [PI] level, in China. As a female scientist in this kind of male-dominant setting, what's the biggest challenge, and what things do you think could be done to improve that?

Dr. Chen: In my Institute, the Ph.D. students are at least 50% female and 50% male, but at the PI level we have only 20% female, maximum. One of the reasons is that many female scientists grow discouraged by the apparent difficulties when they want to have a family. My advice for them: Don't be discouraged by the obvious apparent difficulties. Ask for help. Ask for support. Build up the support network with your families, with your friends, to get through the first several years when the kids are young, when they need a lot of intense attention. In the meantime, I always encourage female scientists: Just be self-motivated. Be self-confident and persistent. If you love science, give it passion. In this way, maybe 10 years later, you're not going to regret: "Oh, I gave up my science."

A Conversation with Caroline Dean

INTERVIEWER: JAN WITKOWSKI

Cold Spring Harbor Laboratory

Dame Caroline Dean, DBE, FRS is a Group Leader and Royal Society Professor in the Department of Cell and Developmental Biology at the John Innes Centre.

Jan Witkowski: Can you give me an overview of your work and how you came to do this?

Dr. Dean: My lab works on how plants know when to flower, so I got into this wonderful world of chromatin and RNA through quite an unintentional route. Many years ago I went back to the U.K., having been a postdoc in the U.S., and decided seasonal timing was really interesting. The place I live in—Norwich—has fairly distinct winters and in spring we have this wonderful bloom. The question is, how is flowering so synchronized?

My lab decided to study the molecular control of flowering, focusing on how plants decide whether to overwinter before flowering, and how they perceive winter. The many genetic routes we used to study these questions led us into the study of one gene. It's a gene that encodes a repressor of flowering. To be able to flower, the plant switches that gene off through a cold-induced Polycomb switching mechanism.

As we dissected this mechanism, we came across a set of antisense transcripts at the repressor locus so we have spent time dissecting how regulatory RNAs affect the chromatin environment of the locus, which affects the transcriptional output, which then affects whether the plants actually need winter and whether they can respond to winter.

Jan Witkowski: And what does this core repressor do?

Dr. Dean: The repressor keeps "off" all the genes in a cell that are required for flowering. Think of it as a brake to flowering: If the protein is expressed, the plant doesn't flower. So, you have to take that away. Winter takes it away. Or you don't make it in the first place, in which case you can flower straight away.

Jan Witkowski: How do the antisense transcripts get turned on? What sensing mechanism does the plant have for perceiving its environment?

Dr. Dean: The antisense transcripts actually completely encompasses the sense transcript. The 5′ end of the antisense starts just upstream of the sense stop site and ends just upstream of the sense start site. The antisense transcripts are made constitutively at a low level but are strongly induced by low temperature. We've looked at how they're cold-induced, and it involves a promoter region that aligns with the end of the sense transcript. This promoter is stimulated by cold temperature in a process involving a chromatin structure called an R-loop, where the antisense RNA invades the chromatin. How that is regulated by temperature, we still don't know.

Jan Witkowski: Are there parallels with regulation in other organisms?

Dr. Dean: The parallel that I like to follow is the RNA-mediated chromatin silencing mechanism from Rob Martienssen's or Shiv Grewal's work on small-RNA-mediated silencing of centromeric repeats. They also study how a chromatin target is selected, how a certain level of expression is determined and then inherited faithfully through mitosis. The question is how do you copy those histone modifications every generation to actually keep that transcription state epigenetically stable through many generations?

Jan Witkowski: How does this work you are doing relate to vernalization?

Dr. Dean: Vernalization is the cold-induced epigenetic silencing of the repressor gene. Plants perceive the long period of cold winter and slowly silence the repressor gene—it's called *FLC* [*Flowering Locus C*] gene—and that is the process of vernalization. We study it in *Arabidopsis* where we can exploit a molecular genetic approach. Unlike rapid cycling varieties that people normally use in their genetics, we use winter annual types that germinate in autumn and overwinter, which synchronizes their flowering in spring. That's the process of vernalization: the ability to perceive the cold, to silence this gene, and then to maintain that silencing through the subsequent warmth and growth.

Jan Witkowski: Can you remind me what Lysenko was doing? He was trying to get spring wheat flowering in the winter?

© 2019 Dean. This article is distributed under the terms of the Creative Commons Attribution-NonCommercial License, which permits reuse and redistribution, except for commercial purposes, provided that the original author and source are credited.

Dr. Dean: Yes. Many plant species show this vernalization requirement. A lot of people will be familiar with winter- and spring-sown varieties of wheat. Winter varieties need to be sown in autumn, and vernalization occurs during the cold temperatures as they overwinter. It's a slightly different mechanism in wheat, but there's an epigenetic memory of that cold exposure and then you get enhanced seed yield. Vernalization gives enhanced fitness to the plant.

Lysenko imagined that once you had that enhanced yield, it would then be stable through plant generations. That was the mistake, because actually vernalization is reset every generation. Each time the seed is sown it needs to experience the cold to get the benefit for yield. Of course, the farmers in Russia didn't realize this. They didn't revernalize each generation, so they didn't get the yield bonus.

Jan Witkowski: Are plants more difficult to work with? Are animal cells more tractable for doing RNA-related work?

Dr. Dean: No, I think plants are a great system to do analyze functions of noncoding RNAs. Plants are very responsive to their environment so need to have mechanisms to integrate noisy signals and remember previous exposure. *Arabidopsis* and other model systems—*Drosophila*, *C. elegans*—are excellent for mechanistic dissection of complex traits.

For example, forward genetics has been a great way to elucidate the complex regulatory hierarchy controlling flowering time. That would have been much harder starting just from -omics data.

Jan Witkowski: But you do have generation time. It takes longer to work with plants.

Dr. Dean: *Arabidopsis* cycles in about 6 weeks, so it's not too bad. The *Arabidopsis* revolution has changed plant biology considerably. It was one of the first eukaryotic genomes to be fully sequenced. We have got knockouts of every gene; we have got natural variants. *Arabidopsis thaliana* grows from the Arctic Circle down to the equator, so there are tremendous resources to actually use for various questions.

Jan Witkowski: There's always the discussion to what extent model organism represents other organisms. Is *Arabidopsis* a true representative of plants in general?

Dr. Dean: Absolutely, and not only plant systems. For example, cellular memory mechanisms are very conserved so what we learn about vernalization is relevant for mammalian silencing. You can most easily dissect mechanism in a model system, and then explore the evolutionary variation.

Jan Witkowski: If I could grant you any wish, what would you like to know about what you work on?

Dr. Dean: My big wish would be to understand how epigenetic memory through Polycomb complexes is faithfully inherited through cell division.

A Conversation with Susan Gasser

INTERVIEWER: JAN WITKOWSKI

Cold Spring Harbor Laboratory

Susan Gasser is a Professor of Molecular Biology and Director of the Friedrich Miescher Institute for Biomedical Research at the University of Basel.

Jan Witkowski: Perhaps we could begin with a synopsis of your work.

Dr. Gasser: We've been studying the mechanisms of heterochromatin repression, of different ways to keep genes silent. Most of that is at the level of transcription through the formation of heterochromatin, but in a screen for derepression of heterochromatin we found three subunits of an RNA-controlling complex. All the other hits in this screen were chromatin modulators as we expected, so these three were a surprise. This was in *C. elegans*, but this complex is conserved from bacteria to yeast to humans. It's called the LSM complex for "like Sm" proteins.

We read up on the LSm complex, and found that it comes in two forms. There's one in the cytoplasm and another in the nucleus. They share six subunits, but the seventh, either subunit 1 or 8, is specific for either cytoplasm or nucleus. The first thing we showed was that the role in silencing was through the nuclear complex. That was good, because it meant that it was probably working at the level of genes.

We then looked at the proteins that are supposedly interacting with this nuclear complex. Its normal role is to bind U6 RNA, catalyze or chaperone splicing, and then help trigger the degradation of the spliced-out intron, through an RNA exonuclease called XRN2. XRN2, but not U6 RNA, was also involved in the heterochromatic silencing of our reporter.

Then we asked, "Which endogenous genes are sensitive to this RNA degradation mode of silencing?" We detected several hundred genes that were derepressed—or up-regulated—in the absence of this LSM complex or of XRN2. We asked, "What's the nature of these genes?" First, we saw that they were all very poorly expressed in wild-type worms. Second, we checked their chromatin state by monitoring enrichment of histone marks, and we found that 95% carried histone H3K27me3, the characteristic methylation deposited by Polycomb.

Polycomb is known to silence genes. It's usually thought to create facultative heterochromatin in a tissue- or cell type–specific manner. It actually poises genes in an "off" state, but such that they can also be switched "on," depending on differentiation. It's quite intriguing that all the genes that were controlled by this LSM2–8 were Polycomb-marked facultative heterochromatin. So we were faced with the question, "How does it work?" Polycomb normally represses by transcriptional repression, by blocking transcription, but here we were looking at RNA degradation. We checked specific genes that were sensitive to this RNA-degrading complex and showed that if you mutate the LSM8 subunit, which is specific for the XRN2-binding nuclear complex, then you stabilize a low level of transcripts from such target genes. Apparently, although genes are repressed by Polycomb, there's a low level of promiscuous transcription possible. Then we asked if the same genes are regulated by XRN2, LSM8, and Polycomb. And indeed, they overlap significantly.

Finally, we asked the question, "Does RNA degradation feed back in any way to the Polycomb mark?" Does the LSM complex and RNA degradation stabilize the transcriptional repression that's mediated by Polycomb?" And indeed it does; we see that the K27 methylation mark is reduced on target genes, but not lost overall, in the *lsm-8* mutant. The LSM8 complex constitutes a backup system that ensures the integrity of the Polycomb repression pathway. This is rather a noncanonical idea, and the Polycomb field seems not very keen on having Polycomb silence by means of something other than the attenuation of Pol II [DNA polymerase II] activity. So it's been an interesting ride to try to publish this.

Jan Witkowski: Does LSM just mop up this low-level transcription?

Dr. Gasser: Right. We believe that it degrades breakthrough transcripts, and usually full-length message. At least, the RNAs we recover are not enriched for the N-termini of the genes. There must be a decapping enzyme that allows the exonuclease XRN2 to degrade the transcript. It's basically reducing the steady state level—often four- to tenfold—from a background level that stems from the incomplete silencing of Polycomb-marked genes. Actually, it is known that Polycomb-marked genes are not 100% "off." Repetitive centromeric heterochromatin, for example, can be more completely "off."

© 2019 Gasser. This article is distributed under the terms of the Creative Commons Attribution-NonCommercial License, which permits reuse and redistribution, except for commercial purposes, provided that the original author and source are credited.

Jan Witkowski: Is the target gene really repressed?

Dr. Gasser: I would say it's quite repressed—or silent—if you have both transcriptional repression and the elimination of promiscuous transcripts. When we knock out this secondary pathway in *C. elegans*, the animals are still alive, but their gonadal development is impaired and they become sterile. So if you ask, "How important is it?," it is likely to be very important, as the organism can't reproduce without it.

Jan Witkowski: Is this the first example of this sort of mopping up?

Dr. Gasser: It's been observed and even characterized very well in *Schizosaccharomyces pombe*—fission yeast —and in plants, where it occurs at centromeric satellite repeats. However, it involves a different pathway, because it depends on a reverse transcriptase that makes a double-stranded RNA that is "diced" and bound by an Argonaute. The Argonaute-dsRNA comes back and helps nucleates H3K9 methylation, which is the canonical repressive mark on constitutive heterochromatin. So, the concept that a promiscuous transcript can feed back and help stabilize repression was already in the field, but it had never been demonstrated or characterized beyond fission yeast and plants. This is the first example in animals, and the first time the Polycomb system and the LSM complex are implicated.

Jan Witkowski: *C. elegans* is a worm. Is it also true of mice and man?

Dr. Gasser: We've just initiated this study. Of course, the complex is conserved and we found that human LSM8 can be bound to chromatin and associates with XRN2. We've just done the RNA-seq to see the genes that are misregulated. We haven't fully characterized the data, but there are genes that both go up and down. Now we're checking to see if targets bear the Polycomb mark.

Jan Witkowski: The subunit 8 of LSM is the one that makes it nuclear, and subunit 1 makes it cytosolic. Are these carrying some sort of signal?

Dr. Gasser: The complex as a whole recognizes poly(A) message in the cytoplasm. In the nucleus, when it recognizes the U6 snRNP [small nucleolar ribonucleoprotein] it recognizes a small poly(U) stretch. We don't really know if there is a signal on these Polycomb-repressed genes somewhere in the 3′ UTR [untranslated region] that's actually recognized by this LSM2-8 complex, but it's clear that you could call this complex "structural," because in both cases—nuclear and cytoplasmic—its job is to bring in an exonuclease. In the cytoplasm, it degrades messages; in the nucleus, it degrades messages and introns. The real link we are missing is the mechanism that targets it to this class of genes. How does it recognize the transcripts of Polycomb-marked genes? We don't know if it's a feature of the RNA or if there is a link to the histone modification. mRNAs could be aberrantly capped or could have some secondary structure or motif. It doesn't seem to be 3′ UTR–specific, as we've swapped 3′ UTRs. Anyway, there are hundreds of genes that are regulated this way, so it's not one single UTR motif. We're looking for the signal.

Jan Witkowski: You used the term "noncanonical" earlier. Is RNA a more interesting molecule than DNA?

Dr. Gasser: Oh, I don't know. The combination is fascinating and I think there are many variants of DNA that we haven't actually explored. There are modifications like methylation, but also sequence motifs, sequence variation, secondary structures, and many selfish repetitive elements, the behavior of which is still not understood. I'm still fascinated with DNA and chromatin, but RNA is interesting, and has unexpected roles. The simple message that we heard about mRNA 50 years ago is pretty much water under the bridge.

Jan Witkowski: Your institute was founded by Ciba-Geigy.

Dr. Gasser: Next year, it's 50 years, and we've been funded by industry for 50 years. First it was Ciba-Geigy and then Novartis; they've been extremely generous. Initially costs were covered 100%, but in the last 15–20 years, about 70%. We compete for grants, which keeps us linked to the academic world. The fabulous thing is that for all that time, they've never told us what to study. We pursue fundamental discovery research, although we are funded by industry. Fortunately, it's one of those rare instances in which the value of open-ended research and the strength of curiosity-driven discovery are recognized.

Jan Witkowski: Is the institute unique in these things?

Dr. Gasser: The Institute of Molecular Pathology in Vienna is funded by Boehringer Ingelheim in a very similar manner. BI even has a second smaller Institute in Mainz, which is near their headquarters. I keep trying to convince other companies that this is the way to go, that they could have a hotline to clever scientists this way. The energy of discovery-driven science is something that keeps a company on its toes, challenging the ideas that sometimes become inbred and narrow within drug development. Very often, they need to be shaken out of their ways of thought, just as basic scientists do. At least in our case, wisdom and vision from the company's leader has kept us alive and hopefully will for another 50 years.

A Conversation with Samie Jaffrey

INTERVIEWER: ANKE SPARMANN

Senior Editor, Nature Structural & Molecular Biology

Samie Jaffrey is the Greenberg–Starr Professor in the Department of Pharmacology at the Weill Cornell Medical College, Cornell University.

Anke Sparmann: Your lab combines tool development using both synthetic and chemical biology approaches and applies these approaches to questions of RNA biology. Can you give us an overview of your work?

Dr. Jaffrey: I'm going to talk about our work on RNA modifications and, in particular, methylated adenosine. Most of the mRNA [messenger RNA] in the cell is composed of A, C, G, and U, but a tiny bit—maybe one in every 300–400 adenosines—is methylated, and when it's methylated on the nitrogen of adenosine, it's called m^6A. We normally don't think of mRNA as containing that many modifications; we think of tRNA [transfer RNA] and other RNAs as having modifications. This was discovered way back in 1974 when scientists were beginning to understand how mRNA capping and other phenomena occur. When they were doing metabolic labeling to study the m^7G cap on mRNA, they ended up inadvertently discovering that there were actually methyl modifications inside mRNA, and with some clever biochemistry they figured out it was m^6A. The people who did this were Dawn Kelley and Robert Perry at Fox Chase and Fritz Rottman and other researchers like Jim Darnell. Some of the pioneers of molecular biology were involved in discovering this modification.

m^6A was known to be low abundance, but even though it was discovered way back then, almost nothing has been known since about what its function could be. It's very appealing to think that it might be a modification like phosphorylation is for proteins. The potential that m^6A has some sort of regulatory function has been overlooked, and it's for many reasons. First, some people didn't even think that m^6A was really in mRNA, because when you purify mRNA you actually have a lot of contaminants and they thought maybe m^6A could have been from tRNA or ribosomal RNA. But people like Jim Darnell did experiments where he did exceptionally high-purity purification, at least to the ability that they had at the time, and they still were able to see m^6A. So, there was belief—but still controversy—about whether it was really in mRNA.

The other thing is, how do you study m^6A? m^6A behaves exactly like adenosine when you do reverse transcription. If you take mRNA and you make cDNA [complementary DNA], which is the reverse transcription step in molecular biology, you're going to lose any methyl marks that were there. A lot of people abandoned the field and quickly moved into splicing because that was discovered in the end of the '70s. A lot of this was lost and it wasn't even in textbooks, but there were a few researchers who stayed on it. Fritz Rottman at Case Western was really the pioneer who cloned the enzyme. It's called METTL3, or methyltransferase-like enzyme 3.

Then, as with so many other things in molecular biology the breakthroughs came from yeast and plants. Some scientists knocked out the enzyme in yeast and they found a very remarkable sporulation defect. In plants they found seeds would go to one stage of development and stop. That was Rupert Fray's paper in 2008. When we saw that paper we were just completely shocked; we never even heard of this modification.

I had a postdoc who started shortly thereafter; her name is Kate Meyer and she's now at Duke. I said, "We need to figure out what is this m^6A. Is it really in mRNA? Where is it? Which mRNA? All mRNAs?" It could have been like the cap, where every RNA has it, or a tail or it could be selected. We had no idea. But that paper—and it related to seed development—told us that the effects were precise. The cells didn't just die. The yeast data, which had come out in 2002, really showed us that there was some connection between developmental and cell-fate decisions like sporulation and seed development; they're all developmental type processes. We wanted to figure out what was going on. That was when we applied for our first NIH grants on this topic.

There were m^6A-binding antibodies that were being developed for other purposes to study DNA methylation in bacteria and things like this. So we got those antibodies and we pulled down mRNA fragments, and it turned out those antibodies bound very specific regions of mRNAs, which are the regions that have m^6A in it. We were able to identify the transcripts in the transcriptome that have it. And it wasn't every transcript; it was very specific transcripts. These transcripts tend to have very unusual features. They tend to have huge exons, and we think those exons are the trigger for the methylation. They tend to occur in transcripts that encode regulators of cell fate and development, just like had been predicted from the work in plants and yeast.

Even though we know now which transcripts have m⁶A based on the original mapping studies from 2012 and some newer methods that we've developed since then to really pinpoint the exact adenosine residue, the functions of m⁶A are still not completely clear. Jim Darnell's great work in the '70s showed the first—and even now the most well-established—function, which is that transcripts that have m⁶A are not as stable as transcripts that don't have it. So, m⁶A is a mark for instability. But besides this function of m⁶A controlling mRNA stability, ways to think about the purpose of m⁶A and how m⁶A transcripts are handled in this cell aren't quite clear.

Anke Sparmann: Is there any idea if the mark is dynamic? Is it removed and then put there?

Dr. Jaffrey: This is the source of a great deal of controversy and debate and a lot of people have different opinions on this. Let me just say that methylation patterns are very similar in almost every tissue. There may be subtle differences but the vast majority of modification seems to be the same, which would go against the idea of some sort of dynamics. The ability to be methylated seems to be hardwired based on gene structure, on the size of the exons. The exons are the same size everywhere because they come from the same DNA, so it doesn't seem that there are dramatic dynamics, but there may be subtle dynamics.

There is evidence for an eraser protein called ALKBH5 that does demethylate m⁶A. It's highly enriched in testes. Animals deficient in ALKBH5 are completely normal, so that does imply that most cell types don't need it, but testes and other germ tissues are abnormal. So, it may have specific roles, but it may not be general. One thing that is interesting is that ALKBH5 is up-regulated in certain cancers. It could be that these patterns of modification—which we call the epitranscriptome—could be shaped to some degree by demethylation, but most of the pattern of m⁶A seems to be constant and fixed. m⁶A therefore marks a set of transcripts that can be regulated in a specific way, but that regulation is going to occur on the same transcripts and in most tissues if those transcripts are expressed. We're trying to understand if there's a logic behind how these transcripts are being regulated.

The core thing that I'm focusing on today is what we discovered by looking at the m⁶A-binding proteins first discovered by Gideon Rechavi and his lab in Israel. They're called the YTH proteins; YTH is the name of the domain that binds the m⁶A. There are five genes in the genome that have YTH domains, but three of them are in the cytosol. The YTH proteins that are in the cytosol are almost identical: they're paralogs. They may have an identical function or a redundant function or a very similar function, but when you look at them, they basically are 450 amino acids or so with a YTH domain and just glutamines and glycines, and pretty much nothing else. Normally, proteins are filled with interesting domains that tell us about what their function is, but this is just filled with these fairly monotonous and uninformative sequences that researchers are starting to realize now are actually biologically meaningful. These are sequences that people call "low-complexity domains" or "intrinsically disordered sequences"; this is what this protein is almost primarily composed of.

We started purifying this protein and looked at it under the microscope. At first it was a clear solution, but then we started to see little droplets form within the buffer. It turns out that the microscope lamp was warming up the buffer and causing a temperature-associated phase transition. These proteins go from a solution state where they're just floating around as, maybe, monomers. But then they come together as droplets within the liquid buffer due to the temperature being changed. And the temperature would get up to 30°C–34°C. Not too warm, but getting near physiological temperatures. We realized that all of these cytosolic m⁶A-binding proteins, which are nearly identical, do this. They undergo this phase separation, and it is regulated by salts and temperatures and other things.

That's interesting, but the more interesting thing was that when we diluted this protein so it did not form these little droplets and then added in methylated RNAs, these proteins quickly formed droplets induced by the RNA. The RNA triggered the phase transition. It turns out that m⁶A's function is to drive these proteins to undergo phase separations. And these proteins—now with RNAs bound to them—partitioned into these naturally occurring phase-separated structures that cells have, like P-bodies [processing bodies] or stress granules, and even other types of structures like the neuronal transport granules. What we find is these structures are highly enriched in m⁶A-modified RNAs, and m⁶A acts as a targeting signal because of this phase transition property because phase-transitioned structures can easily be partitioned into the larger, naturally occurring droplets. They are then driven into these structures. The function of m⁶A—and frankly the reason why some of these genes may have long exons is to encode m⁶A—is to ensure that the transcript then gets handled or triaged to different parts of the cell.

In some cells when they're triaged to P-bodies, these RNAs are probably degraded if P-bodies have a role in degradation. They may not be the actual sites of degradation, but they may be part of the process of the degradation. In other cells where there's stress, you form stress granules which are protein–RNA assemblies. Those are filled with methylated RNAs and the methylated RNAs are disproportionately trafficked to those structures. And in RNA granules in neurons as well, again we see that both the m⁶A-binding proteins and the m⁶A RNAs are enriched in those structures. What our work is telling us is that m⁶A is a signal that directs RNAs to specific structures, and it does the directing through these phase transition pathways. That's how specificity is conferred upon the directionality of these transcripts: By causing these RNAs to phase-separate they can get to their specific target sites like these P-bodies and stress granules, which are themselves a sort of liquid within a cell, liquids that can then fuse with each other and then accumulate and concentrate in specific areas.

Anke Sparmann: Do you think there's any specificity in that? Some go to P-bodies, others might go to stress granules. Is that somehow encoded by the binding proteins or the RNAs themselves?

Dr. Jaffrey: This is going to be the big question: How can this process be regulated? The partitioning behavior of proteins when they first undergo phase separation and then when they partition is something that people don't fully understand. But these low-complexity domains aren't just glutamine/glycine. Some are arginine-rich, some are enriched in aromatic residues like tyrosine. There's actually a code associated with these low-complexity domains that we don't fully understand. These domains can also be phosphorylated and have other modifications that can further change their properties. I think when the YTH proteins bind and they have the low-complexity domain with a certain set of modifications they will partition maybe to P-bodies; but then when phosphorylated or modified in some way, then they maybe partition to other structures. Fundamentally, our ability to understand phase separation principles and the specificity with which these principles allow proteins to go to one compartment versus another will ultimately determine when m^6A has a destabilizing effect or when it has a translational repression effect or other types of effects that had been seen in this cell.

Anke Sparmann: So the specificity comes not from the mark itself?

Dr. Jaffrey: That's definitely something that I think the field is starting to believe: that the specificity is coming from the proteins that recognize the m^6A and how they are regulated rather than the m^6A marks themselves changing. That being said, there may be some transcripts that will have altered methylation but probably at the level of the proteins that bind it, which are sometimes called the "readers" of m^6A. Those readers may be regulated. Now that we had that framework of phase separation to understand how these proteins behave and what I think is their fundamental physical property, we can start to see how these modifications change their partitioning behavior. That might allow us to make predictions about what they would do to the methylated RNA that they're binding to.

A Conversation with Leemor Joshua-Tor

INTERVIEWER: LARA SZEWCZAK

Senior Scientific Editor, Cell

Leemor Joshua-Tor is the W.M. Keck Professor of Structural Biology at Cold Spring Harbor Laboratory and a Howard Hughes Medical Institute Investigator.

Lara Szewczak: You work on a very small protein, Asterix. Remind me what it does.

Dr. Joshua-Tor: We didn't know when we started. We only knew that it was picked up in these piRNA [piwi-interacting RNA] screens where both Greg Hannon's lab and Julius Brennecke's lab were looking for other factors that are involved in the piRNA pathway. This is one that was very clearly giving a strong phenotype, a strong effect. Looking at it, it didn't appear very interesting because it just had these two zinc fingers in it and then nothing much else to go on. But that's something we like to do. We like to go after things that are mysterious. So, we picked that one to try to figure out if we can—maybe using the structure—try to tease out some kind of information as to what exactly it does. We did know where in the pathway it works. We knew that it's in the nucleus. We knew that it doesn't affect piRNA biogenesis. It was pretty clear that it was affecting what's called the "effector" part of the pathway—probably the silencing step itself—but we didn't really know in molecular detail what it actually did.

Lara Szewczak: Now that you've got a handle on it, where do you see the study going?

Dr. Joshua-Tor: We came up with a model where we found that Asterix binds these tRNAs [transfer RNAs]. We're wondering now whether that really is happening. It appears to be the case, and our best guess is that it's binding to the primer binding sites on these retrotransposons because that's where tRNA fits into this whole game. There's a little bit of work still to do to figure out if that all looks good, but we're really excited about it to see if that's really happening.

Lara Szewczak: There are other RNAs that have tRNA-like elements to them. Your data showed a pretty specific size. Have you tried looking at larger things that may have similar structures? I know the model was built around tRNA because of the link to the retrotransposons, but have you thought about other options?

Dr. Joshua-Tor: We were pretty agnostic as to what it would be. It was copurifying with the protein initially when we first looked at it and found that it was a tRNA. The way we expressed the protein was a mouse protein in this insect called armyworm—it's a moth—so we weren't sure that it was relevant at all. That's why we looked for what the mouse protein would pull out in mouse cells that have piRNAs, and in *Drosophila* in ovary cells that also are part of the pathway, to see if it's still doing that. It appears to pick up tRNAs very specifically. It's not fragments. Fragments are all over the place, but this appears to be a full tRNA.

Lara Szewczak: Is the protein restricted to germline?

Dr. Joshua-Tor: Yes, it appears so.

Lara Szewczak: You're a structural biologist. Crystallography is your bread and butter, but you needed to turn to a different approach. How did bringing in NMR [nuclear magnetic resonance] let you answer the question you wanted to?

Dr. Joshua-Tor: We don't shy away from other approaches. We use whatever fits. We got into cryo-EM [electron microscopy] and we're having a lot of fun with that as well. But this is a teensy little guy, so it was just natural to try to do it by NMR. Jon Ipsaro—who's really the hero of the story—wanted to learn how to do it himself, so we chose a collaborator that was nearby.

Lara Szewczak: What kinds of questions do you want to answer now about RNA that you haven't been able to do before because you didn't have crystals or another approach?

Dr. Joshua-Tor: There are a lot of things that you can now do with cryo-EM. You need to make a really good sample that's fairly clean, but it's not quite the same bar as when you want to actually crystallize it. Also, we have a little bit more leeway in terms of conformational flexibility of the molecules, whereas in crystallography you can't do it.

We have another system that we've been working—the TUTases [terminal uridylyl transferases]—with respect to their biogenesis and destruction of an important microRNA called *let-7*. We can now assemble these complexes and they might have conformational flexibility, but we don't have to cut off all the extra loops and things like

© 2019 Joshua-Tor. This article is distributed under the terms of the Creative Commons Attribution-NonCommercial License, which permits reuse and redistribution, except for commercial purposes, provided that the original author and source are credited.

that. We can deal with it because you can computationally purify out the different complexes. It opened up a lot of possibilities for us, so we're interested in that. I started my career in RNA and RNA interference with Argonaute and that also factors into our studies, but the piRNA pathway is probably the most intriguing because there are a lot of factors that have been shown to be a part of this pathway. We know where they fit in and a little bit of what they do, but mechanistically how they do it, we don't. This is an opportunity for us to jump in and figure out some stuff that would be interesting.

Lara Szewczak: To be able to do cryo-EM, you essentially went back to school. What did you have to do to bring the technology on board?

Dr. Joshua-Tor: First of all, we had to get a microscope. Bruce Stillman, who is the president of Cold Spring Harbor, was very generous in raising the money to get the microscope. There are a lot of practical videos online—about 70 h, at least—on the technique, which I watched. We started a cryo-EM course at Cold Spring Harbor, which I was a huge advocate for, and I was Student #1 in the course. I just took it full on, practical and computational and everything, and people in my lab really bought into it. They're all really excited about it. We've been going to courses all around the world, and mostly just doing it and getting advice from people. It's really fun to—at this stage—learn something really new and so powerful, too.

Lara Szewczak: Are you actually going to be in the lab solving a structure, or do you want to be able to just advise?

Dr. Joshua-Tor: I want to be able to advise, to be critical of what people are doing, because if I don't know all the little issues that come along then I don't think I'd be able to do either of those. I wish I could work on the structure, but I don't think the people in my lab would even let me do it.

Lara Szewczak: One of the issues about cryo-EM that's different from crystallography is that the community hasn't yet agreed on set standards, like all the statistics in crystallography that let your peers assess the quality of the structure. Do you think that cryo-EM is going to get to a point where there are those kinds of agreed upon metrics, or just because of the nature of the technique is it always going to be "descriptive?"

Dr. Joshua-Tor: I think it's just a matter of time and development. When I started with crystallography we also didn't have a lot of the quality controls that we have today. We had more primitive ones, but now we're very careful about it. We have all these ways of quality control in assessing structure. It's going to happen faster in cryo-EM, not only because there's more people—the computational techniques are a lot more sophisticated and all that—but also all of us crystallographers are jumping into cryo-EM, so we bring these traditions with us. I think they'll have no choice. This is all going to be developed, and I think not in too long of a time.

Lara Szewczak: What are the kinds of questions where crystallography is going to remain the premier technique, and how are those distinct from the kinds of questions you use cryo-EM or NMR for?

Dr. Joshua-Tor: NMR is very good for looking at dynamics, which we really can't do at all in crystallography, and we can't do as well in cryo-EM. Crystallography is really good at getting very high-resolution structures. The field of drug design and drug development is still going to enjoy what crystallography has to do. It's true that in cryo-EM we're able to look at smaller and smaller things at very high resolution, but those are things that are very well behaved, etc. You also need "well-behaved" samples for crystallography, but it's still so powerful and the technique is so streamlined now—especially for small things and for looking at complexes with drugs—that I think that it still has a place. We can do small things like Argonaute on the cryo-EM, but we know how to crystallize it and it looks so good, so why make our lives more complicated if we can use crystallography?

Lara Szewczak: How does understanding the biology integrate into the process of understanding the structure?

Dr. Joshua-Tor: It's just practical, because we have to know what to do to our sample to coax it in a way that it would like to be in. In many cases it's adding the right partners, whether it's small ligands, but importantly, also big partners. If we don't understand how it associates with them and what pieces of RNA it likes to look at and things like that, our job would be much harder. There's a fantastic description in Barry Werth's book called *The Billion-Dollar Molecule* where he says for crystallization—but I think it's for every sample—that you have to bathe it in the right "amniotic fluids" so that they will "feel good" about themselves.

Lara Szewczak: Are there questions that you wanted to ask about biological systems in the past that you couldn't get around that you now think about going back to?

Dr. Joshua-Tor: With the discussion we were just having with cryo-EM, yes. We've been working on a helicase where it assembles into different states, and each state controls a different part of the maturation into a helicase, from binding double-stranded DNA to melting double-stranded DNA to translocating on DNA. We could get at a couple of these states before, but we couldn't get at one crucial one because we couldn't make enough and we thought it might be too unstable and just not amenable for crystallization. Now, we're going for it. This is part of the piRNA pathway, actually. That's one example, but we actually have a couple of these complexes that are bigger and unwieldy in our freezer that we had very small amounts of and we didn't know what to chop off that would give us more expression that we're taking out of the freezer now. That's fun.

Lara Szewczak: What are the things that students should be aware of when they're getting into structural biology now?

Dr. Joshua-Tor: To me, I don't know how you can live without it, because seeing exactly what things look like is so informative to how they work. That happened very clearly in several cases in my career. I just can't live with blobs in a cartoon. It just doesn't do it for me. You can go back to the blobs, but then you're informed by what the structures are telling you about how things actually work. I'm a chemist at heart and by training, so I always go back to the exact mechanisms. There are a lot of people out there still where that really floats their boat. That's their cup of tea, so they'll come. Also, there's such an aesthetic about structural biology and the techniques that are producing these wonderful images. There's such an aesthetic about the molecules themselves that I hope people are still excited about it.

A Conversation with Alberto Kornblihtt

INTERVIEWER: CARIKA WELDON

Researcher, University of Oxford Wellcome Centre for Human Genetics

Alberto Kornblihtt is Plenary Professor at the Department of Physiology, Molecular and Cell Biology of the Faculty of Exact and Natural Sciences at the University of Buenos Aires and Director of the Institute of Physiology, Molecular Biology and Neurosciences of the National Research Council of Argentina.

Carika Weldon: Could you to tell us about your recent work?

Dr. Kornblihtt: I work on alternative splicing and the coupling with transcription and chromatin, but in particular on recent unpublished results that we obtained in the system of the *SMN2* [survival of motor neurons 2] gene in terms of enhancing exon 7 inclusion in the therapy of spinal muscular atrophy [SMA]. My lab has been working for more than 20 years on the coupling between transcription and splicing, focusing mainly on kinetic coupling. We found that slow elongation can increase inclusion of certain exons in the mature mRNA [messenger RNA]. More recently, we also found that slow elongation can produce skipping of certain exons. Among the exons that are sensitive to elongation, ~80% respond to the first rule—slow elongation increases inclusion—and 20% respond to the second mode in which slow elongation promotes skipping. This is because it gives more time for inhibitors of exon inclusion to bind to the pre-mRNA.

In the case of exon 7 of spinal muscular atrophy, humans have two genes. When *SMN1* is mutated, *SMN2*, which is the backup gene, cannot cope with the lack of SMN protein because it has several mutations that make exon 7 poorly included into the mature mRNA. Adrian Krainer has developed a new drug called Spinraza that is an oligonucleotide that is able to displace the negative factors from the pre-mRNA and allow exon 7 to be more included. This is fantastic, and it has had success in every kind of SMA.

But then we found that exon 7 inclusion also responds to elongation in the second mode, meaning that slow elongation promotes skipping, and we do not want skipping. We reasoned that if we open the chromatin to allow for faster elongation, we could help exon 7 to be included. We tested that in cells in culture and also in mice, and it worked. We now know that chromatin opening with histone deacetylase inhibitors helps the action of Spinraza and that foresees a combined therapy using the oligo and the chromatin-opening drugs.

Carika Weldon: The field of splicing therapeutics has burst into the scene a lot more with Spinraza. It's really exciting to see that we can have a combination to speed up the elongation. What's your next step for this research to try to get to that stage?

Dr. Kornblihtt: One of my graduate students is working here at Cold Spring Harbor in Adrian's lab because we don't have the mouse facilities to perform these experiments. We are very happy that the mice are responding to the combined treatment. It's hard to foresee whether we will do the next step because one of the drugs we are using, valproic acid, is already approved by FDA, but it's used for other neurological diseases. It had been tested for SMA years ago and it didn't work, but now we know that it might work if it is combined with the ASO [antisense oligonucleotide]. I'm not sure how we will proceed. We still have to figure out in more detail the mechanism before we publish. I don't know whether there will be clinical trials or not.

Carika Weldon: Can you go into more detail about exactly how you're slowing down or speeding up the elongation?

Dr. Kornblihtt: You can do it artificially by using a slow mutant of RNA polymerase II [Pol II] to slow down elongation; David Bentley has also tried fast mutants of RNA polymerase II, but you can also use drugs that affect chromatin compaction. For instance, if you treat the cells with camptothecin, which is an inhibitor of topoisomerases, you can generate slow elongation, and you see the effects on elongation and on alternative splicing. On the contrary, if you open the chromatin by inhibiting histone deacetylation with drugs like trichostatin A or valproic acid, you confirm that the chromatin is open and elongation is fast, and then you have changes in splicing.

It seems that the ASO, the oligo itself, apart from displacing the negative factors from the pre-mRNA— HNRNP [heterogeneous nuclear ribonucleoprotein] A1 and A2—is able to create a compact chromatin structure at the site of the chromatin where the pre-mRNA is being made because it promotes histone methylation. That's in agreement with previous results of the so-called transcriptional gene silencing in mammalian cells. We now have evidence that the oligo has two opposite means of action. On the one hand, there's the known displacement of the negative factors. On the other hand, there's an opposite effect that, by creating a more compact chromatin structure and methylating the histones, it would stop polymerase. That would be counter to the positive effect of displacing the negative factors. We show that when we add the histone deacetylase inhibitors like valproic acid, that effect of the oligo is erased: It disappears. It seems that we have a detailed mechanism of why the combined treatment could be successful.

Carika Weldon: When it comes to therapeutics, as always, there are side effects. Finding out that you're actually able to erase one is really encouraging.

Dr. Kornblihtt: The Spinraza oligo is very specific. Adrian showed that it mainly displaces HNRNPA1 and A2 from the actual site of intron 7 that controls SMN2 splicing. On the other hand, the histone deacetylase inhibitors are not specific; they could open the whole chromatin. Depending on the dose, if the dose isn't high, those pleiotropic effects don't seem to be bad for the mice and they haven't been very bad for the humans who have been treated with VPA [valproate] for other diseases.

Carika Weldon: Have you looked into splicing mechanisms in any other diseases?

Dr. Kornblihtt: No, because my lab was reluctant to engage in applied molecular biology. We're always studying very basic science: chromatin transcription, splicing, splicing mechanisms. It was pressure from the families of the diseased children in Argentina who came to my lab and said, "We know Adrian. He will cure the disease, but we want you to do some research in Argentina, and we will fund you." This is great because I always say that the parents of the diseased children know more about the science than sometimes the officials do, because the officials want you to think of applied things. On the contrary, they told me it doesn't matter if you're not going to cure the disease in 2 months or 2 years. We want you to work here in Argentina on this subject. We started about 4 years ago and so far things are going very well.

Carika Weldon: There's another side of science where we have to apply for grants and you always have to try to tie it back to the big picture of "We're trying to help the world." How is it to know that you're actually doing something that is helping a real patient?

Dr. Kornblihtt: At the beginning, only the families from Argentina were supporting this line of research in my lab but then the Cure SMA Association, which is the United States' families, opened a grant call for basic research. We won the first grant, and last year I presented the data at the meeting of the Cure SMA Foundation. They were happy, so they encouraged me to present the second grant and I also got that. So, we are getting money from both the Argentinian families and the United States' families. In the present situation in my country, research funding is very low and the salaries and the Ph.D. fellowships are very low because this government that we've had since 2016 doesn't care very much about supporting science. Having support from abroad changes the way we can perform experiments because we have much more money to do these kinds of experiments than we would have from local sources.

Carika Weldon: We're into an era where we know some mechanisms of certain diseases or of certain genes that cause diseases and now we can actually modulate them. Where do you see the field going in the next 15 years?

Dr. Kornblihtt: I think that the control of splicing is as important as the control of gene expression at the level of transcription. The problem is that not everybody is aware of this and people in many important fields think that the only important thing is whether genes are "on" or "off" and not how many different variants of mRNA and protein genes make. There's still room to study the complexity of alternative splicing regulation because it's key for tissue differentiation, for cell fate, and for counteracting the bad effects that can cause disease.

We also have some people in my lab working on alternative splicing in plants. This is relevant because we just published a paper in *Molecular Cell* in which we show for the first time that kinetic coupling plays a role in a whole organism in response to light and dark, which is an external physiological cue. We showed that in daytime, when there is light, Pol II goes faster and in the dark Pol II goes slower and because of these two different behaviors, alternative splicing changes. This is the first time we saw what we were always seeing in cells in culture. Now we are seeing in a seedling exposed to the light or to the dark the changes in Pol II elongation speed and the change in splicing. The sensor for the light is the chloroplast. In some way through the photosynthetic apparatus it senses light and sends a signal, whose nature we don't know, to the nucleus and affects elongation, which in turn affects splicing.

Carika Weldon: Do you think that could have an effect on agriculture or improving our plant crop growth or anything like that?

Dr. Kornblihtt: I hope so, but I'm not sure. Again, we are interested in the basic problem and we are happy that we could demonstrate that it takes place in a whole organism, not just in cells in culture.

A Conversation with Adrian Krainer

INTERVIEWER: ANKE SPARMANN

Senior Editor, Nature Structural & Molecular Biology

Adrian Krainer is the St. Giles Foundation Professor at Cold Spring Harbor Laboratory.

Anke Sparmann: You were awarded the 2019 Breakthrough Prize in Life Sciences together with Dr. Frank Bennett of Ionis Pharmaceuticals for the development of antisense oligonucleotide drugs to target RNA splicing and the incredible success story of SPINRAZA, the first drug approved for spinal muscular atrophy. Can you start by telling us about this devastating disease and the molecular mechanism underlying it that you discovered?

Dr. Krainer: SMA, or spinal muscular atrophy, is a motor neuron disease. It's very severe, and it mainly affects infants and young children. There are milder forms of the disease, with delayed onset, which affect older patients, including adults. Depending on the type of SMA, it leads to progressive muscle weakness and paralysis, and it can be lethal. It's inherited as an autosomal recessive, Mendelian kind of disorder. The disease was well-characterized, and the responsible gene was identified in 1995. Sometime later—4 years or so—it became clear that a defect in splicing is related to the disease. We began to work on that because our interests in my lab have always been on RNA splicing—both the fundamental science and the relationship to disease.

There are two genes that are closely related. One is missing or defective in patients; the other gene functions as a kind of backup. It can express the correct protein, but in fairly low amounts due to the type of splicing that the transcript undergoes. So, we began to characterize that process. We weren't the ones who described this difference in splicing, but we were interested in that general problem. Between the two genes, there are very few nucleotide differences, but one in particular had been pointed out in the exon that is inefficiently spliced. So, we studied that problem: What is it about that nucleotide? What is normally being recognized in the transcript by various factors? We worked on that for a couple of years, and then we began to think about how to correct the splicing of the *SMN2* [Survival of Motor Neurons 2] RNA in order to allow the gene to produce higher levels of functional protein.

Anke Sparmann: How does this drug that was eventually developed actually work? How does it correct the splicing?

Dr. Krainer: It's a kind of drug called an antisense oligonucleotide. Those come in different modalities or "flavors," if you will. They're synthetic short nucleic acids, single-stranded. They have chemical modifications, and they can be designed to destroy the target RNA. They will home in on an RNA through base-pairing interactions, so they can be very specific. If the chemistry is designed in such a way that the duplex is recognized by endogenous RNase H enzymes, then they destroy the RNA target.

In our case, we use a different type of oligonucleotide design that binds to the RNA target by the same sort of physical chemical interactions, but it doesn't lead to its destruction. Instead, it blocks the binding of RNA-binding proteins. If you place an oligonucleotide in the correct place, then you can block the binding of a protein that affects splicing in some way. In our case, we were looking to block the binding of a splicing repressor, so that the exon that's nearby can now be recognized more efficiently by the splicing machinery. Now splicing looks more like it does in the *SMN1* gene, even though we're targeting the *SMN2* gene that is still present in patients. If you deliver the drug to the right cell types, the cell now knows how to allow this gene to express higher levels of functional protein. That's the molecular mechanism of action of the oligonucleotide.

Anke Sparmann: You were involved from the start figuring out how this works, but then also all the way through to actually drug development. What were the major challenges throughout that whole process?

Dr. Krainer: There were many. This started as a very basic science kind of effort, and we made mechanistic observations that inspired a way to try to correct the defect, and we went through successive modalities for doing that. We learned things along the way. The ultimately successful approach was a bit simpler than the way we had started. Importantly, we began to collaborate with Frank Bennett at Ionis Pharmaceuticals in 2004. We had a lot of discussions and decided to use a particular kind of chemistry and to go with the approach that I just described, which is to find oligonucleotides that will block the binding of a repressor.

We did that very systematically. There was a postdoc who joined the lab at that time, Yimin Hua, who did pretty much all the early work, the key preclinical experiments.

© 2019 Krainer. This article is distributed under the terms of the Creative Commons Attribution-NonCommercial License, which permits reuse and redistribution, except for commercial purposes, provided that the original author and source are credited.

We did have a lot of advice from Ionis people who were doing real pharmacology, but initially we were doing biochemistry, then cell-based experiments. Later, we set up mouse models. This took quite a number of years.

I don't know how to define when exactly we started working on this. We can put the start date when I, and my trainees, began to work on splicing, which is much earlier. On SMA specifically, we began around 2000 or 2001, with the work of a postdoctoral fellow, Luca Cartegni. The antisense screen, as it ultimately was carried out, began in 2004. The SPINRAZA molecule had other names earlier, but we published it in 2008. As things were progressing quite well, particularly when we began to do mouse-model experiments and seeing pretty dramatic results in terms of splicing and protein and phenotype, Ionis got quite serious about undertaking the clinical development and picking among the many different oligonucleotides that were effective, to look for the one that would be most specific and had minimum toxicity at high doses, etc. That's a separate effort in pharmacology, for which they had a lot of experience.

The next step was clinical trials. Those were initially sponsored by Ionis. About a year later, Biogen teamed up with them. The clinical trials were done in a variety of clinical centers and hospitals in several different countries. The key trials—Phases I through III—took about 5 years, which I think all went pretty smoothly. It ended up taking a year less than had been planned, because the results along the way were very encouraging. It was possible to complete the trials so that the patients still remained as part of an open phase extension study, beyond the original clinical trial. There are still ongoing clinical trials, but the ones that were key for obtaining the approval of the drug took about 5 years.

Anke Sparmann: Not that long ago, thinking of RNA as a drug molecule was not really out there. What were the changes that made this possible?

Dr. Krainer: It's all gradual. Like every new modality, there's a concept, and then there's the problems and reducing it to practice. There are always stumbling blocks. Typically, delivery of a new type of drug is something that requires a lot of effort. We were lucky that by the time we started working on this there were already several years of experience with antisense oligonucleotide pharmacology. They had gone through many chemistries. There was clinical experience, not so much with splice modulation, but nevertheless with the related chemistries. A lot of that knowledge—maybe more than 20 years of accumulated knowledge—is what makes these types of things possible. Monoclonal antibodies went through something similar. There was a description and one could see right away the potential, but to turn those into a drug took many years. Now, it's much more routine.

Anke Sparmann: You're looking at other diseases to target. What are you moving onto?

Dr. Krainer: Part of the lab still continues to study the basic fundamental aspects of splicing mechanisms and regulation, because the way we approach the problem is all based on insights about the mechanism. In this case, we were targeting a splicing repressor binding site. A few years earlier, we didn't even know these molecules existed, so one first had to discover that and understand something about how they work. We continue the basics, but we're also pursuing projects in which we try to apply similar or related approaches—blocking splicing components or RNA-binding proteins—in order to change splicing, and also other RNA processing, such as nonsense-mediated mRNA decay. We're exploring several potential targets that could lead to therapeutics for various diseases.

Anke Sparmann: In this basic kind of research, what is the next thing that's going to happen in splicing?

Dr. Krainer: That's moving along on many different fronts. When I started in this field, we were just doing cell-free splicing. It was all biochemistry. I started working on the development of systems for that as a graduate student. There was a lot to do, a lot of biochemistry to identify components. Of course, other labs were using genetic approaches and model organisms. Nowadays, there are many more disciplines that are contributing to understanding the whole process: quantitative approaches, bioinformatics, genomics, transcriptomics. A lot of techniques have been invented since I started in this field, so there are always new ways to revisit an approach. There are many powerful cell biology approaches, as well.

One sees steady progress on many fronts. Every once in a while, there are breakthroughs, so things advance more rapidly. The structural biology approach has had a tremendous impact in recent years with cryo-EM [electron microscopy]. Seeing snapshots of spliceosomes in action felt like the field suddenly moved forward 10 or 20 years. One could appreciate details, some of which were already known, but now you could really see it in real time. It was no longer an indirect demonstration or hypothesis. There's a lot of work that needs to be done using newer approaches like that to get new insights.

I think there will be a lot of surprises, and all these things inform how you might do therapeutics development where splicing underlies the overall approach. There are efforts to develop small molecules. It's not our work, but developments in the field to also modulate splicing, and so we need to understand better how these molecules are actually doing that. What's the mechanism of action? How specific are they, and how applicable is that approach to other targets, other splicing events? Structural insights from the spliceosome can inform those efforts and vice versa.

Anke Sparmann: Are there any problems with off-target effects of these kinds of drugs?

Dr. Krainer: Any drug obviously has that. With antisense oligos, because they're based on base pairing, you obviously have to pick sequences that are not repetitive. SPINRAZA, in particular, binds to a sequence that's unique: It's only present in intron 7 of the *SMN* genes and nowhere else in the genome, at least as a perfect match. This doesn't rule out the possibility that imperfect binding—weaker binding, presumably—with one or more mismatches could occur. The question is, just because it binds somewhere else doesn't mean that it's going to perturb splicing or some other process, but there is that possibility. One has to be very careful about looking for adverse effects of drugs. Obviously, that's part of the whole clinical drug development.

With small molecules, it's a completely different mechanism of action. We need to understand better the few examples that we currently know of and how they're actually eliciting the changes in splicing. We understand a limited number of off-target effects and they appear to be quite specific, but maybe that can be improved. If you move to a different target, is it going to be possible to have similar specificity? That's something that the field is going to learn in the next few years.

A Conversation with Lynne Maquat

INTERVIEWER: RICHARD SEVER

Assistant Director, Cold Spring Harbor Laboratory Press

Lynne Maquat is the J. Lowell Orbison Endowed Chair, Professor of Biochemistry & Biophysics, Founding Director of the Center for RNA Biology, and Founding Chair of Graduate Women in Science at the University of Rochester School of Medicine and Dentistry.

Richard Sever: You've been working on nonsense-mediated decay [NMD], which was something you discovered.

Dr. Maquat: It was toward the end of my postdoc—and also starting my own lab—when we discovered NMD in mammalian cells and began working out mechanism by studying mRNA half-lives using nucleated bone marrow cells from patients with β^0-thalassemia.

Richard Sever: For those who have forgotten molecular biology, what is the problem that's being solved by NMD?

Dr. Maquat: In human diseases of the type that we studied, translation of an mRNA terminates prematurely. The disease-associated mutation is either a frameshift or a nonsense mutation that generates a premature termination codon, which, when recognized by a ribosome, triggers decay of the mRNA. This is a good thing in the sense that if a cell were to make truncated proteins, these proteins could be toxic. The open reading frame of the mutated mRNA is abnormally short, and therefore the encoded protein would be truncated, with the potential to gum up the cellular machine it works in. We figured out the rules for how a cell differentiates a normal termination codon, which generally doesn't trigger mRNA decay, from a premature termination codon, which generally does.

Richard Sever: How does the cell decide it's too short?

Dr. Maquat: Splicing an mRNA precursor, a pre-mRNA, in the nucleus deposits what we called a "mark" that persists until the first round of translation of the mRNA in the cytoplasm. We figured this out not only by studying β^0-thalassemia and other human diseases, but also by generating reporters so we didn't have to rely on patient samples. Using the reporters, we could move introns around, and we could move premature termination codons around. We realized that there must be a mark that persists during that first round of translation, because we knew NMD, at least the type that depends on a mark, depends on the position of the premature termination codon relative to where introns reside within nuclear pre-mRNA, is largely restricted to newly synthesized RNA, and depends on cytoplasmic translation.

Richard Sever: This is a protein complex that sits on a splice site?

Dr. Maquat: That's right. It resides on newly made mRNAs slightly upstream of where splicing occurred to generate the mRNA. When I collaborated with Melissa Moore on our hypothesis of a mark, working together with postdoc, Hervé Le Hir, who started in my lab and then moved to Melissa's lab, we renamed the mark the "exon junction complex" [EJC]. Hervé figured out that this complex of proteins is deposited on newly spliced mRNAs ∼20–24 nt upstream of exon–exon junctions generated by pre-mRNA splicing. We figured out—and Joan Steitz's lab also figured out—that this complex consists of some of the NMD factors that we named UPF [up-frameshift] proteins, after their orthologs in *Saccharomyces cerevisiae*.

So, depending on how many introns are in a pre-mRNA, you can expect to have that many exon junction complexes, one at each exon–exon junction. During the pioneer round of translation, if there is a termination event—it can be premature, or it can be at the normal termination codon—where one or more downstream EJCs remain, then the mRNA will be targeted for NMD.

Richard Sever: Basically, the cell's ordering the "stop" site, and it says "Wait a second, there's one behind where it should be."

Dr. Maquat: That's right. In the process of translation, the ribosome can remove EJCs. We figured out a rule before we even knew where the EJC was deposited. This rule said if translation terminates 50–55 nt or more upstream of an exon–exon junction, then the mRNA will be targeted for NMD. Now that we know where the EJC is deposited, and we know where the leading edge of the terminating ribosome would be situated relative to the termination codon, this rule makes sense.

© 2019 Maquat. This article is distributed under the terms of the Creative Commons Attribution-NonCommercial License, which permits reuse and redistribution, except for commercial purposes, provided that the original author and source are credited.

As I've said, this happens during a pioneer round of translation. But that was another surprise, in the sense that when people at the time were thinking about mRNA translation, they were thinking about mRNA that had eukaryotic initiation factor [eIF]4E at the cap. It is true that mRNAs bound by eIF4E at the cap produce the bulk of cellular proteins. But because we knew NMD is largely restricted to newly synthesized RNA, and we knew that there is a different cap-binding protein that is acquired by the pre-mRNA, I asked postdoc Yasu Ishigaki to test if it was mRNA that has this different cap-binding protein, which is acquired very early on in mRNA biogenesis, that's still there during the pioneer round of translation. The answer turned out to be yes. So we define the pioneer round during which EJC-dependent NMD largely occurs as the translation of mRNA that's bound at the cap not by eIF4E, but by CBP [cap-binding protein] 80, and it's binding partner, CBP20.

It's a matter of timing, because we showed there's a precursor–product relationship between the pioneer round and subsequent steady state rounds of translation in mammals. We think of it as a completely different mRNA in the sense of the proteins that are associated with it when compared to the steady state translation initiation complex. The pioneer translation initiation complex not only has CBP80 and 20 at the cap but it also has the EJCs.

Richard Sever: You mentioned a group of UPFs in those EJCs. What are the proteins there, and how is this coordinated with degradation of the RNA?

Dr. Maquat: We've done a lot of work on understanding how proteins rearrange on an NMD target during the pioneer round of translation and the consequential decay steps. The key NMD factor is an ATP-dependent RNA helicase called UPF1. We find it either transiently or weakly associates with CBP80 at the 5′ cap of the mRNA. To detect that interaction, we have to cross-link proteins before we lyse cells. We found that CBP80 escorts UPF1 to the translation termination complex together with a kinase —SMG1—that phosphorylates UPF1 not as a part of the termination complex, but later when CBP80 escorts UPF1 and its kinase to the EJC. The activation of NMD is when the kinase becomes free to function—that is, it's not inhibited—and it can then phosphorylate UPF1 at the EJC. That's critical.

Then there's a step of translational repression that we found to be important for decay—a feedback to stop more translation initiation once translation terminates and UPF1 becomes phosphorylated. We know the mechanism for that. While ribosomes continue to translate the open reading frame of the mRNA, there's an inhibition of further translation initiation events. Phosphorylated UPF1 goes back and prevents the 60S ribosomal subunit from coming in and joining any 43S translation initiation complex, which includes the 40S ribosome, poised at the translation initiation codon. If you don't have this prevention, you don't get decay.

Richard Sever: Right, and then you get these peptides.

Dr. Maquat: Then you get decay of the mRNA, and we know that decay can be from the 5′ end, from the 3′ end, and/or, endonucleolytic.

Richard Sever: You mentioned that NMD is not the only pathway for this. There are several parallel pathways?

Dr. Maquat: Proteins can multitask in cells, and the way that we discovered other unexpected pathways is by looking to see what else in the cell the RNA helicase, UPF1, interacts with. By yeast two-hybrid analysis, postdoc Yoon Ki Kim found that Staufen interacts with UPF1, and that there is a pathway we named Staufen-mediated mRNA decay (SMD) that's also dependent on UPF1 and translation. The way that UPF1 is recruited to an SMD target at a position situated downstream of what is normally the termination codon is via a Staufen-binding site. When translation terminates—usually normally, but it doesn't have to be—and Staufen isn't removed by the terminating ribosome, it is at that point that I think the SMD and NMD pathways converge. SMD is not restricted to newly synthesized RNA since there is no need for an EJC. SMD targets both the pioneer round and steady state rounds of translation.

Richard Sever: What's it being used for during the steady state rounds of translation?

Dr. Maquat: It is being used to regulate differentiation processes. To make it even more complicated, graduate student Chenguang Gong showed that NMD and SMD are in competition. They're in competition because UPF1 functions in both pathways in a mutually exclusive way. UPF1 can bind to either UPF2, which is important for NMD, or Staufen, which is important for SMD. During myogenesis, the efficiency of SMD goes up. That's important. For example, there's a natural SMD target encoding a protein that keeps myoblasts in an undifferentiated state, so myoblasts eliminate that mRNA by SMD, and thus its encoded protein, to differentiate to myotubes. And by having SMD be more efficient, NMD becomes less efficient. That's important because the expression of myogenin, which derives from a natural NMD target, promotes the differentiation process. The cell has evolved to use these pathways as a way to regulate not only myogenesis but other differentiation processes (e.g., adipogenesis).

Richard Sever: You've also worked on the regulation of apoptosis and the way in which this pathway was used to control that, but it didn't sound like decay was the aim. This wasn't an NMD process, but it was using the same machines. Is that correct?

Dr. Maquat: It is NMD. For apoptosis, postdoc Max Popp found that the cell reduces the efficiency of NMD by the caspase-mediated cleavage of UPF1, which up-regulates natural NMD targets that include those encoding proapoptotic proteins. But, we can ask, why do we have NMD? We constantly make mistakes when we process our pre-mRNAs. We have a lot of alternative splicing, we have a lot of alternative 3′-end formation, And, because of the flexibility in those processes, we make mistakes that are eliminated by NMD. On top of that, I've been talking about natural NMD targets when mentioning differentiation processes and apoptosis. There are ~5%–10% of our messages that are naturally degraded by NMD. In some of these cases, there is an upstream open reading frame; in others there is a splicing event downstream from the normal termination codon, as two possibilities.

Richard Sever: So it's not really nonsense then?

Dr. Maquat: Well, yes it is. I wanted to call Staufen-mediated mRNA decay (Staufen-mediated NMD) and people said "No, 'nonsense' means 'PTC' [premature termination codon]." But actually, of course, it doesn't exclusively. It means "no sense," which applies to both PTCs and normal termination codons, which as we've just discussed can trigger NMD. It's a semantic issue.

Among those mRNAs that are up-regulated during severe DNA damage where it behooves the cell to trigger apoptosis are those that terminate normally. NMD is used as a rheostat by cells to adapt. Now, are all the natural NMD targets proapoptotic? No, but they, too, are up-regulated and translated. They are producing proteins, but it's background gene expression. It's not pertinent to the key process, which also has a lot of other apoptotic caspase-mediated events ongoing that are pertinent. Another way to think about NMD is that it maintains cellular homeostasis until a stress or another change in environment becomes sufficiently severe so that a response should be triggered. In these cases, often the efficiency of NMD is then decreased by one of many possible mechanisms, depending on the cell type and stressor, so as to promote the appropriate response.

A Conversation with Karla Neugebauer

INTERVIEWER: CARIKA WELDON

Researcher, University of Oxford Wellcome Centre for Human Genetics

Karla Neugebauer is a Professor in the Departments of Molecular Biophysics and Biochemistry and of Cell Biology and Director of the Yale Center for RNA Science and Medicine at Yale School of Medicine.

Carika Weldon: Can you give us an overview of your recent work?

Dr. Neugebauer: We're sequencing nascent RNA, which is the very first RNA as it starts to get synthesized by RNA polymerase II. We purify it and sequence the whole length of the RNA. That gives us insight into the whole life history of each individual RNA. We can see where transcription started from the 5' end of the sequence, and then we can see where the elongating polymerase was sitting when we isolated that transcript from the sequence of its 3' end. We can see all the things that might've happened to that RNA in between, like splicing. Splicing is one of my strongest interests.

We're really excited about being able to see the dynamics of nascent RNA. It's this amazing species of RNA that has all these different 3' ends and all of these different sequence features due to the changeability in the body of the RNA. We really feel like this is the tip of the iceberg, that we would be able to associate the progress of transcription with the progress of splicing, but then there's going to be a whole avenue of interesting research that'll be about folding or RNA modifications or changing RNA binding sites as exons become ligated together. I feel sure this is going to be an important regulatory gold mine where we can see what's actually happening to nascent RNA.

Carika Weldon: When you learn about gene regulation, you learn about transcription, then you learn about splicing, and then you learn about translation, and you see them as these three separate events. Can you talk about your research using long-read sequencing that shows this is all happening at once?

Dr. Neugebauer: The wonderful thing about being a biologist is that you go to do an experiment and you think of maybe one, two, or three hypotheses on what the data's going to look like. Then you produce that data and you look at it initially and you're looking for those three expected things... you usually are completely blind to that other thing that you're actually staring at and you just don't even see it. Then some time later you go, "Oh my God, what is *that*?" Then you realize that you're seeing something unexpected, and that's usually the most interesting thing.

The thing that we discovered most recently was a study in *S[chizosaccharomyces] pombe*. We expected to be able to ask if individual introns in the same gene are removed in their order of their synthesis or is it all mixed up due to regulation? It turned out there was evidence for both modes, but the surprise was the presence of transcripts that were completely unspliced, and then the polymerase just read through the poly(A) cleavage site and kept going. We had no reason to suspect that there would be this population, but if you were doing short-read sequencing on nascent RNA, you would never know that because the retained introns would simply be added to all the numbers that you're tallying up for the splicing frequency in each individual intron. You'd never realize that they were all in the same transcript. That really illustrates the power of the long-read sequencing and the ability to link all those splicing events with each other, but then also with poly(A) cleavage: the fact that the polymerase just never stops. In our new study we are also seeing things that we didn't really anticipate.

The thing that's captivating to me is the notion that there's this whole continuum of possibilities. I've tried to illustrate these things, but I feel like when I draw it, it's a little bit too concrete. I can't actually illustrate all of the different possibilities, but I want people to try to imagine all of these dynamics happening at the same time. The fact that when two exons ligate together, they create a new sequence at the junction; they've lost this big intron so now that sequence has gone. Exon 1 is now completely different. I don't know how to illustrate that without making it too concrete. From a public speaking point of view, I'm sure everyone in the audience has had that same experience where it's hard to communicate an idea by drawing it. If you just like look the audience in the eye and go, "You know, I don't even know how to diagram this idea, but maybe you can just imagine it yourself."

Carika Weldon: As a scientist, you have public engagement. Some scientists are really good speakers, some not as great. Do you think that impacts on how science has developed, because other people are not able to "capture the imagination?"

Dr. Neugebauer: It's a definite plus to be a charismatic speaker, to be able to communicate your science. Sometimes there is a speaker who you'll see repeatedly at meetings who's not very charismatic and you're thinking, "Why is this person at every meeting I go to?" It's because their science is so cool. Scientists have a culture where it's not just about selling an idea and being slick; it's about having good ideas and doing beautiful research. The person I was thinking about is a wonderful scientist: very imaginative, and just has amazing data. Even though he's not a charismatic speaker, everybody loves to listen to him and experience the science that he's been producing in his lab. One of the things I like about the scientific community is that it's very inclusive of a lot of different personality types. Even though somebody who's really confident and kind of pushy can get more attention than the person who's more retiring, I do think that the community prioritizes the science and not personality or politics.

Carika Weldon: There used to be a big difference when it came to women in science. We've come a long way with that. What do you think about that?

Dr. Neugebauer: I worked in Europe for 13 years and I was involved in promoting women in science in Europe. Kai Simons, who was the founding director of the Institute [Max Planck Institute for Molecular Cell Biology and Genetics] that I worked at, asked me to start a database of expert women scientists that was meant to show off women who'd actually been achievers, to say, "Look, you can't tell me you can't invite someone to give a seminar, because... what about one of these 800 women in the database?!?!"

In Germany where I was working—and Germany is a wonderful environment for doing science—it's hard for women especially, because there is no tradition of a "career ladder" where you know the expectations for each of the steps and where you don't necessarily have to move cities to get to the next rung on the ladder. I think moving is very stimulatory and it's a wonderful thing to be able to move, but I think that it's frightening for some people who would be excluded. Why should your willingness to move city—move your family, in other words—be a criterion for allowing someone to be a leader in science? That doesn't seem like a rational choice. I feel that in Europe the career structure isn't very helpful, and you can see that because they're really missing women in leadership positions.

Women are better represented in the U.S. Coming back to the U.S. I thought, "There are more women in leadership positions in the U.S., so now I don't have to deal with this issue anymore." Now I realize that we're going to be dealing with this diversity issue for a really long time. First of all, women aren't equally represented in science as you progress through the career track, and there are other diversity issues as well with underrepresented minorities. Clearly, we want to make sure that every person who wants to be a scientist doesn't feel that they don't belong. That's a main goal that I have and I try to express that at Yale, when I teach a lot of undergraduates. I'll have "actions" in my classroom.

Carika Weldon: What is that?

Dr. Neugebauer: The students spoke with me about it. There's a sense of cliquishness in the classroom that some of the students who were not in the major but clearly absolutely capable and credentialed felt excluded. There were a number of different students who felt like they weren't in the mainstream of the classroom and that they were at a disadvantage. At the beginning of the next class, I showed Joan Steitz's Lasker video where she talks about her identity as a scientist: that, "That's who I am. I can define myself any number of ways, but the bottom line for me is that I'm a scientist." I just find that a statement that she makes so moving, so I played that for the class and I also asked them about their identities, too. "Who's an athlete?" It turns out that athletes in college think that they're dumb because they think, "I was recruited because I can catch a football or hit a tennis ball." I was gobsmacked when I realized this because that is so not true. They wouldn't get accepted to Yale University if they weren't capable of doing the work. So that was my action, and some of the students expressed their appreciation for doing that. It's also important to recognize that we can work ourselves into a corner. Anyone—and that can include me at a scientific meeting sometimes—can feel like they don't really belong when we actually do have a right to be there. We do have the credentials.

Carika Weldon: You're really excited about the long-read sequencing and where that could potentially go. Where do you see your field in about 15 years?

Dr. Neugebauer: I'm going to do an exercise instead where, looking back 15 years ago, what am I shocked at now? Back then, I could not possibly have imagined Adrian Krainer's work and the fact that a splicing disease like spinal muscular atrophy can be cured using an antisense morpholino. Also, when Joan Steitz discovered these snRNPs [small nuclear ribonucleoproteins], they were just these small RNAs that assembled little particles and they went off and they "did" splicing. Now, look at how many other small noncoding RNAs are in RNPs [ribonucleoproteins] where the RNA does the business: CRISPR [clustered regularly interspaced short palindromic repeats], telomerase, snoRNPs [small nucleolar ribonucleoproteins], microRNAs in the RISC [RNA-induced silencing complex], Piwi-interacting sRNAs [small RNAs]. And there could be more.

The one thing that's a real black box where I hope we'll go in the future is where there are dynamics, where there's a progression of events that you know must happen to this thing—like splicing and then you make an mRNA—but there are black boxes around what happens if it doesn't do this during a given time period. Does it happen later? Does the thing degrade? We also definitely need to develop new methods that make our understanding of populations of molecules and the quantitative aspect more accessible so that we can interact with physicists and mathematicians and get more quantitative and predictive about RNA biology. That's something that's going on in things like mechanical biology and cell motility and so on. In RNA biology, it's not going on quite as much.

Carika Weldon: Can you remember what inspired you to follow this track to be a scientist?

Dr. Neugebauer: It was that sense of wonder as you discover something, and then that period when you feel chuffed that you've discovered something but then there's all these additional questions. It's the problem-solving that's so much fun. I think that's what all scientists must find so enjoyable, and especially as a young scientist to experience that for the first time is always amazing. I hope young scientists continue to do PhDs and postdoctoral fellowships because with science training, you can go into so many different directions. To actually have the opportunity to discover things is a wonderful life experience.

A Conversation with Nicholas Proudfoot

INTERVIEWER: RICHARD SEVER

Assistant Director, Cold Spring Harbor Laboratory Press

Nicholas Proudfoot, FRS, is Brownlee-Abraham Professor of Molecular Biology in the Dunn School at the University of Oxford.

Richard Sever: You've been working on long noncoding RNA. These are not protein-coding, and there are a bunch of classes. Can you remind us what these are?

Dr. Proudfoot: The concept of the gene is that genes make proteins, so the critical genes in mammalian genomes would be protein-coding genes. Obviously, it was appreciated that there were lots of other transcription units that made structural RNAs like ribosomal RNA or tRNA, and these are actually made by different RNA polymerases. But the RNA polymerase II that makes the protein-coding genes also does a lot of other transcription that doesn't seem to be directly related to producing a messenger RNA and then a protein. When mammalian protein-coding genes were first picked apart, it was clear that a large fraction of the gene is actually noncoding; it's intronic. You get extraordinary situations where 90%–95% of the transcription unit is actually intronic and is removed by splicing and then largely degraded.

Then, once genomic or transcriptomic analysis became possible and you could really get a more complete and higher-resolution profile of transcription across the genome, there was the realization that there are a bunch of transcripts that were entirely noncoding. Initially, the simplest interpretation on the discovery of these noncoding RNAs was that if they exist, they must be there for a purpose. The problem is that, in general, these noncoding RNAs are very unstable and they are rapidly degraded, so the cell doesn't usually take a lot of care over producing very much of them.

Richard Sever: These are distinct from the microRNAs?

Dr. Proudfoot: The microRNAs are really fragments of RNA that come out of other transcription units. Most microRNAs actually come out of these large introns in protein-coding genes. We estimate about 70%–80% of microRNAs come out of the introns of protein-coding genes, but there are some noncoding RNAs that also host transcripts to microRNAs.

Richard Sever: These long noncoding RNAs are where? In the introns, in promoter regions, enhancers, between genes?

Dr. Proudfoot: What's apparent about many long noncoding RNAs is that they're almost made by accident by the protein-coding genes. When RNA polymerase II initiates transcription on the promoter, as well as going in the forward direction to make the pre-messenger RNA, it can also go backward to make a long noncoding RNA. Basically, RNA polymerase II is a fairly naive enzyme complex. It'll bind to anywhere that is accessible, which is a nucleosome-free region, and then it'll just go in either direction. The vast majority of protein-coding gene promoters are bidirectional.

Richard Sever: This produces an antisense transcript?

Dr. Proudfoot: It's not antisense, because it initiates outside the transcriptional region of the protein-coding gene, but it goes into the 5′-flanking region. So there's that type of long noncoding RNA.

Also, enhancers can generate transcripts directly themselves. Protein-coding genes are often regulated by a bunch of really distant promoter elements. You can regard enhancers as being separate promoter elements that come together in 3D by some sort of chromosome looping to form the actual promoter hub. Because these enhancers are regions of open chromatin, RNA polymerase II can also get in there and go in either direction.

Richard Sever: Is that the common feature, that you've got some open chromatin?

Dr. Proudfoot: That's right. Probably in most cell types, the antisense promoter transcripts and the enhancer transcripts will account for well over 50%, if not 75%, of the long noncoding RNAs. Enhancers are very numerous and they're all generating transcripts, usually in both directions.

Richard Sever: Is there such a thing as "junk" RNA, or do these serendipitously generated transcripts actually have a role?

Dr. Proudfoot: Initially, you could regard them as junk, but they are greatly beneficial to evolution. Although they're not used at a particular evolutionary stage, sometimes these transcripts acquire a function that then makes

---

© 2019 Proudfoot. This article is distributed under the terms of the Creative Commons Attribution License, which permits unrestricted reuse and redistribution provided that the original author and source are credited.

them very valuable for the regulation of gene expression in a cell. A classic example is a long noncoding RNA called NORAD [noncoding RNA activated by DNA damage]. It probably began its life as a nonfunctional transcript, but it happened to have a bunch of binding sites for this very interesting protein called Pumilio. Pumilio is very important in development and, indirectly, with DNA damage and other important issues for cells to cope with. I guess evolution then increased the number of Pumilio binding sites, so it has a whole bunch of them. It ends up being a sponge that regulates the levels of Pumilio in the cell.

There's an analogy to introns. Introns are mainly functionless, but evolution has put some introns to good use. You can get microRNAs out of them, and there are other small RNAs that are chopped out of introns like snoRNAs [small nucleolar RNAs]. Of course, the benefit of having introns—or rather, having genes cut up into exons—is that you can have different combinations of exons to give you different proteins. You get a greatly increased proteome through the presence of introns. Maybe long noncoding RNAs also provide potential evolutionary advantage. You certainly see more long coding RNAs the more complex the eukaryote. You see some long noncoding RNAs in S[accharomyces] cerevisiae, but as you go up into mammals you see a lot more.

There are also some long noncoding RNAs that don't seem to be directly connected to an adjacent protein-coding gene, and these are often referred to as long intergenic noncoding RNAs, the lincRNAs. They're really the same thing as eRNAs [enhancer RNAs] and promoter antisense RNAs in that they probably don't code for any proteins and some of them may or may not have a function. Some of them may just be a patch of chromatin in an intergenic region that becomes open for some unknown reason, and then it becomes, potentially, a promoter.

Richard Sever: When we think of mRNA, there's a specific set of signals in the DNA that is used to initiate transcription. What's happening with these sites?

Dr. Proudfoot: Some of them may have evolved binding sites for transcription initiation factors and may become more like bona fide protein-coding promoters. One trick that we think is important for generating a lot of long noncoding RNAs is that when transcripts are made by any polymerase, one of the things they can do is reanneal back to the DNA template. As the polymerase initiates transcription from a promoter, the first thing it does to the DNA template is to unwind it a bit. Because the DNA is partly underwound, and because the polymerase has to expand the nucleosome-free zone as it transcribes the gene, you have a nascent RNA coming out from behind the polymerase that is just lying in wait to get back in. And, indeed, it does. This RNA–DNA hybrid is called an R-loop. The point about the R-loop is that because you displace one of the DNA strands in the double helix, the single-strand DNA is now prone to damage. R-loops are damage-inducing, so the cell has evolved lots of mechanisms to get rid of these awkward structures. There are lots of helicases that unzip the RNA–DNA hybrids. I suspect that virtually all transcriptions will form these RNA–DNA hybrids a lot of the time. If you look at the steady state profile of R-loops across the genome, you see plenty of them, but they're probably there when the formation has outcompeted their removal.

We knew R-loops were particularly associated with the beginnings of the transcription units and also quite often at the end of the transcription units, where the original nucleosome-free zones tend to be in genes. When you form an R-loop, the displaced DNA strand is actually the exact template you need to initiate transcription. The initiation factors pull apart the DNA strands so that RNA polymerase can get into the template strand and start copying it. The R-loops expose the single-strand DNA straight off by forming the hybrid on the other strand. That's now a potential template for initiation of transcription just by itself. The polymerase will go to any open region of chromatin. It'll greatly prefer to go to a single-strand DNA; that's exactly what it needs to start transcribing. What we've shown is that RNA polymerase does indeed bind to these R-loops and transcribes the strand that was displaced. It's making an antisense transcript to the normal sense transcript, which is the exact orientation of many of these long noncoding RNAs. We showed this happens in vitro. We artificially made a plasmid with an R-loop in the middle of it. We stuck it into a nuclear extract and, lo and behold, you generate a transcript that you can see initiates in the single-strand region of the R-loop. We've also spent a lot of time trying to prove this happens in vivo and we think we have.

Richard Sever: How do you do that?

Dr. Proudfoot: It's kind of indirect. You have to show that a long noncoding RNA transcript really is dependent on an R-loop. The simplest way to do that is to overexpress this enzyme that gets rid of the RNA–DNA hybrid, RNase H. If you overexpress RNase H, you can then show that not only do you get rid of R-loops, you also lose a whole set of these long noncoding RNAs, particularly the ones that come from the backward promoters, but also the enhancers. The enhancer transcripts also form R-loops and you can see that if you overexpress RNase H a lot of the enhancer-derived long noncoding RNAs disappear. I think that quite a high fraction of the noncoding transcriptome might well derive from this type of R-loop promoter activity.

Richard Sever: Presumably, as the mRNA is being synthesized, it can either reanneal or interact with the transcriptional processing proteins. Is there some regulation?

Dr. Proudfoot: There clearly are specific cases where this is likely to be regulated, but we have no direct evidence for the regulated formation of R-loops so that they can initiate transcription in a backward direction. It smells more like a "genome accident." Transcription is dangerous, basically. If you make a transcript, you're exposing the DNA to damage. You're exposing it to the formation of these R-loop structures and you generate all these extra transcripts.

What we've shown in my lab is that many of these long noncoding RNAs initiate from this R-loop promoter mechanism, but then the cell has evolved the way to stop them going too far. There's a complex called the "Integrator complex" associated with promoters and enhancers and it appears to have a 3′ end–forming activity. A lot of long noncoding RNAs tend not to be more than several hundred nucleotides. They're long relative to microRNAs, but they're still quite short relative to a pre-messenger RNA. If you knock out a Pol II elongation factor called SPT6 that is important for transcription initiation across long genes in particular, then the regular protein-coding gene transcripts don't work quite so well, but the antisense long noncoding RNAs start working much better. They not only work much better—you get more of them—but they also go much further. We showed that not having SPT6 causes a loss of recruitment of Integrator, and the phenotype of the cells that don't have SPT6 and therefore make a lot more of these long noncoding RNAs is very bad news. The cell cycle grinds to a halt; the cells go into senescence. We showed that there are lots more R-loops being formed by these extended long noncoding RNAs, and lots of DNA damage.

Richard Sever: Would you expect to see more of these at different times in development? And do you have to have additional mechanisms to cope with this?

Dr. Proudfoot: So far, all of our experiments are based on an easy experimental system, the HeLa cell. HeLa cells are great because they grow really fast, and they're very easy to manipulate. With CRISPR, you can do pretty much anything with the HeLa cell. It grows like a yeast, but it's a mammalian cell type, but of course it's very mutated and very weird. It would be nice to look at these transcript mechanisms in a primary cell or in a stem cell. I would guess in stem cells, where there's a lot more open chromatin, you might expect to see more potential for forming noncoding RNAs. Maybe some of these are more likely to be functional. Those experiments end up being very expensive. You have to make a lot of libraries. We've made probably two or three hundred transcriptomic libraries in the HeLa cell so far, so we know an enormous amount. To get that amount of data in a stem cell would involve a lot of work and we haven't yet embarked on it. I would like to do some of these experiments in primary cells or at least in cells that are slightly less weird. HeLa cells, because they've been cultured for so many generations, have acquired lots of polyploidies in the chromosomal regions. They seem to have evolved in culture to cope with the fact that they're loaded with DNA damage. They have a lot of damage factors up-regulated, which is interesting if you want to study DNA damage, but it basically is aberrant.

Richard Sever: Looking through evolution, different organisms have different amounts of these. Are you looking in other species?

Dr. Proudfoot: In simpler organisms like *S. cerevisiae*, the genome is much smaller so there isn't much space to make a separate transcript between the protein-coding genes, but you can still get antisense transcripts. I also suspect unicellular organisms find it harder to energetically afford to make all these weird transcripts that aren't doing anything, so getting rid of them is probably beneficial. Energy is never a very serious issue for a mammalian cell because in an animal you have homeostasis; you guarantee a concentration of ATP. You can make a whole lot of transcripts that may be beneficial in terms of evolution but are not particularly useful to the animal at that point in time.

A Conversation with Oliver Rando

INTERVIEWER: CARIKA WELDON

Researcher, University of Oxford Wellcome Centre for Human Genetics

Oliver Rando is a Professor in the Department of Biochemistry and Molecular Pharmacology at the University of Massachusetts Medical School.

Carika Weldon: Can you give us a brief summary of your work?

Dr. Rando: The general area we work in is paternal effects in mammals: the effects of a male's environment on his offspring. We spend a lot of time working on different things that you can poke dads with that'll affect things like metabolism or anxiety-related behaviors in kids. We're trying to understand how this works: How what a dad sees or experiences affects information in sperm, how sperm information affects early development, and how that might ultimately give rise to a phenotype in the kids.

Carika Weldon: Just to clarify, this is not in humans, correct?

Dr. Rando: We work in mouse entirely.

Carika Weldon: And how exactly does the father's environment affect the child?

Dr. Rando: There are at least three different kinds of epigenetic information that make it into germ cells. There are small RNAs, there's DNA modification like cytosine methylation, and then there are histone modifications. The system we mostly focus on is dads eating a low-protein diet. In that system, we don't think there are many reproducible effects on DNA modification or on chromatin, so we are focused on small RNAs. We don't know for sure that small RNAs in sperm are responsible for programming offspring, although there are several other studies that have reported microinjecting sperm RNAs and getting phenotypes in kids. We haven't tried that yet. We've focused on characterizing the small RNAs that are in sperm and trying to figure out where they come from and what they do when they're in the embryo.

Carika Weldon: You're saying that if the father has a low-protein diet, that then is a negative for the child?

Dr. Rando: In kids, we think it alters the levels of bile acids, so it affects the levels of hepatic cholesterol esters and it affects the levels of hepatic regulation of cholesterol biosynthesis genes. It affects glucose control. All of those things can be good or bad depending on whether or not the kid is seeing a scarce diet or a plentiful diet.

Another system I think is really useful for thinking about is it "good" or is it "bad" is we also studied paternal nicotine. Kids whose fathers consumed nicotine are resistant to a lethal injection of nicotine. Dads experiencing nicotine make kids nicotine-resistant. This is good if you're going to be injected with an LD_{50} of nicotine as a child, but those kids are also bad at glucose clearance. It's a trade-off. It's good if you think the kids are going to be seeing toxic levels of drugs in the environment. It's bad if you think that they're not going to be and they're trying not to get diabetes and metabolic syndrome.

Carika Weldon: Your work has focused on two RNAs: piRNAs [Piwi-interacting RNAs] and tRFs [transfer RNA-derived RNA fragments]. Can you elaborate on those a bit more?

Dr. Rando: In the germline, the piRNAs are a pretty famous class of small RNAs; the testes are filled with them. It depends on the organism, but one of the roles they play is an antitransposon defense system. Sperm are unusual in the RNAs they carry because unlike most cells of the body, they carry some microRNAs—which is the most common small RNA in a typical fibroblast or hepatocyte or something—but they carry much more of these tRNA fragments, which are the 5' ends of mature tRNAs. So, piRNAs are in the male germ cells, but ejaculated sperm carry these totally different tRNA fragments, which is sort of weird. That distinction actually led us to figure out where sperm RNAs come from. They actually aren't made during the process of sperm development in the testes. They're shipped into sperm as they spend a week or 2 weeks going through this long tube called the epididymis. As sperm leave the testes, they somehow eliminate all their piRNAs. As soon as we can get them in the very early epididymis, they've already gained tRNA fragments.

Carika Weldon: Do you know how that happens?

Dr. Rando: We don't know anything about how they lose piRNAs. We've assumed it's by degradation. But the way they gain tRNA fragments is that the epithelium of this tube makes tRNA fragments and we believe it ships them over in little vesicles called epididymisomes.

© 2019 Rando. This article is distributed under the terms of the Creative Commons Attribution-NonCommercial License, which permits reuse and redistribution, except for commercial purposes, provided that the original author and source are credited.

Carika Weldon: But right before they are ejaculated, they gain back some piRNAs?

Dr. Rando: No, there's a third class of small RNAs in the picture. There's a small subset of microRNAs that is present in testicular sperm, absent in sperm in the early epididymis, and then present again in sperm in the late epididymis, which is really weird. It could be that sperm are dumping these RNAs in the early epididymis and then regaining them later, or a mixture of sperm could come out of the testes and the ones that are lacking these RNAs are getting stuck in the early epididymis. Those are the two models for it. But if you just purify sperm from the early and late epididymis, they differ mostly in the levels of a couple microRNAs.

Carika Weldon: In terms of your field, where do you see this going? Is it going toward a therapy?

Dr. Rando: Our interest is in understanding how dads talk to kids. Our model is that the epididymis ships RNA into sperm that will be delivered to the embryo. We also know that the conditions that a dad lives in affect the RNAs that are in the sperm. Putting those together, our model is that the epididymis senses the dad's environment somehow and changes the RNA it makes to give to sperm to give to kids. The million dollar question is "How does the epididymis 'see' the world?" How does its sensing of the world, of amino acid levels, or of leptin, or insulin, or something like that affect which RNAs it makes? That's a big basic question.

Is this something that one might manipulate to clinical ends? It's an interesting idea. It runs into many of the same concerns people might have with editing the genomes of kids; this would be editing the epigenome. That's above my pay grade, in terms of discussing whether society wants to do it or not. But it's certainly reasonable to test the idea that, by changing the RNAs you deliver to sperm, you would alter phenotypes in kids.

Carika Weldon: I recently did a school tour where children asked me questions about genetics. One of the best questions was "Half of you comes from your mom and half from your dad, but can you be more of one than the other?" What do you think, based on your work?

Dr. Rando: Generally speaking, we think of moms having more influence over kids than dads. That's certainly true, because in mammals babies grow inside of mom. Fetal alcohol syndrome is not magic and epigenetic; you're poisoning a baby. Also, women with diabetes have so-called "macrosomic" kids; they're just stewing in sugar and they get huge. In that sense, moms have a huge impact on kids that dads can't hope to replicate. In terms of the molecules in the germ cells, oocytes are where you inherit mitochondria from. Sperm are where you inherit Y chromosomes from. Beyond that, in terms of the small RNAs and the DNA modifications and stuff, it's still the case that probably more of what you inherit comes from the oocyte. Generally speaking, mom tries to erase as much of dad's information as possible.

Carika Weldon: So, when this RNA gets injected into the egg, the egg thinks, "This is not what we want; get rid of it," but it's literally getting rid of information that would have gone to the child?

Dr. Rando: The assumption until 10–15 years ago was that everything except for half a genome came from mom. I think that assumption is now 90% true. Most information is probably still coming through the oocyte, but at least some small amount of information comes through dad. One of the questions we're quite interested in is what the "bandwidth" of sperm is. Exactly how much information does dad manage to get through the mom's screen? Does he just tell kids quality of life: Life is crappy or great? Does he tell his kids about psychic stress, and water supply, and nutrition? Or does he tell his kids extremely detailed information about the world he lived in? This is a really deep question for the paternal effect field. We know some information makes it through, but the question of whether or not it's coarse-grained information or very fine-resolution information has very different implications.

Carika Weldon: Where do you see this field going in the next 10 to 15 years?

Dr. Rando: I'm very interested in how the epididymis sees the world and changes the RNAs it makes. In terms of our own work, the big black box is that we now see changes in gene control in the early embryo depending on which RNAs you give to the zygote. How those changes in gene regulation in an early embryo turn into a cholesterol phenotype in a pup is a question that we have ideas about, but we haven't addressed at all. I think that's the black box that the next decade or two is going to have to take seriously.

Carika Weldon: How did you get into this field? It's not your typical field at RNA meetings.

Dr. Rando: I started in chromatin and transcriptional control. One of the interesting things about chromatin is that, at least in some cases, it seems to carry epigenetic information at cell division. I was just starting my lab, and I was thinking about interesting things that epigenetics could do. Epigenetic information often listens to environmental conditions; the chromatin state changes when you heat-shock yeast or change their conditions. If any of this makes it through to the germ cells, that would allow the environment in one generation to affect the future. It was the kind of thing that people had started to talk about a lot speculatively, because it's reviving a very old question. We set out to test it, and it took a really long time to figure out a useful system for it.

Carika Weldon: How did your colleagues receive what you do in terms of your research?

Dr. Rando: I know there's a lot of skepticism. People don't express it to me personally that often, but when they do I try to take it positively. At this point, simply the observation that a paternal diet affects offspring metabolism has now been repeated in various ways in something like 100 publications. Just by the sheer mass of work, people are starting to believe that idea. I don't believe that we know the mechanisms by which this works in any great detail, so as to whether or not the mechanisms we're working on are going to ultimately be the correct answer for how dads talk to kids, I, myself, am skeptical. Some people believe that the small RNAs we work on might be the answer, and I'm sure lots of people don't. I don't know which of those people is correct.

Carika Weldon: You mentioned that there's some skepticism, and that happens in every area of science. It also happens when there's a big shift, when there's a result that doesn't fit with what everyone thinks. What do you want to say about that, in terms of how we should handle change and how we should handle not ignoring results because it doesn't fit what we think?

Dr. Rando: I don't have anything super-useful to say about that. The whole point of the scientific method is that you have to listen to the evidence, so I think most people ultimately will come around. The flip side of that is that if people have believed something for 100 years, then extraordinary claims require extraordinary evidence. It's very healthy to demand a high level of evidence for something that people suspected wasn't true. On the other hand, it's also the case that ultimately when the preponderance of evidence argues against your deeply held beliefs, you should really be paying attention to the evidence.

Carika Weldon: Your work can seem "out there," but you make it easy to listen to and to understand and accept. When you talk about science, how do you find that balance?

Dr. Rando: It's easy to talk about reproduction, because everybody can relate to reproductive biology one way or another. Personally, I'm actually quite shy. Giving a talk is sort of like acting. I have to feel like it's a performance. That's part of why I'm throwing my hands around and making lots of jokes. It keeps me comfortable. But it is the case that there are people who do amazing science but I just can't watch their talks. I get nothing out of their talks; I'd be better off reading papers than seeing the talk. The delivery means a lot because you don't realize how special a body of work is if people can't communicate that to you. Science communication is super-important for influencing how people think about the world.

A Conversation with Phillip Sharp

INTERVIEWER: CARIKA WELDON

Researcher, University of Oxford Wellcome Centre for Human Genetics

Phillip Sharp is Institute Professor and Professor of Biology at the Massachusetts Institute of Technology and Member of the Koch Institute for Integrative Cancer Research.

Carika Weldon: It's extremely exciting to be talking to the codiscoverer of splicing. How is that for you to be a scientific celebrity?

Dr. Sharp: I try to resist celebrity. It gets in the way of knowing people and doing work. I'm very proud of being able to make that contribution and introducing the world to RNA splicing.

Carika Weldon: When you learn in university about RNA expression and gene regulation, you don't even realize at the time that the person who discovered it is still alive. How is it for you from the other side to see where the field has gone today from where it was when you started?

Dr. Sharp: I was at Cold Spring Harbor for 3 years in '71 to '74. Before that I was at Caltech for 2 years working in bacteria and doing electron microscopy of the DNAs of plasmids and mapping genes on the chromosomes of *E[scherichia] coli*. It was genomics before genomics was called "genomics." I came to Cold Spring Harbor to learn how to work with DNA tumor virus. In those days, the molecular biology community considered working with mammalian cells to be a waste of time. There were a few pioneers: Jim Darnell and a few others. When Jim Watson became director of Cold Spring Harbor, he started the initiative on cancer. It was the period of the "War on Cancer," and I came because I wanted to learn how to work on tumor viruses in mammalian cells. That was the theme that Jim was putting together because these viruses have a few genes that cause tumors in rodents, and we wanted to know how a gene could cause a tumor. So we started working with and studying these viruses.

I began to read about the molecular biology in mammalian cells, and I came across the phenomenon that, in the nucleus of human cells, there was this long RNA called heterogeneous nuclear RNA. It had a poly(A) tail at the 3′ end and a cap at the 5′ end, so it had most of the modifications we knew about in messenger RNA, but it was five to 10 times larger and no one was paying any attention to it. But I knew adenovirus, when infected in human cells, had the same long nuclear RNA but a shorter cytoplasmic one, and that was something I kept my eye on. I thought, "I know how to solve that problem for adenovirus." I didn't know if it would tell us anything in general, but I knew how to solve the issue of what that relationship was.

I left Cold Spring Harbor in '74 to go to MIT and we did the electron microscopy of the RNA from the cytoplasm in the cell compared to the DNA in the nucleus, and there were these three beautiful loops. It told us that the long nuclear RNA was a precursor to the cytoplasmic RNA, and RNA splicing was the process that removed the middle of it. That became a paper that was published in '77. It was discussed here at Cold Spring Harbor in the '77 Symposium—which was about chromatin—but the RNA stole the show.

So, the field has grown. It just exploded after the '77 Symposium, because many labs could begin to clone human genes and genes from mice and other species. Here were these introns between the exons that code for the protein that had to be spliced out, and within 6 months everybody who was literate on the topic knew about this discovery of split genes and RNA splicing. It was absolutely a remarkable time.

Carika Weldon: Were people at the time shocked or were they as excited as you were?

Dr. Sharp: They were stunned. Every major lab had data in their notebooks they had been ignoring because they couldn't explain it. Richard Flavell was working on β-globin and he had this restriction cleavage site in the middle of the cellular gene but it wasn't in the middle of the message, so he couldn't figure out why there was this restriction enzyme site. And I said, "Just listen." And then he went home and within a month he had described that there was an intron; in fact, there are two in globin. Pierre Chambon at Strasbourg and Phil Leder at the National Institutes of Health had similar data in their notebooks, and Susumu Tonegawa and Wally Gilbert had sequenced the v-region of an immunoglobulin gene and there was this little region in the middle of the sequence that coded for amino acids that was meaningless, and they couldn't figure out how it was dealt with. Splicing fixed all that.

We're still struggling with alternative splicing, because it varies from cell to cell. Even though we only have 21,000 genes, every one of these genes are spliced in

---

© 2019 Sharp. This article is distributed under the terms of the Creative Commons Attribution-NonCommercial License, which permits reuse and redistribution, except for commercial purposes, provided that the original author and source are credited.

Published by Cold Spring Harbor Laboratory Press; doi:10.1101/sqb.2019.84.039487

different patterns by alternative splicing, making different proteins in some cells. Until we got the technology—and that's only the last year or so—to do RNA sequencing from a single cell, we really didn't know what type of messenger RNA was expressed in each of the cells. We know that it makes a difference, because if you perturb the alternative splicing, in many cases you see a change in phenotype. So, even though we're in the genomic era, we're still struggling with the complexity of these biological systems.

Carika Weldon: So there's still science to do. Can you tell us about your recent work?

Dr. Sharp: I'm going to talk about a breakthrough now. It's not my breakthrough; it was done by Tony Hyman at Max Planck in Dresden, Germany and Cliff Brangwynne at Princeton. If you look at a cell through a microscope, you see a nucleolus and you see other dense particles. They recognized that many of those dense particles were membraneless subcellular bodies created by what is called a "phase transition." They have a number of molecules that have valency and they coalesce, and they make a concentrated body. In these concentrated bodies, very important reactions occur. In fact, they are critical for these reactions. They make them work at the rate and pace that they need to work. They can be dissociated and reformed. That's a form of regulation. What I've been working on the last 2 years with colleagues at MIT is how that process is central to the transcriptional process that copies RNA from the DNA. That's your major point of regulation.

Every cell in your body has the same DNA, but some cells are skin cells and some cells are blood cells; they express subsets of genes differently. When you get diseases like cancer, the genes change in their expression level and that causes the growth. That regulation of genes, turning them on and off, involves this condensate process. It's a very important insight because now we're going to be able to pinpoint the signals that come into the cell that turn genes up and down. As we learn about this, we will learn how to control gene regulation. That'll help us to treat diseases that we can't treat now, from cancer to Alzheimer's.

It's really surprising that we've been working on cell biology for all these years and it wasn't until about 10 years ago that somebody made the insight that this whole process occurs. Now it sort of tumbles forward and people learn about it. They each relate their own work to it and it's growing and growing in tempo.

Carika Weldon: You've come from the beginning of this field of RNA splicing. We are where we are now. Technology has gotten better: single-cell RNA-seq, long-read RNA sequencing. Where do you think we will be in another 15 years?

Dr. Sharp: I can't answer that question. I know that anything I say is going to be wrong. If I am here 15 years from now, I'll be dramatically surprised by what's happening. I think we're going to see the integration of intense computing into the modeling and understanding of how cells function. It's clear that the complexity of cells is more than the mortal mind can deal with. To make progress now, I can't think gene by gene. I have to think about relationships between genes and relationships between genes and the physiology of the body. That's the challenge we're at now. That's going to require the integration of the capacity to compute models that relate systems of genes to phenotypes and to communicate these insights to each other.

We will make predictions and go in the lab to test them. We will work to expand that knowledge, but we need these tools from systems and computational biology to make the next step. I think that's going to happen in the next 10 years, and it will change our vocabulary. It will change how we even talk at symposiums. We're going to be walking through 3D stereographics saying, "That piece of the cell is dynamically interacting with that piece of the cell over there, and it's doing this." It'll look like *Star Wars*, but it won't be "boom-boom-boom, shoot-'em-up." It'll be proteins and genes interacting and changing how cells become malignant, and explaining and making predictions as to how one could intervene with it. It's a totally exciting field.

Carika Weldon: Can you take us back to 1993? That was a big year for you: the year you won the Nobel prize. What was that like?

Dr. Sharp: Breathtaking. The way prizes work in science is that you get them early in your career if you have done noteworthy work. When you look who had won before and one in four of the people who had gotten *that* prize also got the Nobel prize, you get the feeling that you might be on a list somewhere. I had gotten a number of those prizes with Tom Cech. Tom discovered self-splicing: that the RNA in some introns catalyzes its own excision. He got the Nobel Prize with Sidney Altman in Chemistry in 1989. I assumed that the Nobel committees had recognized splicing.

So, I had sort of relaxed on Nobel Day. I was in bed sleeping in a little late and the phone rings at 6 in the morning. And I said, "Who's calling me at 6?" So I pick up the phone and it's a Swedish colleague who had just come out of the committee meeting and said it had been announced that I was chosen as a Nobel Laureate that year with Richard Roberts. I took a minute to digest this. Then he asked me if I had the telephone number of Rich Roberts, because they didn't have it. He had just moved from Cold Spring Harbor up to New England Biolabs. I tried to find it, but after I called my wife and she responded and jumped out of bed, I said, "You're going to have to find it somewhere else." I had just no chance. I'm not going to be able to do anything.

So, the day just unfolded like that. I get a call from MIT wanting me to be there at 10 in the morning for the press conference. And I go and I have some champagne with my assistant and then I have the press conference, I have champagne. I go to lunch with the Dean, I have champagne. Go to the department. I have champagne. I go home, we have a block party. I have champagne.

Then I go home and the Dean of Science at MIT calls and says, "We want you to talk at 2 in the afternoon about what you did to get a Nobel Prize. And I said, "Who am I going to be talking to?" And he said, "Well, we're going to advertise it to all of MIT." Seventeen thousand people. Actually, about 600 people showed up and I had to put together a very general talk about the subject. It turned out to be one of the most pleasant, exciting moments: to share it with those people who knew me but didn't know the work, but were so excited about MIT and a faculty member getting recognition. Really, a wonderful time. If you get offered a Nobel Prize, say yes.

A Conversation with Maria-Elena Torres-Padilla

INTERVIEWER: ANKE SPARMANN

Senior Editor, Nature Structural and Molecular Biology

Maria-Elena Torres-Padilla is Director of the Institute of Epigenetics and Stem Cells at the Helmholtz Center in Munich.

Anke Sparmann: Your research focuses on transitions of cellular potency and cell fate decisions working with totipotent cells at the very early stages of mammalian development shortly after fertilization. What fascinated you about this cellular system?

Dr. Torres-Padilla: We are interested in understanding how these very early cells of the very early embryo are actually able to establish and maintain the largest plasticity that one can think of. It's quite remarkable. Everybody, at some point, was a single cell. The question is how that single cell is able to generate a new being: not only all the tissues and cells that we have in our body, but really how that single cell builds up the whole program that we call "totipotency." The system is difficult in that we have very limited materials. Obviously, we don't do experiments with humans, but we do use the mouse and other species as a model to understand these transitions. But you don't get a lot of embryos to try to understand biochemically what happens with stem cells and so on. The system is really fascinating, but it is a challenge.

Anke Sparmann: What are you seeing in these very early stages compared to when differentiation occurs?

Dr. Torres-Padilla: There've been quite a lot of things that we did not anticipate when we started to ask how this RNA regulation and chromatin architecture takes place in these early embryos. Every single hypothesis that we put out there was basically wrong. We had a lot of unexpected findings. One is that the retrotransposons, which typically occupy a very large proportion of our genome, are heavily transcribed. In our genomes and in the mouse, these transposons are around 50% of the genome. They're not normal coding genes, like for the proteins of the skin or of the liver. They're really repeats that are half of our genome. These transposons are typically known to be silenced in all your somatic cells. It has been thought that it is important to keep them silent because otherwise they can actually jump. Evolutionarily speaking, they can jump and people have thought that they can originate mutations.

Although this work was pioneered by Barbara Knowles many years ago, in the course of the last decade we have found that there's a large fraction of these transposons that are heavily transcribed. Of course, this has potential for regulation at the chromatin level, but also at the RNA level. What is the RNA that is coded by these transposons? What are these RNAs actually able to do, if anything?

Anke Sparmann: In terms of just the sheer cost of transcribing all those regions, does so much transcription cause a lot of problems for the cell?

Dr. Torres-Padilla: The question has been whether this transcription is actually functional. Is this just a side effect and a waste of energy, or is this really meaningful for the developmental process and for totipotency establishment? There again, we have found quite a few surprises. For example, we've observed recently that the LINE elements [long interspersed nuclear elements] that are very abundant—20% of the genome, roughly—seem to be actually involved in opening up the chromatin structure of the embryo. Rather than a waste of energy, it actually has become part of the developmental program. This is where it becomes interesting. How is it possible that these repeats and these remnants of viral infections that we have in the genome have been co-opted to regulate processing in development?

Anke Sparmann: In terms of the genome being much more open than in a differentiated cell, is that what's causing this transcription, or has it also other effects?

Dr. Torres-Padilla: One of the items that we have to invest quite a lot in understanding transposon function in the genome is, what is cause and what is consequence, or what is just a correlation? There are a few cases where we can really say this is actually causative. In the case of LINE, we actually tried to target transcription factors to them to manipulate their transcription indirectly. At least for a small portion of these transposons, we can probably say that their transcription is causing changes in the chromatin architecture. Whether this is because the genome is just open and is being transcribed, we still have quite a lot of work to do there. There are certain specificities that we don't really understand very well yet.

Anke Sparmann: A recent paper that you published looking at the genome organization showed some of the dynamics that are going on at that stage of differentiation.

Dr. Torres-Padilla: Yeah, we also have been interested in understanding how the nucleus first becomes regionalized. For many years we've known that the genes that tend to be in the internal part of the nucleus in a somatic cell are more prone to activation. The transcriptional activity is higher, whereas the genes that would be a little bit more on the periphery of the nucleus tend to be silenced. The important question was when is this regionalization, which is perhaps functional, first established? That stems from observations that heterochromatin is positioned within the nucleus of the embryo in a very weird manner. It doesn't really have this typical clustering that you see in the somatic cells. Instead, it forms sort of rings around the nucleolic precursors. The question has been whether this atypical nuclear organization is actually important for development as well.

The development of new low-input protocols for looking at the genomic changes in the embryo has made for a change in the field. We teamed up with Jop Kind at the Hubrecht Institute who had set up the DamID [DNA adenine methyltransferase identification] technique to generate molecular mapping of the genome in single cells. We managed to map the regions of the genome that become organized in the nuclear periphery in proximity with nuclear lamina—what we call the lamina-associated domains, or LADs—versus those that could be in the internal regions, the inter-LADs.

There again, there were actually quite a lot of surprises. One thing that we found is that these lamina-associated domains are established very early. A few hours after fertilization the nucleus is already compartmentalized, so there's very clear LAD formation early on. At least in the mouse, 20% of the genome becomes localized in these LADs 3 or 4 h after fertilization, and it remains so for the rest of its life. These are actually regions of the genome that are constantly at the periphery. In a sense, that also could tell you that this very first cell already has kind of a "skeleton" or scaffolding for how the genome is going to be organized.

At the same time, we also found that if you look at the autosomes or at chromosomes in the mouse, we did not detect association with the lamina in the oocyte in the maternal germline before fertilization. That has quite a number of implications. First, that obviously indicates that nuclear organization is established de novo after fertilization and is not inherited. It also gives us the opportunity to try to understand how this nuclear organization is established, mechanistically speaking.

Anke Sparmann: How is that part of the genome partitioned?

Dr. Torres-Padilla: Almost 10 years ago we had done some experiments where we had tethered heterochromatin from the internal part of the nucleus to the periphery. We did those experiments trying to understand whether nuclear organization was important for heterochromatin formation or gene function. What we observed is that heterochromatin becomes derepressed. Basically, we were bringing pericentromeric heterochromatin to the periphery, and instead of having it more silenced like one would expect from what we know from somatic cells, there would be a derepression. Again, that says the embryo does seem to have a bit of a different epigenetic landscape, and that seems to be both from the activation of transposons that we were discussing before, but also the nuclear organization.

Anke Sparmann: You were talking about the female genome. How about the paternal genome?

Dr. Torres-Padilla: Unfortunately, we cannot do these kinds of DamID experiments in male germ cells because there is not really a proper laminar organization in the sperm. Also, except for protamines, their DNA basically goes into the oocyte almost naked during fertilization. It's not very clear what really happens in terms of how the nuclear envelope is formed and what the components are during fertilization. That's a process that we don't know enough about. Regardless, because there is not really a proper nuclear lamina in the sperm, we would anticipate that that organization is also de novo. Of course, we cannot rule out that there's some organization in the sperm or some information that could drive that de novo formation of LADs.

What we do see is that after fertilization when the two pronuclei are still separated the LADs are formed in both pronuclei, not only on the female one. They are slightly different in terms of genomic features in the paternal and the maternal pronuclei, but pretty much 20% of the genome goes to the periphery. Again, that's interesting because it also implies that the two parental genomes are slightly different in terms of nuclear organization, but we found that these differences are resolved by the time the embryo gets to implantation.

Anke Sparmann: It's interesting to see that these cells really are very distinct. They challenge how we think about heterochromatin.

Dr. Torres-Padilla: This is the importance of approaching questions with a very open mind. You launch a hypothesis and then you have to be quite open to see what kind of findings you're going to have, because they might be completely different to what you're used to or what you would expect, which is a cool thing about the biology that we're studying.

A Conversation with Igor Ulitsky

INTERVIEWER: ANKE SPARMANN

Senior Editor, *Nature Structural & Molecular Biology*

Igor Ulitsky is the Sygnet Career Development Chair for Bioinformatics and a Senior Scientist in the Department of Biological Regulation at the Weizmann Institute of Science.

Anke Sparmann: Your research focuses on discovering the functionalities of long noncoding RNAs [lncRNAs] and how these functions are encoded in the sequence, especially what determines their cellular localization. What got you interested in noncoding RNAs in the first place and what are the challenges that you find working with them?

Dr. Ulitsky: I first became interested in noncoding RNAs towards the end of my PhD studies. We were looking at regulatory networks governing transcriptional regulation for protein-coding genes. Then we got interested in microRNAs, so I went for my postdoc to Dave Bartel's lab at the Whitehead Institute. This was around the same time that people began doing large-scale maps and seeing that there is a lot of transcription outside the boundaries of protein-coding genes producing RNAs that seemed to be very similar to mRNAs, but it wasn't clear whether these were functional or what they might be doing.

Together with another postdoc in the lab, Alena Shkumatava, we became interested in studying to what extent long noncoding RNAs are found in different species. We compared zebrafish and human and mouse and we found that there is a lot of turnover, but there is a subset of lncRNAs that is deeply conserved. They're found throughout vertebrates and we could show that two of them were actually functionally important in development of zebrafish and that this functionality was conserved. When I started my own lab about 6 years ago, we decided to focus on understanding what kind of things these long noncoding RNAs are doing and how they're carrying out these functions.

Anke Sparmann: You're using screens, but also computational methods and evolutionary analyses to determine these functions. How do these two methods differ in what you learn from them?

Dr. Ulitsky: My background is in computer science; I did my PhD in computational biology. A lot of what we were doing at the time was based on taking a lot of knowledge that we had on proteins and trying to use networks and similarities and so on to predict protein function. lncRNAs are much more complicated because we don't have much to start with. We don't have any sort of clear "gold standard" where we can say, "We understand *these* lncRNAs, so now let's use that information to learn something about the other lncRNAs," because we don't have enough knowledge to begin with. That's why we have to do more experimental biology to try to build some initial understanding of a few select examples, so we can then go back to the more computational side and generalize it. We've been taking five or six favorite genes and really trying to nail down what they're doing and how they're doing it, while in parallel always trying to think how we can generalize from that.

We screen for fragments of the lncRNA genes or use CRISPR to screen for functionalities of genes, because eventually there are maybe 50,000 different lncRNA genes in the human genome. We don't think that all 50,000 are functional. Maybe there are a few thousand that are functional, maybe just a few hundreds, but that's still a huge diversity. While it's important to study them one at a time, we're not going to get very far very fast if we keep doing that. We always try to understand these examples, but, on the other hand, we're always thinking how we can experimentally or computationally—and ideally, both—try to generalize that and say, "Okay, we're learning something about this gene, but it's also applicable to others." We've made some progress on this, but it still remains the main challenge both for us and for the field in general.

Anke Sparmann: Proteins have functional domains that always look the same and you can predict the function just by the protein having that specific domain. In lncRNA that's not quite how it works.

Dr. Ulitsky: That's still the blueprint, though. In the last 5 years we've tried to find as many of these domains and as many lncRNAs that are behaving like that. I still think that there are some, but today we realize that this is likely a minority. If we think about a "beads-on-a-string" model where you have a long RNA that's built from these functional domains—from "beads"—the way that we like to represent proteins, this would yield certain expectations.

We would expect it to have multiple conserved regions. We'd expect that if we look at its evolution, we're going to see preservation of much of the sequence. Even if the sequence is more flexible because structure is more important than the primary sequence, we would still expect to see that the RNA would maintain at least the general boundaries between, say, human and mouse. There are some lncRNAs that are like that; *Cyrano* and *NORAD* [noncoding RNA activated by DNA damage] behave this way. But if we look at the typical lncRNA and how it evolves and how abundantly it is expressed and what its sequence looks like, it's quite rare that we see such genes.

Again, if we go back to maybe a couple of thousand lncRNAs that are functional, how many of them are actually these RNA "machines" with multiple domains resembling ribosomal RNA? At the extreme, my guess would be that this is a relatively small percentage: maybe 5%–10%. Many of the others might have some functional RNA sequence, but a lot of the functionality is about taking that functional region and expressing it in a particular context. Where the lncRNA is expressed, how it is expressed, how much time it spends on chromatin or in the nucleus, what is happening around it, all seem to be more important than any particular combination of sequences. Still, the field's thinking about this is evolving all the time. If you ask me in another 5 years, maybe everything I just said is wrong.

Anke Sparmann: Is it more the structure or the sequence that determines whether these kinds of the domains might be functional?

Dr. Ulitsky: Thinking about noncoding RNAs is very much shaped by this idea that their structure has to be very important. Other noncoding RNAs—ribosomal RNAs, microRNAs, snoRNAs [small nucleolar RNAs], tRNAs [transfer RNAs]—all act through adopting a particular structure and through interacting with proteins that are recognizing these specific structures. If we look at how these other RNAs evolved, in most cases we can really see that there is a lot of pressure to preserve these particular structures.

If you look at conservation of lcRNAs between different species, there are some cases where we see evidence that evolution has preserved a particular lncRNA structure while changing the rest of the sequence, but these cases are quite rare. For a typical lncRNA, evolutionarily we don't really see evidence for conservation for any particular structure. For the very few that have been interrogated experimentally, there is not a lot of evidence that structure is very important. Of course, our view is very biased: One can say that structure must be important, and if we're not seeing it it's probably because we don't have the right tools.

But it's important to keep in mind that even if you take a random long RNA, it's going to fold into a structure. It's going to be GC-rich; it's going to be a stable structure. That's not to say that lncRNAs don't have a secondary structure; they do, and in many cases it's going to be a stable structure. But the evidence that these structures are actually driving their function or their recognition by other proteins in a way similar to all these other classes of noncoding RNAs is relatively limited. While there is probably some subset where structure is very important, it's likely that this fraction is relatively limited. Evolution-wise, we don't see a lot of evidence for that conservation.

Anke Sparmann: You said earlier that localization of RNAs is very important, and then you found this element that is important for RNA localization overall. How did you find that, and how do you find that it works?

Dr. Ulitsky: We became interested in localization for two reasons. We think that the functionality of many of these RNAs, which in many cases act on chromatin or regulate gene expression in *cis*, will require that the RNA needs to stay in a particular place. There is not a lot known about where RNAs spend their lifetime and what determines the distribution of RNA in the cell. This is relevant for the lncRNAs, but it's increasingly appreciated that it's very important for mRNAs as well. The typical model used to be that RNA is made and then needs to be exported very quickly to be translated in the cytoplasm. Today, from a lot of RNA-seq studies in various different systems, it's clear that while most mRNAs do that, there's also a variety of mRNAs that for different reasons actually stall in the nucleus for quite some time, awaiting a certain signal or stimulus that then allows them to rapidly export an already-made RNA and translate it in the cytoplasm. There was a lot of interest in what determines, for a given RNA, how long it's going to stay on chromatin in the nucleus versus in the cytoplasm.

The other motivation is that, if you're thinking about screening for functional elements in lncRNAs, the ability to do subcellular fractionations efficiently and reproducibly allows for a very easy screening system where we can test different sequences and compare them to see which sequences carry out a certain function: that function being to either keep the RNA in the nucleus or export it to the cytoplasm. Once we've used this screen to find these elements, we can go and look at other more complex, and possibly more interesting, functions.

We decided to take nuclear lncRNAs—long RNAs that we know stay in the nucleus—and then take short fragments of these RNAs and place them in the context of an RNA that in regular conditions is very efficiently exported. For example, we take the long RNA of GFP [green fluorescent protein], which is typically efficiently exported to the cytoplasm, and we test which short fragments of about 100 bases can act as a kind of brake: to see when, once we insert a specific sequence, the RNA is no longer efficiently exported to the cytoplasm and spends more time in the nucleus.

Once we identify these elements, the real power of this approach comes not necessarily in identification of the sequence, but in the fact that you can take that sequence and systematically mutate it. DNA synthesis makes it easy for us to generate thousands of such fragments that are very similar to each other but that differ by only one base or two or 10. This way, we could really narrow it down and identify a specific region that is important, identify the protein that is binding that region, and study what else about the sequence is important.

Once we have identified a functional element, we can look at whether RNA structure is important for that functional element. In this case, we don't really have evidence that structure is important but we can ask questions like, "What happens if, instead of binding in this particular place, we now move the binding site three or four or five or 10 bases sideways," and so on. We can do this reproducibly across hundreds of thousands of different variants. We're also taking this approach to ask other questions. Instead of localization, what if we look at stability? What if we look at functionality? What if we measure an actual phenotype of lncRNA by comparing many different variants at the same time? This also got us interested in studying some other factors that are influencing the distribution of the RNA within the cell.

Anke Sparmann: If you move the element, do you really see differences?

Dr. Ulitsky: Yes. It appears to matter quite a lot, both where it is and its broader context. This particular element is about 40 bases long—and now we can narrow it down to about 20 bases—and appears to be much more effective when it's found in the internal exons of an RNA rather than in terminal exons, which is especially surprising. If you think about the classical models of mRNA localization, typically these elements are found in the 3′ UTR [untranslated region] that we tend to think of as a reservoir of various motifs that are going to influence things like the stability and the localization of the RNA. In this case, we see that this element is actually preferentially found in internal exons. In mRNAs it's typically part of the coding sequence, presumably because it interacts or somehow is influenced by the context of where it is acting.

We're also trying to see what happens if we take that element and put it in a different context altogether. If you take a different mRNA instead of the GFP that we've been using until now, we see some interesting differences. Context matters in this case, presumably because we have an HNRNPK [heterogeneous nuclear ribonucleoprotein K] protein that needs to bind a particular place, but other factors around it are helpful and they can quite dramatically influence the activity of it. We're seeing that within this short sequence element, HNRNPK also needs to bind to a very particular place. There is maybe one alternative place we can find where it binds, but in most other places —even if we take a strong binding site for that protein— it's not going to work. There are only two particular slots where it can be effective.

Anke Sparmann: What do you think will be the next wave of new discoveries in this field?

Dr. Ulitsky: It's hard to say. The field is maturing gradually. We have a large number of lncRNAs for which we have some evidence that they are functional. There are some phenotypes in cells, and an increasing number where there are phenotypes in model organisms. Also increasing, although slowly, is evidence that human diseases are affected. The big challenge now is to figure out how many different mechanisms these are adopting. To what extent are these mechanisms really similar one to another? We have a "box" of the current mechanisms and some of them have been around for quite some time, but I really think that some of these mechanisms are going to be outside of that box. The challenge is to figure out what else is out there in terms of what kinds of things, mechanistically, a long RNA can do within a human cell.

A Conversation with Jeremy Wilusz

INTERVIEWER: JAN WITKOWSKI

Cold Spring Harbor Laboratory

Jeremy Wilusz is an Assistant Professor of Biochemistry and Biophysics at the University of Pennsylvania Perelman School of Medicine.

Jan Witkowski: Could you start with a bit of background on your work?

Dr. Wilusz: I'm interested in unusual looking RNAs. We've known for 50 years what an RNA is supposed to look like: an mRNA [messenger RNA]. The whole point of making an mRNA is to get it to be translated. You need to splice the RNA. You need to add modifications to the 5′ and 3′ ends to stabilize it. That's obviously very important in how many mRNAs look. But do all RNAs look how we think they are supposed to? The analogy I give is microRNAs. Before 20 years ago, microRNAs didn't "exist." Clearly, they did exist; we just didn't know about them because no one thought RNAs that small could be functional. What drives our research is, can we find other unusual classes of RNAs? Can we find RNAs that aren't even linear, that are circular and have covalently closed ends or are processed in unusual ways that we would not expect?

Jan Witkowski: You've set out deliberately to look for these peculiar RNAs?

Dr. Wilusz: Back in my Cold Spring Harbor days, finding unusual RNAs was never our goal. We'd always set out to take loci that we knew were relevant in disease or developmental processes and understand all we can about them. About 10 years ago, we figured out that a very abundant RNA in cells that happens to be noncoding and misregulated in cancer didn't have the canonical poly(A) tail on its 3′ end, which is quite unusual. Once you remove a poly(A) tail from the transcript it should get rapidly degraded, yet here's a transcript that naturally never had a poly(A) tail but it's superabundant. Once we figured out that that existed and understood how its end got stabilized—that was all by accident, to be honest—it said to us, "If we go looking for things, what else can we find?" Nature is very clever and just because we figured out one way that it does its tricks, I'm sure there's others. Some of what we do now is very purposeful to look at high-throughput sequencing data in a different way or in a unique way, or do high-throughput screens to find things that we may not be expecting.

Jan Witkowski: You made a reference to circular RNAs. I would never expect to find a circular RNA.

Dr. Wilusz: Before a few years ago, not many of us thought anything about them. They're made from many genes through canonical splicing processes that were originally found here at Cold Spring Harbor and MIT. In normal splicing you'll take, for example, exon one and join it to exon two and then exon two joins to exon three. With these circles, instead of taking an end of an exon and connecting it to the next one, you'll connect it to the beginning of that exon. On the surface, it's the same exact process of splicing and joining those ends, but in some ways it's really unexpected. If you take a perfectly good protein-coding gene and splice it in that way such that only a single exon is included in the mature RNA, you have to wonder why you would do that because, for example, you may have removed the start codon. Even though this is from a protein-coding gene, there's no way that the mature RNA that you've made can possibly make that protein.

It's still fairly curious why these exist. There are some genes where the dominant thing that's accumulating from that gene is a circle, and it's unclear why. There are examples of circles that can bind specific RNAs or specific proteins and sequester them. There's thoughts that they could be translated, but there's still plenty of these where it's really not clear what it's doing. It's exciting for many reasons: not only understanding what these new RNAs that we really didn't know much about until the last few years are doing, but also how they are regulated. How does the cell decide to make a linear RNA or to make a circle?

Jan Witkowski: When it makes a circle, can it make a circle of the entire RNA?

Dr. Wilusz: Almost everything is possible except for the ends of the gene because at the 5′ end of an RNA you would have a cap structure and at the 3′ end you'd have the poly(A) tail, so you can't covalently close them. But there are clear examples where you'll have many different circles being made from a gene. You might have a single exon form a circle or you might have exons two and three together form a circle. How all that's regulated and what all that means is unclear. Even just identifying it is still a pretty new phenomenon, let alone understanding why cells are doing it or how it would be used in disease processes or in development.

© 2019 Wilusz. This article is distributed under the terms of the Creative Commons Attribution License, which permits unrestricted reuse and redistribution provided that the original author and source are credited.

Jan Witkowski: Presumably, for the ones that are translated, the protein is going to be nonfunctional?

Dr. Wilusz: I don't want to necessarily say it's nonfunctional. What could be really interesting is, for example, if you have a two-domain protein and within your mature circle you might only encode half of that protein because you don't have the other exon. Let's say exon two encodes a DNA-binding domain, and exon three encodes the transactivation domain to promote transcription. If you make a circle that's only exon two and that gets translated, you would just have the DNA-binding domain. You might imagine that as some sort of a repressor or something like that. There aren't really clear examples of that yet, but it's the sort of thing that we're all thinking about how this could be working.

Jan Witkowski: You said that pretty much every gene can produce these circles. Are there exceptions to that?

Dr. Wilusz: It's thought that at least 15% of genes are making circles in a given cell type. We did deep sequencing and looked at hundreds of millions of reads and we often got one or two reads that supported formation of a circle from a gene. It thus still could be that the output of a gene—99.9% of it—is a linear RNA and there's only a little bit of a circle. There's still a debate over if that little bit of circle is functional or not, but there are a lot of genes that are doing this. We've tried to focus on the genes where you're making a lot of the circle—where it's more circle than linear—because it's really confusing why that would be. Why, if this gene encodes a kinase that's a perfectly good protein, would you "waste" an RNA and make a circle?

Jan Witkowski: Are the RNAs that make the circles defective in some way? Is this a way of weeding out messenger RNAs that have some defects?

Dr. Wilusz: It doesn't seem like it. If anything, what would be confusing about a model like that is that once the circles are produced, they're actually very stable because they're covalently closed molecules and most decay happens from the ends of RNAs. In some ways, circles are really clever. Once they're made, they're resistant to the main degradation enzymes. Introns that are spliced out can form lariats, which are a form of circle, but for the most part those are very rapidly debranched and degraded. The cell has figured out that lariats are junk or useless and so they get rid of them very rapidly. If you thought the same thing about these circles coming from exons, you would have thought the cell doesn't want these to accumulate either.

Jan Witkowski: You're working on genes where the circular form predominates. Is there anything particular about that class of genes?

Dr. Wilusz: Not necessarily. A lot of circles are expressed in the brain more than other places and it's unclear why that is. In general, alternative splicing patterns are often more complicated in brain than in other places. Another idea that's out there is that things like neurons are not cycling cells, so you can just simply have them accumulate with time. For example, circles accumulate with aging. What's not clear at the moment is if this is a bad thing. Is this a cause of aging or is it actually just a consequence that is insignificant?

Jan Witkowski: What proportion of the RNA in a neuron is going to be in circles? Presumably, a small fraction?

Dr. Wilusz: Most of the RNAs are ribosomal RNAs and things like that, so it's still a small proportion for any given gene. But others have shown maybe several hundred genes where the circles are more abundant than the linear mRNA from that gene. It's really curious why that could possibly be the case.

Jan Witkowski: If there's a slowing down of pre-mRNA processing, the number of circles goes up. Is that a close linkage?

Dr. Wilusz: It's unclear. What's exciting about this whole field is it's still quite new. We've gone after circles by trying to understand biogenesis: understanding what the sequences are that are important for them and the proteins that are important for all of this. I would have thought—as I think many would—that if you inhibit splicing, all splicing will be inhibited. Instead, we often find that if we knock down core spliceosome components, circles go up when you inhibit general splicing. It seems that cells now shift to making more circles than they do linear RNAs, and it's curious why that would be. It probably has something to do with how the spliceosome is assembled and regulated. It gives a way that the circles can be regulated and, potentially, function differentially depending on the circumstance.

We've also shown for some transcripts where you have what's called "readthrough transcription," where a gene doesn't stop transcribing where it's supposed to but keeps going further and further, that can also lead to circles. In that case, it's unknown whether the circle is actually a very important functional consequence or a way to say, "This polymerase is just going out of control; let's process it in a way so that we can terminate it and get this thing restarted." We like looking for these unusual RNAs because it tells us something about these processes that we didn't know before.

Jan Witkowski: Do you do that by serendipity or do you devise screens that enable you to detect these things?

Dr. Wilusz: It's a little bit of both. In science, you try to be smart about it, but also weird things happen. We did try to do high-throughput screening approaches where we would design reporters and then try to figure out all the factors that are regulating those reporters. Often, the hits that we get have nothing to do with what the original screen was for. For example, we had knocked down, individually, 10,000 different things but when we knocked down this complex called Integrator it caused the levels of our reporter RNA to go up more than anything else, even though our original screen had nothing to do with finding Integrator or anything like it.

Jan Witkowski: What is Integrator?

Dr. Wilusz: Integrator is a complex of 14 subunits. It's actually an endonuclease. It's supposed to cleave snRNAs [small nuclear RNAs] as they're being made and release them from the polymerase so that they can then form a snRNP [small nuclear ribonucleoprotein] and function in splicing. We've found that Integrator also regulates this mRNA that we were studying. We were confused for a while why this would be because we were very much thinking this would be an snRNA effect or something like that. Instead, we find that Integrator can cleave a nascent mRNA just as if it was an snRNA and terminate transcription.

But whereas, when you cleave an snRNA, you release the mature snRNA that functions in splicing, when you cleave an mRNA you're only making a portion of that mRNA. You're not making the whole thing. When that happens, you actually now terminate transcription. It's a way to turn off that gene. What's really interesting there is the question of how is that controlled? It's a really nice way to keep the gene "off" because you're always cleaving the RNA as it's being made but when you do actually want to make this RNA, how do you tell Integrator, "Stop, let's make the whole transcript."

Jan Witkowski: So, making circles is a two-step process. Integrator does this cutting?

Dr. Wilusz: This is unrelated. This has nothing to do with circles. Making circles is all done by the spliceosome. This Integrator thing is completely different. It's interesting, because, to go back to circles for a second, how circles ultimately get degraded is unclear. They can't be degraded from ends, but presumably there's some sort of endonucleases that will cleave them to degrade it.

In total, it's complicated. You take any given gene and a lot of things can happen to it. You can make your normal mRNA. You can also make a circle. You can also never make it even to that stage because you prematurely terminate by having Integrator come in. There's a lot happening there. That's why we go after these different angles. In my lab, we have very diverse interests. In some ways, it becomes hard for me to keep track of all of it because we study transcription, translation, splicing, and all these different stages. But we're still learning a lot.

A Conversation with Feng Zhang

INTERVIEWER: STEVE MAO

Senior Editor, Science

Feng Zhang is the James and Patricia Poitras Professor of Neuroscience at the McGovern Institute for Brain Research, Associate Professor in the Departments of Brain & Cognitive Sciences and Biological Engineering at the Massachusetts Institute of Technology, a Core Member at the Broad Institute of Harvard and MIT, and a Howard Hughes Medical Institute Investigator.

Steve Mao: Would you mind telling us about the new CRISPR [clustered regularly interspaced short palindromic repeats] system that has a great potential to be repurposed as a new tool for genome editing?

Dr. Zhang: Our work looks at the diversity of CRISPR systems. CRISPR is not a single system; there are many different types. This new system is something where a transposable element called Tn7 has, over the course of evolution, co-opted a CRISPR so that it can use the RNA targeting mechanism of CRISPR to spread itself to viruses or plasmids. By studying the molecular mechanism of this, we realized that it's a potentially programmable way to be able to introduce DNA into the genome. One of the major hurdles of gene editing is we can cut DNA, but introducing DNA into the genome in a precise way has been challenging. Using these transposable elements that are RNA-guided, there's the potential to develop a new genome-editing tool.

Steve Mao: Basically, this is a new CRISPR system that is not functioning as an adaptive immune system, but instead it's co-opted by T7-like transposons. It's RNA-guided, and it can insert a large fragment of DNA into a precise location. You've shown that it can be reconstituted in vitro, meaning you only need minimum host factors. You also showed that it can be repurposed in an *E[scherichia] coli* system. Do you think that it will work in mammalian system?

Dr. Zhang: That's a good question. We're now exploring many of these different CRISPR-associated transposase systems, or what we call CASTs. We are very hopeful that we'll have something that can work efficiently in mammalian cells so that we can use it for a broad range of applications.

Steve Mao: What specific application can this system be used for that current systems like Cas9 or Cas12 cannot?

Dr. Zhang: I wanted to have a way to be able to introduce genes into specific positions in the genome so that we can take advantage of endogenous promoters to drive tissue-specific expression. Before working on gene editing, I worked on a system called optogenetics. Optogenetics allows us to stimulate brain cells using light, but the bottleneck of optogenetics is there are so many different types of brain cells. How do we specifically control one type of cell and not other brain cells? One way to do that is if we are able to introduce these light-sensitive channelrhodopsin protein genes into specific promoter regions so that they're only expressed in the cell type of interest. That has been a major challenge. Neurons are postmitotic, so the traditional way of incorporating DNA through homologous recombination is very inefficient. We needed a new way to do it.

One of the potential applications of CRISPR transposase is to introduce genes into specific sites. If you wanted to control parvalbumin interneurons and not perturb excitatory cells at the same time, you can use CAST to introduce channelrhodopsin into the parvalbumin promoter region. If you want to visualize a particular type of cell in the intestine, you can use CAST to introduce GFP [green fluorescent protein] into the promoter region as a unique marker for that cell type. That's one way to use it.

From a therapeutic perspective, there are also exciting applications. Many genetic diseases are caused by single-nucleotide polymorphisms [SNPs]. What that means is that within some exon of an important gene, there is a small mutation. People are working on ways to use gene editing to correct these mutations, but the way that the existing gene-editing systems work is that you have to introduce specific guide RNAs tailored for individual mutations. Even though conceptually it's all doable, from a practical standpoint in terms of developing drugs, you have to have many different compositions to target the same disease. CAST can provide an alternative approach to treat disease because if there are multiple mutations that affect the same exon, rather than fixing an individual mutation we can use CAST to incorporate an intact exon and that can address any mutations affecting the same exon. You end up with one composition that can treat a number

---

© 2019 Zhang. This article is distributed under the terms of the Creative Commons Attribution-NonCommercial License, which permits reuse and redistribution, except for commercial purposes, provided that the original author and source are credited.

of different mutations in the same disease group. These are some of the exciting applications.

In agriculture, it's also very exciting: the ability to be able to introduce genes into the same region so that when you are breeding these crops, the genes don't get segregated. We can significantly increase the pace at which we can develop new crops. These are just some of the applications.

Steve Mao: You mentioned that a lot of these diseases are caused by those SNPs. You are one of the developers of the DNA and RNA base editor systems. Can you tell us something about the Cas13 RNA base editor? What's the current progress in this field and what's the future of this field?

Dr. Zhang: Another thing that we have been working on in the lab is to develop new ways to edit RNA. There are a couple of advantages to editing RNA. First of all, there are diseases that are caused by single genetic mutations. For those, the treatment strategy would be to convert that disease-causing variant back to what is found in the majority of healthy people. If we're able to do it precisely and efficiently at a DNA level, then correcting DNA makes a lot of sense.

But then there are other diseases where you may want to introduce a risk-modifying allele. For those, the thinking is a lot more complicated. Even though we know that it confers either increased or reduced risk for some aspect of a disease, those variations can often have other more complicated interactions that we don't know about. For those, putting in a DNA change is probably less ideal, because what if it causes a catastrophe and you need to reverse it? RNA editing provides that possibility. You can reverse the change.

Another really exciting way to use RNA editing is changing proteins transiently so that we can modulate cellular signaling. A number of studies have shown that if you can modulate the Wnt pathway or the Hippo pathway, you can drive regeneration in liver to get hepatocytes to grow, or we can regenerate photoreceptor cells. When modulating these proteins, you don't want the modulation to be permanent because you will probably end up with a tumor. You actually want it to be just for a short enough period of time so that you get enough regeneration, but no more than that. These are the reasons that we're developing RNA-based systems.

So far we have developed this one system that we call REPAIR [RNA editing for precise A-to-I replacement], which allows us to convert adenosines into inosines. Inosine is an RNA base that basically functions in the same way as a guanosine in splicing and also translation. That means if we can reverse specific adenosines into a guanosine-like behavior, then we can correct the protein product or the splicing result that comes from a specific variant.

We're continuing to work on other types of editors. One of the things that we have been putting the most effort into recently is making a C-to-U editor. C-to-U editors allow us to address a different set of changes, but also are very applicable for modulating protein phosphorylation states.

We decided not to use naturally existing cytosine deaminases because most of the known cytosine deaminases work on single-stranded substrates. RNA is naturally single-stranded. If you have a cytosine deaminase, it will be hard to achieve specificity on a targeted RNA.

Instead, we took a directed evolution approach. We hypothesized that maybe you can turn ADAR [adenosine deaminase, RNA-specific], which normally deaminates adenosine, into a cytosine deaminase. ADAR works on double-stranded RNA and only deaminates adenosine that's mismatched in a bubble, mispaired with a cytosine. That's how you get single-base specificity. We found out that you can actually do that with ADAR. You can turn it into a cytosine deaminase by just having a mispair with a cytosine. You get a C–C bubble. After 16 cycles of directed evolution, we were able to get a CDAR [cytosine deaminase, RNA-specific] that has a similar level of activity as a natural ADAR. We're pretty excited about that.

Steve Mao: A lot of your systems have very cool names, like REPAIR. Do you have a cool name for the C-to-U system?

Dr. Zhang: Yeah. We're calling it RESCUE: RNA editing for specific C-to-U exchange.

Steve Mao: There's another acronym from your lab. It's a system called SHERLOCK [specific high-sensitivity enzymatic reporter unlocking]. That's actually a slightly different system. It's not trying to repair, manipulate, or edit the DNA or RNA. Instead, you're trying to detect the nucleic acids. Can you tell us something about that, especially its potential for diagnosis?

Dr. Zhang: SHERLOCK is a diagnostic system that we developed by taking advantage of a property of Cas13. Cas13 is an RNA-guided RNA nuclease, but unlike Cas9, it doesn't cleave just the target nucleic acid. Once it recognizes the target RNA, it can then also go and cleave many other RNAs. What that means is that there is amplification in the nuclease activity of this enzyme. We thought maybe you can use this as a way to develop an amplifying diagnostic.

One of the latest iterations of the technology we developed is a paper strip test. You can use urine, saliva, or blood, and then you just put in Cas13 protein and the RNA guide that you designed to recognize a Zika virus or Ebola or influenza or a bacterial pathogen. Within the same reaction, there's a shorter reporter RNA that has a biotin and another molecule called FAM attached to the two ends. If Cas13 found the virus or the pathogenic sequence, it will cleave the pathogenic sequence but it will also activate this collateral activity to cleave these reporter RNAs. Then you have biotin and the FAM separated from that RNA linker.

Then you flow this reaction on a paper strip. It's not too different from a pregnancy test strip. It's got two lines. The first line has streptavidin on it, so you'll bind to biotin. Then the other line has an antibody that binds to FAM. When you flow the reaction, if the pathogenic sequence you're trying to find is not there, then the reporter is going to be intact; the biotin stays linked to FAM. When you flow it, biotin will be captured by streptavidin, so you see one line on this paper strip. If the pathogen is present, the reporter will get cleaved. Biotin will get bound by streptavidin, but FAM is now separated from biotin. It will keep flowing and then it will get captured by the antibody. Just by seeing whether you have one line or two lines, you can get a very quick and also low-cost readout for a disease.

Steve Mao: What do you think are the biggest challenges in the genome editing or gene therapy fields?

Dr. Zhang: One of the challenges remaining is how to precisely introduce DNA. Related to that is, how do you precisely delete DNA? A lot of diseases are caused by nucleotide expansion. Huntington's disease is caused by a trinucleotide expansion in the huntingtin gene. There isn't a good way to be able to contract those expanded regions. Generally speaking, new capabilities to manipulate DNA are still very much needed.

The second—and probably even bigger—challenge is how do you deliver these molecules into the body? So far, people have been able to do ex vivo cell manipulation. You take blood cells or immune cells out and they modify them and you can put them back into the patient, but you can't really do that for the vast majority of organs in the body. You can't take out the heart, fix it, and put it back. Ways to be able to deliver a therapeutic agent into the right organ with enough efficiency into enough cells, but also having enough safety so that you're not also causing toxicity in the body, is very important. Those are probably the two major challenges, but there are a lot of other smaller challenges for developing applications for many research needs.

Author Index

A

Adelman, Karen, 253
Aderounmu, Adedeji M., 185
Aguilera, Andrés, 105, 256
Aral, Chie, 115

B

Bartel, David, 259
Basquin, Jérôme, 155
Bass, Brenda L., 185
Benda, Christian, 155
Biswas, Jeetayu, 1
Bonneau, Fabien, 155
Borges, Filipe, 133

C

Cai, Zhiqiang, 115
Carrocci, Tucker J., 11
Chang, Howard Y., 31
Chen, Jia-Yu, 55
Chen, Ling-Ling, 67, 262
Cho, Hana, 47
Conti, Elena, 155
Corbet, Giulia Ada, 203

D

Das, Sulagna, 1
Dean, Caroline, 264
Di, Chao, 115
Donelick, Helen M., 185
Doudna, Jennifer, 245
Dreyfuss, Gideon, 115, 239
Duan, Jingqi, 115

E

Eliscovich, Carolina, 1

F

Falk, Sebastian, 155
Fazal, Furqan M., 31
Fu, Xiang-Dong, 55

G

Gaidatzis, Dimos, 141
Gasser, Susan M., 141, 266
Gerlach, Piotr, 155
Gross, John D., 217

H

Hansen, Sarah R., 185
Hezroni, Hadas, 165
Hirose, Tetsuro, 227

J

Jaffrey, Samie, 268
Joshua-Tor, Leemor, 271

K

Kalck, Véronique, 141
Kornblihtt, Alberto, 274
Krainer, Adrian, 276

L

Lim, Do-Hwan, 55
Lingaraju, Mahesh, 155
Liu, Chu-Xiao, 67
Luna, Rosa, 105
Lynch, Kristen W., 123

M

MacRae, Ian J., 179
Mao, Steve, 262, 302
Maquat, Lynne E., 47, 279
Martienssen, Robert A., 133
Mattout, Anna, 141
Mayr, Christine, 95

N

Nakagawa, Shinichi, 227
Narlikar, Geeta J., 217
Neugebauer, Karla M., 11, 282
Nunez, Leti, 1

P

Parent, Jean-Sebastien, 133
Parker, Roy, 203
Passmore, Lori A., 21
Pawlica, Paulina, 179
Pérez-Calero, Carmen, 105
Perry, Rotem Ben Tov, 165
Pikaard, Craig S., 195
Proudfoot, Nicholas, 285

R

Rambout, Xavier, 47
Rando, Oliver, 288
Rondón, Ana G., 105

S

Salas-Armenteros, Irene, 105
Sanulli, Serena, 217
Schuller, Jan M., 155
Sever, Richard, 259, 279, 285
Sharp, Phillip, 291
Sheu-Gruttadauria, Jessica, 179
Shimada, Atsushi, 133
Singer, Robert H., 1
Singh, Jasleen, 195

So, Byung Ran, 115, 239
Sparmann, Anke, 268, 276, 294, 296
Steitz, Joan A., 179
Szewczak, Lara, 256, 271

T

Tang, Terence T.L., 21
Tatomer, Deirdre C., 83
Thompson, Matthew G., 123
Torres-Padilla Maria-Elena, 294

U

Ulitsky, Igor, 165, 296

W

Weldon, Carika, 274, 282, 288, 291
Wilusz, Jeremy E., 83, 299
Witkowski, Jan, 253, 264, 266, 299

Y

Yamazaki, Tomohiro, 227
Yao, Run-Wen, 67
Yoon, Young J., 1

Z

Zhang, Feng, 302

Subject Index

A

Actin, 240
ADAR, 185, 303
ADAR1, 71
Ago. See Argonaute
AGO4, 135, 195, 198–199
AGO6, 135
AGO9, 135
ALKBH5, 269
ALN, 136
ALS. See Amyotrophic lateral sclerosis
Amyotrophic lateral sclerosis (ALS), 203
ANKRD52, 68
APC, 32
APEX, 6–7, 37–38, 40, 203, 228
AQR, 110
Architectural RNA (arcRNA)
    NEAT1_2
        biogenesis, 229–230
        functional RNA domains, 230–232
        phase-separated paraspeckle nuclear body formation characteristics, 228–229
        mechanisms, 232
    overview, 227–228
    prospects for study, 232–234
arcRNA. See Architectural RNA
Argonaute (Ago), 179–181, 267
ASH1, 3, 35
Asterix, 271

B

BACE1, 173
BBP, 12
BDNF, 173
BIRC3, 97–98
BLM, 110
BRCA, 257
BRCA1, 110
BRCA2, 258

C

Caenorhabditis elegans. See LSM proteins
Caf1, 22, 27–28
Cap-binding complex (CBC), 15, 47–48
CAPRIN1, 100–101
CAST, 302
CBC. See Cap-binding complex
CBP/p300, 48
CBP20, 47, 280
CBP80, 280
    functional overview, 47–48
    PGC-1 interactions, 48–53

CCAT1-L, 67
CCR5, 249
CD47, 97, 99–101
CDAR, 303
CDK12, 120
CDK9, 59
CFIm25, 118–120
CFIm59, 120
CFIm68, 118, 120
ChIP-seq, 56
Chromatin
    RNA-binding proteins
        detection, 56–57
        functional impact on gene expression, 57–58
        heterochromatin formation, spreading, and maintenance, 59–60
        promoters as hotspots for actions, 58–59
        RNA polymerase II interactions, 58
        three-dimensional genome organization, 61–62
    RNA splicing interactions, 15–16
    transcription connection to downstream RNA metabolism events, 60
Circular RNA
    biogenesis, 70–71
    identification of types, 68–70
    innate immunity regulation, 70–71
    localization study prospects, 41
    overview, 299
CLIP, RNA-binding protein studies, 6, 14, 57
CLK1, 128
CLN3, 210
CLSY, 134, 199
CMT3, 198
CNOT6L, 22
CNOT7, 100–101
CPSF, 119, 254
CRISPR, 56, 73–74, 76, 145, 166, 171, 234, 245–249, 263, 283, 287, 296, 302
CRNDE, 173
Csl4, 156
CstF, 119
CstF64, 119
CTCF, 61
CXCR4, 97
Cyrano, 168

D

Dali, 168
DAZAP1, 230

DCL proteins, 133–134
DCL3, 195–198
DDM1, 134
DDX1, 110
DDX5, 110
DDX19, 110–111
DDX19A, 208
DDX21, 59, 73–74, 76, 110–111
DDX23, 110–111
DDX39A, 109
DDX39B, 109
Deadenylation. See Poly(A) tail
DGCR5, 173
DHX9, 110
Dicer
    double-stranded RNA-binding protein modulation of function
        Loquacious, 190
        R2D2, 190–191
        RNase III enzymes, 187–189
        TBRP, 189–190
    functional overview, 185
    helicase domain
        prospects for study, 191
        vertebrate versus invertebrate functions, 186–187
DIS3, 156
DIS3L, 156
DMS3, 136
DNMT1, 59
DNMT3a, 196
DNMT3b, 196
DOGL4, 136
Dorcas Cummings Lecture, 245–250
DRD1, 136, 199
DRM2, 136, 196
DSIF, 59

E

Edc3, 205
EGFP, 71
Egr2, 172
eIF4A, 208
eIF4E, 280
EJC. See Exon junction complex
EMCV. See Encephalomyocarditis virus
Encephalomyocarditis virus (EMCV), 71
Evf2, 168
Exo9, 156–157
Exo10, 157
Exo13, 157, 160
Exo13n, 157
Exo14n, 157
Exon junction complex (EJC), 279–280
Exosome. See RNA exosome

F

FAM, 303–304
FANCM, 110
far-3, 150
FBL, 76–77
Fip1, 119
Firre, 40, 61
FIS2, 137
FLC, 264
FMRP, 32
Frontotemporal lobar degeneration (FTLD), 203
FTLD. *See* Frontotemporal lobar degeneration
FUBP3, 6
FUS, 229–231

G

G3BP1, 205, 208, 210
GAS5, 173
Granules. *See* Ribonucleoprotein granule
grl-23, 150

H

Hand2as, 167
HDA6, 196
HDAC. *See* Histone deacetylase
Heterochromatin. *See* HP1
Histone deacetylase (HDAC), RNA splicing studies, 16
Histone methylation, LSM proteins in heterochromatin silencing, 145–150
HiTS, RNA-binding protein studies, 6
HNRNPA1, 58–59, 231, 274–275
HNRNPA2, 274–275
HNRNPH3, 230
HNRNPK, 229, 298
HNRNPLL, 58
HNRNPU, 33
HP1
  heterochromatin
    assembly modeling
      nucleosome engagement, 222
      octamer deformation, 222
      oligomers versus dimers, 222
      phase separation, 222–223
    chromatin interactions, 219
    features, 217
    histone core disorganization and chromatin compaction into droplets, 220–222
    oligomerization behavior, 218–219
    phase separation, 219–220
Hpr1. *See* THO complex
HSUR1, 179, 182

I

IDN2, 135
IGF2BP1. *See* ZBP1
Influenza A virus
  genome, 123
  NS1, 123, 125–130
  NS2, 123, 125
  RNA splicing
    host splicing regulation by infection, 127–128
    isoforms, 124–125
    M segment splicing regulation, 125–127
    prospects for study, 128–130
Integrator
  development role, 88
  eukaryote distribution, 88
  metallothionein A transcription regulation
    cleavage of transcripts, 85–86
    RNA interference screening, 84–85
  non-MtnA targets of, 87
  small nuclear RNA cleavage, 253–255, 301
  specialized subcomplexes, 88
ITS2, 158, 160

J

JmjC, 196

K

KLF4, 173

L

LINE, 294
lncRNA. *See* Long noncoding RNA
Long noncoding RNA (lncRNA). *See also* Architectural RNA
  classes, 285
  functional overview, 165, 296–297
  nervous system studies
    conceptual roles, 174
    development functions, 167–171
    differential expression, 165–166
    neurological diseases, 173–174
    neuroregeneration role, 171
    peripheral nerve injury, 171–172
    spinal cord injury, 172–173
    subcellular expression, 166
    techniques for study, 166–167
    traumatic brain injury, 173
  snoRNA-ended long noncoding RNA
    identification, 71–73
    Prader–Willi syndrome role, 73
  SLERT, 73–74
Loquacious, 190
LSM proteins
  H3K27me3 level maintenance, 150
  heterochromatic reporter silencing, 144–145
  mRNA reduction from heterochromatic reporters, 141, 143
  overview, 141–142, 267
  polycomb-marked gene targeting, 146–148, 150
  prospects for study, 150–151
  RNA degradation and silencing enzymes in RNA degradation, 148–150
  overview, 148
  silent endogenous heterochromatin maintenance, 145, 147

M

m^6A, 268–270
MALAT1, 33, 67–68, 83–84, 173, 241
Massively parallel reporter assay (MPRA), 39
MB1. *See* THO complex
MEDEA, 137
MEG3, 173
MEG-3, 210
Megamind, 168
Meganuclease, 246
MERFISH, 36
MES-2, 151
MET1, 198
Metallothionein A (MtnA), integrator complex in transcription regulation
  cleavage of transcripts, 85–86
  non-MtnA targets of integrator complex, 87
  RNA interference screening, 84–85
Methylated adenosine. *See* m^6A
Mex67, 106
Miat, 173
MicroRNA (miRNA)
  functional overview, 179, 259–261
  knockdown, 259
  RNA-induced silencing complex, 283
  target-directed microRNA degradation mechanisms, 180–181
  overview, 179–180
  prospects for study, 181–182
miRNA. *See* MicroRNA
Mitochondria, RNA localization, 40–41
Mpp6, 157–158
MPRA. *See* Massively parallel reporter assay
MTF-1, 87
MtnA. *See* Metallothionein A
Mtr4, 157–158, 160
MTREC, 88
Mud2, 12

N

Nab3, 88
NEAT1, 67–68, 77, 83, 173, 207, 230–232, 262
NEAT1_2. *See* Architectural RNA
NELF, 59
NF110, 71
NMD. *See* Nonsense-mediated decay
NONO, 229–231
Nonsense-mediated decay (NMD), 279–281
Nop53, 159
NORAD, 67, 210
NPC. *See* Nuclear pore complex

SUBJECT INDEX

Nrd1, 88
NS1, 123, 125–130
NS2, 123, 125
Nuclear pore complex (NPC), 34
Nucleolus. *See also* Small nucleolar RNA
    ultrastructure, 74–76
NUP160, 50

P

P4R2, 198
PABP-2, 151–152
PABPC1, 21
Pan2, 22–25
Pan3, 22–23
Pantr2, 168
PARN, 182
pax6, 168
PCF11, 88
PCPA, 116–118, 120, 240
Peripheral nerve injury (PNI), long noncoding RNA studies, 171–172
PGC-1α
    CBP80 interactions, 48–53
    RNA accumulation prevention, 50–51
    target gene transcription, 48–49
PGC-1β, 48
Pif1, 110
PINK1, 173–174
piRNA. *See* Piwi-interacting RNA
Piwi-interacting RNA (piRNA), 271, 288–289
PKR, 71, 262–263
PKR185
Plant small RNA
    chromatin reprogramming, 134–136
    generation, 133–134
    germline functions, 136–137
PNI. *See* Peripheral nerve injury
Poly(A) tail
    deadenylation
        Caf1, 22, 27–28
        exonuclease specificity, 22
        tail recognition
            adenine modification studies, 26–27
            deadenylase complexes, 21–22
            Pan2, 23–25
            tail structure importance, 24–25
    intrinsic structure, 23–24
    non-A sequences, 22–23
    overview, 21
    translation role, 28
Polycomb, 265–266
POU3F3, 168, 171
PPIP5K2, 128
Prader–Willi syndrome (PWS), 73
PRC1, 59
PRC2, 59–60
PRDM5, 173
Pre-mRNA, processing, 239–240
P-TEFb, 48, 59
PTEN, 98
Puf3, 33–34
PWS. *See* Prader–Willi syndrome

R

R2D2, 190–191
RAB1, 128
RAG, 171
Rasa1, 173
RBFOX2, 73
RBM14, 229, 231
RBP-12, 151–152
RBP-7, 151–152
RdDM. *See* RNA-directed DNA methylation
RDM1, 136
RDR2, 133, 195–199
REPAIR system, 303
RESCUE system, 303
REXQ5, 110
Ribonucleoprotein granule
    composition
        proteins, 203, 205–206
        RNA, 206–207
    formation, 207–208
    functional overview, 203–205
    ribonucleoprotein
        partitioning into stress granules and P-bodies, 209–210
        recruitment, 208–209
    self-assembly, 240–241
    sorting between multiple granule types, 210
Ribosomal RNA (rRNA)
    RNA exosome in processing
        mechanism, 159–161
        role, 158–159
    translocation and processing, 76–77
R-loop, 107–111, 256–258, 286
RMST, 168
RNA-binding proteins. *See also specific proteins*
    functional overview, 240
RNA-directed DNA methylation (RdDM)
    overview of components, 195–196
    prospects for study, 197–199
    studies
        in vitro, 197–199
        in vivo, 196–197
RNA exosome
    core complex, 155–156
    mechanism of RNA decay, 158
    nuclear cofactors, 157–158
    overview, 155
    ribosomal RNA processing
        mechanism, 159–161
        role, 158–159
RNA localization
    history of study, 31–32
    mechanisms
        active transport, 34
        cis element role, 32–33
        overview, 2
        RNA-binding proteins, 33–34
        splicing, intron retention, and nuclear export, 34–35
    prospects for study
        contribution of translation-dependent localization versus RNA-dependent localization, 39–40
        non-mRNA localization, 41
        organelle RNA localization, 40–41
        RNA influence on genome architecture, 40
        RNA modifications and structure, 40
    techniques for study
        imaging, 35–36
        machine learning, 38–39
        massively parallel reporter assays, 39
        proximity labeling, 37–38
        sequencing
            next generation sequencing, 39
            transcriptome-wide sequencing, 36–37
        single RNA tracking, 35
        ZBP1 and β-actin messenger RNA localization, 2–3
RNA polymerase II
    carboxy-terminal domain modifications, 14–15
    cotranscriptional RNA processing, 240
    cotranscriptional spliceosome assembly, 14
    RNA-binding protein interactions, 58
    transcription connection to downstream RNA metabolism, 60
RNA polymerase IV, 195–199
RNA polymerase V, 195, 199
RNase A, 256
RNase H, 256
RNase III, 187–189
RNA splicing
    chromatin interactions, 15–16
    influenza infection. *See* Influenza A virus
    mechanism, 12–13
    nuclear landscape, 16, 34–35
    overview, 11
    RNA capping, 15
    spliceosome
        assembly, 13–14
        RNA polymerase II modification impact, 14–15
rRNA. *See* Ribosomal RNA
Rrp4, 156, 158
Rrp6, 157–160
RRP6L1, 199
Rrp40, 156
Rrp44, 156–160
Rrp47, 157

S

SAHA, 109
SCI. *See* Spinal cord injury
SDC, 137
SECRETE sequence, 33
Sen1, 88, 109–110
SeqFISH, 36
SET, 100–101
SETX, 109–110
SFPQ, 229, 231

SHERLOCK system, 303
SHH1, 134
SHH2a, 198
SHH2b, 198
Sin3A, 109
SIRLOIN, 39
SIRT7, 68
SIX3, 168
SLE. *See* Systemic lupus erythematosus
SLERT, 73–74
SMA. *See* Spinal muscle atrophy
Small nucleolar RNA (snoRNA), snoRNA-ended long noncoding RNA
  identification, 71–73
  Prader–Willi syndrome role, 73
  SLERT, 73–74
SMG1, 47, 280
SMN, 242, 274
SMN1, 274, 276
SMN2, 274–276
SNHG5, 173
snoRNA. *See* Small nucleolar RNA
SOX2, 168
SPA, 67–68, 73
Spinal cord injury (SCI), long noncoding RNA studies, 172–173
Spinal muscle atrophy (SMA), 242, 263, 274–277
SPINRAZA, 263, 276–277
Spliceosome. *See* RNA splicing
Spp381, 108
SPT5L, 195
SPT6, 287
Staufen, 33–34
Staufen-mediated mRNA decay, 280
Sub2, 106, 108–109, 111
SUV39, 60
SUVH, 196, 198
SUVH2, 136
SUVH9, 136
SWI/SNF, 196, 222, 229–230
Swi6, 219
Systemic lupus erythematosus (SLE), 262

T

T1S11B, 99–100
TALEN, 246
Target-directed microRNA degradation. *See* MicroRNA
TBI. *See* Traumatic brain injury
TBRP, 187, 189–190
TDP-43, 32, 73, 231

Telescripting. *See* U1
TENT4A, 23
TENT4B, 23
TET1, 59
Tex1. *See* THO complex
THO complex
  chromatin distribution, 106
  components, 105
  cotranscriptional role in chromatin, 108–109
  R-loop prevention
    overview, 107–108
    RNA helicase role, 109–111
  structure, 106
  UAP56 role in ribonucleoprotein biogenesis, 109
Tho2. *See* THO complex
Thp2. *See* THO complex
THZ531, 120
TIA-1, 210
TNRC6, 260
TORC1, 70
TP53, 98
Traumatic brain injury (TBI), long noncoding RNA studies, 173
TREX, 35, 106, 120
TRIBE, RNA-binding protein studies, 6
TSA, 109
TUNA, 173
TUT1, 181
TUT4, 181–182
TUT7, 182

U

U1
  inhibition of premature RNA cleavage, 89–90
  recruitment in influenza A virus RNA splicing, 126
  telescripting
    biological roles
      long intron telescripting and gene regulation, 118
      promoter directionality, 117
      short mRNA isoform switching, cell stimulation, and oncogenicity, 117–118
      transcriptome control and proteome diversification, 116–117
    mechanisms, 118–120
    overview, 115–116

U2, 239, 242
U4, 239
U5, 239
U6, 239, 266
UAP56, 106, 108–109, 111, 257–258
UBAP21, 208
3'-Untranslated region
  protein function regulation
    discovery, 97
    mass spectrometry studies, 97–98
    mechanisms, 99–100
    overview, 95
    protein activity versus abundance, 98
    shortening effects on transcript levels, 95–96
UPF proteins, 47, 279–280
URH49, 109
UTPa, 76
UTPb, 76

V

Valproic acid (VPA), 275
VPA. *See* Valproic acid

X

XIST, 230
XRN1, 144, 151, 266
XRN2, 73, 151–152, 266

Y

YB-1, 205
Ym1, 106
YTH proteins, 269–270
YTHDC1, 40
YTHDF, 40
YY1, 61, 172

Z

ZBP1, 240
  dendrite RNA transport, 4–5
  imaging studies, 7
  prospects for study, 7–8
  β-actin messenger RNA
    localization, 2–3, 32
    translation control, 3
ZFN, 246
ZK970.2, 150